Chemical Perspectives of Microelectronic Materials

Chemical Perspectives of Microelectronic Materials

Symposium held November 30-December 2, 1988, Boston, Massachusetts, U.S.A.

EDITORS:

Mihal E. Gross
AT&T Bell Laboratories, Murray Hill, New Jersey, U.S.A.

Joseph M. Jasinski
IBM T.J. Watson Research Center, Yorktown Heights, New York, U.S.A.

John T. Yates, Jr.
University of Pittsburgh, Pittsburgh, Pennsylvania, U.S.A.

| M R S | MATERIALS RESEARCH SOCIETY
Pittsburgh, Pennsylvania

CAMBRIDGE UNIVERSITY PRESS
Cambridge, New York, Melbourne, Madrid, Cape Town,
Singapore, São Paulo, Delhi, Mexico City

Cambridge University Press
32 Avenue of the Americas, New York NY 10013-2473, USA

Published in the United States of America by Cambridge University Press, New York

www.cambridge.org
Information on this title: www.cambridge.org/9781107410848

Materials Research Society
506 Keystone Drive, Warrendale, PA 15086
http://www.mrs.org

First published 1989
First paperback edition 2012

Single article reprints from this publication are available through
University Microfilms Inc., 300 North Zeeb Road, Ann Arbor, MI 48106

CODEN: MRSPDH

ISBN 978-1-107-41084-8 Paperback

This work was supported in part by the Office of Naval Research under Grant Number
N00014-89-J-1302. The United States Government has a royalty-free license throughout the
world in all copyrightable material contained herein.

Acknowledgment is made to the Donors of The Petroleum Research Fund, administered by
the American Chemical Society, for support (or partial support) of this research.

This work was supported in part by the U.S. Army Research Office under Grant Number
DAAL03-89-G-0001. The views, opinions, and/or findings contained in this report are those
of the authors and should not be construed as an official Department of the Army position,
policy, or decision unless so designated by other documentation.

Contents

PART I: CLUSTERS

PART II: I. NOVEL PRECURSORS   COMPOUND SEMICONDUCTORS

*Invited Paper

PART III: COMPOUND SEMICONDUCTOR
DEPOSITION CHEMISTRY

PART IV: SILICON SURFACE CHEMISTRY

*Invited Paper

*Invited Paper

*Invited Paper

PART VIII: LASER-INDUCED CHEMISTRY

PART IX: ION BEAM CHEMISTRY

*Invited Paper

Preface

This volume is a compilation of contributed and invited papers presented at the symposium on "Chemical Perspectives of Microelectronic Materials." This symposium met for the first time at the 1988 Fall Meeting of the Materials Research Society in Boston from November 30 to December 2.

In creating this symposium, the organizers attempted to highlight the crucial role that chemistry plays in the preparation and manipulation of high technology materials used in microelectronics and to create a forum for communication across traditional disciplinary boundaries. Such communication is a vital key to progress in microelectronic materials research. As a result, this symposium contained a diversity of topics and research perspectives. The topics covered and the level of emphasis placed on each were determined at least as much by the participants as by the organizers. This volume presents a cross-section of current research activity at the interface between chemistry and microelectronics. It does not pretend to cover all possible aspects of that interface.

The papers in this proceedings are organized roughly by material system or chemical process, e.g. aluminum deposition, chemical vapor deposition mechanisms, silicon surface chemistry, novel precursors for metallization and compound semiconductors. One common theme throughout is the close relationship between gas phase, surface, and cluster chemistry. Reports on syntheses of novel organometallic precursors containing "tailored" leaving groups to deposit pure metal complemented mechanistic studies of chemical vapor deposition processes that addressed issues of deposit purity. Studies of reactions at clean silicon surfaces related to such microelectronics processing issues as deposition, etching, oxidation, and nitridation. Modelling of deposition reactions and the effect of reactor configuration on the mechanisms were also discussed. Areas such as plasma etching, laser processing, and photolithography, while not specifically excluded from this symposium and certainly heavily dependent on chemistry, are well enough defined specialties that they are subjects of separate symposia.

It is the organizers' hope that the symposium succeeded in making both chemists and non-chemists aware of the relevance of and potential for chemical studies in microelectronics. The enthusiastic response to this symposium presages increased participation by chemists in microelectronic materials research and increased interactions between chemists and materials scientists, physicists, and electrical engineers in all aspects of microelectronics research.

February, 1989

M.E. Gross
J.M. Jasinski
J.T. Yates, Jr.

Acknowledgments

We would like to thank the invited speakers for establishing an exciting framework encompassing the variety of interrelated topics presented in this symposium:

Invited Speakers

Ph. Avouris
B.J. Aylett
B.E. Bent
J.E. Butler
C.M. Friend
K. Gamo
S.M. George
R.E. Smalley
M.L. Steigerwald

and the presenters of contributed oral and poster papers for their enthusiastic participation.

We also thank the session chairpersons for their help in running the symposium:

G. Higashi   B.A. Scott
P. Ho   R.E. Smalley
F.A. Houle   K. Theopold
K.F. Jensen   R.S. Williams

We thank the Office of Naval Research and the Donors of the Petroleum Research Fund, administered by the American Chemical Society, for the principal financial support of this symposium. The support of the Army Research Office, Air Products and Chemicals, Inc., American Cyanamid Co., Matheson Gas Products, and Strem Chemicals is gratefully acknowledged.

ISSN 0272 - 9172

Volume 1—Laser and Electron-Beam Solid Interactions and Materials Processing, J. F. Gibbons, L. D. Hess, T. W. Sigmon, 1981, ISBN 0-444-00595-1

Volume 2—Defects in Semiconductors, J. Narayan, T. Y. Tan, 1981, ISBN 0-444-00596-X

Volume 3—Nuclear and Electron Resonance Spectroscopies Applied to Materials Science, E. N. Kaufmann, G. K. Shenoy, 1981, ISBN 0-444-00597-8

Volume 4—Laser and Electron-Beam Interactions with Solids, B. R. Appleton, G. K. Celler, 1982, ISBN 0-444-00693-1

Volume 5—Grain Boundaries in Semiconductors, H. J. Leamy, G. E. Pike, C. H. Seager, 1982, ISBN 0-444-00697-4

Volume 6—Scientific Basis for Nuclear Waste Management IV, S. V. Topp, 1982, ISBN 0-444-00699-0

Volume 7—Metastable Materials Formation by Ion Implantation, S. T. Picraux, W. J. Choyke, 1982, ISBN 0-444-00692-3

Volume 8—Rapidly Solidified Amorphous and Crystalline Alloys, B. H. Kear, B. C. Giessen, M. Cohen, 1982, ISBN 0-444-00698-2

Volume 9—Materials Processing in the Reduced Gravity Environment of Space, G. E. Rindone, 1982, ISBN 0-444-00691-5

Volume 10—Thin Films and Interfaces, P. S. Ho, K.-N. Tu, 1982, ISBN 0-444-00774-1

Volume 11—Scientific Basis for Nuclear Waste Management V, W. Lutze, 1982, ISBN 0-444-00725-3

Volume 12—In Situ Composites IV, F. D. Lemkey, H. E. Cline, M. McLean, 1982, ISBN 0-444-00726-1

Volume 13—Laser-Solid Interactions and Transient Thermal Processing of Materials, J. Narayan, W. L. Brown, R. A. Lemons, 1983, ISBN 0-444-00788-1

Volume 14—Defects in Semiconductors II, S. Mahajan, J. W. Corbett, 1983, ISBN 0-444-00812-8

Volume 15—Scientific Basis for Nuclear Waste Management VI, D. G. Brookins, 1983, ISBN 0-444-00780-6

Volume 16—Nuclear Radiation Detector Materials, E. E. Haller, H. W. Kraner, W. A. Higinbotham, 1983, ISBN 0-444-00787-3

Volume 17—Laser Diagnostics and Photochemical Processing for Semiconductor Devices, R. M. Osgood, S. R. J. Brueck, H. R. Schlossberg, 1983, ISBN 0-444-00782-2

Volume 18—Interfaces and Contacts, R. Ludeke, K. Rose, 1983, ISBN 0-444-00820-9

Volume 19—Alloy Phase Diagrams, L. H. Bennett, T. B. Massalski, B. C. Giessen, 1983, ISBN 0-444-00809-8

Volume 20—Intercalated Graphite, M. S. Dresselhaus, G. Dresselhaus, J. E. Fischer, M. J. Moran, 1983, ISBN 0-444-00781-4

Volume 21—Phase Transformations in Solids, T. Tsakalakos, 1984, ISBN 0-444-00901-9

Volume 22—High Pressure in Science and Technology, C. Homan, R. K. MacCrone, E. Whalley, 1984, ISBN 0-444-00932-9 (3 part set)

Volume 23—Energy Beam-Solid Interactions and Transient Thermal Processing, J. C. C. Fan, N. M. Johnson, 1984, ISBN 0-444-00903-5

Volume 24—Defect Properties and Processing of High-Technology Nonmetallic Materials, J. H. Crawford, Jr., Y. Chen, W. A. Sibley, 1984, ISBN 0-444-00904-3

Volume 25—Thin Films and Interfaces II, J. E. E. Baglin, D. R. Campbell, W. K. Chu, 1984, ISBN 0-444-00905-1

Volume 50—Scientific Basis for Nuclear Waste Management IX, L. O. Werme, 1986, ISBN 0-931837-15-4

Volume 51—Beam-Solid Interactions and Phase Transformations, H. Kurz, G. L. Olson, J. M. Poate, 1986, ISBN 0-931837-16-2

Volume 52—Rapid Thermal Processing, T. O. Sedgwick, T. E. Seidel, B.-Y. Tsaur, 1986, ISBN 0-931837-17-0

Volume 53—Semiconductor-on-Insulator and Thin Film Transistor Technology, A. Chiang. M. W. Geis, L. Pfeiffer, 1986, ISBN 0-931837-18-9

Volume 54—Thin Films—Interfaces and Phenomena, R. J. Nemanich, P. S. Ho, S. S. Lau, 1986, ISBN 0-931837-19-7

Volume 55—Biomedical Materials, J. M. Williams, M. F. Nichols, W. Zingg, 1986, ISBN 0-931837-20-0

Volume 56—Layered Structures and Epitaxy, J. M. Gibson, G. C. Osbourn, R. M. Tromp, 1986, ISBN 0-931837-21-9

Volume 57—Phase Transitions in Condensed Systems—Experiments and Theory, G. S. Cargill III, F. Spaepen, K.-N. Tu, 1987, ISBN 0-931837-22-7

Volume 58—Rapidly Solidified Alloys and Their Mechanical and Magnetic Properties, B. C. Giessen, D. E. Polk, A. I. Taub, 1986, ISBN 0-931837-23-5

Volume 59—Oxygen, Carbon, Hydrogen, and Nitrogen in Crystalline Silicon, J. C. Mikkelsen, Jr., S. J. Pearton, J. W. Corbett, S. J. Pennycook, 1986, ISBN 0-931837-24-3

Volume 60—Defect Properties and Processing of High-Technology Nonmetallic Materials, Y. Chen, W. D. Kingery, R. J. Stokes, 1986, ISBN 0-931837-25-1

Volume 61—Defects in Glasses, F. L. Galeener, D. L. Griscom, M. J. Weber, 1986, ISBN 0-931837-26-X

Volume 62—Materials Problem Solving with the Transmission Electron Microscope, L. W. Hobbs, K. H. Westmacott, D. B. Williams, 1986, ISBN 0-931837-27-8

Volume 63—Computer-Based Microscopic Description of the Structure and Properties of Materials, J. Broughton, W. Krakow, S. T. Pantelides, 1986, ISBN 0-931837-28-6

Volume 64—Cement-Based Composites: Strain Rate Effects on Fracture, S. Mindess, S. P. Shah, 1986, ISBN 0-931837-29-4

Volume 65—Fly Ash and Coal Conversion By-Products: Characterization, Utilization and Disposal II, G. J. McCarthy, F. P. Glasser, D. M. Roy, 1986, ISBN 0-931837-30-8

Volume 66—Frontiers in Materials Education, L. W. Hobbs, G. L. Liedl, 1986, ISBN 0-931837-31-6

Volume 67—Heteroepitaxy on Silicon, J. C. C. Fan, J. M. Poate, 1986, ISBN 0-931837-33-2

Volume 68—Plasma Processing, J. W. Coburn, R. A. Gottscho, D. W. Hess, 1986, ISBN 0-931837-34-0

Volume 69—Materials Characterization, N. W. Cheung, M.-A. Nicolet, 1986, ISBN 0-931837-35-9

Volume 70—Materials Issues in Amorphous-Semiconductor Technology, D. Adler, Y. Hamakawa, A. Madan, 1986, ISBN 0-931837-36-7

Volume 71—Materials Issues in Silicon Integrated Circuit Processing, M. Wittmer, J. Stimmell, M. Strathman, 1986, ISBN 0-931837-37-5

Volume 72—Electronic Packaging Materials Science II, K. A. Jackson, R. C. Pohanka, D. R. Uhlmann, D. R. Ulrich, 1986, ISBN 0-931837-38-3

Volume 73—Better Ceramics Through Chemistry II, C. J. Brinker, D. E. Clark, D. R. Ulrich, 1986, ISBN 0-931837-39-1

Volume 74—Beam-Solid Interactions and Transient Processes, M. O. Thompson, S. T. Picraux, J. S. Williams, 1987, ISBN 0-931837-40-5

Volume 123—Materials Issues in Art and Archaeology, E.V. Sayre, P. Vandiver,
J. Druzik, C. Stevenson, 1988, ISBN: 0-931837-93-6

Volume 124—Microwave-Processing of Materials, M.H. Brooks, I.J. Chabinsky,
W.H. Sutton, 1988, ISBN: 0-931837-94-4

Volume 125—Materials Stability and Environmental Degradation, A. Barkatt,
L.R. Smith, E. Verink, 1988, ISBN: 0-931837-95-2

Volume 126—Advanced Surface Processes for Optoelectronics, S. Bernasek,
T. Venkatesan, H. Temkin, 1988, ISBN: 0-931837-96-0

Volume 127—Scientific Basis for Nuclear Waste Management XII, W. Lutze,
R.C. Ewing, 1989, ISBN: 0-931837-97-9

Volume 128—Processing and Characterization of Materials Using Ion Beams, L.E. Rehn,
J. Greene, F.A. Smidt, 1989, ISBN: 1-55899-001-1

Volume 129—Laser and Particle-Beam Chemical Processes on Surfaces, G.L. Loper,
A.W. Johnson, T.W. Sigmon, 1989, ISBN: 1-55899-002-X

Volume 130—Thin Films: Stresses and Mechanical Properties, J.C. Bravman, W.D. Nix,
D.M. Barnett, D.A. Smith, 1989, ISBN: 0-55899-003-8

Volume 131—Chemical Perspectives of Microelectronic Materials, M.E. Gross,
J. Jasinski, J.T. Yates, Jr., 1989, ISBN: 0-55899-004-6

Volume 132—Multicomponent Ultrafine Microstructures, L.E. McCandlish, B.H. Kear,
D.E. Polk, and R.W. Siegel, 1989, ISBN: 1-55899-005-4

Volume 133—High Temperature Ordered Intermetallic Alloys III, C.T. Liu, A.I. Taub,
N.S. Stoloff, C.C. Koch, 1989, ISBN: 1-55899-006-2

Volume 134—The Materials Science and Engineering of Rigid-Rod Polymers,
W.W. Adams, R.K. Eby, D.E. McLemore, 1989, ISBN: 1-55899-007-0

Volume 135—Solid State Ionics, G. Nazri, R.A. Huggins, D.F. Shriver, 1989,
ISBN: 1-55899-008-9

Volume 136—Fly Ash and Coal Conversion By-Products: Characterization, Utilization,
and Disposal V, R.T. Hemmings, E.E. Berry, G.J. McCarthy, F.P. Glasser,
1989, ISBN: 1-55899-009-7

Volume 137—Pore Structure and Permeability of Cementitious Materials, L.R. Roberts,
J.P. Skalny, 1989, ISBN: 1-55899-010-0

Volume 138—Characterization of the Structure and Chemistry of Defects in Materials,
B.C. Larson, M. Ruhle, D.N. Seidman, 1989, ISBN: 1-55899-011-9

Volume 139—High Resolution Microscopy of Materials, W. Krakow, F.A. Ponce,
D.J. Smith, 1989, ISBN: 1-55899-012-7

Volume 140—New Materials Approaches to Tribology: Theory and Applications,
L.E. Pope, L. Fehrenbacher, W.O. Winer, 1989, ISBN: 1-55899-013-5

Volume 141—Atomic Scale Calculations in Materials Science, J. Tersoff, D. Vanderbilt,
V. Vitek, 1989, ISBN: 1-55899-014-3

Volume 142—Nondestructive Monitoring of Materials Properties, J. Holbrook,
J. Bussiere, 1989, ISBN: 1-55899-015-1

Volume 143—Synchrotron Radiation in Materials Research, R. Clarke, J.H. Weaver,
J. Gland, 1989, ISBN: 1-55899-016-X

Volume 144—Advances in Materials, Processing and Devices in III-V Compound
Semiconductors, D.K. Sadana, L. Eastman, R. Dupuis, 1989,
ISBN: 1-55899-017-8

Tungsten and Other Refractory Metals for VLSI Applications, R. S. Blewer, 1986; ISSN 0886-7860; ISBN 0-931837-32-4

Tungsten and Other Refractory Metals for VLSI Applications II, E.K. Broadbent, 1987; ISSN 0886-7860; ISBN 0-931837-66-9

Ternary and Multinary Compounds, S. Deb, A. Zunger, 1987; ISBN 0-931837-57-x

Tungsten and Other Refractory Metals for VLSI Applications III, Victor A. Wells, 1988; ISSN 0886-7860; ISBN 0-931837-84-7

Atomic and Molecular Processing of Electronic and Ceramic Materials: Preparation, Characterization and Properties, Ilhan A. Aksay, Gary L. McVay, Thomas G. Stoebe, 1988; ISBN 0-931837-85-5

Materials Futures: Strategies and Opportunities, R. Byron Pipes, U.S. Organizing Committee, Rune Lagneborg, Swedish Organizing Committee, 1988; ISBN 0-55899-000-3

Tungsten and Other Refractory Metals for VLSI Applications IV, Robert S. Blewer, Carol M. McConica, 1989; ISSN: 0886-7860; ISBN: 0-931837-98-7

PART I

Clusters

SURFACE CHEMISTRY ON MICROCLUSTERS: RECENT RESULTS
FROM SILICON AND GERMANIUM CLUSTER BEAMS

J. M. ALFORD AND R. E. SMALLEY
Rice Quantum Institute and Department of Chemistry
Rice University, Houston, Texas 77251

ABSTRACT

    Supersonic cluster beam techniques have recently produced some
fascinating new information as to the surface chemistry and physics of
small (2-100 atom) clusters of various semiconductors. For example, it
appears that silicon clusters exhibit a remarkably pronounced alternation
in reactivity as a function of cluster size. For ammonia chemisorption
certain clusters such as Si_{33}^+, Si_{39}^+, and Si_{45}^+ have been found to be
almost completely inert, while neighboring clusters such as Si_{36}^+ and
Si_{45}^+ chemisorb ammonia readily. Such sharply patterned reactivity
results may provide significant clues as to the detailed nature of
semiconductor surface restructuring and the consequent effects of this
restructuring upon the surface chemistry.

INTRODUCTION

    Over the past few years a new avenue for fundamental research into
the properties of metal and semiconductor surfaces has begun to emerge.
It involves the study of such small pieces of the bulk surface that the
individual atoms may be counted, and explanations for the detailed
surface chemistry and physics may be sought as though these species were
actually just large molecules. With current techniques it is fairly
straight forward to generate cold, fairly intense supersonic beams
containing clusters from two to several hundred atoms in size. Although
these constitute rather large molecules, still most of the atoms are on
the surface, and the properties of this surface have a dominate influence
on all aspects of the cluster chemistry and physics. In a very literal
sense these small atomic clusters are then "molecular models" of real
surfaces.

    As is the case with all simple models, these cluster models should
not be expected to be perfect replicas of the real thing. They will be
useful as models only to the extent that some of the essential new physics
and chemistry we need to learn about real surfaces will also hold
sway in the small cluster. If so, it may be that many of the new insights
will best be gained first on these small clusters. The hope of course is
that many of the powerful techniques we normally associate with small
molecule science can be extended to these cluster models. Particularly
intriguing is the notion that it may soon be possible to calculate the
physical and chemical properties of these small clusters by high level
quantum chemical techniques, and to test the effectiveness of
approximations used in these techniques by direct comparison with detailed
experimental measures on exactly the same cluster in the laboratory.

    The cluster models of surfaces are most likely to be of direct use
when one concentrates on the microscopic details of surface chemistry,
especially when this chemistry occurs on the surface of what is primarily
a covalently bonded material. So a natural early topic of concentrated
cluster research has been carbon, silicon, germanium, gallium arsenide and

a few other main-group semiconductor materials. This short paper presents a few of the recent developments along these lines, emphasizing one of the most promising new techniques, ion cyclotron resonance (ICR) probes of the cluster surface chemistry while the particles are trapped at high vacuum in the high field of a superconducting magnet.

THE SUPERSONIC CLUSTER BEAM ICR APPARATUS

Figure 1 shows a schematic of the supersonic cluster beam FT-ICR apparatus used in this study. Extensive discussion of the design details of this machine have appeared elsewhere [1], along with several early examples of its application to the study of dissociative chemisorption of H_2 on transition metal clusters [2,3]. Here, however, the supersonic cluster beam was arranged to produce clusters of silicon and germanium.

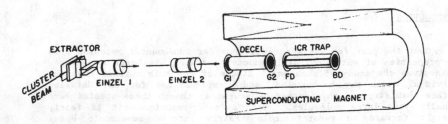

Figure 1. Schematic of supersonic cluster beam FT-ICR apparatus.

These clusters were produced by pulsed laser vaporization (2nd harmonic of a Nd:YAG) of a disc of the corresponding semiconductor in a pulsed supersonic nozzle. Using an arrangement originally developed in our group for the study of gallium arsenide clusters [4], this disc was slowly rotated and translated so that the laser etched the surface in a spiral pattern, smoothly removing material from the surface at a rate of roughly 1000 monolayers per pass. In such a nozzle the hot atoms and small molecules originally produced in the laser-driven plasma are entrained in a fast flow of helium carrier gas at nearly 1 atm pressure, and the resultant mixture is allowed to freely expand into a vacuum, producing an intense supersonic jet which is then skimmed to form a well-collimated beam. Using a second laser (ArF excimer at 1930 angstroms) to irradiate the clusters just as they begin the supersonic expansion it is possible to generate a dense plasma of positively and negatively charged clusters which then receive the full cooling of the supersonic expansion.

Similar techniques have been in use in our group for a few years now in a wide range of experiments with cluster ion beams in the size range from 2 to several hundred atoms per cluster. Although the techniques were originally developed for the spectral and photofragmentation study of the positive cluster ions [5], the negatively charged species have turned out to be particularly useful since they permit the detailed study of the UPS patterns for these clusters as a function of cluster size [6]. Here, however, as shown in Figure 1 the clusters were directed by means of a pulsed extraction field and a pair of einzel lenses down along the central axis of a superconducting magnet operating in the persistence mode at a peak field of 6 Tesla.

Ordinarily, injection of ions into such a strong magnetic field would be extremely difficult owing to the so-called "magnetic mirror" effect. Ions passing into regions of increasing magnetic field begin to turn in response to the $v \times B$ component of the Lorentz force. The effect is to cause the ion trajectories to cross the magnetic field lines with increasing steepness, which in turn increases the Lorentz force further. As a result most incoming trajectories actually never make it into the center of the magnet. Fortunately, as described in an earlier paper from this group [1,2], there is a set of initial trajectories that do succeed in cleanly transiting the fringing field of the magnet. It is those which lie within a narrow cone converging to a point some 20 cm short of the center of the magnet. These trajectories are such that they are nearly tangent to the local magnetic field lines in that region of the fringing field where the Lorentz force is first strong enough to substantially affect the ions motion. There the trajectory begins a slow spiral about the local field line and follows it smoothly into the center of the magnet.

As shown in the figure, the forward momentum of the cluster ion is slowed in two successive deceleration steps as it approaches the ICR trap. This trap is composed of a 15 cm long, 4.8 cm diameter cylinder divided lengthwise into 4 sectors. One opposing set of two of these sectors were used to provide RF excitation in order to coherently pump the cyclotron motion of the clusters. The other set of two sectors were connected to a very sensitive differential amplifier so that the weak image currents generated by the circling cluster ions could be detected, digitized, and the resultant time-dependent waveform submitted to a fast Fourier transform. Drift of the cluster ions out of the trap by motion along the magnetic field lines was prevented by a small repelling potential on two electrodes on either end of the cylinder.

Due to the crude initial pulsed extraction of the cluster ions from the supersonic beam, a considerable energy spread is present in the cluster ion packet as it enters the magnet. As a result it is inefficient to decelerate the cluster ions to less than 5-10 eV as they approach the ICR trap. This and the fact that often 10-100 pulses of the supersonic cluster beam apparatus are necessary to "fill" the ICR cell forces us to use a thermalizing gas in the cell during injection, and to pulse the entrance electrode to the cell down momentarily as each cluster packet arrives. The experiments discussed below used neon as the thermalizing gas held at 1×10^{-5} torr in the ICR trap during the injection of (typically) 100 cluster ion pulses at 10 pulses per second. After an additional 5 seconds to insure the clusters were effectively thermalized, this neon gas was allowed to pump away prior to the RF excitation pulse and subsequent detection of the cyclotron resonance. Once injected and thermalized, cluster ions may be stored in this trap at high vacuum for many minutes. As illustrated in Figure 2 the resultant FT-ICR mass spectrum of clusters trapped in this way can be of superb mass resolution, with excellent signal to noise. This spectrum is due to the very special cluster of carbon, C_{60}^+, which is now widely believed to have the form of a hollow aromatic network with a symmetric bonding pattern identical to that of a modern soccer ball [7]. The smaller peaks to higher mass are due to ^{13}C isotopic variants of the molecule. They appear roughly as expected in accord with the 1.1% normal terrestrial isotopic abundance of ^{13}C.

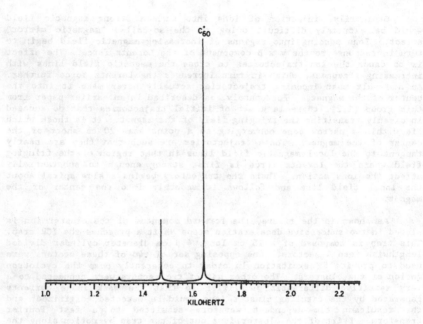

Figure 2. FT-ICR mass spectrum of C_{60}^+ at high resolution in a magnetic trap at 6 Tesla. The horizontal frequency axis is relative to a 125 kHz heterodyne carrier. The lower frequency ICR peaks are due to $^{13}C^{12}C_{59}^+$ and $^{13}C_2^{12}C_{58}^+$, respectively.

Silicon clusters, on the other hand, show no tendency to form a specially stable 60 atom species. Instead the cluster ion distribution from the source is generally found to be a slow monotonically decreasing function of cluster size [8,9]. Figure 3 shows a section of this distribution in the 44-54 atom size range as detected in the FT-ICR apparatus. The fine structure here is due to the several isotopes of silicon in natural abundance.

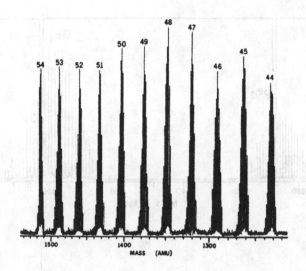

Figure 3.   FT-ICR of positive silicon clusters.

Unlike carbon where one isotope is dominant, most elements in the periodic table have at least two, and often many more, isotopes in major abundance.   As seen here with silicon these isotopes begin to cause somewhat of a problem when one is trying to identify the composition of a cluster by mass alone -- even with the sensational mass resolution of FT-ICR.

A "SWIFT" TECHNIQUE

One of the great virtues of the ICR environment is that each cluster mass has its own cyclotron frequency, so in principle it would appear possible to selectively excite each cluster at resonance until its cyclotron radius became large enough to hit one of the side electrodes, thereby sweeping it from the trap.   If this were cleanly done for all but a single isotopic form of the cluster, one could eliminate the isotopic mass confusion.   For example, the top panel in Figure 4 shows the FT-ICR mass spectrum of germanium clusters in the 11-12 atom mass range.   The broad spread of multiple peaks seen here is due to the 5 major isotopes of germanium in the 70-76 amu mass range.   In order to eject all but a single isotopic variant of these clusters, one would ideally like to arrange the RF excitation such that a uniform RF power was given to all masses shown in the blocked off area to the top panel of this figure.   The problem, of course, is that the desired single mass cluster is exposed to the RF as well and one has to worry that off-resonant excitation of this cluster will leave its cyclotron motion highly excited.

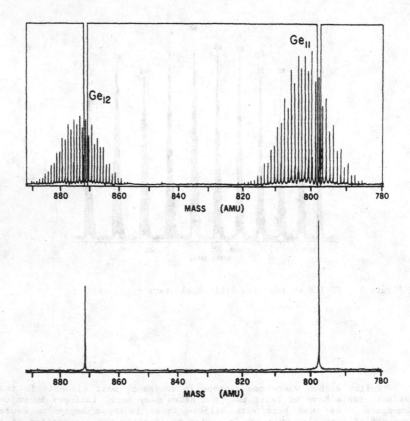

Figure 4. FT-ICR of positive germanium clusters in 11-12 atom size range. The boxed in region of the top panel received uniform RF excitation by the SWIFT technique in order to sweep all but one isotopic variant of the 11 and 12 atom clusters. The bottom panel shows the FT-ICR spectrum of the contents of the ICR trap after this SWIFT ejection.

A beautiful solution to this problem of off-resonant excitation loss of the desired mass is the "Stored Waveform Inverse Fourier Transform" (SWIFT) technique first suggested by Alan Marshal and his coworkers at Ohio State [10].   Here one calculates the discrete inverse Fourier transform of the RF power spectrum necessary eject the unwanted masses, stores this waveform in a large memory, and then uses it to generate an analog time-dependent excitation voltage that is sent to the excitation plates of the ICR cell.   The result is an extremely complicated RF waveform which is guaranteed to minimize off-resonant excitation of the desired cluster. The bottom panel of Figure 4 shows the result of this SWIFT sweeping of all but a single isotopic variant of the 11 and 12 atom germanium clusters from the ICR trap.

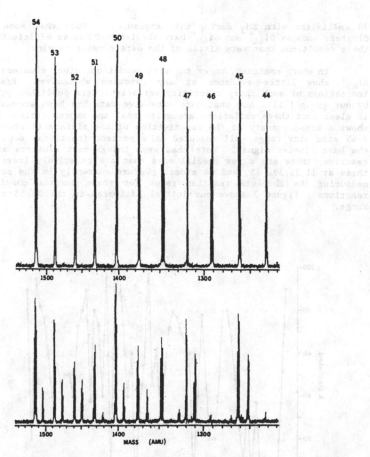

Figure 5. Surface chemistry probe of silicon clusters in the 44-54 atom size range. The top panel shows the (SWIFTed) contents of the ICR cell before reaction. The bottom panel shows the result on this bare silicon cluster distribution of a 5 second exposure to 1 x 10^{-6} torr pressure of ammonia at 300K.

RESULTS FOR THE CHEMISORPTION OF AMMONIA ON SILICON

Using this FT-ICR device it is now possible to begin the study of cluster surface chemistry for a wide variety of materials. One of the most interesting examples of this sort of approach is shown in Figure 5. Here in the top panel one can see the positively charged clusters of silicon in the 44-54 atom size range as they are detected in the ICR trap after injection, thermalization, and SWIFT ejection of most of the isotopic congestion. In the bottom panel one can see the effect on this cluster distribution of a 5 second exposure to 4 x 10^{-7} torr NH$_3$ at near room temperature. A typical silicon cluster ion will experience roughly

70 collisions with NH_3 during this exposure.   Note that some silicon clusters such as Si_{44}^{+3} and Si_{46}^{+} have chemisorbed NH_3 so efficiently under these conditions that very little of the bare cluster remains.

In sharp contrast, under the same conditions other clusters such as Si_{45}^{+} show little evidence of any reaction whatsoever. Preliminary indications of such sharp reactivity variations were published previously by our group [11]. Now that more extensive data has been accumulated it is clear that these variations are quite real and reproducible. Figure 6 shows a broad summary of the reactivities of the silicon clusters in the 3-65 atom size range, all measured as a percent fractional depletion of the bare cluster signal. Note that even though most clusters are quite reactive, there are a few special ones that are relatively inert, namely those at 11,25,39, 45, and 64 atoms. We are currently in the process of measuring the absolute reaction rates for these surface chemisorption reactions. Figure 7 shows our initial estimates in the 39-53 atom size range.

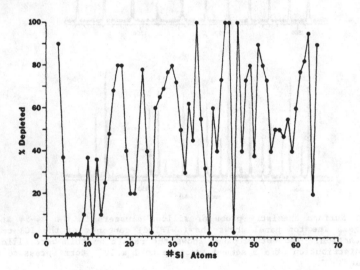

Figure 6. Measured fractional depletion of bare silicon clusters due to chemisorption of NH_3 at 2×10^{-7} torr during a 5 s reaction exposure (roughly 35 NH_3 collisons for each cluster ion).

Such dramatic dependence of reactivity upon cluster size has been one of the hallmarks of the new cluster studies. It is a particularly striking effect in dissociative chemisorption reactions of species like H_2 and N_2 on small transition metal clusters. But it is quite surprising to find it here. First, sharp variations in reactivity are being seen here for very large clusters. In all previous cases the measured reactivity variations largely died out before the cluster size reached 30 atoms, yet here even at 64 atoms special behavior is evident.   Second, it is far easier to produce amorphous silicon by rapid quenching than it is for typical metals. It would not have been surprising to have found these silicon clusters to be essentially amorphous, with each cluster size having many geometrical isomers represented in the ICR trap, each with a distinct reaction chemistry. Yet that clearly cannot be happening. The inertness of clusters like Si_{45}^+ indicates that there is a special structure for this cluster size and that somehow the 45-atom clusters have been able to anneal sufficiently to settle into this structure.

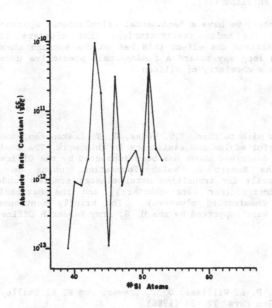

Figure 7. Estimated absolute reaction rate for chemisorption of NH_3 onto the surface of positive silicon clusters as a function of cluster size.

Much work remains to be done to quantify these reaction rates, test for the presence of multiple isomeric forms for some of the more reactive clusters, measure the rates of subsequent reactions after the first chemisorption event, explore other reactants, study the surface chemistry of the negative silicon clusters, etc. But already there is a clear challenge from this early data to theorists interested in understanding silicon surface chemistry. The experimental fact that certain silicon cluster sizes are special is a clue as to how silicon restructures. Somehow these special clusters like Si_{39}^+ and Si_{45}^+ have found a way to restructure all around the surface so that not a single active chemisorption site remains. Yet when even a single atom is added or removed, such a perfect restructuring no longer appears possible.

This is just the sort of intriguing experimental result one hopes to find with this new cluster approach to surface science. Although silicon clusters in this size range still constitute a terrific challenge to electronic structure theorist, it is not a ridiculous challenge. Already extensive ab initio calculations have been published for silicon clusters and their chemistry in the 2-10 atom size range [12-13]. Interestingly, local spin density theory [14] and even elementary tight binding hamiltonian a [15] appear to agree quite well with high level ab initio theory on the structures of the small silicon clusters. So it is conceivable that quite detailed and predictive calculations will be available over the next few years for silicon clusters throughout the 2-100 atom size range. As yet there is no compelling explanation for the new reactivity data based on such fundamental calculations, although some interesting first steps along this road have been published recently [16,17] examining in some detail a structural suggestion originally put forward by J. C. Phillips [18].

When ultimately we have a fundamental calculational approach that is able to predict the unique restructurings that clusters like Si_{45}^{+} undergo, and understand the effect this has on the surface chemistry, we will have come a long way toward a fundamental, predictive understanding of the bulk surface chemistry of silicon.

ACKNOWLEDGEMENT

The authors wish to thank F.D. Weiss, R. T. Laaksonnen, and L. P. F. Chibante for helpful advice and assistance in this work. The semiconductor cluster research described above has been supported by the Office of Naval Research and the Robert A. Welch Foundation, using an apparatus constructed primarily for transition metal cluster studies funded by the Department of Energy (for bare clusters), and the National Science Foundation (for chemisorbed clusters). The briefly mentioned gallium arsenide studies were supported by the U. S. Army Research Office.

REFERENCES

[1] J. M. Alford, P. E. Williams, D. J. Trevor, and R. E. Smalley, Int. J. Mass. Spectrom. Ion Phys. 72, 33 (1986).

[2] J. M. Alford, F. D. Weiss, R. T. Laaksonen, and R. E. Smalley, J. Phys. Chem. 90, 4480 (1986).

[3] J. L. Elkind, F. D. Weiss, J. M. Alford, R. T. Laaksonen, and R. E. Smalley, J. Chem. Phys. 88, 5215 (1988).

[4] S. C. O'Brien, Y. Liu, Q. Zhang, J. R. Heath, F. K. Tittle, R. F. Curl, and R. E. Smalley, J. Chem. Phys. 84, 4074 (1986).

[5] P. J. Brucat, C. L. Pettiette, L. S. Zheng, M. J. Craycraft, and R. E. Smalley, J. Chem. Phys. 85, 4747 (1986).

[6] K. J. Taylor, C. L. Pettiette, M. J. Craycraft, O. Chesnovsky, and R. E. Smalley, Chem. Phys. Lett. 152, 347 (1988).

[7] R. F. Curl and R. E. Smalley, Science 242, 1017 (1988).

[8] L. A. Bloomfield, R. R. Freeman, and W. L. Brown, Phys. Rev. Lett. 54 2246 (1985).

[9] Q. L. Zhang, Y. Liu, R. F. Curl, F. K. Tittle, and R. E. Smalley, J. Chem. Phys. 88, 1670 (1988).

[10] A. G. Marshall, T. C. L. Wang, and T. L. Ricca, J. Am. Chem. Soc. 107, 7893 (1985).

[11] J. L. Elkind, J. M. Alford, F. D. Weiss, R. T. Laaksonen, and R. E. Smalley, J. Chem. Phys. 87, 2397 (1987).

[12] K. Raghavachari, J. Chem. Phys. 84, 5672 (1986).

[13] K. Raghavachari, and C. M. Rohlfing, J. Chem. Phys. 89, 2219 (1988).

[14] P. Ballone, W. Andreoni, R. Car, and M. Parrinello, Phys. Rev. Lett. 60, 271 (1988).

[15] D. Tomanek and M. A. Schluter, Phys. Rev. B 36, 1208 (1987).

[16] D. A. Jelski, Z. C. Wu, and T. F. George, Chem. Phys. Lett. 150, 447 (1988).

[17] J. R. Chelikowsky, Phys. Rev. Lett. 60, 2669 (1988).

[18] J. C. Phillips, J. Chem. Phys. 88, 2090 (1988).

STUDIES OF THE CHEMISTRY OF METAL AND SEMICONDUCTOR CLUSTERS

M. F. JARROLD AND J. E. BOWER

AT&T Bell Laboratories, Murray Hill, New Jersey 07974

ABSTRACT

This article summarizes recent studies of the chemistry of aluminum and silicon cluster ions containing 3-27 atoms.

INTRODUCTION

Atomic clusters are extremely small pieces of bulk material containing 3-~1000 atoms. They are expected to display unique chemical and optical properties due to their hybrid molecular/bulk nature. In particular the chemistry of atomic clusters is expected to be a strong function of the cluster size, and for clusters smaller than some critical size the chemistry is probably quite different from that of the bulk material.

In this article we describe some recent studies of the chemistry of aluminum and silicon clusters. Aluminum is a free electron metal and silicon is a semiconductor. These elements have a relatively simple electronic configuration, and it is possible to perform high quality *ab initio* calculations on quite large clusters (for example, Si_{10} [1]). The experiments described here were performed with positively charged cluster ions. The advantages of working with cluster ions rather than their neutral counterparts are: the clusters can be size selected before the reaction (making it possible to unambiguously determine the reactants and products); and it is relatively easy to vary the cluster ions' kinetic energy, so that studies can be performed over a wide energy range.

EXPERIMENTAL

The basic principle behind the experimental approach [2,3] used to study the reactions of size selected clusters is shown in Fig. 1. Clusters ions are generated by pulsed laser vaporization of an aluminum or silicon rod in a continuous flow of helium buffer gas. After

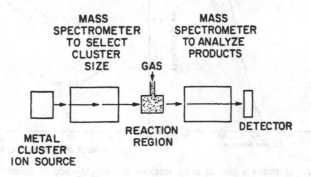

MASS SPECTROMETER TO SELECT CLUSTER SIZE

MASS SPECTROMETER TO ANALYZE PRODUCTS

GAS

REACTION REGION

DETECTOR

METAL CLUSTER ION SOURCE

Figure 1. Schematic diagram of the experimental apparatus.

exiting the source the cluster ions are focussed into a quadrupole mass spectrometer where a particular cluster size is selected. The size selected clusters are then focussed into a low energy ion beam and enter the reaction chamber. There are two possible configurations for the reaction chamber: gas cell or drift tube. In the gas cell configuration the low energy ion beam simply passes through the gas cell which contains a reactant gas at low pressure. Reactions occur under single collision conditions and it is possible to vary the kinetic energy of the cluster ion so that the reactions can be investigated over the collision energy range 0.2-10.0eV. After exiting the gas cell the products and unreacted ions are focussed into a second quadrupole mass spectrometer where they are analyzed, and then detected. In the drift tube configuration the cluster ions are injected into a minature drift tube containing an inert buffer gas (Ne) at a pressure of 0.4torr. The clusters are quickly thermalized and then drift across the drift tube under the influence of a weak electric field. As the ions travel through the drift tube they may react with a reagent diluted in the buffer gas. At the end of the drift tube a representative fraction of the ions exit through a small aperature, they are then mass analyzed and detected. The drift tube can be operated under low field conditions, so that the reactions occur with a thermal energy distribution characterized by the drift tube temperature. The temperature of the drift tube can be varied to investigate the mechanism of the reactions.

RESULTS AND DISCUSSION

Activation barriers for chemisorption of D_2 on Al_n^+ (n=10-27)

The reactions of Al_n^+ with D_2 were investigated using the gas cell configuration [2]. When Al_n^+ with n>10 were passed through the gas cell an $Al_nD_2^+$ adduct was observed. The reactions that form this adduct occur under single collision conditions so the adduct contains enough energy to dissociate back to the reactants. The adduct is metastable and survives long enough (~10μsec) to be detected. The adduct is metastable because of the large number of internal degrees of freedom in the cluster where energy can be distributed. This is the cluster analog of chemisorption on a bulk surface.

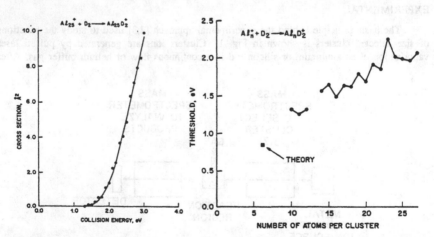

Figure 2. a) shows a plot of cross sections for $Al_{25}D_2^+$ adduct formation against collision energy; and b) shows collision energy thresholds for chemisorption of D_2 on Al_n^+ deduced from data similar to that shown in a).

Fig. 2a shows a plot of the cross sections for metastable adduct formation in the reaction between Al_{25}^+ and D_2. There is a collision energy threshold for adduct formation. The collision energy threshold can be related to the activation barrier for chemisorption of D_2 onto the size selected cluster. The points in Fig. 2a are the experimental data and the line is a simulation which is used to obtain an accurate value of the threshold (it accounts for the broadening of the threshold region). Fig. 2b shows a plot of the collision energy thresholds for chemisorption of D_2 on aluminum clusters with 10-27 atoms. As can be seen from Fig. 2b the activation barriers increase from a little over 1eV upto 2eV with increasing cluster size and then appear to level off. The activation barrier for chemisorption of D_2 on bulk aluminum has not been measured. However, theoretical estimates suggest a value ~1.3eV [4], which is less than the activation barriers measured for the larger clusters. So for clusters with n>27 the activation barrier must eventually fall. Clearly bulk behavior has not been approached with a 27 atom cluster. The point labelled theory in Fig. 2b is from the work of Upton and coworkers [5]. It is for dissociative chemisorption of H_2 on Al_6^+. In these calculations the activation barrier arises from a repulsive interaction between the H_2 molecule and filled orbitals on the Al_6^+ cluster. In order to form the $Al_6H_2^+$ product the cluster must be promoted from its ground electronic state to a state which can accomodate the Al-H bonds. What is not well understood is why the activation barriers increase with increasing cluster size. Since this implies that the promotion energy increases with cluster size, which is the reverse of what would be expected.

Reactions of Al_{25}^+ with CH_4, O_2, CO, N_2 and C_2H_4

In order to get an overview of the chemistry of the aluminum clusters we studied the reactions of Al_{25}^+ with a range of simple molecules [6]. These reactions were also studied using the gas cell configuration.

CH_4 chemisorbs on Al_{25}^+ to give an $Al_{25}CH_4^+$ adduct. At higher collision energies $Al_{24}H^+$ and $Al_{24}CH_3^+$ were observed. The activation barrier for CH_4 chemisorption determined from the collision energy threshold was 3.5eV. This is significantly larger than the activation barrier for D_2 chemisorption (2.0eV) even though the D-D and CH_3-H bond energies are very similar. With O_2 the main product was Al_{21}^+ ($+2Al_2O$). No adduct was observed even down to collision energies as low as 0.2eV. Dissociative chemisorption of O_2 on Al_{25}^+ is very exothermic (~10eV). After the O_2 chemisorbs the adduct rapidly dissociates by loss of $2Al_2O$ molecules. However, despite the large reaction exothermicity, there is apparently an activation barrier associated with the chemisorption of O_2. From our data we estimate the barrier to be ~0.5eV. Since the steps and defects are often more reactive than flat surfaces, we might have anticipated that the clusters would be more reactive than bulk surfaces, (because cluster surfaces are virtually all steps and defects). This is apparently not true. Bulk aluminum is very rapidly oxidized and there is no significant activation barrier. But Al_{25}^+ has an activation barrier of ~0.5eV and would only oxidize slowly if exposed to air at room temperature. Both CO and N_2 chemisorb on Al_{25}^+ to give adducts. With CO the activation barrier is 1.9eV. The $Al_{25}CO^+$ adduct readily dissociates by loss of an Al_2O molecule to yield $Al_{23}C^+$. The activation barrier for chemisorption of N_2 on Al_{25}^+ is 3.5eV, substantially larger than for CO even though CO and N_2 are isoelectronic. At energies significantly above the threshold for adduct formation the $Al_{25}N_2^+$ adduct dissociates by loss of Al_3N to give $Al_{22}N^+$.

Activation barriers measured for chemisorption on Al_{25}^+ are given in Table 1. For D_2, CH_4, O_2, CO, and N_2 the activation barriers show a qualitative correlation with the cluster HOMO (highest occupied molecular orbital) → molecule LUMO (lowest unfilled molecular orbital) promotion energy, suggesting that electron donation from the cluster HOMO to the LUMO of the reactant stabilizes the transition state and lowers the activation barrier [7].

The electronic structure of the metal cluster and the strength of the bonds being broken and formed are probably also important in determining the size of the activation barrier.

Experimental data for the reaction between Al_{25}^+ and C_2H_4 are shown in Fig. 3. It appears that three different types of $Al_{25}C_2H_4^+$ adduct are formed over different collision energy ranges [8]. The collision energy thresholds associated with the formation of the adducts are: ~0.0eV, 0.5-1.0eV and 3.0-3.5eV. Similar results were obtained for other cluster sizes. C_2H_4 is known to give rise to a number of different products on bulk surfaces [9] so it is reasonable to assign the different adducts to different structures of the C_2H_4 molecule on the surface of the aluminum cluster. The measured thresholds are thus related to activation barriers for isomerization between the different structures. Unfortunately our experiments provide no direct structural information so we can not say what the structures are.

Reactions of Si_{25}^+ with D_2, CH_4, O_2, CO, N_2, and C_2H_4

Based on the bulk materials we would expect the bonding in Al_{25}^+ to be metallic with delocalized valence electrons, and the bonding in Si_{25}^+ to involve localized directional covalent bonds. It is thus surprising to find that the chemistry of Si_{25}^+ shows some remarkable similarities to that of Al_{25}^+ [10]. In the reaction between Si_{25}^+ and C_2H_4 three different types of adducts are formed, with similar thresholds to the three processes observed with Al_{25}^+. Table 1 shows a comparison between the activation barriers measured for chemisorption of D_2, CH_4, O_2, CO, N_2, and C_2H_4 on Al_{25}^+ and Si_{25}^+. There is clearly a strong correlation between the size of the activation barriers measured for chemisorption on these two clusters. However, one important difference between the chemistry of Al_{25}^+ and Si_{25}^+ is that Si_{25}^+ often undergoes fission processes similar to the processes observed in the dissociation of the bare clusters [11]. For example, in the reactions with CH_4 the main products observed with Si_{25}^+ are $Si_{25}CH_4^+$ and $Si_{15}CH_4^+$ compared with $Al_{25}CH_4^+$, $Al_{24}H^+$ and $Al_{24}CH_3^+$ which are observed with Al_{25}^+. The $Si_{15}C_2H_4^+$ product can be accounted for by loss of Si_{10} from the $Si_{25}C_2H_4^+$ adduct. Si_{10} is known to be a particularly stable cluster.

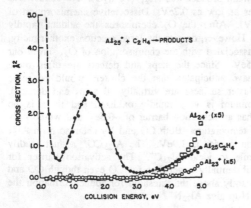

REACTANT	Al_{25}^+	Si_{25}^+
D_2	2.0±0.2	2.0±0.3
CH_4	3.5±0.3	2.9±0.3
O_2	0.5±0.3	0.3±0.3
C_2H_4	~0.0	~0.0
CO	1.9±0.3	3.1±0.5
N_2	3.5±0.3	5.0±0.8

Figure 3. Plot of cross sections against collision energy for the reaction between Al_{25}^+ and C_2H_4.

Table 1. Activation barriers for chemisorption on Al_{25}^+ and Si_{25}^+.

Reactions of Si_n^+ (n=3-24) with C_2H_4

The drift tube configuration was used to study the reactions of Si_n^+ (n=3-24) with C_2H_4 over a temperature range of 143-300K [3]. Si_3^+ dehydrogenates C_2H_4 according to the reaction:

$$Si_3^+ + C_2H_4 \rightarrow Si_3C_2H_2^+ + H_2. \qquad (1)$$

This reaction occurs at close to the collision rate. Clusters with >3 atoms react to form a series of $Si_n(C_2H_4)_m^+$ adducts. Formation of these adducts occurs by an association reaction which under the conditions employed is a two step process:

$$Si_n^+ + C_2H_4 \rightleftarrows Si_nC_2H_4^{+*} \qquad (2)$$

$$Si_nC_2H_4^{+*} + Ne \rightarrow Si_nC_2H_4^+ + Ne. \qquad (3)$$

The Si_n^+ and C_2H_4 collide and form a metastable adduct which either dissociates back to reactants or is stabilized by collisions with the Ne buffer gas. Fig. 4 shows some kinetics data for the reactions of Si_5^+ and Si_{10}^+. The Figure shows a plot of the log of the relative abundance of the reactant ion against C_2H_4 pressure in the drift tube. The data for Si_5^+ shows a simple exponential decay which is expected for pseudo first order kinetics. The data for Si_{10}^+, however, shows non-exponential behavior. These results suggest that there are two components of Si_{10}^+ which react at different rates. The line in Fig. 4 is a simulation where we assume that there are two components of Si_{10}^+: the dominant component (85%) which reacts at a rate of $1.6 \times 10^{-10} cm^3 s^{-1}$ and a minor component (15%) which reacts at 1/4 of the rate of the dominant component. The two components of Si_{10}^+ are probably different structural isomers. In their recent calculations Raghavachari and Rohlfing [1] found two low energy structures for Si_{10}: a tetracapped octahedron and a tetracapped biprism.

Fig. 5 shows a plot of the rate constants for the adsorption of C_2H_4 on Si_n^+ (n=4-24) at room temperature and 143K. It is clear that there are enormous changes in the reactivity with cluster size. Si_{13}^+ and Si_{14}^+ are particularly inert. At the lower temperature all the rate constants have increased and a significant fraction of the clusters now react at the collision rate. There are two reasons why the overall association reaction (Eqns. 2 and 3 above) can

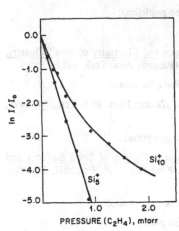

Figure 4. Plot of the relative abundance of the reactant ion against C_2H_4 pressure.

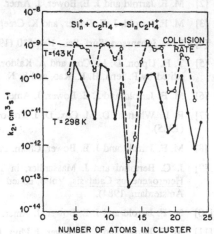

Figure 5. Plot of the rate constants for chemisorption of C_2H_4 on Si_n^+ for n=4-24 at 143K and 298K.

occur at less than the collision rate (sticking probability < 1). Either there is an activation barrier associated with chemisorption of C_2H_4 on Si_n^+ or a significant fraction of the $Si_nC_2H_4^{+*}$ metastable adduct dissociates before it can be stabilized. The observed temperature dependence of the reactions indicates that there is no significant activation barrier for chemisorption of C_2H_4 on Si_n^+. So the measured rate constants reflect the lifetimes of the $Si_nC_2H_4^{+*}$ metastable adducts. Two main factors influence the lifetimes: the number of internal degrees of freedom and the binding energy of the C_2H_4 molecule to the cluster. Thus the large oscillations in the rate constants shown in Fig. 5 reflect changes in the binding energy of the C_2H_4 molecule to the Si_n^+ cluster and the reason Si_{13}^+ and Si_{14}^+ appear to be particularly inert is that C_2H_4 is particularly weakly bound to these clusters. Statistical phase space theory can be used to model the dissociation of $Si_nC_2H_4^{+*}$ and provide an estimate of the Si_n^+-C_2H_4 binding energies. For Si_4^+-Si_{10}^+ the binding energies were found to be 0.8-1.9eV - too large to be accounted for by ion-dipole interactions and too small to be due to strong di-σ bonding. So the interaction between Si_n^+ and C_2H_4 is probably best described as a π bonding interaction. Si_4^+, Si_6^+ and Si_{10}^+ are known to be particularly stable clusters but there is nothing special about these clusters in Fig. 5. Apparently cluster stability does not influence the Si_n^+-C_2H_4 binding energies.

CONCLUSIONS

In this article we have described the results of several recent studies of the chemistry of size selected aluminum and silicon clusters. The chemistry of clusters in the 3-27 atom size range changes dramatically with cluster size, and surprisingly the larger clusters studied appear to be less reactive than the bulk materials. Many aspects of the reactivity of the clusters are not yet well understood.

REFERENCES

[1] K. Raghavachari and C. M. Rohlfing, J. Chem. Phys., 89, 2219 (1988).

[2] M. F. Jarrold and J. E. Bower, J. Amer. Chem. Soc., 110, 70 (1988).

[3] M. F. Jarrold, J. E. Bower, and K. Creegan, (to be published).

[4] P. K. Johansson, Surf. Sci., 104, 510 (1981).

[5] T. H. Upton, D. M. Cox, and A. Kaldor, in Physics and Chemistry of Small Clusters, edited by P. Jena, B. K. Rao, and S. N. Khanna (Plenum, New York, 1987).

[6] M. F. Jarrold and J. E. Bower, J. Amer. Chem. Soc., (in press).

[7] R. L. Whetten, D. M. Cox, D. J. Trevor, and A. Kaldor, Phys. Rev. Lett., 54, 1494, (1985).

[8] M. F. Jarrold and J. E. Bower, Chem. Phys. Lett., (in press).

[9] J. C. Bertolini and J. Massardier, in The Chemical Physics of Solid Surfaces and Heterogeneous Catalysis, Vol 3, edited by D. A. King and D. P. Woodruff, (Elsevier, Amsterdam, 1984).

[10] M. F. Jarrold and J. E. Bower, J. Amer. Chem. Soc., (in press).

[11] M. F. Jarrold and J. E. Bower, J. Phys. Chem., (in press).

METAL ATOM ROUTES TO METAL-BASED CLUSTERS IN POLYMERS

MARK P. ANDREWS, MARY E. GALVIN, AND SHARON A. HEFFNER
AT&T Bell Laboratories, 600 Mountain Avenue, Murray Hill, NJ 07974

ABSTRACT

Past syntheses of polymer composites have largely evolved from chemical reduction or thermal decomposition of organometallic or inorganic precursor molecules in polymers, or plasma and thermal co-deposition of metal vapors and carbonaceous free radicals. Our approach involves the site-specific capture of metal atoms deposited in vacuum to give isolated, high energy mononuclear organometallic centers within a polymer film. These centers can be converted at ambient or sub-ambient temperatures (ie, below the polymer glass transition temperature) to, for example, metal oxide microclusters.

We describe the results of our studies of a prototypical system involving chromium atoms and their conversion to corundum-type oxide microclusters in arene-functionalized polymer films. Thus Cr was deposited into 150 K liquid tetrahydrofuran solutions of polystyrene or poly(styrene-isoprene-styrene) triblock, spun *in vacuo* as thin films on the surface of a rotating glass cryostat. Evidence from epr spectrscopy shows that the resulting polymer-anchored (inter/intra-chain) bis(arene)Cr sandwich complex is locally mobile in the macroscopically rigid film at room temperature. The Cr atom is discharged from the rings by subsequent reaction with oxygen diffused into the film. Although α-Cr_2O_3 is a classic two-sublevel antiferromagnet that is not epr active above 308 K, we observe an intense signal even at 77 K in these films. Cr_2O_3 microclusters are indicated, and these are confirmed by *in situ* measurements of the oxidation and aggregation process.

The metal atom methodology has also been used to synthesize silver microsphere/polymer composites. With quadratic electrooptic phase modulation, these composites were found to show a third order susceptibility enhanced by coupling the dipolar surface plasmon mode of the particles with incident light.

INTRODUCTION

Other lectures in this Symposium have provided some elegant demonstrations of the reactivity and unusual electronic structure of gas phase cluster molecules. For materials applications, these studies also raise questions as to whether or not cluster science will mature into its own (quantum) technology, be subsumed under the developing cluster ion beam processing technologies, or contribute in more subtle ways by mediating our understanding of the role that size quantization plays in electron transport, magnetic, and optical processes. Although cluster science is an emerging discipline, such questions are only premature if we insist on extrapolating *directly* from the gas phase to current trends in technology, or linking too literally, technology's material manifestations in processing and devices. Cluster science may ultimately drive new discoveries, conceptual and material, in areas like structural composites (fibers, whiskers), nanocomposites,[1] intermetallics, electro-optics, high temperature ceramic superconductors, microstructure processes and properties, protective films and corrosion.

In the short term, one approach to utilizing clusters is to explore their properties more or less compounded with those of a host, such as a polymer. Of course, certain information available from a cluster in the perturbation-free environment of a vacuum is complicated by the presence of interfaces in the condensed phase. Nevertheless, polymers are attractive hosts because, under appropriate conditions, they can supply the kinetic barrier necessary to prevent diffusion and aggregation of the highly reactive cluster species.[2] As the chemistry of polymer molecules achieves a sophistication sufficient for us to develop molecular architectures by chemical design, we can envisage locating clusters in spatially well defined ways within the polymer microstructure. This raises the possibility of fabricating composites ("crystals") with desirable anisotropies. Moreover, the fact that many polymers are processible satisfies an important condition for device fabrication.

In this article we report new results of the use of metal atoms to synthesize molecular clusters, sub-colloids and colloids in organic, polymeric media. The use of metal atoms as reagents for chemical synthesis is well established.[3] The chemistry of transition metal atoms in polymer science has been the subject of a review,[4] where the first experiments that yielded polymer-supported π-sandwich complexes of a variety of transition metals are described. These complexes consist of a single metal atom sandwiched between two co-planar arene (benzene derivative, ϕ) rings. Bonding is accomplished through a synergic "push-pull" flow of charge density between the d-orbitals of the metal and the π-orbitals of the rings. Similar compounds now form the point of departure for the work reported in this paper.

Atomic chromium was chosen for initial study because of the ease with which it sublimes, as compared with the more vigorous evaporation conditions required to generate atomic Ti, V, Mo, Nb, Ta or W. This makes feasibility studies easier to conduct. Chemical selectivity is ensured because the metal atom undergoes a low activation energy, simple orbital mixing process yielding a sandwich complex (Scheme I). The resulting chromium/polymer system in this instance should be viewed as a dispersion of atoms interacting strongly with the support (phenyl substituents), but poised for subsequent conversion in the macromolecular host. The polymer thus carries the latent reactivity of the atom in a high energy state. Chromium also carries useful spectroscopic information about its location in the macromolecular host. The oxidized, polymer-bound $\phi_2 Cr^+$ complex is paramagnetic and epr active at room temperature and below. Cr^+ therefore spin-labels the polymer. Consequently, information can be acquired concerning the degree of dispersion of the chromium atoms in the polymer, their state of aggregation, and local polymer chain dynamics. Lastly, oxidation under mild conditions (room temperature, 1 atm air) converts the chromium to clusters of Cr_2O_3 in solid films of the polymer. Bulk Cr_2O_3 is diamagnetic below its Neel temperature (308 K); however, minute clusters of the material are paramagnetic and can therefore be detected by epr spectroscopy.

We conclude with an application of the metal atom methodology to prepare silver microsphere/polymethylmethacrylate composites. These experiments demonstrate for the first time a useful procedure for synthesizing stable liquid organosols of Ag from which solid polymer films can be cast. Spatial confinement in the small silver particles modifies the nonlinear optical properties of silver and confers a resonantly enhanced third order susceptibility on the composite medium.

EXPERIMENTAL

All materials were intitially prepared by quantitative resistive evaporation of metal atoms into tetrahydrofuran (THF) solutions of the target polymer (2-5 % w/v), spun *in vacuo* at 145 K in a rotating glass cryostat.[3] Polystyrene, poly(styrene-isoprene-styrene) triblock and polymethylmethacrylate (PMMA) were purchased from Aldrich chemical. The triblock is 35% styrene, 38%*cis*-1,4-isoprene, 17% *trans*-1,4-isoprene and 10% 3,4-isoprene, as determined by ^{13}C NMR. The triblock molecular weight was determined against a polystyrene standard to be M_n = 27000, with a polydispersity of 1.3. The polystyrene and PMMA molecular weights were, respectively, 200,000 and 172,500. All solvents were distilled from living polystyrene anion and stored under argon. Manipulations of materials were performed anaerobically, as required. Details regarding the synthesis and analysis of these and similar materials can be found elsewhere.[5]

Nascent silver particles were formed by condensation of atomic silver into 200 ml volumes of 2-5% w/v solutions of PMMA in THF, cooled to 150 K in an evacuated rotating glass cryostat. Particles grow to a maximum size of 200 Å by diffusion and aggregation in the liquid organic medium. Typically, enough metal is evaporated to give metal loadings in the PMMA of less than 0.1%. The resulting golden-yellow silver organosol is filter-cannulated through a column of Celite A (4x5 cm) at 150 K under an argon back pressure. A 500 ml Schlenk flask maintained at 200 K acts as a receiver. Solvent is removed via a rotary evaporator over a period of 15 min, while the flask warms gradually to 283 K. An intensely colored orange to red-orange solid polymer film eventually deposits on the walls of the Schlenk tube. The PMMA/Ag composite is removed from this container in an argon-filled (< 5 ppm O_2) dry-box. The composite can be reversibly dissolved in THF with no evidence of decomposition or aggregation of the metal particles. Prolonged dissolution in propylene glycol methyl ether acetate (PGMEA) causes gradual aggregation and precipitation of massive silver particles.

Electronic absorption spectra were collected from solid films or solutions of the composite in organic solvents. For the third order measurements discussed in this paper, a robust sample cell was made. Solid PMMA/Ag composite is dissolved in PGMEA to give a final concentration of 15% w/v.[6] An aliquot of the solution is filtered by syringe through a 0.2 to 4.0 μm membrane and deposited onto a glass plate patterned with transparent indium-tin-oxide (ITO) electrodes. A 0.5 μm thick film is spun from this solution and baked at 373 K to remove PGMEA. Two square sections are cut from the plate, overlapped and compressed in an oven programmed to cycle the sample through a temperature ramp from 298 K to well above the glass transition of the polymer (403 K). This thermal treatment bonds the ITO plates into a sandwich, providing an hermetic environment for the polymer-metal composite.

The third order nonlinear optical susceptibility was determined with electrooptic phase modulation. The sample is placed in one arm of a Mach-Zehnder interferometer with the film perpendicular to the laser beam. The electric field polarization is in the plane of the film while the modulating field is applied with the transparent ITO electrodes perpendicular to the film plane. The experimental layout has been previously discussed.[7] The phase modulation results in an intensity modulation of the light out of the interferometer. An apertured silicon detector collects the light and the output is passed through a lock-in amplifier. A computer controlled interface is used to measure the amplitude of the modulated beam at both the modulating frequency and at twice the modulating frequency as a function of phase difference between the two arms of the interferometer. The relationship between the measured interferograms and the third order susceptibility is described in detail elsewhere.[8]

RESULTS AND DISCUSSION

Electronic Absorption and EPR Spectroscopy of Chromium Atoms Supported on Polystyrene and Poly(styrene-isoprene-styrene) Triblock

The unsaturation present in the isoprene portion of the triblock is potentially troublesome because chromium atoms are known to react with monoenes and alkadienes.[9] We observed no tendency for Cr to attack the C=C bonds in the styrene-isoprene triblock. The formally zerovalent polymer-supported bis(η^6-arene)Cr (d^6) complex has a singlet A_1 electronic ground state. The electronic structure of this kind of sandwich molecule has been fully discussed in the literature.[10] [11] The electronic absorption spectrum of the polymer-bound bis(arene)Cr compound in the region of the $^1A_1 \leftarrow {}^1A_1$ transition at 318 nm agrees with published data.[10] Mild oxidation of the complex results in the removal formally of one electron, giving a compound of $^2A_1$ symmetry. The unpaired electron responsible for the observed epr signal in Figure 1 is confined primarily to the Cr d_z^2 orbital. Partial spin delocalization gives rise to proton hyperfine interactions which can be seen in both the liquid and glassy state. The satellite lines at low field (Figure 1(a)) are due to the ^{53}Cr isotope. The g-values and hyperfine constants (see Figure 1) are identical to those reported for other mono-substituted bis(arene)Cr$^+$ molecules.[12]

There is important information to be gained through a qualitative reading of the spectra in Figure 1. We turn our attention first to spectrum (a) for bis(arene)Cr$^+$ supported on homopolystyrene in solution. The polymer contains 2% Cr w/w bound up in the complex. By definition the formation of such a complex involves intra- and/or interchain crosslinking through the intermediary of the phenyl substituents (Scheme 1). The fact that the signal is not exchange-narrowed indicates that the Cr$^+$ centers are not aggregated in the polymer. The signal is also characteristic of a bis(arene)Cr$^+$ complex in the rapid tumbling regime where the anisotropic **g** tensor and hyperfine coupling interaction are almost averaged. In spite of the inter-/intramolecular crosslinks, the probe ion is substantially free to move. It is rather surprising that a macromolecular system, crosslinked to the extent of 2%, appears soluble. A possible explanation is that a sufficient portion of the crosslinking is intramolecular in nature. This spectrum should be contrasted with that in Figure 1 (b) for a solution of styrene-isoprene triblock. Here the signal is modulated by motional damping. The Cr(d^5) centers are clearly sensing the more restricted chain motions. The triblock of styrene-isoprene is also loaded

SCHEME 1

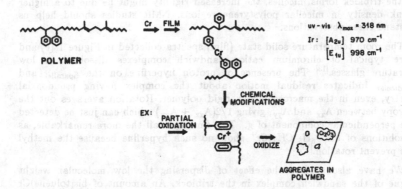

POLYMER

Cr FILM

uv–vis λ_{max} = 318 nm

Ir: $[A_{2u}]$ 970 cm^{-1}
$[E_{1u}]$ 998 cm^{-1}

CHEMICAL MODIFICATIONS

EX: PARTIAL OXIDATION

Cr$^+$

OXIDIZE

AGGREGATES IN POLYMER

Polystyrene

<g> = 1.9856 ± 0.0005
a_H = 3.6 ± 0.1

$g_\parallel$ = 2.0030 ± 0.0005
$g_\perp$ = 1.9764 ± 0.0005
a_{zH} = 3.1 ± 0.1 G

(a)

(c)

Cr$^+$

Poly(styrene–isoprene–styrene) Triblock

<g> = 1.9856 ± 0.0005
a_H = 3.6 ± 0.1

$g_\parallel$ = 2.0028 ± 0.0005
$g_\perp$ = 1.9764 ± 0.0005
a_{zH} = 3.1 ± 0.1 G

(b)

(d)

LIQUID SOLUTION

SOLID FILM

20 G
H

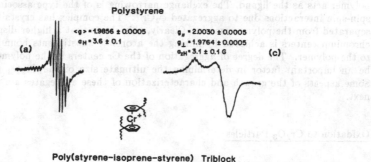

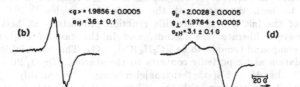

Fig. 1. Liquid phase (a, b, in THF) and solid phase (c, d, film) epr spectra of polymer-bound (arene)$_2$Cr$^+$ complexes prepared from atomic Cr and the homo- or triblock polymer. All spectra were collected at room temperature.

with approximately 2% Cr. The effective Cr concentration must, however, be higher because of the lower proportion of styrene monomer units in the chains. Since the triblock forms micelles, the increased rigidity might be due to a higher crosslink density in micellar polystyrene regions. NMR studies should help us understand these observations.

The room temperature solid state (film) spectra collected in Figure 1 (c) and (d) are typical of chromium cation sandwich complexes dissolved in low temperature glasses.[13] The presence of proton hyperfine on the $g_{parallel}$ and $g_{perpendicular}$ indicates residual motion about the complex having pseudoaxial symmetry, even in the macroscopically rigid polymer. Rotation averages out the anisotropy between A_{xx} and A_{yy}, giving $1/2(A_{xx} + A_{yy})$ which can just be detected on the perpendicular component of g. This finding is all the more remarkable, as solid solutions of bis$(C_6H_5$-$CH_3)Cr^+$ show no such hyperfine because the methyl groups prevent rotation.[13]

We have also studied the effect of dispersing the low molecular weight analogue of the sandwich complex in the triblock. An amount of bis(toluene)Cr equivalent to 2% w/w Cr was co-dissolved with the polymer in toluene. A film of the polymer containing the dispersed complex was obtained by casting from solution and allowing the solvent to evaporate undisturbed in an argon-filled drybox. Portions of the film were cut into 1 mm x 10 mm filaments, packed into an epr tube, and sealed under argon. The sample was initially diamagnetic. Mild oxidation of the film produced a paramagnetism detectable as a Lorentzian line (Figure 2), indicating the presence of an exchange interaction. The signal displayed in Figure 1 differs strikingly from that exhibited by the sample in which the polymer acts as the ligand. The exchange narrowing is of the type associated with spin-spin interactions due to aggregated ϕ_2Cr^+. The complex has crystallized and separated from the polymer matrix. Clearly, this shows that a higher dispersity of chromium centers is achieved by linking the atoms to substituents bound directly to the polymer. The degree of aggregation of the Cr centers in the polymer should be an important factor in determining the ultimate size of the Cr_2O_3 particles. Some aspects of the growth and characterization of these aggregates are taken up next.

Oxidation to Cr_2O_3 Particles

The oxidation of bis(arene)Cr compounds has been reported to be first order with respect to both dioxygen and the sandwich complex;[14] [15] however, identification of the initial organometallic reaction products is at best sketchy. Exposure to oxygen liberates the ligands, and in the case of bis(benzene)Cr, precipitates a compound thought to be $[(C_6H_6)Cr]_2CrO_4$. This material is unstable to further oxidation and reportedly converts to the species, $5Cr_2O_4.2C_6H_6$.[15] On the other hand, bis(ethylbenzene)chromium apparently deposits $[(C_6H_5C_2H_5)_2]CrO_4$ on reaction with O_2. Likewise, this molecule is unstable to further oxidation, but the final product is unknown. Our own data gathered from extensive epr studies, infrared and electronic absorption spectroscopy, indicate that the polymer supported complexes convert to Cr_2O_3. This is similiar to studies which showed that polymer-supported cyclopentadienyl Fe complexes will form iron oxide on exposure to O_2 whereas the small molecule analogs will not.[16]

Chromium incorporated as Cr^{3+} in clusters of point defect Cr_2O_3 undergoes changes in its electromagnetic properties as the volume of the corundum crystal

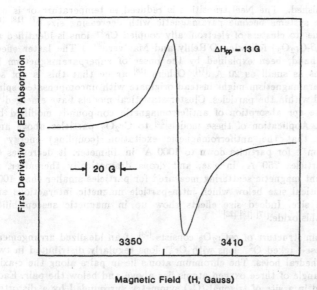

Fig. 2. Epr spectrum of bis(toluene)Cr$^+$ in a film of poly(styrene-isoprene-styrene) triblock at room temperature. Bis(toluene)Cr was dispersed in the polymer to give a concentration of 2 weight % in Cr. The exchange-narrowed signal results from aggregation of bis(toluene)Cr molecules in the polymer matrix. Compare Figure 1 c and d.

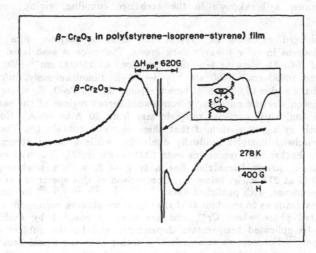

Fig. 3. Epr signal from minute β-Cr$_2$O$_3$ particles in a poly(styrene-isoprene-styrene) film. The spectrum was recorded 30 K below the Neel temperature of bulk chromia.

structure is diminished. The Neel transition is reduced in temperature or is not observed, and the systems become paramagnetic with decreasing size.[17] [18] [19] (The epr signal due to clusters of electronically coupled Cr^{3+} ions is identified as the β-resonance (β-Cr_2O_3) following O'Reilly and MacIver.[20]) The latter effect has, on the one hand, been explained by the onset of superparamagnetism in individual particles as small as 20 $\overset{\circ}{A}$.[17] Others [18] argue that this is not so, suggesting that paramagnetism might instead originate with uncompensated spins at the surface and within the particles. Cluster statistical models have emerged[19] [21] to describe the epr absorption of antiferromagnetic compounds modified by magnetic dilution. Application of these models[22] to Cr_2O_3 particles above and below T_N shows that the antiferromagnetic excitation (coupling) energy is constant at 430 cm^{-1} for particles down to 1000 $\overset{\circ}{A}$ in diameter. It decreases to 260 cm^{-1} for particles 750 $\overset{\circ}{A}$ in size, and drops dramatically thereafter. A decrease in coherent magnetic scattering measured for particles smaller than 1000 $\overset{\circ}{A}$ indicates a critical size below which intra-particle magnetic interactions are affected by grain size. Indeed size effects show up in magnetic susceptibility measurements of this oxide.[17] [18] [23]

The corundum structure of α-Cr_2O_3 consists [24] of an idealized arrangement of hexagonally close-packed O^{2-} ions with Cr^{3+} ions regularly distributed in two thirds of the octahedral holes. The chromium atoms lie in pairs along the c-axis. An equilateral triangle of three oxygen atoms lies above and below the pair. Each Cr atom is located in a site of trigonal (C_3) symmetry surrounded by a distorted octahedron of oxygen atoms. Ideally, every cation is coupled to 13 others through the intervening octahedra. The largest superexchange interactions occur between the Cr-Cr pairs lying along the c-axis. [25] Direct t_{2g} orbital overlap results in strong antiferromagnetic coupling between them. Weaker exchange couplings are experienced throughout the remainder of the lattice. Vacancies in any of the Cr positions will cause a breakdown in the exchange coupling, giving rise to paramagnetism.

After prolonged exposure to oxygen, the triblock polymer film with complexed Cr deepens in color towards dark green. The color is associated with the formation of β-Cr_2O_3 showing tell-tale absorptions at 25 000 cm^{-1} (400 nm, $^4T_2 \leftarrow {}^4A_2$) and 16950 cm^{-1} (590 nm $^4T_1 \leftarrow {}^4A_2$). Simultaneously, infrared bands emerge in the region of 650 cm^{-1} (broad, strong) due to Cr-O, E_u stretching modes. Transmission electron microscopy from two different regions of the sample reveals that the individual particles range in size from 20 $\overset{\circ}{A}$ to 90 $\overset{\circ}{A}$. Energy dispersive analysis by x-rays confirmed that these particles contain Cr. The epr signal for the sandwich complex gradually diminishes while a broad absorption emerges from the background, resonating over 4000 G (Figure 3). The absorption consists of a single, nearly Lorentzian line near g = 2, with a peak-to-peak linewidth of 620 G at 278 K, ie, below T_N. Comparison of this spectrum, and its temperature dependence, with published data [17] [18] [20] [21] [22] [23] [26] for minute β-Cr_2O_3 particles, leads us to conclude that the signal we observe corresponds to a Cr^{3+} concentrated phase where Cr^{3+} ions are strongly coupled by exchange interaction. Its complicated temperature dependence will be the subject of a forthcoming paper.[27] For now we summarize by noting that the signal broadens and weakens somewhat as the temperature is lowered. This indicates that a portion of the resonance does not occur in the ground state, but in an excited state whose population decreases (but only slightly in this instance) with decreasing temperature. This is the case for Cr^{3+} ions coupled antiferromagnetically by exchange interaction.[20] Clusters of Cr_2O_3 which are antiferromagnetic will, of course, not be detected by epr.

Silver Microsphere/Polymer Composites

The electron density profile and the breaking of translational symmetry at a particle surface have decisive effects on the optical properties of sub-100 $\overset{\circ}{A}$ (Rayleigh limit) metal particles. It is now known that the cubic nonlinear dielectric susceptibility of the effective medium (metal spheres diluted in a dielectric host) is resonantly enhanced when the dielectric constant of the host and metal spheres obey the condition $Re[\epsilon_{sphere}(\omega)] = -2\epsilon_{host}$. This relation defines the dipolar surface plasmon mode in small metallic spheres. Here, both the metal nonlinearity and the dielectric host nonlinearity near the metal inclusions dominate the material response. To date experiments with colloids have been confined to aqueous solutions and a precipitate glass.[28] Further study of the phenomenon, with a view to evaluating its potential in implementing some device ideas, would be helped by immobilizing the particles in a suitable host.[29] Andrews and Kuzyk [29] showed that sub-100 $\overset{\circ}{A}$ microspheres of silver could be grown from silver atoms deposited into 145 K solutions of PMMA in THF. Solid films of the composite could be spun in thicknesses ranging from 1 to 1000 μm on conventional substrates. The measurement of the third order susceptibility uses Mach-Zehnder interferometry, which requires high electric fields to modulate the refractive index of the sample. Since conventional methods for producing silver colloids by chemical reduction generate adventitious charged species, such procedures cannot be used to produce the composites. The intrinsically cleaner metal vapor methodology avoids this problem by giving liquid, silver organosols from which Ag/PMMA composites can be obtained directly. PMMA was selected because it is optically transparent in the region of interest in this experiment, it is commonly used to make optical devices, and it is a macroscopically rigid host suitable for stabilizing silver particles against agglomeration.

Figure 4 shows a linear absorption spectrum of the silver spheres embedded in a thin film of the polymer. The absorption profile is characteristic of the dipolar surface plasmon mode of small silver particles in a dielectric host.[30] The arrow positioned at 633 nm locates the frequency of the He-Ne laser used in the electrooptic modulation experiment. Transmission electron microscopy of films containing 0.045 weight % silver revealed that the microspheres collect into colonies of 200-500 non-overlapping spheres, some of which exhibit dark bands from twin dislocations and/or stacking faults. The colonies are separated over distances greater than a thousand angstroms. The mean particle size before processing between the ITO patterned plates is 79 $\overset{\circ}{A}$ with a standard deviation of 36 $\overset{\circ}{A}$. After processing we observe that both the average particle size and also the number density of particles have increased. The colloid is now more evenly distributed throughout the PMMA matrix. This suggests diffusion and growth of spheres smaller than 20 $\overset{\circ}{A}$ in diameter (not visible under the microscope), when the medium is annealed above the polymer glass transition temperature.

The film thickness in the sandwich was $d = 6.8 \times 10^{-6}$m and the light intensity at the detector at twice the modulating frequency ($\Omega = 4.0$kHz) is shown as a function of phase difference in Figure 5 for an R.M.S. modulating voltage of $V_{RMS} = 83.4$V. The arrows show the phase difference where the modulating efficiencies are expected to be the greatest as determined from the phase difference dependence of the light output from the interferometer with no modulation.

The measured value of the quadratic electrooptic coefficient of the composite is $s_{1133} = 2.0(\pm0.5) \times 10^{-22}$m^2/V^2. A similar measurement of pure PMMA shows that the coefficient is smaller than $s_{1133} = 4.2(\pm0.5) \times 10^{-23}$m^2/V^2. The laser

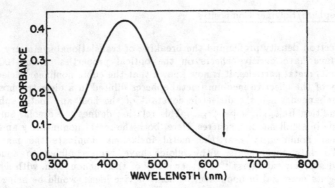

Fig. 4. Linear absorption spectrum in the region of the dipolar surface plasmon mode for small silver particles embedded in PMMA. The film was cast from THF onto sapphire. The arrow marks the region of the absorption spectrum probed by the He-Ne laser in the electrooptic experiment.

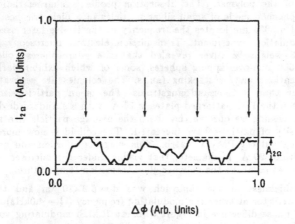

Fig. 5. Electrooptic modulation amplitude as a function of phase difference between the two arms of the interferometer. The arrows show the positions where the modulation efficiency is expected to be maximum, as determined from the phase difference dependence of the interferometer without modulation.

wavelength of both measurements was at $\lambda = 633$nm, located away from the absorption maximum at $\lambda = 435$nm (see Figure 4). Although the enhancement is expected to be much smaller here than for a resonant measurement, the third order susceptibility of the composite at low levels of loading (0.045 weight % or a volume fraction of 4.5×10^{-4}) was at least 5 times larger than that of undoped PMMA.

The third order susceptibility calculated from the quadratic electrooptic coefficient is $\chi^{(3)}_{1133} = 1.6 \times 10^{-14}$esu. Using optical phase conjugation techniques,[28] a third order susceptibility, enhanced at the maximum of the plasmon absorption, was determined to be $\chi^{(3)}_{1133} = 2.4 \times 10^{-9}$ esu. The enhancement factor, as determined by the quotient of the phase conjugation measurement and our measurement, is 8.5×10^{5} when the difference in concentration between the two samples is taken into account. This compares favorably with the predicted enhancement factor of 3.6×10^{6}.[28] It is currently thought that the nonlinear response is dominated by the electrons in the metal particles.[29] [28]

CONCLUSIONS

We have shown that Cr atoms can be attached selectively to the pendant phenyl substituents of polystyrene and a poly(styrene-isoprene-styrene) triblock by vaporizing the metal into liquid solvent solutions of the polymer in a rotating cryostat. Evidence from epr spectroscopy shows that the atoms are anchored to the phenyl substituents in the form of bis(arene)Cr complexes, and that they do not interact electronically. We have demonstrated that at room temperature the chromium atom is discharged from the rings when oxygen is diffused into solid polymer films. Although Cr_2O_3 is a classic two-sublattice antiferromagnet that is not epr active below its Neel temperature (308 K), we observe an intense signal even at 77 K. β-phase Cr_2O_3 is indicated, and this is confirmed by TEM measurements of the aggregates.

We have introduced a new technique for making polymer/Ag colloid composites directly from atomic silver. These composites are in a form that is convenient for leisurely spectroscopic examination and fabrication into thin films. The composite shows a third order susceptibility, measureable by quadratic electrooptic modulation.

ACKNOWLEDGMENT

We thank Debbie Fish for help in sample preparation and T. T. Sheng for transmission electron microscopy (Ag/PMMA studies). Stimulating discussions with A. M. Lyons, M. A. Marcus and C. W. Dirk deepened our curiosity in the physical properties of minute structures.

REFERENCES

1. See Symposium G, Materials Research Symposium meeting, Boston, Nov. 1988.

2. M. P. Andrews and G. A. Ozin, J. Phys. Chem., **90**, 2929 (1986).

3. M. P. Andrews, "Using Metal Atoms and Molecular High Temperature Species in New Materials Synthesis", in Experimental Organometallic Chemistry, A. L. Wayda and M. Darensbourg, eds., ACS Symposium Series **357**, ch. 7, p. 158 (1987); M. P. Andrews and G. A. Ozin, Chemistry of Materials,in press.

4. M. P. Andrews, in Encyclopedia of Polymer Engineering and Science, 2nd ed., vol.9 (John Wiley and Sons, New York, 1987).

5. M. P. Andrews, M. E. Galvin and S. A. Heffner, J. Am. Chem. Soc., submitted for publication.

6. Improved colloid stability is achieved by first dissolving the Ag/PMMA in THF. PGMEA (to give a 35:65 v/v THF:PGMEA ratio) is then added at the final step, and mixed vigorously. Films must be spun immediately from this solvent combination.

7. K.D. Singer, M.G. Kuzyk, W.R. Holland, J.E. Sohn, S.J. Lalama, R.B. Comizzoli, H.E. Katz and M.L. Schilling, Appl. Phys. Lett., to be published.

8. M.G. Kuzyk and C.W. Dirk, unpublished.

9. P. S. Skell, D. L. Williams-Smith and M. J. McGlinchey, J. Am. Chem. Soc., **95**, 3337 (1973).

10. Andrews, M.P.; Mattar, S.M.; Ozin, G.A.O. *J. Phys. Chem.* **1986**, *90*, 1037; Cloke, F.G.N.; Dix, A.N.; Green, J.C., Perutz, R.N.; Seddon, E. *Organometallics* **1983**, *2*, 1150; Weber, J.; Geoffrey, M.; Goursot, A.; Penigault, E. *J. Am. Chem. Soc.* **1978**, *100*, 3995; Wittmann, G. T.W.; Krynauw, G. N.; Lotz, S. Ludwig, W. *J. Organomet. Chem.* **1985**, *293*, C33.

11. K. D. Warren in: Structure and Bonding, **27**, 45 (1976).

12. Ch. Elschenbroich, R. Mockel, U. Zennneck, and D. W. Clack, Ber. Bunsenges. Phys. Chem. **83**, 1008 (1979).

13. R. Prins, and F. J. Reinders, Chem. Phys. Lett. **3**, 45 (1969).

14. Y. Aleksandrov, V. M. Fomin, and A. V. Lunin, Kinetics and Catalysis **531**, 531 (1975).

15. V. M. Fomin, Y. A. Aleksandrov, and V. A. Umilin, J. Organomet. Chem. **61**, 267 (1973).

16. L. F. Nazar, Ph.D. Dissertation, University of Toronto, 1985.

17. K. G. Srivastava and R. Srivastava, Nuovo Cimento **39**, 71 (1965).

18. L. Pintschovius and W. Gunsser, Zeitsch. Phys. Chem. **100**, 83 (1976).

19. K. Drager, Z. Naturforsch. **38a**, 1223 (1983).

20. D. E. O'Reilly and D. S. MacIver, J. Phys. Chem. **66**, 276 (1962).

21. K. Drager, Ber. Busenges. Phys. Chem. **79**, 996 (1975); K. Drager and R. Gerling, Phys. Stat. Sol. **38a**, 547 (1976).

22. K. Drager and R. Gerling, Surf. Sci. **106**, 427 (1981).

23. A. Ellison and K. S. Sing, J. Chem. Soc. Faraday Trans. **74**, 2807 (1978).

24. R. E. Newnham and Y. M. de Haan, Z. Kristall. **117**, 235 (1962).

25. F. S. Stone and J. C. Vickerman, Trans. Faraday Soc. **67**, 316 (1971).

26. C. P. Poole, W. L. Kehl and D. S. MacIver, J. Catal. **1**, 407 (1962).

27. Andrews, M. P.; Galvin, M. E.; Heffner, S. A. to be published.

28. F. Hache, D. Ricard and C. Flytzanis, J. Opt. Soc. Am. B. **3**, 1647 (1986); D. Ricard, Ph. Roussignol and C. Flytzanis, Opt. Lett., **10**, 511 (1985).

29. M. P. Andrews and M. J. Kuzyk, Appl. Phys. Lett., submitted for publication.

30. U. Kreibig in Contribution of Clusters Physics to Materials Science and Technology, J. Devenas and P.M. Rabette, eds., NATO ASI Series **104**, Martinus Nijhoff Publ., Dordrecht (1986).

I. Novel Precursors
Compound Semiconductors

SYNTHESES OF METAL CHALCOGENIDES USING ORGANOMETALLIC METHODS

MICHAEL L. STEIGERWALD
AT&T Bell Laboratories, 600 Mountain Avenue, Murray Hill,
New Jersey, 07974

ABSTRACT

The precursor method is being used increasingly in the preparation of solid state inorganic materials. Use of this general method allows the isolation of otherwise inaccessible phases and the preparation of known phases under much milder conditions. One approach which holds promise is the use of organometallic precursors for the preparation of both thin films and bulk samples of inorganic materials. In this paper I describe our syntheses of several metal chalcogenides from organometallic reagents.

INTRODUCTION

The synthesis of inorganic solid state compounds is most typically achieved by combination of the proper stoichiometries of the elements as solids [1]. To insure complete interdiffusion of the solid reagents these reactions are usually conducted at high temperature. This can be a severe limitation. Lower processing temperatures allow the preparation of metastable phases and give wider latitude in the fabrication of complicated physical structures (heterostructures, quantum wells, etc.) which are unstable at higher temperatures. In part to avoid such harsh reaction conditions there has been increasing interest in the use of precursor methods [2]. The general technique here is not to combine the elements, but rather to combine molecular precursors to the elements such that when the precursors are heated or otherwise chemically treated the ancillary components are removed as the solid state compound is formed. Since the elements are intimately mixed in the "molecular' stage of the process the problem of the interdiffusion of the elements is removed and the low temperature synthesis is facilitated.

Organometallic molecular compounds have features which make them attractive as precursors to solid state compounds. The most important of these is that they are the elements masked in molecular form, which can be dispersed molecularly in a variety of innocent solvents. This obviates elemental interdiffusion since suitable precursors for each element can be dispersed in the same solvent. When the solvent is subsequently removed the different precursors are left mixed on the molecular level. A second advantage is that there exists an extensive list of ligand systems in organometallic chemistry. The creative use of these ligand systems will allow molecular control over processing. As an example, with the proper choice of ligands precursor compounds can be made volatile. This is crucial for the application of organometallic methods to the preparation of thin films by vapor phase epitaxy.

In this manuscript I will describe some of our recent work on the preparation and use of organometallic precursors in the synthesis of solid state compounds.

II-VI COMPOUNDS

A serious impediment to the growth of thin-films of HgTe and CdTe by organometallic vapor phase epitaxy (OMVPE) has been the high reaction temperature (400°C) which is required for the thermal decomposition of the traditional tellurium source compound, diethyltellurium (DET). At such a high temperature the evaporation and diffusion of Hg make the preparation of HgTe/CdTe superlattices not only inconvenient but impossible [3]. Since the need for the high reaction temperature was the stability of the Te source, we sought a volatile organotellurium compound which could be relied upon to provide elemental Te at considerably lower temperatures than those required by DET.

If OMVPE is assumed to proceed via the unimolecular decomposition of the source compounds to give the required element in gas-phase atomic form (equation 1), then the minimum temperature required by the precursors is directly related to the activation energy for the rate-limiting step in their pyrolytic decomposition. In the case of DET, the rate-limiting step is the homolysis of the first Te-C covalent bond to give the ethyl- and the ethyltelluryl-radicals. The growth temperature could therefore be reduced by the direct use of the ethyltellury radical as the source compound. Clearly this is not possible since the radical dimerizes to give diorganoditelluride, but this reasoning suggests the use of diorganoditellurides. The Te-Te bond in ditellurides is weak by comparison to the Te-C bond and therefore the activation barrier and consequently the growth temperature for the OMVPE process should be lowered. We have found [4] that dimethylditelluride (DMDT) can be used to prepare films of CdTe at temperatures at least as low as 250°C.

$$R - Te - R \xrightarrow{-R\cdot} R - Te\cdot \xrightarrow{-R\cdot} Te_{(atom)}$$
$$\searrow$$
$$CdTe_{(solid)} \quad (eq\ 1)$$
$$R' - Cd - R' \xrightarrow{-R'\cdot} R' - Cd\cdot \xrightarrow{-R'\cdot} Cd_{(atom)} \nearrow$$

In order to gain more insight into the mechanism of the reactions involved in II-VI film growth by this technique, we studied some similar reactions in solution. It is known that at fairly low temperatures dialkyl compounds of both Hg and Cd decompose to give elemental Hg and Cd respectively; we also found [4] that DMDT is stable at 250°C (under OMVPE conditions); and we therefore studied the reaction of diphenylditelluride (DPDT) with elemental Hg [5, 6]. The reaction of DPDT with an excess of elemental Hg at room temperature in an inert solvent quickly gives the complex Hg(TePh)$_2$, 1, as an insoluble oligomer. We have found that upon heating, this compound reversibly eliminates Hg and in an independent reaction eliminates diphenyltellurium with the concomitant production of HgTe (equation 2). In this pair of reactions the yields are quite high and there are no observed by-products.

$$RTe - TeR + Hg \rightleftharpoons Hg(TeR)_2 \xrightarrow{\Delta} HgTe + TeR_2 \quad (eq.2)$$

This simple model reaction indicates an important feature in precursor design. The organic ligands which are required to keep the precursors molecular must be removed in the pyrolysis, and it is important to design a "leaving group" feature into the precursor molecules and processes. In the present study "extra" Te is added to the reaction mixture, but it is required in order to give the organic ligands an exit route.

The model in equation 1 implies that another low-temperature route to films of metal tellurides would be available if gas-phase atomic Te could be used directly in the OMVPE process. This suggests that phosphine tellurides (R_3PTe, R=organic radical) might be valuable Te source compounds. We were led to phosphine tellurides by the original observation [7] that trialkylphosphine tellurides reversibly deposit elemental Te when warmed.

In our initial studies [8] of phosphine tellurides we found that: (1) Trimethylphosphine can be used to chemically transport Te; (2) Triethylphosphine telluride reacts very quickly with liquid Hg to give HgTe; and (3) Triethylphosphine telluride reacts with either diethylmercury or diphenylmercury to give HgTe with the elimination of diorganotellurium "leaving groups". If this last reaction is very rich in Te (9 equiv/Hg) a significant amount of diphenylditelluride is seen in addition to diphenyltellurium. This observation leads us to speculate that an intermediate such as 1, arising from the insertion of Te directly into the Hg-C bonds, is reasonable.

MANGANESE TELLURIDE

Manganese telluride MnTe is a valuable target for precursor-based synthesis for at least two reasons. Firstly, solid solutions of MnTe and CdTe have been widely studied as examples of dilute magnetic semiconductors. Secondly, based on our initial results with phosphine tellurides, a potentially useful synthesis route presented itself: if the phosphine tellurides did give direct insertion of Te into the Hg-C bond in the diorganomercury compounds described above, it seemed likely that they would give insertion of Te into the Mn-Mn bond of manganese carbonyl, $Mn_2(CO)_{10}$, and the attendant formation of Te-Mn covalent bonds would represent the first step in the formation of bulk MnTe.

We have found [9] that triethylphosphine telluride does react with manganese carbonyl to give insertion of Te into the Mn-Mn bond. When two equivalents of phosphine telluride are allowed to react with manganese carbonyl in toluene in the presence of excess phosphine the crystalline complex $\{(Et_3P)_2(CO)_3MnTe\}_2$, 2, can be isolated in good yield (equation 3). The structure of this complex was determined by X-ray crystallography which established the Mn-Te-Te-Mn connectivity of an organometallic ditelluride. We have found further that when heated to approx. 300°C this complex first melts, evolves CO and Et_3P, and finally forms MnTe as the solid pyrolyzate. The yield in this process is high, and MnTe is the only observed solid state product, i.e., no Te, Mn or other Mn/Te phases are seen.

$$(CO)_5 Mn - Mn(CO)_5 + 2\ Et_3PTe + 2\ Et_3P \xrightarrow[\text{reflux}]{\text{toluene}}$$

(eq. 3)

$$(Et_3P)_2(CO)_3Mn \overset{Te}{\diagup} \overset{}{\diagdown} \overset{Te}{\diagdown} \overset{}{\diagup} Mn(CO)_3(Et_3P)_2$$

These results emphasize that phosphine tellurides are useful in the preparation of solid state compounds, and that some of the typical ancillary ligands of organometallic chemistry (viz., CO and phosphines) can be valuable "leaving groups" in materials synthesis. In this regard it is worthwhile to note that the supporting ligands in 2 are stable compounds themselves, not organic radicals (such as the methyl radicals in a typical Group III source such as trimethylgallium), and can be removed from the pyrolysis reaction intact, thereby being much less likely to be incorporated in the final solid product.

IRON TELLURIDES

An important question facing the development of the precursor method is that of selectivity. For example, it is not clear in general whether the solid phase which is formed from the pyrolysis of a given precursor is determined by the nature of the precursor, by kinetic features of the pyrolysis reaction(s) or by the thermodynamics of the product phase diagram. While it is to be expected that all of the above contribute to the outcome of a precursor pyrolysis, it is valuable to catalog some examples of selective reactions in order to see if some patterns will emerge. We became interested in the iron-tellurium system for two reasons. Firstly, it gave the chance to extend the utility of phosphine tellurides; and secondly it presented the question of reaction selectivity. Aside from the two end points, Fe and Te, there are two Fe/Te phases which are stable at room temperature and pressure, FeTe and FeTe$_2$[10]. We sought precursor syntheses for each, exclusive of the other.

The (cyclopentadienyl) (dicarbonyl) iron dimer (hereinafter, Fp$_2$) is a common starting material in the organic chemistry of iron. The reactivity intrinsic to the Fe-Fe bond led us to examine the reactions of this compound with phosphine tellurides [11]. When Fp$_2$ is treated with two equivalents of triethylphosphine telluride and excess triethylphosphine in refluxing toluene the compound, di[(cyclopentadienyl)(carbonyl) (triethylphosphine)iron]ditelluride, 3, is formed and may be isolated in high yield as a crystalline solid (equation 4). Compound 3 has been characterized by IR and NMR spectroscopies which show: (1) only terminal CO ligands; (2) cyclopentadienyl and triethylphosphine ligands (one each) on each Fe; (3) a simple P-Fe-Te array; and (4) the presence of two diastereomers. The simplest structure which is consistent with these features is 3.

$$Fp_2 + Et_3PTe + Et_3P \longrightarrow \qquad + \qquad \text{(eq 4)}$$

When Fp$_2$ is treated with only one equivalent of triethylphosphine telluride under conditions identical to those above, the compound di[(cyclopentadienyl) (carbonyl) (triethylphosphine)iron]telluride, 4, is formed (equation 5). Spectroscopic parameters of 4 are quite similar to those of 3, but the two compounds are readily distinguished. It is significant that the mono- and di-telluride compounds can be interconverted. Thus when 4 is treated with phosphine telluride 3 results. (This is evidence that the organoiron telluride complexes are formed by simple insertion reactions, atomic tellurium (stabilized as the phosphine complex) being inserted directly into the Fe-Fe and Fe-Te bonds). Similarly, when 3 is treated with triethylphosphine and Fp$_2$ 4 results. This is crucial because if it were not the case then 4 would not be isolable, it being irreversibly carried on to 3. The equilibration between 3 and 4 is also important to the solid state reactions described below.

$$F_{p_2} + 2 \, Et_3PTe \longrightarrow \quad (eq.5)$$

In both complexes 3 and 4 the cyclopentadienyl ligands are tightly bound to the iron atoms, and unlike the phosphine and carbonyl ligands they are organic radicals, not stable, closed-shell molecules; therefore it might be expected that complexes 3 and 4 would not be valuable precursors to FeTe$_x$. One might expect that the radical ligands would remain in the solid product after the pyrolysis of these compounds. This is not the case. Pyrolysis of 3 at approx. 275°C in the solid state under vacuum (scheme 1) gives polycrystalline FeTe$_2$ with the elimination of CO, triethylphosphine and ferrocene (dicylcopentadienyliron). In a similar way, the pyrolysis of 4 gives FeTe, CO, triethylphosphine and ferrocene. The production of ferrocene in these reactions is, perhaps, not surprising in view of the particular stability of this compound, but the fact that it would be formed in essentially quantitative yield (97% and 96%, resp.) was not anticipated. This very high yield indicates that little if any carbon impurity in the iron tellurides arises from the cyclopentadienyl ligands. This is encouraging since the use of this ligand is very common in traditional organometallic chemistry.

It is noteworthy that the pyrolyses are selective with respect to products, 3 giving FeTe$_2$ and 4 giving FeTe. Different precursors lead to different solid state products, i.e., the product distribution from a given precursor reaction is not dominated solely by the thermodynamics of the solid state product phase diagram. It is also noteworthy that the separate precursors, 3 and 4, need not be isolated and purified in advance of pyrolysis. The (readily available) starting materials may be used directly. As an example, since Fp$_2$ itself pyrolyzes to give Fe, and 3 pyrolyzes to give FeTe$_2$, we were curious as to the results of a co-pyrolysis of the two. When the two crystalline complexes were ground together into a powder and subsequently pyrolyzed, only FeTe was observed in the solid product. There was no evidence for either Fe or FeTe$_2$. Since the activation energies for the extrusion/insertion of Te out of/into complexes 3, 4 and Fp$_2$ are lower than those for "decomposition" to the solid state compounds, the comproportionation of Fp$_2$ and 3 to give 4 must occur prior to the production of the solid. This sequence emphasizes the importance of knowing the molecular reactions which are open to the organometallic precursors. Our present knowledge of these is summarized in scheme 1.

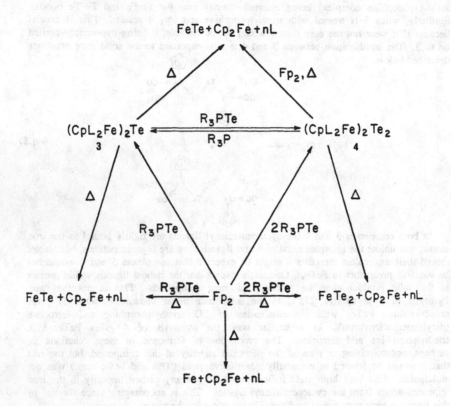

Scheme 1: Organoirontelluride reactions discussed in the text.

CONCLUSION

In this manuscript I have outlined several of our syntheses of solid state inorganic materials which are based on the thermal reactions of organometallic compounds. These preparations have several advantages. Firstly, they all require only modest reaction temperatures. Secondly, the starting materials for each are all readily available, and the subsequent preparative reactions are simple. Thirdly, the "intermediate precursors", i.e., 1, 2, 3 and 4 need not be isolated (although their isolation may give more direct control over the ultimate products in terms of stoichiometry, etc.). In addition the pyrolytic reactions have some interesting chemical features: a wide variety of ancillary ligands may be used, including the large cyclopentadienyl radical and the reactions and reagents can also be made selective with respect to products.

ACKNOWLEDGEMENTS

I would like to gratefully acknowledge D. W. Kisker, T. Y. Kometani and K. S. Jeffers for their collaboration in the use of ditellurides as OMVPE sources; C. E. Rice for her determination of the structure of complex 2; and C. R. Sprinkle for his assistance in the preparation and characterization of complex 1 and related materials.

REFERENCES

[1] See, for example, (a) Schafer, H., Angew. Chem. (Int. Ed. Engl.), 1971, 10, 43-50. (b) Wold, A., J. Chem. Educ., 1980, 57, 531-6.

[2] (a) West, A. R., "Solid State Chemistry and Its Applications:, Wiley and Sons, 1984, N.Y. p. 16-31. (b) Rao, C. N. R., Gopalakrishnan, J. "New Directions in Solid State Chemistry:, Cambridge University Press, 1986, Cambridge, UK, p. 116-24. (c) Foise, J., Kim, K., Covino, J., Dwight, K., Wold, A., Chianelli, R., Passaretti, J. Inorg. Chem. 1984, 23, 872-4. (e) Jensen, J. A., Gozum, J. E. Pollina, D. M., Girolami, G. S., J. Amer. Chem. Soc., 1988, 110, 1643-4. (f) Nanjundswamy, K. S., Vasanthacharaya, N. Y., Golapakrishnan, J., Rao, C. N. R., Inorg. Chem., 1987, 26, 4286-8.

[3] Tunnicliffe, J., Irvine, S. J. C., Dosser, O. D., Mullin, J. B., J. Cryst. Growth, 1984, 68, 245.

[4] Kisker, D. W., Steigerwald, M. L., Kometani, T. Y., Jeffers, K. S., Appl. Physi. Lett. 1987, 50, 1681-3.

[5] Steigerwald, M. L.; Sprinkle, C. R., J. Amer. Chem. Soc. 1987, 109, 7200.

[6] (a) Okamoto, Y.; Yano, T. J. J. Organomet. Chem. 1971, 29, 99-103. (b) Dance, N. S.; Jones, C. H. W. J. Organomet. Chem. 1978, 152, 175-85.

[7] Zingaro, R. A.; Stevens, B. H.; Irgolic, K. J. Organomet. Chem. 1965, 4, 320-3.

[8] Steigerwald, M. L.; Sprinkle, C. R., Organometallics, 1988, 7, 245.

[9] Steigerwald, M. L.; Rice, C. E., J. Amer. Chem. Soc. 1988, 110, 4228-31.

[10] Gronvold, F.; Haraldsen, H.; Vihode, J. Acta Chem. Scand. 1954, 8, 1927-42. The designation of the β phase of Fe/Te as FeTe is a misnomer. The stability range of this phase (as prepared from the direct combination of the elements) is reported to be from FeTe0.8 to FeTe0.9. In the work described in this manuscript the phase we report as FeTe shows the X-ray diffraction pattern of this β phase.

[11] Steigerwald, M. L. Chem. Mat., 1989, 1, in press.

THE USE OF TRIS(TRIMETHYLSILYL)ARSINE TO PREPARE AlAs, GaAs AND InAs. THE X-RAY CRYSTAL STRUCTURE OF (Me₃Si)₃AsAlCl₃·C₇H₈.

RICHARD L. WELLS, COLIN G. PITT, ANDREW T. McPHAIL, ANDREW P. PURDY, SOHEILA SHAFIEEZAD, AND ROBERT B. HALLOCK.
Department of Chemistry, Duke University, Durham, NC 27706.

ABSTRACT

The reactions of $(Me_3Si)_3As$ with group III halides have been utilized to prepare AlAs, GaAs and InAs. The adduct $(Me_3Si)_3AsAlCl_3$ has been isolated as an intermediate in the formation of AlAs from $(Me_3Si)_3As$ and $AlCl_3$. The crystal structure of its toluene solvate has been determined.

INTRODUCTION

The OMCVD growth of arsenic-containing III-V alloys typically involves the use of AsH_3; however, concerns regarding the use of AsH_3 have prompted a number of researchers to evaluate other sources of arsenic [1]. Recent research in our laboratories has demonstrated that dehalosilylation reactions are a facile route to the formation of the Ga-As covalent bond and, as a result, we have prepared and characterized a number of new Ga-As compounds [2]. We now report the first examples of the use of dehalosilylation reactions to prepare AlAs, GaAs and InAs [3]. Reactions between $(Me_3Si)_3As$ and MX_3 (M = Al, X = Cl; M = Ga, X = Cl or Br; M = In, X = Cl) proceed according to equation 1. Elimination of

$$(Me_3Si)_3As + MX_3 \longrightarrow MAs + 3Me_3SiX \qquad (1)$$

Me_3SiCl from the $(Me_3Si)_3As/AlCl_3$ reaction mixture occurs less readily than in the Ga or In systems, thereby allowing for isolation of the intermediate adduct $(Me_3Si)_3AsAlCl_3$. At higher temperatures, $(Me_3Si)_3AsAlCl_3$ undergoes dehalosilylation to yield AlAs.

EXPERIMENTAL [4]

Reaction of (Me₃Si)₃As and GaCl₃.

Upon combining $(Me_3Si)_3As$ [5] (0.343g, 1.16 mmol) and $GaCl_3$ (0.205 g, 1.16 mmol) in ligroin at room temperature a white precipitate formed which quickly changed color to yellow then to orange. After stirring for 3 days at 75 °C, 2.89 mmol of Me_3SiCl were evolved. The resulting brown solid was heated in the absence of solvent to 185 °C and 0.39 mmol of Me_3SiCl were collected (total Me_3SiCl

= 3.28 mmol, 94% theoretical). Further heating to 380 °C resulted in the elimination of a trace of yellow liquid beginning at 330 °C. The nonvolatile black solid (0.136 g, 81% yield) was shown to be GaAs of about 95% purity by elemental analysis and comparison of its X-ray powder diffraction pattern with that of a bona fide sample of GaAs. Anal. Calcd (Found) for GaAs: C, 0.00 (1.35); H, 0.00 (0.98); Cl, 0.00 (2.36); Si, 0.00 (0.45). In a separate experiment, $GaCl_3$ (0.411 g, 2.33 mmol) and $(Me_3Si)_3As$ (0.688 g, 2.33 mmol) were combined in pentane and 4.47 mmol of Me_3SiCl were isolated after the mixture had stirred overnight at room temperature. Further elimination of Me_3SiCl from the reaction mixture occurred as follows: 1.42 mmol, overnight at 90 °C in toluene; 0.60 mmol, overnight at 140-180 °C neat (total Me_3SiCl = 6.49 mmol, 93% theoretical). The solid was heated in vacuo with a cool flame (propane, no oxygen; ~400-500 °C) for 2 h. A trace of slightly volatile yellow liquid was formed and GaAs (0.294 g, 87% yield) of about 96% purity was isolated. Anal. Calcd (Found) for GaAs: C, 0.00 (1.68); H, 0.00 (0.30); As, 51.80 (50.08); Cl, 0.00 (2.51); Ga, 48.21 (45.96); Si, 0.00 (0.51); (Ga:As mole ratio = 1.00:1.01).

Reaction of $(Me_3Si)_3As$ and $GaBr_3$.

Upon mixing $(Me_3Si)_3As$ (0.350 g, 1.19 mmol) and $GaBr_3$ (0.367 g, 1.19 mmol) in benzene a precipitate formed as in the $GaCl_3$ reaction. Stirring the mixture overnight at 47 °C led tó the formation of 1.39 mmol of Me_3SiBr; an additional 1.16 mmol of Me_3SiBr was isolated after heating overnight at 75-85 °C. Further heating of the resultant reddish-brown solid in the absence of solvent to 410 °C did not lead to clean elimination of additional Me_3SiBr, rather the formation of an unidentified slightly volatile yellow liquid was observed and GaAs of 88% purity was isolated. Anal. Calcd (Found) for GaAs: C, 0.00 (3.58); H, 0.00 (0.19); Br, 0.00 (7.21); Si, 0.00 (<0.7).

Reaction of $(Me_3Si)_3As$ and $InCl_3$.

Indium trichloride (0.314 g, 1.42 mmol) and $(Me_3Si)_3As$ (0.415 g, 1.41 mmol) were combined in pentane. A salmon colored precipitate formed immediately upon mixing and rapidly changed color to dark brown. The formation of Me_3SiCl occurred as follows: 2.33 mmol, 3 days at 25 °C in pentane; 0.99 mmol, 4 days at 70-75 °C in benzene; 0.32 mmol, 18 h at 150 °C neat, 0.05 mmol, 15 min heating with a cool flame (total Me_3SiCl = 3.69 mmol, 87% yield). The resulting black solid (0.153 g, 57% yield) was shown to be InAs of 98% purity by elemental analysis and comparison of its X-ray powder diffraction pattern with that of a bona fide sample of InAs. Anal. Calcd (Found) for InAs: C, 0.00 (0.56); H, 0.00 (0.17); As, 39.49 (38.32); Cl, 0.00 (0.80); In, 60.51 (59.21); Si 0.00 (<0.1); (In:As mol ratio = 1.00:0.99).

Reaction of (Me₃Si)₃As and AlCl₃.

Equimolar amounts of $(Me_3Si)_3As$ (0.380 g, 1.29 mmol) and $AlCl_3$ (0.172 g, 1.29 mmol) were combined in pentane and an off-white solid remained insoluble. In contrast to the reactions of $GaCl_3$ and $InCl_3$, only a trace (0.03 mmol) of Me_3SiCl was formed after stirring the mixture overnight at room temperature. After removal of pentane, toluene was distilled onto the solid reaction mixture and a clear pale yellow solution formed upon heating to 40 °C. Additional Me_3SiCl (0.42 mmol) was isolated after heating overnight at 115 °C. When the solid reaction product was heated for 10 min with a cool flame, a white solid sublimed as the remainder of the material became brown, and 1.61 mmol of Me_3SiCl were isolated (total Me_3SiCl = 2.06 mmol, 53% theoretical). The volatile solid was resublimed at 120 °C (10^{-4} mm) and shown to be the adduct $(Me_3Si)_3AsAlCl_3$. Anal. Calcd (Found) for $C_9H_{27}AlAsCl_3Si_3$: C, 25.27 (25.46); H, 6.36 (6.27); Cl, 24.86 (24.20); ^1H NMR (C_6D_6, ref. 7.15 ppm) 2.72 ppm (s, Me_3Si). The non-volatile brown solid was heated with a cool flame for an additional 30 min without further evolution of Me_3SiCl. The brown solid was shown to be AlAs of about 92% purity by elemental analysis and comparison of its X-ray powder diffraction pattern with that of the isostructural compound GaAs. Anal. Calcd (Found) for AlAs: C, 0.00 (1.53); H, 0.00 (0.46); Al, 26.48 (26.14); As, 73.52 (67.58); Cl, 0.00 (3.73); Si, 0.00 (2.44); (Al:As mole ratio = 1.00:0.93).

Preparation of (Me₃Si)₃AsAlCl₃.

Equimolar amounts of $(Me_3Si)_3As$ (0.411 g, 1.40 mmol) and $AlCl_3$ (0.186 g, 1.40 mmol) were combined in toluene and heating of the mixture to 60 °C resulted in a yellow solution. Upon cooling slowly to room temperature, large colorless solvated crystals formed. These crystals readily lost solvent at room temperature and became opaque. A first crop of crystals was isolated by decanting the solution from the crystals grown at room temperature. This solution was then cooled to -28 °C for several days and a second crop of crystals was isolated. Toluene was completely removed from the product by evacuation of the solvated crystals at room temperature for 30 min. The total yield of solvent-free $(Me_3Si)_3AsAlCl_3$ was 0.501 g (84% theoretical). Anal. Calcd (Found) for $C_9H_{27}AlAsCl_3Si_3$: C, 25.27 (25.46); H, 6.36 (6.14); Cl, 24.86 (24.57); ^1H NMR (C_6D_6 ref. 7.15 ppm) 2.70 ppm (s, Me_3Si).

<u>Crystal Structure of the Toluene Solvate of (Me₃Si)₃AsAlCl₃</u>.

For X-ray data collection, a crystal of dimensions 0.33 x 0.39 x 0.80 mm was sealed inside a thin walled glass capillary. *Crystal data*: $(Me_3Si)_3AsAlCl_3 \cdot C_7H_8$, $M = 519.98$, orthorhombic, space group $Pbca(D_{2h}^{15})$, $a = 17.642(3)$ Å, $b = 22.564(4)$ Å, $c = 13.824(3)$ Å, $V = 5503.0$ Å³, $Z = 8$, $D_{calcd.} = 1.255$ g cm⁻³, μ(Cu-$K\alpha$ radiation, $\lambda = 1.5418$ Å) = 60.3 cm⁻¹. One octant of intensity data (5649 reflections), recorded on an Enraf-Nonius CAD-4 diffractometer [Cu-$K\alpha$ radiation, incident-beam graphite monochromator; ω-2θ scans; scanwidth $(1.15 + 0.14\tan\theta)°$; $\theta_{max.} = 75°$], yielded 2517 reflections [$I > 3.0\sigma(I)$] which were retained for the analysis. The intensities of two control reflections, monitored every 2 h during data collection, showed no significant variation. In addition to the usual Lorentz-polarization corrections, an empirical absorption correction ($T_{max.}:T_{min.} = 1.00:0.73$) was also applied to the data. The crystal structure was solved by direct methods [6]. Initial Al, As, Cl, and Si positions were derived from an E-map. Carbon atoms were located in difference Fourier syntheses. Full-matrix least-squares refinement of non-hydrogen atom positional and anisotropic temperature factor parameters, with Me₃Si hydrogen atoms included at their calculated positions, converged at $R = 0.043$ ($R_w = 0.069$) [7]. A view of the (Me₃Si)₃AsAlCl₃ molecule and a Newman projection, viewed down the Al-As bond, are provided in Figure 1. The packing arrangement in crystals of the toluene solvate is illustrated in Figure 2.

<u>Preparation of AlAs from (Me₃Si)₃AsAlCl₃</u>.

A sample of (Me₃Si)₃AsAlCl₃ (0.211 g, 0.49 mmol) was heated at 300 °C for 2 h, then with a cool flame for 15 min. to afford 1.12 mmol of Me₃SiCl (76% theoretical) and a brown nonvolatile powder. Some of the adduct sublimed into the trap with the Me₃SiCl rather than decomposing. The brown powder was heated again with a cool flame for 15 min to give AlAs of about 91% purity as shown by elemental analysis and comparison of its X-ray powder diffraction pattern with that of the isostructural compound GaAs. Anal. Calcd (Found) for AlAs: C, 0.00 (1.63); H, 0.00 (0.21); Al, 26.48 (24.52); As, 73.52 (67.63); Cl, 0.00 (2.60); Si, 0.00 (4.78); (Al:As mole ratio = 1.00:0.99).

RESULTS AND DISCUSSION

This work demonstrates that dehalosilylation reactions provide a viable route to the formation of AlAs, GaAs and InAs. These III-V semiconductors were isolated in 88-98% purity from very crude experiments with no attempts to optimize the reaction conditions. Tris(trimethylsilyl)arsine is a liquid with a

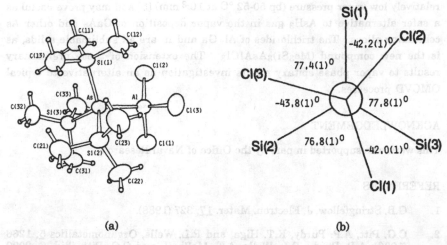

(a)   (b)

Figure 1. (a) Structure of (Me₃Si)₃AsAlCl₃; small circles represent hydrogen atoms. Selected distances and angles follow: As-Al 2.463(2) Å, As-Si 2.374(2) - 2.381(2) Å, Al-Cl 2.116(2) - 2.123(2) Å; Al-As-Si 108.19(6) - 109.58(6)°, Si-As-Si 109.04(6) - 110.38(6)°, As-Al-Cl 107.7(1) - 107.9(1)°, Cl-Al-Cl 110.0(1) - 111.6(1)°. (b) Newman projection viewed down the Al-As bond.

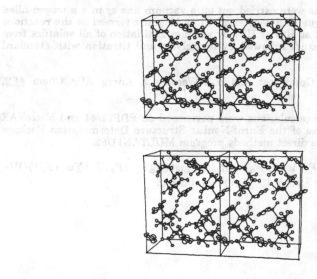

Figure 2. Stereoscopic view of the packing arrangement in crystals of (Me₃Si)₃AsAlCl₃·C₇H₈.

relatively low vapor pressure (bp 50-52 °C at 10^{-3} mm) [5] and may prove useful as a safer alternative to AsH_3 gas in the vapor deposition of GaAs and other As containing alloys. The trichlorides of Al, Ga and In are all sublimable solids, as is the new compound $(Me_3Si)_3AsAlCl_3$. The extension of these preliminary results to vapor phase epitaxy merits investigation as an alternative to typical OMCVD processes.

ACKNOWLEDGEMENT

This work was supported in part by the Office of Naval Research.

REFERENCES

1.   G.B. Stringfellow, J. Electron. Mater. 17, 327 (1988).

2.   C.G. Pitt, A.P. Purdy, K.T. Higa, and R.L. Wells, Organometallics 5, 1266 (1986); A.P. Purdy, R.L. Wells, A.T. McPhail, and C.G. Pitt, ibid. 6, 2099 (1987); R.L. Wells, S. Shafieezad, A.T. McPhail, and C.G. Pitt, J. Chem. Soc., Chem. Commun. 1987, 1823; R.L. Wells, A.P. Purdy, A.T. McPhail, and C.G. Pitt, J. Organomet. Chem. 354, 287 (1988).

3.   The reactions of $(Me_3Si)_3As$ with $GaCl_3$, $GaBr_3$ and $InCl_3$ are described in more detail in the following: R.L. Wells, C.G. Pitt, A.T. McPhail, A.P. Purdy, S. Shafieezad, and R.B. Hallock, Chemistry of Materials (in press).

4.   All manipulations were carried out on a vacuum line or in a nitrogen-filled glovebox. The quanitities of Me_3SiCl or Me_3SiBr formed in the reactions were determined as HCl or HBr by vacuum distillation of all volatiles from the reaction mixture followed by hydrolysis and titration with standard NaOH solution.

5.   V.G. Becker, G. Gutekunst, and H.J. Wessely, Z. Anorg. Allg. Chem. 462, 113 (1980).

6.   Crystallographic calculations were performed on PDP11/44 and MicroVAX computers by use of the Enraf-Nonius Structure Determination Package incorporating the direct methods program MULTAN11/82.

7. $R = \Sigma ||F_o| - |F_c|| / \Sigma |F_o|$; $R_w = [\Sigma w(|F_o| - |F_c|)^2 / \Sigma w(|F_o|)^2]^{1/2}$; $w = 1/\sigma^2(|F_o|)$; $\Sigma w(|F_o| - |F_c|)^2$ minimized.

ORGANOMETALLIC CHEMICAL VAPOR DEPOSITION OF GaAs USING NOVEL ORGANOMETALLIC PRECURSORS

R. A. JONES*, A.H. COWLEY*, B. L. BENAC*, K. B. KIDD*, J. G. EKERDT**, AND J. E. MILLER**
Departments of Chemistry* and Chemical Engineering**, The University of Texas at Austin, Austin, Texas 78712

ABSTRACT

The goals of the research are the design and synthesis of a new class of precursor compounds for III/V compound semiconductor materials, growth of films with these precursors and developoment of an understanding of the relationships between precursor structure, film growth reactions and film properties. Conventional OMCVD of III/V compound materials has a number of inherent safety and processing problems associated with the group III alkyl and group V hydride sources. Our approach to these problems is the synthesis of a single precursor with a fixed III:V stoichiometry and a direct two-center, two-electron sigma III-V bond. These compounds have the general formula $[R_2M(R_2E)]_2$ and $R_2M(R_2E)_2M R_2$ (M,M' = Al, Ga, In; E=P,As; R,R' = alkyl, aryl). The III-V bond in these compounds is stronger than the other bonds and the minium deposition temperature can be controlled by employing subsituents that undergo facile hydrocarbon elimination.

A typical example is the use of $[Me_2Ga(\mu-t-Bu_2As)]_2$ as the single source for GaAs films. The organometallic precursor is a solid crystalline powder which is maintained at 130°C to generate enough vapor for OMCVD. Typical film growth conditions involve the use of H_2 or He as the carrier gas, substrate temperatures of 500 to 700°C, and a total system pressure of 0.0002 Torr. GaAs(100), Si(100) (As-doped 3° off toward (011) and quartz have been used as substrates. Film composition has been established with XPS. The Ga 3d, As 3d, and C 1s signals at 18.8, 40.9, and 284.6 eV, respectively, reveal the films to be 1:1 Ga:As and void of carbon. The carbon levels are less than 1000 ppm. X-ray diffraction and SEM results suggest polycrystalline GaAs on quartz and epitaxial GaAs on GaAs(100) and Si(100). (2 K) photoluminescence measurements on GaAs, grown on semi-insulating GaAs(100) and Si-doped GaAs(100) at 570 °C. produce PL signals indicating that crystalline domains are present,the measurements indicate degeneratively n-doped material and show that good Ga:As ratios and low levels (ca. 1 ppm) of impurities are present. Growth rates:~ 1.0 mm/hour.

INTRODUCTION

Group III/V semiconductors have received much attention due to their usefulness in high speed digital circuits, microwave devices, and optoelectronics [1-26]. Several techniques have been employed for the preparation of thin films of these materials, including organometallic chemical vapor deposition

(OMCVD) and molecular beam epitaxy (MBE). The OMCVD method is often preferred for larger scale processes and typically involves the reaction of a group III trialkyl such as Me_3Ga, with a group V hydride such as AsH_3 or PH_3 at elevated temperatures (600-700°). A carrier gas such as H_2, He or N_2 is used to sweep volatile species through the reaction chamber. The most studied reaction is that of Me_3Ga and AsH_3 to produce GaAs.

There are a number of important disadvantages to this conventional OMCVD method. Alkyls of the group III elements tend to be pyrophoric and group V hydrides such as AsH_3 are extremely toxic. Apart from the potential environmental, safety and health hazards of handling the reagents, the conventional OMCVD methodology also suffers from several other drawbacks. These include stoichiometry control, impurity incorporation, and unwanted side reactions. For GaAs grown from Me_3Ga and AsH_3 the incorporation of carbon has been a persistent problem [25-29]. Moreover, the high temperatures involved in some processes can promote interdiffusion of layers and dopants which prevents sharp heterojunctions from being achieved.

Attempts to grow superior films by modifications of the OMCVD process include low pressure-, [30-35] plasma enhanced-, [36] rapid thermal-, [37] and hybrid MBE-OMCVD systems[38]. Several groups of workers have investigated the use of alternative sources of both the group III and group V components. Group V sources which have been investigated include Me_3P, Et_3P, $t-BuPH_2$, $i-BuPH_2$, [39] $t-BuAsH_2$, Et_2AsH, [40-42] Me_3As, and Et_3As [43]. Group III alkyls have generally been limited to trimethyl and triethyl derivatives [44,45].

In addition, various adducts such as $Me_3In.PEt_3$ and $Me_3In.NMe_3$ have been explored [46-51]. Although adducts have a III:V stoichiometry of 1:1 and less sensitivity to air and moisture, they do not have ideal chemical and physical properties for optimum OMCVD processes and using them to provide both the group III and V elements has met with limited success. The donor-acceptor dative type bonds are generally considered to be relatively weak compared to the other bonds (such as Ga-C or As-H) present in the adducts [52]. Ultimately, these have to be broken in order to form the III-V material. Dissociation of the adduct and loss of stoichiometry can occur, and typically excess PH_3 or AsH_3 is required for the production of good quality films [48,49,53].

RESULTS AND DISCUSSION

One chemical solution to the problems outlined above is to employ a single source precursor which contains both the group III and group V elements in the same molecule in the correct ratio (1:1). One would then cause the III-V bond to be as strong as, or stronger than, the other bonds in the molecule. Thus, under the reaction conditions required for film growth, the bonds between the group III and group V elements remain intact while the other bonds are broken. One way to do this would be to make the III-V bond a direct 2-center, 2-electron sigma (σ) bond instead of a dative donor-acceptor type.

In order to test this hypothesis, our initial studies focussed on the design and synthesis of suitable organometallic molecules which feature direct σ-bonding between the group III and group V elements [54]. Examples of such compounds which we have synthesized and characterized in our laboratories include dinuclear complexes of the general formula $[R_2M(\mu-t-Bu_2E)]_2$ (M= Al, Ga, In; E=P, As). Several preliminary publications describing our work in this area have appeared in the literature [54-56].

The use of bulky groups attached to the group V element prevents oligomerization and mononuclear and dinuclear complexes may be isolated conveniently and in high yields. There are relatively few other examples of compounds of these types. Pioneering work by Coates established that compounds of the the type $(Me_2EMMe_2)_3$ may be produced from the interaction of Me_3M and $HEMe_2$ (M= Ga, In; E=P, As) [57,58]. However these compounds do not appear to be appreciably volatile . More recently Wells and Beachley have reported compounds of similar stoichiometry which are dinuclear as well as related compounds [59-64].

FILM GROWTH STUDIES

We have concentrated our initial studies on the compounds of general formula $[R_2M(\mu-t-Bu_2E)]_2$ since they have a III:V stoichiometry of 1:1 and are extremely safe and easy to handle [65]. These compounds are hydrocarbon soluble, non-corrosive, non-toxic, essentially air stable and are sufficiently volatile for the OMCVD of III/V films. Unlike the conventional OMCVD process which required a large excess of the group V source, we have grown GaAs films using the methyl or ethyl derivatives $[Me_2Ga(\mu-t-Bu_2As)]_2$ (1) and $[Et_2Ga(\mu-t-Bu_2As)]_2$ (2) as the sole sources of both Ga and As. Films have been grown using a cold-wall, vertical tube reactor built to permit growth of films over the pressure range 760 to 10^{-6} torr and at temperatures up to 780 °C.

MATERIALS CHARACTERIZATION STUDIES

The most extensive characterization of films has been done on films grown with the methyl derivative (1). We have employed: X-ray diffraction to establish crystallinity, XPS to determine chemical composition and gross carbon contamination, SIMS to identify impurities, SEM to observe the morphological properties, and low temperature (2 K) photoluminescence (PL) to explore the band edge.

For the methyl derivative (1) growth rates of 0.7 to 1.0 μm/hr have been realized for the methyl derivative when the compound was heated to 130 °C in the saturator. XPS results have been obtained after sputtering the surface for 10 min. with Ar^+ ions and match those for a single crystal substrate. No carbon was detected by XPS. The Ga:As ratios were 1:1 to within the limits of the XPS experiment.

Carbon impurity levels were tested with SIMS because XPS cannot detect less than 1000 ppm of carbon. We did not have calibration standards when the SIMS profiles were taken; however, SIMS studies of the 570 °C films showed that the films did not contain any more carbon than the GaAs substrates. SIMS did identify the presence of metallic impurities, Na, K, Al, Si, In, Ti, Cr, and Fe. The SIMS results are significant and demonstrate that, as expected, precursor ligand loss was complete and that carbon incorporation into the films did not occur, however, simple chemical measures alone are inappropriate to establish composition for electronic applications. Most of the films have been found to be polycrystalline by X-ray diffraction; however, of key significance are low temperature (2 K) photoluminescence measurements on films of GaAs, grown on semi-insulating GaAs(100) and Si-doped GaAs(100) at 570 °C. These materials do produce PL signals indicating that crystalline domains are present, and although the measurements indicate degeneratively n-doped material they also show that we have achieved very good Ga:As ratios and very low levels (ca. 1 ppm) of impurities. We have so far made no attempts to rigorously purify the precursor compounds in order to prepare films of ultra-high electrical quality. However we believe that this will be ultimately possible. Films have been grown in the presence of a carrier gas to assist in sweeping the compound from the saturator. A reactive carrier gas is not required to facilitate ligand loss. Both He and H_2 have been used with equal success indicating the potential these compounds may have in chemical beam epitaxy (CBE) [66].

To our knowledge there have been only two reports of OMCVD studies which are related to our work. The use of the trimer $[Et_2GaPEt_2]_3$ as a precursor for GaP has been described. However relatively high levels of carbon were observed in the polycrystalline films obtained [67,68]. In addition Bradley, Factor and coworkers have recently achieved the OMCVD of InP using $[Me_2In(\mu-t-Bu_2P)]_2$ [69]. Our preliminary results indicate that the single source precursors are viable alternatives to current OMCVD approaches.

MECHANISTIC CONSIDERATIONS

The mechanism of decomposition of the precursor molecules either at or on the surface of the growing layer is a key issue which will determine the temperature of layer formation and ultimately the carbon content of the grown film. Since our compounds are single source molecules, study of the pathways of decomposition and ultimately of layer growth should be considerably easier than with the conventional systems. One clear requirement which has emerged from our studies so far is for a low energy kinetic pathway which permits the facile expulsion of hydrocarbon ligands. Our preliminary studies indicate that this may be possible via the well known β-hydrogen elimination pathway.

Preliminary thermal decomposition studies on $[Me_2Ga(\mu-t-Bu_2As)]_2$ (1) show that methane (CH_4) and isobutylene ($Me_2C=CH_2$) are the main hydrocarbons produced during layer growth. This, plus the solid state crystal structure of the compound which indicates a relatively short H...Ga interaction of a tert-butyl-As hydrogen atom lead us to tentatively propose a decomposition pathway involving transfer of this H atom to the Ga and expulsion of CH_4 [65]. This process, accompanied by a cleavage of the As-C bond producing isobutylene would then result in the clean removal of the hydrocarbon groups from the inner Ga_2As_2 core.

(part of **1**)

Initial studies using the ethyl and n-butyl analogues [$Et_2Ga(\mu-t-Bu_2As)]_2$ (**2**) and [$(n-Bu)_2Ga(\mu-t-Bu_2As)]_2$ (**3**) suggest that an alternative β-hydrogen elimination pathway, which has an even lower kinetic barrier, may be the main process occuring during layer growth. The presence of β-hydrogen atoms on the group III alkyl permits the expulsion of the corresponding alkene and isobutane from the t-Bu-As units.

REFERENCES

1. Nakanisi, T. J. Cryst. Growth. **1984**, 68, 282.
2. Reep, D. H.; Ghandi, S. K. J. Electrochem. Soc.; Solid-State Sci. and Tech. **1983**, 103, 675.
3. Nishizawa, J.; Kurabayashi, T. J. Electrochem. Soc. Solid-State Sci. and Tech. **1983**, 103, 413.
4. Bediar, S. M.; Tischler, M. A.; Katsuyama, T. Appl. Phys. Lett. **1986**, 48, 30.
5. Atsutoshi, D.; Yoshinobu, A.; Namba, S. Appl. Phys. Lett, **1986**, 49, 785.
6. Morris, B. J. Appl. Phys. Lett. **1986**, 48, 867.
7. Yoshidia, M.; Watanaba, H.; Uesugi, F. J. Electrochem. Soc.; Solid State Sci. Tech. **1985**, 132, 676.
8. Schlyer, D. J.; Ring, M.A. J. Electrochem. Soc.; Solid-State Sci. Tech. **1977**, 124, 569.
9. Leys, M. R.; Veenvliet, H. J. Cryst. Growth **1981**, 55, 145
10. Stringfellow, G. B. J. Cryst. Growth **1983**, 62, 225.
11. Haigh, J.; O'Brein, S. J. Cryst. Growth **1984**, 67, 75.
12. Koppitz, M.; Vestavik, O.; Pletchen, W.; Mircea, A.; Heyen, M.; Richter, W. J. Cryst. Growth **1984**, 68, 136.
13. Stringfellow, G. B. J. Cryst. Growth **1984**, 70, 133.
14. Seki, H.; Koukitu, A. J. Cryst. Growth **1986**, 74, 172.
15. Koukitu, A.; Suzuki, T..; Seki, H. J. Cryst. Growth **1986**, 74, 181.
16. Kasai, K.; Komeno, J.; Takikawa, M.; Nakai, K.; Ozeki, M. J. Cryst.Growth **1986**, 74, 659.
17. Stringfellow, G. B. J. Cryst. Growth **1986**, 75, 91.
18. Larsen C. A.; Stringfellow, G. B. J. Cryst. Growth **1986**, 74, 181.
19. Arens, G.; Heinecke, H.; Putz, N.; Luth, H.; Balk, P. J. Cryst. Growth **1986**, 76, 302.
20. Huelsman, A. D.; Reif, R.; Fonstad, C. G. Appl. Phys. Lett. **1987**, 50, 206.
21. Butler, J. E.; Bottaka, N.; Sillmon, R.S.; Gaskill, D. K. J. Cryst. Growth, **1986**, 77, 163.
22. Monteil, Y.; Berthet, M. P.; Favre, R.; Hariss, A.; Bouix, J.; Vaille, M.; Gibart, P. J. Cryst. Growth **1986**, 77, 172.
23. DenBaars, S. P.; Maa, B. Y.; Dapkus, P. D.; Danner, A. D.; Lee, H. C. J. Cryst. Growth **1986**, 77, 188.
24. Chen, K.; Mortazvi, A. R. J. Cryst. Growth **1986**, 77, 199.

25.Heinecke, H.; Brauers, A.; Luth, H.; Balk, P. J. Cryst.
Growth 1986, 77, 241.
26.Akiyama, M.; Kawarada, Y.; Ueda, T.; Nishi, S.; Kaminishi, K.
J. Cryst. Growth 1986, 77, 490.
27.Manasevit, H. M. Appl. Phys. Letters, 1968, 12, 156.
28.Makita, Y.; Nomura, T.; Yokota, M.; Matsumori; T.; Izumi, T.;
Takeuchi, T.; Kudo, K. Appl. Phys. Lett. 1985, 47, 623.
29.Balk, P.; Heinecke, H. in "Physical Problems in
Microelectronics" Ed Kassabov, J. (World Sci. Publ.) P.190.
30.Manasevit, H. M.; Simpson, W. I. J. Electrochem. Soc., 1969,
116, 1725. See also papers on the First International
Conference on MOCVD, J. Cryst. Growth. 1986, 551, 1 and
refs. therein.
31.Shastry, S. K.; Zemons, S.; Oren, M. J. Cryst. Growth. 1986,
77, 503.
32.Norris, P.; Black, J.; Zemon, S.; Lambert, G. J. Cryst.
Growth. 1984, 68, 437.
33.Duchemin, J. P.; Bonnet, M.; Koelsch, F.; Huyghe, D. J.
Electrochem. Soc. 1978, 43, 181.
34.Duchemin, J. P.; Bonnet, M.; Koelsch, F. J. Electrochem. Soc.
1978, 125, 637.
35. Duchemin, J. P.; Bonnet, M.; Beuchet, G.
J. Vacuum Sci. Technol. 1979, 16, 1126.
36.Huelsman, A. D.; Reif, R.; Fonstad, C G. Appl. Phys. Lett.
1987, 50, 206.
37.Reynolds, S.; Vook, D. W.; Gibbons, J. F. Appl. Phys. Lett.
1986, 49, 1720.
38.Fraas, L. M.; McLeod, P. S.; Partain, L. D.; Weiss, R. E.;
Cape, J. A. J. Cryst. Growth 1986, 77, 386.
39.Larsen, C. A.; Chen, C. H.; Kitamura, M.; Stringfellow, G.
B.; Brown, D. W.; Roberston, A. J. Appl. Phys. Lett. 1986,
48, 1531.
40.Lum, R. M.; Klingert, J. K. Lamont, M. G. Appl. Phys. Lett.
1987, 50, 284.
41.Chen, C. H., Cao, D. S., Stringfellow, G. B., J. of
Electronic Materials, 1988, 17, 67.
42.Chen, C.H., Larsen, C.A., Stringfellow, G. B., Appl. Phys.
Lett., 1987, 50, 218 .
43.Lum, R. M., Klingert, J. K., Wynn, A. S., Lamont, M.G., Appl.
Phys. Lett., 1988, 52 , 1475 .
44.Ludowise, M. J., J. Appl. Phys. 1985 , R31 58 .
45.Duchemin, J. P., Bonnet, M., Beuchet, G., Koelesch, F., Inst.
Phys. Conf. Ser. 45, Inst. of Phys., 1979, 10 .
46.Moss, R. H. J. Cryst. Growth. 1984, 68, 78.
47.Bradley, D. C. Factor, M. M. White, E. A. D. Frigo, D. M.
Young, K. V. Chemtronics 1988, 3, 50.
48.Bass, S. J.; Skolnick, M. S.; Chudzynska, H.; Smith, L., J.
Cryst. Growth, 1986, 75, 221.
49.Zaouk, A.; Salvetat, E.; Sakaya, J.; Maury, F.; Constant, G.
J. Cryst. Growth, 1981, 55, 135.
50.Maury, F.; El Hammadi, A.; Constant, G. J. P. J. Cryst.
Growth, 1984, 68, 88.
51.Chatterjee, A. K.; Faktor, M. M.; Moss, R. H.; White, E. A.
D. J. de Physique, 1982, C5, 491.
52.Zaouk, A.; Salvetat, E.; Sakaya, J.; Maury, F.; Constant, G.
J. Cryst. Growth, 1981, 55, 135.
53.Haigh, J.; O'Brien, S. J. Cryst. Growth, 1984, 68, 550.
54.Arif, A. M. ; Benac, B. L.; Cowley, A. H.; Geerts, R. L;

Jones, R. A.; Kidd, K. B.; Power, J. M.; Schwab, S. T. J. Chem. Soc. Chem. Comm., **1986**, 1543.

55. Heaton, D. E., Jones, R. A., Kidd, K. B. Cowley, A. H., Nunn, C. M. Polyhedron in press.

56. Arif, A. M., Benac, B. L., Cowley, A. H., Jones, R. A., Kidd, K. B., Nunn, C. M. New J. Chem. **1988**,*12*, 553.

57 Coates, G.E.; Graham, J. J. Chem. Soc. **1963**, 233.

58. Beachley, O. T. ; Coates, G. E. J. Chem. Soc. **1965**, 3241.

59. Wells, R. L.; Purdy, A. P.; McPhail, A. T.; Pitt, C. G. J. Organomet. Chem. **1986**, *308*, 281.

60. Beachley, O. T.; Kopasz, J. P.; Zhang, H.; Hunter, W. E.; Atwood, J. L. J. Organomet Chem.**1987**, *325*, 69

61. Pitt, C. G.; Purdy, A. P.; Higa, K. T.; Wells, R. L. Organometallics **1986**, *5*, 1266.

62. Pitt, C. G.; Higa, K. T.; McPhail, A. T.; Wells, R. L. Inorg. Chem. **1986**, *25*, 2483.

63. Wells, R. L.; Purdy, A. P.; Higa, K. T.; McPhail, A. T.; Pitt, C. G. J. Organometallic Chem. **1987**, *325*, C7.

64. Purdy, A. P.; Wells, R. L. ; McPhail, A. T.; Pitt, C. G. Organometallics, **1987**, *6*, 2099.

65. Cowley, A. H. Benac, B. L. Ekerdt, J. G. Jones, R. A. Kidd, K. B. Lee, J. Y. Miller, J. E. J. Amer. Chem. Soc. **1988**, *110*, 6248.

66. Chiu, T. H.; Tsang, W. T., Ditzenberger, J. A.; Tu, C. W.; Ren, F.; Wu, C. S J. Electronic Matls., **1988**, *17*, 217.Chiu, T. H.; Tsang, W. T.; Cunningham, J. E., Robertson, Jr. A. J. Appl.Phys. Lett. **1987**, *62* 2302.Tsang, W. T., Appl. Phys. Lett. **1984**, *45*, 1234.

67. Maury, F.; Combes, M.; Constant, G.; Carles, R.; Renucci, J. B. J. de Physique **1982**, *45*, C1-347.

68. Maury, F.; Constant, G. Polyhedron, **1984**, *3*, 581.

69. Bradley D. C.-personal communication. See also Andrews, D., Davies, G.J., Bradley, D. C. Frigo, D. M., White, E. A. D.Appl. Phys. Lett. in press.

ORGANOMETALLIC PRECURSORS FOR III-V SEMICONDUCTORS

ERIN K. BYRNE, TREVOR DOUGLAS, AND KLAUS H. THEOPOLD
Department of Chemistry, Baker Laboratory, Cornell University, Ithaca, New York 14853

ABSTRACT

Organometallic molecules containing covalently linked gallium and arsenic or indium and phosphorus have been synthesized and characterized spectroscopically and by X-ray diffraction. These precursors can be transformed into the corresponding III-V materials in a chemical reaction proceeding at ambient temperature. The compound semiconductors prepared in this way are obtained as amorphous powders. During the reaction, quantum size effects may be observed by UV-VIS spectroscopy as the particles grow.

INTRODUCTION

III-V semiconductors (i.e. compounds consisting of an element from group III and one from group V in a 1:1 stoichiometry, e.g. gallium arsenide, indium phosphide etc.) are materials of great technological interest. The high electron mobility in GaAs as well as the band gaps of these compounds (in the visible and infrared light region) make them indispensable for the design of very fast integrated circuits and optoelectronic devices. The manufacture of these devices requires the deposition of very thin films used to form quantum wells or twodimensional electron gases, and the deposition of highly pure semiconductors on various substrates. Currently many device quality films are grown by "Organometallic Vapor Phase Epitaxy" (OMVPE), a process in which volatile molecules containing the desired elements are flowed over a heated substrate. The technology associated with carrying out these reactions in a controlled fashion has evolved to the point that films of near atomic thickness (~3 Å) can be produced.

Despite this apparent success several problems remain to be solved before the full potential of III-V semiconductors can be realized. Arsine (AsH_3) is one of the most toxic gases known (inhalation of as little as 0.5 ppm may be dangerous), making its large scale use in plants unattractive. The nonstoichiometric nature of the gas mixtures is responsible for deviations from the desired 1:1 ratio of group III element and group V element in the solid, thus introducing anti-site defects (e.g. As occupying Ga sites) into the semiconductor film. The reactant chemistry usually governs the range of acceptable deposition temperatures and some heterostructures can not be formed in this range of temperatures, as they are destroyed by thermal diffusion. Finally, and probably most serious of all, the OMVPE process in its current form does not lend itself to commercialization of devices based on III-V semiconductors because of the complexity of apparatus which is required to achieve dimensional control and uniformity of the semiconductor films. The solutions to these problems lie in the design of novel and more sophisticated precursors for the preparation of III-V semiconductors. Some time ago we began work involving the synthesis of III-V molecules and a study of their reactivity. It is our belief that investigations of the reaction pathways and mechanisms in III-V chemistry are critical in designing useful precursors.

RESULTS AND DISCUSSION

Among the various arsinogallanes we have prepared to date is the first example of a monomeric molecule of this type (i.e. containing only one group III and group V atom each) .[1] Figure 1 shows the result of a crystal structure determination of $Cp^*_2Ga\text{-}As(SiMe_3)_2$ (1, $Cp^* = \eta^5\text{-}C_5Me_5$), which was prepared by a reaction of $[Cp^*_2GaCl]_2$ [2] with $Li(THF)_2As(SiMe_3)_2$ [3]. The bulky nature of the alkyl substituents prevents dimerization of the molecules. The design of 1 had included a provision for the ultimate removal of all substituents on the binuclear core of the molecule. In this scenario, the basicity of the gallium bound carbon was to be used to introduce another substituent, which in turn would form a strong bond to silicon, thus driving the formal reduction to GaAs. Accordingly, reaction of 1 with tert-butanol in pentane yielded 2 equivalents each of pentamethylcyclopentadiene and ^{t}butyl(trimethylsilyl)ether (see Scheme 1). The only other product of this reaction was a reddish powder, which precipitated from the reaction solution. Elemental analysis as well as scanning electron microprobe (SEM) analysis of this powder showed it to be gallium arsenide with some organic impurities.

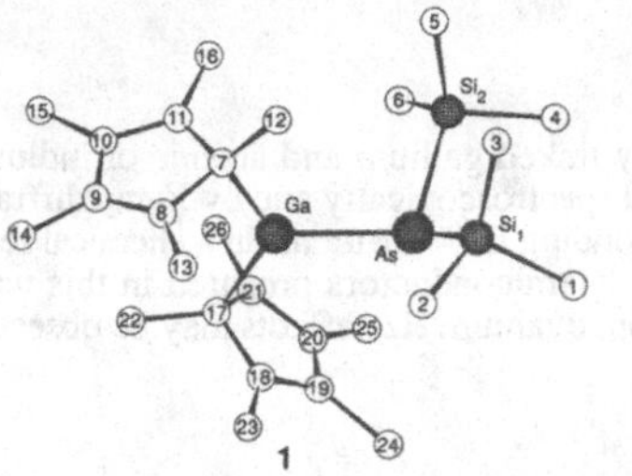

$$1 / 2 \, [(C_5Me_5)_2GaCl]_2 \qquad Li(THF)_2As(SiMe_3)_2$$

$$\downarrow 2 \textit{ tert}-Butanol$$

$$GaAs + 2 \, C_5Me_5H + 2 \textit{ tert}-BuOSiMe_3$$

Figure 1. The molecular structure of **1**.   **Scheme 1**

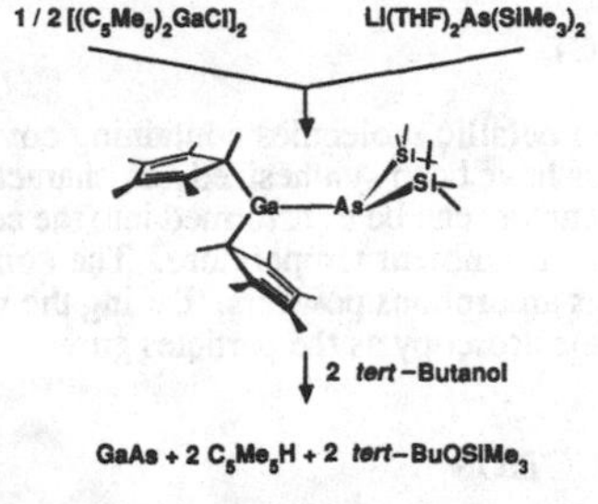

The mechanism of this superficially simple reaction is complicated and involves catalysis by a chloride containing trace impurity in the starting material. Based on several observations we have identified a compound of the composition $[Cp*(Cl)Ga-As(SiMe_3)_2]_n$ (**2**) as the most likely candidate for this impurity. A possible catalytic role for this molecule might consist of its reaction with tert-butanol to form a gallium alkoxide and release HCl, which in turn could cleave a Ga-Cp* bond in the precursor, thereby reforming the catalyst and so forth. We have obtained some circumstantial evidence for the above hypothesis. First, the reaction of $Cp*_2Ga-As(SiMe_3)_2$ with tert-butanol was sped up in the presence of small amounts of HCl (generated by addition of a small amount of Me_3SiCl to the reaction mixture). Second, addition of the base K^tOBu to the reaction mixture slowed the reaction dramatically (presumably by removal of the HCl). However, more direct proof came from the independent synthesis of said impurity, and the demonstration that it can serve as a catalyst for the formation of gallium arsenide from the precursor $Cp*_2Ga-As(SiMe_3)_2$. Reaction of $Cp*_2Ga-As(SiMe_3)_2$ with 1.0 equivalent of HCl yielded a new compound, which has been fully characterized and to which we assign the composition $[Cp*(Cl)Ga-As(SiMe_3)_2]_n$ (n = 1,3 in solution). It reacted instantaneously with tert-butanol to yield gallium arsenide, and more importantly, addition of a small amount of it to a reaction mixture consisting of $Cp*_2Ga-As(SiMe_3)_2$ and tert-butanol dramatically increased the rate of formation of gallium arsenide.

More recently, we have extended our synthetic efforts to include precursors for indium phosphide also. Some representative reactions and precursors thus produced are shown in Scheme 2.

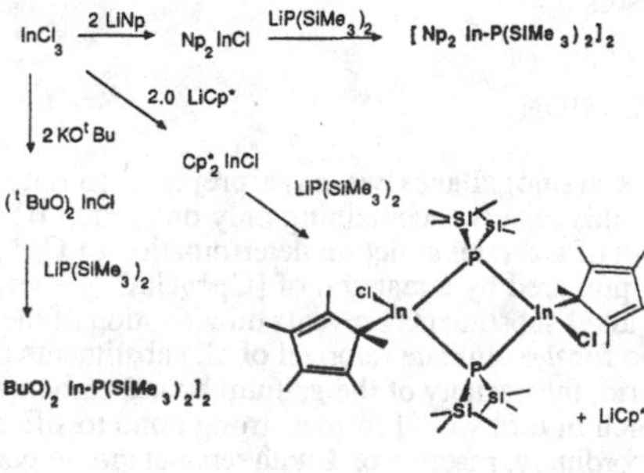

Scheme 2

Surprisingly, [Cp*(Cl)In-P(SiMe₃)₂]₂ (3) was the product of the reaction of Cp*₂InCl with LiP(SiMe₃)₂. Apparently the pentamethylcyclopentadienyl ligand is a better leaving group than chloride in this reaction. **3** is also a direct analogue of the "catalytic impurity" **2** in the arsinogallane **1**. Its crystal structure was determined by X-ray diffraction and the result of this study is shown in Figure 2. Due to the lesser steric demand of the chloride substituent (as compared to a second Cp*-group) the molecule is a dimer in the solid state. That this structure is also retained in solution was indicated by the observation of virtual coupling of the protons of the trimethylsilyl groups to two equivalent phosphorus nuclei in the ¹H NMR spectrum.

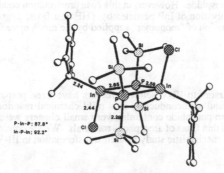

P-In-P: 87.8°
In-P-In: 92.2°

Figure 2. The molecular structure of **3** with selected bond distances in Ångstrom.

Phosphinoindane **3** reacted rapidly with tert-butanol to yield pentamethylcyclopentadiene (Cp*H, 1 equivalent), tert-butyl(trimethylsilyl)ether (2 equivalents) and a yellow powder, which precipitated from the solution. This powder is insoluble in standard organic solvents, it does not melt below 300 °C, and it exhibits a broad diffraction maximum in the X-ray powder pattern centered about the strongest reflection of indium phosphide (InP). Based on these observations, and in analogy to the formation of gallium arsenide in the reaction of **1** with tert-butanol, we believe this material to be very finely divided and amorphous indium phosphide.

We have previously reported the observation of quantum size effects in the absorption spectra of very small particles of gallium arsenide. When the reaction of **1** with tert-butanol was carried out in the donor solvent THF (tetrahydrofuran), the solution remained homogeneous for extended periods of time and changed color from yellow over orange and red to dark brown. During the same time monitoring of this solution by UV-VIS spectra revealed a continuous red shifting of the onset of absorption (see Figure 3a).

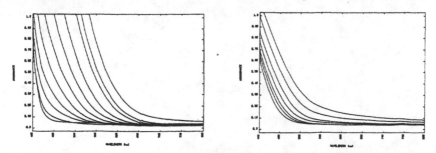

Figure 3. Evolution of UV-VIS spectra of small III-V semiconductor particles in THF solution with time. (a) **1** + ᵗBuOH, duration 14 hours. (b) **3** + ᵗBuOH, duration 30 minutes.

We believe that the formation of the III-V semiconductor begins with the liberation of single molecules of GaAs. These aggregate to clusters and then colloidal particles until they finally reach a size which induces precipitation. During this aggregation process the bandgap of the small semiconductor particles changes, shifting to lower energies as the size of the particles increases. Calculations on the dependence of the bandgap of gallium arsenide on particle size [4] are consistent with an average cluster size of 60 Å in the solution which exhibits the most red shifted spectrum of Figure 3a.

Figure 3b shows absorption spectra of a THF solution of 3 after addition of tert-butanol. Once again a red shift of the absorption onset with time was observed, indicating the growth of small particles of indium phosphide. However, in this case precipitation occurred much earlier. This may reflect weaker solvation of InP particles by THF or a faster aggregation due to the higher steady state concentration of "monomer" supplied by the much more facile reaction of 3 with the alcohol.

CONCLUSION

Organometallic precursors to III-V semiconductors have been prepared which can be transformed to the compound semiconductors in a mild chemical reaction. "Solutions" of gallium arsenide and indium phosphide containing very small clusters were generated, which may allow the deposition of thin films of amorphous materials. We are continuing the synthesis of novel precursors and the mechanistic study of their transformation to III-V alloys in various processes.

REFERENCES

1. E. K. Byrne, L. Parkanyi, and K. H. Theopold, Science 241, 332 (1988).

2. G. Becker, G. Gutekunst, H. J. Wessely, Z. Anorg. Allg. Chem. 113, 462 (1980).

3. O. T. Beachley, Jr., R. B. Hallock, H. M. Zhang, J. L. Atwood, Organometallics 4, 1675 (1985).

4. L. Brus, J. Phys. Chem. 90, 2555 (1986).

EFFECTS OF THE SELENIUM PRECURSOR ON THE GROWTH OF ZnSe BY METALORGANIC CHEMICAL VAPOR DEPOSITION

KONSTANTINOS P. GIAPIS, LU DA-CHENG, AND KLAVS F. JENSEN
Department of Chemical Engineering and Materials Science
University of Minnesota, Minneapolis, MN 55455

ABSTRACT

The growth of ZnSe on GaAs substrates by metalorganic chemical vapor deposition was investigated in a specially designed vertical downflow reactor. Dimethylzinc was used as the Zn source while different Se source compounds (hydrogen selenide (H_2Se), diethylselenide and methylallylselenide) were employed to determine the effect of different source combinations on morphology, thickness uniformity, growth rate, electrical properties and photoluminescence (PL) characteristics of the grown ZnSe films. The H_2Se was produced in situ by reaction of H_2 and Se followed by distillation to control the amount of H_2Se entering the reaction zone. H_2Se produced very high mobility films with good PL spectra but poor surface morphology. Diethylselenide led to layers of good morphology and PL characteristics but the films were highly resistive. Unusual surface features were observed for methylallylselenide.

INTRODUCTION

High quality epitaxial layers of ZnSe, that can be doped controllably, have a well-recognized potential use in blue light emitting diodes and injection lasers. Metalorganic chemical vapor deposition (MOCVD) is a promising technique for reproducible growth of high purity ZnSe films. Different source compounds have been investigated over the past 10 years to improve material properties and to gain insight into the underlying growth chemistry. It was recognized early that the use of hydrogen selenide (H_2Se) with zinc alkyls led to premature gas phase reactions, resulting in poor film morphology, thickness uniformity, and inefficient source utilization [1,2]. This fact, in combination with the high toxicity of H_2Se has led to investigations of alternative organometallic Se sources [3-7].

In this paper, three different selenium sources: H_2Se, diethylselenide (DESe), and the new methylallylselenide (MASe) are investigated. ZnSe films, grown by MOCVD with dimethylzinc (DMZn) as the common Zn source, are evaluated by their morphology, thickness uniformity, growth rate, electrical properties and photoluminescence characteristics. The results are reported in terms of the best material grown at optimal conditions for each source combination.

EXPERIMENTAL

The ZnSe films were grown in a loadlock-equipped, vertical stagnation point flow MOCVD reactor. A mass spectrometer system, with a sampling port immediately downstream of the susceptor, was used to monitor stable reaction by-products. DMZn (Texas Alkyls) was used as the Zn source while three different Se sources were tried : MASe (American Cyanamid), DESe (Alfa Products) and H_2Se. The H_2Se was produced *in-situ* from Pd-purified hydrogen and 99.9999 % pure selenium (Osaka Asahi Metals) in a small reactor unit, separated from selenium and partially condensed, so that a controlled amount of H_2Se in H_2 could be introduced into the reaction chamber. The ZnSe films were grown epitaxially on semi-insulating (100) GaAs substrates, misoriented 2° towards (110). The substrates were degreased and etched according to standard procedures [1] before being placed on a Mo susceptor inside the loadlock. The reactor was pumped down to a base pressure of 1×10^{-8} Torr and then purged overnight with H_2 at low pressure prior to growth. After transferring the substrate into the main reactor chamber, the deposition sequence was initiated with native oxide desorption in 1 SLM H_2 flow at 650°C and a total reactor pressure of 300 Torr for 10 minutes. During the adjustment to the growth temperature (T_G), the Se source diluted in H_2 was switched into the reactor and, after establishing the desired T_G and reactor pressure, DMZn in H_2 was introduced to initiate growth.

The films were grown under a constant DMZn flux and Se rich conditions. The [VI/II] ratio was 2 for the organometallic Se sources and 10 for the H_2Se; the later high ratio was needed to achieve Se rich growth conditions because of the inlet nozzle employed to avoid premature reactions. The Se rich condition was necessary for producing high quality ZnSe, as judged by PL spectra and it was verified by two observations: 1) An increase in the Se-source flux did not result in a variation of the growth rate and 2) an increase in the Zn-source flux resulted in an increase of the growth rate. Typical flow rates of the reactants as well as other experimental growth conditions for each source combination are listed in Table I.

Table I. Optimal conditions for growth of ZnSe with the best photoluminescence spectra and electrical properties for each source combination. T_G is the growth temperature, P_G is the reactor pressure, R_G is the growth rate, μ_{77} and μ_{300} are the mobility at 77 K and 300 K, respectively. The total flow rate (in balance H_2) was 1 slm for all cases, and the Se and Zn source molar flow rates are given (in μmol/min) under the ratio [VI/II].

Source combination	T_G (°C)	P_G (Torr)	[VI/II]	R_G (μm/hr)	μ_{77} (cm^2/Vsec)	μ_{300} (cm^2/Vsec)	Morphology
H_2Se/DMZn	325	30	150/15	24.0	8000-9000	450-500	Hillocks
DESe/DMZn	500	300	40/20	3.6	high ρ	high ρ	Mirror-like
MASe/DMZn	480	300	40/20	3.6	350-400	200	Hazy(gaps)

After growth, the sample thickness was measured by Scanning Electron Microscopy (SEM) for thin films and by an optical microscope (calibrated for measuring line thicknesses on VLSI masks) for thick films. The surface morphology was examined by SEM. All layers investigated in the present study were single crystalline, as verified by their electron channelling patterns and electro-optical characteristics.

Photoluminesence (PL) spectra at 9 K were obtained by using the ultraviolet output of an Ar$^+$-ion laser at low power densities (< 30 mW/cm^2) in an unfocussed beam of diameter 2 mm, and analyzed by using a 0.85-m SPEX 1403 double monochromator and photon counting electronics. Spectral resolution in the range of interest was better than 0.5 cm^{-1}.

Electrical properties were evaluated by measurements of resistivity and Hall effect. Ohmic contacts were made by pressing small pieces of In on the freshly grown ZnSe surface, followed by annealing in purified H_2 for 5 min at 300°C. Formation of ohmic contacts was verified by using a curve tracer. PL spectra recorded before and after annealing were identical in sharpness and intensity of peaks, indicating that annealing at the above conditions had no significant effect on the properties of the films.

RESULTS AND DISCUSSION

Growth rate

Figure 1(a) shows the measured growth rates as function of reciprocal growth temperature for growth from DESe and MASe . The growth rates become constant and equal at temperatures above 500°C and 480°C for growth from DESe and MASe, respectively. Combined with the observation that the growth rate is independent of an increase in the Se flow rate but linearly dependent on the DMZn flow rate, this strongly suggests that the growth rate at the temperature plateau is limited by mass transfer of the Zn-alkyl to the deposition surface. The apparent activation energies for the kinetically limited regimes are 13 kcal/mole for MASe and 18 kcal/mole for DESe. The lower activation energy for growth with MASe is consistent with an expected lower decomposition temperature of MASe compared to DESe because of the more stable allyl radical relative to the ethyl radical. Similar observations have been made for organometallic Te compounds [8].

Mitsuhashi et al. [7] have reported apparent activation energies of 22 kcal/mole for the growth of ZnSe from DMSe and DESe under similar conditions to those used in this study. The higher activation energy of DMSe is in line with the above bond strength arguments since the methyl-Se bond is stronger than the ethyl-Se bond. However, the reported value for DESe is larger than that obtained in this study. Mitsuhashi et al. also observed significantly different mass transfer limited growth rates for the two Se compounds, which is surprising since the diffusion coefficients differ by less than 10%. The low values for the activation energies obtained in this and the previous study indicate that the growth process is limited not only by the gas phase decomposition of the precursors but also by surface reactions.

The growth rate of ZnSe from H_2Se and DMZn, shown in Figure 1(b), varies only slightly with respect to changes in temperature and shows no clear kinetic and mass transfer limited regimes. These films were grown with a higher VI/II ratio and at a lower pressure than was used for the organometallic Se sources (cf. Table I) in order to reduce effects of premature gas phase reactions. Therefore, it is not possible to make a direct comparison between the growth rates in Figures 1(a) and (b). Nevertheless, it is clear that ZnSe growth from H_2Se at optimal conditions is much faster than growth from organometallic Se sources. Furthermore, the efficiency of Se incorporation is 75% higher for growth with H_2Se than with the Se organometallic sources. This and the different temperature behavior of the growth rates suggest that disparate mechanisms control the deposition of ZnSe from Se hydride versus from Se organometallic compounds.

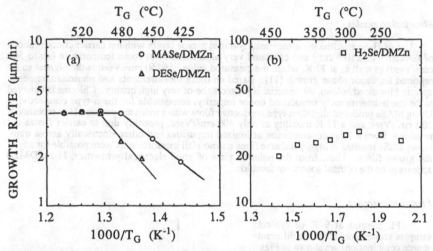

Figure 1. The effect of growth temperature on the growth rate of ZnSe deposited from three different source combinations: (a) MASe/DMZn and DESe/DMZn and (b) H_2Se/DMZn; the other growth variables were set according to Table I.

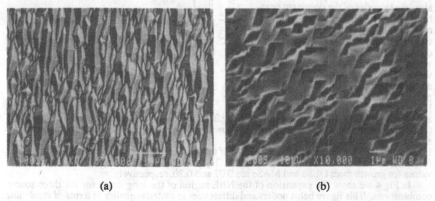

Figure 2. SEM photomicrographs of single crystalline films of ZnSe grown from (a) H_2Se/DMZn and (b) MASe/DMZn under conditions given in Table I.

Morphology

Among the ZnSe films grown on (100) GaAs substrates from the different source combinations, only those using DESe resulted in thick (4-5 μm) films with mirrorlike surfaces and with no features observable by SEM even under very high magnification. H_2Se/DMZn produced well-documented but poorly understood [9,10] ridge-shaped hillock structures parallel to the (110) cleavage plane over the entire wafer area, as illustrated in Fig.2(a). This rough surface morphology is undesirable in optoelectronic applications. It was found that hillock formation could be minimized by a reduction in the growth rate or by the addition of an adduct forming reagent (allylamine), which blocks the parasitic gas phase reaction. These phenomena are currently under investigation. Using MASe led to a surface that appears hazy to the naked eye and that has unusual surface structures when observed by SEM. The surface, shown in Fig.2(b), appears to have oddly shaped gaps between patches of smooth surface. To our knowledge these features have not been observed before in the growth of ZnSe.

Electrical properties

Use of H_2Se resulted in unintentionally doped n-type layers with net carrier concentrations of 6.4×10^{14}-1.5×10^{16} cm^{-3} and exhibited very high mobility at room temperature (up to 500 cm^2/Vsec) as well as at 77 K, where the measured value of 9250 cm^2/Vsec is the highest so far reported for vapor phase growth [11]. Based on these measurements and photoluminescence spectra (discussed below), the material is judged to be of very high quality. Chlorine is believed to be the unintentionally introduced donor impurity, responsible for the n-type conductivity. Using MASe produced slightly n-type conductive films with a room temperature Hall mobility of 200 cm^2/Vsec and a 77 K mobility of only 400 cm^2/Vsec, probably due to carrier freeze-out resulting from a large concentration of ionized impurities and other electrically active traps. Using DESe resulted in highly resistive films and no Hall measurements were possible for any of the grown films. Thus, from the point of view of good electrical properties, H_2Se/DMZn appears to be the optimal source combination.

Photoluminescence

PL spectra at 9 K of typical samples grown from the three different source combinations are shown in Figs. 3 and 4. In Fig.3, we compare the long scans with particular emphasis in the magnitude of the deep level (DL) emission ($h\nu < 2.5$ eV) as well as the absolute intensity of the dominant near-band-edge (NBE) emission. Both are measures of the relative quality of the corresponding film. All spectra are composed of a strong NBE emission, while the DL emission is broad and weak. However, in terms of the absolute intensity of the dominant NBE peak, the quality of ZnSe grown from H_2Se is the best (x1), while the DESe and MASe produce ZnSe with weaker (x40) and much weaker (x100) NBE emission intensity, respectively.

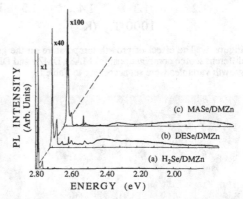

Fig.3. Comparison of the PL spectra of ZnSe films grown from the three different source combinations.

In terms of deep levels, we compare the ratio of the maximum DL peak over the corresponding free excitonic emission. The ZnSe grown from H_2Se with ratio 0.01 is again the best, while the values for growth from DESe and MASe are 0.07 and 0.20, respectively.

In Fig.4 we show the expansion of the NBE region of the long scans for the three source combinations. This figure helps understand differences in material quality in terms of resolvable peaks related to shallow donor or acceptor impurities. We proceed with analysis of the first spectrum and then compare it and discuss the differences with the other ones. Using H_2Se

produces films with NBE spectra shown in Fig.4(a), which are dominated by an intense and narrow peak at 2.7954 eV, identified as the commonly observed I_x [12]. This may be assigned to $I_3 \cdot Cl$, as described by Dean at al. [13], which would explain the n-type conductivity of these films. A second dominant peak, referred to as I_{20}, is clearly defined at 2.7970 eV. This has been attributed to recombination of excitons bound to shallow extrinsic neutral donors. There are also two peaks due to free exciton recombination: the very intense lower branch polariton, denoted E_x^L at 2.8002 eV and the upper branch polariton, E_x^U at 2.8026 eV. The peak on the low-energy side of the I_x at 2.7938 eV is the I_3 line, presumably due to excitons bound to an ionized donor [12]. One also distinguishes a low I_v peak at 2.7768 eV and the phonon replica of the free exciton at 2.7711 eV. Very weak donor-acceptor pair (DAP) emission is also observed (Fig.3), which means that the concentration of acceptor-like impurities or defects is considerably reduced in the films grown from $H_2Se/DMZn$. As compared to other PL results in the literature [12,14], this PL spectrum is distinguished by its sharp, narrow and distinct peaks in the NBE region and the weakness of its I_1^d line, usually appearing at 2.782 eV, attributable to Zn vacancies [14]. In fact, since there is no discernible I_1^d line in the PL spectra of all the samples discussed in this paper, we believe that our material is very close to being stoichiometric.

Using DESe results in films with NBE spectra shown in Fig.4(b). Important differences with the previous spectrum are: The dominant NBE peak is now the free exciton E_x^L, whereas I_{20}, I_x and I_3 are relatively weak. A new unidentified peak, marked as I_1, appears at 2.7920 eV, which may be due to a shallow acceptor impurity. Dean at. al. [15] have identified a peak occuring at this energy as due to the nitrogen acceptor bound exciton (I_1^N). However, since it is difficult to incorporate nitrogen in ZnSe and we do not have any indication of nitrogen containing impurities in the DESe, this possibility seems unlikely. Although deep levels are higher for ZnSe from DESe than when using H_2Se, as discussed above, the dominance of the free excitonic emission in the NBE spectrum indicates that the intrinsic properties of this material are good enough for n- or p-type doping to be attempted. The lack of any dominant shallow donor or acceptor bound peak is consistent with the high resistivity of the film.

Using MASe results in films with NBE spectra shown in Fig.4(c). The picture is different here, with I_x and I_1 as the dominant NBE peaks. I_1 appears at 2.7896 eV i.e. 2.4 meV lower than the I_1 of Fig.4(b). It is therefore believed to be due to a different shallow acceptor impurity, possibly Li since it is very close to the experimentally observed peak for Li [16]. However, the Li bound exciton usually appears as a doublet and gives rise to a fairly intense DAP, which are not observed in the PL spectrum, therefore cannot be assigned unambiguously to any of the reported acceptor impurities. Nevertheless, it is a peak of nature different from the I_1 of Fig.4(b)

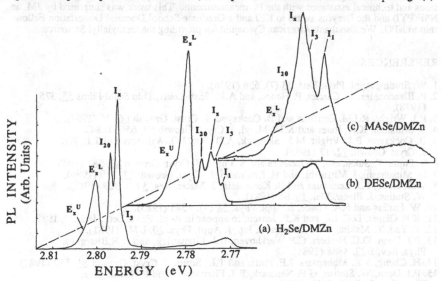

Fig.4. Expanded NBE spectra of ZnSe grown from the three source combinations.

and it stems from an impurity in the MASe. This material is judged to be of inferior quality as compared to the films with the spectra of Fig.4(a) and 4(b). Thus, the new MASe source needs additional purification to remove impurities, and further evaluatation as a potential Se source.

CONCLUSIONS

Comparison of growth rates, surface morphology, electrical and optical properties for ZnSe grown heteroepitaxially on GaAs from DMZn and three different Se sources shows that each source combination has potential advantages as well as disadvantages. $H_2Se/DMZn$ appears to be the best choice in terms of electrical properties, growth rate and photoluminescence characteristics. Another advantage of this source combination is the low growth temperature, close to 325°C for optimal material properties or as low as 225°C for single crystalline material. This low growth temperature has important consequences for the growth of high purity ZnSe, relatively free from the presence of stoichiometric defects, which form deep centers in ZnSe, compensating shallow electrically active centers. However, the persistent problem of premature reactions complicates reactor design and causes poor surface morphology, which may be minimized by lowering the growth rate. Therefore, if this premature reaction could be inhibited, $H_2Se/DMZn$ would appear to be the optimal source combination, since it already has led to undoped material with the best electrical properties so far reported for vapor phase growth [11]. The ultimate proof, of course, would be the ability to dope the material p-type at high enough concentrations for device applications.

DESe/DMZn produces material with acceptable PL characteristics, excellent surface morphology and film thickness uniformity. However, in terms of electrical properties it appears that the films are highly compensated. Again, doping experiments will be necessary to test the viability of this source combination for growth of device quality ZnSe. The new source MASe needs to be further purified and additional growth experiments performed before any conclusions can be made. However, it has already led to ZnSe of better quality than other Se alternative sources such as selenophene [5] or methylselenol [6] and, in this respect, is a very promising source.

ACKNOWLEDGEMENTS

The authors are grateful to J.E. Potts, H. Cheng, J.M. DePuydt and G. Haugen for discussions and technical assistance with the PL measurements. This work was supported by 3M, an NSF(PYI) and the Dreyfus award to KFJ and a Graduate School Doctoral Dissertation Fellowship to KPG. We also thank American Cyanamid for providing the methylallyl Se source.

REFERENCES

1. W. Stutius, Appl. Phys. Lett. 33 (7), 656 (1978).
2. P. Blanconnier, M. Cerclet, P. Henoc, and A.M. Jean-Louis, Thin Solid Films 55, 375 (1978).
3. P.J. Wright, R.J.M. Griffiths and B. Cockayne, J. Cryst. Growth 66, 26 (1984).
4. S. Sritharan, K.A. Jones and K.M. Motyl, J. Cryst. Growth 68, 656 (1984).
5. B. Cockayne, P.J. Wright, M.S. Skolnick, A.D. Pitt, J.O. Williams and T.L. Ng, J. Cryst. Growth 72, 17 (1985).
6. S. Fujita, T. Sakamoto, M. Isemura and S. Fujita, J. Cryst. Growth 87, 581 (1988).
7. H. Mitsuhashi, I. Mitsuishi, and H. Kukimoto, J. Cryst. Growth 77, 219 (1986).
8. W.E. Hoke, P.J. Lemonias and R. Korenstein, J. Mater. Res. 3 (2), 329 (1988).
9. W. Stutius, J. Electr. Mat. 10, 95 (1981).
10. W. Stutius and F.A. Ponce, J. Appl. Phys. 58 (4), 1548 (1985).
11. K.P. Giapis, D.C. Lu, and K.F. Jensen, to appear in Appl. Phys. Lett., Jan. 23, 1989.
12. T. Yao, Y. Makita, and S. Maekawa, Jap. J. Appl. Phys. 20, L741 (1981).
13. P.J. Dean, D.C. Herbert, C.J. Werkhoven, B.J. Fitzpatrick, and R.N.Bhargava, Phys. Rev. B23, 4888 (1981).
14. H. Cheng, S.K. Mohapatra, J.E. Potts, and T.L. Smith, J. Cryst. Growth 81, 512 (1987).
15. P.J. Dean, W. Stutius, G.F. Neumark, B.J. Fitzpatrick, and R.N.Bhargava, Phys. Rev. B27 (4), 2419 (1983).
16. H. Cheng, J.M. DePuydt, J.E. Potts, and T.L. Smith, Appl. Phys. Lett. 52 (2), 147 (1988).

NEW ORGANOTELLURIUM PRECURSORS FOR THE PYROLYTIC AND PHOTOLYTIC DEPOSITION OF $Hg_{1-x}Cd_xTe$

ROBERT W. GEDRIDGE, JR.,* KELVIN T. HIGA, AND ROBIN A. NISSAN
Chemistry Division, Research Department, Naval Weapons Center, China Lake, CA 93555

ABSTRACT

Organometallic precursors with low decomposition temperatures are essential in the fabrication of high performance mercury cadmium telluride ($Hg_{1-x}Cd_xTe$) infrared detectors by pyrolytic and photolytic metal-organic chemical vapor deposition (MOCVD). Film growth temperature is governed by the relative stability and/or reactivity of the organotellurium precursor, which is determined by the strength of the Te-C bonds. Since the rate-determining step in the pyrolysis of organometallic compounds involves bond breaking and free radical formation, we have concentrated on the synthesis of a variety of organotellurium precursors with substituents that possess low activation energies for the formation of hydrocarbon free radicals. The synthesis, characterization, and properties of methylallyltelluride, ethylallyltelluride, isopropylallyltelluride, tertiarybutylallyl-telluride, methylbenzyltelluride, and methylpentadienyltelluride are reported. These unsymmetrical tellurides were characterized by ^1H, ^{13}C, and ^{125}Te NMR spectroscopy. The potential applicability of these organotellurium precursors to lower film-growth temperatures in MOCVD is discussed.

INTRODUCTION

Metal-organic chemical vapor deposition (MOCVD) is an increasingly valuable technique used to prepare high quality mercury cadmium telluride ($Hg_{1-x}Cd_xTe$) semiconductor films. The development of pyrolytic and photolytic MOCVD growth of mercury cadmium telluride superlattices and focal plane arrays is essential in the fabrication of high performance infrared detectors [1]. Low substrate temperature is the critical requirement for growth of complex structures that contain sharp heterojunction interfaces. However, conventional MOCVD of mercury cadmium telluride with elemental mercury, dimethylcadmium, and diethyltelluride between 390-450°C [2] results in film degradation due to mercury evaporation and diffusion. Lower growth temperatures can be achieved by a variety of methods which include the use of alternative organotellurium source compounds [3-7] and photo-assisted MOCVD [1,8].

Lowering the growth temperature of mercury cadmium telluride is governed by the relative stability and/or reactivity of the organotellurium source compound, which is determined by the strength of the Te-C bonds. The rate-determining step in the pyrolysis of organometallic compounds is homolytic cleavage of the metal-carbon bond, which is generally the weakest bond in the molecule. The relative thermal stability of organotellurium precursors parallels the activation energy for breaking the C-H bond in the parent hydrocarbon and formation of a hydrocarbon free radical [9]. Lower activation energies for breaking the C-H bond result when the hydrocarbon free radical is stabilized by resonance. Thus allyl (C-H = 87 kcal/mol), benzyl (88 kcal/mol), and pentadienyl (80 kcal/mol) tellurides should be less stable than ethyl (98 kcal/mol) or tertiarybutyl (92 kcal/mol) tellurides [10]. Our research has focussed on developing alternative organotellurium source compounds with reduced activation energy for Te-C bond cleavage and sufficient vapor pressure for MOCVD.

Diallyltelluride was used to grow HgTe films at 180°C at 1 μm/h, which is the lowest reported film growth temperature of HgTe by pyrolytic MOCVD [4]. Recently methylallyltelluride was reported as an alternative source compound used to grow CdTe films at 290°C at 30 μm/h in an unassisted pyrolytic MOCVD process [7]. With these encouraging results, other allylic and unsaturated organotellurium derivatives have been investigated which may prove to be even better organotellurium source compounds.

RESULTS AND DISCUSSION

Synthesis

The unsymmetrical diorganyl tellurides were prepared by two synthetic methodologies. In the first, tellurium metal reacted with an alkyl lithium reagent by inserting into the Li-C bond to give the lithio(alkyl)telluride (LiTeR) which was not isolated [11]. Lithio(alkyl)telluride was then immediately allowed to react with an alkyl halide (R'X) to give the desired unsymmetrical telluride (eq 1) [12].

$$LiR + Te \xrightarrow{THF} LiTeR \xrightarrow{+R'X} RTeR' + LiX$$

$$R = CH_3, (CH_3)_3C \quad R'X = CH_2=CHCH_2I$$

$$R = CH_3 \quad R'X = C_6H_5CH_2Br, \quad CH_2=CHCH=CHCH_2Br \tag{1}$$

The methylallyltelluride, tertiarybutylallyltelluride, methylbenzyltelluride, and methylpentadienyltelluride compounds were prepared by this synthetic route. In an alternative synthetic route, tellurium metal reacts with a Grignard reagent to yield alkyltelluro magnesium chloride (RTeMgCl), which was immediately treated with an alkyl halide to form the desired unsymmetrical telluride (eq 2) [13,14].

$$RMgCl + Te \xrightarrow{THF} RTeMgCl \xrightarrow{+R'X} RTeR' + MgClX$$

$$R = CH_2CH=CH_2$$

$$R'X = CH_3I, CH_3CH_2I, (CH_3)_2CHBr \tag{2}$$

The methylallyltelluride, ethylallyltelluride, and isopropylallyltelluride compounds were prepared by this method.

Generally, the unsymmetrical organotellurium compounds prepared from lithium alkyl reagents were isolated in better yields than those prepared from Grignard reagents. The unpurified products from the Grignard reactions contained significant amounts of the undesired symmetrical diorganyl telluride. Complex mixtures of organotellurium compounds resulting from unwanted side reactions have been reported when Grignard reagents were used to alkylate tellurium [15].

Properties

The unsymmetrical organotellurium compounds were isolated as malodorous yellow liquids. With the exception of methylbenzyltelluride, these compounds are extremely air-, light-, and heat-sensitive. These compounds were purified by several fractional vacuum distillations under an inert atmosphere in the dark and stored cold in opaque containers. The extreme light-sensitivity of methylallyltelluride and methylpentadienyltelluride suggests that they are potential precursors to photo-assisted MOCVD. Though the new allyl and pentadienyl compounds developed do not possess vapor pressures as high as diethyltelluride, their vapor pressures are sufficient to be practical for MOCVD (Table I). Unfortunately, methylbenzyltelluride possessed a low vapor pressure, minimal light-sensitivity, and greater thermal stability, suggesting that it would not be a practical candidate for pyrolytic or photolytic MOCVD.

Table I. Experimental Data for the Unsymmetrical Tellurides

Compound	Method of Preparation	Boiling Point ° C (mmHg)	%Yield Crude
(Methyl)Te(Allyl)	LiR / RMgCl	41 (13)	50-60
(Ethyl)Te(Allyl)	RMgCl	53 (13)	50-80
(i-Propyl)Te(Allyl)	RMgCl	58 (12)	50-70
(t-Butyl)Te(Allyl)	LiR	60 (12)	80-90
(Methyl)Te(Benzyl)	LiR	61 (0.15)	80-90
(Methyl)Te(Pentadienyl)	LiR	66 (6)	80

Characterization

Characterization involved ^{1}H, ^{13}C, and ^{125}Te NMR spectroscopy. The chemical shifts of the protons of the desired unsymmetrical diorganyl tellurides were similar to those of the symmetrical tellurides. The ^{1}H NMR spectra of the crude products were further complicated since the olefinic protons of the allyl and pentadienyl substituents appeared as complex multiplets due to spin-spin coupling.

The ^{13}C NMR spectra of the allylic tellurides revealed that the chemical shifts of the three carbon atoms of the allyl group were unchanged by different alkyl substituents on tellurium. The chemical shifts of the corresponding carbon atoms of the symmetrical tellurides were very similar to those of the unsymmetrical tellurides. These spectroscopic observations suggested that ^{1}H and ^{13}C NMR are not effective methods of measuring the relative amount of the symmetrical telluride impurities.

^{125}Te chemical shifts are sensitive to concentration, temperature, solvent, and substituents [16,17]. The ^{125}Te chemical shift of ditertiarybutyltelluride (999 ppm) [17] is significantly downfield in comparison to diallyltelluride and tertiarybutylallyltelluride by greater than 600 ppm and 280 ppm, respectfully (Table II). The absolute sensitivity of ^{125}Te NMR, which is a product of the relative sensitivity and the natural abundance, is an order of magnitude greater than that of ^{13}C NMR. Therefore, ^{125}Te NMR experiments require less time than ^{13}C NMR experiments and the described considerations suggest that ^{125}Te NMR is a valuable probe for the identification of organotellurium impurities during purification of the telluride.

Table II. ^{125}Te NMR Chemical Shifts of Diorganyl Tellurides

Compound	Chemical Shift (ppm)[*]
(Methyl)Te(Allyl)	182
(Ethyl)Te(Allyl)	382
(Allyl)Te(Allyl)	391
(i-Propyl)Te(Allyl)	550
(t-Butyl)Te(Allyl)	717
(Methyl)Te(Benzyl)	314
(Methyl)Te(Pentadienyl)	215

[*] Spectra were recorded from 1 mol/L telluride samples in C_6D_6 and referenced to neat dimethyltelluride whose chemical shift is 0 ppm.

CONCLUSIONS

The unsymmetrical allyl and pentadienyl tellurides described are potential candidates for pyrolytic and photolytic MOCVD. The extreme light-sensitivity of methylallyltelluride and methylpentadienyltelluride infers that they are potential precursors to photo-assisted MOCVD. Unfortunately, the low vapor pressure and greater observed stability of methylbenzyltelluride suggests that this compound would not be a suitable candidate for MOCVD. Preparation of these compounds by utilizing organolithium reagents gave products in high yields with less symmetrical telluride impurities than observed when Grignard reagents were used. It was imperative that purification of these compounds involved fractional vacuum distillation under an inert atmosphere in the dark at reduced temperatures. These compounds must be stored in the dark at low-temperatures. The significant deviations in chemical shifts among organotellurium compounds as well as the greater sensitivity of ^{125}Te NMR over ^{13}C NMR, implies that ^{125}Te NMR is the spectroscopic probe of choice in identifying organotellurium compounds.

ACKNOWLEDGMENTS

The authors gratefully acknowledge financial support from the Office of Naval Research and the postdoctoral fellowship from the American Society for Engineering Education/Office of Naval Technology for Dr. Robert W. Gedridge, Jr.

REFERENCES

1. W.L. Ahlgren, E.J. Smith, J.B. James, T.W. James, R.P. Ruth, E.A. Patten, R.D. Knox, and J.-L. Staudenmann, J. Cryst. Growth 86, 198 (1988).
2. W.E. Hoke, P.J. Lemonias, and R. Traczewski, Appl. Phys. Lett. 44, 1046 (1984).

3. W.E. Hoke and P.J. Lemonias, Appl. Phys. Lett. 46, 398 (1985); 48, 1669 (1986).
4. R. Korenstein, W.E. Hoke, P.J. Lemonias, K.T. Higa, and D.C. Harris, J. Appl. Phys. 62, 4929 (1987).
5. L.S. Lichtmann, J.D. Parsons, and E.-H. Cirlin, J. Cryst. Growth 86, 217 (1988).
6. D.W. Kisker, M.K. Steigerwald, T.Y. Kometani, and K.S. Jeffers, Appl. Phys. Lett. 50, 1681 (1987).
7. J.D. Parsons and L.S. Lichtmann, J. Cryst. Growth 86, 222 (1988).
8. S.J.C. Irvine, J.B. Mullin, H. Hill, G.T. Brown, and S.J. Barnett, J. Cryst. Growth 86, 188 (1988).
9. W.E. Hoke, P.J. Lemonias, and R. Korenstein, J. Mater. Res. 3, 329 (1988).
10. Handbook of Chemistry and Physics, edited by R.C. Weast, 67th ed. (CRC, Boca Raton, Fl, 1986), p. F-178.
11. K.J. Irgolic, The Organic Chemistry of Tellurium, (Gordon and Breach Science Publishers, New York, 1974), p. 26.
12. J.L. Piette and M. Renson, Bull. Soc. Chim. Belges. 79, 353 (1970); 79, 367 (1970).
13. K. Bowden and A.E. Braude, J. Chem. Soc. 1952, 1068.
14. G. Pourcelot, C. R. Acad. Sci., Paris 260, 2847 (1965).
15. N. Petragnani and M. de Moura Campos, Chem. Ber. 96, 249 (1963).
16. N.P. Luthra and J.D. Odom, The Chemistry of Organic Selenium and Tellurium Compounds, edited by S. Patai and Z. Rappoport (John Wiley & Sons, New York, 1986), vol. 1, pp. 221-233.
17. D.H. O'Brien, N. Dereu, C.-K. Huang, K.J. Irgolic, and F.F. Knapp, Jr., Organomet. 2, 305 (1983).

ASSESSMENT OF ORGANOTELLURIUM COMPOUNDS FOR USE AS MOVPE PRECURSORS

J.E. HAILS, S.J.C. IRVINE, J.B. MULLIN, D.V. SHENAI-KHATKHATE* AND D. COLE-HAMILTON*

Royal Signals & Radar Establishment, St. Andrews Road, Malvern, Worcs. WR14 3PS. UK

*   University of St. Andrews, St. Andrews, Fife, UK

ABSTRACT

In order to reduce the growth temperature of (Hg,Cd)Te by MOVPE below 350–400°C, alternative organometallic precursors will be required which either decompose at a lower temperature than existing precursors or which absorb strongly at a suitable wavelength in the ultraviolet. The features required for a programme of assessment of organometallics are discussed. UV absorption spectra for dimethyltelluride, dimethylditelluride, diethyltelluride, di-iso-propyltelluride and diallyltelluride are presented and their usefulness as photolytic MOVPE precursors discussed.

INTRODUCTION

Efforts are currently in progress to reduce the temperature at which the infra-red detector material (Hg,Cd)Te can be grown. This reduction in temperature is necessary in order to reduce and control the equilibrium mercury vacancy concentration which in turn would allow greater control of the electrical properties. In addition, the lowering of the growth temperature would also reduce the interdiffusion between substrates and epitaxial layers thus enabling more complex structures to be grown. The lowest temperature at which pyrolytic growth can occur is dependent upon the decomposition temperature of the more stable of the organometallic precursors. Among the most commonly used precursors for the MOVPE growth of (Hg,Cd)Te are Me_2Cd, Et_2Te and Pr^i_2Te and it is the stability of the tellurium alkyl that limits the growth temperature to around 410°C for Et_2Te and 350°C for Pr^i_2Te. In order to reduce the growth temperature a less stable organotellurium compound is required. An alternative method of growth is by photolysis which also offers the potential for photo-patterning and an organotellurium compound which strongly absorbs in the UV at a wavelength corresponding to a laser or arc lamp emission, for example, 193, 248, 254 or 257nm would be invaluable in photolytic growth.

For any organometallic compound to be suitable as an alternative to the existing ones it must possess several other properties. It must be stable at room temprature, under an inert atmosphere if necessary, for a period of at least several months and preferably indefinately. It must also be sufficiently volatile to be transported around the MOVPE system, ideally with a vapour pressure of >1 Torr at room temperature. Equally importantly an alternative organometallic should not undergo any undesirable reactions such as polymerisation or premature vapour phase reaction with the other organometalics in the system. It must also be obtainable chemically pure.

Several techniques are likely to be of use in the assessment of organotellurium compounds for use as MOVPE precursors and these include UV spectroscopy, vapour pressure measurements, study of decomposition products and the growth of HgTe, CdTe and (Hg,Cd)Te. Several alternatives have been tried recently with varying degrees of success (see table 1). In the published data the growth of layers with alternative precursors seems to have been limited largely to HgTe and CdTe and measurements of their properties to vapour pressure at one or two temperatures.

EXPERIMENTAL

We now have in operation an MOVPE system specifically designed to investigate organotellurium compounds for suitability as MOVPE precursors. This system includes facilities for sampling the gas stream to look at the UV spectrum, measure the vapour pressure of the organometallics, investigate decomposition products and grow HgTe, CdTe and (Hg,Cd)Te. A schematic diagram of the system is shown in figure 1. In order to

Table 1 : Literature survey of organotellurium
compounds which have been used as MOVPE precursors

Organometallic	UV Spectra	Vapour Pressure	HgTe μm/hr	HgTe °C	CdTe μm/hr	CdTe °C	(Hg,Cd)Te μm/hr	(Hg,Cd)Te °C	Reference
Me_2Te		7.97−1865/T (a)	0.6	325					1
Et_2Te	✓	7.99−2093/T (a)	25	410	20	350	20	410	2, 3, 4
Pr^n_2Te		2.0τ @ 30°C							
Pr^i_2Te		5.6τ @ 30°C	16–18	350	8–10	350	15	350	5, 6
Bu^t_2Te		4τ @ 40°C	1.6	230	1.6	250			7
C_4H_6Te(b)		Solid @ 0°C			4	250			8
$Me(CH_2=CH-CH_2)Te$(c)			15	320	24	290	4	350	9
$(CH_2=CH-CH_2)_2Te$(d)		3τ @ 45°C	1	180	2	270			10
Me_2Te_2		6.94−2200/T (a)			<4	250			11

(a) Vapour pressure equation log P(mmHg)
(b) 2,5-dihydrotellurophene
(c) Methylallyltelluride
(d) Diallyltelluride

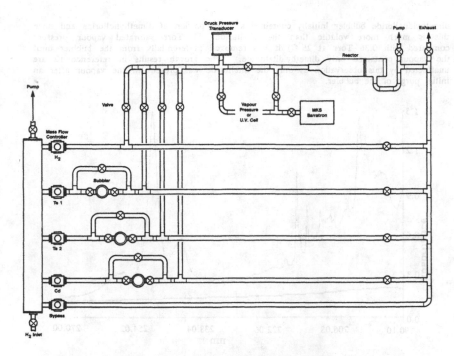

Fig 1. Schematic illustration of the MOVPE system designed to investigate metalorganic precursors.

measure the vapour pressure of an organotellurium compound the carrier gas is passed through the required bubbler where it picks up the organometallic. The gas stream is then passed through the vapour pressure cell held at liquid nitrogen temperature to condense out the precursor. When sufficient material has been collected the cold cell is evacuated, the pressure measured, the cell isolated and allowed to warm up to a set temperature. The pressure is measured again and the vapour pressure calculated from the two pressure measurements. The vapour pressure equation can be calculated from several vapour pressure measurements at different temperatures. For ultra-violet spectroscopic measurements a 4cm cell was connected to the system, evacuated and sealed off. The gas stream containing the organometallic was allowed to stabilise over 5 to 10 minutes before the valve to the UV cell was opened. The sealed cell was transferred to the Perkin-Elmer Lambda 15 spectrometer and the UV spectrum recorded.

RESULTS AND DISCUSSION

The UV spectrum initially obtained from the Me_2Te_2 bubbler and published in reference 12 is illustrated in figure 2. After the Me_2Te_2 bubbler had been in use for some time a different spectrum, shown in figure 3, was reproducibly obtained and which we now believe to have arisen from the Me_2Te_2. The UV spectrum obtained from a sample of vapour taken from a Me_2Te bubbler is shown in figure 4. Comparison of the spectra in figures 2, 3 and 4 and absorbances at 222 and 249nm shows that our original sample from the Me_2Te_2 bubbler consisted of a mixture of 46% Me_2Te and 54% Me_2Te_2. The peaks arising from the dimethyltelluride are clearly visible in figure 2 while the presence of the dimethylditelluride is revealed in the increased absorption between 220 and 236nm compared with the dimethyltelluride spectrum. It seems that our dimethylditelluride bubbler initially contained a small portion of dimethyltelluride and since this is much more volatile than the ditelluride (51 Torr saturated vapour pressure compared with 0.36 Torr at 25°C) it was removed preferentially from the bubbler until the vapour contained only dimethylditelluride. The growth results in reference 12 are unaffected by this observation as only the ditelluride was present in the vapour after an initial purge of the bubbler.

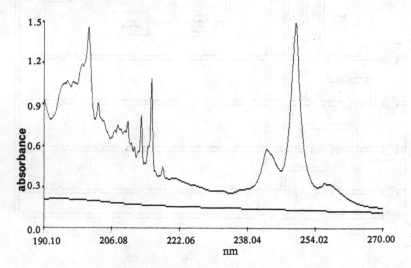

Fig 2.　UV spectrum of supposed Me$_2$Te$_2$ from reference 12.

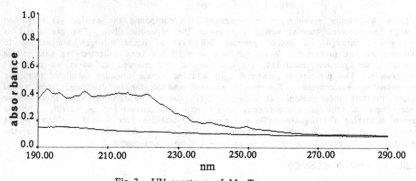

Fig 3.　UV spectrum of Me$_2$Te$_2$.

The UV spectra of di–iso–propyltelluride, diethyltelluride and diallyltelluride are shown in figures 5, 6 and 7 respectively.　Table 2 lists absorption cross–sections at the peaks and at some other wavelengths useful for photo–MOVPE calculated from

$$\sigma = \frac{kT}{LP} \ln I_0/I$$

k = Boltzmann constant
P = Partial pressure
L = UV cell length
T = temperature

Diallytelluride is not included in this table because we have not yet established its vapour pressure reliably.   At 257nm the two precursors with the stronger absorption cross-sections are Me_2Te and Et_2Te and these could be useful precursors for use with a frequency doubled argon ion laser.   At 248nm Me_2Te, Et_2Te and Pr^i_2Te all have very large cross-sections and (Et_2Te especially) absorb strongly. Me_2Te, Et_2Te and Pr^i_2Te have similar cross-sections at 254nm and would therefore be expected to have a similar performance in photolytic growth of CdTe, HgTe and (Hg,Cd)Te using a mercury arc lamp. Me_2Te_2 absorbs UV light very poorly by comparison but does exhibit a tail in absorption out to 285nm.

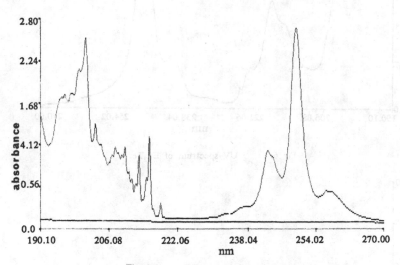

Fig 4.   UV spectrum of Me_2Te.

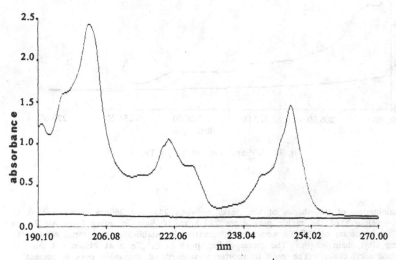

Fig 5.   UV spectrum of Pr^i_2Te.

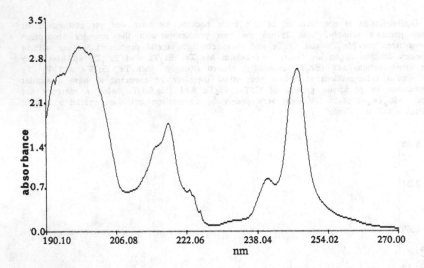

Fig 6.   UV spectrum of Et$_2$Te.

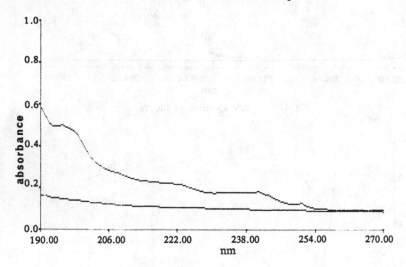

Fig 7.   UV spectrum of (allyl)$_2$Te.

Examination of the shapes of the peaks between 235 and 260nm shows a shift to shorter wavelength of the major peak from 249.3 to 246.9nm in going from Me$_2$Te to Et$_2$Te. A spectrum of Pr$^n_2$Te would be informative in establishing if a trend exists with increasing alkyl chain length. The corresponding peak in Pr$^i_2$Te is at 249nm but this has a branched alkyl chain. The peak to shorter wavelength of the main peak is present in all 3 spectra at 242.6, 240.2 and 242.9nm for Me$_2$Te, Et$_2$Te and Pr$^i_2$Te respectively. To longer wavelength of the main peak is another smaller peak in the Me$_2$Te spectrum, a similar one is just discernable in the tail of the Et$_2$Te absorption at 259nm. In contrast the absorption by Pr^i_2Te drops away sharply above 254nm giving rise to the differing cross-sections at 257nm. A shoulder is observed in Pr^i_2Te at 246nm. Further spectra from other organotellurium compounds will be required before further conclusions can be drawn.

The absorbance of the diallyltelluride is disappointing as it was anticipated that the carbon-carbon double bond of the allyl group would absorb strongly. It seems that this is not the case and the spectrum shows broad absorption only.

Precursor	λ max/nm	σ	σ at wavelength		
			257nm	254nm	248nm
Me_2Te	256.3 249.3 242.6	1.21 7.37 2.70	1.18	1.00	4.04
Me_2Te_2	215.6	1.79	0.19	0.18	0.27
Et_2Te	246.9	10.7	0.91	1.47	9.85
Pr^i_2Te	249.1	5.45	0.15	1.08	4.83

Table 2: Absorption cross sections (σ) for some organotellurium precursors at different wavelengths x $10^{17}cm^{-2}$

CONCLUSIONS

The factors which need to be taken into account when assessing organometallics for use as MOVPE precursors have been discussed. Not least of these is the ability to obtain pure precursors as exemplified by the masking of the Me_2Te_2 spectrum by the much more volatile Me_2Te impurity. Of the organotellurium compounds examined Me_2Te and Et_2Te are possible precursors for photo-MOVPE at 257, 254 and 248nm and Pr^i_2Te at 254 and 248nm.

REFERENCES

1. T.F. Kuech and J.O. McCaldin, J. Electrochem. Soc. 128, 1142 (1981).
2. S.J.C. Irvine, J.B. Mullin, D.J. Robbins and J.L. Glaser, Mat. Res. Soc. Symp. Proc. 29, 253 (1984).
3. S.J.C. Irvine, J.B. Mullin and A. Royle, J. Crystal Growth 57, 15 (1982).
4. J.B. Mullin, S.J.C. Irvine and J. Tunnicliffe, J. Crystal Growth 68, 214 (1984).
5. J. Thompson, P. Mackett and L.M. Smith, Materials Letters 5(3), 72 (1987).
6. W.E. Hoke and P.J. Lemonias, Appl. Phys. Lett. 46(4), 398 (1985).
7. W.E. Hoke and P.J. Lemonias, Appl. Phys. Lett. 48(24), 1669 (1986).
8. L.S. Lichtmann, J.D. Parsons and E.H. Cirlin, J. Crystal Growth 86, 217 (1988).
9. J.D. Parsons and L.S. Lichtmann, J. Crystal Growth 86, 222 (1988).
10. R. Korenstein, W.E. Hoke, P.J. Lemonias, K.T. Higa and D.C. Harris, J. Appl. Phys. 62 (12), 4929 (1987).
11. D.W. Kisker, M.L. Steigerwald, T.Y. Kometani and K.S. Jeffers, Appl. Phys. Lett. 50(23), 1681, (1987).
12. S.J.C. Irvine, H. Hill, O.D. Dosser, J.E. Hails, J.B. Mullin, D.V. Shenai-Khatkhate and D. Cole-Hamilton, Materials Letters 7, 25 (1988).

THE SYNTHESIS OF InP FROM INDIUM HALIDES
AND TRIS(TRIMETHYLSILYL)PHOSPHINE

MATTHEW D. HEALY, PAUL E. LAIBINIS, PAUL D. STUPIK AND
ANDREW R. BARRON[*]
Department of Chemistry, Harvard University, Cambridge, MA 02138,

ABSTRACT

InP has been prepared by the reaction of InX_3 (X = Cl, Br, I) with $P(SiMe_3)_3$. The intermediates and product have been characterized by XPS, elemental analysis and X-ray powder diffraction.

INTRODUCTION

The technological importance of the III-V semiconductor compound indium phosphide, InP, has led to a renewed interest in the chemistry of compounds containing covalent bonds between indium and phosphorus.[1] One such class of compounds with the empirical formula $(R_2InPR'_2)_n$ was originally prepared by Coates et al. 25 years ago by the reaction of indium trialkyl compounds with a secondary phosphine (e.g., Reaction 1).[2] Recently, Jones et al. have shown that lithium dialkyl-phosphides will react with indium halides to yield indium phosphido compounds. (Equation 2).[3]

$$3InMe_3 + 3HPMe_2 \longrightarrow (Me_2InPMe_2)_3 + 3MeH \qquad (1)$$

$$InCl_3 + 4LiPPh_2 \longrightarrow Li[In(PPh_2)_4] + 3LiCl \qquad (2)$$

Recent research in our laboratories has demonstrated that the dehalosilylation reaction is a facile route to the formation of In-P bonds (Equation 3).[4]

$$2Me_2InCl + 2Ph_2P(SiMe_3) \longrightarrow (Me_2InPPh_2)_2 + 2Me_3SiCl \qquad (3)$$

Mat. Res. Soc. Symp. Proc. Vol. 131. ©1989 Materials Research Society

The relatively mild conditions required for this reaction, often below room temperature prompted us to investigate the use of $P(SiMe_3)_3$ for the preparation of polycrystalline indium phosphide.[5]

RESULTS AND DISCUSSION

Addition of $P(SiMe_3)_3$[6] (5% molar excess) to a rapidly stirred toluene suspension of $InCl_3$ at $-78°C$, results in the formation of a pale yellow solid. As the reaction is warmed to room temperature the reaction mixture darkens. Removal of all volatiles by vacuum distillation, at room temperature yields a bright orange powder which is insoluble in organic solvents. The insolubility of this material suggests that it may be a polymeric species. Based on XPS and elemental analysis this orange powder is proposed to be $[Cl_2InP(SiMe_3)_2]_x$ (1a).

Heating 1a at $550°C$ for 1 h yields InP (2a) as a black powder. No chlorine or silicon is detected by XPS or elemental analysis. A typical survey spectrum of the InP is shown in Figure 1. In addition to peaks due to the In and P, peaks arising from O and C are observed. They are possibly due to the native oxide and some inadvertent carbon contamination due to air exposure. The X-ray powder diffraction confirms the product to be primarily poly-crystalline indium phosphide.[6]

The thermogravometic analysis of 1a is consistent with the loss of two equivalent of Me_3SiCl. Initial decomposition occurs at ca $110°C$, the maximum decomposition rate is above $410°C$. The conversion of 1a to InP is complete at $560°C$.

The reaction of $P(SiMe_3)_3$ with $InBr_3$ proceeds analogously. A dark orange intermediate (1b), formulated as $[Br_2InP(SiMe_3)_2]_x$, decomposes completely to give InP (2b) at $600°C$ (2 h). No bromine or silicon is detected by XPS or elemental analysis, however, the quantity of carbon incorporated (ca 1%) is larger than that found in the InP prepared from $InCl_3$.

The addition of $P(SiMe_3)_3$ to a toluene solution of InI_3 at $-78°C$, results in the initial formation of a yellow solution from which a dark solid eventually precipitates (1c). Although the elemental analysis indicates the composition of 1c to be $[I_2InP(SiMe_3)_2]_x$, the compound decomposes rapidly in air during sample preparation for XPS, which procludes the measurement of a spectrum.

<u>Figure 1</u>. XPS Survey spectrum of InP (2a).

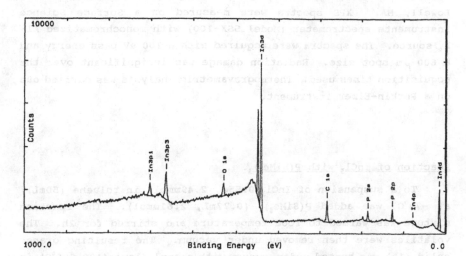

Compound **1c** is stable to 260°C at which temperature decomposition
is initiated. No more volatile products are obtained upon heating
above 540°C.   The InP produced (**2c**) is contaminated with significant quantities of iodine and carbon.

CONCLUSION

We have demonstrated that InP of 99%+ purity can be synthesized from InX_3 (X = Cl, Br) and $P(SiMe_3)_3$. The use of InI_3
results in incorporation of iodine. It is interesting to note that
no Si, a <u>n</u>-type dopant, could be detected by XPS. We are investigating methods to optimize the purity of the InP produced as well
as exploring the $InX_3/P(SiMe_3)_3$ systems for the growth of InP thin
films.

EXPERIMENTAL

All manipulations were carried out under a nitrogen atmosphere. Toluene was dried over sodium under nitrogen, and degassed
prior to use. $InCl_3$, $InBr_3$ and InI_3 (Strem Chemicals) were used

as received. $P(SiMe_3)_3$ was prepared according to published procedures.[5] Microanalysis were performed by Multichem Laboratories,
Lowell, MA.   XPS spectra were measured on a Surface Science
Instruments spectrometer (Model SSX-100) with monochromatized Al-
K_α source.   The spectra were acquired with a 100 eV pass energy and
a 600 μm spot size.   Radiation damage was insignificant over the
acquisition times used.   Thermogravametric analysis was carried out
on a Perkin-Elmer instrument.

Reaction of InCl_3 with P(SiMe_3)_3.

To a suspension of $InCl_3$(0.55g, 2.49mmol) in toluene (50mL),
at -78°C, was added $P(SiMe_3)_3$ (0.77mL, 2.61mmol).   The reaction
mixture was warmed to room temperature and stirred for 2h.   The
volatiles were then removed under vacuum.   The resulting orange
solid (1a) was heated under vacuum with a cool flame (400-500°C) in
the absence of solvent for 2h, to give a black solid (2a).

Compound 1a

XPS, eV (%); P 2p, 125.7; Cl 2p, 194.9(60) 196.5(40); In $3d_{5/2}$,
441.6; In $3d_{3/2}$, 449.1.   Atomic ratio In:P:Cl; 1:1:2. Anal. Calcd.
for $C_6H_{18}Cl_2InPSi_2$: C,19.8; H,4.99; Cl,19.5; P,8.52.   Found: C,20.0;
H,5.02; Cl,19.3; P,8.51.

Compound 2a

XPS, eV (%); P 2p, 128.4; In $3d_{5/2}$, 444.9; In $3d_{3/2}$ 452.5.
Atomic ratio In:P; 1:1.   Anal. Calcd. for InP: P,21.2. Found:
P,20.9; C,0.01.

The reaction of $P(SiMe_3)_3$ with $InBr_3$ and InI_3 was carried out
under the same conditions as described for the reaction using
$InCl_3$.

Compound 1b

XPS, eV (%); P 2p, 125.1; Br 3s, 255.2 , In$3d_{5/2}$, 447.0, In
$3d_{3/2}$, 451.5.   Atomic ratio In:P:Br; 1:1:2.   Anal. Calcd. for
$C_6H_{18}Br_2InPSi_2$: C,15.9; H,3.98; Br,35.4; P,6.86.   Found: C,15.7;
H,3.95; Br,36.0; P,6.88.

Compound 2b

XPS, eV (%); P 2p 128.5; In $3d_{5/2}$, 444.9; In $3d_{3/2}$, 452.5. Atomic ratio In:P; 1:1. Anal. Calcd. for InP: P, 21.2. Found: P,20.8; C,1.00.

Compound 1c

Anal. Calcd. for $C_6H_{18}I_2InPSi_2$: C,13.2; H,3.29; I,46.5; P,5.68. Found: C,12.9; H,3.20; I,46.9; P,5.70.

Compound 2c

XPS, eV (%); P 2p 128.5; In $3d_{5/2}$, 444.6, In $3d_{3/2}$, 452.2. I 4d, 48.9. Atomic ratio In:P:I, 1:1:0.1. Anal. Calcd. for InP: P,21.2. Found: P,20.0; C,1.03.

ACKNOWLEDGMENT

We thank Dr. W. Rees for assistance with the thermogravemetric measurements.

REFERENCES

1. (a) F. Maury and G. Constant, Polyhedron, 3, 581 (1984).
   (b) O.T. Beachley, Jr., J.P. Kopasz, H. Zhang, W.E. Hunter and J.L. Atwood, J. Organometal. Chem., 325, 69 (1987).   (c). C.J. Carrano, A.H. Cowley, D.M. Giolanda, R.A. Jones, C.M. Nunn and J.M. Power, Inorg. Chem., 27, 2709 (1988).
2. G.E. Coates and J. Graham, J. Chem. Soc., 1963, 233.
3. A.M. Arif, B.L. Benac, A.H. Cowley, R. Geertz, R.A. Jones, K.B. Kidd, J.M. Power and S.T. Schwab, J. Chem. Soc., Chem. Commun., 1986, 1543.
4. A.M. Arif and A.R. Barron, Polyhedron, in press.
5. M.D. Healy, P.E. Laibinis, P.D. Stupik, and A.R. Barron, J. Chem. Soc., Chem. Commun., in press.
6. G. Becker and W. Holderich, Chem. Ber., 108, 2484, (1975).

PART III

Compound Semiconductor Deposition Chemistry

PART II

COHERENT ANTI-STOKES RAMAN SCATTERING MEASUREMENTS OF GROUP V HYDRIDE AND TRIMETHYLGALLIUM DECOMPOSITION IN ORGANOMETALLIC VAPOR PHASE EPITAXY

R.LÜCKERATH*, H.J.KOSS*, P.TOMMACK*, W.RICHTER** and P.BALK***
* I. Physik. Inst., RWTH Aachen, Sommerfeldstr., Turm 28,
D-5100 Aachen, Federal Republic of Germany
** Inst. f. Festkörperphysik, TU Berlin, Hardenbergstr. 36,
D-1000 Berlin 12, Federal Republic of Germany
*** Inst. of Semicond. Electronics, RWTH Aachen, D-5100 Aachen,
Federal Republic of Germany

ABSTRACT

The thermal decomposition of AsH_3 and TMG is measured in-situ under different experimental conditions. Simultaneously the production of H_2, CH_4 and C_2H_6 is observed. The data indicate a situation where AsH_3 is only partially decomposed at the GaAs surface. The hydrogen released removes additional CH_3 groups from the trimethyl-gallium (TMG) molecule, enhances the decomposition of TMG, and thereby forms methane.

INTRODUCTION

The group V hydrides together with methyl or ethyl compounds of the group III elements are the primary reactants for metalorganic vapor phase epitaxy (MOVPE) of III-V semiconductors such as GaAs, InP, InGaAs and GaAsP [1]. Although there has been remarkable progress in growing device quality material, the fundamental chemical reactions and the related kinetic parameters determining the growth process are only poorly known. Such information, on the other hand, is essential in order to develop new growth procedures such as selective growth and to minimize byproducts which degrade the quality of the material. In addition, kinetic rate data are also needed to formulate detailed transport-reaction models for the design of MOVPE reactors with uniform growth properties over large area substrates.

A number of investigations concerned with MOVPE chemistry have been been reported in recent years using various spectroscopic techniques [3-16]. Among those coherent anti-Stokes Raman scattering (CARS) [14-16] has certain advantages. It is first of all an in-situ technique and one has not to consider possible postsampling gas-gas or gas-surface reactions. Secondly it can detect a large number of species with a good spatial resolution needed for the large gradients appearing for all quantities in MOVPE reactors. Finally as compared to spontaneous Raman scattering the detectivity and spectral resolution is very much improved since CARS is a resonant four wave mixing process.

Mat. Res. Soc. Symp. Proc. Vol. 131. ©1989 Materials Research Society

We have previously reported on the application of CARS to in-situ MOVPE diagnostics [14-16]. Here we present additional data on the thermal decomposition of AsH_3 and $Ga(CH_3)_3$ under various experimental conditions.

EXPERIMENTAL

In order to simplify the interpretation of the data an isothermal tubular reactor (Fig. 1) was used in the CARS measurements presented here, although, for comparison, measurements were also performed in a standard growth reactor. The

Fig. 1. Experimental configuration used for CARS measurements of MOVPE reactants in a tubular reactor. The inserts show different surface configurations applied.

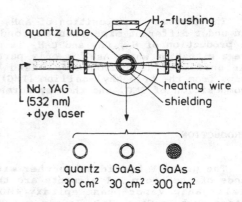

heated length of the reactor was 90 mm with a diameter of 10 mm. Typical operation conditions were: total pressure 50...200 mbar, flow rates 5...50 sccm, average velocity 1...20 cm/s and partial input pressures of the reactants in the mbar range. Temperature was measured by thermocouples, and was additionally controlled by analysing the rotational structure in the CARS spectra of N_2 and AsH_3.

We used the collinear CARS spectroscopic arrangement which has been described in detail elsewhere [14]. The molecular vibrations used for monitoring the concentrations of reactants and products were selected either on account of their scattering strength or because their frequency position was convenient to the range of the dye-laser. Concentrations of reactants were determined by normalizing the measured total intensity at temperature T to that one at 300 K, taking into account thermal expansion, and assuming that no decomposition takes place at room temperature. Concentrations of products (H_2, CH_4, C_2H_6) were determined by measuring the intensities originating from decomposed reactants, and comparing these results with intensities obtained from measurements, where those gases were directly introduced into the reactor under controlled conditions.

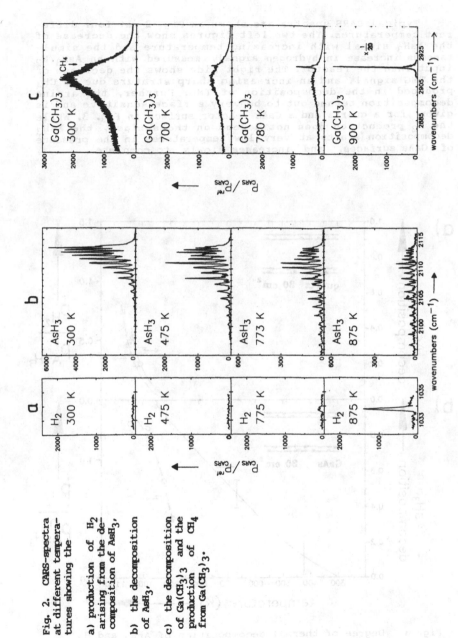

Fig. 2. CARS-spectra at different temperatures showing the

a) production of H_2 arising from the decomposition of AsH_3,

b) the decomposition of AsH_3,

c) the decomposition of $Ga(CH_3)_3$ and the production of CH_4 from $Ga(CH_3)_3$.

RESULTS

    Typical CARS spectra are shown in Fig. 2 at four diffe-
rent temperatures. The two left figures show the decrease of
the AsH_3 signal with increasing temperature and the simultaneous increase in hydrogen signal, measured with an AsH_3/N_2
input into the reactor. The right side shows the decrease of
the TMG signal, and an increasing sharp struture due to CH_4
produced in the decomposition of TMG. Further, the arsine
decomposition turned out to be very surface sensitive and is
given for a quartz and a GaAs reactor surface in Fig. 3. While
the H_2 production does not depend on the surface, the AsH_3
decomposition starts at very low temperatures in the presence
of GaAs surfaces, and increases slowly to complete

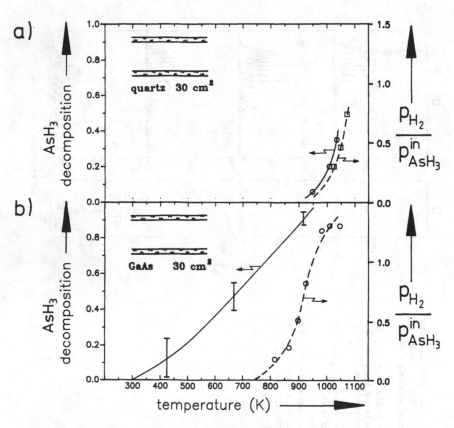

Fig. 3. Degree of thermal decomposition of AsH_3 and H_2
production over (a) quartz and (b) GaAs versus temperature.

conversion at around 900 K. This behaviour turned out to be independent of the carrier gas (N_2, H_2) and the presence of TMG. On the other hand, as shown in Fig. 4, the H_2-production decreases with increasing amount of TMG present.

On the other hand the TMG decomposition increased strongly by choosing H_2 instead of N_2 as carrier gas. This observation is in agreement with [4,6]. Furthermore, in N_2 the decomposition also increased by adding AsH_3. In addition to CH_4 we also observed C_2H_6 (but no C_2H_4) as a product from the TMG decomposition. In N_2, around 900 K, the C_2H_6-concentration was one third of the amount possible from all CH_3-grounps of TMG. By using H_2 as carrier gas, only a few percent C_2H_6 were observed. Detailed data will be published later.

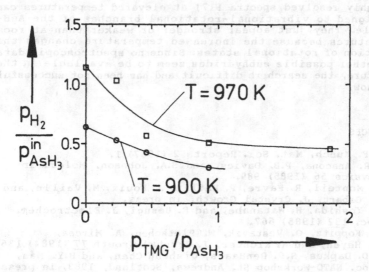

Fig. 4. H_2 production at two different temperatures versus TMG partial pressure relative to AsH_3 partial pressure (8 mbar) in the presence of GaAs surface.

DISCUSSION

The following experimental facts stand out from these results: (i) the catalytic effect of the GaAs surface on the AsH_3 decomposition, (ii) the different temperature behaviour of AsH_3 decomposition and H_2 production over GaAs, (iii) that at fixed temperature by addition of TMG the H_2 production from AsH_3 is decreased but the AsH_3 concentration remains constant, (iv) the TMG concentration as well as the CH_4 production show a strong decrease respectively increase in the presence of AsH_3, at least in N_2, and finally (v) the difference between ex-situ and in-situ data observed previously [14-16].

The possible consequences of these points are that AsH_3 decomposes only partially at low temperatures on the GaAs surface (i, ii). Subhydrides must desorb and be able to recombine at lower temperatures (v). By adding TMG the hydrogen originating from the AsH_3 is captured by CH_3 groups released from the TMG, thus reducing the H_2 production. Simultaneously, the TMG decomposition and the CH_4 production are accelerated, at least in N_2 (iii, iv).

A remaining experimental problem is to prove the existence of group V subhydrides. For that reasons we have very carefully investigated the frequency region from 1800 to 2300 cm^{-1} [17], expecially from 2050 to 2105 cm^{-1}, where frequency values have been assigned for the totally symmetric vibration of AsH in the electronic ground state [2,3]. However, all features present in the highly resolved spectra [17] at elevated temperatures can be assigned to vibrational-rotational branches of the AsH_3 molecule. They just appear stronger or weaker than at room temperatures because the increased temperature changes the occupation of rotational states. Since no spectroscopic data about other possible subhydrides seem to be available in the literature, the search is difficult and has been not successful up to now.

REFERENCES

1. T.F. Kuech, Mat. Sci. Reports 2 (1987) 1.
2. J.R. Anacona, P.B. Davies, and S.A. Johnson, Molecular Physics 56 (1985) 989.
3. Y. Monteil, R. Favre, P. Raffin, J. Bouix, M. Vaille, and P. Gibart, J. Crystal Growth, in press.
4. M. Yoshida, H. Watanabe, and F. Uesugi, J. Electrochem. Soc. 132 (1985) 667.
5. M. Koppitz, O. Vestavik, W. Pletschen, A. Mircea, M. Heyen, and W. Richter, J. Crystal Growth 77 (1984) 136.
6. P.D. Dapkus, S.P. DenBaars, Qisheng Chen, and B.Y. Maa, Proc. NATO-Workshop St. Andrews, Scotland, 1988, in press.
7. C.A. Larsen, N.I. Buchan, and G.B. Stringfellow, Appl. Phys. Lett. 52 (1988) 480.
8. S. D. DenBaars, B. Y. Maa, P.D. Dapkus, A.D. Danner, and H.C. Lee, J. Crystal Growth 77 (1986) 188.
9. M.R. Leys, Chemtronics 2 (1987) 155.
10. J. Nishizawa and T. Kurabayashi, J. Electrochem. Soc. 130 (1983) 413.
11. J.E. Butler, N. Bottka, R.S. Sillmon and D.K. Gaskill, J. Crystal Growth 77 (1986) 163.
12. P.W. Lee, T.R. Omstead, D.R. McKenna, and K.F. Jensen, J. Crystal Growth 85 (1987) 165.
13. V.M. Donnelly, R.F. Karlicek, J.Appl.Phys. 53 (1982) 6399.
14. R. Lückerath, P. Balk, M. Fischer, D. Grundmann, A. Hertling, and W. Richter, Chemtronics 2 (1987) 199.
15. R. Lückerath, P. Tommack, A. Hertling, H. J. Koß, P. Balk, K.F. Jensen, and W. Richter, J.Crystal Growth 93(1988)151.
16. R. Lückerath, W. Richter, K.F. Jensen, Proc. NATO-Workshop St. Andrews, Scotland, 1988, in press.
17. P. Tommack, R. Lückerath, H.J. Koß, W.Richter, to be publ.

GAS PHASE INTERACTIONS BETWEEN
TRIETHYLINDIUM AND TRIMETHYLGALLIUM

P. D. AGNELLO AND S. K. GHANDHI
Electrical, Computer and Systems Engineering Department,
Rensselaer Polytechnic Institute, Troy, NY. 12180

ABSTRACT

A study of the room temperature gas-phase interactions between gallium and indium alkyls was undertaken using a mass spectrometer sampling system, mounted on a low pressure organometallic vapor phase epitaxial reactor. Mixtures of triethylindium with triethylgallium or trimethylgallium were investigated. Both combinations formed addition compounds; moreover, the triethylindium-trimethylgallium mixture underwent alkyl exchange. Both admixtures showed reduced reactivity towards arsine. A structure for the addition compound is proposed.

INTRODUCTION

$Ga_{0.47}In_{0.53}As$ films, lattice matched to InP, are important because of their suitability for the fabrication of optoelectronic devices. Device requirements for ultra-thin layers, abrupt changes in doping and composition, and uniform growth over large areas, make organometallic vapor phase epitaxy (OMVPE) ideally suited [1] for these applications. However, the growth of indium containing compounds by OMVPE is complicated by a room temperature, parasitic reaction between the indium bearing species and the group V hydride [2]. A lack of understanding of these reactions has prevented this growth technique from achieving its full potential.

EXPERIMENTAL

A description of the sampling system has been given [3]. In summary, the sample was taken by a 150 μm I.D. silica capillary which reached into the system from the gas inlet end by means of flexible stainless steel bellows. The gases exiting the capillary impinged on a 100 μm dia. orifice in a skimmer chamber. This orifice connected directly to a mass spectrometer in a bakeable UHV system. The transit time of the sample from the capillary to the ionizer of the mass spectrometer was less than 0.2 sec.

The reactor was held at a pressure of 152 torr with a hydrogen carrier gas flow of 3 slm. Organometallic reactants were Et_3In, Me_3Ga and Et_3Ga, which were studied separately and as mixtures. Experiments were carried out over flow conditions which include the range of pressure and flow conditions over which device quality $Ga_{0.47}In_{0.53}As$ has been grown by us. The organometallic reactant flows were premixed and transported in hydrogen gas to the reaction chamber, at which point the arsine was introduced, when required. Data was taken at room temperature, which we have shown [3] to be essentially the same as the gas temperature right up to the susceptor during growth.

RESULTS AND DISCUSSION

In one set of experiments, we have monitored the fragments due to 0.02 torr of Et_3In, Et_3Ga, and the mixture of these organometallic compounds, in the absence of AsH_3. The sampling probe position was 2.5 cm from the reactor inlet for these measurements. Figure 1(a) shows the mass spectrum for Et_3Ga alone over the range from 65 to 130 AMU, where the fragments ranging from Ga^+ to Et_2Ga^+ are located. Likewise, Fig. 1(c) shows the mass spectrum for Et_3In alone over the range from 110 to 175 AMU, where the fragments ranging from In^+ to Et_3In^+ are located. The spectrum for the mixture is shown in Figs. 1(b) and (d) over the entire 65-175 AMU range.

When Et_3Ga was added to the Et_3In, the Et_2In^+ signal doubled. In addition, the Et_2Ga^+ signal from the mixture was about a factor of two lower when compared to an equivalent amount of Et_3Ga alone. The mass spectra for the mixture of Et_3In and Et_3Ga contains no peaks which are not also observed in the spectra of the individual organometallics. However, because the spectrum for the mixture is not a superposition of the individual spectra (i.e., the peak ratios have changed) we conclude that weak bonding or interaction has occurred between the organometallic species. A detailed examination of the spectra, before and after mixing, has been made on a fragment by fragment basis. By subtracting the spectra of the individual species from the spectrum of the mixture, we calculate that a maximum of 65% of the initial reactants remain in the mixture. Thus, at least 35% of the end products are in the form of addition compounds.

When Me_3Ga and Et_3In are mixed together prior to their introduction into the reactor, they also form an addition compound. Moreover, some alkyl exchange occurs. Figures 2a-d show scans for 0.13 torr of each of the individual reactants Me_3Ga and Et_3In and for the mixture of the two, in the absence of AsH_3 gas. Again, we note that the ion pattern of the mixture is not a superposition of the individual ion peaks. For the mixture, the appearance of a series of Me-In and Et-Ga peaks, accompanied by a reduction in the Me-Ga and Et-In peaks is observed. As expected, the spectrum of pure Et_3In contains some methyl fragments due to ionization. However, the spectrum of pure Me_3Ga does not contain ethyl fragments as their creation during the ionization process would be highly improbable.

The signals at 113 and 115 AMU could result from both indium and gallium fragments. However, the separate contributions can be evaluated by taking into account the natural isotopic proportions of Ga^{69} to Ga^{71}, which is 3:2, and those of In^{113} to In^{115} (4:96). The spectrometer data shows that, upon mixing Et_3In and Me_3Ga, there is an increase in the Me_2In^+ signal with a concurrent decrease in the Et_2In^+ peak. The behavior of the gallium signals shows an opposite trend, i.e., a decrease in the Me_2Ga^+ signal and a concurrent increase in the Et_2Ga^+ signal.

The parent ion signal of the exchange product Et_3Ga^+ falls at masses 156/158. Although weak, this signal is clearly observed, since it is not obstructed by other signals. Parent ion signals of the exchange products $EtMe_2Ga^+$ and $MeEt_2Ga^+$ fall at 128/130 and 142/144 AMU respectively, which are close to (or coincide with) other strong fragment signals. Again, some of these exchange products can be identified if we take into account the isotopic proportions of Ga and In. Thus, we can show that the signal at

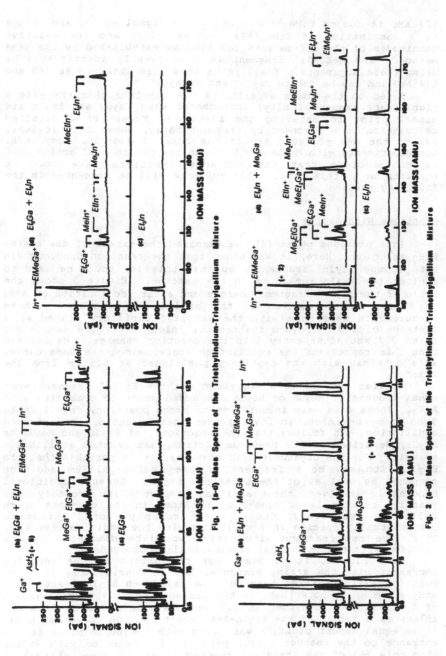

Fig. 1 (a-d) Mass Spectra of the Triethylindium-Triethylgallium Mixture

Fig. 2 (a-d) Mass Spectra of the Triethylindium-Trimethylgallium Mixture

128 AMU is due to $EtMe_2{}^{69}Ga^+$ alone. The signal at 130 AMU is due to a combination of $EtMe_2{}^{71}Ga^+$ and $Me^{115}In^+$, and the relative magnitudes of these fragments can also be established by the same method. The $MeEt_2Ga^+$ fragment is also clearly identifiable by using these arguments. Small peaks were also observed at 160 and 174 AMU and may be due to Me_3In^+ and $EtMe_2In^+$.

Based on the above experiments, we conclude that there is a significant amount of alkyl interchange when Me_3Ga and Et_3In are mixed, prior to entering the reactor. Moreover, a detailed calculation on a fragment by fragment basis, shows that at least 69% of the end products are in the form of addition compounds. Our experiments with Me_3Ga and Et_3In in the presence of arsine lead us to conclude that the exchange reactions have come to equilibrium at this time. This evidence will be presented in the following section.

Reactions With Arsine

In a previous paper [3], we examined the extent of the Et_3In-AsH_3 reaction. Here, it was shown that the reaction products did not produce Et_2In^+ fragments, so that this ion could be used to monitor the Et_3In remaining in the reactor. Figure 3 shows the ratio of the Et_2In^+ fragment, before and after the addition of AsH_3 to Et_3In, as a function of the arsine overpressure. These measurements were made with the sampling capillary placed at a distance of 9.0 cm beyond the reactor inlet, i.e., the sample was taken 0.7 sec after entry into the reaction chamber. We believe that this represents the equilibrium state, since the same curves were obtained with the capillary positioned at 6.5 cm from the inlet.

Similar sets of data are shown for the Et_2In^+ fragment when equal amounts of Me_3Ga or Et_3Ga are added prior to combining with AsH_3. These data were insensitive of probe position, from 1 cm to 9 cm from the inlet, indicating that these reactions had reached equilibrium. It follows that the reactions, of Et_3In and Me_3Ga or Et_3Ga, described in the previous section, had reached equilibrium as well. For the organometallic mixtures, it is possible that the Et_2In^+ ion may be a fragment of the indium-gallium addition compound as well as of the unreacted Et_3In. Because additional bonding is involved, these addition compounds are probably less reactive towards Lewis bases, as outlined in the comments which follow. Moreover, there are probably some unreacted organometallic species in equilibrium with the addition compound. Thus, the net reaction with arsine would be dominated by the reaction with the individual organometallic compounds.

From Fig. 3, it is seen that the extent of the reaction increases with the arsine pressure, but is considerably less for both the alkyl mixtures than for the case when Et_3In reacts with AsH_3 alone. We note that, in the presence of 3.04 torr AsH_3, all of the Et_3In was consumed in 0.7 seconds (i.e., 9.0 cm from the inlet) as a result of the Et_3In-AsH_3 reaction. On the other hand, if an equal amount of Et_3Ga was mixed with the Et_3In prior to its entrance to the reactor (i.e., prior to its reaction with AsH_3), then only 60% of the Et_3In was reacted after the same interval of time. In effect, the presence of Et_3Ga has reduced the extent of the reaction between Et_3In and AsH_3. This is further evidence that the two organometallic species form some type of addition compound. Moreover, analysis of the the data shows that at least 40% of the Et3In is tied up in this way, and is less susceptible to reaction with arsine in this form. The Me3Ga-Et3In mixture is even less reactive with AsH3 than the Et3Ga-Et3In mixture. Here, the Et2In+ signal is reduced by less than 40% upon the addition of AsH3, indicating that at least 60% of the Et3In is in the form of an addition compound.

We propose that the gallium-indium association is similar to the gas phase dimerization [4] of Me3Al, which involves two bridging alkyl groups as shown in Fig. 4. Studies of mixtures of Me3Tl and Et3Tl have shown [5] alkyl exchange via a structure of the type proposed here. The structure of Fig. 4 would be less acidic and therefore less likely to react with AsH3. Further, the dynamic nature of the structure would provide a mechanism for alkyl exchange of the type observed in the mass spectrometer data. We postulate a similar structure for the Et3Ga-Et3In mixture. Such structures, with three centered bond arrangements, would be unlikely to react with arsine (unless they dissociate), because indium has a coordination of four. Such structures are likely, since indium has a coordination of four in solid Me3In

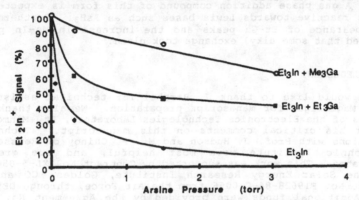

Fig. 3 Steady State Diethylindium Ion Signal as a Function of Arsine Pressure for Gas Mixtures

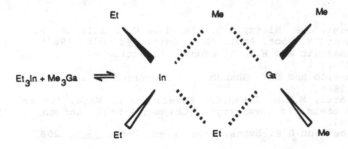

Fig. 4 Proposed Triethylindium-Trimethylgallium

Though there is clear evidence that Me_3Ga, Et_3Ga, Me_3In and Et_3In do not form stable dimers in the gas phase, it has been suggested that a slight change in the thermodynamics might allow for the formation of stable dimers of Me_3Ga, Et_3Ga, Me_3In and Et_3In [4]. Certainly more detailed studies are required, perhaps by techniques which could explore the bonding in such structures, to obtain more direct evidence that these compounds exist. The fact that some reaction between arsine and Et_3In still occurs with these structures is evidence that either some of the Et_3In is dissociated from the gallium species or that the arsine is still somewhat reactive towards the indium, even when it is bonded to the gallium, because a closed shell electronic configuration has not been achieved.

CONCLUSIONS

Both Et_3Ga and Me_3Ga form a weak association with Et_3In, and can reduce the gas phase reaction with arsine when mixed prior to introduction into the epitaxial reactor. The association is probably similar to the well-established addition compound form of Me_3Al. A gas phase addition compound of this form is expected to be less reactive towards Lewis bases such as AsH_3. Furthermore, the appearance of Et-Ga peaks and the increase in Me-In peaks indicated that some alkyl exchange took place.

ACKKNOWLEDGEMENT

We would like to thank J. Barthel for technical assistance and P. Magilligan for manuscript preparation. We also thank Dr. K. Jones of the Electronics Technologies Laboratory, Ft. Monmouth, NJ, for his critical comments on this manuscript. Technical discussions with Prof. J. Hudson and Mr. P. Chinoy of Rensselaer Polytechnic Institute were most helpful, and are greatly appreciated. This work was supported by Contract No. XL-5-05018-2 from the Solar Energy Research Institute, Golden, CO and by Contract No. F19628-84-C-0066 from the Air Force, through DEVCOM, Inc. Additional funds were provided by the Agreement No. 900-ERER-ER-87 from the New York State Energy Research and Development Authority (NYSERDA). This support is hereby acknowledged.

REFERENCES

1. M. Razeghi, J.P. Hirtz, U.O. Ziemelis, C. Delalande, B. Etienne and M. Voos, Appl. Phys. Lett., 43, 585 (1983).
2. H.M. Manasevit and W.I. Simpson, J. Electrochem. Soc., 120, 135 (1973).
3. P.D. Agnello and S.K. Ghandhi, J. Electrochem. Soc., 135, 1530 (1988).
4. G.E. Coates, M.L.H. Green, P. Powell and K. Wade, "Principles of Organometallic Chemistry", (Chapman & Hall, London, 1977), pp. 38-40.
5. J.P. Maher and D.F. Evans, Proc. Chem. Soc., 1961, 208.

GAS PHASE AND SURFACE REACTIONS IN MOCVD OF GaAs FROM TRIETHYLGALLIUM, TRIMETHYLGALLIUM, AND ORGANOMETALLIC ARSENIC PRECURSORS

Thomas R. Omstead, Penny M. Van Sickle and Klavs F. Jensen
Department of Chemical Engineering and Materials Science, University of Minnesota,
Minneapolis, Minnesota 55455

ABSTRACT

The growth of GaAs from triethylgallium (TEG) and trimethylgallium (TMG) with tertiarybutylarsine (tBAs), triethylarsenic (TEAs), and trimethylarsenic (TMAs), has been investigated by using a reactor equipped with a recording microbalance for *in situ* rate measurements. Rate data show that the growth with these precursors is dominated by the formation of adduct compounds in the gas lines, by adduct related parasitic gas phase reactions in the heated zone, and by the surface reactions. A model is proposed for the competition between deposition reactions and the parasitic gas phase reactions. Model predictions are in very good agreement with experimental data for all combinations of precursors except for TEG/TMAs where extensive gallium droplet formation is observed at low temperatures. Growth of reasonable quality GaAs with Hall mobilities of 7600 cm^2/Vs at 77 K using TEG and tBAs is reported for the first time.

INTRODUCTION

Although the use of arsine in organometallic chemical vapor deposition (MOCVD) of GaAs has been demonstrated to give excellent electrical properties [1 and references within], its use has a number of disadvantages. Since arsine is supplied in pressurized cylinders there is the possibility for an accidental large release of toxic gas, particularly during the changing of cylinders. Tertiarybutylarsine (tBAs), triethylarsenic (TEAs), and trimethylarsenic (TMAs) are alternative, liquid organometallic arsenic sources to arsine with the potential for growing good quality GaAs films [2,3] at relatively low V/III ratios.

The gas phase chemistry associated with the use of TMAs, TEAs, and tBAs in conjunction with trimethylgallium (TMG) and triethylgallium (TEG) is complex, involving free radical reactions and possible adduct formation. A mass spectroscopy study of gas phase decomposition reactions and related possible adduct formations is described elsewhere [4,5]. Here we report on deposition rate measurements at 1 torr to demonstrate the relative influence of adduct formation, parasitic gas phase reactions and surface kinetics on the growth of GaAs by using organometallic precursors with TMG and TEG. The low pressure conditions were selected to minimize gas phase reactions and reduce mass transfer limitations that would otherwise complicate the data interpretation. For the same reasons, as well as the growth of abrupt interfaces and improvement in uniformity, there is a trend toward low pressure operation in device processing [1].

EXPERIMENTAL

A radiantly heated, microbalance equipped, MOCVD reactor was employed as the primary means of investigating the growth rate of organometallic arsenic precursors with trimethylgallium (TMG) and triethylgallium (TEG). This reactor is shown in Figure 1 and it is described further elsewhere [5]. It consists of a Cahn Microbalance from which a double-side polished GaAs wafer is suspended by a thin molybdenum wire. The wafer is held perpendicular to a 2000 Watt water and air cooled quartz-halogen lamp by a small sample-holder made out of a single piece of molybdenum wire.

The GaAs substrates were heavily doped with zinc ($\sim 10^{18}$ cm^{-3}) to make them absorbing in the near infrared region, where the lamp has the majority of its power output. The high doping also gave the GaAs wafer a high and constant emissivity in the 4.9 to 5.5 micron range of our optical pyrometer due to valance band transitions [6]. This allowed for accurate temperature measurement in an optical region not interfered with by the radiation from the quartz lamp.

Wafers were degreased and etched before being loaded into the reactor after which the system was baked and purged overnight before taking data. During experiments the wafer was held at a given temperature for about ten minutes to obtain a growth rate measurement except for conditions with very low growth rates. In those cases, up to 30 minutes were required to obtain an accurate reading above the noise caused by minor pressure fluctuations in the system. The surface morphology of the sample was monitored and if it had deteriorated significantly after a set of runs the data from the entire run were discarded.

The total system pressure was usually kept constant at 1 torr while the V/III ratio was varied between 3 and 10. Typical feed rates of TMG and TEG corresponded to 0.01 torr partial pressure of the alkyl in the system. Hydrogen flow rates were varied from 30 sccm to 45 sccm to evaluate the effect of residence time.

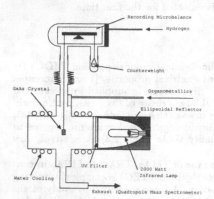

Figure 1. Schematic of microbalance low pressure MOCVD reactor.

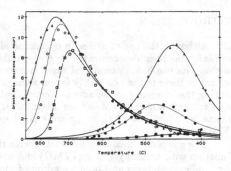

Figure 2. Variation in the growth rate of GaAs from TEG+tBAs (solid points) and TMG+tBAs (open points) with substrate temperature and carrier gas (H$_2$) flow rate, □ 30 sccm, △ 37.5 sccm, o 45 sccm. P$_{total}$ = 1 torr, P$_{TEG}$ = P$_{TMG}$ = 0.01 torr, V/III = 5.

RESULTS

Growth of GaAs using tBAs and TMG or TEG

The kinetic behavior of the TMG/tBAs system was studied as a function of temperature and residence time (total flow-rate) in the reactor while holding the partial pressure of TMG and tBAs constant at 0.01 torr and 0.05 torr, respectively. The results of this study, shown in Figure 2, shows three distinct regions of growth. At low temperatures the deposition rate is independent of flow-rate; this indicates that growth in this regime is controlled by surface (heterogeneous) reactions. In the mid-temperature region, at temperatures above about 610°C, a significant dependence of the growth rate on the total flow rate is observed.

Two possible mechanisms for the dependence of growth-rate on flow-rate may be proposed, either mass transfer limitations or parasitic gas phase reactions. Because of the high diffusivity of the precursors at 1 torr, mass transfer limitations are unlikely. Estimates assuming diffusion limited growth predict rates an order of magnitude higher than the experimentally observed rate. Gas-phase depletion caused by parasitic reactions is a more likely explanation for the observed flow-rate dependence on growth. Growth efficiency (moles gallium in the growing film/moles entering the reactor) is very high (41% at 700°C), thus the loss of precursor from the gas-phase will also directly affect the growth rate. Comparison of growth rates obtained when the reactants (TMG and tBAs) are mixed in the reactor to those obtained when reactants combined in the gas handling system, as illustrated in Figure 3, suggest that the parasitic reaction is related to adduct formation. Since the growth rates for split and combined streams are nearly equal at low temperatures the parasitic reaction does not take place in the lines but requires higher temperatures to occur.

The temperature dependence of the growth rate of GaAs from TEG and tBAs at low pressures is similar to that observed for the TMG+tBAs, except that the growth is shifted to lower temperatures as illustrated in Figure 2. The maximum growth rate occurs around 750°C for TMG while it takes place around 450°C for TEG. This is consistent with the larger bond strength of the methyl group compared to the ethyl compound. The deposition rate again shows a strong dependence on the residence time, which is an indication that a parasitic gas phase reaction also occurs in the TMG+tBAs system. Unlike the situation for TMG, a surface kinetic controlled regime, where the growth rate is independent of flow rates, is not apparent at low temperatures (400-600°C) as it is for the TMG system.

To investigate the electrical properties of growth with the TEG/TBAs system, which has not been previously investigated, films were grown on undoped GaAs using doped GaAs as a susceptor in the microbalance reactor. Table 1 lists 300 K and 77 K mobilities for films grown at four different V/III ratios. The conversion to p-type at high V/III ratios is most likely due to carbon incorporation. The best 77 K mobility of 7600 cm^2/Vs was obtained at 650°C and a V/III ratio of 5. Future experiments will concentrate on the optimization of this system for electrical properties.

TABLE 1. Electrical Properties for GaAs Grown using tBAs and TEG.

| Precursor | T | P | V/III | Carrier Conc. | Type | Mobility (300 K) | Mobility (77 K) |
	C	torr		cm^{-3}		cm^2/Vs	cm^2/Vs
TEG/tBAs	650	1	3	$3 \cdot 10^{16}$	n	1522	–
TEG/tBAs	650	1	4	$4 \cdot 10^{16}$	n	1952	1913
TEG/tBAs	650	1	5	$6 \cdot 10^{15}$	n	4170	7592
TEG/tBAs	650	1	7.5	$1 \cdot 10^{15}$	n	1323	1643
TEG/tBAs	650	1	10	$3 \cdot 10^{17}$	p	110	187

Growth of GaAs using TMAs and TEAs

The use of TMAs and TEAs as arsenic precursors results in drastically reduced growth rates due to their enhanced tendency to form parasitic gas-phase reactions as shown in Figures 4 and 5, respectively. The TEAs/TMG and TMAs/TMG data show the same qualitative behavior as the tBAs data with variations in temperature and flow rates. Growth with TEG/TEAs was found to be impossible under these conditions due to the formation of a non-volatile adduct. Growth with TMAs/TEG was adversely affected by the formation of Ga droplets most likely due to the much higher decomposition temperature of TMAs relative to TEG [5]. The apparent increase in adduct formation

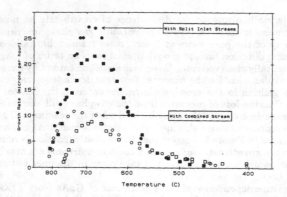

Figure 3. Variation in the growth rate of GaAs from TMG and tBAs with substrate temperature and H_2 carrier gas flow rate for split streams (solid points) and combined streams (open points); (□ 30 sccm, (○ 37.5 sccm, $P_{TMG} = 0.01$ torr, $P_{total} = 1$ torr, V/III = 5.

and parasitic reactions with TEAs and TMAs relative to tBAs is consistent with the electron donating nature of the alkyl groups. This implies that the lone electron pair on the arsenic atom is more negatively charged and thus, more likely to form addition compounds.

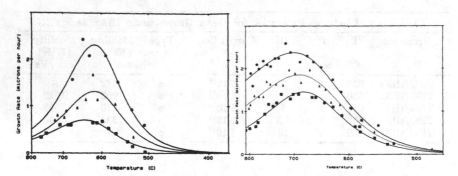

Figure 4. Fit of the model to the variation in the growth rate of GaAs from TMG and tMAs for changing substrate temperature and carrier gas (H_2) flow rate. (Split streams) ■ 30 sccm, △ 37.5 sccm, ● 45 sccm, P_{total} = 1 torr, P_{TMG} = 0.01 torr, V/III=5.

Figure 5. Fit of the model to the variation in the growth rate of GaAs from TMG and tEAs for changing substrate temperature and carrier gas (H_2) flow rate. (Split streams) ■ 30 sccm, △ 37.5 sccm, ● 45 sccm, P_{total} = 1 torr, P_{TMG} = 0.01 torr, V/III=5.

DISCUSSION

The described complex behavior of the growth rates for the organometallic arsenic and systems with residence time and temperature suggests that adduct-based parasitic gas-phase reactions among the organometallic precursors play an important role in the deposition mechanism. Based on the following simplified picture of the chemistry, it is possible to formulate a model that is consistent with the experimental observations.

$$\text{organometallic precursors} \xrightarrow{k_1} \text{complex} \xrightarrow{} \text{polymeric deposits}$$
$$\downarrow k_s \qquad\qquad \underset{k_r}{\rightleftharpoons} \qquad k_d$$
$$\text{GaAs} \qquad\qquad \text{(adduct)}$$

Each step obviously involves several elementary reactions, but there are not sufficient data to provide additional mechanistic detail at this stage. A similar model has been proposed earlier [7]. Here we relax assumptions in that model and derive more accurate rate expressions. Because of the low pressures (i.e. large diffusion coefficients), the hot reaction zone around the substrate may be considered to be a *well mixed reactor* with residence time τ. A steady state balance over the gallium alkyl then takes the form:

$$P_{G0} - P_G - \tau k_f P_G P_A + \tau k_r P_C - \tau(S/V)R_{surface} = 0 \qquad (1)$$

Here P_A, P_G, and P_C represents the partial pressures of the arsenic precursor, the gallium precursor, and the adduct complex respectively. The first two terms gives the difference between gallium precursor in and out flow in the reacting gas volume. The next two terms represent the net reversible formation of adduct, while the last term gives the loss of gallium alkyl in deposition reactions. A similar balance over the adduct complex (C) gives:

$$P_{C0} - P_C + \tau k_f P_G P_A - \tau k_r P_C - \tau k_d P_C = 0 \qquad (2)$$

The first term represents adduct formed in the gas handling system. For the case of a low boiling adduct as in the TEG experiments or split streams as in the TMG experiments this term is essentially zero. Since the activation energy for adduct formation is small, we may assume that k_r is much larger than k_d and we have:

$$P_C = \frac{\tau k_f P_G P_A}{[1 + \tau k_r + \tau k_d]} \simeq K_c P_G P_A \qquad (3)$$

Based on observations for atomic layer epitaxy, the surface reactions of Ga species appears to be self-limiting [8]. Furthermore, for large V/III ratios the surface reaction will be saturated with respect to arsenic, while gallium surface coverages are expected to be low. Hence, the surface reactions may be approximated by the first order reaction expression:

$$R_{surface} = k_s K_G P_G \qquad (4)$$

Since the residence time is inversely proportional to the flow rate, F, one obtains the following simplified expression for the growth rate as a function of H_2 carrier gas flow rate (F) and temperature (T):

$$R_{surface} = \frac{(k_s'/k_1)F P_{G0}}{1 + F/k_1} \qquad (5)$$

where the quantities:

$$k_s' = k_s K_G S = k_{s0}' \, \exp[E_{As}'/RT]$$
$$k_1 = [k_d K_c P_A V + k_s'] = k_{10} \, \exp[E_{A1}/RT]$$

represent loss of precursor from the gas-phase by parasitic reaction and by film growth.

The unknown constants (k_{s0}, E_{As}, EA1, k_{10}) were found by nonlinear regression of Equation (5) with E_{As} for TMG obtained from a linear regression fit of the surface controlled region of the TMG/TBAs data. The resulting constants are summarized in Table 2. The values represent averages of the individual curve fits. For low Ga utilization (i.e. for the TMAs and TEAs cases) k_1 reflects the product of rate constant k_d and the equilibrium constant K_c associated with the parasitic reactions. The good fits support the qualitative features of this simple model, but we emphasize that additional kinetic analysis remains in order to identify the individual steps.

TABLE 2. Rate parameters determined by fit of the model to experimental data.

	Figure #	k'_{so} (cm/torr s)	E_{As} (kJ/mol)	k_{10} (sccm)	E'_{A1} kJ/mol
TMG/tBAs	2	690	123	$5 \cdot 10^{19}$	280
TEG/tBAs	2	170	98	10^{18}	193
TMG/TEAs	5	70	123	$2 \cdot 10^{12}$	191
TMG/TMAs	4	130	123	$4 \cdot 10^{13}$	204

CONCLUSIONS

The low pressure growth of growth of GaAs from triethylgallium (TEG) and trimethylgallium (TMG) with tertiarybutyl-arsine (tBAs), triethylarsenic (TEAs) and trimethylarsenic (TMAs) has been investigated in a MOCVD recording microbalance for *in situ* rate measurements. The observed variation in rate data with variations in substrate temperature and H_2 carrier gas flow rate are consistent with a model describing the competition between deposition reactions and the parasitic gas phase reaction. This parasitic reaction is dependent on the formation of an apparent adduct. In agreement with experimental data, the model predicts the growth efficiency may be improved by reducing the residence time of the precursors and by decreasing the total pressure in the reactor. With these precautions it possible to deposit single crystalline GaAs of reasonably high mobility at low pressure (1 torr) using TEG and tBAs which show promise for further investigation of alternative As sources.

ACKNOWLEDGEMENTS

The authors are grateful to the National Science Foundation and Air Products and Chemicals for support of this work. They would also like to thank Michelle Hoveland for help with the electrical characterization work.

REFERENCES

1. T.F. Kuech, *Mat. Sci. Reports* **2**, 1 (1987).
2. C.H. Chen, C.A. Larsen and G.B. Stringfellow, *Appl. Phys. Lett.* **50**, 218 (1987).
3. R.M. Lum, J.K. Klingert and M.G. Lamont, *Appl. Phys. Lett.* **50**, 284 (1987).
4. P.W. Lee, T.R. Omstead, D.R. McKenna and K.F. Jensen, *J. Crystal Growth* **93**, 134-142 (1988).
5. P.W. Lee, T.R. Omstead, D.R. McKenna and K.F. Jensen, *J. Crystal Growth* **85**, 165 (1987).
6. A.S. Jordan, *J. Appl. Phys.* **51**(4), 1980.
7. T.R. Omstead, P.M. Van Sickle, P.W. Lee and K.F. Jensen, *J. Crystal Growth* **93**, 20-28 (1988).
8. J. Nishizawa, T. Kurabayashi and H. Abe, *Surface Science* **185**, 249 (1987).

AN EVALUATION OF SIMPLIFIED MODELS FOR SURFACE KINETICS IN MOVPE PROCESSES

MAX TIRTOWIDJOJO AND RICHARD POLLARD
Department of Chemical Engineering, University of Houston, Houston, TX 77204

ABSTRACT

A general MOVPE model has been used to assess the applicability of simplified representations for surface kinetics. With the general model, predictions for GaAs deposition on (111)Ga using trimethylgallium and arsine show excellent agreement with observed growth rates. However, if Langmuir-Hinshelwood kinetics is assumed, the model only matches the deposition rates over a narrow range of operating conditions, even when several rate-limiting steps are included. This limitation arises because combinations of equilibrium constants and local partial pressures often do not give reasonable approximations for the surface concentrations of reactive intermediates. The form of the Langmuir-Hinshelwood relation(s) and the parameter values can be fitted empirically to experimental data, but this could lead to erroneous conclusions concerning process behavior and the model would have limited predictive capabilities. An alternative approach is to use surface reaction probabilities, but they can only be applied in an empirical fashion and their magnitudes depend on gas flow rate, inlet composition, and reactor pressure as well as surface temperature.

INTRODUCTION

A mathematical model for metalorganic vapor phase epitaxy (MOVPE) of GaAs has recently been developed [1,2]. The analysis considers multicomponent heat and mass transport, the kinetics of reactions in the gas phase and at the deposition surface, and fluid flow for impinging-jet and rotating-disk reactors. A unique feature of the model is that it considers many plausible reactions, and the rate constants for each elementary process are estimated from statistical mechanics, transition-state theory, and bond dissociation enthalpies. With this approach, dominant reaction pathways are predicted rather than assumed, and species compositions and deposition rates are determined without adjusting the value for any kinetic parameter. The theoretical predictions show quantitative agreement with available experimental data [1].

A total of 79 species and 347 elementary processes are included in the model for homoepitaxial formation of GaAs from $Ga(CH_3)_3$ (TMG) and AsH_3 on (111)Ga substrates [1,2]. However, theoretical calculations over a wide range of operating conditions indicate that many of the intermediate species and their reactions do not have an appreciable effect on the system behavior. Therefore, the process can be described accurately with only 19 species and 25 reactions [1,2].

In the model, the forward and backward rates of each surface reaction include the fractional occupancies θ_{im} of reactant and product species adsorbed on active sites of type m [1]. The finite rates of adsorption and desorption processes are also treated, and steady-state material balances are used to determine the values for θ_{im}. This approach can be contrasted with previous models of epitaxial chemical vapor deposition (CVD) systems which traditionally have assumed Langmuir-Hinshelwood (LH) kinetics [3,4] or have used a surface reaction probability (SRP) to represent the deposition rate [5,6]. In this paper, the applicability of simplified rate expressions for the surface processes in MOVPE of GaAs is evaluated. In particular, the fundamental assumptions incorporated into LH kinetic models are tested by making quantitative comparisons with results from the general MOVPE model. Also, the limitations of the SRP approach are discussed.

Mat. Res. Soc. Symp. Proc. Vol. 131. ©1989 Materials Research Society

RESULTS AND DISCUSSION

Langmuir-Hinshelwood kinetics

With an LH model, a number of the elementary processes involved in the major reaction pathways are assumed to be at equilibrium [7]. Consequently, the surface concentrations of adsorbed species (that appear in the kinetic expressions for the rate-determining surface reactions) can be replaced by a combination of partial pressures of gaseous species and equilibrium constants. In general, several simultaneous reactions may be regarded as rate-limiting, provided that the number of relationships that can be obtained by setting the remaining major reactions to equilibrium is sufficient to eliminate all concentrations of adsorbates from the rate equations. A set of LH rate expressions developed in this way for deposition of GaAs on a (111)Ga substrate is presented in Table I. The five reactions were chosen on the basis of the pathways and rate-limiting steps identified using the general MOVPE

Table I. Rate expressions for the reactions used in the Langmuir-Hinshelwood model. Subscripts with the prefix S refer to reaction numbers in the list of major surface processes given in Table II. The quantities $k_{j,b}$ and K_j are the backward rate constant and the equilibrium constant calculated for elementary step j [2]. The equilibrium constant $K_{O,I}$ for reaction I is evaluated using the stoichiometries (i.e., exponents) and thermochemical properties of the species appearing in the square bracket of the rate equation for that reaction.

I.
$$r_I = k_{S11,b}\,\alpha_1^{\,2}\left[K_{O,I}\,p_{As_2}p_H^{\,6} - \left(p_{AsH_3}\right)^2\right]\frac{\theta_{vs}^{\,2}}{p_H^{\,6}}$$

II.
$$r_{II} = k_{S12,b}\,K_{S4}K_{S5}\left[K_{O,II}\,p_{CH_4} - p_{CH_3}p_H\right]\theta_{vs}\,\theta_{vl}$$

III.
$$r_{III} = k_{S14,b}\,\alpha_1\,K_{S13}\left[K_{O,III}\,a_{GaAs}\,p_{CH_3}p_H^{\,3} - p_{GaCH_3}p_{AsH_3}\right]\frac{\theta_{vs}\,\theta_{vl}}{p_H^{\,3}}$$

IV.
$$r_{IV} = k_{S15,b}\,\alpha_2\,K_{S13}\left[K_{O,IV}\,a_{GaAs}\,p_{CH_3}p_H^{\,3} - p_{GaCH_3}p_{AsH_3}\right]\frac{\theta_{vs}\,\theta_{vl}}{p_H^{\,2}}$$

V.
$$r_V = \frac{k_{S17,b}}{K_{S16}}\left[K_{O,V}\left(p_{CH_3}\right)^2 p_{GaCH_3} - p_{TMG}\right]\frac{\theta_{vl}}{p_{CH_3}}$$

where

$$\alpha_1 = \frac{K_{S3}}{\left(K_{S4}\right)^3 K_{S8}K_{S9}K_{S10}} \qquad \alpha_2 = \frac{K_{S3}}{\left(K_{S4}\right)^2 K_{S8}K_{S9}}$$

$$\left(\theta_{vl}\right)^{-1} = 1 + K_{S13}\,p_{GaCH_3} + \left(\frac{K_{S4}\,K_{S9}}{K_{S18}}\right)p_H + \frac{1}{K_{S16}}\frac{p_{TMG}}{p_{CH_3}}$$

$$\left(\theta_{vs}\right)^{-1} = 1 + \alpha_1\frac{p_{AsH_3}}{p_H^{\,3}} + \alpha_2\frac{p_{AsH_3}}{p_H^{\,2}} + \left(\frac{K_{S3}}{K_{S4}\,K_{S8}}\right)\frac{p_{AsH_3}}{p_H} + K_{S5}\,p_{CH_3} + K_{S4}\,p_H$$

Table II. Major elementary surface processes in homoepitaxial deposition of GaAs on (111)Ga substrates. The reactants are TMG and AsH_3. Vacant sites on planar surfaces and ledges are denoted by v and l, respectively, and species adsorbed on these sites are specified by (a_s) and (a_l).

S1.	AsH	+	v	⇔	$AsH(a_s)$		
S2.	AsH_2	+	v	⇔	$AsH_2(a_s)$		
S3.	AsH_3	+	v	⇔	$AsH_3(a_s)$		
S4.	H	+	v	⇔	$H(a_s)$		
S5.	CH_3	+	v	⇔	$CH_3(a_s)$		
S6.	$AsH(a_s)$	+	$H(a_s)$	⇔	AsH_2	+	2 v
S7.	$AsH_2(a_s)$	+	$H(a_s)$	⇔	AsH_3	+	2 v
S8.	$AsH_2(a_s)$	+	$H(a_s)$	⇔	$AsH_3(a_s)$	+	v
S9.	$AsH(a_s)$	+	$H(a_s)$	⇔	$AsH_2(a_s)$	+	v
S10.	$As(a_s)$	+	$H(a_s)$	⇔	$AsH(a_s)$	+	v
S11.	As_2	+	2 v	⇔	$2 As(a_s)$		
S12.	CH_4	+	2 v	⇔	$CH_3(a_s)$	+	$H(a_s)$
S13.	$GaCH_3(a_l)$			⇔	$GaCH_3$	+	l
S14.	GaAs(s)	+	$CH_3(a_s)$ + l	⇔	$GaCH_3(a_l)$	+	$As(a_s)$
S15.	GaAs(s)	+	$CH_3(a_s)$ + $H(a_l)$	⇔	$GaCH_3(a_l)$	+	$AsH(a_s)$
S16.	$DMG(a_l)$	+	CH_3	⇔	TMG	+	l
S17.	$GaCH_3(a_l)$	+	CH_3	⇔	$DMG(a_l)$		
S18.	$AsH_2(a_s)$	+	$H(a_l)$	⇔	$AsH_3(a_s)$	+	l

model. The effective rate constants for these reactions are combinations of equilibrium constants and rate constants for elementary steps, and their magnitudes have been calculated previously from first principles [2].

The general model demonstrates that, even under a set of operating conditions where there is one major rate-controlling reaction, there are invariably other elementary processes that are significantly removed from equilibrium. For example, at low values of the surface temperature T_s, decomposition of $Ga(CH_3)_2$ (DMG) at ledge sites (S17, see Table II) is important but it is still necessary to consider the rate of formation of CH_4 (reaction S12) since it also affects the occupancy of ledges by DMG. Similarly, at high temperatures, reactions S11, S12, S14, and S15 are far from equilibrium. This factor, coupled with the gradual nature of the shifts between the various rate-limiting regimes make it impossible to set any of the five reactions in Table I to equilibrium without automatically precluding agreement between calculated results and observed trends in behavior. On the other hand, if additional reactions are regarded as rate-determining, the number of equilibrium relationships remaining would be insufficient to eliminate the surface concentrations.

In Fig. 1, deposition rates r_s calculated using the LH treatment of the surface kinetics are compared with those obtained from the general MOVPE model. At atmospheric pressure, the LH model shows the observed trends in behavior, but it overestimates r_s in the range $780 < T_s(K) < 950$ because kinetic limitations for decomposition of TMG (reaction S16) invalidate the use of an equilibrium relationship to represent the fractional coverage $\theta_{DMG,l}$ at ledge sites. At lower values of T_s, the curves labelled A and A_{LH} converge because DMG covers most of the ledge sites [1]

112

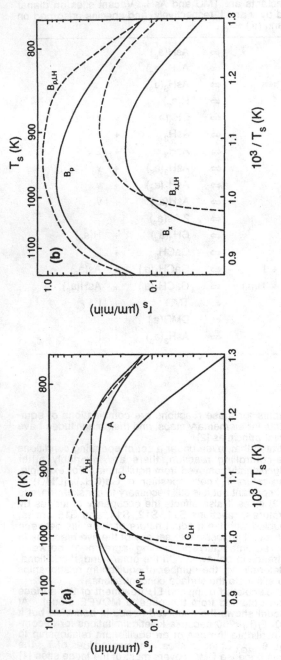

Figure 1. Comparison of GaAs deposition rates predicted using Langmuir-Hinshelwood (LH) kinetics and using the general MOVPE model [2]. The deposition rates are calculated for growth on a (111)Ga substrate in an impinging jet reactor. The operating characteristics are as follows: carrier gas = H_2; inlet gas temperature T_∞ = 298 K; ratio of partial pressures of reactants, $p^o_{AsH_3}/p^o_{TMG}$ = 20; total reactor pressure p_∞ (atm): A - 1.0, B - 0.1, and C - 0.01; p^o_{TMG} (10^{-4} atm): A, B_p - 3.40 and B_x, C - 0.34; gas flow rates (slpm) A, B_x - 1.5 and B_p, C - 0.15. Curves A_{LH}, $B_{x,LH}$, $B_{p,LH}$, and C_{LH} are the corresponding results from the LH model described in Table I. Curve A^o_{LH} is the same as A_{LH} except that gas-phase reactions are neglected. (a) Results for 1 atm and 0.01 atm; (b) results for 0.1 atm.

and, hence, overestimation of the TMG decomposition rate does not alter $\theta_{DMG,I}$ significantly. For $T_s > 950$ K, the LH model gives a reasonable prediction since surface decomposition of TMG is replaced by formation of $GaCH_3$ in the gas phase and the reactions between adsorbed $GaCH_3$ and As or AsH control the deposition rate (reactions S14 and S15). However, the success of the LH model for $T_s > 950$ K is contingent on inclusion of the gas-phase reactions (contrast curves A and A^o_{LH}). Note that, with smaller gas flow rates, homogeneous reactions would become even more important and their influence would extend to values of T_s below 1000 K.

At low reactor pressures, the results from the LH model are less satisfactory because the equilibrium assumptions that form the cornerstone of the LH approach are less appropriate. For example, at low values of T_s, removal of H atoms from ledge sites, decomposition of TMG, and adsorption of CH_3 radicals are more kinetically-controlled at 0.01 atm than at 1 atm. As a result, the LH model, which assumes these reactions to be at equilibrium, significantly overestimates the deposition rate (see curves C_{LH} and C). At higher values of T_s, $GaCH_3$ desorbs, but the rate of this process is dictated by kinetics rather than by thermodynamic constraints. Therefore, under these operating conditions, the LH calculations give lower deposition rates than the complete MOVPE model. The curves in Fig. 1a demonstrate that the applicability of LH kinetics diminishes as the reactor pressure is reduced. However, it should be emphasized that the other operating conditions also affect the validity of the simplified kinetics. For instance, less agreement is obtained for case B_x than for case B_p (see Fig. 1b) even though both examples are for a total pressure of 0.1 atm. Case B_x has lower inlet partial pressures of the reactants than case B_p, and this tends to give process behavior which is closer to that for example C [2].

Overall, Fig. 1 shows that even a sophisticated LH model, developed with the knowledge of the rate-limiting steps and rate constants from the detailed MOVPE calculations, only gives a reasonable description of process behavior over a limited range of conditions. Note that a separate LH model, specifically designed for low reactor pressures, would still fail because several consecutive surface processes are substantially removed from equilibrium. This factor prevents any equilibrium relation from giving a realistic value for the fraction of sites covered by a given species.

Another difficulty with LH kinetics is the need to know the rate-controlling reactions in advance. Development of LH expressions using theoretical estimates of the rate constants but without the insight gained from the general model would be extremely difficult because there are many equilibrium relations and rate-determining reactions that could be chosen which would not yield meaningful results. For this reason, one might be tempted to choose reaction stoichiometries and LH rate equations on an empirical basis and to use fitted kinetic parameter values. For example, a rate expression involving the partial pressures of the reactants has been proposed for atmospheric-pressure MOVPE of GaAs [3], and different kinetic parameters are chosen for high and low values of T_s in order to match the observed dependence of the deposition rate on the inlet gas composition. However, procedures of this type are likely to give a misleading indication of the rate-limiting factors for the process. For instance, strong adsorption of AsH_3 was postulated to explain why the deposition rate is independent of the inlet partial pressure of AsH_3. However, the detailed MOVPE model predicts the observed behavior with a relatively low surface coverage of AsH_3, and it shows that deposition would be far too slow if arsine filled the surface sites [8]. In addition, desorption of $GaCH_3$ is proposed as an explanation for the decline in r_s at high temperatures, whereas the general model [1] shows that kinetic limitations for reactions S14 and S15 are responsible together with desorption of arsenic and a lower thermodynamic driving force for adsorption of $GaCH_3$.

The limitations of LH kinetics, demonstrated here for MOVPE of GaAs, can also be expected in other CVD systems. At low pressures in particular, it is unlikely that sufficient elementary processes will be at equilibrium to allow the coverages of active sites to be represented by local partial pressures. Furthermore, unless the critical reaction pathways are established in advance, it will be necessary to use empirical LH relations, and one cannot expect these equations either to be correct from a fundamental standpoint or to predict observed behavior outside the range of operating conditions used to fit the parameter values.

Surface reaction probabilities

The sticking coefficient s is commonly used in the analysis of surface processes in molecular beam epitaxy (MBE) systems [9]. This coefficient is defined by a flux balance at the surface, i.e., the fraction of the total incoming flux of an atomic species that is incorporated into the solid deposit. The relationship used to describe the surface flux is based on the kinetic theory of gases and it is applicable in high-vacuum systems, as encountered in MBE.

Some previous models [5,6] of transport and kinetics in CVD have assumed that the deposition rate can be described using surface reaction probabilities which are set equal to sticking coefficients. However, this approach has three fundamental problems: (i) the species flux relationship used to define s has questionable validity for CVD since, in these systems, continuum mechanics is usually applicable and the effects of diffusion, thermal diffusion, and convection need to be taken into consideration, (ii) the effects of several elementary processes are represented by lumped parameters that are equivalent to rate constants for irreversible first-order reactions, and (iii) the rate expression uses local partial pressures of reactive species (containing atoms that form the deposit) and it does not consider the fractional coverages of adsorbates. The SRP models [5,6] have assumed further that a single reaction probability can be used to describe the deposition rate, i.e., that it is only necessary to consider the surface flux of one gas-phase species.

The SRP approach is empirical rather than fundamental. Nevertheless, its applicability to MOVPE of GaAs was tested by representing the deposition rate as the sum of products of reaction probabilities (values fitted by matching theory to data) and partial pressures of gallium-containing species adjacent to the surface. The calculations indicate that the approximation is reasonable for the limited range of operating conditions where the deposition rate is found to be linear in the inlet partial pressure of TMG. However, even under these circumstances, the SRP's have a significant dependence on reactor pressure and gas flow rate as a result of changes in the reaction pathways and rate-limiting reactions. Therefore, a wide range of coefficients would be needed to match deposition rates at a fixed surface temperature. For example, $GaCH_3$ is adsorbed at 1000 K and 1 atm [1] (i.e., the conditions of case A in Fig. 1) but it desorbs at 1000 K and 0.01 atm [2] (case C in Fig. 1), and this could only be accounted for by changing both the sign and the magnitude of the SRP. At high temperatures, the observed drop in the deposition rate would require the use of reaction probabilities with negative activation energies. Also, in this operating regime, the deposition rate tends to have a slightly superlinear dependence on p^o_{TMG} [2], but the SRP model gives a sublinear relation. Furthermore, the sublinear dependence of r_s on p^o_{TMG} observed at low temperatures [3] could not be matched using a single set of values for the SRP's.

CONCLUSIONS

Langmuir-Hinshelwood (LH) kinetics or surface reaction probabilities only provide a reasonable representation of MOVPE of GaAs on (111)Ga substrates over a limited range of operating conditions. The sticking coefficient approach is purely empirical and would involve fitting values of the coefficient(s) for each combination of pressure, flow rate, inlet gas composition, and temperature. The LH approach can be applied using calculated rate constants if dominant pathways for homogeneous and heterogeneous processes and the rate-controlling reactions have been identified previously using a general MOVPE model. However, even with this advantage and with inclusion of several rate-limiting steps, there are many sets of operating conditions for which the LH model is inadequate. Erroneous predictions using LH kinetics arise primarily because the compositions of reactive intermediates at surface sites or ledge sites cannot be described in terms of their local partial pressures and equilibrium constraints, and this problem is particularly acute at low reactor pressures. Furthermore, application of a single LH equation over a small temperature range can only match experimental data on a purely empirical basis since it has been found that, at any set of processing conditions, more than one reaction is significantly displaced from equilibrium.

ACKNOWLEDGMENT

This work was supported by the National Science Foundation (Grant No. CBT-845112).

REFERENCES

1. M. Tirtowidjojo and R. Pollard, J. Cryst. Growth 93, 108 (1988).
2. M. Tirtowidjojo and R. Pollard, J. Cryst. Growth, submitted for publication.
3. D.H. Reep and S.K. Ghandhi, J. Electrochem. Soc. 130, 675 (1983).
4. S.K. Shastry, in Initial Stages of Epitaxial Growth, edited by R. Hull, J.M. Gibson, and D.A. Smith (Mat. Res. Soc. Symp. Proc., 94, Pittsburgh, PA 1987) pp. 267-272.
5. M.E. Coltrin, R.J. Kee, and J.A. Miller, J. Electrochem. Soc. 133, 1206 (1986).
6. H.K. Moffat and K.F. Jensen, J. Electrochem. Soc. 135, 459 (1988).
7. C.G. Hill, Jr., An Introduction to Chemical Engineering Kinetics and Reactor Design (J. Wiley & Sons, New York, 1977).
8. M. Tirtowidjojo and R. Pollard, to be published.
9. J.R. Arthur, Surf. Sci. 43, 449 (1974).

A KINETIC MODEL FOR METALORGANIC CHEMICAL VAPOR DEPOSITION OF GaAs FROM TRIMETHYLGALLIUM AND ARSINE

TRIANTAFILLOS J. MOUNTZIARIS AND KLAVS F. JENSEN
Department of Chemical Engineering and Materials Science
University of Minnesota, Minneapolis, MN 55455

ABSTRACT

A kinetic model for metalorganic chemical vapor deposition (MOCVD) of GaAs from trimethylgallium and arsine is presented. The proposed mechanism includes 15 gas-phase species, 17 gas-phase reactions, 9 surface species and 29 surface reactions. The surface reactions take into account different crystallographic orientations of the GaAs substrate. Sensitivity analysis and existing experimental observations have been used to develop the reduced mechanism from the large number of reactions that might in principle occur. Rate constants are estimated by using thermochemical methods and reported experimental data. The kinetic mechanism is combined with a two-dimensional transport model of a hot-wall tubular reactor used in experimental studies. Model predictions of gas-phase composition and GaAs growth rates show good agreement with published experimental studies. In addition, the model predicts reported trends in carbon incorporation.

INTRODUCTION

Metalorganic chemical vapor deposition (MOCVD) has been used successfully to grow a wide variety of III-V and II-VI compound semiconductors on a laboratory scale [1]. However, to obtain uniform layers with controlled impurity incorporation, needed for large scale production of electronic and optical devices, the combined chemical reactions and transport phenomena underlying MOCVD must be understood. Modelling of MOCVD has typically focused on predicting complex flow and heat transfer phenomena for different reactor configurations, while simple overall rate expressions have been used to describe the growth rates [2,3]. Detailed kinetic models must be included in reactor descriptions to go beyond growth rate simulations and to predict film composition and impurity incorporation. As an example of this approach, a kinetic model of silane pyrolysis with mass transfer limited growth of GaAs has been used to simulate Si doping of GaAs from silane and disilane [4]. In this paper we propose a detailed kinetic model for epitaxial growth of GaAs from trimethylgallium (TMG) and arsine (AsH_3) to describe kinetically limited growth conditions and carbon incorporation.

Although MOCVD of GaAs is perhaps the simplest and most understood compound semiconductor growth process, the development of a kinetic model is a challenging task. The model development is complicated by a large number of gas-phase and surface reactions, incomplete experimental data for gas-phase species, and almost complete lack of mechanistic information and kinetic data for the surface reactions for the various crystallographic orientations. The kinetics of GaAs deposition have recently been addressed through estimates of thermochemical data for intermediate species and equilibrium computations in the gas phase [5] and subsequently by employing finite rate expressions of elementary reactions in reactor models [6,7]. In addition to predicting growth rate measurements, as done in the previous studies, the model also predicts recent gas phase composition data, as well as trends in carbon incorporation versus temperature and V/III ratio.

Kinetic models contain rate constants that are often only known within an order of magnitude. Furthermore, conditions and reactor geometries vary considerably in reported studies making direct comparison of growth rates difficult. Therefore, in this paper we emphasize the prediction of trends in data reported in different studies rather than exact agreement with one particular set of data.

GAS PHASE AND SURFACE REACTION MECHANISM FOR GaAs GROWTH

Gas-phase reactions in the $TMG/AsH_3/H_2$ system are complex and not well understood. The pyrolytic decomposition of TMG was first studied in a toluene carrier system and found to be limited by the loss of the first methyl radical [8]. More recent kinetic studies based on steady state downstream composition measurements [9,10] indicate that TMG decomposition in H_2 or D_2 is not affected by adding or changing the surface area. However, under steady state conditions the surface will rapidly be covered by polymeric deposits of the type discussed in [8] and it will not play a measurable role in the decomposition. However, in an actual deposition process new sites are continuously generated which could serve as TMG decomposition sites. Thus, kinetic parameters reported from steady state pyrolysis studies [8-10] reflect homogeneous decomposition reactions of TMG and the role of an active growth surface remains uncertain. TMG pyrolysis proceeds faster in H_2 than in N_2 due to hydrogen radical attack on TMG [9,11]. $GaCH_3$ is expected to be the most stable gas phase species [8-10].

The decomposition of AsH_3 is considered to be a heterogeneous reaction [12], proceeding through adsorption and subsequent loss of H to the surface. Recent *in situ* studies [13] indicate that the mechanism may be more complex. The rate of AsH_3 decomposition depends strongly on the type of surface and it is enhanced by the addition of TMG [9,10,13]. The enhancement and preferential CH_4 (rather than CH_3D) production in a D_2 carrier gas [9] may in principle be explained by the reaction of methyl radicals with AsH_3 [14]. However, because of the slow homogeneous decomposition rate of TMG at low temperatures, there are not enough methyl radicals to yield the observed acceleration. This is a clear indication that TMG will either be directly involved in heterogeneous reactions with surface arsenic species, or it will react with AsH_3 to form adducts [9], which can subsequently rearrange through the loss of CH_4. Equilibrium computations [5], however, indicate that adducts are too unstable to play a major role in the growth chemistry.

The gas phase and surface reaction mechanism summarized in Table 1 reflects the above experimental observations. The surface reaction sequences involve two kinds of sites (As and Ga) and their relative positions according to the surface orientation. Reaction rate constants for the mechanism were obtained from published gas phase pyrolysis studies, combustion rate data [15], and estimated by thermochemical methods [16]. Further details will be presented elsewhere along with discussion of the underlying assumptions [17].

The carbon incorporation is proposed to occur through the formation of carbene containing gallium species. These compounds are formed via hydrogen abstraction by methyl radicals. They are shortlived and only exist in small concentrations which explains the relatively low carbon incorporation in GaAs films under standard growth conditions ($\sim 10^{15}$ atoms/cm^3) [1]. The reactions involved are unimportant for the growth rate predictions.

TRANSPORT MODEL

A two dimensional model of a tubular hot-wall reactor is used to solve the flow and heat transfer problem. Subsequently, the mass transfer problem is solved by using collocation in the radial direction and integrating the resulting nonlinear differential-algebraic equations with the program DASSL [18]. This reactor configuration was chosen over the commonly used vertical and horizontal reactor configurations in order to efficiently simulate large systems of chemical reactions and carry out sensitivity analysis. Accurate transport predictions for actual deposition require time consuming computations that would unnecessarily confound the kinetic model development. Furthermore, a number of steady state decomposition studies have been performed in hot-wall tubular reactors [e.g. 8,9,11]. Ultimately, the detailed kinetic and transport have to be combined to predict process performance, as exemplified in the case of Si doping of GaAs [4].

RESULTS AND DISCUSSION

As a first test of the gas phase kinetics, model predictions and reported steady state data for TMG decomposition in H_2 and N_2 as a function of temperature [9,11] are compared in Figure 1. The kinetic parameters used for the homogeneous decomposition of TMG are those reported in [8]. The agreement between model predictions and experimental observations is excellent for TMG decomposition in N_2 and falls within the range of experimental data for decomposition in H_2. More accurate prediction of the H_2 data will require a better understanding of the reactions between hydrogen radicals and gallium species. As a second example, the decomposition of TMG and AsH_3 in D_2 carrier gas is considered. Figure 2 shows the observed [9] and predicted

Table 1. Gas phase and surface reaction mechanism for GaAs growth by MOCVD

Gas-Phase Reactions

Pyrolytic Decomposition of TMG :

$$Ga(CH_3)_3 \leftrightarrow Ga(CH_3)_2 + CH_3\cdot$$
$$Ga(CH_3)_2 \leftrightarrow GaCH_3 + CH_3\cdot$$

Methane Formation :

$$CH_3\cdot + H_2 \leftrightarrow CH_4 + H\cdot$$
$$CH_3\cdot + AsH_3 \leftrightarrow AsH_2 + CH_4$$

Free Radical Recombination :

$$H\cdot + H\cdot + M \leftrightarrow H_2 + M$$
$$CH_3\cdot + H\cdot + M \leftrightarrow CH_4 + M$$
$$CH_3\cdot + CH_3\cdot \leftrightarrow C_2H_6$$

Gallium-Carbene Formation :

$$Ga(CH_3)_x + CH_3\cdot \leftrightarrow Ga(CH_3)_{x-1}CH_2 + CH_4 \quad x=1,2,3$$
$$Ga(CH_3)_xCH_2 \leftrightarrow Ga(CH_3)_{x-1}CH_2 + CH_3\cdot \quad x=1,2$$
$$Ga(CH_3)_xCH_2 + H\cdot \leftrightarrow Ga(CH_3)_{x+1} \quad x=1,2$$

Decomposition of TMG in H_2 carrier gas :

$$Ga(CH_3)_x + H\cdot \leftrightarrow Ga(CH_3)_{x-1} + CH_4 \quad x=2,3$$

Surface Reactions

Adsorption / Chemisorption :

$$H\cdot + S_x \leftrightarrow H_x^* \quad x=A,G$$
$$CH_3\cdot + S_x \leftrightarrow (CH_3)_x^* \quad x=A,G$$
$$Ga(CH_3)_x + S_G \leftrightarrow Ga(CH_3)_x^* \quad x=1,2,3$$
$$AsH_x + S_A \leftrightarrow AsH_x^* \quad x=1,2,3$$
$$Ga(CH_3)_{x-1}CH_2 + S_A + S_G \leftrightarrow Ga(CH_3)_{x-1}CH_2^{**} \quad x=1,2,3$$

Recombination :

$$H_x^* + H_y^* \leftrightarrow H_2\uparrow + S_x + S_y \quad x,y=A,G$$
$$(CH_3)_x^* + H_y^* \leftrightarrow CH_4\uparrow + S_x + S_y \quad x,y=A,G$$
$$(CH_3)_x^* + (CH_3)_y^* \leftrightarrow C_2H_6\uparrow + S_x + S_y \quad x,y=A,G$$

Decomposition of the Precursors :

$$AsH_x^* + S \leftrightarrow AsH_{x-1}^* + H^* \quad x=1,2,3$$
$$Ga(CH_3)_x^* + S \leftrightarrow Ga(CH_3)_{x-1}^* + CH_3^* \quad x=1,2,3$$

Hydrogen Abstraction :

$$CH_3 + H^* \leftrightarrow CH_4\uparrow + S$$
$$CH_3 + AsH_x^* \leftrightarrow AsH_{x-1}^* + CH_4\uparrow \quad x=1,2,3$$

Growth :

(111), (110) surfaces:

$$Ga(CH_3)_x^* + AsH_y^* \leftrightarrow GaAs + (x-1)CH_3^* + (y-1)H^*$$
$$+ CH_4\uparrow + (2-x)S_G + (2-y)S_A \quad x,y=1,2$$

(100) surface:

$$Ga(CH_3)_x^* + S_G \leftrightarrow GaAs + xCH_3^* \quad x=0,1,2$$
$$AsH_x^* + S_A \leftrightarrow GaAs + xH^* \quad x=0,1,2$$

Carbon Incorporation :

$$Ga(CH_3)_{x-1}CH_2^{**} + S \leftrightarrow GaC + (x-1)CH_3^* + H_2\uparrow \quad x=1,2,3$$

Recombinative Desorption :

$$2\,AsH^* \leftrightarrow As_2\uparrow + H_2\uparrow + 2\,S_A$$
$$2\,As^* \leftrightarrow As_2\uparrow + 2\,S_A$$

S_A, S_G : free sites available for As-species and Ga-species adsorption, respectively.

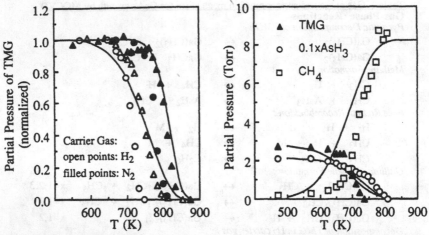

Figure 1. Comparison between theory and experiments (o:[9], Δ:[11]) for TMG decomposition.

Figure 2. Decomposition of TMG and AsH₃. Model: solid line. Data: from [9].

variations in partial pressures of major stable species with temperature. The observed trends are predicted and the increased CH_4 at low temperatures relative to the experimental observations is related to the differences between model and experiments in the H_2 data shown in Figure 1. The simulations show that CH_4 is primarily produced by growth reactions between methylated gallium species and adsorbed AsH_x. The reaction of methyl radicals with D_2 produces detectable amounts of CH_3D only at high temperatures, in accordance with experimental observations [9]. The homogeneous reaction of AsH_3 with methyl radicals contributes less than 5% over the whole temperature range.

To explore the ability of the model to predict actual growth data, simulations are made for the conditions reported by Reep [19], and Reep and Ghandhi [20]. Since these authors use a cold-wall horizontal reactor with sloped susceptor, it is not possible to directly compare data controlled by diffusion, but it is still possible to compare trends in the growth rate with variations in temperature and reactant partial pressures. To preserve the relative importance of gas phase and surface reactions, the computations are done using the same residence time as in the original experiments. As illustrated in Figure 3, the model predicts the correct variation of growth rate with respect to temperature. The apparent activation energy for the low temperature, surface reaction controlled regime, of 25 kcal/mole compares well with the observed value of 24.5 kcal/mole. The high temperature

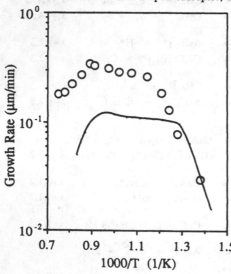

Figure 3. Growth rate of GaAs as a function of reciprocal temperature. Surface orientation: (110) Partial pressure of AsH_3 = 2.51 Torr Partial pressure of TMG = 0.14 Torr Model : Solid line Data : from Reep [19]

reduction in the rate is reproduced by the model and is caused by increased desorption of film precursors. A diffusion limited plateau is also predicted in qualitative agreement with the data. As mentioned, different reactor geometries preclude a direct comparison of the diffusion limited rates, but the model shows quantitative agreement with data in the kinetically limited regime.

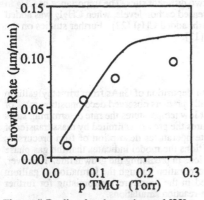

Figure 4. Predicted and experimental [20] growth rate versus partial pressure of TMG. T = 500°C ; p AsH₃ = 2.51 Torr.

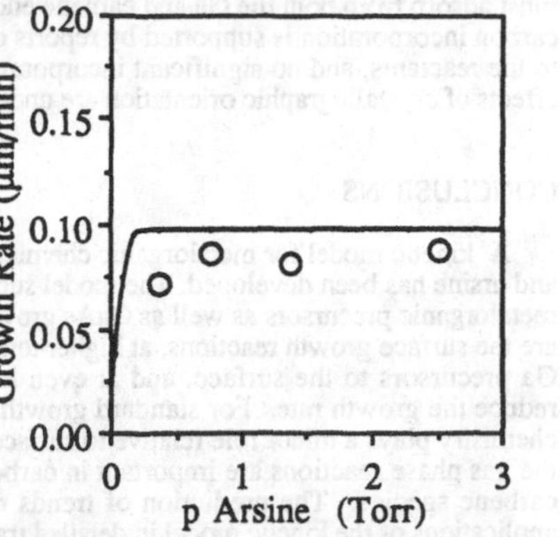

Figure 5. Predicted and experimental [20] growth rate versus partial pressure of AsH₃. T = 500°C ; p TMG = 0.14 Torr.

Figures 4 and 5 show experimental and predicted growth rate variations with TMG and AsH₃ partial pressures, respectively, for growth at 500°C. The model simulations demonstrate the saturation of the growth rate at high partial pressures and are in good agreement with the experimental data. Exact agreement could be obtained by fitting the rate constants of the surface reactions, but the emphasis here is to demonstrate that trends in data from different sources can be reproduced by the mechanism.

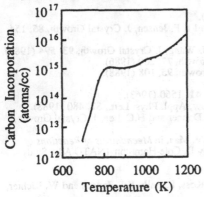

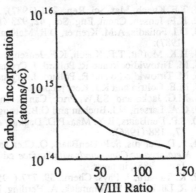

Figure 6. Carbon incorporation versus T. p TMG = 0.14 Torr ; p AsH₃ = 2.51 Torr.

Figure 7. Carbon incorporation versus V/III ratio. T=700°C ; p AsH₃ = 2.51 Torr.

Relatively simple rate models are often able to reflect observed growth rate variations. However, detailed gas phase and surface chemistry models are needed to predict impurity incorporation levels, as illustrated for Si doping of GaAs in [4]. Here, we consider carbon incorporation occurring via the formation of carbene containing Ga species. Predicted variations in carbon incorporation with temperature and V/III ratio are shown in Figures 6 and 7 for similar conditions to those used in the growth rate comparisons. The simulations show the correct order of magnitude and experimentally observed trends, i.e. increasing carbon incorporation with increased temperature and reduced V/III ratio [21]. The carbene mechanism explains the much lower than expected carbon incorporation, if methyl radicals were directly responsible. The gallium carbene species will rapidly abstract hydrogen from other surface species, in which case carbon incorporation is not likely to occur. Thus, for C to replace an As, the carbene species must adsorb from both the Ga and carbene ends on two different sites. The importance of CH_2 in carbon incorporation is supported by reports of increased carbon levels, when CH_2I_2 was added to the reactants, and no significant incorporation with added CH_3I [22]. Further studies on the effects of crystallographic orientation are underway [17].

CONCLUSIONS

A kinetic model for metalorganic chemical vapor deposition of GaAs from trimethylgallium and arsine has been developed. The model successfully predicts observed decomposition rates of metalorganic precursors as well as GaAs growth. At low temperatures the rate determining steps are the surface growth reactions, at higher temperatures the growth is limited by mass transfer of Ga precursors to the surface, and at even higher temperatures desorption of film precursors reduce the growth rate. For standard growth conditions the model indicates that the gas phase chemistry plays a minor role relative to surface reactions in predicting the growth rate. However, the gas phase reactions are important in carbon incorporation through the formation of gallium carbene species. The prediction of trends reported in the literature is promising for further applications of the kinetic model in detailed transport-reaction simulations.

ACKNOWLEDGEMENTS

This work was supported by the National Science Foundation (DMR-8704355) and the Minnesota Supercomputer Institute.

REFERENCES

1. T.F. Kuech, Mat. Sci. Rep., **2**, 1 (1987).
2. K.F. Jensen, Chem. Eng. Sci., **42**, 923 (1987).
3. D.I. Fotiadis, A.M. Kremer, D.R. McKenna and K.F. Jensen, J. Crystal Growth, **85**, 154 (1987).
4. H.K. Moffat, T.F. Kuech, K.F. Jensen and P.J. Wang, J. Crystal Growth, **93**, 594 (1988).
5. M. Tirtowidjodjo and R. Pollard, J. Crystal Growth, **77**, 200 (1986).
6. M. Tirtowidjodjo and R. Pollard, J. Crystal Growth, **93**, 108 (1988).
7. M.E. Coltrin and R.J. Kee, preprint.
8. M.G. Jacko and S.J.W. Price, Can. J. Chem., **41**, 1560 (1963).
9. C.A. Larsen, N.I. Buchan and G.B. Stringfellow, Appl. Phys. Lett., **52**, 480 (1988).
10. S.P. DenBaars, B.Y. Maa, P.D. Dapkus, A.D. Danner and H.C. Lee, J. Crystal Growth, **77**, 188 (1986).
11. P. D. Dapkus, S. P. DenBaars, Q. Chen and B.Y. Maa, in *Mechanisms of Reactions of Metalorganic Compounds with Surface*, edited by D. Cole-Hamilton (NATO Adv. Study Inst., 1988), in press.
12. K. Tamaru, J. Phys. Chem., **59**, 777 (1955).
13. R. Lückerath, P. Tommack, A. Hertling, H.J. Koss, P. Balk, K.F. Jensen and W. Richter, J. Crystal Growth, **93**, 151 (1988).
14. J.E. Butler, N. Bottka, R.S. Sillmon and D.K. Gaskill, J. Crystal Growth, **77**, 163 (1986).
15. W. Tsang and R.F. Hampson, J. Phys. Chem. Ref. Data, **15**, No. 3, 1087 (1986).
16. S.W. Benson, *Thermochemical Kinetics,* 2nd ed. (Wiley, New York, 1976).
17. T.J. Mountziaris and K.F. Jensen, in preparation.
18. L.R. Petzold, Report #SAND82-8637, Sandia National Laboratories, Livermore, CA, 1982.
19. D.H. Reep, PhD Thesis, Rensselaer Polytechnic Institute, 1982.
20. D.H. Reep and S. K. Ghandhi, J. Electrochem. Soc., **130**, 675 (1983).
21. T.F. Kuech and E. Veuhoff, J. Crystal Growth, **68**, 148 (1984).
22. T.F. Kuech, personal communication.

EQILIBRIUM ANALYSIS OF THE TMGa-TMAl-AsH₃-H₂ MOCVD EPITAXIAL GROWTH SYSTEM

HYUK J. MOON AND THOMAS G. STOEBE
Department of Materials Science and Engineering, University of Washington,
Seattle, WA 98195
BRIAN K. CHADWICK
United Epitaxial Technologies, Inc., 19545 N.W. Von Neumann Drive,
D4/210, Beaverton, OR 97006

ABSTRACT

The thermodynamic equilibrium state of the Ga-Al-As-C-H system was determined theoretically by means of an iterative equilibrium constant method. This method of calculation is presented and discussed. With very little operator input, the program is capable of computing the partial pressures of the gas-phase species present in the equilibrated system.

In these calculations the system was considered to be saturated with solid-phase AlGaAs and included 58 plausible gas-phase intermediates which evolved from the initially present gas species; trimethylgallium, trimethylaluminum, arsine, and hydrogen. Temperature and total system pressure ranges investigated were 750-1100 K and 0.1 atm-1.0 atm, respectively. The effects of temperature and pressure variations, in addition to effects caused by changes in the appropriate atom ratios, have been delineated. The properties of this equilibrated system are compared with those from recent thermodynamic research efforts on AlGaAs systems consisting of only gaseous constituents.

INTRODUCTION

Several important compound semiconductor epitaxial layer growth techniques have been introduced, some of which are currently being employed for the production of device quality materials; among these are Liquid Phase Epitaxy (LPE), Vapor Phase Epitaxy (VPE), Metalorganic Chemical Vapor Deposition (MOCVD), Molecular Beam Epitaxy (MBE), etc. MOCVD has demonstrated the ability to grow high quality epitaxial layers, however, understanding and controlling the numerous parameters associated with the MOCVD processes is difficult. Thermodynamic analyses provide useful information needed for additional understanding of the MOCVD growth process. It has been shown that thermodynamic analyses can help predict alloy compositions and maximum growth rates in MOCVD systems [1,2]. Although epitaxial growth by MOCVD is generally regarded as a kinetically controlled process, since the rates of chemical reactions at the growth surface are high compared to the arrival rate of reactants to the growth surface, we may assume that a near-thermodynamic equilibrium is established at the growth surface.

This study is intended to delineate the chemical thermodynamic equilibrium state of an MOCVD system when saturated with undoped AlGaAs solid. Han and Rao [3] have used the iterative equilibrium constant method to compute the equilibrium state of the Ga-As-H-Cl VPE growth system. Rao [4] and Chadwick [5] have described the iterative approach in detail for other systems. Chadwick [6] has used the iterative equilibrium constant method to describe the equilibrium state of a system with undoped GaAs

Mat. Res. Soc. Symp. Proc. Vol. 131. ©1989 Materials Research Society

solid. A similar analysis was performed by Tirtowidjojo and Pollard [7] using a different computational technique than that used here. The thermochemical values given by Tirtowidjojo [8] are used here along with data for two additional species, GaAs-solid [9] and AlAs-solid [10]. AlAs solid-phase heat capacity, as a function of temperature, was obtained by interpolating values reported by Kagaya [11]. Solid phase $Al_xGa_{1-x}As$ ($x=0.5$) is assumed to be an ideal solid solution. The relative stabilities of the 58 gas-phase species, considered as intermediates in this system, are deduced in terms of their partial pressures.

ANALYSIS

When determining which species should be chosen, it is important to first include the species which occur in non-negligible concentrations. The thermochemical data should also be available for each species chosen (if not, the thermochemical values must be computed).

The chemical species chosen to be present in this system are identical to those chosen by Tirtowidjojo and Pollard [7] with one exception; AlGaAs solid phase, which is taken to be in its standard state, has been included. The following list of species are envisioned to exist within the temperature range (750 - 1100 K) investigated:

AlGaAs(s), As, As_2, As_3, As_4, AsH, AsH_2, AsH_3, $As(CH_3)_3$, $As(CH_3)_2$, $As(CH_3)$, $HAs(CH_3)_2$, $HAs(CH_3)$, $H_2As(CH_3)$, C, CH_3, CH_4, C_2H_6, C_2H_4, C_2H_5, C_2H_6, H_2, H, Ga, GaH, GaH_2, GaH_3, Ga_2H_6, GaAs, $(GaAs)_{3,L}$, $(GaAs)_{5,L}$, $(GaAs)_{3,C}$, $(GaAs)_{5,C}$, $Ga(CH_3)_3$, $Ga(CH_3)_2$, $Ga(CH_3)$, $HGa(CH_3)_2$, $HGa(CH_3)$, $H_2Ga(CH_3)$, $AsGa(CH_3)H$, $H_3AsGa(CH_3)_3$, $HAsGa(CH_3)$, $AsGa(CH_3)$, $AsGa(CH_3)_2$, Al, AlH, AlH_2, AlH_3, Al_2H_6, AlAs, $Al(CH_3)_3$, $Al_2(CH_3)_6$, $Al(CH_3)_2$, $Al(CH_3)$, $HAl(CH_3)$, $H_2Al(CH_3)$, $AsAl(CH_3)_2$, $AsAl(CH_3)$, $HAsAl(CH_3)$, and $HAl(CH_3)_2$.

The present equilibrium system is seen to consist of 2 phases (1 solid and 1 gas), 59 species, and 5 kinds of atoms - As, Ga, Al, C, and H. We must determine the equilibrium partial pressures of 58 gaseous species. The maximum number of independent reactions needed to describe the system are 54 and one possible set of 54 reactions chosen to represent this system is shown in Table I.

The degrees of freedom was found to be 5 and are satisfied by specifying total system pressure, temperature, and three atom ratios (R1=H/C, R2=(As-(Ga+Al))/H, R3=(Ga-Al)/H). The partial pressures of the majors are modified in each successive iteration by using the following convergence formulae;

$$(P_{H_2})_{new} = (P_{H_2})_{old}(R1E)^{0.1}$$

$$(P_{AsH_3})_{new} = (P_{AsH_3})_{old}(P_T/P_{T(new)})^{0.15}(R2ExR3E)^{-0.05}$$

$$(P_{TMGa})_{new} = (P_{TMGa})_{old}(P_T/P_{T(new)})^{0.5}(R2E)^{0.05}$$

$$(P_{TMAl})_{new} = (P_{TMAl})_{old}(P_T/P_{T(new)})^{0.5}(R3E)^{0.05}$$

where,

$$R1E = R1/R1_{new}$$
$$R2E = As/(R2 \times H + Ga + Al) \quad \text{when R2 is positive}$$

Table I. An independent reaction set for the TMGa-TMAl-AsH₃-H₂ system.

Reaction		Reaction	
$GaAs = Ga + As$	(R 1)	$(GaAs)_{5,L} = (GaAs)_{5,C}$	(R28)
$GaAs = Ga + 1/2As_2$	(R 2)	$GaAs + 2CH_3 = AsGa(CH_3)_2$	(R29)
$GaAs = Ga + 1/3As_3$	(R 3)	$Ga(CH_3)_3 = Ga(CH_3)_2 + CH_3$	(R30)
$GaAs = Ga + 1/4As_4$	(R 4)	$Ga(CH_3)_2 = Ga(CH_3) + CH_3$	(R31)
$As + 1/2H_2 = AsH$	(R 5)	$Ga(CH_3) + 1/2H_2 = HGa(CH_3)$	(R32)
$As + H_2 = AsH_2$	(R 6)	$HGa(CH_3) + 1/2H_2 = H_2Ga(CH_3)$	(R33)
$AsH_2 + 1/2H_2 = AsH_3$	(R 7)	$H_2Ga(CH_3) + HGa(CH_3) = HGa(CH_3)_2 + GaH_2$	(R34)
$As + CH_3 = As(CH_3)$	(R 8)	$Ga(CH_3) + AsH = AsGa(CH_3)H$	(R35)
$As + 2CH_3 = As(CH_3)_2$	(R 9)	$Ga(CH_3) + AsH = HAsGa(CH_3)$	(R36)
$As(CH_3)_2 + CH_3 = As(CH_3)_3$	(R10)	$H_3AsGa(CH_3)_3 = 3/2H_2 + GaAs + 3CH_3$	(R37)
$As(CH_3)_2 + 1/2H_2 = HAs(CH_3)_2$	(R11)	$Ga(CH_3)_3 = Ga + 3CH_3$	(R38)
$HAs(CH_3)_2 = HAs(CH_3) + CH_3$	(R12)	$AsH_3 + 1/2 Al(CH_3)_3 + 1/2 Ga(CH_3)_3 =$	(R39)
		$\quad 3CH_4 + Al_{0.5}Ga_{0.5}As$	
$HAs(CH_3) + 1/2H_2 = H_2As(CH_3)$	(R13)	$Al(CH_3)_3 = Al(CH_3)_2 + CH_3$	(R40)
$C_2H_4 = C + CH_4$	(R14)	$Al(CH_3)_2 = Al(CH_3) + CH_3$	(R41)
$CH_4 = CH_3 + 1/2H_2$	(R15)	$Al(CH_3) + 1/2H_2 = HAl(CH_3)$	(R42)
$CH_3 + CH_4 = C_2H_6 + 1/2H_2$	(R16)	$HAl(CH_3) + 1/2H_2 = H_2Al(CH_3)$	(R43)
$C_2H_6 = C_2H_5 + 1/2H_2$	(R17)	$Al(CH_3)_3 = Al + 3CH_3$	(R44)
$C_2H_6 = C_2H_4 + H_2$	(R18)	$Al + H = AlH$	(R45)
$GaAs + CH_3 = AsGa(CH_3)$	(R19)	$AlH + H = AlH_2$	(R46)
$H_2 = 2H$	(R20)	$AlH_2 + H = AlH_3$	(R47)
$Ga + H = GaH$	(R21)	$2AlH_3 = Al_2H_6$	(R48)
$GaH + H = GaH_2$	(R22)	$HAl(CH_3) + Al(CH_3) = HAl(CH_3)_2 + Al$	(R49)
$GaH_2 + H = GaH_3$	(R23)	$As(CH_3) + AlH = HAsAl(CH_3)$	(R50)
$2GaH_3 = Ga_2H_6$	(R24)	$HAsAl(CH_3) = AlAs + CH_4$	(R51)
$3GaAs = [GaAs]_{3,L}$	(R25)	$AlAs + CH_3 = AsAl(CH_3)$	(R52)
$(GaAs)_{3,L} = (GaAs)_{3,C}$	(R26)	$AsAl(CH_3) + CH_3 = AsAl(CH_3)_2$	(R53)
$(GaAs)_{3,L} + 2GaAs = (GaAs)_{5,L}$	(R27)	$2Al(CH_3)_3 = Al_2(CH_3)_6$	(R54)

$$R2E = (As - R2 \times H)/(Ga + Al) \quad \text{when R2 is negative}$$
$$R3E = Ga/(R3 \times H + Al) \quad \text{when R3 is positive}$$
$$R3E = (Ga - R3 \times H)/Al \quad \text{when R3 is negative.}$$

The following 4 conditions must be satisfied or the program will perform another iteration:

$$|(P_T\text{-}P_T(\text{new}))/P_T|<0.001$$
$$|1.0\text{-}R1E|<0.001$$
$$|1.0\text{-}R2E|<0.001$$
$$|1.0\text{-}R3E|<0.001$$

RESULTS

Temperature Effect

Figure 1(a) shows the effect increasing temperature has on the first 12 species at constant total pressure and constant $(As-(Ga+Al))/H$, $(Ga-Al)/H$, and H/C ratios. Nearly all species exhibit increasing pressures as higher temperatures are encountered; As_4, GaH_3, $H_2Ga(CH_3)$, and Ga_2H_6 behave oppositely. H_2 is always observed to be nearly equal to the total pressure and has been omitted from the figures for clarity purposes. As the system is subjected to higher temperatures the dimer form of gaseous arsenic becomes more stable at the expense of the tetramer molecule. Note also that the pressures of As_3 and As increase at higher temperatures. For the temperature range of interest, the most stable species is CH_4 (except for H_2) and the major Ga containing species are $Ga(CH_3)$ and GaH_2. GaH_3 becomes more important at lower temperatures while GaH and Ga become more important at high temperatures. Monatomic Ga becomes more stable than AsH_3 above 950 K. The major Al-containing species are AlH_3, AlH_2, and $Al(CH_3)$ and the equilibrium partial pressures of these are more than seven orders of magnitude smaller than those of major Ga-containing species; they are even less stable than $Ga(CH_3)_3$ below 850 K.

Hydrogen to Carbon Ratio Effect

Figure 1(b) provides a comparison between different H/C ratios when viewed in conjunction with Figure 1(a). Decreasing H/C ratio seems to have the effect of shifting some curves higher and some lower, but the general trend followed by the species as the temperature is increased is preserved. The relative stabilities also remain unchanged for the most part.

Total Pressure Effect

Figure 1(c) shows the effect of total system pressure on the individual partial pressures at constant temperature and constant $(As-(Ga+Al))/H$, $(Ga-Al)/H$, and H/C ratios. As the system pressure is decreased the dimer molecule becomes increasingly more stable than the tetramer form of gaseous arsenic, this agrees with the work of Chadwick [6]. One obvious inconsistency between this work and the work of Chadwick [6] is in the behavior of the gaseous Ga. In the latter study the pressure of Ga drops gradually as the total system pressure is increased. An important feature at a pressure of 0.1 atm, which is a common setting for the low-pressure MOCVD epitaxial growth, is the increased stability of Ga over AsH_3 and the

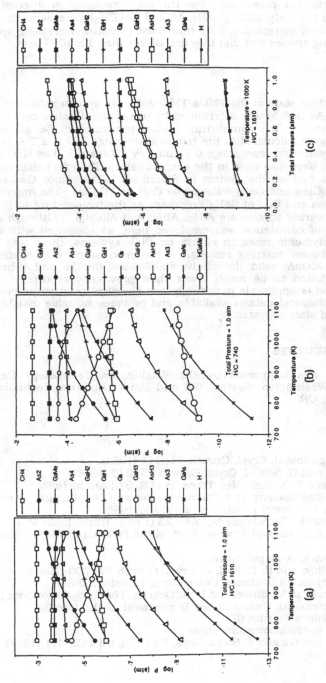

Figure 1. (a),(b) Effect of temperature and (c) total system pressure on the equilibrium partial pressures of the first 12 gas-phase species in the TMGa-TMAl-AsH3-H2 system. The legend to the right of each plot lists (in decreasing order) the relative stabilities at T=1000 K for (a) and (b) and P$_T$=1.0 atm for (c). The partial pressure of H$_2$ is near unity and has been omitted for clarity purposes.

increased fall in As_4 pressure. For the pressure range of interest, the pressures of all but AlH_2, $Al(CH_3)$, AlH, Al, and $HAl(CH_3)$ increase as the total system pressure is increased. Note that those pressure-decreasing species are all Al-bearing species and that they are all less than 1×10^{-13} atm.

CONCLUSION

The equilibrium state of the TMGa-TMAl-AsH_3-H_2 system saturated with solid $Al_xGa_{1-x}As$ (x=0.5) was determined by use of the iterative equilibrium constant method. The equilibrium partial pressures of 58 gas-phase intermediates were calculated in the temperature range 750 K $\leq$ T $\leq$ 1100 K and the total system pressure range 0.1 atm $\leq P_T \leq$ 1.0 atm. Two H/C ratios (1610 and 740) were also used in the computation. The effect that each of these variables had on the system is described. The major Ga-bearing species are $Ga(CH_3)$ and GaH_2 followed by GaH and GaH_3. The importance of GaH increases and that of GaH_3 decreases as the temperature increases. The major Al-bearing species are AlH_3, AlH_2, and $Al(CH_3)$. Although a different method of calculation was employed here, as compared with other recent thermodynamic research efforts on III/V systems, there was good agreement between existing results. However, since this system is saturated with AlGaAs solid, the relative stabilities of the Ga- and Al-bearing species were found to be much lower than previously reported. The iterative method is applicable to virtually any chemical reaction process for which thermochemical data is available and provides valuable insight into the equilibrated state of systems.

ACKNOWLEDGEMENTS

This work was supported by the Washington Technology Center, University of Washington, Seattle, WA and United Epitaxial Technologies, Inc., Beaverton, OR.

REFERENCES

1. G. B. Stringfellow, J. Cryst. Growth **68**, 111 (1984).
2. A. Koukitu and H. Seki, J. Cryst. Growth **76**, 233 (1986).
3. H. G. Han and Y. K. Rao, Met. Trans. B **16B**, 97 (March, 1985).
4. Y. K. Rao, *Stoichiometry and Thermodynamics of Metallurgical Processes* (Cambridge University Press, New York, NY, 1985).
5. B. K. Chadwick, TMS Paper No. A86-25 (1986); Trans. Instn Min. Metall.(Sect. C: Mineral Process. Extr. Metall.) **97**, C143 (September 1988).
6. B. K. Chadwick, to be published.
7. M. Tirtowidjojo and R. Pollard, J. Cryst. Growth **77**, 200 (1986).
8. M. Tirtowidjojo, Ph.D. thesis, University of Houston, 1988.
9. L. B. Pankratz, J. M. Stuve, and N. A. Gokcen, *Thermodynamic Data for Mineral Technology*, United States Department of the Interior, Bureau of Mines Bulletin 677.
10. R. Pollard (private communication).
11. H. Matsuo Kagaya and T. Soma, Phys. Stat. Sol. (B) **142**, 411 (1987).

ADSORPTION AND DECOMPOSITION OF TRIMETHYLARSENIC ON GaAs(100)

J. R. CREIGHTON
Sandia National Laboratories, Division 1126, P.O. Box 5800, Albuquerque,
NM 87185

ABSTRACT

Alkylated arsenic compounds have shown some promise as alternatives to arsine as the group-V source gas for GaAs MOCVD. However, little is known about the fundamental chemical interactions of these compounds with the GaAs surface. We have investigated the adsorption and reactivity of trimethylarsenic (TMAs) on GaAs(100) using temperature programmed desorption (TPD), Auger electron spectroscopy, and LEED. For the exposures and temperatures studied, TMAs did not pyrolytically decompose on the GaAs(100). TPD results indicate that TMAs chemisorbs, apparently non-dissociatively, and desorbs $\approx$ 330 K. Multilayers of TMAs desorb $\approx$ 140-160 K. Exposure of adsorbed TMAs to 70 eV electrons results in irreversible decomposition of the molecule. After electron irradiation, TPD shows that methyl radicals desorb at 660 K, which corresponds to a desorption activation energy of $\approx$ 40 kcal/mol. At higher temperatures, As_2, H_2, C_2H_2, and a smaller amount of methyl radicals desorb, and a small coverage of carbon remains on the surface.

INTRODUCTION

The hazard posed by the group-V hydride source gases (i.e. AsH_3 and PH_3) commonly used for III-V MOCVD has spawned interest in the search for safer alternatives [1,2]. Most of work to date has focused on alkylated derivatives of arsine such as trimethyl, triethyl, or tertiarybutylarsine. High levels of carbon incorporation has been a major problem in the development of growth processes utilizing the alkylated source gases. It is generally believed that carbon incorporation is the result of some intrinsic chemical reaction involving the alkyl groups of the source gas. As part of our overall effort to investigate the heterogeneous chemistry relevant to MOCVD, we have chosen to examine the surface chemistry (i.e. reaction pathways, carbon incorporation mechanisms) of the simple alklylated arsenic compounds. In this paper we report some of our initial results on the chemical interactions of trimethylarsenic (TMAs) with the GaAs(100) surface. The true pyrolytic behavior of TMAs will be compared to that observed when adsorbed TMAs is irradiated with 70 eV electrons.

EXPERIMENTAL

Experiments were performed in a multilevel vacuum system equipped with a cylindrical mirror analyzer for Auger Spectroscopy, a quadrupole mass spectrometer for temperature programmed desorption (TPD), and rear-view LEED optics (see Fig. 1). The chamber was pumped by turbomolecular pumps and liquid-nitrogen cold traps to a typical working base pressure of 5 X 10^{-10} Torr. A doser, with a liquid-nitrogen-cooled shutter, was used to dose TMAs in the upper chamber level. The absolute flux of TMAs impinging on the sample was not known, but it was normally adjusted to produce about 1 ML of adsorbed TMAs every 5-30 seconds. The sample was a 1.8-cm x 0.8-cm rectangular piece of semi-insulating GaAs(100) on which a 2-μm-thick conductive epilayer (Si doped, n-type) of GaAs was deposted by MBE. Tantalum clips held the GaAs sample to the manipulator stage and allowed for direct resistive heating to above 600°C. The sample could be cooled to $\approx$ 105 K by adding LN_2 to the

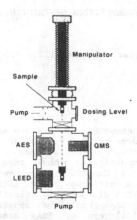

Figure 1. Schematic of experimental apparatus.

manipulator stage. A sacrificial GaAs sample mounted in the upper chamber level could be heated to above 700°C and served as a source of excess As_2 [3] to vary the initial surface stoichiometry of the primary GaAs(100) sample. For temperature measurement, a chromel-alumel thermocouple was spot-welded to a small tantalum clip which was attached to the top edge of the sample with a small drop of high-temperature cement (Aremco #516). The thermocouple consistently gave somewhat lower tempertures than our optical pyrometer (IRCON mod #2000). For instance, with the sample heated to a thermocouple reading of 550°C, the pyrometer read 587°C. Since we do not yet know which device is more accurate, we report temperature using uncorrected thermocouple values, realizing that these values may be low by a few tens of degrees at the highest temperatures.

TPD was performed with a multiplexed mass spectrometer, typically monitoring 10 masses, and a sample heating rate of 5 K/sec. The sample was normally placed about two inches in front of the mass spectrometer ionizer. To prevent electron irradiation, the sample could be biased -135 V from ground. If the sample was simply grounded, an electron flux of ≈3 $\mu A/cm^2$ (E = 70eV) impinged on the surface. A liquid-nitrogen-cooled cryoshroud surrounded the ionizer portion of the mass spectrometer and this served to reduce the scattered component of the TPD signals (for species that condense on the cryoshroud), thus enhancing the detection of the line-of-sight component. Unfortunately, reduction of the scattered component also decreases the overall sensitivity. The cryoshroud also appears to quench the coversion of methyl radicals to methane, which apparently occurs via radical-wall collisions.

RESULTS AND DISCUSSION

The GaAs(100) sample was cleaned by ion bombardment (4keV Ar^+), and subsequent annealing to 520-590°C, normally resulting in a (4 X 6) LEED pattern and a low As/Ga surface stoichiometry [4,5], as determined by Auger spectroscopy. The C/Ga ratio was always ≤ 0.005 after the sputter-anneal cycle. Other reconstructed surfaces, in particular the c(8 X 2), c(6 X 4), and c(2 X 8), could be formed by dosing the surface with As_2 at 400°C and annealing [6].

TMAs was adsorbed onto a GaAs(100)-(4 X 6) surface at ≈105 K and then desorbed in a TPD experiment in which the sample was heated to 848 K at 5 K/sec. Desorption of TMAs was monitored by measuring intensities at several m/e ratios that are normally seen in the ion fragmentation pattern for the

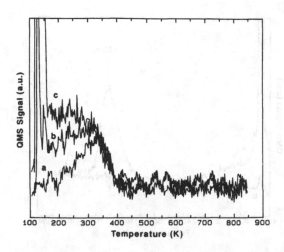

Figure 2. TPD spectra of TMAs/GaAs(100) monitoring m/e = 105
for increasing exposures; (a) 5 sec, (b) 10 sec, (c) 30 sec.

parent molecule. TPD results for three different TMAs exposures monitoring
m/e = 105 (As(CH$_3$)$_2^+$) are displayed in Fig. 2. Other ions known to originate
from TMAs (i.e. As$^+$, AsCH$_3^+$, and As(CH$_3$)$_3^+$) show similar TPD spectra with signal
level ratios consistent with the fragmentation pattern we measured by
backfilling with TMAs. At the lowest exposure, TMAs desorption exhibits a
peak at ≈330 K and as the exposure is increased (curves b-c), a multilayer
peak appears ≈150 K. Since the quality of the data does not warrant a
detailed kinetic analysis, we used a simple Redhead desorption analysis [7] to
calculate the activation energy for desorption (E$_d$). Using the peak
temperature measured (330 K) for low initial coverages and assuming a
preexponential factor of 10^{13}/sec, we obtain a value for E$_d$(TMAs) ≈ 20
kcal/mol. This value should only be considered as a reasonable estimate and
only has validity in the low coverage range. It should be noted that a
binding energy of ≈20 kcal/mol does fall in the range of typical III-V donor-
acceptor bond strengths [8].

Surface decomposition of TMAs can also be probed with TPD if products
desorb at temperatures below 850 K (the highest temperature normally ramped
to). The CH$_3^+$ and CH$_4^+$ TPD signals arising from an electron irradiated
multilayer of TMAs are displayed in Fig. 3. Note the large CH$_3^+$ peak at 660 K
and the absence of a corresponding CH$_4^+$ peak. Since methane gives a CH$_3^+$/CH$_4^+$
ratio of ≈0.8 in our mass spectrometer, we have multiplied the CH$_4^+$ signal by
this ratio to facilitate comparison. The difference between curves (a) and
(b) is then attributed to the portion of the CH$_3^+$ signal that arises from
sources other than methane. An extensive search yielded no evidence for ions
that would be formed by C$_2$H$_x$, C$_3$H$_x$, As(CH$_3$)$_x$, and Ga(CH$_3$)$_x$ species that might
also give rise to the 660 K CH$_3^+$ peak. We therefore attribute this peak to the
desorption of methyl radicals from the surface. We believe the small rise in
the CH$_4^+$ signal around 600 K is due to the coversion of a small amount of the
desorbed methyl radicals into methane upon collision with a chamber wall,
rather than CH$_4$ desorption from the GaAs surface. The peak temperature of the
methyl radical desorption feature remains constant as the initial coverage is
varied, which is consistent with a first-order desorption process [8]. If we
assume a preexponential factor of 10^{13}/sec and use Redhead's analysis [8], we
calculate an activation energy for methyl radical desorption of ≈40 kcal/mol.
This value is appreciably less that the mean carbon-metal bond strength in
gas-phase TMAs (57 kcal/mol) or TMGa (60 kcal/mol) [9].

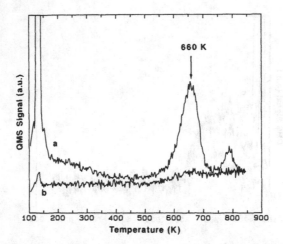

Figure 3. TPD spectrum for electron irradiated TMAs/GaAs(100) monitoring; (a) m/e − 15, CH_3^+, and (b) m/e − 16, CH_4^+ X 0.8.

Before going into to a more detailed discussion, we should state that it is apparently difficult to pyrolyze TMAs on GaAs(100) under our experimental conditions and that some other source of excitation, such as electrons from the mass spectrometer ionizer, is required to initiate decomposition. In fact, the methyl radicals shown in Fig. 3. thermally evolve from a product formed primarily from the electron induced dissociation of TMAs in some reaction sequence as follows;

$$As(CH_3)_3(s) + e^- \longrightarrow As(CH_3)_x(s) + (3-X)CH_3(s) \xrightarrow{\Delta} 3CH_3(g)$$

For the data reported in Fig. 3, the sample was grounded and, as described in the experimental section, an appreciable electron flux was incident on the sample. Even though the TMAs multilayers and monolayer were exposed to the electrons for less than 100 seconds, this time was sufficient for significant decomposition to occur. If the sample was biased -135 V, a current less than 1 nA flowed to the sample when it was placed in the TPD position. When the sample was biased -135 V, the methyl radical desorption peak from a multilayer dose of TMAs was attenuated by about a factor of 10 relative to the results for the unbiased sample, demonstrating that decomposition is primarily due to the impinging electron flux. A better way of determining if any TMAs decomposes pyrolytically on GaAs(100) is outlined below. A sample covered with multilayers of TMAs was left in the upper chamber and heated to 480 K at the same rate used for the TPD experiments (5 K/sec). This thermal ramp desorbs all of the physisorbed and chemisorbed TMAs. Any products from the pyrolytic decomposition of TMAs should remain adsorbed on the sample at this temperature. The sample was then lowered in front of the mass spectrometer and TPD performed. This procedure produced no measurable amount of methyl radicals, which indicates that when they were observed (as in Fig. 3) they were not produced by a pyrolytic process. The residual amount of methyl radical desorption observed when the sample is biased negatively (preventing electron irradiation) and not preheated in the upper chamber level may be due to bombardment of the adsorbed TMAs by positive ions created in the ionizer (although this seems unlikely because of the low current measured, <1 nA) or perhaps it is due to irradiation by UV photons or soft X-rays also created in the ionizer. This is obviously a source of concern for future experiments.

After electron irradiation of adsorbed TMAs, we did not detect any TMAs desorption at high temperatures that might have arisen from the recombination of As$(CH_3)_x$ and CH_3 fragments. It therefore appears that the electron induced decomposition of TMAs is irreversible. In fact, upon heating, the adsorbed fragments produce excess surface arsenic which eventually desorbs as As$_2$(g), as illustrated in Fig. 4. Over this same temperature range, a clean GaAs(100)-(4 X 6) surface evolves As$_2$ and Ga at rates such that their mass spectrometer signals are virtually superimposable. The As$_2^+$ signal between 760-830 K in Fig. 4 is appreciably larger than the Ga$^+$ signal and represents the desorption of the excess surface arsenic. Interestingly, the TPD spectrum in Fig. 4. closely resembles spectra taken for the arsenic-rich GaAs surfaces prepared by dosing with As$_2$ [6].

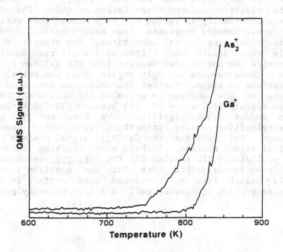

Figure 4. TPD spectrum for electron irradiated TMAs/GaAs(100) monitoring m/e = 150 (As$_2^+$) and m/e = 69 (Ga$^+$).

The lack of evidence for the pyrolysis of TMAs on GaAs(100) suggests that the chemisorbed TMAs has a larger barrier to dissociation than it does to desorption. Therefore, upon heating, the TMAs preferentially desorbs rather than dissociates. We have also performed experiments where the GaAs surface was heated in the presence of gas-phase TMAs to see if dissociation would proceed. For substrate temperatures of 400, 570, and 700 K and an estimated TMAs exposure of ≈100 Langmuirs, TPD showed no products (methyl radicals or excess As$_2$) expected from TMAs decomposition. From this observation we conclude that the reactive sticking coefficient of TMAs must be ≤0.001 for the conditions studied.

In addition to the evolution of methyl radicals at 660 K, other products were observed when electrons were allowed to irradiate the adsorbed TMAs. Note in Fig. 3 that there is also a small CH$_3^+$ peak at ≈795 K. In this case we do observe a C$_2$ hydrocarbon species desorbing coincident with this CH$_3^+$ peak, but by analyzing the intensity ratios of m/e = 25, 26, 27, 28, and 29, the species can clearly be identified as acetylene (C$_2$H$_2$), which obviously has no CH$_3^+$ component in its ion fragmentation pattern. As was the case with the 660 K CH$_3^+$ TPD peak, we found no evidence for a higher mass species that could produce the 795 K CH$_3^+$ peak so we attribute it to another reaction pathway giving rise to methyl radical desorption. Hydrogen (H$_2$) is also observed desorbing nearly coincident with the 795 K CH$_3^+$ peak. This suggests that the three species (CH_3, H_2, C_2H_2) are products of surface reactions occurring at $\approx$795 K which involve a common surface intermediate. The desorption of H_2 and C_2H_2 may indicate that some of the surface methyl groups have decomposed, either pyrolytically or by electron irradiation, into CH_2 or CH groups and surface hydrogen.

Heating the GaAs surface to 850 K during the TPD experiments removes the excess arsenic, as described above, and normally restores the initial surface stoichiometry and (4 X 6) periodicity. However, there is a residual carbon deposit left on the surface after each experimental cycle which increases the C/Ga Auger ratio by about 0.01. We suspect that this carbon may be the end product of the methyl group decomposition reactions which also may have produced the hydrogen and acetylene which desorbed $\approx$795 K.

It is interesting to speculate on the origin of the methyl radicals which desorb at 660 K. Do they arise from surface $Ga(CH_3)_x$ or $As(CH_3)_x$ species created by the electron induced dissociation of TMAs? The behavior as the initial surface stoichiometry is varied may help answer this question. As the As/Ga surface stoichiometry increases, less methyl radical desorption occurs for otherwise identical experimental conditions. For example, the amount of methyl radicals that desorb $\approx$660 K from an arsenic-rich GaAs(100)-c(2 X 8) surface is only $\approx$1/2 the amount that desorbs from the gallium-rich GaAs(100)-(4 X 6) surface. These results strongly suggest that the methyl radicals were bonded to the surface Ga atoms. Further evidence that some kind of $Ga(CH_3)_x$ fragments exists on the surface is revealed directly with TPD. We measure small desorption signals (for m/e = 69, 99) between 530-650 K that arise from dimethylgallium and/or trimethylgallium. We have not yet quantified the amount of dimethylgallium and/or trimethylgallium which desorbs in this temperature range, but note that the $Ga(CH_3)_2^+$ signal maximum is about 2% of the methyl radical signal maximum. Perhaps more striking evidence is that some of our preliminary TPD results for the pyrolytic decomposition of TMGa are similar to results described here for the electron-irradiated TMAs. Methyl radicals desorb from a TMGa-dosed GaAs surface in a peak $\approx$660 K, preceded by the desorption of dimethylgallium and/or trimethylgallium [10].

SUMMARY

We have used TPD to investigate the surface chemistry of TMAs adsorbed on GaAs(100). TMAs desorbs from the GaAs(100)-(4 X 6) $\approx$330 K with an activation energy of $\approx$20 kcal/mole (low-coverage limit). The activation energy for decomposition of the adsorbed TMAs is apparently greater than the activation energy for desorption because upon heating all of the molecules desorb and no decomposition occurs. Attempts to decompose TMAs on hot (400-700K) GaAs at exposures of $\approx$100 Langmuirs were unsuccessful, which places an upper limit on the reactive sticking coefficient of 0.001.

Irradiation of the adsorbed TMAs with electrons from the mass spectrometer ionizer (E = 70 eV, σ = 3 μA/cm^2, t $\leq$ 100 sec) leads to irreversible dissociation. Upon heating, the products of the irradiated TMAs evolve methyl radicals in a peak at $\approx$660 K with a desorption activation energy of $\approx$40 kcal/mole. Excess surface arsenic is also formed by heating the irradiated TMAs layer. As$_2$ desorbs between 760-830 K, restoring the initial (4 X 6) periodicity and stoichiometry. At $\approx$795 K additional methyl radicals, hydrogen, and acetylene desorb. A small amount of carbon is left on the surface after heating to 850 K.

The amount of methyl radical desorption at $\approx$660 K decreases as the initial surface stoichiometry becomes more arsenic rich, suggesting that these methyl groups were bonded to the surface gallium atoms. Small TPD signals arising from dimethylgallium and/or trimethylgallium are detected between 530-650 K, indicating that a $Ga(CH_3)_x$ species existed on the surface. The similarity of these results with our preliminary results for the pyrolytic decomposition of TMGa on GaAs(100) [10] suggests that some of the same surface species may be formed in both cases.

ACKNOWLEDGEMENTS

The author thanks T. Brennan and E. Hammon for growing the MBE epilayer, Gary Karpen for technical support, and Kevin Killeen for enlightening discussions. This work performed at Sandia National Laboratories supported by the US Department of Energy under contract #DE-AC04-76DP000789.

REFERENCES

1. R.K. Lum, J.K. Kingert, and M.G. Lamont, J. Cryst. Growth 89, 137 (1988).
2. G.B. Stringfellow, J. Electron. Mater. 17, 327 (1988).
3. C.T. Foxon, J.A. Harvey, and B.A. Joyce, J. Phys. Chem. Solids 34, 1693, (1973).
4. R.Z. Bachrach, R.S. Bauer, P. Chiaradia, and G.V. Hansson, J. Vac. Sci. Technol. 18, 797 (1981).
5. P. Drathen, W. Ranke, and K. Jacobi, Surface Sci. 77, L162 (1978).
6. J.R. Creighton, manuscript in preparation.
7. P.A. Redhead, Vacuum 12, 203 (1962).
8. R.H. Moss, J. Cryst. Growth 68, 78 (1984).
9. P.J. Barker and J.N. Winter, in The Chemistry of the Metal-Carbon Bond, Vol 2., ed. by F.R. Hartley and S. Patai (Wiley, New York, 1985), p. 151.
10. J.R. Creighton, unpublished results.

APPLICATION OF X-RAY SCATTERING TO THE
IN SITU STUDY OF ORGANOMETALLIC VAPOR PHASE EPITAXY

P.H. FUOSS*, D.W. KISKER*, S. BRENNAN** and J.L. KAHN***
*AT&T Bell Laboratories, Holmdel, N.J.   07733
**Stanford Synchrotron Radiation Laboratory, Stanford, CA.   94305
***Physics Department, Stanford University, Stanford, CA.   94305

1. ABSTRACT

Despite their importance, the detailed surface reactions and rearrangements which occur during chemical vapor deposition remain largely undetermined because of the lack of suitable experimental probes. In principle, x-ray scattering and spectroscopy techniques are well suited to studying these near atmospheric pressure processes but advances in this area have been limited both by the lack of suitable x-ray sources and by the difficulty of integrating the growth and measurement experiments. We have developed equipment and techniques to perform *in situ* x-ray scattering studies of the structure of surfaces during organometallic vapor phase epitaxial (OMVPE) growth using the extremely bright undulator radiation from the PEP electron storage ring. In this paper, we describe our initial experimental results studying cleaning and subsequent reconstruction of GaAs (001) surfaces in a flowing H_2 ambient. These results demonstrate the excellent surface sensitivity, low background and high signal levels necessary to study the dynamic processes associated with semiconductor growth using OMVPE.

2. INTRODUCTION

Chemical vapor deposition processes have become well established for preparing thin films of III-V and II-VI semiconductors such as GaAs and ZnSe. Despite the technological importance of these materials, little is known about the detailed mechanisms of these growth processes. In addition to chemical reactions which occur in the vapor phase, growth involves reactions and diffusion which occur on the surface of the growing material as well as structural changes induced in the material as growth proceeds.

In constrast, UHV techniques such as molecular beam epitaxy (MBE) and chemical beam epitaxy (CBE) have been carefully studied using *in situ* electron based analytical techniques including reflection high energy electron diffraction (RHEED), Auger electron spectroscopy and x-ray photoemission spectroscopy (XPS). Unfortunately, due to the high gas pressures used in CVD growth processes, *in situ* electron based techniques cannot be employed. In spite of this lack of probes, great progress has been made in understanding CVD processes by phenomenological studies of relationships between growth parameters and material properties. A detailed understanding of the microscopic processes occuring during CVD will enhance our understanding and control of these processes.

In this work we describe a new approach which couples x-ray scattering and spectroscopy techniques with the brightest x-ray synchrotron radiation sources to produce a powerful technique for *in situ* analysis of CVD.

3. EXPERIMENTAL APPROACH

Our initial experiments will use the grazing incidence x-ray scattering (GIXS) approach to study the growth of ZnSe epitaxial films grown on GaAs by organometallic vapor phase epitaxy (OMVPE), using diethylselenium and diethylzinc as source compounds. With the

Mat. Res. Soc. Symp. Proc. Vol. 131. ©1989 Materials Research Society

GIXS geometry, x-rays impinge on the surface at a shallow grazing angle (ϕ) and are observed at a second grazing angle (ϕ') (see Figure 1). The use of GIXS to study *in situ* growth processes is well established for ultra-high vacuum monolayer growth[1] and for analysis of surface and interface structures of grown films.[2] GIXS can analyze 1) the crystal structure of the surface and thin film, 2) study size distributions of islands and 3) analyze defect structures on the surface and in the thin, growing film.

In addition, analysis of crystal truncation rods can be used to study the roughness of 1) ordered regions surrounded by amorphous regions or 2) regions of one phase surrounded by another phase.[3] Mochrie has recently established that crystal truncation rod analysis is very suitable for studying roughening transitions.[4] Finally, x-ray reflectivity measurements can provide information about surface roughness.[5]

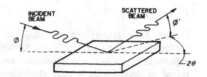

Figure 1: The grazing incidence x-ray scattering geometry. X-rays are incident at a grazing angle ϕ and are detected at a second grazing angle ϕ'.

Thus, the application of x-ray scattering techniques offers a powerful tool to study *in situ* growth. However, there are several concerns associated with extending these techniques from the UHV environment to growth processes close to atmospheric pressure. First, how large is the background from the ambient vapor? We estimate the background from 1 atmosphere of H_2 buffer gas and a 10^{-4} partial pressure of $ZnEt_2$ and $SeEt_2$ to be an order of magnitude smaller than the signal.

The second concern is that data can be collected rapidly enough to enable reasonable growth rates, enabling analysis of the non-equilibrium structures that may be present during growth. Such growth rates might be of order 1 micron/hour or approximately 1 bilayer/second. Thus, we would like measurement times on the order of 1 second for appropriate data to be collected. Such collection times generally require the brightest synchrotron x-ray sources currently available.

4. SYSTEM DESIGN

The goal of performing in situ x-ray scattering while growing material using OMVPE techniques requires a unique blending of techniques and a very special design of both the OMVPE reactor and the x-ray diffractometer. Because x-ray scattering samples a relatively large spatial area, high growth uniformity must be achieved. Since the experiment is not permanently stationed on a synchrotron radiation beamline, it must be portable. Due to the high radiation levels associated with the measurements, all routine processes must be remotely controlled. Finally, the most troublesome problem is providing a large aperature window which allows the full range of structures to be studied and which is protected from CVD deposition. Even a small amount of deposits would rapidly degrade the transmission of the incident and diffracted beams.

The resultant reactor system is shown in Figure 2. The basic unit of construction is a 4" OD stainless steel tube with 6" OD ultra-high vacuum flanges. Discussing the outer shell of the reactor from the top down, the first component is the gas inlet system which consists of a double conflat flange arrangement to capture the internal quartz nozzle (to be discussed later) and a spool piece for internal space.

The beryllium windows are mounted on a 5/16" wall stainless steel pipe. This pipe has a large cutout and o-ring groove for mounting the beryllium window for the diffracted beam. The diffracted beam window consists of 0.5mm thick beryllium brazed to a monel flange. A separate beryllium window for the incident beam is mounted on the 2 3/4" flange which comes perpendicularly out of the window spool piece. These windows are configured so that diffracted beams can be observed over an angle of 120° in the plane of the substrate and 40° normal to the substrate.

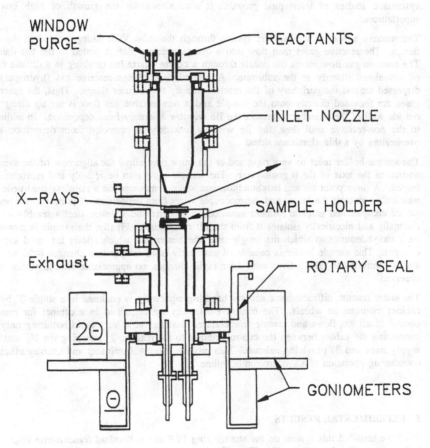

Figure 2: An overall drawing of our special OMVPE reactor. The details are described in the text.

The final section is a teflon sealed, differentially pumped rotary seal which allows rotary motion (θ) to be precisely coupled to the sample. In our application, we pump the space between the second and third seal with our roughing pump and supply a small overpressure of nitrogen to the first pumping port. This insures that any reactive gases will be pumped through a charcoal scrubbing system and also minimizes the possibility that any oxygen will be fed into the reactor. This rotary seal and the sample holder are directly coupled to the θ goniometer through an adapter flange.

The specially constructed, remote controlled, gas handling system (CVD Equipment, Inc.) allows for three organometallic sources. All of the switching valves are interfaced to an external, programable controller which also interacts with the central data acquisition computer. This, coupled with dynamic pressure balancing between the vent/run lines, allows systematic studies of interrupted growth. It also allows for the growth of high quality superlattices.

The reactive gases are fed into the system through the axial VCR fitting welded to the top flange. The reactive gases then flow into a quartz nozzle which is sealed to the top flange. The reactive gas flow leaves the nozzle through a coarse quartz frit resulting in a diffuse flow of gas aimed directly at the substrate. A second flow of non-reactive gas (hydrogen) is dispersed around the periphery of the reactor tube by the diffuser flange. Thus, the reactive gases are focussed directly onto the sample and a non-reactive gas flow is set up along the outside wall of the chamber to protect the Be window assembly from deposition. In addition to the non-reactive wall flow, the Be window assembly is protected from deposition and overheating by a thin aluminum shield.

The sample holder must be very rigid and at the same time allow for alignment of the sample normal to the axis of the θ goniometer. The sample must also be reliably and reproducibly heated. A three point tilt and height adjustment stage is mounted on a stainless steel pedestal which also protects the heater and thermocouple wires from deposition. The heater is a boron nitride encapsulated graphite element mounted on a quartz and stainless steel assembly which thermally and electrically isolates it from the tilt mechanism. Finally, the sample is mounted on a molybdenum cup which fits snugly over the heater and which allows for rapid sample changes. This sample holder is capable of extremely rapid temperature changes ($\approx$50 K/sec.) and is controlled to ±0.5°. Even with these rapid changes, no appreciable sample motion was observed.

The entire reactor, diffractometer and gas handling equipment is enclosed in a single 3' by 6' cabinet mounted on wheels. The control electronics are contained in a cabinet for remote control of all gas flows and sample temperature. Installation is very rapid requiring only 1) connecting the cables between the control cabinet to the reactor, 2) attaching the N_2 and H_2 supply lines and 3) providing exhaust. This process takes $\approx$30 minutes and enables efficient timesharing operation of a synchrotron beamline.

5. EXPERIMENTAL RESULTS

We have installed this system on the storage ring PEP at the Stanford Synchrotron Radiation Laboratory. The beamline used, PEP 5B, is equipped with a 26 pole undulator which produces maximum intensity at $\approx$12 KeV with the typical storage ring operating energy of 14 GeV. A 10 KeV photon beam was monochromatized with a two crystal, Si(111) parallel setting monochromator. While the beamline is equipped with a focussing mirror, the data shown were taken without the use of the mirror. Photon fluxes of 1.6×10^{10} photons/second were observed into a 0.4mm×4mm entrance slit.

We used the limited amount of beamtime available ($\approx$24 hours) to study the surface structure of GaAs (001) surfaces prepared in flowing H_2. Following sample installation and alignment, the GaAs crystal was heated to 585°C in 300 Torr of flowing H_2 for 10 minutes. This treatment is similar to that typically used in MBE processes to thermally desorb any oxide present after wet chemical sample preparation. If complete oxide removal occurs, a so-called "streaky" RHEED pattern would be observed, indicative of the presence of surface reconstruction. We have attempted to observe evidence of a similar result in this system.

After rapid cooling to room temperature ($\approx$5 minutes), the diffraction pattern shown in Figure 3a was obtained. This diffraction scan, taken along a (110) direction, clearly shows an eight-fold reconstruction of the surface. In contrast, Figure 3b shows a diffraction scan along the ($1\bar{1}0$) direction. This data shows indications of an eight-fold reconstruction but also shows extra, extremely sharp diffraction peaks at $(1.2,\bar{1.2}0)$ and $(1.6,\bar{1.6},0)$; and a series of peaks around $(1.48,\bar{1.48},0)$.

After dosing the surface with 5% air to re-establish an oxide, only the $(1.5,\bar{1.5},0)$ and the $(1.875,\bar{1.875},0)$ diffraction peaks remained along this azimuth. In addition, there was a diffuse peak located near $(1.5,\bar{1.5},0)$ We interpret this diffuse component and the multiple diffraction peaks seen before dosing as indication of an oxide remaining on the surface. As the sample was annealed in H_2 at slowly increasing temperatures ($\approx$50°C/hour), the diffuse peak slowly disappeared and at 400°C, a sharp diffraction peak at $(1.25,\bar{1.25},0)$ appeared.

Because beamtime was limited, we were unable to complete this study of the GaAs surface reconstruction. As a result, our data leaves many unanswered questions. However, it is clear that we can obtain information very similar to that obtained by RHEED analysis during MBE growth but with much higher resolution and without the complexities of multiple scattering. We are actively pursuing this problem and expect to greatly extend these results in the near future.

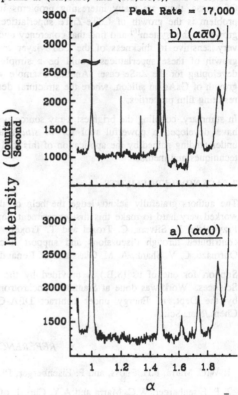

Figure 3: Data taken along the a) [110] and b) [1$\bar{1}$0] directions from a GaAs(001) surface prepared by heating in 300 Torr of H_2 at 585°C for 10 minutes.

6. CONCLUSIONS AND FUTURE POSSIBILITIES

We have established the feasibility of using x-ray based analytical tools for the *in situ* analysis of CVD systems such as OMVPE processes. Even our preliminary results on oxide removal from GaAs indicate that we have excellent surface sensitivity, low backgrounds, and high signal levels so that we should be able to complete these studies with the next available beamtime. Then, having such an analytical tool at our disposal, we can proceed to the study of more challenging systems, such as the growth of epitaxial films. Starting with ZnSe, we expect to be able to monitor growth transients using reflected beam intensities (cf. RHEED oscillations), establish the presence of layer by layer growth processes if they exist, and characterize the extent of surface roughness as a function of growth parameters such as temperature, pressure, growth rate, source compounds, etc. In addition, because the ZnSe/GaAs system is slightly lattice mismatched, we will be able to monitor the development of strain in the overlayer, which will then be relieved by the formation of misfit dislocations at the substrate interface. The *in situ* study of this process should give valuable insight into the kinetics of lattice relaxation in mismatched systems.

Of course, there are many interesting problems in other systems besides ZnSe. One such problem is the growth of CdTe-ZnTe superlattices. We have made an extensive study of growth in this system[6] and find that coherency and defect structure of the final superlattice is very sensitive to thickness of the CdTe layer and to growth conditions. Analysis of the growth of these superlattices would be a simple extension using the capabilities we are developing for the ZnSe case. Another example would be the study of the initial stages of growth of GaAs on silicon, where the structural details of the interface apparently control the resulting film properties.

In summary, coupling the brightest x-ray source with proven x-ray scattering techniques, we have developed a powerful tool for *in situ* analysis of CVD systems. We expect the understanding gained by the application of this tool will lead to the development of new CVD techniques and methods.

7. ACKNOWLEDGEMENTS

The authors gratefully acknowledge the help of the large number of SSRL personnel who worked very hard to make this first experiment happen. They include J. Cerino, D. Day, H. Przybylski, R. Silvers, C. Troxel and T. Troxel. In addition, several Bell Labs colleagues contributed through discussions and support, including L.J. Norton, R. D. Feldman, A. Ourmazd, C. V. Shank, A. M. Glass and G. Renaud.

Support for one of us (S.B.) is provided by the Dept. of Energy Office of Basic Energy Sciences. Work was done at Stanford Synchrotron Radiation Laboratory which is supported by the Dept. of Energy under contract DEA-C0382ER13000, Office of BES, Div. of Chem./Mat. Sci.

REFERENCES

1. W.C. Marra, P.H. Fuoss, and P. Eisenberger, Phys. Rev. Lett., **49**, 1169 (1982).

2. P. Eisenberger, W.C. Marra and A.Y. Cho, J. of Applied Physics, **50**, 6927 (1979).

3. I.K. Robinson, Phys. Rev. B, **33**, 3830 (1986).

4. S.G.J. Mochrie, Phys. Rev. Lett., **59**, 304 (1987).

5. R.W. James, "The Optical Principles of the Diffraction of X-rays", Oxbow Press, Woodbridge, Conn., (1982).

6. D.W. Kisker, P.H. Fuoss, J.J. Krajewski, P. Armithiraj, S. Nakahara, and J. Menendez, J. of Crystal Growth, **86**, 210 (1987).

INVESTIGATION OF Ga DIFFUSION IN (001) AND (111)
CdTe LAYERS GROWN ON (001) GaAs

J.J. DUBOWSKI[*], J.M. WROBEL[*], J.A. JACKMAN[**] AND P. BECLA[+]
[*] Laboratory for Microstructural Sciences, NRCC, 100 Sussex Dr., Ottawa, Ont., Canada K1A 0R6
[**] Metals Technology Laboratories, CANMET, 568 Booth St., Ottawa, Ont., Canada K1A 0G1
[+] Francis Bitter National Magnet Laboratory, MIT, Cambridge, Ma 02139, USA

ABSTRACT

A secondary ion mass spectroscopy study of Ga diffusion in CdTe grown on (001) GaAs was carried out. The layers were grown by pulsed laser evaporation and epitaxy. Two characteristic regions with increased Ga concentration were found. The first was the CdTe/GaAs interface where the concentration of Ga decreases rapidly to the detection limit of ~ 8 x 10^{14} cm^{-3}. This region was usually less than 300 nm wide. The second was a surface region with a Ga accumulation of up to ~ 10^{17} cm^{-3}. Ion imaging revealed that in the (001) CdTe layers, Ga accumulates near the surface at localized spots, up to about 8 μm in diameter. This feature is less apparent in the (111) CdTe layers. Annealing at 500 °C for 1 h increased the Ga concentration in the whole layer to above 10^{16} cm^{-3}. We also observe thermal annealing induced precipitation of Ga at the surface of bulk CdTe samples which were originally uniformly doped with Ga.

INTRODUCTION

Significant attention has been paid to the use of Si, GaAs and InSb substrates in the growth of CdTe and HgCdTe thin films. This interest has been stimulated by the considerable cost of commercial CdTe wafers. Also, the size and concentration of defects, together with their tendency to continue in the grown epilayers, make CdTe wafers difficult to handle, for example, in the growth of planar detector structures.

GaAs substrates are an attractive alternative in the growth of CdTe films. Depending on the substrate treatment and growth conditions, it is possible to produce epitaxial layers of both (001) and (111) CdTe on (001) GaAs, as well as (111) CdTe on (111) GaAs. However, there have been reports of Ga migration into CdTe layers[1 - 4]. Such migration may strongly affects the electronic properties of CdTe, leading to limited application of GaAs substrates in CdTe technology. Thus, the problem of Ga migration in CdTe/GaAs heterojunctions should be addressed in more detail.

In this paper, we describe a secondary ion mass spectroscopy (SIMS) study of Ga behaviour in (001) and (111) CdTe layers grown on (001) GaAs by pulsed laser evaporation and epitaxy (PLEE)[5].

EXPERIMENTAL DETAILS

The epitaxial layers of (001) and (111) CdTe were grown on Si doped (001) GaAs substrates (Sumitomo Electric, Inc.). The growth temperature for most of the samples studied was 300 °C and in some cases it was 260 °C. One sample was deposited on a substrate held at 25 °C and it had a polycrystalline structure. The growth of (111) oriented layers was carried out on Ne$^+$ ion cleaned substrates, and (001) layers were grown on chemically etched and vacuum baked substrates[6 - 7]. The structural quality of the (001) CdTe layers was studied with a single crystal x-ray diffractometer[8]. The full width at half maximum (FWHM) for the [400] reflection decreased sharply as the thickness of the film increased. For a film 6.2 μm thick the FWHM for this reflection was ~ .07 °. The preliminary results for the (111) CdTe layers showed that they were inferior (layers with thicknesses up to ~ 2 μm were studied) compared to the

(001) CdTe layers, and sometimes (001) oriented CdTe was incorporated in (111) CdTe.

The photoluminescence (PL) study confirmed observations concerning the crystallographic quality of PLEE layers. For ~ 2 μm thick (111) CdTe layers, either the donor bound exciton line at 1.593 eV or the acceptor bound exciton line at 1.583 eV, were very clear in the PL spectra. But, strong bands in the near edge emission, from 1.44 to 1.57 eV, were also present. The line at 1.583 eV was always accompanied by a broad defect band at ~ 1.42 eV. The FWHM of excitonic peaks at 1.593 and 1.583 eV was ~ 7 meV. The spectra of the (001) CdTe layers[9] which were thinner than ~ 3 μm were dominated by transitions at 1.593, 1.585 and 1.581 eV. A broad defect band about 1.42 eV always accompanied these spectra. For thicker samples, the dominating transition was at 1.589 eV which is ascribed to an acceptor bound exciton emission. The FWHM for this peak was 1.7 meV which is comparable to the corresponding value observed in bulk CdTe. In a sample 6.2 μm thick, the intensity ratio of this excitonic peak to the most intense feature associated with free to bound or donor acceptor pair recombination was about 15, and defect band at 1.42 eV was not observed.

The SIMS study was carried out using a Cameca IMS 4f system. An O_2^+ ion beam operating at 1.5 keV (impact energy per ion) was used for sputtering the samples at a rate of ~ 60 nm/min. The primary beam was scanned over an area of 250 μm x 250 μm. The size of the analyzed area was 150 μm in diameter for ion imaging. It should be noted that because sputtered Ga tends to redeposit around the crater, the choice of limited field of view has some influence on the SIMS signal. Therefore, for in-depth profiling we always used small diameter field aperture, limiting the size of the analyzed area to 62 μm in diameter. Calibration of the system was performed with the use of a (001) CdTe crystal implanted with a dose of 5×10^{14} cm^{-2} Ga ions having an energy of 180 keV. The SIMS profile of Ga in the implanted sample is shown in Fig. 1. It can be seen that a rapid change of the Ga concentration from 4×10^{19} cm^{-3} to 5×10^{14} cm^{-3} is detected within ~ 1 μm depth. Similar SIMS profiles were redone whenever a new series of measurements began. This allowed us to establish the Ga detection limit as being not worse than 8×10^{14} cm^{-3}. Mass Ga^{69} was used for tracing the presence of Ga. In some cases, additional measurements of the ratio of the SIMS signal for Ga^{69} and Ga^{71} were taken to verify the results. This ratio was always constant. A complementary study of an As profile and its distribution in the films was carried out with a Cs$^+$ primary beam.

RESULTS AND DISCUSSION

For each of the as-grown thin films studied, two characteristic regions in the SIMS profiles were always seen, showing the presence of Ga, both at the CdTe/GaAs interface and at the CdTe surface. Near the interface, the concentration of Ga decreased rapidly as the distance from the interface increased. A transition region, where the change in Ga concentration from that at the surface of GaAs to the detection limit in the CdTe layer, usually had a width of < 300 nm. It appeared that reducing the growth temperature resulted in an even narrower transition region. We note, however, that a broad transition region recorded by SIMS may also be the result of pits created near the interface which act as an additional source of Ga in SIMS readings. This problem will be

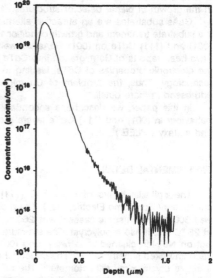

Figure 1 The SIMS Ga profile in bulk CdTe implanted with 180 keV Ga ions (dose 5×10^{14} cm^{-2}).

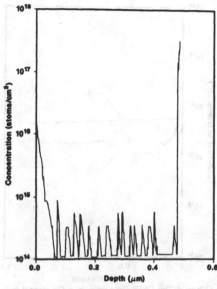

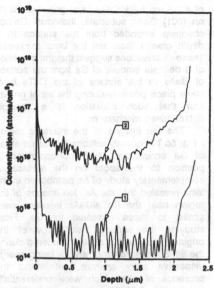

Figure 2 The SIMS Ga profile in a (001) CdTe epilayer grown on (001) GaAs.

Figure 3 The SIMS Ga profile in a (111) CdTe as-grown layer (1) and in the same layer after annealing at 500 °C for 2 h (2).

discussed below. An example of a Ga depth profile in a 0.5 μm thick (001) CdTe sample which was grown at 260 °C is shown in Fig. 2. Besides the interface transition region, which in this case is < 25 nm, a significant accumulation of Ga can be seen, up to ~ 10^{16} cm^{-3}, at the surface of the grown epilayer. The concentration of Ga decreases below the detection limit within ~ 80 nm from the layer surface. There were no significant differences between Ga profiles for the (001) and (111) CdTe layers, and two Ga rich regions were always observed.

We studied the Ga profiles in epilayers thermally annealed in the temperature range 500 - 700 °C. This treatment always leads to intense diffusion of Ga into the layers. The profiles of Ga in the (111) CdTe as-grown film and in the same film after annealing at 500 °C for 2 h are shown in Fig. 3. It can be seen that the concentration of Ga in the whole annealed layer (curve 2) increased to above 10^{16} cm^{-3}. An accumulation of Ga is also indicated in the near surface region, and a substantial broadening of the interface is seen. We also observed an effect of Ga surface accumulation in bulk CdTe as a result of thermal annealing. This is illustrated in Fig. 4, where we compare a Ga profile in Bridgman grown, uniformly doped CdTe to a profile in a sample which was annealed at 800 °C for 48 h. An intense accumulation of Ga on the surface occurs at the expense of its reduced concentration in the bulk of the annealed sample. The gettering of Ga at the surface of CdTe kept at elevated temperatures is similar to the gettering of Li, Na, K and Al observed in bulk CdTe[10].

A Ga ion imaging study revealed that migration of Ga from the GaAs substrate into (001) CdTe films is a strongly nonuniform process. 'Hot spots' of Ga, against the more uniform Ga background, were always seen at the surface of (001) CdTe films, even for the thickest films studied. The diameter of an individual spot ranged from 2 μm to 8 μm and their concentration was in the range (5 - 10) x 10^6 cm^{-2}. Examples of Ga images obtained for a (001) CdTe film 6.2 μm thick are shown in Fig. 5. It can be seen that the spots disappeared as the distance from the surface increased. Typically, the spots decayed within 300 - 400 nm. This thickness is slightly larger than the distance from the surface at which the measured Ga concentration usually decreased below the detection limit. We also observed Ga accumulation at the surface of a 1.6 μm thick CdTe layer grown at 25 °C on (001) GaAs substrate. Individual Ga-rich channels extended from the surface to a depth greater than half the layer thickness. These observations suggest that the migration of a definate amount of Ga from the surface of GaAs to the surface of the CdTe layer takes place primarily during the initial growth and that such migration is a strongly defect-driven mechanism.

The ion images of the interface region (Fig. 5c) frequently exhibited localized areas of Ga enrichment which corresponded in position to the spots on the surface. A complementary study of As distribution in the films revealed that the As⁻ ion images of the region near the CdTe/GaAs interface were similar to those observed for Ga. From studying the surface morphology of the original films and the craters sputtered during the SIMS measurements, and by using energy dispersive x-ray analysis, we found the presence of pits which were preferentially sputtered in CdTe. The development of pits in CdTe as a result of chemical etching or vacuum sublimation[11] is well known, and it is strongly related to the concentration of various

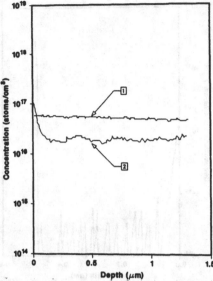

Figure 4 The SIMS Ga profile in a Ga doped bulk CdTe: (1) as-grown sample, (2) after annealing at 800 °C for 48 h.

defects. Thus, it seems reasonable to link the presence of sputtered pits with the presence of defects in CdTe films as well. An example of a typical pit developed on the surface of a sputtered (001) CdTe film is shown in Fig. 6. Up to ~ 10^5 cm⁻² of such pits, typically 3 μm x 5 μm in size, were observed on the surface of the sputtered films. Additional pits began to develop when the CdTe/GaAs interface was approached.

We conclude from these observations that the sputtered pits pierce the GaAs substrate in advance of the main part of the crater, producing images near the interface which, sometimes, closely resemble the surface images. This conclusion does not support suggestions given in Ref. 4 that Ga spots observed near the interface result from a particular treatment of GaAs substrate prior to the deposition of CdTe. The precipitations of Ga micro-droplets like those, for example, observed at the surface of the annealed GaAs wafer[12], even if present in our samples would not be seen since the sputtered pits are larger and they occur before the CdTe layer is completly sputtered. Since we frequently observe a correlation between the

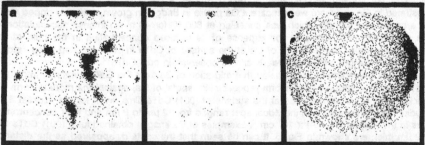

Figure 5 The Ga ion image of the (001) CdTe film: (a) as-grown surface, (b) surface after sputtering of a layer - 300 nm thick, (c) surface near the interface. The diameter of the shown field is ~ 150 μm.

location of Ga-rich spots at the interface (caused by sputter induced pits) and at the surface where Ga enrichment is genuine, we suggest that some defects, which we suspect are responsible for the formation of sputtered pits in the layer, extend through the whole layer in the direction perpendicular to the CdTe/GaAs interface.

A typical Ga image observed for (111) CdTe layers is shown in Fig. 7. There is evidence for a Ga accumulation at the surface (Fig. 7a) but, in contrast to (001) CdTe films, no Ga hot spots were found at the surface. In the interface region (Fig. 7c), however, Ga activated spots were observed. They were also the result of pits sputtered

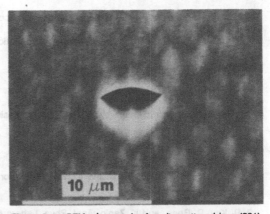

Figure 6 An SEM micrograph of a pit sputtered in a (001) CdTe epilayer near the CdTe/GaAs interface.

during the SIMS measurements. This preliminary study suggests that the migration of Ga in CdTe grown on GaAs may be an orientation dependent process.

CONCLUSIONS

The SIMS study of Ga incorporation in (001) and (111) CdTe films grown on (001) GaAs shows that: (1) There are two characteristic regions with increased Ga concentration. The first is the CdTe/GaAs interface where the concentration of Ga decreases to the detection limit of ~ 8 x 10^{14} cm^{-3}, usually, within less than 300 nm. However, the accurate width of this region is difficult to estimate since the pits wich are formed during SIMS measurements may act as an additional source of Ga. The second is a surface region with a Ga accumulation of up to ~ 10^{17} cm^{-3}. (2) For (001) CdTe films, even thicker than ~ 6 μm, an accumulation of Ga in localized spots, against the more uniform Ga background, is observed near the surface of the grown film. Spots with diameters of up to ~ 8 μm are found. (3) For (111) CdTe films, no surface spots of Ga are found, but a surface concentration of Ga up to ~ 10^{17} cm^{-3} is observed. (4) The intensive migration of Ga from the GaAs to the CdTe film takes place during the early growth stage and it seems to be a strongly defect-driven mechanism. (5) Characteristic pits are formed during SIMS analysis. The concentration of pits is up to

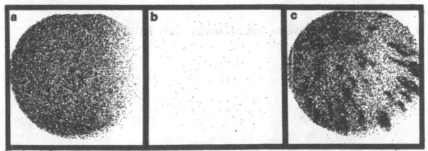

Figure 7 The Ga ion image of the (111) CdTe film: (a) as grown surface, (b) surface after sputtering of a layer ~ 300 nm thick, (c) surface near the interface. The diameter of the shown field is ~ 150 μm.

$\sim 10^5$ cm^{-2}, and they are responsible for the appearance of enhanced SIMS Ga and As signals near the CdTe/GaAs interface. (6) Thermal annealing increases diffusion of Ga to the CdTe films grown on GaAs, and it also induces a precipitation of Ga at the surface of bulk CdTe uniformly doped with Ga.

ACKNOWLEDGMENTS

The authors wish to thank Mr. S. Rolfe for his assistance in the SIMS measurements.

REFERENCES

1. J. Giess, J.S. Gough, S.J.C. Irvine, G.W. Blackmore, J.B. Mullin, and A. Royle, J. Cryst. Growth 72, 120 (1985).

2. R. Kay, R. Bean, and K. Zanio, Appl. Phys. Letters 51, 2211 (1987).

3. R. Korenstein and B. MacLeod, J. Cryst. Growth 86, 382 (1988).

4. B.K. Wagner, J.D. Oakes, and C.J. Summers, J. Cryst. Growth 86, 296 (1988).

5. J.J. Dubowski, Chemtronics 3, 66 (1988).

6. J.J. Dubowski, D.F. Williams, P.B. Sewell, and P. Norman, Appl. Phys. Letters 46, 1081 (1985).

7. J.J. Dubowski, D.F. Williams, J.M. Wrobel, P.B. Sewell, J. LeGeyt, C. Halpin, and D. Todd, Can J. Phys. (to be published).

8. J. Noad (unpublished results).

9. J.M. Wrobel, J.J. Dubowski, and P. Becla, presented at the 1988 U.S. Workshop on the Physics and Chemistry of Mercury Cadmium Telluride, Orlando, FL, 1988 (to be published in J. Vac. Sci. Technol.).

10. L.O. Bubulac, J. Bajaj, W.E. Tennant, P.R. Newman and D.S. Lo, J. Cryst. Growth 86, 536 (1988).

11. J.J. Dubowski, J.M. Wrobel, D.F. Mitchell, and G.I. Sproule, J. Cryst. Growth 94, (to be published).

12. M.G. Lagally and D.G. Welkie, Surf. Int. Analysis 3 (1), 8 (1981).

ELECTROCHEMICAL PHOTOVOLTAIC CELL WITH 6.9% EFFICIENCY USING POLYCRYSTALLINE CdSe GROWN BY A SIMPLE LIQUID METAL-VAPOUR REACTION

MARCUS F. LAWRENCE[*], ZHITSING DENG[*] AND LOUIS GASTONGUAY[**]
[*]Concordia University, Dept. of Chemistry, 1455 de Maisonneuve Blvd. West, Montreal, Quebec, Canada H3G 1M8.
[**]I.N.R.S.-Energie, C.P. 1020, Varennes, Quebec, Canada J0L 2P0

INTRODUCTION

Research on II-VI semiconducting compounds during the last two decades has been motivated by possible device applications such as thin film transistors, photodetectors and solar energy converters. In particular, the direct gap n-type semiconductor CdSe, has remained the subject of studies aimed at developing efficient photovoltaic and photoelectrochemical cells. For the low-cost production of polycrystalline CdSe layers, many noteworthy methods have been employed, such as: vacuum deposition, spray pyrolysis, chemical bath deposition and electrodeposition [1-5]. The best reported performances of photoelectrochemical cells using CdSe films obtained by these methods, in contact with an aqueous polysulfide electrolyte and under solar or simulated solar radiation, have varied between 5 and 7% [6,7].

More recently, studies based on the work of Iwanov and Nanev concerning the direct synthesis of epitaxial II-VI films on single crystal metal substrates [8], have led to the development of another promising technique for the low-cost production of polycrystalline CdSe layers. This method, referred to in the past as the "tarnishing reaction" [9] or as the "gas-solid process" [10], involves the reaction of the chalcogen vapour (Se) with the surface of the heated metal substrate (Cd), under a constant argon flow. The results presented here show that, under the proper experimental conditions, the liquid metal-vapour reaction enables the synthesis of polycrystalline CdSe semiconductor layers with a 6.9% conversion efficiency when in contact with a 1 M polysulfide electrolyte and under 80 mW cm^{-2} of white light illumination. The highly textured surface of the samples thus obtained, would seem to be responsible for the high photovoltaic efficiency.

EXPERIMENTAL METHODS

Liquid metal-vapour reaction:

The liquid metal-vapour reaction system is illustrated in Fig. 1. Cadmium substrates (2 x 2 cm^2) were cut from a 1 mm thick plate of Cd

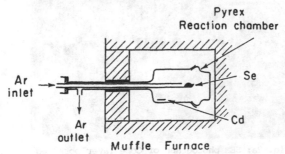

Fig. 1: Liquid metal-vapour reaction apparatus.

(Ventron, 99.999%). The substrates were etched for 5 sec in a 5:4:1 mixture of $H_2O:HCl:HNO_3$, thoroughly rinsed with doubly deionized water, and then quickly dried with nitrogen before being introduced in the reaction chamber. To remove oxygen, the reaction chamber containing a Cd substrate is purged for 30 min (100 cm^3 min^{-1}) with ultra-high purity argon (99.999%) which has passed through an Oxysorb oxygen removal unit (SPECTREX). The argon is introduced via a 1/4 inch glass tube which acts simultaneously as the gas inlet and as the spoon containing the selenium (Fluka, >99.999%). At first, the reaction chamber is placed into the muffle furnace (Pyradia, equipped with a programmable REX P-100 microprocessor temperature control), and the argon flow is reduced to 35 cm^3 min^{-1}. The Cd substrate is then heated to 320^0C (within 30 min), and the reaction proceeds as the Se vapour is brought in contact with constant argon flow, with the liquid Cd surface. After 4 h, the reaction is stopped and the CdSe sample is allowed to regain room temperature under argon atmosphere.

Characterization of the CdSe layers

The CdSe layers were analyzed by scanning electron microscopy and EDAX (Hitachi Model S-570, SEM), and by X-ray diffractometry (Siemens D-500, Cu Kα : 1.5406 Å). Electrochemical and photoelectrochemical measurements were performed under potentiostatic control with a three electrode cell using a saturated calomel electrode as reference and a platinum foil (6 cm^2) as counter electrode. The electrolyte was an aqueous polysulfide solution of composition $Na_2S(1M)/S(1M)/NaOH(1M)$. Measurments of the semiconductor/electrolyte interface impedance were made to obtain the donor density and flatband potential values according to the classical Mott-Schottky technique. The capacitance-voltage curves were obtained with a PAR 173 potentiostat and a 5206 lock-in amplifier, both controlled by an Apple IIe computer. A 10 mV peak to peak, 10 kHz ac voltage was usually used to get the resistance and capacitance values.

The cell could be illuminated either with white light (80 mW cm^{-2}) or with monochromatic light from a Xe lamp (Bausch and Lomb, 150 W) coupled to a monochromator. Light intensities were measured with a radiometer (Optikon model 88 X LC, with model 400 sensor head).

RESULTS AND DISCUSSION

Fig. 2a presents a SEM photograph of a typical CdSe film obtained after 4 h of reaction time. To the naked eye the surface of such a sample

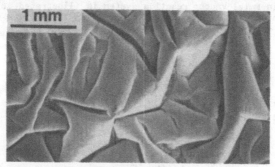

Fig. 2a: SEM photograph of CdSe layer obtained after 4 h of reaction time.

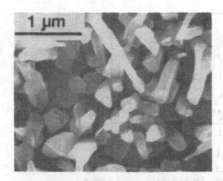
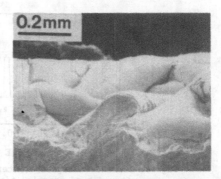

Fig. 2b: Surface of CdSe layer shown   Fig. 2c: Cross-section of CdSe
     in Fig. 2a.   layer of Fig. 2a.

appears heavily textured and pitch black. Etching off the CdSe layer
reveals a Cd substrate with the same topography which may indicate that
this crumpled texture is induced by stress during the cooling of the
sample. At a magnification 1,000 times greater than that of Fig. 2a, the
layer surface appears as an agglomeration of crystallites with diameters
ranging from approximately 0.3 to 1 μm (Fig. 2b). The EDAX measurements
indicate that the CdSe forming these layers is rich in Cd, with the
following atomic percentages: Cd = 53 % and Se = 47%. This rather high
off-stoichiometry, however, might not be representative of the actual
layer composition. As revealed in Fig. 2c, the precise thickness of the
CdSe layer is difficult to assess and the coverage of the Cd substrate may
be somewhat non-uniform.

X-ray diffraction performed on these polycrystalline samples
indicates that the positions (2θ degrees) of the observed diffraction
peaks correspond reasonably well to those reported in the JCPDS file for
hexagonal type (wurtzite) CdSe but that their normalized integrated
intensities, however, show little correlation (Fig. 3). In particular,
the diffraction peak of maximum intensity given in the JCPDS file for
hexagonal CdSe corresponds to diffraction by the (100) planes, whereas it
is the (110) planes which dominate in the diffraction pattern of CdSe
grown by the liquid metal-vapour reaction. The probable cause of this may
be that some orientations are favoured over others during formation of the
CdSe layer and evidence for this is given in Fig. 2b which shows the
columnar aspect of the crystallites.

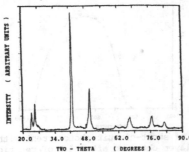

Fig. 3: X-ray diffraction pattern of CdSe sample obtained
after 4 h of reaction time.

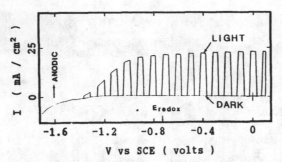

Fig. 4: Current-voltage characteristic of CdSe layer obtained
after 4 h of reaction time, in the dark, and under
chopped white light illumination of 80 mW cm⁻².

The current-voltage characteristics shown in Fig. 4 were obtained
in a three electrode cell under potentiostatic control using a CdSe
specimen as the working electrode in contact with a 1M Na₂S/1M S/1M NaOH
aqueous solution. Under white light illumination of $I = 80$ mW cm⁻², the
cell yields a short-circuit photocurrent density of $J_{sc} = 22$ mA cm⁻², open-
circuit photovoltage of $V_{oc} = 0.56$ V, and a fill factor of ff = 0.45, for
an overall photovoltaic conversion efficiency of $\eta = 6.9\%$ as calculated
according to the expression

$$\eta = \frac{J_{sc} \ V_{oc} \ ff \ 100}{I} \qquad (1)$$

This value rates amongst the highest efficiencies measured under
similar conditions with polycrystalline CdSe in a photoelectrochemical
cell [6,7] and, furthermore, it should be emphasized that in this case no
post-etching or post-annealing treatments were required. The maximum
efficiency of 6.9% refers to the performance of fresh electrodes since
they undergo the usual degradation caused by photocorrosion of the CdSe
layer. Under constant white light illumination of 80 mW cm⁻², J_{sc}
typically decreased by approximately 20% in 3 h. Nevertheless, the loss
in performance experienced during the electrochemical and photoelectrochemical measurments used to characterize these electrodes is negligeable.
The photocurrents are anodic which indicates that this CdSe is n-type, and
extrapolation of the values obtained from the measurements of photocurrent
versus wavelength in the near infrared, yields a bandgap energy of 1.65
eV for this material (Fig. 5). This value is slightly lower than the room

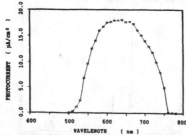

Fig. 5: Variation of short-circuit photocurrent with wavelength
for CdSe layer obtained after 4 h of reaction time
(adjusted for a constant photon flux).

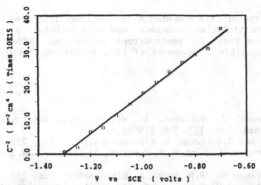

Fig. 6: Mott-Schottky plot for CdSe layer obtained after 4 h of reaction time (10 mV peak to peak, 10 kHz ac modulation).

temperature value of 1.73 eV reported in the literature for wurtzite type single crystal CdSe [11], but it coincides with the bandgap energies of 1.69 and 1.71 eV found by Coblow et al [12] for polycrystalline CdSe films that were obtained by electrodeposition on titanium and nickel, respectively.

Fig. 6 presents the variation of space charge layer capacitance, C, as a function of applied potential, V_A, which enables the determination of the semiconductor's majority charge carrier density, N, and flatband potential, V_{fb}, using the well known Mott-Schottky relation

$$C^{-2} = \frac{2}{E E_0 \, q \, N \, s^2} \left(V_A - V_{fb} - \frac{kT}{q} \right) \qquad (2)$$

where q is the electronic charge, E is the semiconductor's dielectric constant (E = 10 for CdSe [13]), E_0 = 8.86 x 10^{-14} F cm^{-2}, k is the Boltzmann constant, T is the absolute temperature, and s is the sample area in contact with the electrolyte. From the intercept and the slope of the Mott-Schottky plot, values of V_{fb} = -1.31 V vs SCE and N = 2.6 x 10^{17} cm^{-3} (calculated using the sample's apparent (geometrical) surface area) were obtained. These results were found to be frequency independent in the range of 5 to 20 kHz. It is important here to recognize that the evaluation of N is based on the attribution of equal "geometrical" and "actual" surface areas to the samples. Clearly, as shown in Figures 2a and 2c, the actual surface area of the CdSe layers obtained by the liquid metal-vapour reaction is greater than the geometrical area. If one estimates that the actual area of these electrodes is roughly 3 times the geometrical value, a value of N = 3 x 10^{16} cm^{-3} is found.

CONCLUSION

The low-cost liquid metal-vapour reaction enables the synthesis of polycrystalline CdSe thin films giving a 6.9% conversion efficiency when used in an electrochemical photovoltaic cell configuration in contact with 1M polysulfide electrolyte and under 80 mW cm^{-2} of white light illumination.

The highly textured surface of the samples produced in this study is suspected of being mainly responsible for the good photovoltaic efficiency. Further studies are needed to evaluate how the actual surface area of this type of film compares with the geomtrical area (possibly by the Brunauer–Emmett–Teller technique) and also to obtain a precise value of their quantum yield for photocurrent generation. Efforts will also focus on the possible enhancement of this efficiency through voluntary doping.

REFERENCES

1.   M.A. Russak, J. Reichman, H. Witzke, S.K. Deb, S.N. Chen, J. Electrochem. Soc. 127, 725 (1980).
2.   C.J. Liu and J.H. Wang, J. Electrochem. Soc. 129, 719 (1982).
3.   M. Tomkiewicz, I. Ling, W.S. Parsons, J. Electrochem. Soc. 129, 2016 (1982).
4.   R.A. Boudreau and R.D. Rauh, J. Electrochem. Soc. 130, 513 (1983).
5.   J.P. Szabo and M. Cocivera, J. Electrochem. Soc. 133, 1247 (1986).
6.   B. Parkinson, J. Chem. Education 60, 338 (1983).
7.   S.M. Boudreau, R.D. Rauh, R.A. Boudreau, J. Chem. Education 60, 498 (1983).
8.   D. Iwanov and C. Nanev, Acta Physica Academiae Scientiarum Hungaricae 47, 83 (1979).
9.   J.S. Curran, R. Philippe, M. Roubin, L. Mosoni, Solar Energy Materials 9, 329 (1983).
10. M.F. Lawrence, N. Du, R. Philippe, J.P. Dodelet, J. Cryst. Growth 84, 133 (1987).
11. R. Ludeke, J. Vac. Sci. Tech. 8, 199 (1971).
12. K. Colbow, D.J. Harrison, B.L. Funt, J. Electrochem. Soc. 128, 547 (1981).
13. H. Gerisher, J. Electoanal. Chem. 150, 553 (1983).

PART IV

Silicon Surface Chemistry

STUDIES OF Si SURFACE CHEMISTRY AND EPITAXY USING SCANNING TUNNELING MICROSCOPY AND SPECTROSCOPY

PHAEDON AVOURIS AND ROBERT WOLKOW
IBM Research Division, T. J. Watson Research Center, Yorktown Heights, NY 10598.

ABSTRACT

We apply scanning tunneling microscopy (STM) and spectroscopy (STS) to study the reaction of NH_3 with Si(111)-(7x7), and the epitaxial growth of CaF_2 on Si(111). By a combination of topographs and atom-resolved spectra we can follow the spatial distribution of the reaction and changes in electronic structure with atomic resolution. We find that there are strong site-selectivities for the NH_3 reaction on the 7x7 surface. We also observe the initial stages of the CaF_2 deposition and even are able to image insulating multi-layer CaF_2 films.

INTRODUCTION - WHY THE STM IS NEEDED IN SURFACE CHEMISTRY.

Surface chemistry is a local phenomenon. Reaction can take place at different sites of a crystal surface or at defect sites. Different sites may interact, in the sense that reaction at one site may influence reaction at a neighboring site. Moreover, understanding chemistry implies that one can make a correlation between local electronic structure and reactivity. From the above it is clear that in studying surface chemistry ideally we need an experimental technique (or techniques) that will allow us to follow the spatial distribution of a reaction as well as the local electronic structure with atomic resolution. Conventional topographical and electronic structure techniques, however, average over an area defined by the probe (light, electron, . . .) beam. This is usually larger than $\sim 10^{11}$ atomic sites.

In the following, we will try to demonstrate that STM [1] and STS can, under the appropriate conditions, be used to achieve the above goals. Specifically, we will present results on the application of STM/STS in : (A) the study of surface electronic structure and reactivity using the reaction of Si(111)-(7x7) with NH_3 as an example, and (B) in the study of epitaxy and thin-film growth processes using the epitaxy of CaF_2 on Si(111) as an example.

THE STRUCTURE OF THE CLEAN Si(111)-(7x7) SURFACE.

The nature of the 7x7 reconstruction has been the subject of intense study for about 30 years. Currently, the model proposed by Takayanagi et al. [2] is generally accepted. This model for the 7x7 unit cell is shown in Fig. 1 (top). There are two triangular subunits, each surrounded by nine Si dimers. In addition, there is a stacking-fault in the left triangle. On the surface there are six triply-coordinated Si atoms, (labeled A and B in Fig. 1), so-called restatoms. The top layer is composed of twelve Si adatoms (black circles), and finally, at the corners of the unit cell there are vacancies usually referred to as corner-holes. We further separate the adatoms in two groups: the ones located next to a corner-hole are termed corner-adatoms while the other six are called center-adatoms. The most important chemical effect of the reconstruction is a severe reduction in the number of surface dangling-bonds (dbs). While on the unreconstructed Si(111) surface there are 49 dbs, only 19 survive in the 7x7 unit cell. Of these, 12 are located on the two types of adatoms, 6 on the restatoms and one on the atom at the bottom of the corner-hole. Thus, there are a variety of chemically-active surface sites which would allow us to study the role of local structure on reactivity.

On the bottom of Fig. 1 we show an STM constant-current-topograph of the occupied states of the 7x7 unit cell obtained with the sample biased at -3V. Adatoms, restatoms and the stacking fault are clearly seen.

PROBING ELECTRONIC STRUCTURE AND REACTIVITY WITH ATOMIC RESOLUTION.

The Si(111)-(7x7) +NH₃ Reaction

In addition to being able to image the constant electron density contours of a surface (see Fig. 1, bottom) STM can be used to obtain electronic spectra of the occupied and unoccupied states of the surface with atomic resolution. In our lab we accomplish this while scanning a constant current topograph, by periodically freezing the tip-surface distance and ramping the sample bias. In this way we obtain a topograph and a set of points on this topograph for which we have a complete I-V curve[3]. The distance between such points is variable, typically ~1Å. The quantity $(dI/dV)/(I/V)$ is roughly proportional to the density-of-states[4]. Thus, a plot of $(dI/dV)/(I/V)$ vs. the applied sample bias gives the density of occupied states (negative sample bias) and unoccupied states (positive bias) of the sample.

In Figure 2A (top) we show a topograph of the unoccupied states of clean Si(111)-(7x7) obtained with the sample biased at +2V. Under these conditions, the 12 adatoms in the unit cell are prominent. In the bottom of Fig. 2A we show atom-resolved tunneling spectra obtained above a restatom site (A), a corner-adatom site (B), and over a center-adatom site. Restatoms are characterized by an occupied db-state peaked at ~0.8 eV below the Fermi level. Given that the Coulombic repulsion between two electrons at a Si db-site is only ~0.4 eV [5], the restatom db-state should be fully occupied. The adatom dbs (B and C), however, clearly are only partly occupied. In particular, the occupation of the center-adatom db appears to be the lowest with the intensity of the unoccupied band at +0.5 eV correspondingly stronger. We have interpreted the differences between corner and center-adatoms seen in the above spectra and also in the topograph of Fig. 1 as due to adatom-restatom interactions and adatom to restatom charge-transfer[3]. Center-adatoms have two restatom neighbors to interact with, while corner-adatoms have only one.

When the Si(111)-(7x7) surface is exposed to NH₃ its constant current topographs change drastically. In Fig. 2B (top) we show a topograph (bias +2V) of the unoccupied states of the 7x7 surface after exposure at 300K to ~2L of NH₃. Roughly half of the adatoms have reacted and have become dark. If on the same surface we probe spectroscopically the restatom sites we do not find the characteristic spectrum of the db-state (Curve A in Fig. 2A) but instead we obtain a featureless spectrum (Curve A in Fig. 2B). It is clear that while about half of the adatoms remain unreacted, all restatoms have reacted. Indeed with systematic studies we find that restatoms are much more reactive

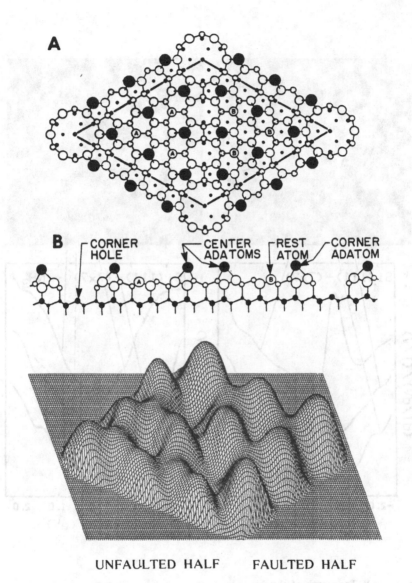

UNFAULTED HALF   FAULTED HALF

Fig. 1 Top: (A) Top view of the Takayanagi et al.[2] model of the Si(i111)-(7x7) unit cell. Atoms at increasing distances from the surface are indicated by circles of decreasing size. (B) Side view along the long diagonal of the unit cell. The dangling bond-bearing restatoms, adatoms and corner hole are labeled. In the left half of the unit cell the stacking sequence is faulted. Bottom: A topograph of the occupied states of the 7x7 surface obtained with a sample bias of -3V.

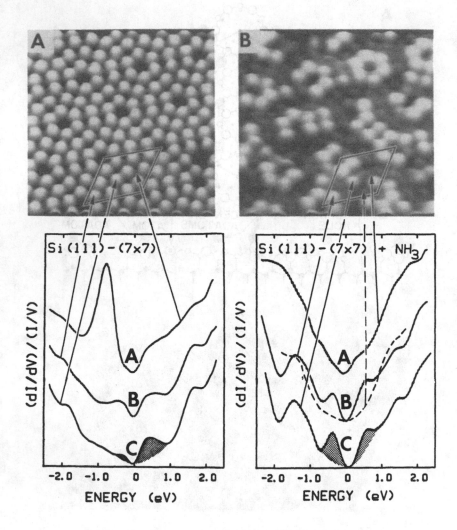

Fig. 2. (A): Topograph of the unoccupied states of the clean 7x7 surface (top), and atom-resolved tunneling spectra (below.) Curves A, B and C give the spectra over restatom, corner-adatom and center-adatom sites, respectively. (B): Topograph of the unoccupied states and tunneling spectra of an NH_3-exposed surface. Curves A and B (dashed line) give the spectra obtained over reacted restatom and adatom sites, respectively. B (solid line) and C give the spectra obtained above unreacted corner- and center-adatom sites, respectively.

than adatoms and thus upon exposure to NH_3 they react first [3]. The spectrum obtained over a dark adatom site (B, dotted line) is also featureless, indicating the elimination of the corresponding db-states by the reaction. By inspection of the topograph in Fig. 2B, we see there are about four times more reacted center-adatoms than corner-adatoms, indicating that center-adatoms are more reactive than corner-adatoms.

Thus, STM allows us to see directly the spatial distribution of the surface reaction. We find that the local environment indeed plays an important role on db-reactivity. Moreover, through STS we can detect interactions between sites on the clean and reacted surface; compare adatom spectra B and C on the clean surface and after the reaction of restatoms[3].

In addition to being able to determine which Si surface atoms have reacted and which have not, we also are able to see the products of the reaction. In Fig. 3 we show a topograph (-3V) of the adatoms surrounding a corner-hole on the partially reacted surface. Sites B, D, E and F are unreacted adatoms, while sites A and C represent products of the reaction. From vibrational [6] and photoemission [7] studies we know that the products of the Si(111)-(7x7) +NH_3 reaction are Si- NH_2 and Si-H groups. Given the relative contributions of these two groups to the density-of-states in the range E_F to E_F -3eV as measured by the corresponding intensities of the valence photoemission bands, feature A is most likely an Si- NH_2 group and C is an Si-H group.

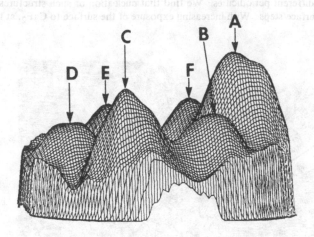

Fig. 3. Topograph of the adatom sites surrounding a corner hole on an NH_3-exposed Si(111) surface. Sample bias =-3 V. Features B, D, E and F are unreacted adatoms, while features A and C are reaction products.

EPITAXY OF CaF$_2$ ON Si(111) - STM IMAGING OF INSULATORS.

Here we will discuss an STM study of a different type of process that is of interest in microelectronics: an epitaxial film-growth process. As an example, we will consider the growth of CaF$_2$ on Si(111). CaF$_2$ is a very good insulator with a band-gap of about 12 eV. Its lattice constant is the same as that of Si(111) to within one percent. Since CaF$_2$ evaporates as a molecule, the problem of achieving the correct stoichiometry in the epitaxial process is automatically solved. CaF$_2$ therefore appears to be a very promising material for 3-dimensional devices. CaF$_2$ films have been studied by a variety of structural [8] and electronic [10] techniques. Our interest is to use the STM to study the initial phases of CaF$_2$ epitaxy, when sub-mono layer amounts of CaF$_2$ are deposited on Si, and understand how these surface structures nucleate, and how a multilayer film grows. Moreover, we would like to test the possibility of imaging multilayer films despite the fact that these films are insulating. The conventional point of view has been that STM is not able to image insulating materials, and that atomic force microscopy is needed for this purpose.

We find that a variety of surface structures are formed at sub-mono layer coverages, depending on the temperature of the deposition process. On the other hand, we find that good films are grown only at temperatures in the range of 750° to 800°C. Under these growth conditions, the predominant structures are row-structures such as those shown on figure 4. This figure shows a Si step, and growing out of that step we have two row structures with different periodicities. We find that nucleation of such structures invariably occurs at surface steps. With increasing exposure of the surface to CaF$_2$, at the sub-

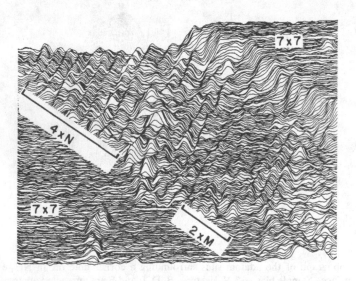

Fig. 4. Topograph (bias = +2V) showing a Si surface step and growing out of it two different calcium fluoride row-structures.

mono layer regime, the dominant row structure becomes the 2x3 structure shown in figure 5. These row structures can be imaged not only in the unoccupied states mode but also in the occupied states mode with a sample bias in the order of -1V. This indicates that the Ca atom cannot be in the Ca^{2+} state as in bulk CaF_2 because the lowest binding energy occupied states of Ca^{2+} lie at about 26 eV below the Fermi energy, and thus are inaccessible to STM. This observation, along with detailed spectroscopic studies [11], indicates that upon interaction of CaF_2 with the Si surface a chemical reaction occurs in which one fluorine is removed from the CaF_2 molecule. In this way, the Ca atom acquires an electron

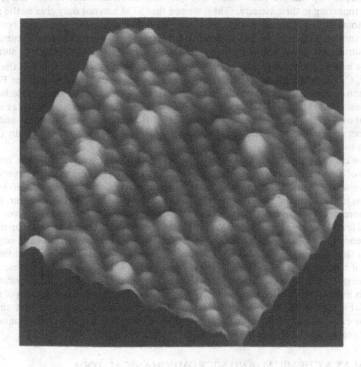

Fig. 5. Topograph (140Å x 140Å) of the unoccupied states of the calcium fluoride 2x3 surface structure. Bias = +2V.

from the fluorine that now resides in the 4s level of Ca. This 4s electron is used to interact with the dangling bonds of surface Si atoms to form the Ca-Si bond. Our STM spectroscopic studies, in agreement with photoemission [9] have shown that the resulting bonding level is peaked at about 1.5 eV below E_F. A corresponding anti-bonding level is found at about 1.2 eV above E_F. As we increase the amount of CaF_2 deposited on the surface to the point where a monolayer of CaF_2 is deposited, we observe a transition from the 2x3 row-structure to the 1x1 structure. This structure has very low electronic corrugation and it appears in the STM topographs as a very smooth surface. However, depending on the deposition conditions and annealing, this surface can be dotted by holes which give it the appearance of a pincushion, as seen in figure 6(top). We ascribe these holes to negatively charged sites which, through their Coulomb fields reduce the proba- bility for tunneling in their vicinity. Thus, we see that STM can not only give us the posi- tion of atoms at the surface, but also the distribution of local fields. In Fig. 6 (bottom) we show another picture of the 1x1 surface, and on top of it we see some rather symmetric, round protrusions. These are CaF_2 molecules deposited on top of the 1x1 CaF_2 surface. The Ca 4s level in these molecules is totally unoccupied, while as we saw above in the first 1x1 layer, Ca exists in the Ca^{1+} state. The 4s energy level of Ca^{2+} is closer to the Fermi energy than that of Ca^{+1}. Thus, we find that with a positive bias on the sample below +2 V, these individual CaF_2 ad-molecules are clearly seen. However, as the bias is in- creased above +2 V, the resonance with the Ca^{1+} 4s level becomes important and the contrast between the ad-molecules and the 1x1 underlayer is reduced, and eventually these molecules become invisible.

If we continue to deposit CaF_2, multilayers are formed. Such multilayers have a band-gap of about 12 eV and therefore they are insulating. According to the conventional point of view, we should not be able to image such multilayers. Indeed, if we try to image these films in the occupied states mode, we find that this is not possible. However, if we look in more detail at the energy levels of the Si-CaF_2 system [9], we see that given that the Fermi level in this system is pinned at the top of the Si valence band, and the top of the CaF_2 valence band lies at ~8.5 eV below E_F, it follows that the conduction band edge of the CaF_2 film lies only about 3.5 eV above E_F. Thus it is accessible to the tunneling electrons if the sample is biased with a positive bias of 3.5 V or higher. Under these con- ditions, tungsten electrons can tunnel into the conduction band of CaF_2. To demonstrate that this is indeed feasible, we show in Fig. 7 a topograph of the unoccupied states of such a 15Å CaF_2 film on Si(111). Consideration of other insulators indicates that similar conditions exist in many other systems and that STM-imaging of insulators by tunneling into their conduction bands should be feasible.

THE STM AS A CHEMICAL AND MICROMECHANICAL TOOL.

The use of STM in surface chemistry is not limited to being a tool for the study of atomic geometry and electronic structure. The STM also could be used to induce surface chemical or morphological modifications of electronic materials on a nanometer scale. As we have shown recently, electron-beam-excitation [11] can be used to induce a variety of surface reactions useful to microelectronics and lead to local material deposition or etch- ing. By using the STM tip as the source of the excitation electrons and operating in the field-emission mode, surface chemistry can be induced. Recent reports show that this is indeed feasible [12].

Besides using the electrons emitted from the tip, one could also use the high electric fields that can be produced in the vicinity of the tip to induce surface modifications. Thus, gas phase molecules can be field-ionized and the ions driven to the surface to deposit or

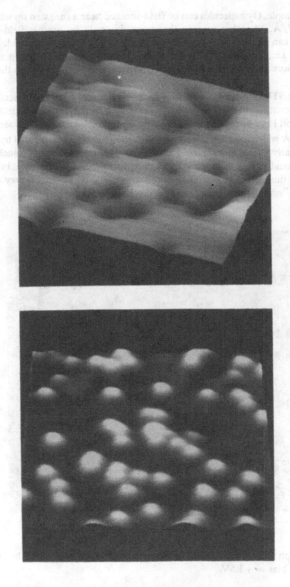

Fig. 6. Top: Topograph (140Å x 140Å) of the unoccupied states of the 1x1 calcium fluoride monolayer (sample bias = +2 V). Charged sites appear as depressions. Bottom: Imaging of the unoccupied states of individual CaF_2 molecules deposited on top of the 1x1 monolayer (sample bias = +3.5 V).

implant. For example, O_2 molecules can be field-ionized near a tungsten tip when the field becomes ~1.5 V/Å. Another process that possibly could be used is field-desorption. Atoms on the tip can be desorbed as positive ions above a certain critical field. This critical field can be low, i.e., less than 1 V/Å for weakly bound species on the tip and at small tip-surface distances so that image and possibly chemical forces can act on the desorbing species.

Finally, the STM tip can be used as a micromechanical tool to physically modify a surface. We demonstrate this in Figure 8 which shows a trench produced mechanically on a NH_3-exposed Si(111) surface by the STM tip and then imaged using the same tip. The trench is only 20Å wide (less than ten atoms) with the atomic debris formed by the digging of the trench clearly visible at its sides. This is probably the narroweast trench ever to be produced intentionally. It is most remarkable that the tip was not destroyed in the process so it could image the surface modification with atomic resolution. In this way the STM tip can be used as a "surgical tool" to operate on device structures.

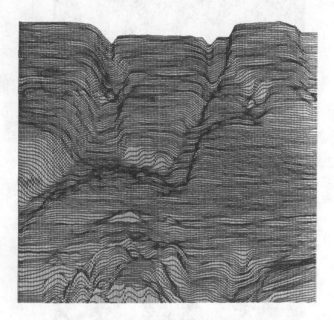

Fig. 7. Topograph (350Å x 300Å) of the unoccupied states of a 15Å film of CaF_2 on Si(111). Sample bias = +3.5V.

CONCLUSIONS

In conclusion, we found that by using STM topographs and atom-resolved spectra we can study surface electronic structure and reactivity with atomic resolution. In the case of the reaction of Si(111)-(7x7) with NH_3 we have been able to determine the relative reactivity of the various dangling-bond sites and selectively image unreacted sites and the

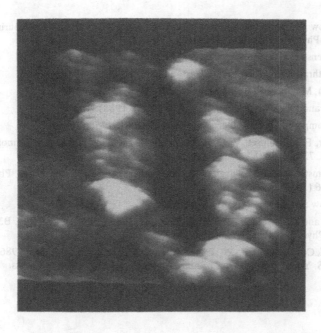

Fig. 8. Trench (~20Å wide) dug mechanically by the STM tip on an NH_3-exposed Si(111) surface. The same tip was used to image with atomic resolution the modified surface.

reaction products. In the case of CaF_2 epitaxy we found that when sub-monolayer amounts of CaF_2 are deposited on Si(111), row-like structures are formed. The 1x1 monolayer has small corrugation and allows the observation of charged sites, as well as individual CaF_2 molecules deposited on top of it. More importantly, we can image insulating CaF_2 multi-layers by tunneling into their conduction bands. Finally, we showed that the STM can be used to manipulate surfaces with nanometer resolution.

REFERENCES

1. For reviews see: G. Binnig and H. Rohrer, Rev. Mod. Phys. **59**, 615 (1987); P. K. Hansma and J. Tersoff, J. Appl. Phys. **61**, R1 (1987).

2. K. Takayanagi, Y. Tamishiro, M. Takahashi, H. Motoyoshi, and K. Yagi, J. Vac. Sci. Technol. **A3**, 1502 (1985).

3. R. Wolkow and Ph. Avouris, Phys. Rev. Lett. **60**, 1049 (1988); Ph. Avouris and R. Wolkow, Phys. Rev. B, to be published.

4. R. M. Feenstra, J. A. Stroscio and A. P. Fein, Surf. Sci. **181**, 295 (1984).

5. J. E. Northrup, Phys. Rev. Lett. **57**, 154 (1986).

6. S. Tanaka, M. Onchi and M. Nishijima, Surf. Sci. **L756**, (1987).

7. F. Bozso and Ph. Avouris, Phys. Rev. **B38**, 3937 (1988).

8. R. M. Tromp and M. C. Reuter, Phys. Rev. Lett. **61**, 1756 (1988).

9. D. Rieger, F. J. Himpsel, V. O. Karlsson, F. R. McFeely and J. A. Yarmoff, Phys. Rev. **B34**, 7295 (1986).

10. M. A. Olmstead, R. I. G. Uhrberg, R. D. Bringans, and R. Z. Bachrach, Phys. Rev. **B35**, 7526 (1987).

11. R. Wolkow and Ph. Avouris, to be published.

12. F. Bozso and Ph. Avouris, Appl. Phys. Lett. **53**, 1095 (1988); Phys. Rev. **B38**, 3943 (1988); Phys. Rev. Lett. **57**, 1185 (1986).

13. M. A. McCord and R. F. W. Pease, J. Vac. Sci. Technol. **B4**, 86 (1986); E. E. Ehrichs, S. Yoon, and A. L. deLozanne, Appl. Phys. Lett. (1988); in press.

OXIDATION KINETICS OF SILICON SURFACES: REACTIVE STICKING COEFFICIENT, APPARENT SATURATION COVERAGE AND EFFECT OF SURFACE HYDROGEN

S.M. GEORGE, P. GUPTA, C.H. MAK AND P.A. COON
Dept. of Chemistry, Stanford University, Stanford, Calif. 94305

ABSTRACT

The kinetics of the initial oxidation of silicon surfaces by O_2 were studied using laser-induced thermal desorption (LITD), temperature-programmed desorption (TPD) and Fourier Transform Infrared (FTIR) spectroscopy. The LITD results showed that the oxidation of Si(111)7x7 by O_2 was characterized by two kinetic processes: an initial rapid oxygen uptake followed by a slower growth that asymptotically approached an apparent saturation oxygen coverage. The initial reactive sticking coefficient of O_2 on Si(111)7x7 decreased with surface temperature. In contrast, TPD experiments on Si(111)7x7 and FTIR studies on porous silicon demonstrated that the apparent saturation oxygen coverage increased as a function of surface temperature. Experiments with preadsorbed hydrogen also revealed that silicon oxidation was inhibited as a function of increasing hydrogen coverage on the Si(111)7x7 surface.

Introduction

The reaction of oxygen with silicon surfaces is of great fundamental and technological interest. Not only is oxidation a model silicon surface reaction, but this reaction also produces dielectric isolation in silicon devices. As the dimensions of integrated circuitry are progressively reduced in the submicron regime, the understanding of this basic reaction becomes increasingly important. In the nanometer limit, surface-to-volume ratios are extremely high, and the kinetics of silicon surface oxidation are critical to comprehending and controlling this essential chemical processing step.

In this study, we report on laser-induced thermal desorption (LITD),temperature programmed desorption (TPD) and Fourier Transform Infrared (FTIR) experiments that were used to investigate the kinetics of both the fast and slow steps in the oxygen adsorption process. These studies allowed the initial reactive sticking coefficient of oxygen on the Si(111)7x7 surface to be measured as a function of surface temperature. In addition, the kinetics of the slow adsorption step were studied on Si(111)7x7 and porous silicon as a function of surface temperature. Moreover, in order to study the dependence of oxidation rate on surface dangling bond density, the initial reactive sticking coefficient and apparent saturation oxide coverage on Si(111)7x7 were measured as a function of preadsorbed hydrogen.

The experimental apparatus for LITD and FTIR experiments has been described in detail previously [1,2]. Briefly, the ultrahigh vacuum (UHV) chamber used for the LITD studies was pumped by a 300 1/sec ion pump and a titanium sublimation pump. These pumps maintained background

Mat. Res. Soc. Symp. Proc. Vol. 131. ©1989 Materials Research Society

pressures of approximately 4×10^{-10} Torr during these experiments. This chamber was also equipped with a low energy electron diffraction (LEED) spectrometer, a cylindrical mirror analyzer for Auger electron spectroscopy (AES), and a quadrupole mass spectrometer for LITD and TPD studies. Likewise, the UHV chamber for the FTIR studies was pumped by dual turbomolecular pumps and has been described in an earlier publication [2].

The laser induced thermal desorption (LITD) experiments utilized a TEM-00 Q-switched Ruby laser with a temporal width (FWHM) of 80-100 nsec and a Gaussian spatial profile [1]. The laser pulses had energies of 3.3 mJ before entering the chamber. These pulses were focused onto the silicon sample by a 1.0 m lens. The focused laser beam had a Gaussian beam profile at the focus of the lens with a width of 260 μm (FWHM). The laser pulses were directed onto the silicon sample at an angle of approximately 54° with respect to the surface normal. Under such conditions, the desorption spot on the crystal was elliptical. The spot size measured using the auto-correlation method [3] was roughly 420 μm in diameter along the major axis and 250 μm in diameter along the minor axis.

Temperature programmed desorption (TPD) spectra obtained after silicon oxidation experiments on Si(111)7x7 were carried out using a linear heating rate of 9 K/sec. Before the TPD experiment, the crystal was rotated to a position approximately 5 cm from the ionizer of the quadrupole mass spectrometer. The only species observed in the TPD spectra was SiO. Desorption of SiO was observed only when the silicon crystal was directly in front of the ionizer of the mass spectrometer.

Initial Sticking Coefficient

The initial oxidation of Si(111)7x7 was studied as a function of surface temperature using LITD techniques [4]. In these experiments, the clean Si(111)7x7 surface was maintained at a constant surface temperature. The surface was then oxidized by backfilling the UHV chamber with a constant pressure of O_2. During the oxidation of the surface, the oxygen coverage on the crystal was measured as a function of time by monitoring the SiO LITD signals. By measuring SiO LITD signals from different positions at different times during oxygen exposure, the oxygen surface coverage could be determined in real time.

The results of the LITD oxidation measurements on Si(111)7x7 performed with an oxygen background pressure of 5×10^{-9} Torr for three different surface temperatures are shown in Fig. 1. The initial oxidation rate is proportional to the initial slope of the oxygen uptake curves displayed in Fig. 1. The solid lines shown in Fig. 1 are least-square fits to the initial slopes of the oxygen uptake curves. Notice that the initial oxidation rate clearly decreases as a function of surface temperature.

At all oxygen background pressures investigated, the initial oxidation rate on Si(111)7x7 decreased with increasing surface temperature. In particular, the initial oxidation rate was reduced by approximately a factor of four when the surface temperature was raised from 200 to 600 K. A similar decrease in the initial oxidation rate as a function of surface temperature on Si(100) was also observed recently using molecular beam techniques [5].

The initial oxidation rate on Si(111)7x7 was found to be directly proportional to oxygen background pressure at a fixed surface

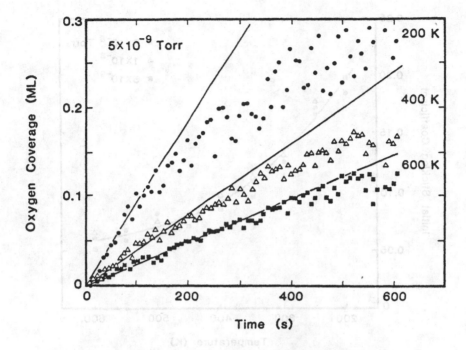

Fig. 1 LITD measurements of the oxygen coverage on Si(111)7x7 as a function of time. These LITD measurements were performed at three different temperatures at a background O_2 pressure of P= 5×10^{-9} Torr.

temperature. This proportionality indicates that silicon surface oxidation is first-order in oxygen pressure. The initial sticking coefficients, S_o, were obtained by dividing the initial oxidation rate by the oxygen flux which is simply proportional to the oxygen pressure. The results for the initial sticking coefficient on Si(111)7x7 are shown in Fig. 2.

The observed temperature dependence of the initial oxygen sticking coefficient on Si(111)7x7 is consistent with a precursor-mediated adsorption model [6]. In the precursor model, the first stage of adsorption involves trapping O_2 in a precursor state. This precursor species may either react to form the chemisorbed oxygen or desorb into the gas phase. The model is defined by the following steps:

$$O_2(g) + n * \xrightarrow{k_a} O_2(a) \tag{1}$$

$$O_2(g) + n * \xleftarrow{k_d} O_2(a) \tag{2}$$

$$O_2(a) + m * \xrightarrow{k_r} 2O(a) \tag{3}$$

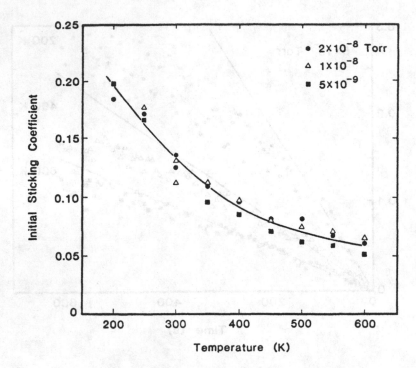

Fig. 2. Initial reactive sticking coefficient for O_2 on Si(111)7x7 as a function of temperature. The different symbols represent oxidation rate measurements at three different O_2 background pressures.

where $O_2(a)$ is the molecular precursor state, $O(a)$ is the chemisorbed state, and * represents an empty site. k_a is the adsorption rate constant and k_d and k_r are the desorption and reaction rate constants of the precursor state, respectively. The adsorption rate constant is defined such that $k_a O_2(g) = \alpha\phi$ where α is the trapping coefficient of the precursor state and ϕ is the flux of oxygen molecules impinging on the surface.

Using the precursor-mediated adsorption model, the temperature-dependent sticking coefficient can be fit using the ratio of the preexponential factors (k_d^0/k_r^0) and the difference in the activation barriers E_d-E_r between the desorption and the reaction pathways. With a value of $\alpha = 0.5$, k_d^0/k_r^0 and E_d-E_r were determined to be approximately 15 and 1 kcal/mole, respectively. The solid line in Fig. 2 shows the initial oxygen sticking coefficients calculated according to the precursor-mediated adsorption model using these parameters.

Apparent Saturation Coverage

The utility of LITD techniques for measuring oxygen coverages was limited to oxygen coverages $\theta < 0.3$ ML [4]. Consequently, the TPD

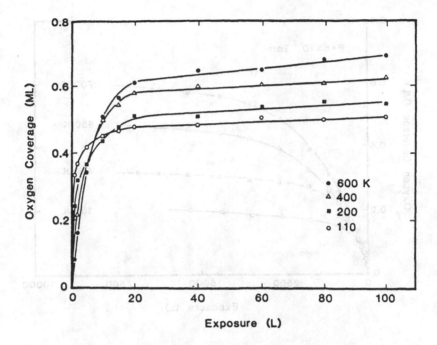

<u>Fig. 3.</u> TPD measurements of oxygen coverage on Si(111)7x7 as a function of oxygen exposure for several temperatures.

spectra of SiO were employed to study the adsorption of oxygen on Si(111)7x7 at higher oxygen coverages. The oxygen exposures were performed using an oxygen background pressure of 1×10^{-7} Torr at a variety of surface temperatures from 110 K to 600 K. These TPD measurements of oxygen uptake are shown in Fig. 3.

The initial oxidation behavior revealed by the SiO TPD measurements shown in Fig. 3 is similar to the initial oxidation rates determined by the LITD measurements shown in Fig. 2. The oxygen uptake as a function of time is given by the SiO TPD area. The slope of the oxygen uptake curves in the limit of zero coverage defines the initial sticking coefficient. Notice that the slope of the oxygen uptake curves decreases with increasing surface temperature.

At intermediate oxygen surface coverages, the silicon oxidation rate decreases dramatically. This decrease leads to a bend in the oxygen uptake curves. This bending over or turnover point occurs at smaller oxygen exposures for lower surface temperatures. In particular, Fig. 3 shows that the oxygen uptake curve at 110 K bends over after an oxygen exposure of 1 L. In contrast, the turnover point occurred at roughly 10 L of oxygen exposure at a surface temperature of 600 K. Consequently, higher apparent saturation oxide coverages occur at higher surface temperatures. Although the initial oxidation rate was faster at lower surface temperatures, the final oxygen coverage was higher for higher surface temperatures.

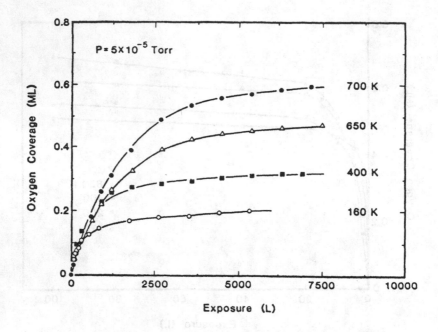

Fig. 4. FTIR measurements of the oxygen coverage on porous silicon surfaces as a function of oxygen exposure for several temperatures.

FTIR studies of the oxidation of porous silicon surfaces also revealed that the apparent saturation oxide coverages were temperature dependent [7]. For these FTIR experiments, the infrared absorbance by the asymmetric Si-O-Si stretch was employed as a measure of the oxide coverage. FTIR measurements of the oxide coverage on porous silicon surfaces as a function of oxygen exposure are shown in Fig. 4.

The temperature dependence of the apparent saturation oxide coverage suggests that the slow oxygen uptake step is an activated process with a rate characterized by a coverage-dependent activation barrier. This coverage dependent activation barrier may result from stress induced by the distortion of the silicon lattice during oxidation [8] or electrostatic repulsion between the partial negative charges on the oxygen adatoms [9].

As a result of the induced stress or electrostatic repulsion, the activation barrier for the oxygen reaction is likely to increase as a function of the number of Si-O-Si units on the surface. An increasing activation barrier with increasing oxygen coverage would cause the oxygen adsorption rate to decrease as a function of oxygen coverage. At higher temperatures, higher oxide coverages could be obtained before the higher activation barrier would cause the oxygen adsorption rate to decrease.

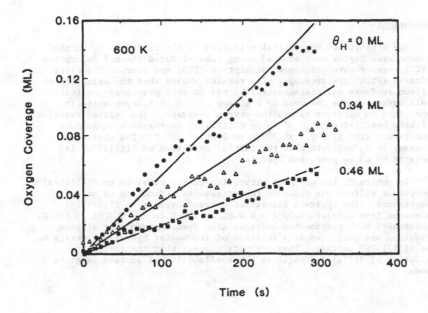

Fig. 5. LITD measurements of the effect of hydrogen on the initial oxidation of Si(111)7x7 at 600K.

Effect of Preadsorbed Hydrogen

Figure 5 shows the LITD oxidation measurements during initial oxidation at 600 and 200 K with an oxygen pressure of 5×10^{-9} Torr for Si(111)7x7 surfaces preadsorbed with hydrogen. The corresponding measurement for a clean Si(111)7x7 surface is also displayed for comparison. Figure 5 clearly demonstrates that the initial oxidation rate decreases as a function of increasing hydrogen coverage on Si(111)7x7.

Hydrogen is known to tie up surface dangling bonds on silicon surfaces. Previous EELS and FTIR experiments [2,10] have shown that silicon-monohydride and dihydride species are formed on hydrogen-covered silicon surfaces. These hydride species reduce the surface dangling bond density on silicon surfaces and are expected to inhibit surface reactions [11]. The decrease in the initial sticking coefficient with preadsorbed hydrogen coverage suggests that the initial adsorption of oxygen requires the availability of surface dangling bonds. Consequently, a reduction in the dangling bonds caused by the preadsorbed hydrogen atoms results in a decrease in the initial oxygen reactive sticking coefficient.

Conclusions

The kinetics of the initial oxidation of Si(111)7x7 by O_2 in the submonolayer regime were studied using laser-induced thermal desorption (LITD), temperature-programmed desorption (TPD) and Fourier Transform Infrared (FTIR) spectroscopy. The results showed that the oxidation of silicon surfaces was characterized by two kinetic processes: an initial rapid oxygen uptake followed by a slower growth that asymptotically approached an apparent saturation oxygen coverage. The initial reactive sticking coefficient (S_o) of O_2 on Si(111)7x7 decreased with surface temperature from S_o = 0.2 at 200 K to 0.06 at 600 K. The observed decrease in S_o suggests that the initial oxidation of Si(111)7x7 is mediated by an O_2 precursor species.

In contrast, the apparent saturation oxygen coverage on Si(111)7x7 and porous silicon was observed to increase as a function of surface temperature. The apparent saturation oxygen coverage on Si(111)7x7 increased from approximately θ = 0.4 ML at 110 K to θ = 0.7 ML at 600 K. Experiments with preadsorbed hydrogen also demonstrated that silicon oxidation was inhibited as a function of increasing hydrogen coverage on the Si(111)7x7 surface. These results suggest that the initial oxidation of Si(111)7x7 requires the availability of surface dangling bonds.

Acknowledgments

This work was supported by the Office of Naval Research under contract N00014-86-K-545. Some of the equipment utilized in this work was provided by the NSF-MRL program through the Center for Materials Research at Stanford University. CHM gratefully acknowledges the National Science Foundation for a graduate fellowship.

References

[1] B.G. Koehler, C.H. Mak, D.A. Arthur, P.A. Coon and S.M. George, J. Chem. Phys. 89, 1709 (1988).

[2] P. Gupta, V.L. Colvin and S.M. George, Phys. Rev. B37, 8234 (1988).

[3] S.M. George, A.M. DeSantolo and R.B. Hall, Surf. Sci. 159, L425 (1985).

[4] C.H. Mak, P. Gupta, P.A. Coon and S.M. George, (in preparation).

[5] M.P. d'Evelyn, M.M. Nelson and T. Engel, Surf. Sci. 186, 75 (1987).

[6] W.H. Weinberg, in Kinetics of Interface Reactions, edited by M. Grunze and H.J. Kreuzer (Springer Verlag, New York, 1987), p.94.

[7] P. Gupta, V.L. Colvin and S.M. George, (in preparation).

[8] E.A. Lewis and E.A. Irene, J. Electrochem. Soc. 134, 2332 (1987).

[9]   J.A. Stroscio, R.M. Feenstra and A.P. Fein, Phys. Rev. Lett. 58, 1668 (1987).

[10] R. Butz, E.M. Oellig, H. Ibach and H. Wagner, Surf. Sci. 147, 343 (1984).

[11] F. Bozso and Ph. Avouris, Phys. Rev. Lett. 57, 1185 (1986).

SILANE ADSORPTION AND DECOMPOSITION ON Si(111)-(7X7)

S.M. Gates, C.M. Greenlief and D.B. Beach

I.B.M. T.J. Watson Research Center, Yorktown Heights, N.Y. 10598.

ABSTRACT

The reactive sticking coefficient, S^R, of SiH_4 on the Si(111)-(7X7) surface has been studied as a function of hydrogen coverage (Θ_H) in the surface temperature (T_S) range 80-500°C. At 400°C, evidence is seen for two adsorption regimes which are proposed to correspond to minority and majority surface sites. On the minority sites ($\Theta_H = 0$ to 0.08), S^R is approximately independent of T_S. On the majority sites ($\Theta_H > 0.08$), S^R is a complicated function of T_S and Θ_H. After SiH_4 exposure, surface SiH is the dominant *stable* decomposition intermediate from 80 to 500°C, with detectable populations of SiH_2 and SiH_3 present at the lowest temperatures.

Thermal decomposition of silane, $SiH_4(g)$, leads to silicon film growth and evolution of $H_2(g)$ under appropriate conditions. Surface reaction steps in the overall film growth reaction (1) include SiH_4 adsorption, diffusion of adsorbed SiH_x species, decomposition of these species, Si-Si bond formation and $H_2(g)$ evolution.

$$SiH_4(g) \rightarrow Si(s) + 2 H_2(g) \qquad (1)$$

Measurements of kinetic parameters that characterize the net reaction (1) are extensive and have been recently compared and summarized {1}. Kinetic data pertaining to the individual heterogeneous reaction steps in reaction (1) has been limited to H_2 desorption from surface SiH {2,3} and SiH_2 {2,4}.

The first heterogeneous reaction step in the process of Si film growth from silane is examined here in the temperature range 80-500°C using an atomically clean Si(111)-(7X7) surface and controlled UHV conditions. The reactive sticking coefficient, S^R, is approximately 10^{-5} near $\Theta_H = 0$. The dependence of the rate of silane adsorption on hydrogen surface coverage (Θ_H) and surface temperature (T_S) is discussed, and the stable surface hydride species after SiH_4 exposure are identified.

The experiments are performed in a stainless steel UHV chamber equipped with a differentially pumped quadrupole mass spectrometer (QMS), a calibrated effusive doser, an ion gun and Low Energy Electron Diffraction (LEED) optics. The stable surface intermediates formed are identified using static secondary ion mass spectrometry (SSIMS) as described elsewhere {5,6}.

Temperature programmed desorption (TPD) is used to quantify surface hydrogen, and hence the amount of chemisorbed silane, following calibrated silane exposures as described in detail previously {7,8}.

The number of silane molecules adsorbed on the Si(111)-(7X7) surface expressed relative to the silicon atom density per cm.$^{-2}$ {9} is plotted versus SiH$_4$ exposure in Fig. 1 for three surface temperatures (T$_S$). The predominant stable surface species detected in these experiments is SiH (see Fig. 2). The right ordinate indicates the resulting H atom coverage (Θ_H) in H/Si {9}. At T$_S$ = 400°C, Θ_H rigorously equals the coverage of surface SiH species. At lower T$_S$, Θ_H approximates the SiH coverage, as small amounts of SiH$_2$ and SiH$_3$ are also present (see Fig. 2).

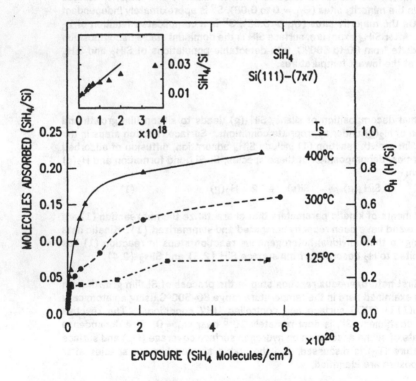

Fig. 1. Adsorption of silane on the Si(111)-(7X7) surface at 400°C (triangles), 300°C (dots), and 125°C (squares). Y axis: SiH$_4$ Adsorbed per 1st layer Si {9}, as measured by H$_2$ TPD area. X axis: Molecules Exposure cm.^{2}, from calibrated effusive gas doser. Gas temperature equals 25°C.

The inset to Fig. 1 contains data for the lowest SiH_4 exposures at $T_S = 400^{\circ}C$, emphasizing adsorption on the minority sites (solid line) which are about 2% of the surface sites. The slope of the solid line is S^R near $\Theta_H = 0$, and equals 10^{-5}. A distinct decrease in S^R occurs when about 0.019 SiH_4/Si are adsorbed ($\Theta_H \approx 0.08$). The unit cell of the (7X7) structure contains one deep corner hole (essentially a vacancy defect) per unit cell {10}, with a coverage in vacancies per 1st layer Si atom of 1/49, or 0.020. We speculate that silane adsorption at $\Theta_H \approx 0$ with $S^R \approx 10^{-5}$ occurs at the deep corner holes of the (7X7) structure, although the agreement in coverage could be fortuitous. When T_S is varied at a constant SiH_4 exposure corresponding to minority site adsorption, S^R is essentially independent of T_S {8}. Comparison with an ion bombardment damaged (7X7) surface and with the Si(100)-(2X1) surface (data not shown) confirms that S^R depends on the surface structure, as shown here in the comparison of minority and majority sites.

The data of Fig. 1 concern adsorption on the remaining 98% of the surface (majority sites). Both S^R and the maximum observed Θ_H increase with increasing T_S, consistent with an activated chemisorption process, and in contrast to the minority site chemisorption. The data of Fig. 1 have been analyzed using a finite differential approximation to S^R {8}. At $T_S = 125^{\circ}C$, S^R drops below 10^{-7} for $\Theta_H > 0.1$, while at $T_S = 300^{\circ}C$, S^R smoothly decreases and only approaches 10^{-7} for $\Theta_H \approx 0.6$. Data at $400^{\circ}C$ are the most relevant to CVD film growth. At $T_S = 400^{\circ}C$, S^R remains well above 10^{-6} until ≈ 0.12 SiH_4/Si are adsorbed ($\Theta_H \approx 0.5$). Finally, S^R drops to almost 10^{-7} when Θ_H exceeds 0.65 at $400^{\circ}C$.

The reactive sticking coefficient, S^R, is about 10^{-5} on the clean surface, and decreases as function of hydrogen coverage. The functional dependence of S^R on Θ_H is, in turn, dependent on T_S, so that accurate models or analytical expressions for S^R require both variables. Enhanced reaction at minority sites indicates that the rate of SiH_4 adsorption during CVD film growth will be a sensitive function of the exact structure of the growth interface.

Figure 2 shows the SSIMS spectra of the Si(111) surface for three different conditions. The top spectrum is of the clean Si surface and Si^+ is the only secondary ion observed between 28 and 34 amu. The middle spectrum is of a surface exposed to SiH_4 at $T_S = 125^{\circ}C$. The exposure of 3×10^{20} SiH_4 cm.$^{-2}$ (plateau region for squares, Fig. 1) resulted in 0.08 SiH_4 molecules adsorbed per surface Si atom. SiH^+ is observed in SSIMS as the dominant secondary hydride ion. Small amounts of SiH_2^+ and SiH_3^+ are also detected (multiplied by ten in figure). The bottom spectrum is of a hydrogen saturated surface. Three secondary hydride ions, SiH^+, SiH_2^+, and SiH_3^+, are observed.

Our previous SSIMS work {5,6} involving the surface populations of silicon hydrides allows one to identify and determine the surface coverage of SiH_x species present. The hydrogen saturated surface (bottom, Fig. 2) contains three hydrides, SiH, SiH_2, and SiH_3. SiH is the most abundant with an average of 1 SiH/Si. SiH_2 and SiH_3 are also present in small amounts, 0.08 SiH_2/Si and 0.04 SiH_3/Si. Turning to the silane exposed surface we see that SiH is the majority species while the SiH_2 and SiH_3 concentrations are small. The SSIMS results indicate that surface SiH groups are the dominant *stable* decomposition

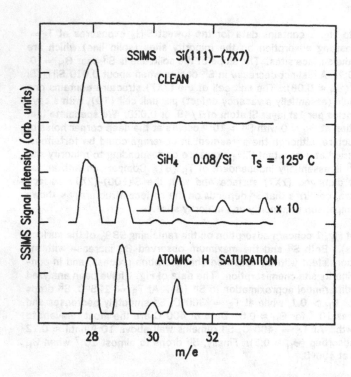

Fig. 2. SSIMS spectra for the clean Si(111)-(7x7) surface (top), for Si exposed to silane (middle), and for a Si surface saturated with atomic hydrogen (bottom). $T_S = 125°C$ for each experiment.

intermediate from SiH_4 adsorption. The SSIMS signal for SiH_2^+ is small, and corresponds to a coverage of ≈ 0.04 SiH_2/Si, and the SiH_3 coverage is < 0.01 SiH_3/Si. These data support the proposal that SiH decomposition is the rate controlling step in the net reaction (1).

We have contrasted the dominant adsorption process which depends on T_S and occurs for $\Theta_H > 0.08$ with adsorption on minority sites which may be the corner holes of the (7X7) structure. The latter process is essentially independent of T_S {8}. We are seeking further evidence for the proposal that SiH_4 adsorption occurs by two distinct pathways at different classes of surface sites. The results presented here show that the initial heterogeneous reaction step leading to Si film growth is dependent on surface temperature, hydrogen coverage, and surface structure. Silane decomposes on the initially clean Si surface to form SiH as the predominant stable intermediate. Investigation of the surface species linking $SiH_4(g)$ with surface SiH is in progress.

The authors gratefully acknowledge Dr. B.A. Scott for many helpful discussions, and P.A. Holbert for experimental work.

REFERENCES

1. R.J. Buss, P. Ho, W.G. Breiland and M.E. Coltrin; J. Appl. Phys. 63 2808 (1988).

2. G. Schulze and M. Henzler; Surface Science 124 336 (1983).

3. B.G. Koehler, C.H. Mak, D.A. Arthur, P.A. Coon and S.M. George; J. Chem. Phys. 89 1709 (1988).

4. P. Gupta, V.L. Colvin and S.M. George; Phys. Rev. B37 8234 (1988).

5. C.M. Greenlief, S.M. Gates and P.A. Holbert; J. Vac. Sci. Tech., submitted.

6. C.M. Greenlief, S.M. Gates and P.A. Holbert; Phys. Rev. Lett., submitted.

7. S.M. Gates; Surface Science 195 307 (1988).

8. S.M. Gates, C.M. Greenlief, D.B. Beach and R.R. Kunz; Chem. Phys. Lett., submitted.

9. A bulk terminated (111) plane with 7.8×10^{14} Si cm.$^{-2}$ is used as the reference to report Θ in H or SiH_4 per 1st layer Si atom.

10. K. Takayanagi, Y. Tanishiro, M. Takahashi, and S. Takahashi; J. Vac. Sci. Tech. A3 1502 (1985).

CHEMISORPTION OF HF ON SILICON SURFACES

S. A. JOYCE, J. A. YARMOFF, A. L. JOHNSON,[a] and T. E. MADEY[b]
Surface Science Division, National Institute of Standards and Technology,
Gaithersburg, MD 20899
(a) Present address: Department of Chemistry, Liverpool University, Liverpool,
L69 3BX, UK
(b) Present address: Department of Physics, Rutgers, The State University
Piscataway, NJ 08855

ABSTRACT

We have investigated the interaction of gaseous hydrofluoric acid (HF) with single crystal silicon surfaces using soft x-ray photoemission spectroscopy and Electron Stimulated Desorption Ion Angular Distributions (ESDIAD). Examination of the Si(2p) core level for surfaces saturated with HF shows the formation of silicon-fluoride bonds indicating the dissociative chemisorption of HF on both Si(111) and Si(100) surfaces. Inspection of the F(2s) and F(2p) valence levels at saturation coverage indicate that only one-half monolayer of fluorine bonds to the silicon. The primary ion desorbed by electron bombardment of these surfaces is F^+ with only a minor contribution from H^+. ESDIAD images from a saturation coverage of HF on stepped Si(100) surfaces reveal F^+ desorption primarily along the direction of the terrace dimers. The ESDIAD patterns from HF adsorbed on Si(111) are characterized by strong normal F^+ emission with a weak background component of off-normal emission. These results are consistent with the dissociative chemisorption of HF where the ion emission direction is determined by the Si-F bond directions.

INTRODUCTION

Hydrofluoric acid (HF) is a commonly used reagent in the production of many semiconductor devices. While most frequently employed in its aqueous solution phase for the initial removal of SiO_2 from silicon wafers,[1] gaseous HF is also present in significant quantities in many processing environments (for example CF_4/H_2 and CF_3H plasmas) as a major gas phase reaction byproduct.[1] Given its importance, it is rather surprising that only recently have the fundamental interactions of HF with silicon and silicon dioxide surfaces been investigated.[2-4] While it is generally agreed that the etching of SiO_2 with aqueous HF leaves behind a silicon substrate which is only hydrogen terminated,[2,3] the authors know of no previous work which has studied the interaction of gaseous HF with clean, well characterized silicon surfaces. We have employed the techniques of soft X-ray photoemission and electron stimulated desorption ion angular distributions (ESDIAD) to determine the chemical and geometric structures of the overlayers formed after exposure of the Si(111) 7x7 and Si(100) 2x1 surfaces to gas phase hydrofluoric acid.

EXPERIMENTAL

These experiments were conducted in two separate ultra-high vacuum (UHV) instruments. The photoemission studies were carried out on beamline UV-8b at the National Synchrotron Light Source. Well oriented, planar Si(100) and Si(111) crystals were cleaned by annealing to ~950°C in vacuum. Surface cleanliness was determined by measuring the characteristic surface states for Si(111) 7x7 and Si(100) 2x1 at the Si(2p) core level and in the valence region.[5] The surfaces were exposed to HF in a separate reaction chamber (base pressure = 1 x 10^{-9} torr) and then transferred under UHV to the spectrometer chamber (base pressure = 5 x 10^{-11} torr). Si 2p core level and valence region photoemission were measured using 130 eV and 90 eV photons, respectively. The typical overall energy resolution of the incident radiation and the

Mat. Res. Soc. Symp. Proc. Vol. 131. ©1989 Materials Research Society

ellipsoidal mirror analyzer was 0.1 eV.[6]

The ESDIAD experiments were performed in a UHV chamber equipped with ESDIAD/LEED optics, a CMA for Auger electron spectroscopy, and a quadrupole mass spectrometer. The ESDIAD optics and the video data acquisition system are described in detail elsewhere.[7] The planar Si(111) samples were cleaned by heating in vacuum and checked for cleanliness with Auger and the appearance of a sharp 7x7 LEED pattern. For the ESDIAD studies of HF on the Si(100) surface, a vicinal surface (cut 5° towards the (011) direction) was used so that only a single majority domain of the 2x1 reconstruction was present. This stepped surface was cleaned by argon ion bombardment, followed by annealing in vacuum and checked for cleanliness with Auger and the appearance of a sharp, single domain 2x1 LEED pattern.

RESULTS AND DISCUSSION

Representative Si $2p_{3/2}$ photoemission spectra for a Si(100) 2x1 surface before and after exposure to a saturation dose of HF are shown in figure 1. The small shoulder to lower binding energy seen in the clean Si spectrum is indicative of the surface state present on a well ordered Si(100) 2x1 surface.[5] Adsorption of HF results in the disappearance of this state and the growth of peaks at higher binding energies. The new peaks are indicative of the formation of silicon fluorides, thus demonstrating that the adsorption of HF on silicon surfaces is dissociative.

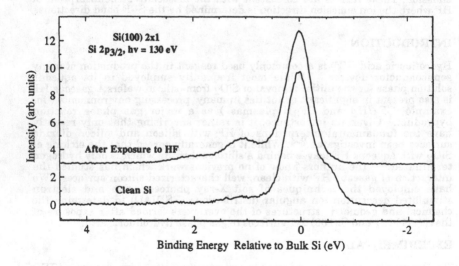

Figure 1. The Si 2p spectra of a) a Si(100) surface exposed HF and b.) a clean Si(100) surface.

Earlier photoemission studies of the interaction of XeF_2, a fluorinating agent, with silicon surfaces showed that the various silicon fluorides (SiF_x, x = 1-3) yield characteristic Si 2p binding energy shifts of ~1 eV for each fluorine atom bound to a silicon atom.[8] It can be seen in figure 1 b. that HF adsorption forms predominantly Si-F species with only small amounts of the higher fluorides. Similar Si 2p spectra were obtained after exposing the Si(111) 7x7 surface to HF. As with the (100) surface, only the formation of the monofluoride species was observed with only small amounts of higher fluorides present. It should be noted that the adsorption of HF on both surfaces saturates as demonstrated by the fact the intensity of the shifted Si 2p features does not increase with respect to the bulk peak even with exposures of up to 10000L of HF.

In order to determine the fluorine coverage on these surfaces we have measured the intensities of the F 2s pseudo-core level in the valence region at saturation coverage. Figure 2 shows the valence survey spectra for a saturation coverage of HF on Si(111) and for comparison, similar survey spectra for a Si(111) surface first exposed to 50L of XeF_2 and then annealed to 600K. Earlier work has shown that a 50L XeF_2 exposure yields a Si(111) surface with slightly more than 1 monolayer of fluorine in the form of SiF and SiF_3 species.[9]

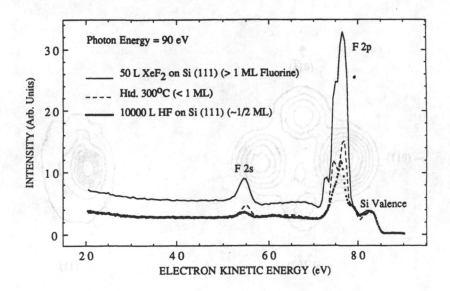

Figure 2. Comparison of the valence band photoemission for a Si(111) surface exposed to 50L XeF_2 at room temperature, the same sample annealed at 600K, and exposed to a saturation coverage of HF.

Annealing this surface to ~300°C removes the higher fluorides and results in a surface terminated solely by SiF species with an overall fluorine coverage of slightly less than one monolayer. As can be seen in figure 2, the F 2s intensity from a HF saturated surface is ~1/2 that of annealed XeF_2 exposed surface, thus indicating that the fluorine coverage saturates at ~1/2 monolayer. Since the fluorine does not cover the entire surface, it is proposed that the hydrogen occupies the remaining surface sites.

Some of the results of our ESDIAD studies of HF adsorption on the stepped Si(100) surface have been published previously,[10] where more details of the experimental results may be found. The ESDIAD patterns from saturation coverages of HF on the stepped Si(100) and the planar Si(111) surfaces are shown in figure 3. For both surfaces, the primary desorbing ion is F^+, with only a minor contribution from H^+. The minority of H^+ is based on experiments where a QMS was used to monitor the masses of the desorbing ions and from separate experiments where the clean surfaces were exposed to H atoms from a hot filament source and little or no H^+ desorption was seen, suggesting that the cross-section for H^+ desorption from Si is very small.

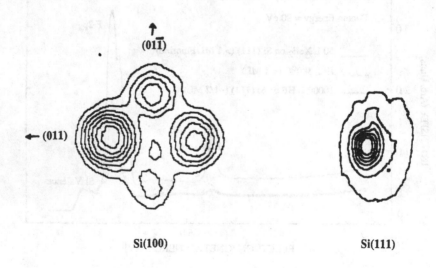

$(01\bar{1})$

← (011)

Si(100)   Si(111)

Figure 3. Contour plots of the ESDIAD pattern of F^+ desorption from saturation coverages of HF on a) a stepped Si(100) surface and b.) a planar Si(111) surface. The electron impact energy is ~180 eV, including a crystal bias of +80 eV.

For both surfaces, coverages of HF up to saturation do not qualitatively change the LEED, suggesting that the (100)2x1 and (111)7x7 reconstructions are *not* significantly perturbed upon chemisorption. Both ESDIAD patterns are stable upon heating to 600K; upon further heating, the intensities of the images decrease and eventually disappear. This is consistent with the dissociative chemisorption of HF to form silicon fluorides.

For the stepped (100) surface (figure 3a), the most intense emission of F^+ is along the (011) azimuths and therefore along the Si dimer direction. The weaker F^+ emission along the (011) direction is mostly due to the residual minority domains. The polar angle for the F^+ desorption is $29^\circ \pm 3^\circ$. Since the polar ion emission angle is not too far from the surface normal and the F^+ kinetic energy is sufficiently high (~2 eV),[9] distortions of the polar ion emission angle by image charge and reneutralization effects should be small.[11] This means that the polar Si-F bond angle is approximately equal to the polar angle of 29° along which the ions desorb.

In comparison, the ESDIAD pattern of F^+ from the Si(111) surface (figure 3b) is markedly different. In this case the predominant F^+ emission is along the surface normal with only small amounts of off-normal desorption, and no observable azimuthal ordering. This result indicates that the majority of of the Si-F bonds are oriented along the surface normal.

The photoemission and ESDIAD/LEED results are consistent with a model for a dissociative chemisorption of HF at the available dangling bonds on the silicon surfaces. On both surfaces, there is never more than one dangling bond per surface atom. The retention of the higher order reconstruction spots in LEED and the presence of mainly monofluoride species seen in the photoemission results implies that little or no Si-Si bond breaking has occurred. The ESDIAD results are consistent with SiF bond directions which are similar to the dangling bond directions of the clean silicon surfaces: directed off-normal and ordered azimuthally on the Si(100) 2x1 surface and directed along the surface normal for the Si(111) 7x7 surface.

CONCLUSIONS

We have shown that the chemisorption of HF on Si(100) and Si(111) surfaces is dissociative, producing Si-F and Si-H bonds. It is proposed that the HF reacts only with the available dangling bonds on the reconstructed silicon surfaces and does not break the Si-Si bonds to form substantial amounts of the higher fluorides. The ESDIAD results show that on the (100) surface, the majority of the Si-F bonds are oriented 29° away from the surface normal, while on the (111) surface most of the Si-F are pointing along the surface normal.

ACKNOWLEDGEMENTS

SAJ and ALJ acknowledge the NBS/NRC Research Associateship Program for postdoctoral support. The authors wish to acknowledge Mary Walczak(Iowa Satate) for her assistance in the intital phases of this work, Yves Chabal(Bell Labs) for the donation of the stepped Si(100) sample, and the staff of the IBM U8 beamline. This work was funded in part by the Office of Basic Energy Sciences of the U.S. Department of Energy. This work was performed in part at the National Synchrotron Light Source, which is supported by the Department of Energy, Department of Material Sciences.

REFERENCES

1. E. Kay, J. Coburn, and A. Dilks, Topics Curr. Chem. **94**, 1 (1980).

2. V.A. Burrows, Y.J. Chabal, G.S. Higashi, K. Raghvarachi, and S.B.

Christman, Appl. Phys. Lett. **53**, 998 (1988).

3. J.A. Yarmoff, F.J. Himpsel, and B. Meyerson, (unpublished results).

4. B.R. Weinberger, G.G. Peterson, T.C. Eschrich, and H.A. Krasinski, J. Appl. Phys. **60**, 3232 (1986).

5. F.J. Himpsel, D.E. Eastman, P. Heimann,B. Reihl, C.W. White, and D.M. Zehner, Phys. Rev. B **24**, 1120 (1981).

6. D.E. Eastman, J.J. Donelon, N.C. Hein, and F.J. Himpsel, Nucl. Instrum. Methods **172**, 327 (1980).

7. A.L. Johnson, R. Stockbauer, D. Barak, and T.E. Madey, in M. Knotek and R. Stulen, eds. Proceedings of DIET III, Springer Series in Surface Science 13 (1988) p. 130.

8. F.R. McFeely, J.F. Morar, N.D. Shinn, G. Landgren, and F.J. Himpsel, Phys. Rev. B **30**, 764 (1984).

9. J.A. Yarmoff, A. Taleb-Ibrahimi, F.R. McFeely, and Ph. Avouris, Phys. Rev. Lett. **60**, 960 (1988).

10. A.L. Johnson, M.M. Walczak,and T.E. Madey, Langmuir **4**, 277 (1988).

11. Z. Miskovic, J. Vukanic, and T.E. Madey, Surf. Sci. **141**, 285 (1984); **169**, 405 (1986).

INFRARED SPECTROSCOPY OF Si(111) AND Si(100) SURFACES AFTER HF TREATMENT: HYDROGEN TERMINATION AND SURFACE MORPHOLOGY

Y. J. CHABAL, G. S. HIGASHI AND K. RAGHAVACHARI
AT&T Bell Laboratories, Murray Hill, NJ 07974

ABSTRACT

The methodology for extracting structural information from surface infrared spectra is exemplified by considering the silicon-hydrogen stretching modes of flat and vicinal Si(111) and Si(100) surfaces obtained after HF treatment.

INTRODUCTION

Chemical preparation of semiconductor surfaces is a necessary first step in device technology. In fact the influence of chemical treatments on the *electronic* properties of semiconductor surfaces was pointed out as early as 1958.[1] Recently, surface recombination measurements have revealed that chemically oxidized and subsequently HF-stripped Si surfaces (RCA cleaning technique[2]) are nearly perfect electronic interfaces[3], terminated mostly by atomic hydrogen.[3-6]

Despite the importance of understanding the microscopic nature of these chemically prepared surfaces, little is known because few probes can investigate the surfaces in situ (e.g., in the solution and after treatment at atmosphere pressure). Partial information (e.g. after introducing the sample into a UHV chamber) may not describe the original state well because contamination may occur during the pumping process. As a result, there is a need for several experimental probes. Among them, optical techniques have the advantage of being non-destructive and compatible with investigations at atmospheric pressure. The purpose of this paper is to outline the methodology involved with one of such optical spectroscopies: surface infrared spectroscopy. The details of the analysis have already been reported in refs. (5) and (6).

EXPERIMENTAL

An important aspect of infrared spectroscopy of semiconductor surfaces is the use of a polarized multiple internal reflection (MIR) geometry.[7] The *internal* reflection provides sensitivity to vibrational components *parallel* to the surface, multiple reflections enhance the overall sensitivity and polarized radiation separates all components of the spectra.[8]

To illustrate this point, we consider the geometry used in our experiment: 3.8 cm $\times$ 1.9 cm $\times$ 0.05 cm silicon plates with 45° bevels for input and output coupling. The infrared radiation normally incident on the input bevel, is reflected a total of 75 times on both sides. For this 45° *internal* incidence angle, the electric field intensities on the vacuum side of the vacuum/Si interface (see lower right part of Fig. 1) are:[7,8] $|E_z/E°|^2 = 2.4$ and $|E_x/E°|^2 = 2.0$ for p-polarized radiation (electric field in the plane of incidence) and $|E_y/E°|^2 = 2.2$ for s-polarized radiation (electric field orthogonal to the plane of incidence). Thus, all components (x, y and z) of the adsorbate vibrations are probed with equal sensitivity within 20%. This is not the case for *external* reflection (or for dipole EELS) because the field intensities parallel to the surface are comparatively much smaller, resulting in one order of magnitude higher sensitivity to the perpendicular component. That is, both external reflection IR spectra and EEL spectra will be dominated by the components of vibrations *perpendicular* to the surface. Note that non-specular EELS (impact scattering regime) will not be more sensitive to *dipole* modes oriented parallel to the surface since impact scattering is much weaker (over an order of magnitude) than dipole scattering.

The sensitivity enhancement of MIR over a single external reflection can be calculated by considering the focussed spot size, the silicon reflectivity and the number of internal reflections (N = 75 in our case). For 45° incidence and 2.5 mm diameter spot, we find[9]:

$$\Delta R|_{MIR} \approx 18\ \Delta R|_{1\ int.refl.} \approx 50\ \text{to}\ 500\ \Delta R|_{1\ ext.refl.}$$

depending on whether a parallel or normal component is considered. Here $\Delta R|_{MIR} \equiv I_0 - I$ where I_0 and I are the total throughput intensities of the reference and sample MIR Si plates, respectively. Notice that we are comparing ΔR (and not $\Delta R/R$) because the experiments are usually detector noise limited and therefore the S/N is proportional to the overall magnitude of R, the total signal intensity. The use of MIR has made it possible to observe vibrational absorptions as small as $\Delta R/R = 10^{-6}$ per reflection, corresponding to 1% of a H monolayer on Si.[5]

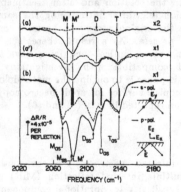

Fig. 1. Polarized IR spectra of silicon-hydrogen stretching vibrations for three isotopic concentrations: (a) 10%H:90%D, (a') 25%H: 75%D, and (b) 100%H. The dashed spectra correspond to s-polarization (electric field parallel to the surface only) and the solid spectra to p-polarization (both // and ⊥ components). The arrows labelled M, M', D and T indicate the experimentally measured lines in the "isolated" spectra. Ab-initio cluster calculations indicate that the isolated frequency of the relaxed dihydride, D, is 24 cm^{-1} higher than M, and T is 25 cm^{-1} higher than D. The vertical bars in bold represent the calculated coupled mode splittings from the measured isolated frequencies (M = 2077 cm^{-1}, D = 2111 cm^{-1}, and T = 2137 cm^{-1}).

The motivation for using stepped samples [cut vicinal to the (111) and (100) planes] is to study the influence of steps on the morphology of the HF-prepared Si(111) and Si(100) surfaces. If the chemically treated surfaces display a regular array of double layer steps, i.e. a high step density, then 9° miscut samples should give a detectable contribution to the spectra arising from H at steps and a high enough step density to relieve possible strain on the adjacent terraces.

METHODOLOGY

Data acquisition

The collection of reference spectra is important for weak spectra because the ultimate signal to noise ratio depends on how well the spectra of interest can be ratioed to the reference spectra. Ideally, the sample is not moved between runs to maintain the alignment. However, for HF-prepared samples, it is not possible to chemically etch the samples in the spectrometer chamber. Chemically prepared samples are usually mounted in air and installed in the chamber within a few minutes, and polarized infrared spectra (I) are recorded. The samples are then removed, chemically oxidized and returned to the chamber to obtain reference spectra (I_o).

The transmittance spectra, I/I_o, yield the absorptions of the HF-prepared surfaces *relative* to that of the oxidized surfaces. Although the oxidized surfaces have no vibrational features in the Si-D and Si-H stretch frequency regions, they display some OH and CH stretch absorptions in the 3200-3600 cm^{-1} and 2800-3000 cm^{-1} regions due to water and hydrocarbon contamination, respectively. These absorption bands can be quantified by flashing the surfaces in vacuum an recording a different set of reference spectra.

Isotopic Mixture Experiments

When complex spectra of polyatomic molecules are considered with several normal modes in some small frequency region, isotopic mixture experiments are particularly useful to simplify the spectra. For instance, "isolated" spectra are obtained by mixing a small amount of one isotope (5-10%) with another and measuring the spectrum of the minority species. In such a spectrum, each chemical species (eg, Si-H, SiHD, $SiHD_2$, ...) is characterized by only 1 stretch vibration as illustrated in Fig. 1(a) and 1(a'). The relative spectral positions of these "isolated" bands is the first piece of information necessary for proper assignment.

Next, the "pure" spectra associated with both isotopes (eg. H and D) are measured and compared with the isolated spectra. In the pure spectra, the interactions between similar atomic motions (eg. stretching) are turned on resulting in several absorption lines associated with each chemical species (e.g. 3 lines for SiH_3, 2 for SiH_2 and 1 for SiH).

Assignment of Modes

In the case of atomic adsorption (eg. H) on semiconductor surfaces, the assignment relies on ab-initio molecular orbital calculations that are reliable in the prediction of vibrational parameters.[8-11] Such calculations require model clusters (e.g. $H_3Si-SiH_2-SiH_3$) that are good representations of the surface structure, including reconstruction.[10] Often, the frequencies associated with the normal modes of a given species are very characteristic, making it possible to give an unambiguous assignment irrespective of the orientation of the species. for example, the dihydride exhibits an asymmetric stretch frequency 4 cm^{-1} *higher* than its isolated mode frequency and symmetric stretch frequency 5 cm^{-1} *lower* than its isolated mode frequency, i.e. $\nu_{as} - \nu_{ss} = 9$ cm^{-1}. In comparison, dideuteride is characterized by $\nu_{as} - \nu_{ss} = 22$ cm^{-1}. This large increase in splitting is the best evidence for the presence of dihydride.[10,5]

Polarization Considerations

Once the mode assignment is performed without ambiguity, then the relative absorption strengths of the two different polarizations (s and p-pol) can be used to extract the orientation of each species. The results of such considerations are illustrated in Fig. 2, where all the possible hydride species on the (111) surface are shown.

The trihydride (T), for example, is oriented normal to the surface (termination of (111) plane) because its asymmetric stretch modes [at 2139 cm^{-1} in Fig. 1b] are excited only by $E_{//}$. Explicitly, the intensity in of the s-polarized spectrum in Fig. 1b is ~ 1.1 that of the p-polarized spectrum, indicating that there is no absorption associated with $E_\perp$ at that frequency.

The two dihydride absorptions (as and ss), on the other hand, can be observed in both polarizations (Fig. 1b). Instead of a ss normal to the surface (only detectable with p-polarization) and an as parallel to the surface (both in s- and p-polarizations),[10] both modes are present in s-pol and the ss is only slightly stronger than the as in p-pol., indicating a substantial tilt of the dihydride axis with respect to the surface normal. The configuration marked D in Fig. 2 is prevalent over that marked D' because the as is as intense in p-pol. as it is in s-pol (ie. mostly parallel to the surfaces).

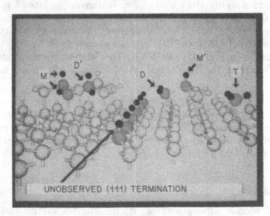

Fig. 2. Schematic representation of possible surface structures and hydride terminations, some of which accounting for the observed silicon-hydrogen vibrations.

Similar considerations are applied to the spectra associated with monohydrides to complete the schematic picture given in Fig. 2. Note that the presence of simple monohydride termination of the (111) plane is ruled out on the basis of the frequency associated with the Si-H mode normal to the surface.[5]

Stepped Surfaces

The study of stepped surfaces provides additional information both on the manner in which the surface accommodates the miscut, and the influence of steps on the surface microscopic arrangement. For a 9° stepped (111) surface, the miscut results in the formation of a regular array of double layer steps[6] after HF treatments. As a result, the adsorption of H at steps can be studied. In addition, the H spectra of hydrogen on the terraces are modified indicating that the steps influence the terrace atomic arrangement.[6]

CONCLUSIONS

The overall approach to the IR study of chemically modified semiconductor surfaces has been outlined. In the case of HF-prepared silicon surfaces, the spectroscopy has shown that H terminates the surfaces that tend to be microscopically rough. It is particularly interesting that H terminates the surface, rather than F. Simple thermodynamic arguments would predict that Si-F would be the stable phase. Furthermore, very recent work by Joyce and coworkers[12] indicates that gaseous HF dissociates on both Si(100) and Si(111), leaving a half monolayer of fluorine on the surface.

The silicon-fluorine vibrations cannot be studied using MIR because the silicon bulk absorption is too strong in the frequency region of interest. On the other hand, the Si-F stretch mode may be strong enough to be detected by a single internal reflection using a thin sample. Clearly, a lot more work, involving several complementary techniques, is required to further our understanding of chemically modified semiconductor surfaces.

ACKNOWLEDGMENTS

We are grateful to V. A. Burrows for her initial contribution to this work and S. B. Christman and E. E. Chaban for technical support.

REFERENCES

(1)   T. M. Buck and F. S. McKim, J. Electrochem. Soc. *105*, 709 (1958).

(2)   W. Kern, Semicond. Int. *94* (April 84).

(3)   E. Yablonovitch, D. L. Allara, C. C. Chang, T. Gmitter, and T. B. Bright, Phys. Rev. Lett. *57*, 249 (1986).

(4)   M. Grundner and R. Schultz, in *Deposition and Growth Limits for Microelectronics*, edited by G. W. Rubloff, AIP Conf. Proc. No 167 (American Institute of Physics, New York, 1988) pp 329-337; Am. Vac. Soc. Ser. No4 (AVS, Anaheim, 1988).

(5)   V. A. Burrows, Y. J. Chabal, G. S. Higashi, K. Raghavachari, and S. B. Christman, Appl. Phys. Lett. *53*, 998 (1988).

(6)   Y. J. Chabal, G. S. Higashi, K. Raghavachari, and V. A. Burrows, J. Vac. Sci. Technol (May/June 1989).

(7)   N. J. Harrick, *Internal Reflection Spectroscopy* (Wiley, N.Y., 1967) (Second printing: Harrick, Ossining, 1979).

(8)   Y. J. Chabal, Surf. Sci. Repts *8*, 211 (1988).

(9)   ibid, p. 265 and Y. J. Chabal in *Chemistry and Physics of Solid Surfaces*, edited by R. Vanselow and R. F. Howe, Springer Series in Surfaces Sciences, vol. 10 (Springer, Berlin, 1988) pp. 109ff.

(10)   Y. J. Chabal and K. Raghavachari, Phys. Rev. Lett. *53*, 282 (1984); ibid *54*, 1055 (1985).

(11)   K. Raghavachari, J. Chem. Phys. *81*, 2717 (1984).

(12)   S. A. Joyce, J. A. Yarmoff, A. L. Johnson, and T. E. Madey, MRS Proceedings (This volume).

ADSORPTION OF SILICON TETRACHLORIDE ON Si(111) 7×7

P. GUPTA, P.A. COON, B.G. KOEHLER AND S.M. GEORGE
Dept. of Chemistry, Stanford University, Stanford, Calif. 94305

Abstract

The kinetics of $SiCl_4$ adsorption on Si(111) 7×7 were studied using laser induced thermal desorption (LITD) and temperature programmed desorption (TPD) techniques. The initial reactive sticking coefficient of $SiCl_4$ on Si(111) 7×7 was observed to decrease with increasing surface temperature. This decrease was consistent with a precursor-mediated adsorption model. Both LITD and TPD experiments monitored $SiCl_2$ as the main desorption product. These results suggest that $SiCl_2$ may be the stable chlorine species on the Si(111) 7×7 surface.

Introduction

The adsorption of chlorosilane molecules on silicon surfaces plays an essential role in silicon epitaxial growth. The kinetics of $SiCl_4$ adsorption are critical to understanding the $SiCl_4 + 2H_2 \longrightarrow Si + 4HCl$ silicon CVD growth reaction. Unfortunately, the adsorption kinetics and the reaction products of simple chlorosilane molecules such as $SiCl_4$, $SiCl_2H_2$ and Cl_2 on silicon surfaces have not been explored.

In this study, laser induced desorption (LITD) and temperature programmed (TPD) techniques were used to probe the Si(111) 7×7 surface following $SiCl_4$ adsorption. The LITD experiments measured the kinetics of the initial reaction of $SiCl_4$ with Si(111) 7×7. These studies allowed the initial reactive sticking coefficient of $SiCl_4$ on Si(111) 7×7 to be determined as a function of surface temperature.

The experimental apparatus for LITD experiments has been described in detail previously (1). Briefly, the ultrahigh vacuum (UHV) chamber used for the LITD studies was pumped by a 300 l/sec ion pump and a titanium sublimation pump. These pumps maintained background pressures of approximately 4.0×10^{-10} Torr during these experiments. This chamber was also equipped with a low energy electron diffraction (LEED) spectrometer, a cylindrical mirror analyzer for Auger electron spectroscopy (AES) and a quadrupole mass spectrometer for LITD and TPD studies.

The laser induced thermal desorption (LITD) experiments utilized a TEM-00 Q-switched Ruby laser with a temporal width (FWHM) of 80-100 nsec and a Gaussian spatial profile (1). The laser pulses had energies of 5.0 mJ before entering the chamber. The pulses were focused on the silicon sample by a 1.0 m lens. The focused laser beam had a Gaussian beam profile at the focus of the lens with a width of 260 μm (FWHM). The laser pulses were directed onto the silicon sample at an angle of approximately 54° with respect to the surface normal. Under such conditions, the desorption spot on the crystal was elliptical. The spot size using the auto-correlation method (2) was roughly 350 μm in diameter along the major axis and 280 μm in diameter along the minor axis.

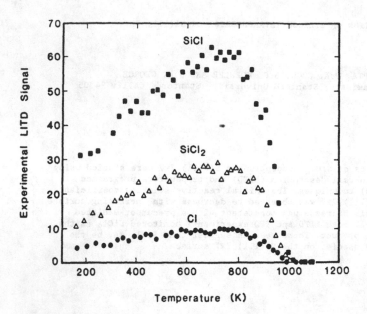

Fig.1 Temperature dependence of Cl, SiCl, $SiCl_2$ LITD signals following a saturation $SiCl_4$ dose.

Laser-Induced Thermal Desorption (LITD) Studies

$SiCl_2$, SiCl and Cl were the only species observed by the mass spectrometer from the LITD yield following a saturation dose of $SiCl_4$ on Si(111) 7×7 at 160 K. No $SiCl_3$ or $SiCl_4$ LITD signals were observed. Fig. 1 shows the temperature dependence of these LITD signals. For this temperature programmed LITD experiment, the silicon surface was heated linearly at 2 K/sec.

Fig. 1 shows a rise in the $SiCl_2$, SiCl and Cl LITD signals with surface temperature. This behavior is caused by the temperature dependence of the LITD signals and should not be confused with an actual increase in the surface coverage. When the surface is at a higher temperature, the Gaussian laser pulses can raise the temperature of a larger area of the surface above the threshold desorption temperature (3). Consequently, the LITD desorption spot size increases with surface temperature and the LITD desorption signal increases. The decrease of the $SiCl_2$, SiCl and Cl LITD signals between 900-950 K were consistent with the thermal desorption of $SiCl_2$, SiCl and Cl, respectively.

The LITD signals may result from the cracking of one silicon chloride species in the electron impact ionizer of the mass spectrometer. Consequently, the $SiCl_2$, SiCl and the Cl signals were scaled to determine if all the signals were proportional to each other. This scaling revealed that all the LITD signals could be fit to one another and suggested that the $SiCl_2$, SiCl and Cl LITD signals all emanate from $SiCl_2$.

The absence of $SiCl_3$ and $SiCl_4$ LITD signals argues against the presence of $SiCl_3$ or $SiCl_4$ species on the Si(111) 7×7 surface. The $SiCl_2$

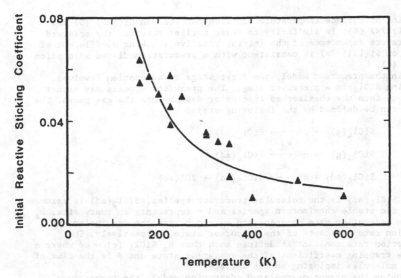

Fig. 2 Initial reactive sticking coefficient for $SiCl_4$ on Si(111) 7×7.

LITD signal and the proportionality of the $SiCl_2$, SiCl and Cl LITD signals suggest that the stable chemisorbed chlorine species is $SiCl_2$. $SiCl_2$ has been postulated earlier to be the stable chemisorbed species after Cl_2 adsorption on Si(111) 7×7 (4). However, spectroscopic evidence will be required to determine if $SiCl_2$ is the stable chemisorbed chlorine species on the Si(111) 7×7 surface.

Initial Reactive Sticking Coefficient

The kinetics of $SiCl_4$ adsorption on Si(111) 7×7 were studied as a function of surface temperature using LITD techniques. In these experiments, the clean Si(111) 7×7 surface was maintained at a constant surface temperature. The pressure in the UHV chamber was then held at a constant pressure of $SiCl_4$. During the adsorption of $SiCl_4$ on the Si(111) 7×7 surface, the surface coverage was measured as a function of time by monitoring the LITD signals. By measuring the SiCl LITD signals from different positions on the surface at different times during $SiCl_4$ exposure, the surface coverage could be determined in real time.

The LITD adsorption measurements were performed with a $SiCl_4$ pressure of 1.0×10^{-7} Torr for various surface temperatures. The initial adsorption rate was proportional to the initial slope of the SiCl LITD signal versus time. The initial adsorption rate was observed to decrease as a function of surface temperature. The initial reactive sticking coefficient, S_o, was obtained by dividing the initial adsorption rate by the $SiCl_4$ flux. The results for the initial reactive sticking coefficient of $SiCl_4$ on Si(111) 7×7 are shown in Fig. 2.

Initial sticking coefficients that decrease as a function of temperature are a fairly general observation. For example, a similar decrease in the initial reactive sticking coefficient of O_2 as a function of surface temperature has been reported on Si(100) (5) and Si(111) 7×7 (6). In similarity to these earlier results, the observed temperature dependence of the initial reactive sticking coefficient of $SiCl_4$ on Si(111) 7×7 is consistent with a precursor-mediated adsorption model (7).

In the precursor model, the first stage of adsorption involves trapping $SiCl_4$ in a precursor stage. The precursor species may either react to form the chemisorbed species or desorb into the gas phase. The model can be defined by the following steps:

$$SiCl_4(g) \; + \; n* \; \xrightarrow{\; k_a \;} \; SiCl_4(ad)$$

$$SiCl_4(g) \; + \; n* \; \xleftarrow{\; k_d \;} \; SiCl_4(ad)$$

$$SiCl_4(ad) \; + \; m* \; \xrightarrow{\; k_r \;} \; SiCl_2(ad) \; + \; 2Cl(ad)$$

where $SiCl_4(ad)$ is the molecular precursor species, $SiCl_2(ad)$ is assumed to be the stable chemisorbed species and * represents an empty site. k_a is the adsorption rate constant and k_d and k_r are the desorption and reaction rate constants of the precursor state, respectively. The adsorption rate constant is defined such that $k_a SiCl_4$ (g) ~ $\alpha\phi$ where α is the trapping coefficient of the precursor state and ϕ is the flux of $SiCl_4$ molecules impinging on the surface.

Using the precursor-mediated adsorption model, the temperature dependent reactive sticking coefficient can be fit using the ratio of the preexponential factors (k_d/k_r) and the difference in the activation barriers E_d-E_r between the desorption and reaction pathways. With a value of α ~ 0.13, k_d/k_r and E_d-E_r were determined to be approximately 20 and 1 Kcal/mole, respectively. The solid line in Fig. 2 shows the initial $SiCl_4$ sticking coefficients calculated according to the precursor-mediated adsorption model using these parameters.

Temperature Programmed Desorption Studies

Temperature programmed desorption (TPD) spectra obtained after $SiCl_4$ adsorption on Si(111) 7×7 were performed using a linear heating rate of 9 K/sec. Before the TPD experiments, the crystal was rotated to a position approximately 5 cm from the ionizer of the quadrupole mass spectrometer. The only species present in the TPD spectra following $SiCl_4$ adsorption from Si(111) 7×7 were $SiCl_2$, SiCl and Cl. These species were detected at a surface temperature of approximately 900-950 K.

The TPD spectra were used to study the adsorption of $SiCl_4$ on Si(111) 7×7 at higher surface coverages. A glass capillary array doser was utilized for these $SiCl_4$ exposures. The exposures were performed at a variety of surface temperatures from 160 to 200 K. The TPD spectra were recorded and the areas under the TPD spectra were measured as a function of exposure. These TPD measurements of $SiCl_4$ uptake utilizing the $SiCl_2$ TPD signals are shown in Fig. 3.

The initial adsorption observed by the TPD measurements shown in Fig. 3 is similar to the initial reactive sticking coefficient determined by the LITD measurements. The slope of the $SiCl_4$ uptake curve in the limit of low coverage defines the initial sticking coefficient. Notice that the slope of the uptake curves decreases with increasing surface temperature.

At higher coverages, the adsorption rate decreases dramatically. At larger exposures, the TPD signals saturate at the same level for the three surface temperatures. This surface coverage at saturation defines θ_s, the saturation coverage on Si(111) 7×7. The existence of this saturation coverage argues that $SiCl_4$ can adsorb onto Si(111) 7×7 until all the available surface dangling bonds are occupied.

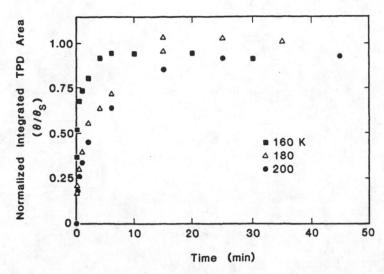

Fig. 3 Initial adsorption of $SiCl_4$ on Si(111) 7×7 measured by integrating the area under the $SiCl_2$ TPD spectra.

Acknowledgments

This work was supported by the Office of Naval Research under contract N00014-86-K-545. Some of the equipment utilized in this work was provided by the NSF-MRL program through the Center for Materials Research at Stanford University. BGK gratefully acknowledges the National Science Foundation for a graduate fellowship.

References

1. B.G. Koehler, C.H. Mak, D.A Arthur, P.A. Coon and S.M. George, J. Chem. Phys. <u>89</u>, 1709 (1988).
2. S.M. George, A.M. Desantolo and R.B. Hall Surf. Sci., <u>159</u>, L425 (1985).
3. J.L. Brand and S.M.George, Surf. Sci. <u>167</u>, 341 (1986).
4. J.V. Florio and W.D. Robertson, Surf. Sci. <u>18</u>, 398 (1969).
5. M.P. d'Evelyn, M.M. Nelson and T. Engel, Surf. Sci. <u>186</u>, 75 (1986).
6. S.M. George, P.Gupta, C.H. Mak, P.A. Coon, (this journal).
7. W.H. Weinberg in <u>Kinetics of Interface Reactions,</u> ed. by M. Grunze and H. Kreuzer (Springer Verlag NY 1987) p 94.

SURFACE CONDITION IN THE PLASMA-CVD OF a-Si:H,F FROM SiF$_4$ AND H$_2$

A. Maruyama, D.S. Shen, V. Chu, J.Z. Liu, J. Jaroker, I. Campbell, P.M. Fauchet
and S. Wagner
Department of Electrical Engineering, Princeton University, Princeton, N.J. 08544

ABSTRACT

We present a detailed study of the growth of a-Si:H,F from SiF$_4$ and H$_2$. The growth surface appears to have a high density of surface states. These surface states can be thermally relaxed by keeping the films at growth temperature after the termination of growth, suggesting that the states were created during film growth. When frozen in, the surface state density is found to depend on the conditions during film growth. The density is related to the sharpness of the valence band tail as measured by the Urbach Energy. We believe that a reaction on the growth surface resulting in fluorine elimination creates these surface states and also affects the formation of the Si-network.

INTRODUCTION

The deposition of amorphous hydrogenated and fluorinated silicon (a-Si:H,F) from SiF$_4$ and H$_2$ was first reported by Ovshinsky and Madan [1]. Since then, fluorinated source gases have been employed in the growth of high quality a-Si,Ge:H,F alloys[2], especially with low optical gaps (down to 1.2 eV)[3], and of low-temperature epitaxial Si [4,5]. Although the growth mechanism from fluorides and hydrogen is not well understood, a reductive reaction between fluorinated gas and hydrogen must occur either in the gas phase or on the growth surface. This reaction should affect the formation of the three-dimensional network.

a-Si:H,F and a-Si,Ge:H,F contain only about 1at.% of fluorine [6]. This means that most of the fluorine atoms introduced with the source gas must be eliminated during film growth. No film grows when precursors (SiF$_n$), produced in a glow discharge, are mixed with molecular H$_2$ [4]. This suggests that the hydrogen radicals are needed to eliminate the fluorine atoms. Does this elimination reaction affect the structure of the films?

We have studied the growth of a-Si:H,F from SiF$_4$ and H$_2$ by varying the growth conditions to provide a wide range of film properties. The density of surface states N$_{ss}$ was found to be highly sensitive to the conditions of growth termination. N$_{ss}$ is correlated with the characteristic Urbach Energy, E$_u$, of the grown film. In this paper, we describe the identification of surface states, the relation between this density and the Urbach Energy, and the nature of the surface states.

EXPERIMENTAL PROCEDURES

Sample preparation

We used a film growth system employing a DC-excited glow discharge in a triode reactor, which has been described elsewhere [7]. The area of the circular electrodes is 120 cm^2, the distance between the cathode grid and the substrate is normally fixed at 1 cm. The deposition parameters were: gas ratio SiF$_4$/H$_2$ = 1:4 to 8:1, growth pressure P = 0.30 to 0.65 torr, a fixed total gas flow rate = 35 sccm, DC power = 34 W and substrate temprature T$_s$ = 250 C. The growth conditions for the best bulk properties are: SiF$_4$/H$_2$ = 2:1, P = 0.65 torr. The density of surface states was adjusted by the procedure for growth termination. A low density is observed by holding the film at growth temperature for 10 min. after the glow discharge has been turned off. Rapid cooling establishes a high density of surface states [8].

For a determination of the depth profile of the surface states we removed thin layers by RF plasma etching in a separate station [9]. The etching conditions were: gas flow rates CCl$_2$F$_2$/O$_2$

= 43/15 sccm, process pressure = 190 mtorr, forward power = 100 W, DC self bias = 90 V, substrate temperature = 18 C, resulting etching rate = 551 A/min.

Evaluation techniques

The film structure was investigated by Raman Scattering, X-ray diffraction and infrared absorption. There was no sign of micro-crystallization in any of the films discussed here. Optical properties were measured by visible/near-IR transmission spectroscopy, sub-band gap absorption with the constant photocurrent method (CPM) and photothermal deflection spectroscopy (PDS). The CPM spectra were evaluated for the characteristic Urbach Energy E_u. E_u is commonly taken as an indication of the Si-Si bond angle distortion, such that a high E_u reflects strong distortion. As described later, CPM and PDS were also used to determine the density of surface states. As for electronic transport evaluation, we measured dark and photoconductivity as functions of temperature. The surface and bulk compositions, including fluorine concentration, were determined by secondary ion mass spectroscopy (SIMS).

RESULTS AND DISCUSSION

By varying the ratio of SiF_4 and H_2 gas flow and growth pressure, the optoelectronic properties and composition of the grown films can be made to change over a wide range. Two properties, the dark conductivity and the sub-gap absorption measured by PDS, appeared to be highly sensitive to the growth conditions. These two properties are correlated. Fig.1 shows the dark conductivity as a function of the temperature of measurement for films with different intensities of sub-gap absorption, obtained from PDS. Two distinct branches can be identified. The high-temperature branch has a high activation energy of about 0.8eV, the low-temperature branch shows a low activation energy of about 0.15eV. The intensity of PDS sub-gap absorption increases with the dark conductivity in the low-temperature branch. The sub-gap absorption intensity measured by PDS reflects the sum of the contributions by surface and bulk states, while the absorption measured by CPM reflects only the bulk states [10]. Since the intensities of sub-gap absorption from CPM are essentially the same for all films shown in Fig.1, the dark conductivity in the low temperature region originates in surface states.

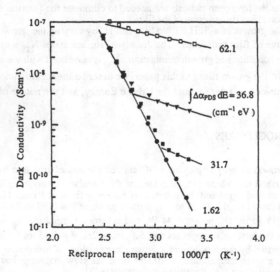

Fig.1 Dark conductivity as a function of temperature for four a-Si:H,F films with a range of integrated sub-gap absorption intensities from PDS.

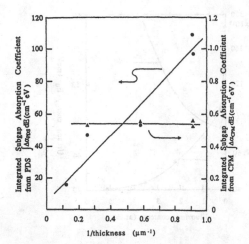

Fig.2 The integrated subgap absorption coefficient measured by PDS and by CPM versus reciprocal film thickness.

The existence of a surface layer with a high density of defect states is supported by the thickness dependence of the sub-gap absorption intensity in PDS [10]. In Fig.2, we show the PDS and CPM results of a series of samples with a range of thicknesses. The films shown in Fig.2 were grown under conditions that provide good bulk properties (low Urbach Energy and low defect density). Following growth, we applied the growth termination procedure which provided the highest density of surface states. While the integrated sub-gap absorption coefficient from CPM remains constant in Fig.2, the values from PDS decrease with increasing sample thickness. This result shows that the high density of states measured by PDS is located in a thin layer either on the top surface or the bottom interface. However, because the states can be established by appropriate termination of growth, and because of the result of surface etching described later, this thin layer must be located on the top surface.

The density of surface states N_{ss} can be determined from the difference of the integrated sub-gap absorption between PDS and CPM [10]. Since the conversion factors from integrated sub-gap absorption coefficient to number of defect states are 1.9×10^{16} for CPM and 7.9×10^{15} for PDS, respectively, N_{ss} is determined by the following equation.

$$N_{ss} \ (cm^{-2}) = \left(7.9 \times 10^{15} \int \Delta\alpha_{PDS} \ dE - 1.9 \times 10^{16} \int \Delta\alpha_{CPM} \ dE \right) \times film \ thickness$$

The N_{ss} and the Urbach Energy E_u determined from the CPM spectra correlate with the fluorine concentration in the films. In Fig.3, opposite tendencies with fluorine content are observed for E_u and N_{ss}. While N_{ss} increases with decreasing fluorine content, E_u is smallest at a fluorine content of close to zero. In a-Si:H,F, the Urbach Energy, E_u, from CPM indicates the sharpness of the valence band tail.

Growth termination procedures also affect the density of surface states. In Fig.4, we show the PDS spectra of two films, which were grown under the same deposition condition but different termination procedures. The difference in those two termination procedures are the length of the period for which the films were kept at the growth temperature after the glow discharge had been cut off. While film A67 was cooled down immediately after growth, A57 was kept at the growth temperature for 10 min before cool-down. A large difference in the PDS spectra is observed. After subtraction of the corresponding CPM intensities, these PDS spectra translate to densities of surface states of 1×10^{14} cm^{-2} for A67 and 4×10^{12} cm^{-2} for A57. Thus the 10 min. anneal just after growth reduces the density of surface states by more than one order of magnitude.

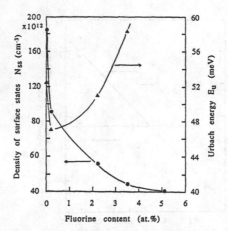

Fig.3 Surface state density N_{SS} and Urbach energy E_u plotted against the fluorine concentration in the a-Si:H,F films.

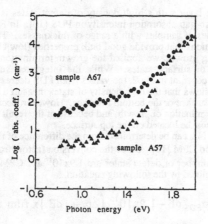

Fig.4 PDS spectra of two films with immediate (A67) and 10 min delayed (A57) cooling after growth termination. Cooling rate is 7 C/min for both samples.

Three films with different total density of surface states were etch-profiled in a reactive plasma of CCl_2F_2 and O_2. After each etching step, the density of surface states was determined from the difference between the PDS and CPM spectra. The density of surface states vs.etch depth is shown in Fig.5. Note that the etching procedure did not raise N_{SS} in the sample with a low initial N_{SS} (A57). This proves that the etching procedure did not introduce new states. The surface states were reduced at the etch depth of about 1000 A, though it is not clear why the density of surface states remains constant beyond 1000A. The depth profile of the density of surface states is shown in Fig.6.

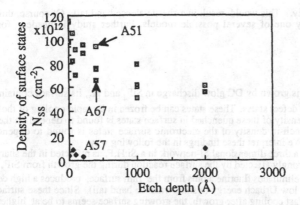

Fig.5 Nss vs. thickness of removed layer (etch depth) for two samples with high initial Nss (A51 and A67) and one sample with low initial Nss (A57). The values for A57 show that the plasma etching dose not introduce new surface states.

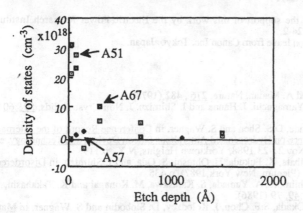

Fig.6 Depth profile of the surface states determined by PDS conbined with reactive plasma etching.

The results of Fig.3 to 5 provide a clue about the growth mechanism of a-Si:H,F. According to the result shown in Fig.3, we can assume that the fluorine elimination is accompanied by creation of surface states, and that this elimination also affects the formation of the Si-network, the network being manifest in the sharpness of the valence band tail (E_u). The results in Fig.4 suggest that the surface states are created during film growth and are thermally relaxed during some extended time period. With SIMS we could not find any evidence of surface contamination. Therefore, the surface states probably originate at Si dangling bonds, suggesting that fluorine elimination may take place by the following reaction.

$$Si\text{-}F + H^* \longrightarrow Si \text{ dangling bond} + HF$$
$$H^* : \text{hydrogen radical}$$

We surmise that the role of the fluorine atoms in the growth of a-Si:H,F is to provide the growth surface with a high density of Si dangling bonds. Because of this high density of dangling bonds, the surface layer may become so flexible that the rearrangement of the Si-Si network occurs easily. This would result in a sharp valence band tail. Of course, this growth mechanism is only one of several possible models. Further study is needed for a firm conclusion.

CONCLUSIONS

When a-Si:H,F is grown by DC glow discharge in SiF_4 and H_2, its surface contains a high density of electronic defect states. These states can be frozen in by rapid cooling of the samples after growth. The density of these quenched in surface states is found to depend on the growth condition. The quench-in density of the electronic surface states is found to depend on the growth conditions. We interpret these findings in the following way.

The formation of a three dimensional Si-network in a-Si:H,F, as reflected in the sharpness of the Urbach tail, is strongly affected by the surface reaction during the growth from SiF_4 and H_2. This reaction, which eliminates fluorine atoms from the film surface, produces a high density of surface states and a low Urbach energy (a sharp valence band tail). Since these surface states can be frozen-in by fast cooling after growth, the growing surface seems to be at higher energy than the bulk. The surface relaxes during a time period which is associated with the rearrangement of the Si-network.

ACKNOWLEDGEMENT

We appreciate the support of this work by the Electric Power Research Institute under Contract No. RP2824-2.
Akio Maruyama is on leave from Canon Inc. Tokyo, Japan.

REFERENCES

[1] S. Ovshinsky and A. Madan, Nature, 276 , 482 (1978)
[2] K. Nozawa, Y. Yamaguchi, J. Hanna and I. Shimizu, J. Non-Cryst. Solids, 59&60 533 (1983).
[3] V. Chu, J.P. Conde, D.S. Shen and S. Wagner, in Conference Record of the International Topical Conference on Hydrogenated Amorphous Silicon Devices and Technology, edited by Jerzy Kanicki, Nov.21-23 1988, Yorktown Heights, N.Y., p.119
[4] J. Hanna, N. Shibata, K. Fukuda, H. Ohtoshi, S. Oda, and I. Shimizu, in Disordered Semicondutors, (Plenum, New York,1987), p.435.
[5] S. Nishida, T. Shiimoto, A. Yamada, S. Karasawa, M. Konagai and K. Takahashi, Appl.Phys.Lett. 49, 79 (1986).
[6] R. Schwarz, S. Okada, S.F. Chou, J. Kolodzey, D. Slobodin and S. Wagner, in Materials Issues in Amorphous Silicon Technology, edited by D. Adler, A. Madan, and M.J. Thompson (Mater. Res. Soc. Proc. 70, Pittburgh, PA 1985), pp.283-288.
[7] D. Slobodin, S. Aljishi, R. Schwarz, and S. Wagner, in Materials Issues in Applications of Amorphous Silicon Technology, edited by D. Adler, A. Madan, and M.J. Thompson (Mater. Res. Soc. Proc. 48, Pittburgh, PA 1985), pp.153-158.
[8] A. Maruyama, D.S. Shen, V. Chu, J.Z. Liu, and S. Wagner, to be published
[9] P.H. Singer, Semiconductor International 11, 68 (1988).
[10] Z E. Smith, V. Chu, K. Shepard, S. Aljishi, D. Slobodin, J. Kolodzey, T.L. Chu, and S. Wagner, Appl. Phys. Lett. 50, 1521 (1987).

EFFECTS OF POTASSIUM ON THE ADSORPTION AND REACTIONS OF NITRIC OXIDE ON SILICON SURFACE

Z. C. YING AND W. HO
Laboratory of Atomic and Solid State Physics and Materials Science Center
Cornell University, Ithaca, New York, 14853

ABSTRACT

The adsorption, thermoreactions, and photoreactions of NO coadsorbed with K on Si(111)7×7 at 90 K have been studied and compared with the results obtained from the K-free surface. The experiments were performed under ultra-high vacuum conditions using high resolution electron energy loss spectroscopy, work function change measurements, and mass spectrometry. NO adsorbs both molecularly and dissociatively on the K-free surface. Two molecular N–O stretching modes are observed at 188 and 225 meV. The concentration of these NO molecules on the surface decreases as the K exposure increases and vanishes at high K exposures. A new N–O stretching mode, attributed to adsorption of NO molecules on K clusters, is observed at 157 meV. After thermal heating or photon irradiation, the surface is covered with atomic O and N. The surface is more oxidized in the presence of K. A steady decrease in the photodesorption cross section is observed as the K exposure increases and is attributed to K-induced band structure changes.

INTRODUCTION

The adsorption of alkali metals on semiconductor surfaces has been observed to cause a decrease in the work function and to induce changes in the chemical and physical properties of the surface. Recent studies have focussed on the determination of the electronic properties of K chains on Si(100) surface[1] and the promotion of chemical reactions, such as oxidation[2] and nitridation[3] of the silicon surface by submonolayer coverages of K. The presence of alkali metals is also expected to promote or suppress photochemical reactions on solid surfaces. Studies of photoreactions on surfaces preadsorbed with alkali metals are expected to lead to an increase in our general understanding of the fundamental mechanisms of photo-induced reactions on solid surfaces.[4,5] Furthermore, results from these studies can also have technological implications for microelectronic materials processing.

The adsorption, thermoreactions, and photoreactions of NO coadsorbed with K on Si(111)7×7 at 90 K have been studied in detail and the main results are reported in this paper. The present study extends our previous investigation of NO on the K-free Si(111)7×7 surface.[6,7]

EXPERIMENTAL ARRANGEMENT

The experiments were performed under ultra-high vacuum (UHV) conditions, using high resolution electron energy loss spectroscopy (HREELS), thermal desorption spectroscopy (TDS), photon induced desorption spectroscopy (PIDS), and work function change ($\Delta\phi$) measurements. An n-type (5.5×10^{14} cm^{-3}) Si(111) sample was used. The sample was cleaned by repeated cycles of sputtering and annealing. The clean surface, cooled down to 90 K, was exposed to K from a heated K getter. The surface was then exposed to a saturation exposure of NO. The coverage of K deposited on the surface was achieved reproducibly by controlling the heating time of the getter. No attempt was

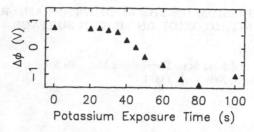

Figure 1. Work function change of the Si(111)7×7 surface at 90 K coadsorbed with K and saturation exposure of NO for different K exposure times.

made to calibrate the absolute coverage of K. Detailed description of the apparatus and the data collection procedures have been published elsewhere.[8]

RESULTS

Changes in the work function of the Si(111)7×7 surface coadsorbed with NO and K were investigated for various K exposures. The work function changes, referenced to the clean Si(111)7×7 surface, are shown in Fig. 1. For saturation exposure of NO on the K-free surface a 0.7 eV increase in the work function is observed, as shown in Fig. 1 at 0 s of K exposure time. The work function decreases with K coadsorbed on the surface, which is clearly observed with K exposures longer than 30 s. (There are no significant changes in the work function for K exposure times shorter than 30 s, indicating that insignificant amounts of K are released from the getter due to the finite time required to heat the getter to the evaporation temperature.) A maximum decrease in the work function is observed at an 80 s K exposure; the work function decreases by 1.4 eV compared to the clean surface, or 2.1 eV compared to the K-free surface with saturation exposure of NO. The work function of the surface coadsorbed with NO and a 45 s K exposure is approximately the same as that of the clean Si(111)7×7 surface, i.e., the effects of NO and K are compensated at this K exposure. The 45 s exposure will be used as a dividing point for describing low and high K exposures.

Figure 2 shows HREEL spectra of NO on the K-free and K preadsorbed Si(111)7×7 surface at 90 K. Spectrum (a) corresponds to saturation exposure of NO on the K-free surface. A detailed discussion of the assignment of the loss peaks shown in the spectrum is published elsewhere.[7] The major results are summarized as follows. NO adsorbs on Si(111)7×7 at 90 K both molecularly and dissociatively. The observed N–O vibrational stretching modes at 188 and 225 meV lie in the range corresponding to bridge and atop site bonding, respectively. The presence of dissociated N and O atoms are deduced from the lower frequency peaks at 103 and 115 meV, which correspond to vibrations of N and O against the Si surface, respectively. In addition, N_2O molecules are observed on the surface, with the N–O and N–N stretching modes observed at 154 and 277 meV, respectively. These physisorbed N_2O molecules are produced from reactions of NO on the surface during adsorption.

HREEL spectra (b)–(d) in Fig. 2 were recorded for the Si surface preadsorbed with different amounts of K followed by saturation exposures of NO. The surface becomes very disordered after K exposures, especially at low K exposures, which results in a much reduced signal to noise ratio in the HREEL spectra. The major peaks observed in the HREEL spectrum of NO adsorbed on the K-free surface [spectrum (a)] are also present

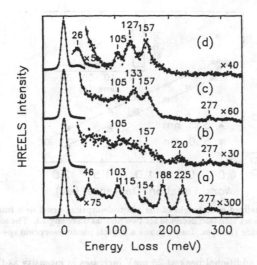

Figure 2. HREEL spectra of saturation exposure of NO on Si(111)7×7 at 90 K for different initial conditions: (a) without K; coadsorbed with (b) 35 s, (c) 45 s, and (d) 80 s K exposures.

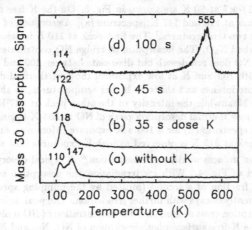

Figure 3. Thermal desorption spectra of NO coadsorbed with different K pre-exposures on Si(111)7×7 at 90 K. A linear temperature rise with a heating rate of 2.0 K/s was used.

in the spectrum of NO coadsorbed with a 35 s K exposure [spectrum (b)]. In addition, a new peak at about 157 meV is observed. This peak is attributed to an N–O stretching mode of molecular NO adsorbed on the metallic K, as will be discussed in the next section. As the K exposure increases to 45 s, the two N–O stretching modes originally at 188 and 225 meV are no longer resolvable [spectrum (c)]. Three loss peaks at 105, 133, and 157 meV, which are attributed to vibrations of atomic N and O against the surface and the new N–O stretching mode, respectively, begin to dominate the spectrum. At

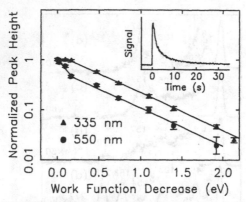

Figure 4. Normalized peak height of the photodesorption signal as a function of the K-induced work function decrease for two photon wavelengths. The solid lines are drawn to guide the eyes. Insert shows a typical photodesorption spectrum.

high K exposures, an additional peak at 26 meV increases in intensity as the K exposure increases and becomes very intense for an 80 s K exposure, as shown in spectrum (d) in Fig. 2.

Thermal desorption spectra monitoring mass 30 for NO coadsorbed with different K exposures on Si(111)7×7 at 90 K are shown in Fig. 3. On the K-free surface thermal desorption of NO occurs at 110 and 147 K [spectrum (a)]. Desorption of N_2 and N_2O is also observed at these two temperatures.[7] The first peak at 110 K is associated with the desorption of physisorbed N_2O. The desorption of bridge NO contributes to the second peak at 147 K. (Atop NO does not desorb but dissociate between 200 and 300 K.) On the surface coadsorbed with NO and K at low exposures the first thermal desorption peak at 110 K becomes more intense and shifts to higher temperatures, as shown in spectra (b) and (c) in Fig. 3. Meanwhile the intensity of the other peak at 147 K decreases. At higher K exposures a new thermal desorption peak of NO at 555 K begins to increase in intensity, as shown in spectrum (d) for a 100 s K exposure. Moreover, an additional N_2 thermal desorption peak at 315 K is observed at high K exposures (not shown).

Photon irradiation induces nonthermal desorption.[6] A typical desorption spectrum is shown in the insert of Fig. 4. With cw irradiation the desorption signal reaches a maximum in a small fraction of a second (limited by the pumping speed of the UHV chamber) and subsequently decays with further irradiation. The peak height is a function of both the photodesorption cross section and the concentration of NO molecules adsorbed on the surface. On the K-free surface photodesorption of NO, N_2, and N_2O is observed with approximately the same cross sections.[7] The desorption occurs over a wide range of wavelengths from the UV to IR (to at least 830 nm, the longest wavelength used in the present experiments). An enhancement in the photodesorption cross section is observed for photons of higher energy than the first direct band gap transition for Si at 3.37 eV (370 nm). This observation reveals that the photodesorption is initiated by photogenerated electron-hole pairs.[6]

The peak height of the photodesorption signal decreases on the surface preadsorbed with K. The magnitude of the decrease for the desorption of NO and N_2O is about the same but is larger than that of N_2. Figure 4 shows the peak height of the photodesorption of NO as a function of the K-induced work function decrease for two wavelength regions.

One region is in the UV, centered at 335 nm and with a bandwidth of 75 nm; the other is in the visible, centered at 550 nm and with a bandwidth of 65 nm. The data shown in Fig. 4 have been normalized to the values obtained from the K-free surface. The peak height decreases by nearly two orders of magnitude with K adsorption. Furthermore, a larger decrease is observed with visible irradiation compared with UV irradiation.

HREELS was used to identify the species on the surface after thermal heating or photon irradiation. On the K-free surface, molecularly adsorbed NO is desorbed and dissociated after either thermal heating or photon irradiation, leaving the surface covered only with atomic N and O.[6,7] Similar results are observed on the surface preadsorbed with low K exposures; however, the intensity ratio of the N peak to the O peak in HREELS deceases as the K exposure increases. At high K exposures the intensity of the NO peak at 157 meV decreases (but not to zero) and the intensity of the O peak at about 130 meV increases after either photon irradiation or thermal heating to a temperature below 555 K. After heating the surface above 555 K, the peaks at 26 and 157 meV disappear, and the surface is covered only with atomic O and N.

DISCUSSION AND CONCLUSIONS

Adsorption of NO on Si(111)7×7 at 90 K is modified by K coadsorption. On the K-free surface N–O stretching modes observed at 188 and 225 meV lie in the range nominally attributed to bridge and atop site bonding, respectively. On the surface coadsorbed with K and NO, the intensity of these two modes decreases as the K exposure increases and are completely absent at high K exposures. Similarly, the intensity of the thermal desorption peak at 147 K for bridge NO diminished to zero with K. These observations indicate that the adsorption of K on Si(111)7×7 competes with molecular NO adsorption at both bridge and atop sites.

A new peak at 157 meV is observed in HREELS of NO coadsorbed with K. A natural assignment of this peak is the stretching mode of NO molecules adsorbed on metallic K. The energy of the peak is lower than the two N–O peaks observed on the K-free surface, which is consistent with the expectation that NO molecules are strongly bonded with K. It is interesting to note that this peak is observed even at low K exposures and that its energy shows no apparent shift as the K exposure increases. It is therefore suggested that K clusters are formed on the surface even at low K exposures, and the observed peak at 157 meV is the N–O stretching mode of molecular NO adsorbed on the K clusters. These NO molecules contribute to the high temperature desorption peak at 555 K observed at high K exposures, since the 157 meV peak in the HREEL spectrum disappears only after heating the surface above 555 K. However, there is no observable desorption at 555 K with low K exposures even though the 157 meV peak is clearly observed in HREELS. Moreover, with high K exposures, the intensity of the 157 meV peak decreases with thermal heating. It is possible that NO molecules adsorbed on K clusters may migrate away from the clusters during thermal heating and dissociate and/or desorb before the temperature reaches 555 K. At high K exposures larger clusters are formed so that only a fraction of NO molecules are able to migrate away from the clusters before the surface temperature reaches 555 K, where the desorption of the remaining NO molecules from the K clusters occurs. More experiments are necessary to confirm this hypothesis. An additional peak at 26 meV is also observed in HREELS for the surface coadsorbed with NO and high K exposures. A similar feature has been previously observed for NO adsorbed on Ag(111).[9] A possible explanation for the low energy peak is a phonon mode associated with the metallic K.

The surface is covered with atomic O and N after either thermal heating or photon irradiation. (An exception occurs at high K exposures, where molecular NO remains on the surface even after long time UV irradiation.) HREELS shows that the intensity ratio of the O peak to the N peak increases as the K exposure increases. Moreover, the magnitude in N_2 desorption during either thermal heating or photon irradiation increases with increasing K exposures. These observations indicate that the surface is more oxidized in the presence of K.

The photodesorption peak height decreases as the K exposure increases. Part of this decrease can be explained by a K-induced decrease in the coverages of bridge and atop NO. (NO molecules adsorbed on the K clusters are only weakly perturbed by the photons; the intensity of the 157 meV peak is only slightly reduced after irradiation.) Most significantly, a larger decrease occurs with visible irradiation than UV irradiation, which can not be explained by the decrease in the molecular NO coverage. This additional decrease is consistent with the mechanism of the photodesorption involving photogenerated carriers. The photodesorption is initiated by hot charge carriers, in particular holes, generated by photons.[6,10] The surface adsorbed with NO shows a 0.7 eV increase in the work function compared with the clean surface, which gives rise to a band bending up near the surface.[11] As a consequence, the photogenerated holes experience an acceleration potential towards the surface, which gives rise to favorable reaction cross sections. When the surface is coadsorbed with K, the work function decreases and the band bending is diminished and eventually becomes downward, depending on the K exposure. Therefore, holes moving towards the surface are accelerated in a weaker potential or even deaccelerated, which causes the decrease in the cross section. The K-induced effects are especially pronounced with visible or lower energy photon irradiation where photogenerated holes have lower energies.

ACKNOWLEDGMENTS

Support of this research by the Office of Naval Research Under Grant No. N00014-81-K-0505 is gratefully acknowledged. One of us (ZCY) would like to thank the Materials Research Society for a Graduate Student Award.

REFERENCES

[1] T. Aruga, H. Tochihara, and Y. Murata, Phys. Rev. Lett. **53**, 372 (1984).

[2] P. Soukiassian, T.M. Gentle, M.H. Bakshi, and Z. Hurych, J. Appl. Phys. **60**, 4339 (1986); E.M. Oellig, E.G. Michel, M.C. Asensio, and R. Miranda, Appl. Phys. Lett. **50**, 1660 (1987).

[3] P. Soukiassian, M.H. Bakshi, H.I. Starnberg, Z. Hurych, T.M. Gentle, and K.P. Schuette, Phys. Rev. Lett. **59**, 1488 (1987).

[4] T.J. Chuang, Surf. Sci. Rep. **3**, 1 (1983).

[5] W. Ho, Comments Cond. Matter Phys. **13**, 293 (1988).

[6] Z. Ying and W. Ho, Phys. Rev. Lett. **60**, 57 (1988).

[7] Z. Ying and W. Ho, J. Vac. Sci. Technol. A **7**, 000 (1989).

[8] Z. Ying and W. Ho, Surf. Sci. **198**, 473 (1988).

[9] S. K. So (private communication).

[10] E. Ekwelundu and A. Ignatiev, Surf. Sci. **179**, 119 (1987).

[11] A. Many, Y. Goldstein, and N.B. Grover, *Semiconductor surfaces* (North-Holland, Amsterdam, 1965).

KINETIC AND X-RAY PHOTOELECTRON SPECTROSCOPIC STUDIES
OF THE THERMAL NITRIDATION OF Si(100)*

C.H.F. PEDEN, J.W. ROGERS, JR., D.S. BLAIR, AND G.C. NELSON
Sandia National Laboratories, Albuquerque, NM 87185-5800

ABSTRACT

The thermal nitridation of Si(100) by NH_3 and N_2H_4 has been studied by X-ray photoelectron (XPS) and Auger electron (AES) spectroscopies. The pressure dependence of the rates as a function of reaction time has been measured. It has been found that the growth of the first monolayer (ML) of nitride is mediated by a surface reaction step. For subsequent growth, diffusion of one or more of the reacting species becomes an important process in determining the rate of reaction. Such species may be substrate Si diffusing to the vacuum/Si_3N_4 interface, or possibly network nitrogen diffusing into the Si substrate. The independence of the reaction rate on NH_3 or N_2H_4 pressure at long reaction times rules out a mechanism involving molecular diffusion of the nitriding gas to the Si_3N_4/Si interface in a manner similar to the oxidation of Si by O_2 or H_2O. Careful analysis of the Si(2p) XPS spectra reveals the presence of a unique Si species with a Si(2p) binding energy intermediate between elemental Si and Si in Si_3N_4. Further, the relative intensity of the Si(2p) features due to this species, and the angular dependence of the XPS peaks indicate that they result from a ML of Si at the outermost surface layer, on top of the growing Si_3N_4 film.

INTRODUCTION AND EXPERIMENTAL

Future scaled VLSI devices will require very thin (≤ 100 Å) gate insulators of high quality. The particular material requirements for these thin insulators have been discussed recently by Moslehi and Saraswat [1]. In this paper, these authors also review the preparation and properties of thin Si_3N_4 films which are a promising alternative to SiO_2 as device dimensions are reduced. Generally it has been found that the growth of Si_3N_4 thin films by the thermal nitridation of Si with dinitrogen (N_2) or ammonia (NH_3) is initially rapid, followed by a marked decrease in rates until growth nearly stops at a maximum thickness of about 40 Å [1-5]. This behavior can, in fact, be useful in controlling the thickness of insulator films based on Si_3N_4 [1]. To understand the growth and chemical state of such films in more detail, we have performed kinetic and spectroscopic studies of the thermal nitridation of Si(100) using hydrazine (N_2H_4) or NH_3 as the gas source for nitrogen. (We have previously found [6] that initial growth with N_2H_4 is significantly more rapid than with NH_3.) In this short report, we present the results of XPS studies and briefly describe some of the kinetic data [7]. In addition, we compare and contrast the kinetic results for thermal nitridation with those previously measured for the thermal oxidation of Si under "dry" (O_2) and "wet" (H_2O) conditions in order to compare the mechanism of reaction for oxide and nitride film growth.

Both n- and p-type, doped Si(100) single crystals (20Ω-cm resistivity) from commercially available sources (Monsanto, Westinghouse) were used in this study. Sample mounting and cleaning, and nitriding experimental procedures have been described previously [8,9]. The experiments were performed in two ultra-high vacuum (UHV) apparatuses, both operated at base pressures less than 5×10^{-10} Torr. The first system was equipped for Auger electron spectroscopy (AES) with a Physical Electronics single-pass

---

*This work, performed at Sandia National Laboratories, was supported by the U.S. Department of Energy under contract number DE-AC04-76DP00789.

cylindrical mirror analyzer containing an internal, coaxial electron gun. X-ray photoelectron spectroscopy (XPS) was performed with a VG ESCALAB 5 surface spectrometer. The photoemission was excited with a Mg(Kα) X-ray source operated at 240 W (12 kV, 20 mA), and analyzed with a hemispherical electron energy analyzer operated in the constant pass energy mode at a total instrument resolution (including broadening due to the X-ray source) of 1.2 eV. Binding energies (BEs) were reproducible to ±0.1 eV, and were referenced to the Fermi level of gold ($Au(4f)_{7/2}$ = 84.0 eV).

The kinetic studies to be described were conducted in order to determine the pressure dependence of reaction. This dependence can be described as an "order of reaction" according to Eq. 1:

$$\text{Reaction Rate} = \text{const.} \times [\text{reactant pressure}]^m. \tag{1}$$

Thus, we determine here the exponent, m (the order of reaction), directly from the slope of a log/log plot of the reaction rate versus the reactant pressure [10].

RESULTS AND DISCUSSION

Pressure Dependence for First Monolayer Growth

<u>Ammonia:</u> Fig. 1a shows the growth of the N(KVV) Auger transition near 375 eV as a function of time during the thermal nitridation with NH_3 at 823 K. Two straight line segments can be drawn through the data suggesting layer-by-layer film growth for the first several monolayers (MLs). (Similar behavior has been observed previously for Si_3N_4 film growth with NH_3 on Si(100) [4]). Thus, the slope of the first line segment is a measure of the growth rate for the first ML with the second straight line slope being proportional to the rate of growth of the second ML. The slopes of such lines for the first ML at several NH_3 pressures are plotted in Fig. 1b using a log/log format in order to abstract the order of reaction. This data was taken at a lower temperature than that shown in Fig. 1a in order to conveniently measure the reaction order at higher pressures. Within experimental error, the growth of the first ML is first order in NH_3 pressure.

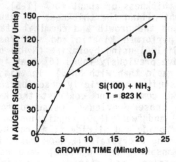 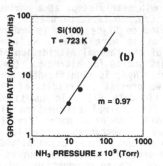

FIGURE 1: (a) N(KVV) signal strength (peak-to-peak height of feature near 375 eV) plotted versus exposure time during the thermal nitridation with NH_3 at 823 K (P(NH_3) = 2 x 10^{-8} Torr). (b) Log/log plot of the growth rate (obtained from the slope of the first straight line segment in Fig. 1a) versus NH_3 pressure.

Hydrazine: We have previously measured [9] the pressure dependence for growth of the first ML in N_2H_4 by exposing the crystal to various background pressures of N_2H_4 for a fixed length of time (5 minutes). These experiments were performed at lower temperature than the NH_3 experiments due to the initially higher reactivity of N_2H_4 relative to NH_3 [6]. In this case, the reaction rate for first ML growth was found to be one-half order in N_2H_4 pressure [9].

For both ammonia and hydrazine, the measured order of reaction is consistent with a process mediated by a surface reaction step. For example, Blanc [11] has recently analyzed the kinetics of reaction for the oxidation of Si by O_2 at early times where the pressure dependence has been measured to be 0.5-1.0 [12,13]. Blanc derives a kinetic expression which gives a reaction order in O_2 of 0.5 (which he assumes to be the correct value) for early times from a mechanism involving the equilibration of chemisorbed O atoms (O_{ads}) and gas-phase O_2 molecules. The slow step of the oxidation process at early reaction times is the oxidation of the Si network by O_{ads} which involves the rupture of Si-Si bonds. The $P^{1/2}$ dependence of oxidation then comes from the fact that two atoms of O_{ads} are formed for each molecule of O_2 reacted [11]. Thus, the difference in the pressure dependence observed for the two nitrogen containing reactants during thermal nitridation may be simply a reflection of the different number of N-atoms in the molecules; two for N_2H_4 and one for NH_3. A surface reaction limited process is expected since there is no need to diffuse either reactant (Si or N-containing species) in order to form the first layer. Interestingly, recent work by George, et al. [12] found the initial (first ML) oxidation rate to be first order in O_2 gas pressure. Analogously to the situation with N_2H_4, one might expect a $P^{1/2}$ dependence on O_2 gas pressure as assumed by Blanc [11] and found by others [13] for early reaction times (not necessarily first ML growth). However, the results of George and coworkers [12] also implicate an adsorbed molecular O_2 precursor to explain the decrease in the initial sticking coefficient with temperature between 200 and 600 K, a species ignored by Blanc [11] in his analysis. Since it is these early-time kinetics that will be most important in growing very thin dielectric films, additional kinetic studies of Si oxidation and nitridation are warranted to clarify the surface chemical and physical processes leading to thin film growth on Si.

Pressure Dependence for Subsequent Layer Growth

Hydrazine: The pressure dependence of reaction in N_2H_4 to form films with thicknesses greater than 7 Å is markedly reduced from the $P^{1/2}$ behavior for the first ML as described above. In addition, the order of reaction continues to decrease with film thickness (longer reaction times). For example, we found that the growth of films approximately 7 Å and 12 Å in thickness (data not shown) gave orders of reaction, m, of 0.30 and 0.24, respectively. Thus, it appears that growth is becoming independent of N_2H_4 pressure (zero-order) as the film thickness increases.

Ammonia: In a recent publication [4], Glachant and Saidi show curves similar to Fig. 1a, taken at several NH_3 gas pressures, in which clear breaks are seen after completion of the first ML. In addition, second breaks are evident in the data (Fig. 5, Ref. 4) indicating the completion of the second ML and the onset of growth of the third ML. From this data, we calculated the rate of formation of the first and second MLs as a function of NH_3 pressure. As with our data (Fig. 1), these results suggest that the reaction is first order in NH_3 pressure for the first ML. However, in the second ML the order of reaction in NH_3 pressure already is becoming smaller (order of reaction = 0.75). This trend apparently continues until the reaction becomes essentially independent of pressure as we found with N_2H_4. Evidence for this trend comes from the work of Hayafuji and Kajiwara [2] who measured the change in thickness of Si_3N_4 films grown at 1173-1373 K for 5 hours. With NH_3 pressure changes of over three orders of magnitude (0.005 < $P(NH_3)$ < 50 Torr), film thickness increased at most only 20%.

It is instructive to compare the kinetics observed here for "long-time" nitride film growth with the kinetics of oxidation of Si. For both wet (H_2O) and dry (O_2) oxidation, oxide film growth becomes first-order in oxidant pressure as film thickness increases [13]. In both cases, oxidation becomes rate-limited by the diffusion of the oxidant molecule (H_2O or O_2) through the growing SiO_2 film in order for reaction to occur at the Si/SiO_2 interface. At early times, some have found the reaction to be less than first order in O_2 pressure with values for the order of reaction ranging from 0.5 to 1.0 [13]. While this "early time" reaction regime is not well understood [14], it is clear that oxidation kinetics are quite different than nitridation kinetics with respect to the tendancy for Si_3N_4 film growth to become independent of nitriding gas pressure.

Underscoring the differences in the mechanisms of nitridation and oxidation of Si are the results from isotopic tracer studies [5,13] in which oxide and nitride films are grown consecutively in light and heavy O and N isotopes, respectively. In the case of NH_3, the last isotope used to grow the film resides at the surface of the nitride film. To account for these results, the authors propose two mechanisms, both of which are also consistent with the zero-order pressure dependence we and others have observed for "long-time" nitride film growth. The first mechanism involves the rate-limited diffusion of Si from the substrate, through the growing nitride film, to the surface where reaction with the N-containing source gas can take place. The second possibility involves the one-directional (into the substrate), step-by-step motion of charged network nitrogen atoms. The vacancies left by the migrating atoms would then be refilled from behind, eventually by N-atoms from the gas-phase source which have reacted with Si-atoms at the surface. If the refilling process is much faster than the initial migration (rate-limiting step), then again zero-order pressure dependence would be expected. Simple diffusion of N-atoms in the network would be in both directions and would contradict the observed lack of intermixing among the two N isotopes in the tracer studies [5]. Thus if the second model is correct, the movement of charged N-species in the network is constrained to be one-directional into the substrate, perhaps by a strong electric field arising from charge-transfer processes between Si and N at the surface [5].

Si(2p) XPS Results

XPS spectra of the Si(2p) region near 100 eV binding energy are shown in Fig. 2 for Si_3N_4 films thermally grown for various lengths of time in N_2H_4. Similar results to these were found for the direct thermal nitridation using NH_3 suggesting that the characteristics of the film (e.g., N/Si stoichiometry, see below) are also similar. Fig. 2a shows the spectra obtained from the clean Si surface. Also shown is the decomposition into the $Si(2p)_{1/2}$ and $Si(2p)_{3/2}$ spin-orbit split components after removal of a Shirley-type background [15]. The decomposition was accomplished in the following manner. The spectral features of the two components were constrained such that i) the energy splitting was 0.6 eV [16], ii) the relative intensity ratio was 2:1 as required by the degeneracy of the 3/2 and 1/2 levels, and iii) the relative lineshapes (Gaussian/Lorentzian product function [17]) and the relative full-width at half-maxima (FWHM) were equivalent. The _absolute_ energies, intensities, lineshapes (Gaussian/Lorentzian ratios), and FWHM were then varied to obtain the best fit to the data as determined by minimizing chi-squared using a standard fitting algorithm (VG 5000 software based on Ref. 17). The FWHM of the elemental Si(2p) spin-orbit split features obtained from this fitting procedure were 1.2 eV. This observed value is dominated by and is in excellent agreement with the known

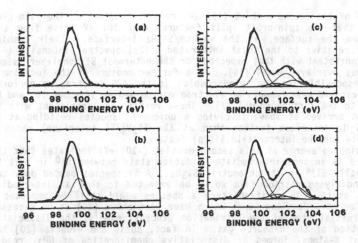

FIGURE 2: XPS Si(2p) spectra as a function of N_2H_4 exposure for the direct thermal nitridation of Si(100) at a substrate temperature of 973 K. (a) Spectrum of clean Si(100). (b-d) Spectra obtained after reaction in 5 x 10^{-8} Torr N_2H_4 for (b) 100 seconds, (c) 2000 seconds, and (d) 7000 seconds. Also shown are the results of a fit to the spectra consisting of several spin-orbit split pairs of peaks as described in the text. For clarity, the doublet near 101 eV is shaded. Intensities are shown in arbitrary units.

instrument broadening expected for our experimental operating conditions. This observation, along with the magnitude of the chi-squared values, as well as the excellent visual fit to the data validates the use of this curve fitting method.

As the first ML of nitride grows, a new Si(2p) spin-orbit split feature appears with a BE centroid near 101 eV as shown in Fig. 2b. Film growth beyond the first ML regime results in a third doublet whose BE centroid occurs near 102 eV, and whose intensity continues to increase with nitridation time (Figs. 2c-2d). These latter features clearly arise from Si in the growing Si_3N_4 film. (Note: attempts to satisfactorily fit the spectra shown in Figs. 2c and 2d, and numerous other spectra obtained in the course of these studies, with two sets of doublets near 99 and 102 eV were unsuccessful.) The FWHM for the components of the 101 and 102 eV doublets was 1.65 eV which is in excellent agreement with that observed by others for Si_3N_4 [3] and SiO_2 [16] when corrected for differences in analyzer operating conditions. This increased breadth relative to the width of the elemental features is generally attributed to phonon broadening in the nitride or oxide film [18].

The relative intensity of the second doublet (101 eV) as a function of nitridation time was observed to rise at the one ML exposure level to a value of about 7% of the total integrated Si(2p) intensity and remain constant upon further exposure to N_2H_4 [7]. In contrast, the features due to elemental Si are attenuated by the growing Si_3N_4 overlayer while the intensity of the Si features from Si_3N_4 near 102 eV increase monotonically as the film thickens. The fact that the doublet at 101 eV is observed prior to the appearance of any intensity for the doublet at 102 eV, and that the intensity of the second doublet (101 eV) remains constant as the film grows suggests that the spin-orbit split features near 101 eV arise from a Si species on the surface, at the vacuum/Si_3N_4 interface. Their combined intensity relative to the total integrated Si(2p) spectral intensity (~7%) is also consistent with that expected for the outermost Si monolayer using a homogeneous overlayer model [19]. As a further measure of the location of the Si responsible for the third doublet in the XPS spectra, we performed angular resolved measurements at electron escape angles of about 15° and 85° relative to the surface normal [7]. These results are consistent with the conclusion arrived at above involving a unique Si species residing at the vacuum/Si_3N_4 interface (rather than at the Si_3N_4/Si interface) which is responsible for the intermediate Si(2p) features.

The binding energy of this second doublet (~101 eV) indicates that this Si species is in some intermediate oxidation state between Si^0 in bulk Si, and nominally Si^{+4} in stoichiometric Si_3N_4. A Si species bonded only to N in the underlying nitride film would be expected to show an intermediate oxidation state. In addition, such a species would still be expected to have one or more dangling bonds which have been shown [20] to be necessary for the initial step of the nitridation process; namely, the dissociative chemisorption of the N-source gas. In fact, Bozso and Avouris [20] have found that N-atoms, formed by dissociative chemisorption of NH_3, readily migrate under the outer Si layer even at 90 K.

Similar intermediate Si(2p) features were observed recently by Soukiassian, et al. [21] upon formation of a Si_3N_4 film prepared by codeposition of Na and N_2 onto Si(100) and subsequent removal of Na by annealing. In fact, the relative intensity of the intermediate features was significantly greater in their spectra owing to the use of low energy (130 eV) excitation photons from a synchrotron source. This provided greater surface sensitivity since the exiting photoelectrons had low kinetic energy with a correspondingly low mean-free path. These authors assigned the intermediate features to a substoichiometric silicon nitride species at the Si/Si_3N_4 interface indicating that the interface was not an abrupt one. However, our results described above clearly indicate that the intermediate features observed in films grown by thermal nitridation are due to a Si species located at the surface of the growing nitride film (vacuum/Si_3N_4 interface). Thus, it seems likely that the intermediate BE feature observed by Soukiassian, et al. [21] is also due to a surface Si species, although we cannot rule out the possibility that the different method of Si_3N_4 film preparation used in their study may result in a different Si_3N_4/Si interfacial structure than that obtained by thermal nitridation as used in our work. Angle-resolved XPS measurements on films prepared by Na promotion, similar to those described above, would resolve this question.

Finally, we calculated the N/Si ratio in the Si_3N_4 film using the total integrated intensity of the N(1s) transition near a BE of 398 eV and the intensities of the features for Si in the growing nitride near 101 and 102 eV. The calculated ratio stayed essentially constant with film thickness giving a value of 1.35±0.05; very close to the ratio expected for stoichiometric Si_3N_4 (1.33). Considering the assumptions of the model [19], and the uncertainties in the experimental data, this result suggests that the films are stoichiometric Si_3N_4 and that the Si_3N_4/Si interface is abrupt. (However, it should be noted that our XPS spectra are not particularly sensitive to the composition of such a buried interface. Synchrotron photoemission studies, similar to those described recently for the thermal oxidation of Si [16], are needed.)

ACKNOWLEDGMENTS

The authors would like to thank the Department of Energy, Office of Basic Energy Sciences, Division of Materials Sciences for funding this work, and

W.J. Choyke (Westinghouse R&D Center, Pittsburgh, PA) for providing the p-type Si(100) single crystals used in this study. The technical assistance of R. A. McWilliams and S. B. Van Deusen is also gratefully acknowledged.

REFERENCES

1. M.M. Moslehi and K.C. Saraswat, IEEE Trans. Elect. Devices, ED-32, 106 (1985).

2. Y. Hayafuji and K. Kajiwara, J. Electrochem. Soc. 129, 2102 (1982).

3. C. Maillot, H. Roulet and G. Dufour, J. Vac. Sci. Tech. B2, 316 (1984).

4. A. Glachant and D. Saidi, J. Vac. Sci. Tech. B3, 985 (1985).

5. C. Maillot, H. Roulet, G. Dufour, F. Rochet and S. Rigo, Appl. Surf. Sci. 26, 326 (1986), and references therein.

6. J.W. Rogers, Jr., D.S. Blair and C.H.F. Peden, in Deposition and Growth: Limits for Microelectronics, edited by G.W. Rubloff (American Institute of Physics, New York, 1988), American Vacuum Society Series 4, p. 133.

7. C.H.F. Peden, J.W. Rogers, Jr. and D.S. Blair, in preparation.

8. D.S. Blair and G.L. Fowler, J. Vac. Sci. Tech. A6, 3164 (1988).

9. C.H.F. Peden and S.B. Van Deusen, J. Vac. Sci. Tech. A5, 2024 (1987).

10. J.W. Moore and R.G. Pearson, Kinetics and Mechanism, 3rd ed. (John Wiley & Sons, New York, 1981), p. 16.

11. J. Blanc, Appl. Phys. Lett. 33, 424 (1978).

12. S.M. George, P. Gupta, C.H. Mak and P.A. Coon, in Chemical Perspectives of Microelectronic Materials, edited by M.E. Gross, J.M. Jasinski and J.T. Yates, Jr. (Materials Research Society, Pittsburgh, 1989), in press.

13. S. Rigo, in Instabilities in Silicon Devices, Vol. 1, edited by G. Barbottin and A. Vapaille (North Holland, Amsterdam, 1986), p. 5.

14. E.A. Irene, in The Physics and Chemistry of SiO_2 and the Si/SiO_2 Interface, edited by B.E. Deal (Plenum Press, New York, 1988) in press.

15. A.D. Shirley, Phys. Rev. B 5, 4709 (1972).

16. F.J. Himpsel, F.R. McFeeley, A. Taleb-Ibrahimi, J.A. Yarmoff and G. Hollinger, Phys. Rev. B 38, 6084 (1988).

17. R.O. Ansell, T. Dickinson, A.F. Povey and P.M.A. Sherwood, J. Electroanal. Chem. 98, 79 (1979).

18. G. Hollinger and F.J. Himpsel, Appl. Phys. Lett. 44, 93 (1984).

19. C.S. Fadley, Prog. in Surf. Sci. 16, 275 (1984).

20. F. Bozso and Ph. Avouris, Phys. Rev. Lett. 57, 1185 (1986).

21. P. Soukiassian, M.H. Bakshi, H.I. Starnberg, Z. Hurych, T.M. Gentle and K.P. Schuette, Phys. Rev. Lett. 59, 1488 (1987).

CATALYTIC OXIDATION OF SILICON NITRIDE THIN FILMS USING POTASSIUM*

J. W. ROGERS, JR., D. S. BLAIR, and C. H. F. PEDEN
Sandia National Laboratories, Albuquerque, NM 87185-5800

ABSTRACT

Thin silicon nitride films on a Si(100) substrate have been oxidized using potassium in a low thermal budget process. The presence of potassium on the Si_3N_4 surface greatly lowers the temperature-time requirements for oxidation as compared with direct thermal oxidation.

Introduction

Silicon nitride (Si_3N_4) thin films are used extensively in the semiconductor industry for diffusion masks, surface passivation layers, and masking layers for selective oxidation. Si_3N_4 is used in this broad range of applications because of its chemical stability and desirable electrical properties.

However, Si_3N_4 thin films are known to oxidize at elevated temperatures and surface oxidation has been exploited to produce silicon oxynitride whose electronic properties can be tailored by judicious control of the oxygen content of the film. These films show potential for fabrication of radiation hardened microelectronics. Unfortunately the high temperatures necessary to thermally oxidize the Si_3N_4 will produce dopant diffusion and resulting degradation of device performance.

One goal of an ongoing program in our laboratory is to find alternate low temperature synthetic routes to the formation of Si_3N_4 and silicon oxynitride thin films. Several recent studies suggest [1-3] that certain alkali metals (Cs, Na, and K) act as "catalysts" and enhance the oxygen uptake during direct oxidation of Si to SiO_2. We have investigated the use of an alkali catalyst to oxidize thin films of Si_3N_4 on Si at low temperature. A comparison has been made between the low temperature catalytic oxidation of Si_3N_4 and direct thermal oxidation at elevated temperatures (1125 to 1275 K).

Experiment

Si_3N_4 thin films were prepared by low-pressure, direct thermal nitridation of a Si(100) crystal at 1025 K using hydrazine as the nitridant gas [4]. This procedure yielded 5-25 Å thick films of pure, stoichiometric Si_3N_4. Unlike silicon nitride prepared by chemical vapor deposition methods, these films were free from oxygen contaminants.

Potassium (K), from a heated alkali metal source, was then deposited at 300 K in submonolayer to multilayer quantities onto the Si_3N_4 film for a predetermined period of time in an ambient of 1×10^{-6} Torr of O_2. The substrate was then heated to 900 K for less than two minutes to facilitate removal of the potassium and oxidation of the Si_3N_4. X-ray Photoelectron Spectroscopy (XPS) was used with standard operating conditions [4] to follow the extent of oxidation and the chemical state of surface intermediates after various treatments.

*This work, performed at Sandia National Laboratories, was supported by the U.S. Department of Energy under contract number DE-AC04-76DP00789.

Results and Discussion

Following deposition of K onto the Si_3N_4 surface, examination of the surface by XPS suggests that K was present in two distinct chemical forms. These surface species were characterized by O(1s) binding energies (BEs) of 530.0 and 530.8 eV, respectively. The stoichiometry, determined from K(2p) and O(1s) peak areas (adjusted for atomic sensitivity), indicate that both are $(KO)_x$ species. The BEs of the O(1s) transition for these species suggest that they are present as peroxide in which the oxygen-oxygen bond is intact. Subsequent annealing to 900 K for less than two minutes resulted in complete removal of K from the surface by thermal desorption and left a partially to fully oxidized Si_3N_4 film depending upon the amount of K originally deposited. Preliminary temperature programmed desorption results suggest that K and K_2O were desorbed during this annealing procedure. Ninety percent of the oxygen initially deposited remained on the surface after annealing. The O(1s) region from the oxidized film which remained after annealing was characterized by a single peak whose BE was 532.8 eV which is characteristic of that observed from silicon oxynitride.

The Si(2p) region from a heavily oxidized silicon nitride film (several K-deposition/anneal cycles) is shown in the figure. The original Si_3N_4 film was ~25 Å thick. This spectrum was curve resolved using standard techniques [5] into five distinct peaks. In order to obtain a qualitative understanding of the oxidation process, it was not necessary to further decompose each feature into its spin-orbit split doublets [5]. Starting from the lowest BE feature, these peaks were derived from i) elemental Si (99.6 ± 0.1 eV), ii) a monolayer of Si at the film/vacuum interface (101.2 eV) [5], iii) Si bonded only to nitrogen in the oxynitride film (102.3 eV), iv) Si bonded to both nitrogen and oxygen ligands in the oxynitride film (103.5 eV), and v) Si bonded only to oxygen in SiO_2 (104.2 eV). Before oxidation, only peaks i, ii, and iii were present and their BEs did not change as the film was oxidized. As oxidation proceeded, peak iv appeared and its intensity increased monotonically as the number of oxidation cycles increased, until

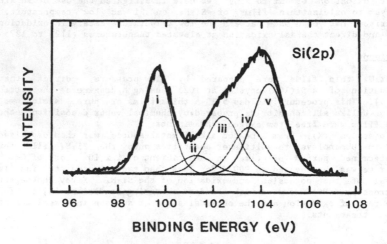

The Si(2p) region for an oxidized film of Si_3N_4. The 25 Å-thick Si_3N_4 film was initially grown on a Si(100) substrate by direct thermal nitridation using hydrazine. The film was oxidized by repeated cycles of K-deposition in an ambient of O_2 followed by a brief anneal (~2 minutes) at 900 K. The intensity is shown in arbitrary units.

its area nearly equaled that of peak iii. Further oxidation resulted in oxidation of Si to SiO_2 at the Si/silicon oxynitride interface as evidenced by the appearance and angular dependence of peak v and a concomitant attenuation of the underlying elemental Si (peak i) due to the thickening overlayer.

In contrast to the catalytic oxidation of Si_3N_4 thin films, direct thermal oxidation required more extreme processing conditions. Si_3N_4 films were oxidized at 1125 to 1275 K in an ambient of 5×10^{-7} Torr of O_2 for periods of up to one hour. The maximum extent of oxidation was only about twenty percent of that achieved using the K catalyst.

Conclusions

Si_3N_4 can be oxidized by room temperature deposition of K in an ambient of O_2, followed by a brief anneal to moderate temperature. The oxidizing agents appear to be potassium oxides which thermally decompose during the anneal releasing atomic oxygen which readily reacts with the Si_3N_4 surface to form silicon oxynitride [6]. Potassium is not a true catalyst because its chemical state changes during the anneal. However, it appears to be very effective in weakening the oxygen-oxygen bond and allowing oxidation of the surface under relatively mild conditions.

Acknowledgements

The authors would like to thank the Department of Energy, Office of Basic Energy Sciences, Division of Materials Sciences for funding this work, and W. J. Choyke (Westinghouse R&D Center, Pittsburgh, PA) for providing the Si(100) single crystals.

References

1. P. Soukiassian, T. M. Gentle, M. H. Balshi, and Z. Hurych, J. Appl. Phys. 60, 4339 (1986).
2. A. Franciosi, P. Soukiassian, P. Philip, S. Chang, A. Wall, A. Raisanen, and N. Troullier, Phys. Rev. B35, 910 (1987).
3. M. C. Asensio, E. G. Michel, E. M. Oellig, and R. Miranda, Appl. Phys. Lett. 51, 1714 (1987).
4. J. W. Rogers, Jr., D. S. Blair, and C. H. F. Peden, AIP Monograph - Proceedings of the Topical Conference on Deposition and Growth, 34th National AVS Symposium, Nov. 2-6, 1987, Anaheim, CA.
5. C. H. F. Peden, J. W. Rogers, Jr., D. S. Blair, and G. C. Nelson, in Chemical Perspectives of Microelectronic Materials, edited by M. E. Gross, J. M. Jasinski, and J. T. Yates, Jr. (Materials Research Society, Pittsburgh, PA, 1989) this volume.
6. E. G. Michel, E. M. Oellig, M. C. Arensio, and R. Miranda, Surf. Sci. 189/190, 245 (1987).

GAS-SURFACE REACTION STUDIES RELEVANT TO SiC CHEMICAL VAPOR DEPOSITION

C.D. Stinespring, A. Freedman, and J.C. Wormhoudt
Center for Chemical and Environmental Physics, Aerodyne Research, Inc.,
45 Manning Road, Billerica, MA 01821

ABSTRACT

Reactions of C_2H_4, C_3H_8, and CH_4 on Si(111) and C_2H_4 on Si(100) have been investigated for surface temperatures in the range of 1062 K to 1495 K. These studies used x-ray photoelectron spectroscopy to identify the reaction products, characterize the solid state transport process, determine the nucleation mechanism and growth kinetics, and assess orientation effects. The results are used to provide insight into the mechanisms of SiC CVD processes.

INTRODUCTION

The use of epitaxial β-SiC as a high temperature semiconductor has motivated a number of studies of hydrocarbon reactions on Si. During the late 1960's, the reactions of C_2H_2 and C_2H_4 on Si surfaces were investigated by Kahn and Summergrad.[1] These and similar studies performed during this time frame focused on thin film morphology and characteristics. More recently, Yates and coworkers[2-4] and the authors of this paper[5] have used surface sensitive techniques such as x-ray photoelectron spectroscopy (XPS) to gain a more fundamental understanding of hydrocarbon-Si surface chemistry and its relationship to the growth process. The surface studies described in this paper were performed to provide a basis for understanding the SiC chemical vapor deposition (CVD) process developed at the NASA Lewis Research Center.[6]

RELATIONSHIP OF GAS-SURFACE REACTION STUDIES TO SiC CVD

The NASA CVD procedure is a two step process. In the first step, an initial layer of SiC approximately 10 nm thick is deposited by flowing a dilute mixture of C_3H_8 in H_2 over the Si substrate as its temperature is ramped (70 K s^{-1}) to 1673 K. The second step involves homoepitaxy of SiC on the initial deposit using a dilute mixture of C_3H_8 and SiH_4 in H_2 at 1673 K. Deposition during the initial step is found to be essential if high quality epitaxial SiC is to be deposited during the second step.

The work described here focused primarily on the initial deposition step. In defining the surface studies relevant to this process, it is important to note that the species which interact with and on the surface to form SiC need not be the input species, but may also include non-equilibrium thermal decomposition and reaction products. In gas phase chemical kinetics calculations reported by the authors,[7] this was found to be the case for C_3H_8. At low temperatures in the ramp (i.e., <1050 K), C_3H_8 is in fact the dominant hydrocarbon species available to react on the surface. CH_4, the species predicted by equilibrium considerations, is not formed as a result of kinetic limitations in the decomposition process. For temperatures greater than 1050 K, unsaturated hydrocarbons such as C_2H_2 and C_2H_4 are the dominant carbon containing species, again as a result of kinetic limitations in the reaction to form CH_4. Only at 1673 K is CH_4 the major hydrocarbon species, and even then, significant amounts of unsaturated hydrocarbons are present. Consequently our surface studies included C_3H_8, C_2H_4, and CH_4 as representatives of the species available during the course of the CVD process.

Mat. Res. Soc. Symp. Proc. Vol. 131. ©1989 Materials Research Society

EXPERIMENTAL APPROACH

The apparatus used in this work has been described elsewhere.[5] It consists of a UHV surface analysis system and reaction cell interfaced directly by a gate valve. Samples were transferred between the chambers by a magnetically coupled linear motion feedthrough and were heated from behind by electron bombardment. The sample temperature was monitored using a disappearing filament optical pyrometer. Experiments involved exposing a heated Si surface to a known fluence of a specific hydrocarbon species in the reaction cell. This was accomplished using a simple molecular beam arrangement or by backfilling the chamber to pressures on the order of 10^{-7} to 10^{-4} torr. After a specified reaction was completed, the sample was transferred into the analysis chamber where XPS was used to characterize the effects of the reaction. The samples were electronic grade (p-type) single crystal Si with (111) and (100) orientations. These were cleaned using Ar ion bombardment followed by annealing at 1100 K to restore crystal order. The hydrocarbon gases had a certified purity of 99.999%.

RESULTS AND DISCUSSION

a) Reaction Products

The reactions of C_3H_8, C_2H_4, and CH_4 on the Si(111) and (100) surfaces were studied as a function of exposure for surface temperatures in the range of 1062 K to 1495 K. For each temperature, the course of the reaction was monitored by following the evolution of the Si 2p and C 1s XPS peaks and the Si 2s plasmon loss peak. Figure 1 shows the Si 2s spectral region for a clean, annealed Si(111) surface before and after exposure to increasing fluences of C_2H_4 at 1062 K. For comparison, the Si 2s spectral region for SiC is also shown. The main spectral feature is the Si 2s peak. The less intense feature at about 17 eV is the bulk silicon plasmon loss peak. As the C_2H_4 exposure increases, the intensity of this peak decreases, while that of a second plasmon loss peak at about 22.5 eV increases. This latter feature corresponds to the bulk SiC plasmon and indicates the formation of bulk SiC from only gas-phase C_2H_4 and solid Si. This remarkable process raises questions concerning transport mechanisms which will be discussed in the following section.

Figure 2 shows the C 1s peak for successively greater C_2H_4 exposures at 1062 K. The binding energy for each of these peaks is 282.5 eV (± 0.25 eV) which is in good agreement with that reported for SiC (282.6 eV).[3,4] Estimates of the reaction product surface coverage corresponding to each exposure may be made using the C 1s peak intensities, the corresponding Si 2p intensities, and a numerical technique developed by Madey et al.[8] On this basis, the reaction product formed by C_2H_4 exposures of 1.4 x 10^{17} cm^{-2} and 1 x 10^{18} cm^{-2} may be shown to correspond to ~0.5 and ~1 monolayer, respectively. The deposit produced by the largest exposure of 4 x 10^{19} cm^{-2} consists of multiple layers and is characterized by the bulk SiC plasmon.

A major point established by the spectroscopy of the substrate and adspecies is that the reaction product for both the Si(111) and (100) surfaces is essentially carbidic in nature throughout the course of the reaction for the range of temperatures investigated. Given the spectral resolution of the conventional XPS technique used here, however, it is not possible to determine if the adspecies are hydrogenated.

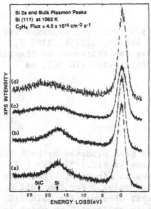

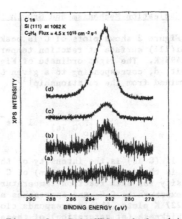

Figure 1. Si 2s XPS peak and associated plasmon loss features for (a) clean annealed silicon and silicon exposed at 1062 K to (b) 1 x 10^18 C_2H_4 cm^-2, (c) 4 x10^19 C_2H_4 cm^-2, and (d) SiC.

Figure 2. C 1s XPS peak for (a) clean annealed silicon and silicon exposed at 1062 K to (b) 1.4 x 10^17 C_2H_4 cm^-2, and (c) 1 x 10^18 C_2H_4 cm^-2, and (d) 4 x 10^19 C_2H_4 cm^-2.

Additional insight into the nature of C adspecies bonding is provided by the Monte Carlo simulations performed by Tiller and coworkers.[9] Their calculations show that C adspecies on Si(111) and (100) surfaces are well contracted into the surface, and each C atom is coordinated by Si much as it would be at the SiC surface. Although this picture should possibly be modified to account for at least the partial hydrogenation of the C adspecies, it suggests that minimal changes observed in the chemical state of C might be anticipated as the reaction proceeds from the submono-layer to the multilayer regime.

b) **Transport Mechanisms**

The observation of the multilayer SiC deposits indicates that signifi-cant diffusion of C or Si or both occurs during the reaction. Either process is quite feasible in Si since the diffusivities of both C and Si in Si are quite high for the temperatures of interest.[10] This contrasts with the situation for SiC where the diffusivities of C and Si are small,[11] and a continuous SiC deposit of only several layers would act as a diffusion barrier.[12]

To explore the issue of C in-diffusion for the Si(111) and (100) sur-faces, a number of experiments were performed. These spanned the submono-layer through the multilayer range and reaction temperatures of 1062 K through 1495 K. As discussed in detail elsewhere,[5] no evidence for C in-diffusion could be found for the Si(111) surface. Thus, because the formation of multilayer reaction products requires diffusional mixing, these results indicate that Si out-diffusion rather than C in-diffusion must be the key transport mechanism in the SiC growth process for the Si(111) and (100) surfaces. This is consistent with the simulation calcula-tions of Tiller and coworkers[9] which indicate that C adspecies penetra-tion into the Si surface is not energetically favored. We believe similar calculations would show that hydrogenation of the C adspecies, if it occurs, also limits C in-diffusion.

c)   Nucleation Mechanism and Growth Kinetics

Figure 3 shows plots of C 1s peak intensity versus C_2H_4 exposure for the Si(111) surface at reaction temperatures of 1062 K, 1227 K, 1395 K, and 1495 K. The right ordinate of Figure 3 shows the thickness of the SiC deposit, d, corresponding to a given C 1s peak intensity, I(C 1s), as determined from the relationship[13]

$$\frac{I(C\ 1s)}{I_o(C\ 1s)} = 1 - e^{-d/\lambda} \qquad (1)$$

where I_o (C 1s) is the intensity of the C 1s XPS peak produced by bulk SiC and λ is the free path (~3.2 nm) of C 1s electrons in SiC.[14] The C 1s intensity for each reaction temperature, as shown in Figure 3, increases with exposure to C_2H_4 and approaches a saturation value. For the 1062 K and 1227 K plots, this corresponds closely to that observed for bulk SiC. The thickness (at saturation) of these deposits is at least 9 nm which is comparable to the conversion layer thickness observed by Addamiano and Sprague[15].

The C 1s intensity versus exposure data for 1062 K and 1227 K are shown fitted to an equation having the same functional form as Equation 1. In this case, however, the thickness of the deposit, d, has been taken as the growth rate times the ratio of the C_2H_4 exposure to the C_2H_4 flux. (The ratio of the exposure to the flux is the reaction time.) Obtained as a fitting parameter, a growth rate of 2.9×10^{-2} nm s^{-1} or a reaction efficiency of 1.6×10^{-3} may be determined from the data for 1062 K and 1227 K. As discussed by Rhead et al[16] and Bauer and Poppa,[17] this functional form is representative of Frank-van der Merwe or layer-by-layer growth involving a two dimensional (2D) nucleation process. The data for 1395 K and 1495 K are also initially described by the same growth mechanism and reaction efficiency. For C_2H_4 exposures greater than 2×10^{18} cm^{-2} or a thickness of about 0.8 nm (i.e., ~2 atomic layers), however, apparent deviations are observed. Specifically, the plots for these reaction temperatures saturate at lower C 1s intensities than those for 1062 K and 1227 K.

A growth mechanism consistent with these data may be summarized as follows. For submonolayer to monolayer thick deposits, the 2D SiC nuclei grow and coalesce using readily available surface Si and adsorbed hydrocarbon species to form SiC. Once the deposit exceeds a monolayer thickness, however, Si must be supplied by out-diffusion. Because even one or two layers of SiC acts as a barrier to bulk diffusion, this most likely involves Si out-diffusion along the boundaries between adjacent nuclei followed by surface diffusion of Si and hydrocarbon adspecies to a 2D growth site on the SiC deposit. Although we have no direct structural information for the deposits formed at 1062 K and 1227 K, we suspect that the nuclei/grains are epitaxial based on estimates of surface mobility at this temperature.[9]

The saturation of SiC growth for temperatures of 1395 K and 1495 K may be ascribed to two competing processes, diffusion and sublimation, which limit the Si adspecies coverage on the SiC surface. At higher growth temperatures the grain size increases and the density of grain boundaries decreases[16,17] because faster surface diffusion favors growth over nucleation. Because bulk diffusion in SiC is extremely slow, a reduction in the density of grain boundaries results in a reduction in the major Si diffusion channels, and, consequently, the supply of Si at the

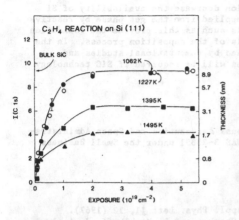

Figure 3. C 1s intensity produced as a result of C_2H_4 reaction on the Si(111) surface and plotted as a function of C_2H_4 exposure for surface temperatures of 1062 K, 1227 K, 1395 K, and 1495 K.

surface becomes transport limited. At the same time, the Si sublimation rate increases substantially at the higher growth temperatures. This has been observed experimentally in XPS characterization studies of SiC surfaces,[3] and it is consistant with expectations based on Si-SiC bond strengths for Si adspecies on SiC.[9] For the deposit thickness (exposure) at which the Si transport and sublimation rates are effectively equal, the SiC growth rate becomes zero since Si is no longer available on the SiC surface, and the C 1s peak due to SiC saturates. The saturation of the C 1s peak due to carbon in any form indicates that in the absence of Si adspecies either the reaction of C_2H_4 on the SiC surface has a low probability or it "passivates" the surface with respect to further C_2H_4 adsorption.[5]

Similar studies of the reaction of CH_4 and C_3H_8 on the Si(111) were also performed, and the initial reaction efficiencies for CH_4 and C_3H_8 were found to be 5×10^{-5} and 2×10^{-5}, respectively. Given the generally low C 1s peak intensity produced by the reaction of either CH_4 and C_3H_8, however, it was difficult to unambiguously identify the nucleation and growth mechanisms for these gases. Within the experimental uncertainty, the data could be described by either 2D nucleation/layer-by-layer growth or 3D nucleation and growth process.[16,17]

Studies of the C_2H_4 reaction on the Si(100) surface showed the nucleation and growth mechanism to be quite similar to that for the Si(111) surface, and a reaction efficiency of ~ 7×10^{-4} was observed. This value is a factor of three lower than that reported by Bozso and Yates[2] for the reaction of C_2H_4 on the Si(100) surface, but, given the complexities of the experiment, this represents reasonably good agreement.

CONCLUSIONS

Based on the experimental results and previously reported modeling studies,[7] a physical model for the initial step of the β-SiC CVD process may be proposed. The key features of this model are: i) During the initial temperature ramp, kinetic limitations in the gas phase decomposition of C_3H_8 lead to the formation of reactive species such as C_2H_4. ii) These reactive species allow the initiation of SiC growth at a low temperature (e.g., ~1000 K) during the ramp which prevents substantial evaporative losses of Si from the substrate. The initiation of growth at low temperatures also leads to a larger density of nucleation sites which results in a larger density of grain boundaries when the nuclei coalesce. iii) Si required for multilayer growth during the ramp is supplied by out-diffusion from the substrate through defects and grain boundaries in the SiC thin film. iv) Continued growth as the temperature increases produces an epitaxial microcrystalline surface, upon which SiC homoepitaxy occurs during the crystal process at 1673 K. v) At high temperatures (e.g., ≥ 1300 K) during the ramp and crystal growth at 1673 K, enhanced Si sublimation and decreased out-diffusion decrease the availability of Si from the substrate, and Si must be supplied from the gas phase by reactive species such as SiH$_2$. Physical models such as this provide the foundation for more accurate computational models of the deposition process. In the future, experimental CVD studies guided by computational studies and complemented by additional UHV studies will be required if SiC technology is to mature.

ACKNOWLEDGEMENTS

This research supported by National Aeronautics and Space Administration Contract No.'s NAS 3-24531 and NAS 3-23891 under the Small Business Innovative Research program.

REFERENCES

1. I.H. Kahn and R.N. Summergrad, Appl. Phys. Lett $\underline{11}$, 12 (1967).
2. F. Bozso, J.T. Yates, Jr., W.J. Choyke, and L. Muehlhoff, J. Appl. Phys. $\underline{57}$, 2771 (1985).
3. L. Muehlhoff, W.J. Choyke, M.J. Bozak, and J.T. Yates, Jr., Appl. Phys. Lett. $\underline{60}$, 2842 (1986).
4. F. Bozso, L. Muehlhoff, M. Trenary, W.J. Choyke, and J.T. Yates, Jr., J. Vac. Sci. Technol. $\underline{2}$, 1271 (1984).
5. C.D. Stinespring and J.C. Wormhoudt, J. Appl. Phys. (to appear).
6. S. Nishino, A.J. Powell, and H.A. Will, Appl. Phys. Lett. $\underline{42}$, 460 (1983).
7. C.D. Stinespring and J.C. Wormhoudt, J. Cryst. Growth, $\underline{87}$, 481 (1988).
8. T.E. Madey, J.T. Yates, Jr., and N.E. Erickson, Chem. Phys. Lett. $\underline{19}$, 487 (1973).
9. T. Takai, T. Halicioglu, and W.A. Tiller, Surface Sci. $\underline{164}$, 327 (1985).
10. R.C. Newman and J. Wakefield, Solid State Physics in Electronics and Telecommunications, $\underline{1}$, 318 (1960).
11. J.D. Hong, R.F. Davis, and D.E. Newbury, J. Mat. Sci. $\underline{16}$, 2485 (1981).
12. M.A. Taubenblatt and C.R. Helms, J. Appl. Phys. $\underline{59}$, 1992 (1986).
13. T.A. Carlson, Photoelectron and Auger Spectroscopy (Plenum Press, New York 1978) p. 261.
14. J. Sazajman, J. Liesegang, J.G. Jenkin, and R.C.G. Leckey, J. Electron Spectrosc. $\underline{23}$, 97 (1981).
15. A. Addamiano and J.A. Sprague, Appl. Phys. Lett. $\underline{44}$, 525 (1984).
16. G.E. Rheed, G.-M.Barthes, and C. Argile, Thin Solid Films $\underline{82}$, 201 (1981).
17. E. Bauer and H. Poppa, Thin Solid Films $\underline{12}$, 167 (1972).

LASER STUDIES OF THE SiH RADICAL/SURFACE INTERACTION DURING DEPOSITION OF A THIN FILM

PAULINE HO, RICHARD J. BUSS, AND WILLIAM G. BREILAND
Sandia National Laboratories, Albuquerque, NM 87185-5800

ABSTRACT

This paper presents a new method for studying the interaction of radicals with the surface of a depositing film using a combination of laser spectroscopy and molecular beam techniques. The reactivity of SiH molecules with the surface of a depositing a-Si:H film is measured to be at least 0.95, with no strong dependence on rotational state.

INTRODUCTION

Understanding the interactions of molecules with surfaces is important for the development and control of many thin-film materials processing technologies, including chemical vapor deposition (CVD), plasma-enhanced CVD, and plasma etching. These processes are often controlled by the chemical reactions that occur when gas-phase molecules collide with the substrate. In deposition systems, the starting compounds themselves may be relatively unreactive and the process can depend to a major extent on the generation of gas-phase species that are more reactive with the surface. For example, silane, which is used in both thermal and plasma-enhanced CVD processes, is relatively inert; reactive sticking coefficients for silane on hot silicon of 10^{-2}–10^{-5} have been observed [1,2]. However, decomposition of the starting compounds and subsequent gas-phase reactions produce radical species which can be much more reactive at the surface. Unfortunately, little information is available on the reactivity of such intermediate species with surfaces.

In this paper, we introduce a technique that can directly monitor the surface reactivity of an intermediate species from within a complex mixture of molecules and ions colliding with a surface upon which deposition or etching is occuring. The work in this paper will be described in more detail elsewhere [3]. The technique is illustrated by measurements of the interaction of the free radical SiH with the surface of depositing silicon. Using laser-excited fluorescence of molecules in a molecular beam, we are able to measure the reactivity of SiH with the surface of amorphous hydrogenated silicon (a-Si:H) during plasma deposition.

EXPERIMENT

Our method for studying the reactivity of a radical with a surface combines molecular beam and laser spectroscopy techniques. An effusive molecular beam produced from a silane plasma impinges on a silicon sample at 30 degrees from the surface normal. This beam contains virtually all the neutral and ionized species present in the plasma. A laser beam, directed parallel to the surface and propagating in the principal scattering plane, is tuned to an absorption frequency of a particular molecular species, in this case SiH, and thus selects one chemical component of the beam for study. The laser simultaneously excites both the molecules in the incident beam and molecules scattered from the surface. The spatial dependence of the fluorescence excited by the laser beam is measured by imaging light from the interaction region on an optical multichannel analyzer (OMA). The diode array is oriented with its long axis parallel to the laser beam. Measurements are made with the surface positioned a known distance from the laser beam and also with the surface removed. The difference between the spatial distribution of fluorescence with the surface in and out of the molecular beam gives information on the surface reactivity of SiH with the depositing film.

Only a general description of the experimental apparatus will be given in this paper–a more detailed description will be published elsewhere [3]. The experiments were carried out in a vacuum chamber which was pumped to a base pressure of 1×10^{-7} Torr with an oil diffusion pump and rotary roughing pump. The source for the molecular beam was a cylindrical glass chamber containing a silane plasma. A 10 mm diameter orifice connected the plasma tube to the scattering chamber. The pressure in the tube was 15 mTorr with a silane flow rate of 10 sccm during plasma operation. RF power (13.56 MHz, 30 W) was inductively coupled into the plasma through a coil wrapped around the glass tube.

Two slits were used in front of the plasma tube to define the molecular beam spatially. The use of a rectangular molecular beam rather than a round beam had two advantages. First, it allowed higher spatial resolution in the horizontal direction. Second, it increased the sensitivity to molecules scattered/desorbed from the surface because molecules emanating from a line source decrease in density as $1/r$ rather than the $1/r^2$ resulting from a point source. The sample was a 1×2 cm section of a silicon wafer mounted so that it could be reproducibly rotated in and out of the path of the molecular beam.

An excimer laser (XeCl, 15 ns, 150–220 mJ) was used to pump a dye laser at a repetition rate of 100 Hz. Laser light at ~410 nm was produced with pulse energies of 1–5 mJ. The laser beam was steered into the chamber through a quartz window with prisms and focused above the sample surface with a 350 mm f.l. lens. The laser beam intersected the molecular beam 4.1 cm from the orifice in the plasma tube. Fluorescence was collected at a right angle to the principal scattering plane and focused on the OMA (a doubly intensified photodiode array with 1024 active channels) through an interference filter (450 nm, 40 nm bandpass). This optical arrangement provided a slightly (~20%) magnified image of the interaction region. The OMA was gated on after the laser pulse to avoid interference from scattered laser light. Because SiH has a fluorescence lifetime of 534 ns [4], an OMA gate width of 1–2 μs was used to collect the fluorescence. Signals were collected and analyzed with the microcomputer in the OMA system. The signal was generally collected for 10^4 laser pulses before being stored. Data with the surface in and out of the molecular beam path were taken alternately. Background signals resulting from continuous emission from the plasma and pixel noise were also measured and subtracted from the SiH signals.

RESULTS AND DISCUSSION

In these experiments, the spectral selectivity of the laser-excited fluorescence technique was used to identify SiH and study it separately from the other species in the molecular beam. The spectroscopy of SiH is well known [5,6]. Figure 1 shows the fluorescence excitation spectrum obtained with laser wavelengths between 407 and 415 nm, along with calculated line positions for lines up to J=15.5 for the SiH $A^2\Delta - X^2\Pi$ transition. Although not visible in the figure, the lambda doubling of the rotational lines was just resolved. This spectrum is not corrected for variations in laser power and was obtained using a broader molecular beam than was used for the reactivity measurements to improve the signal levels.

The curve in Fig. 2a was obtained by spatially resolving the laser-excited fluorescence when the surface was out of the path of the molecular beam. The laser was tuned to 411.58 nm to excite only molecules in the J=5.5 state. The SiH in the molecular beam can be clearly seen. The upper curve in Fig. 2b shows the signal obtained when the surface was in the path of the molecular beam, 1.5 mm from the laser beam. This signal is dominated by fluorescence from SiH in the molecular beam, but there is a

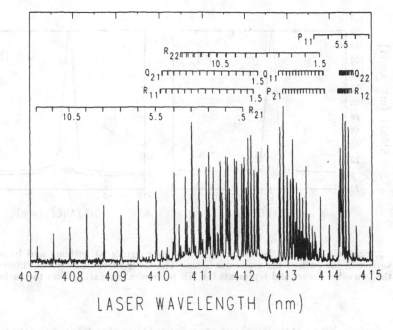

1. Fluorescence excitation spectrum of SiH in the molecular beam.

small additional component. Subtracting the signal obtained without the surface from that obtained with the surface yields the lower curve in Fig. 2b, which is the signal from SiH molecules emanating from the surface. The fact that the signals from the scattered/desorbed SiH molecules are only a few percent of the signals from the SiH in the molecular beam indicates that most of the SiH molecules react at the surface. A more quantitative analysis of the desorbing SiH required a numerical model of the experiment, described below. Note that these experiments measure the spatial distribution of state-specific SiH coming from the surface but do not distinguish SiH molecules which adsorb/desorb from the surface without reaction from SiH molecules produced by the surface reaction of some other species in the molecular beam, such as SiH_2 or SiH_3.

Interpreting data such as that shown in Fig. 2 requires a quantitative model of the experiment. The calculation is divided into three parts: 1) simulation of the spatial distribution of SiH number density at the intersection of the molecular beam and the laser beam, 2) calculation of the spatial distribution of flux of SiH molecules at the substrate surface, and 3) simulation of the spatial distribution of desorbed SiH number density along the line defined by the laser beam. The calculations are outlined below, but a more detailed description will be published elsewhere [3].

The flux of molecules incident on an element of the surface was obtained by integrating $1/r^2$ differential contributions from the orifice in the plasma tube. The integration was limited by line-of-sight calculations to eliminate contributions from regions of the orifice that were blocked by the slits. A similar procedure was used to calculate the distribution of SiH in the molecular-beam/laser-beam intersection. The laser-excited fluorescence technique measures number density, rather than flux, so the flux through a

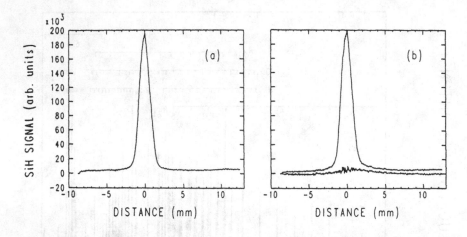

2. Spatially dispersed laser-excited fluorescence signals a) without and b) with a surface in the path of the molecular beam. The lower curve in b) is the difference of the two other curves and represents the SiH molecules emanating from the surface.

given area element at the laser beam was converted to number density using the average velocity of molecules in the beam assuming room temperature. For the conditions used in this particular experiment, the translation of excited molecules out of the viewing region during the observation time is not significant.

After the molecular beam has been turned on, the reactions at the surface reach a steady-state. If X% of the incident molecules react at the surface, then the flux of molecules emanating from an element of the surface is (100-X)% of the molecules incident on that surface element. In the case of direct elastic scattering, the surface acts as a "mirror" for the molecules, so the density distribution of the scattered molecules at the laser beam is the same as that for the molecular beam at a larger distance from the orifice, i.e. the image of the beam in the "mirror". Thus, the density distribution for specular scattering can be calculated using the procedure described for the molecules in the beam. In the case of adsorption/desorption, each area element of the surface becomes a source for desorbing molecules. The spatial distribution of the desorbing molecules will thus depend not only on the spatial distribution of incident molecules, but will also depend on the dynamics of the desorption. For simplicity, we have assumed that the molecules do not migrate and that they desorb with a cosine angular distribution. The density of desorbed molecules in a particular volume element of the laser beam is thus obtained by adding the contributions from the various area elements of the surface. As was the case for the incident beam, number density is obtained by dividing by the average velocity. For purposes of the calculation, the desorbing molecules were assumed to have a thermal velocity distribution at the surface temperature.

The width and shape of the molecular beam calculated from the measured slit parameters agree well with the experimental measurements shown in Fig. 2a, although the experimental results exhibit some extra intensity at the wings. The broadening in the wings was accounted for by including two small effects that result from distortion in the light collection optics and from the finite width of the laser which causes an apparent broadening due to the laser/molecular beam crossing angle.

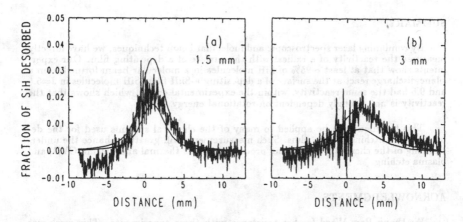

3. SiH signals at a) 1.5 and b) 3 mm laser–surface distances. Smooth curves give calculated signals for desorption of 4 and 8% of the incident molecules.

Fig. 3 shows the signals from desorbed SiH molecules obtained using laser–surface distances of 1.5 and 3 mm, along with calculated curves for desorption of 4 and 8% of the SiH in the molecular beam. These data show no evidence for specular scattering of SiH molecules from the surface. The experiment is very sensitive to elastically scattered SiH and less than 2% undergoes specular reflection. Rather, the width of the residual peak suggests that the molecules come from the surface with a broad angular distribution. Within the substantial noise, the shapes of the experimental curves agree with the shapes of the calculated curves. The experimental curves show the increasing peak width and the decreasing peak height with increasing laser–surface distance expected for molecules desorbing with a cosine distribution.

For the data shown in Figs. 2 and 3, the laser was tuned to the $R_{11}(5.5)$ line of SiH (411.58 nm). Thus, the results of these experiments rigorously pertain only to SiH molecules in J=5.5. To investigate the possibility that the reactivity of SiH varies with rotational state, experiments were also done using the $R_{21}(9.5)$ spectral line (408.30 nm). Although the lower signal levels observed using the $R_{21}(9.5)$ line lead to a larger experimental uncertainty, the reactivity obtained is the same as the results obtained for J=5.5. This indicates that there is not a strong dependence of the reactivity on rotational energy, and that the reactivity measurements obtained using specific rotational lines can be reasonably extrapolated to SiH molecules as a whole. Combining all the data indicates that at least ~95% of the SiH molecules incident on the surface react.

There is some information in the literature on the surface interactions of the $SiH_n(n \leq 3)$ species. From deposition rate measurements and kinetic modeling of silane plasmas, Perrin [7] and Miyazaki, et al. [8] estimated average reactive sticking coefficients for SiH_n (n≤3) of 0.7–0.8 and near 1, respectively. By analyzing deposition rates in masked and unmasked areas in a silane plasma, Perrin and Broekhuizen [9] concluded that SiH_3 had a reactive sticking coefficient of 0.1–0.2 in the 40–350°C temperature range. Although they are not directly comparable, these results are all compatible with our determination of a reactivity close to unity for SiH at room temperature. As mentioned in the introduction, reactive sticking coefficients for silane are on the order of 10^{-2}–10^{-4}, substantially lower than observed for the $SiH_n(n \leq 3)$ species.

SUMMARY

By combining laser spectroscopic and molecular beam techniques, we have directly measured the reactivity of a radical with the surface of a depositing film. Our experiments show that at least ~95% of SiH molecules in a molecular beam formed from a silane discharge react at the surface of a depositing a-Si:H film. SiH molecules in J=5.5 and 9.5 had the same reactivity, within the experimental errors, which shows that the reactivity is not strongly dependent on rotational energy.

This technique can be applied to many of the chemical systems used for the deposition and etching of thin films. Such measurements will greatly enhance the understanding of the chemistry underlying processes such as thermal and plasma CVD, and plasma etching.

ACKNOWLEDGMENTS

We thank Pam Ward for her assistance with these experiments. This work was supported by the U.S. Department of Energy under contract No. DE-AC04-76DP00789.

REFERENCES

1. R.J. Buss, P. Ho, W.G. Breiland and M.E. Coltrin, J. Appl. Phys. **63**, 2808 (1988).

2. S.M. Gates, Surf. Sci. **195**, 307 (1988).

3. P. Ho, W.G. Breiland and R.J. Buss, in preparation.

4. W. Bauer, K.H. Becker, R. Düren, C. Hubrich and R. Meuser, Chem. Phys. Lett. **108**, 560 (1984).

5. L. Klynning and B. Lindgren, Ark. Fysik **33**, 73 (1966).

6. K.P. Huber and G. Herzberg, *Molecular Spectra and Molecular Structure, IV. Constants of Diatomic Molecules,* (Van Nostrand Reinhold, New York, 1979).

7. J. Perrin, PhD Thesis, as cited in J. Perrin and T. Broekhuizen, Appl. Phys. Lett. **50**, 433 (1987).

8. S. Miyazaki, H. Hirata, S. Ohkawa, and M. Hirose, J. Non-Cryst. Solids **77** and **78**, 781 (1985).

9. J. Perrin and T. Broekhuizen, Appl. Phys. Lett. **50**, 433 (1987).

INTERACTION OF In ATOM SPIN-ORBIT STATES WITH Si(100) SURFACES

DOEKE J. OOSTRA, RUSSELL V. SMILGYS AND STEPHEN R. LEONE[†]
Joint Institute for Laboratory Astrophysics, University of Colorado and
National Institute of Standards and Technology, and Department of Chemistry
and Biochemistry, University of Colorado, Boulder, CO, 80309
[†]Staff member, Quantum Physics Division, National Institute of Standards
and Technology.

ABSTRACT

Scattering and desorption of In from Si(100) is investigated. Laser
induced fluorescence is used to probe the desorbing and or scattered
species. Auger Electron Spectroscopy is used to study the composition on
the surface. The results show that at surface temperatures below 820 K a
two dimensional layer of In desorbs by a half order mechanism. This is
explained by assuming two dimensional In islands on the surface. Above 820
K, desorption takes place by a first order mechanism. The desorption parameters appear to be spin-orbit state specific. The desorption energy for In
$^2P_{3/2}$ is 2.8 ± 0.4 eV and for In $^2P_{1/2}$ 2.5 ± 0.2 eV. The difference is
equal to the difference in the spin-orbit energy. So far no specular
scattering of In is observed, suggesting that the sticking coefficients are
unity.

INTRODUCTION

The interaction of molecular beams of group III elements with Si surfaces is being investigated extensively. The possibility to grow high
quality GaAs layers on Si has stimulated detailed studies of adsorption,
desorption and growth of Ga on Si(111) and Si(100) surfaces [1-5]. Al, Ga
and In are also used as dopants in Si-device technology. However, these
metals show strong surface segregation during MBE processing, which leads to
distorted dopant depth profiles [6,7]. This has motivated some studies on
the interaction of In with Si surfaces [8].

In this paper we investigate adsorption and desorption of In on Si(100)
for both spin-orbit states $^2P_{1/2}$ and $^2P_{3/2}$ (upper state). This is done for
two reasons: Firstly, at the Si surface In is chemically bound [8] suggesting that there is no spin-orbit character present. In vacuum the energy
between the two spin-orbit states, ΔE, is 0.27 eV. Here, we examine whether
this energy difference, ΔE, shows up as a difference in the desorption
energy between the two states. This information will provide insights into
the microscopic mechanisms of thermal desorption and the influence of the
electronic states on adsorption and/or desorption.

EXPERIMENTAL

The experiments are performed in a previously described UHV chamber
with a base pressure of 5×10^{-11} Torr [3-5,9]. A differentially pumped
chamber houses commercially obtained Knudsen evaporation sources. A silicon
(100) sample is mounted on a manipulator. By simple rotation the surface is
positioned in front of the In beam from a Knudsen source or is examined by
low energy electron diffraction (LEED) or by auger electron spectroscopy

Mat. Res. Soc. Symp. Proc. Vol. 131. ©1989 Materials Research Society

(AES). The In beam comes in at an angle of $\approx 20°$ from the surface normal. Laser induced fluorescence (LIF) probing of gas species is measured parallel to the surface. The laser can probe the incoming In beam, In desorbing from the surface, or the specular reflected beam.

Cleaning of the silicon surface is done by a sputter and anneal method. After throroughly rinsing the silicon surface with trichloroethane and ethanol, the surface is mounted on the sample holder. In vacuum the surface is heated resistively. After cooling down the sample, the surface is sputtered by 1.5 keV Ar$^+$ ions and subsequently annealed at 900°C. This always leads to sharp 2x1 LEED patterns and Auger spectra without any trace of contaminants. Knall et al. show that the growth mechanism of In on Si(100) 2x1 is very sensitive to carbon contamination [8]. They find that In grows with a 2x2 LEED pattern at room temperature only if less than 0.01 ML of carbon is present. After a day of experimenting we always find the 2x2 pattern indicating that contamination of the surface is not of importance in our experiments.

RESULTS AND DISCUSSION

A. Adsorption

Sticking of In to the surface is investigated for surface temperatures up to 1000 K. No specularly scattered In atoms are observed by LIF for either spin-orbit state above the noise level. This means that less than 5% of the In $^2P_{1/2}$ and In $^2P_{3/2}$ are specular scattered as $^2P_{1/2}$ and $^2P_{3/2}$ respectively. Thus the sticking coefficient of the ground state In $^2P_{1/2}$ is higher than 0.95. A possibility exists that there is relaxation of the $^2P_{3/2}$ to $^2P_{1/2}$ during specular scattering. This could not be determined accurately because of the low initial population of the $^2P_{3/2}$ state and the relatively high $^2P_{1/2}$ background. However, we estimate, that the sticking coefficient of In $^2P_{3/2}$ is $\gtrsim 0.6$. This is examined more thoroughly in a forthcoming paper [9].

B. Desorption order

Approximately one monolayer of In is deposited on the Si(100) surface. The desorption energy and the preexponential factor of In are determined by isothermal desorption measurements at different surface temperatures using LIF-probing and AES. AES measures the In coverage N as a function of time, t, whereas LIF-probing measures the desorption rate (-dN/dt) of each spin-orbit state specifically as a function of t. The desorption kinetics are described by the equation: $-dN/dt = kN^n$, where k is the desorption rate constant and n is the desorption order.

LIF-probing of both In spin-orbit states is performed at surface temperatures ranging from 790 K to 870 K. An example of the log of the LIF $^2P_{1/2}$ signal versus time is shown in Fig. 1. For first order desorption this presentation yields a straight line. The plot shows two contributions. First the curve shows a log behavior. Then after a distinct break the curve shows a linear behavior. Comparable to Ga on Si(100), these features are explained as follows: Initially three dimensional In islands are present, from which desorption takes place. After the break point these islands have desorbed completely and a two dimensional layer is left, from which desorption takes place. This explanation is in agreement with the results of Knall et al [8]. They conclude from LEED, AES and SEM mesurements that In grows two dimensionally up to 0.5 ML. At higher coverages three dimensional island formation takes place. We concentrate here on the desorption of the two dimensional layer of In.

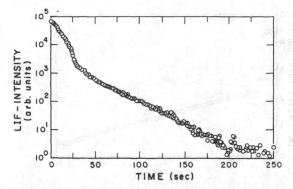

Fig 1: Log of the LIF-intensity of desorbing In $^2P_{1/2}$ as a function of time. Surface temperature is 830 K.

Figure 2 shows LIF measurements of the desorption rate of In $^2P_{1/2}$ for the two dimensional layer at different surface temperatures. In the log-log presentation the measurements yield linear plots with a slope equal to the desorption order n. Figure 2 shows that for surface temperatures above 820 K the desorption order is 1.0, and below 820 K the desorption order changes to 0.5. This conclusion is confirmed by AES. AES of the In coverage as a function of time is performed at surface temperatures between 745 K and 805 K. At all temperatures a desorption order n - 0.5 is found. An example of one Auger measurement is shown in Fig. 3. In the figure the square root of the In Auger signal is plotted versus time, showing a linear plot for this half order desorption. The plot shows two components, just as in the case for the LIF spectra. First the signal from the three dimensional islands is observed, and after a distinct break the two dimensional In layer desorbs.

A half order desorption mechanism can be explained by assuming that the layer of In consists of two dimensional islands [10], and that desorption takes place predominantly from the perimeters of the islands because those atoms have a lower coordination number. The atoms from the perimeter can move away from the island, are mobile on the surface, and desorb. If the

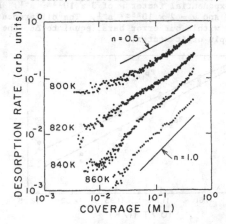

Fig 2: Log of the LIF-intensity of desorbing In $^2P_{1/2}$ as a function of log of the In coverage at different surface temperatures. n - 1.0 and n - 0.5 indicate slopes for first and half order desorption respectively. The vertical offsets between the curves have no physical meaning.

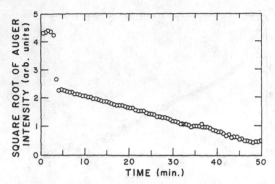

Fig 3: Square root of the Auger intensity of In coverage as a function of time. Surface temperature is 760 K.

total In coverage is N, the number of atoms in the perimeters is roughly equal to $N^{1/2}$. This leads to the half order desorption form of the kinetics. At higher surface temperatures the desorption mechanism changes to first order. That means that every atom acts individually in the desorption. The change in the order can be explained by assuming that the two dimensional island structures break up at these higher temperatures.

C. Desorption energies

The LIF signals from In desorption for surface temperatures above 820 K are fitted to a first order desorption mechanism. The desorption rate constants are obtained using a weighted non-linear least squares fitting routine. The rate constants obtained for both spin-orbit states are shown in Fig. 4. In this Arrhenius plot the slope of the line yields the activation energy for desorption and the intercept yields the preexponential factor. The plots show different desorption energies for the $^2P_{1/2}$ state and the $^2P_{3/2}$ state. Fitting the data yields for In $^2P_{1/2}$ a desorption energy U of 2.5 ± 0.2 eV and a preexponential factor ν of $3 \times 10^{13\pm1}$ s^{-1}. For the In $^2P_{3/2}$: $U = 2.8 \pm 0.4$ eV and $\nu = 2 \times 10^{15\pm2}$ s^{-1}. The difference between the desorption energies is, within our error bars, equal to ΔE, the difference in energy between the spin-orbit states.

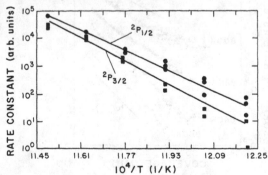

Fig 4: Log of the desorption rate constants as measured by LIF as a function of the inverse of the surface temperature. All data points relate to first order desorption.

Microscopic models of thermal desorption describe the surface-atom potential by a Morse oscillator [11,12]. By phonon interaction the system steps up and down the ladder of states. For In there are two outgoing channels: The $^2P_{1/2}$ state and, 0.27 eV higher, the $^2P_{3/2}$ state. Our results can be explained as follows: For In on the surface, in the bottom of the potential well, no spin-orbit character is present. The spin-orbit character is developed when the system is highly excited. The atom can escape into the vacuum into the $^2P_{1/2}$ state or the system can be excited further up the ladder, and the atom can escape into the $^2P_{3/2}$ state. This model has been tested for Ga desorption from Si(100) [13]. For both Ga states the same desorption energy is observed. When the atoms have no spin-orbit character at the surface, the desorption rates for both states are expected to be equal to the total desorption rate. Therefore we expect to find the same desorption energy for each of the states [13]. Hence, the observation of the energy difference, ΔE, in case of In is surprising. It may indicate that the spin-orbit character of the desorbing atoms is already determined on the surface, in the bottom of the potential well. There can be different desorption sites on the surface. If the In atoms do not rapidly interchange between the sites, the results can be explained by assuming that the final spin-orbit state of an In atom depends on its desorption site. On the other hand, the finding of the energy difference, ΔE, may very well be fortuitous, because of the large error bars. More detailed experiments are necessary to settle this.

The $^2P_{3/2}$ state has both a higher desorption energy and a higher pre-exponential factor. This higher value of ν might be somewhat artificial. The fitting routines used have a strong tendency to increase the intercept when increasing the slope. On the other hand there may be physical reasons for this behavior. For fully mobile surface atoms $\nu \approx 10^{13}$ s^{-1}, whereas for fully immobile surface atoms $\nu \approx 10^{16}$ s^{-1}. Above it is argued that different states might be desorbing from different sites. Therefore a higher ν, in case of $^2P_{3/2}$, might be related to higher mobilities. Reordering of the surface during desorption can also lead to changes in ν [2,10]. When an adatom is removed from the surface, the surface may relax to a particular reconstruction. This process may be influenced by the spin-orbit character of the leaving In atom.

The desorption parameters for the half order desorption mechanism can also be obtained from the Auger data. A thorough discussion of the Auger results is presented in a forthcoming paper [9].

CONCLUSIONS

Desorption of two dimensional layers of In from Si(100) is investigated using LIF and AES. The measurements show that at surface temperatures below 820 K, In desorbs by a half order desorption mechanism. This is explained by assuming that In is present on the surface in two dimensional islands and that desorption takes place from the perimeters of the islands. For surface temperatures above 820 K In desorbs by a first order mechanism. This indicates that there is no collective interaction between the In atoms at the higher temperatures. This is explained by assuming that above 820 K the two dimensional islands break up. In the first order desorption region the desorption energy and preexponential factor are determined for both In spin-orbit states. The results suggest that the difference in desorption energy of the two spin-orbit states is equal to ΔE, which is the difference in energy between the two spin-orbit states of In in vacuum. Thus far no direct scattering of In from the Si surface has been observed.

244

ACKNOWLEDGMENTS

This work is generously supported by the Air Force Office of Scientific Research, Electronic and Material Sciences.

REFERENCES

1. K. L. Carleton, B. Bourguignon, R. V. Smilgys, D. J. Oostra, and S. R. Leone, presented at the 1988 MRS Spring Meeting, Reno, NV, 1988.
2. M. Zinke-Allmang and L. C. Feldman, Surf. Sci. 191, L749 (1987).
3. B. Bourguignon, R. V. Smilgys, and S. R. Leone, Surf. Sci. 204, 473 (1988).
4. B. Bourguignon, K. L. Carleton, and S. R. Leone, Surf. Sci. 204, 455 (1988).
5. K. L. Carleton and S. R. Leone, J. Vac. Sci. Technol. B5, 1141 (1987).
6. Y. Ota, J. Appl. Phys. 51, 1102 (1980).
7. J. C. Bean, Appl. Phys. Lett. 33, 654 (1978).
8. J. Knall, J.-E. Sundgren, G. V. Hansson, and J. E. Greene, Surf. Sci. 166, 512 (1986) and ref. 1 and 2 cited herein.
9. D. J. Oostra, R. V. Smilgys, and S. R. Leone, (in preparation).
10. R. Kern, G. Le Lay, and J. J. Metois, in Current Topics in Materials Science, Vol 3, edited by E. Kaldis, (North Holland, Amsterdam, 1979) p.
11. G. De Sarkar, U. Landman, and M. Rasolt, Phys. Rev. B21, 3256 (1980).
12. G. Korzeniewski, E. Hood, and H. Metiu, J. Vac. Sci. Technol. 20, 594 (1982).
13. K. L. Carleton, B. Bourguignon, and S. R. Leone, Surf. Sci. 199, 442 (1988).

REACTIVE ATOM-SURFACE SCATTERING
THE ADSORPTION AND REACTION OF ATOMIC OXYGEN
ON THE Si(100) SURFACE

J. R. ENGSTROM, M. M. NELSON AND T. ENGEL
Department of Chemistry, University of Washington, Seattle, WA 98195

ABSTRACT

The adsorption and reaction of atomic oxygen on the Si(100) surface has been examined by employing supersonic beam techniques, X-ray photoelectron spectroscopy and mass spectrometry. Atomic oxygen adsorbs with a unit probability of adsorption on the clean Si(100) surface. The probability of adsorption decreases monotonically with increasing coverage. At surface temperatures above approximately 1000 K, the adsorption of atomic oxygen results in the gasification of the substrate, producing SiO(g). In comparison to molecular oxygen for this reaction, where two surface intermediates were implicated, only one surface intermediate is formed from the reaction of atomic oxygen with the Si surface.

INTRODUCTION

Unlike the interaction of molecules and unreactive (i.e., noble gas) atoms with solid surfaces, very little is known about the interaction between gas phase radicals and surfaces. This is despite the fact that radical-surface chemistry plays a fundamental role in many developing technologies, such as plasma processing. We have examined in detail the interaction of atomic oxygen with the Si(100) surface employing X-ray photoelectron spectroscopy (XPS), mass spectrometry and ion scattering spectroscopy. Although the results of these studies will undoubtedly have a practical impact on various technologies (e.g., microelectronic device fabrication), it is also hoped that a more fundamental understanding of the differences between the reactivity of atomic and molecular species on solid surfaces will emerge.

EXPERIMENTAL PROCEDURES

The experiments were conducted in a molecular beam apparatus that has been described in detail previously [2]. Briefly, the apparatus consists of a stainless steel UHV chamber, which contains facilities for XPS, Ar^+ ion sputtering, and mass spectrometric analysis of the desorbing gas phase reaction products. The Si(100) crystal is mounted on a liquid-nitrogen-cooled sample holder which may be rotated to permit variation of the incident angle of the molecular beam. The supersonic oxygen atom beam is generated by a radio frequency glow discharge, similar to that described previously [3]. Once formed, the beam passes through two intermediate differential pumping stages before striking the sample. In the first intermediate chamber the beam passes through a 10 cm long, 3 kV-cm^{-1} deflection field, which removes any ions present in the beam, whereas in the second intermediate chamber the beam may be chopped by a 50% duty cycle rotating blade chopper. Typically,

Mat. Res. Soc. Symp. Proc. Vol. 131. ©1989 Materials Research Society

the level of dissociation achieved in the beam was 40-70%, with corresponding fluxes of atomic oxygen impinging on the sample of 0.01-1.0 ML-s^{-1} (1 monolayer = 1 ML ≡ 6.8 x 10^{14} atoms-cm^{-2}).

RESULTS AND DISCUSSION

Previous work involving the interaction of molecular oxygen with the Si(100) surface [2] has identified two major reaction pathways, the relative "selectivity" depending on substrate temperature. Below approximately 1000 K, the interaction involves the dissociative chemisorption of molecular oxygen, leading eventually to the formation of an amorphous SiO$_2$ overlayer. On the other hand, above approximately 1000 K, and at sufficiently low pressures, the interaction is characterized by adsorption and reaction leading to the formation of SiO(g). The former interaction is termed "passive" oxidation, whereas the latter is termed "active" oxidation. It is of interest to compare and contrast the behavior of atomic and molecular oxygen in regard to these two major reaction pathways.

We have examined the kinetics of the adsorption of atomic oxygen on the Si(100) surface by employing XPS. The coverage-exposure relationships for both atomic and molecular oxygen are displayed in Fig. 1. Several points can be made from the data displayed in Fig. 1. First of all, for all

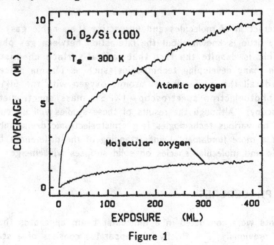

Figure 1

The coverage-exposure relationships for both atomic and molecular oxygen on the Si(100) surface. In both cases, the substrate temperature was 300 K and the angle of incidence of the molecular (or atomic) beam was 73° from the surface normal. Average translational energies of the beam constituents were 22 kcal-mol^{1} for atomic oxygen and 9 kcal-mol^{-1} for molecular oxygen. Coverages have been corrected for attenuation of the O(1s) photoelectrons by utilizing standard procedures (4), and employing an inelastic mean free path, $\lambda_{O(1s)}$, of 26Å.

coverages, the reaction probability (given by the slope of these curves) of atomic oxygen is much greater than that of molecular oxygen. This fact, for example permits us to essentially ignore the presence of nondissociated molecular oxygen in the beam of atomic oxygen, which greatly simplifies the analysis. Secondly, two stages of adsorption of atomic oxygen are readily apparent—the initial, "fast" stage of adsorption reaches completion near 3-4 ML. A second, much slower stage of adsorption continues above coverages of 3-4 ML, and does not saturate even at coverages of 10 ML and greater. This behavior, i.e., two stages of adsorption, is qualitatively similar to that observed for molecular oxygen, however, for molecular oxygen the "fast" stage of adsorption reaches completion near 1 ML [2].

Curves such as those displayed in Fig. 1 can be utilized to derive the initial (zero-coverage) probability of adsorption. Separate experiments [1] conducted at differing substrate temperatures (120-800K), average translational energies (4-22 kcal-mol^{-1}) and angles of incidence (23°-73° from normal) indicate that for all reaction conditions considered the initial probability of adsorption of atomic oxygen is equal to 1.0 ± 0.2. These observations are consistent with facile, "direct" adsorption of atomic oxygen. In contrast, the initial probability of adsorption of molecular oxygen on the Si(100) surface depends in a complex fashion on both average beam energy and substrate temperature, and varies between approximately 0.01 and 0.05 [2]. A similar investigation of the coverage-exposure relationship for atomic oxygen indicates that the adsorption kinetics at finite coverages are essentially independent of both beam energy and angle of incidence. The "fast", initial stage of adsorption has also been found to be essentially independent of substrate temperature. However, in the "slow" stage of adsorption, the rate of oxide growth has been found to increase by approximately a factor of two as the substrate temperature is increased from 300 to 800 K. Thermally activated diffusion across the oxide film could play a part in explaining this temperature dependence.

We have also examined the kinetics of SiO(g) formation from the reaction of atomic oxygen with the Si(100) surface. The transition from "passive" to "active" oxidation has been shown recently to depend strongly on the oxygen partial pressure (i.e., flux) [5]. For our flux conditions, the transition occurs near 1000 K. Previous work [2,6] has identified the following reaction sequence for the reaction of molecular oxygen on the Si(100) surface:

$$O_2(g) \xrightarrow{S_{O_2}} I_1 \xrightarrow{k_1} I_2 \xrightarrow{k_2} SiO(g) \tag{1}$$

i.e., the reaction pathway involves two intermediates, which are produced sequentially, with the further reaction of each intermediate following first-order kinetics. Note that the previous analysis [2,6], which identified a "fast" and a "slow" step, could not unambiguously assign these two measured rate coefficients as to their order of occurrence in the mechanism given by Eq. (1). Likewise, the nature of the intermediates, I_1 and I_2, was also not entirely clear.

Modulated molecular beam techniques have been applied to evaluate the kinetics of the reaction $O(g) + Si(s) \rightarrow SiO(g)$. A Fourier analysis of the desorbing $SiO(g)$ product waveforms, presented elsewhere [1], indicates that the reaction of atomic oxygen follows a single-step, first-order process, involving a single intermediate. Thus, for atomic oxygen the reaction mechanism is given by

$$O(g) \xrightarrow{S_0} I' \xrightarrow{k'} SiO(g) \qquad (2)$$

The first-order rate coefficients derived from this analysis and representing a number of different temperatures are plotted in Fig. 2 versus the reciprocal temperature.

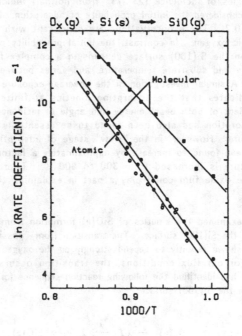

Figure 2

Reaction rate coefficients for the formation of $SiO(g)$ from the reaction of the Si(100) surface with atomic oxygen (open circles) and molecular oxygen (solid symbols). A single reaction step (i.e., intermediate) was found for atomic oxygen, whereas a sequential, two step process was found for molecular oxygen.

From these data we derive a first-order preexponential factor of 1.3×10^{19} s^{-1} and an activation energy of 79 ± 3 kcal-mol^{-1}. More importantly, however, is a comparison of these data to those found for the reaction with molecular oxygen. Lines representing least-squares fits to the "fast" and "slow" rate coefficients deduced for the molecular oxygen reaction are displayed also in Fig. 2. As may be seen, the rate coefficients k' reported here for atomic oxygen are *identical*, within experimental uncertainty, to those found for the "slow" reaction step involving molecular oxygen.

The most reasonable interpretation of the results displayed in Fig. 2 is the following. The adsorption of atomic oxygen results directly in the formation of the intermediate designated I_2 in Eq. (1). Consequently, the rate coefficient measured for atomic oxygen, k', is identical to that measured for the "slow", second step for molecular oxygen, i.e., k_2. A likely candidate for the intermediate I_2 is an adsorbed diatomic-like monoxide species, Si-(Si=O)-Si [7,8]. This intermediate is formed directly from the attachment of atomic oxygen to the dangling bond states present on the Si(100) surface. Thus, the rate coefficient measured here for atomic oxygen, which is equivalent to the "slow" step for molecular oxygen, represents the desorption rate coefficient for this adsorbed monoxide species. The preferential, direct formation of this intermediate from atomic oxygen, as compared to molecular oxygen, would be consistent with recent work employing SEXAFS [9], which suggests that the predominant species formed at room temperature from molecular oxygen on the Si(100) surface is a bridging atomic species, i.e., Si-O-Si.

CONCLUSION

In conclusion, both quantitative (e.g., reaction probabilities) and qualitative differences (e.g., reaction pathways) have been observed concerning the reactivity of atomic and molecular oxygen. It can be reasonably expected that similar behavior will be observed for other systems involving comparisons of atomic and molecular reactivity on solid surfaces.

ACKNOWLEDGMENT

This research was supported by the AFOSR under grant 87-0166.

References

1. J.R. Engstrom, M.M. Nelson, and T. Engel, to be published.

2. M.P. D'Evelyn, M.M. Nelson, and T. Engel, Surface Sci. 186, 75 (1987).

3. S.J. Sibener, R.J. Buss, C.Y. Ng and Y.T. Lee, Rev. Sci. Instrum. 51, 167 (1980).

4. M.P. Seah and W.A. Dench, Surface Interface Anal. 1, 2 (1979).

5. R.E. Walkup and S.I. Raider, Appl. Phys. Lett. 53, 888 (1988).

6. M.L. Yu and B.N. Eldridge, Phys. Rev. Lett. 58, 1691 (1987).

7. A.J. Schell-Sorokin and J.E. Demuth, Surface Sci. 157, 273 (1985).

8. G. Hollinger, J.F. Morar, F.J. Himpsel, G. Hughes and J.L. Jordan, Surface Sci. 168, 609 (1986).

9. L. Incoccia, A. Balerna, S. Cramm, C. Kunz, F. Senf and I. Storjohann, Surface Sci. 189/190, 453 (1987).

ATOM- AND RADICAL-SURFACE STICKING COEFFICIENTS MEASURED USING RESONANCE ENHANCED MULTIPHOTON IONIZATION (REMPI)

ROBERT M. ROBERTSON[*] AND MICHEL J. ROSSI[**]
[*]Present Address: Applied Materials Corporation, Santa Clara CA 95054
[**]Department of Chemical Kinetics, SRI International, Menlo Park CA 94025

ABSTRACT

Sticking coefficients γ of neutral transient species at ambient temperature were measured using in situ Resonance Enhanced Multiphoton Ionization (REMPI) of the transients in a Knudsen cell. γ for I and $CF_3I^{\ddagger}$ on a stainless steel surface were 0.16 and >0.5, respectively, whereas γ for CF_3 on the same surface was measured to <0.01; γ of SiH_2 on a growing carbon containing amorphous silicon surface was 0.11; this value increased to 0.15 for interaction of SiH_2 with a "pure" growing silicon-hydrogen surface, and γ of $SiH_2^{\ddagger}$ on both types of surfaces was found to be >0.5.

INTRODUCTION

Due to the ever shrinking dimensions of thin film devices and VLSI chips, it is essential to control the different processes involved in VLSI chip production. In order to achieve control over these processes, fundamental understanding in terms of elementary reactions occurring in the gas phase and on the solid substrate is required. Our approach involves selective generation of one and only one atomic or molecular species using photolytic methods and to investigate its interaction in situ with a given surface under experimental conditions that come as close as possible to process conditions.

In this paper we present results on the sticking coefficients of several atomic and molecular species on surfaces of practical interest. The use of the term "sticking" is ambiguous in the case of wall deactivation of vibrationally excited species, because the process may not correspond to mass accomodation but rather to energy transfer or energy accomodation. We use this term only to describe the disappearance of the detected species due to heterogeneous interaction with the surface of interest.

EXPERIMENTAL

The experimental apparatus has been described in detail elsewhere [1], so that only a brief description will be given here. The Very Low Pressure Photolysis (VLPΦ) reactor is a Knudsen cell, which is part of a flowing gas experiment. The pressure inside the VLPΦ reactor is in the mTorr range (give or take a decade), so that the molecules undergo predominantly gas-wall collisions. The molecules of interest are generated from appropriate precursors either via IR-MPD using a pulsed CO_2 TEA laser or via single photon UV photolysis using a pulsed excimer laser.

The measurement principle rests on the fact that one compares the rate of the process of interest with the rate of a "reference" process. By calibrating this "reference" process for the atom or free radical of interest, one can then put this relative comparison of rates on an absolute basis. The Knudsen cell used in our investigations uses the gas-wall collision rate as a reference process for decay processes on the tens of μs time scale, whereas slow decay processes are measured against the effusion of the species out of the Knudsen cell. Consequently, fast transient decay processes are monitored in situ using time dependent resonance enhanced multiphoton ionization (REMPI) detection of the atom or free radical, whereas slow processes are followed either in steady state or in a time resolved (ms) manner using quadrupole mass spectrometry.

The Knudsen cell is displayed schematically in Figure 1. It is a stainless steel six-way cross coated with gold on the inside. The cell is equipped with an inner tube that is oriented coaxially to the beam of the photolysis laser. The pulsed laser generates the neutral transients initially in a fraction of the volume of the inner tube in a collisionless manner. Subsequently, the transient species fill the volume of the inner tube homogeneously by molecular diffusion on a μs time scale (Table I). This process of diffusion is followed by gas-wall collisions taking place on a longer time scale (see Table I). The decay of the REMPI signal (k/s^{-1}) of the transient is compared to the gas-wall collision frequency ω, hence a sticking coefficient γ is calculated ($\gamma=k/\omega$), which describes the number of wall collisions required to remove the transient species irreversibly from the gas-phase.

The subsequent fate of the radical is to escape from the inner tube into the plenum of the Knudsen cell, and finally escape into the vacuum chamber on a progressively longer time scale (Table I). In experiments with CF_3 radical, the inner tube was removed. In this case mass spectrometric observation of the loss of reactant and build-up of stable product molecules was used to determine an accurate numerical value of γ [2].

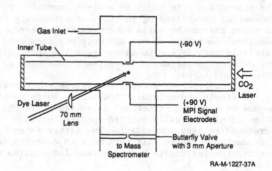

Figure 1

The stainless steel VLPΦ reactor for measurement of sticking coefficients.

RA-M-1227-37A

RESULTS AND DISCUSSION

The sticking coefficient of I atoms and highly vibrationally excited CF_3I molecules on stainless steel at ambient temperature was determined using the following generation scheme:

$$CF_3I + nh\nu \rightarrow CF_3 + I \tag{1}$$
$$\rightarrow CF_3I^{\ddagger} \tag{2}$$

At fluences below 2 J cm^{-2}, IR-multiphoton excitation causes only partial decomposition of CF_3I with a sizable fraction of the irradiated CF_3I in

TABLE I

CHARACTERISTIC RATE CONSTANTS FOR THE VLPΦ/REMPI EXPERIMENT

T/K, M/amu, v_{th}/cm s^{-1}, r_B/cm, A/cm^2, V/cm^3

Process		Rate Constant (s^{-1})	Time Constant (for SiH_2)
Production			$2\mu s$
Dilution	k_d	$\sim v_{th}/r_B$ 30000$(T/M)^{1/2}$	$10\mu s$
Wall collision	ω	$v_{th}A_T/4V_T$ 9000$(T/M)^{1/2}$	$35\mu s$
Escape tube	k_{eT}	$v_{th}A_{eT}/4V_T$ 135$(T/M)^{1/2}$	2.3 ms
Escape reactor	k_{eR}	$v_{th}A_{eR}/4V_R$ 0.24$(T/M)^{1/2}$	1.3 s

high lying vibrational states. The REMPI spectrum of IR-multiphoton excited CF_3I in the range of 460 to 490 nm shows sharp resonances due to both I in the ground state $(5p^5$ $^2P_{3/2})$ and I in the upper excited spin-orbit state $(^2P_{1/2})$, that is I^* [3]. The origin of I^* can be traced back to the presence of highly excited CF_3I, that is $CF_3I^{\ddagger}$, which undergoes single photon dissociation to $CF_3 + I^*$ in high quantum yield [4]. IR-MPD of CF_3I cannot lead to excited I in the upper spin state. Both I and I^* are ionized through a [3 + 1] REMPI process, and thermochemical arguments confer a lower limit of 5900 cm^{-1} of internal energy to $CF_3I^{\ddagger}$, roughly corresponding to one quantum of vibrational energy in each of the normal modes of CF_3I [3].

The time dependence of the REMPI signals for I monitored at 474.5 nm and for $CF_3I^{\ddagger}$ (via REMPI of I^*) monitored at 477.8 nm is displayed in Figure 2. The rapid decay of the REMPI signal in the first 30 μs corresponds to homogeneous filling of the inner tube by I or $CF_3I^{\ddagger}$ with a rate constant given in Table I.

Figure 2

Time dependence of the
total REMPI signals
at 474.5 nm (I, circles)
and 477.8 nm
($CF_3I^{\ddagger}$ squares).

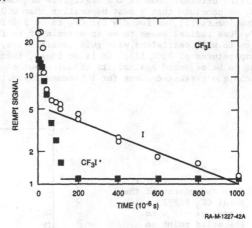

RA-M-1227-42A

The I REMPI signal decays with a unimolecular rate constant of 2.2×10^3 s^{-1} leading to a sticking coefficient $\gamma=0.16$ with ω from Table I. The relevant surface is the inner surface of the stainless steel tube that was exposed to CF_3I, CF_3 and I. The decay of the REMPI signal monitoring the time dependent density of $CF_3I^{\ddagger}$ is merged with the molecular diffusion of that species within the inner tube, so that the two time scales can not be separated. This situation is presented in Figure 2, where the REMPI signal at 477.8 nm rapidly falls to a constant level with a unimolecular rate constant of 2.2×10^4 s^{-1}, thus yielding a lower limit of 0.5 for γ.

Kinetic data on wall deactivation of atoms and free radicals are sparse, and even more so for vibrationally excited species. The sticking coefficient for atomic oxygen on single crystal silicon has been found to be essentially unity [6], whereas atomic bromine and chlorine hardly interact at all with a Teflon or halocarbon wax surface [7]. The present data indicate a rapid deactivation of I on an oxidized metal surface, which is perhaps surprising in view of the expected weak bonding of I on an oxide surface. Highly vibrationally excited CF$_3$I with approximately 5900 cm^{-1} of excess energy transfers energy upon essentially every collision, a finding that is not unexpected for those internal energies. Similarly, the fate of I after a deactivating collision with the VLPΦ vessel wall is not addressed in the present experiments. We observe molecular I$_2$ by mass spectrometry as a result of I interaction with the vessel walls and subsequent recombination of I on the walls of the reaction vessel.

The sticking coefficient of CF$_3$ free radical was measured on the stainless steel surface of the inner tube of the VLPΦ vessel using hexafluoroacetone, CF$_3$COCF$_3$, as a precursor. The IR-MPD of CF$_3$COCF$_3$ has been discussed before [1] and is briefly presented in equation (3):

$$CF_3COCF_3 + nh\nu \rightarrow CF_3 + COCF_3 \qquad (3)$$

The REMPI spectrum of CF$_3$ was studied in detail by Duignan et al [8]. It is due to [3+1] ionization in the wavelength range of 420 to 490 nm. The time dependence of the REMPI signal was monitored at 455.6 nm and is presented in Figure 3. The REMPI signal decreases rapidly in the first 100 μs and reaches a constant level thereafter. On a much longer time scale (90 ms), Figure 3 shows a slight decrease of the REMPI signal (filled circle). Due to the negligible CF$_3$ decay on the time scale of a ms, we conclude that γ must be smaller than 0.01. In fact, from other experiments [1], we found γ for CF$_3$ to be 2.2×10^{-5} on a gold surface. CF$_3$ free radical seems to be an example of an inert radical in relation to wall collisions with gold, stainless steel and SiO$_2$ up to temperatures of 720K [3]. It is so inert, that its deactivation rate has to be measured against its effusion out of the Knudsen cell and the technique presented above for I becomes inapplicable. In contrast,

Figure 3

Time dependence of the total CF$_3$ REMPI signal at 455.6 nm. The solid point on the right ordinate corresponds to the REMPI signal at 90 ms.

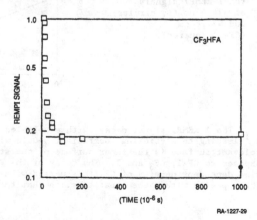

RA-1227-29

Winters finds γ to be between 0.08 and 0.75 for CF_3 on clean silicon under UHV conditions [9]. This is almost certainly the consequence of the presence of dangling bonds on the Si surface, which become saturated through the interaction with CF_3 free radical.

The sticking coefficient of SiH_2 and $SiH_2^{\ddagger}$ on an amorphous silicon surface was studied using n-$C_4H_9SiH_3$ and Si_2H_6 as precursors in IR-MPD schemes presented in equations (4), (5) and (6):

$$C_4H_9SiH_3 + nh\nu \rightarrow C_4H_9SiH + H_2 \qquad (4)$$

$$C_4H_9SiH + mh\nu \rightarrow SiH_2 + C_4H_8 \qquad (5)$$

$$Si_2H_6 + nh\nu \rightarrow SiH_2 + SiH_4 \qquad (6)$$

The REMPI spectrum of SiH_2 was not known prior to our study, so that we had to confirm the presence of SiH_2 and the assignment of the resulting REMPI spectrum by using different precursors. Pertinent details are discussed in [10]. Briefly, the REMPI spectrum of SiH_2 consists of two kinds of transitions: one consists of a series of congested lines corresponding to [3+1] REMPI via molecular Rydberg states of SiH_2, and the other consists of single intense sharp peaks corresponding to [2+1] REMPI Si($^1D_2$). This excited silicon state is generated by single photon dissociation of highly vibrationally excited $SiH_2^{\ddagger}$ to the molecular A state of SiH_2, which predissociates into Si($^1D_2$) + H_2. The REMPI signal at 487.9 nm probes $SiH_2^{\ddagger}$ with an internal energy content of approximately 7000 cm^{-1}, whereas similar sharp lines in the REMPI spectrum around 363 to 366 nm probe the ground state of SiH_2.

Figure 4 presents the time dependent REMPI signal for SiH_2 monitored at 500 nm and for $SiH_2^{\ddagger}$ monitored at 487.9 nm. The sticking coefficient for SiH_2 on an Si-C surface was measured to be 0.11, whereas γ for $SiH_2^{\ddagger}$ is larger than 0.5, in analogy to the results for $CF_3I^{\ddagger}$ discussed above. We note the excellent dynamic range of the REMPI signal over almost four decades of signal strength. Due to the overlap of the REMPI spectrum via the atomic resonance in Si($^1D_2$) and via molecular Rydberg states in SiH_2, the REMPI signal at 487.9 nm has a slower decay component that corresponds to the decay rate constant of the REMPI signal at 500 nm. At this wavelength of 487.9 nm the REMPI signal probes both the fast decaying $SiH_2^{\ddagger}$ as well as the slower decaying ground state of SiH_2.

Figure 4

Time dependence of the total REMPI signals at 487.9 nm ($SiH_2^{\ddagger}$, x) and at 500.0 nm SiH_2 (triangles).

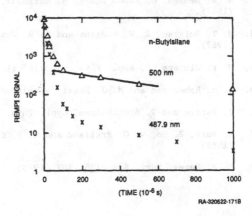

RA-320522-171B

The sticking coefficient for SiH_2 on a silicon surface containing carbon (by virtue of using n-butylsilane as the SiH_2 precursor) is 0.1, whereas γ for a "pure" silicon surface using disilane as the SiH_2 precursor is approximately 50% larger, on the order of 0.15. This difference in γ could be due to the fact that the former surface contains C-H bonds that have low reactivity towards SiH_2 insertion. This value seems low in view of the fact that modeling efforts "require" values on the order of 0.7 to 1.0. However, these modeling efforts relate to deposition conditions, for surface temperatures between 150 to 250 C. The present results, however, refer to a surface at ambient temperature. For comparison purposes we note that the sticking coefficient for the closely related SiH_3 is on the order of 0.1 for a surface at higher temperature [11], whereas the sticking coefficient for SiH_4 is negligible in comparison, even at high temperatures [12,13].

ACKNOWLEDGEMENT

This research was supported by AFOSR under Contract No. F49620-85-K-0001.

REFERENCES.

1. R. M. Robertson, D. M. Golden and M. J. Rossi, J. Phys. Chem., 92, 5338 (1988); J. Vac. Sci. Technol. A5, 3351 (1987).

2. N. Selamoglu, M. J. Rossi and D. M. Golden, J. Chem. Phys. 84, 2400 (1986).

3. R. M. Robertson, D. M. Golden and M. J. Rossi, J. Chem. Phys. 89, 2925 (1988).

4. W. P. Hess, S. C. Kohler, H. K. Haugen, and S. R. Leone, J. Chem. Phys. 84, 2143 (1986); W. P. Hess and S. R. Leone, ibid., 86, 3773 (1987).

5. S. C. Saxena and R. K. Joshi, in Thermal Accomodation and Adsorption Coefficients of Gases, McGraw-Hill/CINDAS Data Series on Material Properties, Vol. II-1, McGraw-Hill Book Co., 1981.

6. J. R. Engstrom, M. M. Nelson and T. Engel, This Symposium (E: Chemical Perspectives of Electronic Materials).

7. S. W. Benson, O. Kondo and R. M. Marshall, Int. J. Chem. Kinet., 19, 829 (1987).

8. M. T. Duignan, J. W. Hudgens and J. R. Wyatt, J. Phys. Chem. 86, 4156 (1982).

9. H. F. Winters, J. Appl. Phys. 49, 5165 (1978).

10. R. M. Robertson and M. J. Rossi, submitted to J. Chem. Phys.

11. J. Perrin and T. Broekhuizen, Appl. Phys. Lett. 50, 433 (1987).

12. R. Buss, P. Ho, W. G. Breiland and M. E. Coltrin, J. Appl. Phys. 63, 2808 (1988).

13. S. M. Gates, Surf. Sci. 195, 307 (1988).

PART V

Chemical Vapor Deposition

IN-SITU DIAGNOSTICS OF DIAMOND CVD

J.E. Butler, F.G. Celii, P.E. Pehrsson, H.-t. Wang, H.H. Nelson
Chemistry Division, Naval Research Laboratory, Washington, DC 20375-5000.

ABSTRACT

The deposition of diamond, a metastable crystalline form of carbon, from low pressure gases poses intriguing questions about the mechanisms of growth. Tunable IR Diode Laser Absorption Spectroscopy, Laser Multi-Photon Ionization Spectroscopy, and Laser Induced Fluorescence were used to characterize the gaseous environment in the Chemical Vapor Deposition growth of diamond films. The quality of the deposited material was examined by optical and SEM microscopies, and Raman, Auger, and XPS spectroscopies. When a reactant mixture of 0.5% methane in hydrogen, was passed across a hot Tungsten filament (2000 C), C_2H_2, C_2H_4, H and CH_3 were detected above the growing diamond surface, and concentration limits for undetected species were determined. These results are discussed in terms of simple models for species formation and consumption, as well as the implications for the diamond growth mechanism.

Introduction

Diamond is a material which possess many superlative physical properties, eg. hardness, wear and chemical resistance, thermal conductivity, optical transparency, large bandgap and high carrier mobilities.[1] The growth of diamond by chemical vapor deposition (CVD) is now a well established phenomena[2] and is receiving considerable attention because of the desire to control the shape and impurity distributions for the production of devices that would be difficult to form or mold out of natural diamonds. Less understood are the mechanisms of growth and impurity incorporation of this thermodynamically metastable material. It is hoped that by understanding the gaseous and surface chemistry of diamond CVD, a deeper insight into CVD processes will be attained and this will aid in developing various technical applications.

Experimental and Results

The CVD technique chosen for this study was one reported and confirmed by various laboratories[3,4,5,6]: filament-assisted CVD. Growth was from a reactant mixture of methane in hydrogen (0.1 to 2.0% methane, 20 to 40 torr, 100 sccm total flow) passed over a hot tungsten filament (2000 to 2300 C) suspended above a substrate (ca. 900 C). The presence and quality of diamond deposited on the substrate was determined using optical and scanning electron microscopies, x-ray diffraction, AES, XPS, and Raman scattering.

The gaseous species present during diamond growth were determined using tunable IR diode laser spectroscopy[7], resonance enhanced multiphoton ionization spectroscopy (REMPI)[8], and laser induced fluorescence(LIF). Using diode laser probing of the chamber, the species detected (from highest to lowest concentration) were CH_4, C_2H_2, CH_3, and C_2H_4. No other hydrocarbons were detected within the sensitivity limits of the technique. The REMPI experiments confirmed the presence of CH_3[9], and detected atomic H using a three photon resonance at 364.7 nm with the 1s to 2p Lyman-alpha transition[10]. The suppression of the REMPI signal using traveling wave excitation by the third harmonic light (generated by the atomic hydrogen) was overcome by using a standing wave geometry[11].

Both the methyl radical and atomic hydrogen signals were observed to increase with the filament temperature. At high methane concentrations, ca. 3.0%, the atomic hydrogen concentration was suppressed at filament temperatures below 2200 C, and the carbon deposited under these conditions becomes more amorphous and graphitic.

Discussion

Comparison of these observations with a thermodynamic equilibrium calculation[12] of the concentration of the species present at near the filament temperature gives qualitative, but not quantative agreement. Thus a simple picture of the diamond deposition environment is that of the hot filament dissociating the methane and hydrogen, followed by simultaneous gaseous reactions and species transport. The flow dynamics are such that the gas has insufficient time to achieve the full thermodynamic species concentrations. The actual growth species are still undetermined, but by combining the observed growth rate (0.3 to 1.0 microns per hour) and species flux derived from these observations and the calculations of Harris, et al[13], the potential growth species can be limited to CH_4, C_2H_2, C_2H_4, CH_3, and possibly C_2H_6, C_2H, and C_3H. The detection of atomic hydrogen in the diamond growth environment, as well as the observation of more non-diamond carbon deposition when the H atom concentration is suppressed, lends support to the surface role of H and the model of diamond CVD as dynamic balance of deposition and etching of all forms of carbon. In this model, the role of the atomic hydrogen is to etch the various forms of carbon, diamond at a rate slower than its deposition rate, while graphite, etc. at rates exceeding their deposition rates.

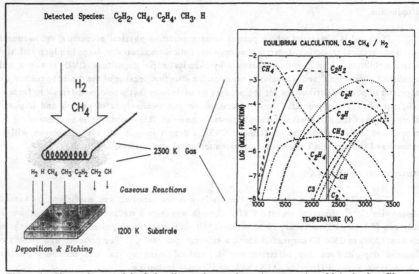

Figure 1: Shown is a model of the diamond growth environment in which the hot filament pyrolyzes the gas to approach a local thermodynamic equilibrium and the less reactive species are transported to the cooler surface on which solid carbon is grown and etched.

One implication of the oversimplified thermodynamic picture of the filament role is that the CVD of diamond should be relatively insensitive to the nature of the hydrocarbon reactar:. This has been observed by employing various organic compounds and reactants, including alcohols[4], acetone[4], and, in this lab, digester gas generated of the treatment of sewage sludge. In this experiment, the methane reactant was replaced by sewage gas (ca. 65% methane, 34% CO_2, 1% other gases) in the filament-assisted CVD reactor, and a finely grained diamond film was deposited on a silicon substrate.

Acknowledgements

This work was supported in part by the Office of Naval Research.

REFERENCES

1. J.E. Field, The Properties of Diamond, Academic Press, New York, 1979.

2. R.C. DeVries, Ann. Rev. Mater. Sci. 17, 161 (1987).

3. S. Matsumoto, Y. Sato, M. Kamo, and N. Setaka, Jpn. J. Appl. Phys. 21, L183 (1982).

4. Y. Hirose, and Y. Terasawa, Jpn. J. Appl. Phy. 25, L519 (1986).

5. B. Singh, Y. Arie, A.W. Levine, and O.R. Mesker, Appl. Phys. Lett. 52, 451 (1988).

6. T.D. Moustakas, J.P. Dismukes, L. Ye, K.R. Walton, and J.T. Tiedje, Proc. of the 10th International Conference on Chemical Vapor Deposition, The Electrochemical Society, Inc., 1164 (1987).

7. F.G. Celii, P.E. Pehrsson, H.-t. Wang, and J.E. Butler, Appl. Phys. Lett. 52, 2043 (1988).

8. F.G. Celii and J.E. Butler, Appl. Phys. Lett., submitted for publication.

9. J.W. Hudgens, T.G. DiGuiseppe and M.C. Lin, J. Chem. Phys. 79, 571 (1983).

10. P.J.H. Tjossem and T.A. Cool, Chem. Phys. Lett. 100, 479 (1983).

11. D.J. Jackson and J.J. Wynne, Phys. Rev. Lett. 49, 543 (1982); M.G. Payne, W.R. Garrett and W.R. Ferrell, Phys. Rev. A 34, 1143 (1986).

12. S. Gordon and B.J. McBride, "Computer Program for Chemical Equilibrium Calculations," NASA-Lewis Research Center, NASA SP-273, March, 1973.

13. S. J. Harris, A. M. Weiner, and T. A. Perry, Appl. Phys. Lett. 53, 1605 (1988).

CHEMICAL REACTION PATHWAYS FOR THE DEPOSITION OF AMORPHOUS SILICON-HYDROGEN ALLOYS BY REMOTE PLASMA ENHANCED CVD

GN PARSONS, DV TSU and G LUCOVSKY
Department of Physics, NC State Univ, Raleigh, NC 27695-8202

ABSTRACT

We have grown thin films of a-Si:H alloys by Remote Plasma Enhanced CVD (Remote PECVD) and have studied the deposition process by mass spectrometry. We find that the concentration of silane fragments (SiH$_x$ x=0 to 3) and higher silanes (e.g. disilane Si$_2$H$_6$) in the gas phase is below our detection limit of ≈0.5%. Bias experiments and a comparison of the a-Si:H deposition rate with the known concentration of silane in the gas phase suggest that in Remote PECVD, silane molecules, SiH$_4$*, vibrationally excited in the gas phase or on the deposition surface may lead dirtectly to a-Si:H film deposition.

INTRODUCTION

Amorphous silicon-hydrogen (a-Si:H) alloy films have been deposited by a variety of thermal and plasma enhanced techniques. These techniques include: 1) reactive magnetron sputtering [1-2] (MS), in which silicon atoms are sputtered in the presence of hydrogen from a silicon target using an argon plasma ; 2) conventional glow discharge [3] (GD) or Direct PECVD, in which the silane reactant gas is directly plasma excited and the substrate is immersed in the plasma glow; 3) triode glow discharge [4,5], where a grid separates the substrate from the silane plasma; 4) ECR Plasma deposition, where a magnetic field confines the silane plasma in a region removed from the substrate surface [6]; and 5) Remote PECVD [7], where helium flows through a plasma excitation region and mixes with silane introduced downstream in a region removed from the plasma glow, resulting in film deposition on a substrate outside of the helium plasma.

The specific plasma processes, reaction pathways and gas phase precursors leading to film deposition have not been unambiguously determined for the Direct PECVD process. Although the SiH$_3$ radical has been suggested, and is often "accepted" as the dominant precursor leading to high quality a-Si:H films in Direct PECVD [8,9], its role has recently been questioned [10]. The objective of the research described here is to determine the nature of the precursor species and the reaction pathways leading to the deposition of a-Si:H in the Remote PECVD process. This process is "easier" to study and analyze than Direct PECVD because the silane reactant gas is not directly plasma excited. We observe that this reduces the number of active precursor species and therefore reduces the number of reaction pathways leading to film deposition.

REMOTE PLASMA a-Si:H DEPOSITION

To produce Remote PECVD a-Si:H films, helium is excited in an inductively coupled rf plasma (13.56 MHz) separated by more than 10cm from a heated substrate. Excited helium species (e.g., He metastables and ions) and electrons flow from the plasma region into the deposition chamber. Silane (diluted with 90% Ar or He) is introduced *outside the plasma region* through a 'showerhead' gas

Mat. Res. Soc. Symp. Proc. Vol. 131. ©1989 Materials Research Society

dispersal ring 5cm from the substrate surface. Mixing of excited helium species with neutral silane produces reactant species that undergo a heterogeneous CVD reaction at the substrate surface producing the a-Si:H thin film. Pressure and flow conditions in the chamber are sufficient to prevent significant diffusion of silane back into the plasma glow region. The rf power supplied to the plasma is typically 25W resulting in a growth rate of about 0.15Å/s which is independent of substrate temperature (T_s) between 38°C and 325°C.

The hydrogen concentration, [H], is shown in Fig. 1 versus substrate temperature for a-Si:H films deposited by Remote PECVD and conventional glow discharge [5]. For T_s between 100°C and 400°C, the hydrogen content in Remote PECVD films is substantially lower than in GD [5] or other [1,2] films deposited at the same T_s. Details of the electrical and optical propeties of the Remote PECVD a-Si:H films are presented elsewhere [7].

MASS SPECTROMETRY RESULTS

We have used mass spectroscometry in a specially designed deposition/analysis chamber [11] to investigate the silane species present in the gas phase near the substrate (i.e. downstream from the silane introduction ring) during Remote PECVD of high quality a-Si:H films. Figure 2 shows the mass spectrometer signal intensity for 28amu to 32amu plotted versus input rf power between 0 and 60W. With the He plasma off (zero Watts), the mass peaks from 28 to 32 amu reflect the cracking pattern for neutral (unexcited) silane as it is fragmented and ionized in the mass spectrometer. The peaks from 28 to 31 are predominantly due to Si, SiH, SiH_2 and SiH_3 ions, respectively, while the measured peak intensity at 32 and 33 amu relates directly to published isotope fractions for silicon. The intensity of all silane fragments decreases exponentially with increasing rf power. At 60W the mass signal intensity is about 30% of its original value, indicating that a large fraction of the silane introduced into the chamber has been consumed in deposition. Under these conditions in our analysis chamber, the a-Si:H deposition rate is greater than 1Å/sec.

Figure 3 shows the mass spectrometer signal intensity for mass in the range from (a) 25 to 35 amu; and (b) 55 to 65 amu for an rf power of 40W supplied to helium. The lack of signal greater than 55 amu shows that disilane species (Si_2H_y , y = 0 to 6) are not present in the gas phase (sensitivity limit < 0.5%). This result is in contrast with mass spectrometry studies of silane glow discharges where the mass signal intensity for disilane is comparable to that for silane [4].

We have investigated the presence of neutral silane fragments using the appearance potential technique described by Robertson and Gallagher [8] in our Remote PECVD analysis system. The results are shown in Fig. 4 where we have plotted the derivative of the mass signal versus the energy of the electrons responsible for species ionization. Neutral SiH_2 and SiH_3 species will ionize with 2 to 4 eV lower energy than that needed for SiH_4 dissociative ionization. With an electron energy distribution of ≈2eV in our ionizer, the shape of the derivative signal does not change when rf power is increased from 0W to 40W. Therefore, within a detection limit of ≈0.5%, neutral fragments (SiH_2, SiH_3 etc.) are not observed in the gas phase during film deposition.

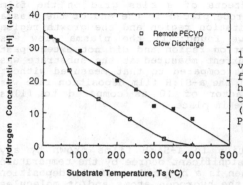

1) Hydrogen concentration versus substrate temperature for amorphous silicon-hydrogen alloys produced by conventional glow discharge (Direct PECVD) and by Remote PECVD.

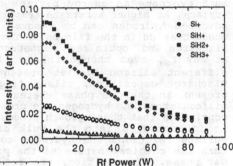

2) Mass spectrometer signal intensity for various silane fragments (ions) versus rf power supplied to the plasma. The fragment distribution at 0Watts (plasma off) is the cracking pattern for neutral (unexcited) silane in our mass spectrometer.

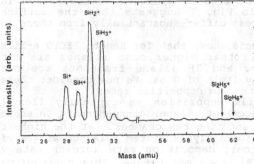

3) Mass spectrometer signal intensity for mass in the range from 25 to 35 amu and 55 to 65 amu. Rf power supplied to helium was 40W.

4) Derivative of the mass spectrometer signal intensity versus ionizer electron energy. Three conditions are shown: a) plasma excitation off; b) helium is plasma excited with 40W rf power; and c) helium and oxygen ($[He]/[O_2] = 90\%$) are plasma excited with 40W.

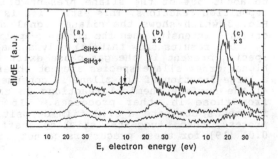

We have studied the effects of a bias grid on the film deposition. When a mesh electrode placed in the end of the plasma tube between the plasma excitation region and the growth region is electrically isolated, we found: 1) the plasma glow was confined to the plasma excitation region and did not extend past the mesh electrode; 2) the current measured at the substrate was reduced by a factor of two as compared to that measured without the electrode in place; and 3) the a-Si:H film deposition rate was reduced by more than a factor of 10 as compared to films deposited without the electrode in place.

DEPOSITION REACTIONS

The hydrogen concentration in a-Si:H films grown by any technique is determined to a significant degree by the temperature of the growth surface. Hydrogen is a by-product of the deposition of silicon from silane, and the hydrogen atoms and/or molecules must overcome an energy barrier to be removed from the growth surface. At higher substrate temperatures a larger fraction of the adsorbed hydrogen can be removed resulting in a lower hydrogen concentration in the film bulk. The hydrogen concentration in glow discharge and optimized magnetron sputtered films is comparable at any T_s, even though the two techniques have substantially different silicon species present in the gas phase. For glow discharge deposition, silane fragments and hydrogen atoms are present in the gas phase whereas for magnetron sputtering only silicon atoms and hydrogen are present. This leads to an important conclusion that these growth processes, including the way hydrogen is incorporated into the bulk of the film, are dominated by surface reactions and hydrogen coverage of the surface, rather than the chemical nature of the precursor species formed in the gas phase. In addition, the reduced hydrogen concentration in Remote PECVD films shown in Fig. 1 suggests that the surface reactions in the Remote process differ substantially from those in other deposition processes.

Mass spectrometry analysis shows that for Remote PECVD a-Si:H deposition: 1) disilane or other higher mass silanes are not observed in the gas phase; and 2) silane fragments are not observed within a sensitivity limit of 0.5%. We now consider these observations as they relate to film deposition rates.

During a-Si:H thin film deposition we typically flow 1 standard cubic centimeter (sccm) of silane, corresponding to about 2×10^{19} molecules per minute entering the chamber. At the highest growth rates in the deposition/analysis system (>1Å/sec at 60W rf power), we estimate the atomic deposition rate (growth rate x effective area) to be 10^{19} Si atoms per minute. This corresponds to about 50% of the silane present being consumed by the film deposition reactions. This estimation is confirmed by the data in Fig. 3 which shows the silane signal decreasing to 30% of its original intensity when the rf power applied is 60W.

Our results show that: 1) only a small fraction of the silane species present in the gas phase exist as neutral fragments; 2) higher order silane species are not observed in the gas phase; 3) a large fraction of the silane is consumed in film deposition; and 4) the surface chemical reactions in the Remote process differ from those in other processes. If we suppose that silane fragments dominate a-Si:H film deposition in the Remote PECVD process, then suggested sticking coefficients of approximately 0.01 [19] for these precursors cannot account for both the large silane consumption factor and the low concentration of silane fragments observed in the gas phase. We conclude, therefore, that non-fragmented SiH_4 molecules with additional internal (i.e. vibrational) energy (denoted here as SiH_4^*), may lead to a-Si:H deposition in the Remote PECVD process. In this picture, the fragmentation of the silane takes place predominantly *on the deposition surface* and may require additional surface activation. This surface activation may result from impact of additional energetic species from the gas phase to produce chemically activated adsorbtion sites [14,17].

Further evidence supporting our conclusion that vibrationally excited SiH_4 molecules are important reaction intermediates for a-Si:H deposition is found from: 1) bias grid experiments and the electron energy distribution in the plasma; 2) electron impact cross sections for vibrational excitation of silane [12]; 3) other studies establishing that vibrational internal energy enhances both homogeneous and heterogeneous reaction rates [13,14]; and 4) the reduced hydrogen concentration in Remote PECVD a-Si:H films.

Bias experiments discussed above suggest that charge species emanating from the plasma glow play a critical role in the film deposition reaction. In the plasma glow, electrons respond easily to the rf field and acquire translational energies of the order of a few eV. These electrons carry the majority of the energy out of the plasma region and hence are good candidates for the species responsible for neutral silane excitation. For a wide range of electron energies, electron impact excitation and ionization cross sections for silane have been determined [12]. The cross section for vibrational excitation is on the order of 10^{-15} cm^2, independent of electron energy up to 8 eV then falls to 10^{-18}cm^2 for electron energy above 10eV. The neutral dissociation cross section is negligible for electron energy below 8 eV and rises to 10^{-16} cm^2 at 10 eV and above. Considering the large cross section for vibrational excitation and the density of silane in the gas phase, a significant concentration of vibrationally excited silane molecules is possible under our conditions of deposition. Vibrationally excited silane has been observed in other systems, e.g., Knights and co-workers [15] observed vibrationally excited silane using infrared emission and absorption in a silane/hydrogen ($[SiH_4]/[H_2]$= 1/3) plasma at 200 mTorr.

It is known that internal (rotational, vibrational or electronic) energy can contribute to overcoming an activation energy barrier in homogeneous [13] and heterogeneous [14] reactions. In specific experiments, internal vibrational energy in methane molecules has been shown to enhance surface adsorbance of the molecules on clean metal surfaces [16]. Although the reactivity of unexcited silane molecules on relatively cool substrates (< 500°C) is known to be very low, internal vibrational or translational energy corresponding to internal temperatures near 800K [15] may enhance the surface reactivity of silane molecules.

Energetic molecular silane (SiH_4^*) may enhance the desorption of hydrogen molecules and limit the hydrogen available at the a-Si:H deposition surface. Enhanced hydrogen desorption reactions may lead to the observed lower hydrogen content in Remote PECVD a-Si:H films as compared to GD films.

Following work of other groups developing models for a-Si:H deposition from a silane glow discharge, we speculated that SiH_3 radicals may be the dominant gas phase precursors leading to high quality film deposition in the Remote PECVD process. Veprek et. al. have recently suggested that in glow discharge deposition, SiH_3 species cannot account for film deposition, and that disilane produced by the insertion of SiH_2 into a SiH_4 molecule is a more viable candidate for the a-Si:H precursor [10]. We suggest that vibrationally excited silane may react directly with the deposition surface to produce a-Si:H in the Remote PECVD process.

CONCLUSIONS

We believe that the differences in Remote PECVD a-Si:H films and those deposited by other techniques can be accounted for by differences in the reactions on the deposition surface and on the nature of the precursor species involved in film deposition. We suggest that SiH_4^* adsorbed directly on the deposition surface may lead, through additional surface activation, to a-Si:H films in the Remote PECVD process. This conclusion is based on: 1) the observed concentration of silane fragments in the gas phase as compared to the observed deposition rates; 2) the silane depletion factor; and 3) reduced hydrogen in Remote PECVD a-Si:H films. We find that reactions at the deposition surface in Remote PECVD differ from those in other techniques, and that Remote PECVD is a highly efficient process that consumes a significant fraction of the silane introduced into the reaction chamber. Our conclusions are consistent with electron impact cross sections for vibrational excitation of silane, the reduced hydrogen concentration and the distribution of hydrogen in mono- and poly-hydride bonding groups in the Remote PECVD films.

REFERENCES

[1]   T.D.Moustakas in "Semiconductors and Semimetals" Vol.21 Part A, ed.by J.I.Pankove (Academic, New York 1984).
[2]   R.A.Rudder, J.W.Cook Jr., and G.Lucovsky, Appl.Phys.Lett.45, 887(1984).
[3]   H.Fritzsche,Solar Energy Mat.3,447 (1980).
[4]   A.Matsuda and K.Tanaka, Thin Solid Films 92,171 (1982).
[5]   K.Tanaka and A.Matsuda, Mat.Sci.Reports 2,141 (1987).
[6]   S.R.Mejia et.al.,J.Non-Cryst.Solids 77&78,765(1985).
[7]   G.N.Parsons,D.V.Tsu and G.Lucovsky, J.Non-Cryst.Solids 97&98,1375 (1987).
[8]   R.Robertson and A.Gallagher, J.Appl.Phys. 59, 3402 (1986).
[9]   P.A.Longeway, R.D.Estes and H.A.Weakliem J.Phys.Chem 88, 73(1984).
[10] S.Veprek, M.Heintze, F.A.Sarott, M.Jurcik-Rajman and P.Willmott, MRS Symp. Proc. Vol.118 (1988) in press.
[11] D.V.Tsu, G.N.Parsons and G.Lucovsky, J.Vac.Sci.Technol. A 6, 1849 (1988).
[12] A.Garscadden, G.L.Duke, and W.F.Bailey, Appl.Phys.Lett. 43, 1012 (1983).
[13] J.W.Moore and R.G.Pearson, "Kinetics and Mechanism", (Wiley, New York) 1981.
[14] S.G.Brass and G.Ehrlich, Phys.Rev.Lett. 57, 2532 (1986).
[15] J.C.Knights, J.P.M.Schmitt, J.Perrin and G.Guelachvili, J.Chem.Phys.76, 3414 (1982).
[16] C.T.Rettner, H.E.Pfnur and D.J.Auerbach, Phys.Rev.Lett. 54, 2716 (1985).
[17] J.C.Knights, Mat.Res.Symp.Proc.Vol.38,371,(1985).

AMORPHOUS TO CRYSTALLINE STRUCTURAL TRANSITION DURING GROWTH OF SILICON BY PLASMA ENHANCED CHEMICAL VAPOR DEPOSITION

W.J. VARHUE*, S. KRAUSE**, J. DEA** and C.O. Jung**
*University of Vermont, Department of Electrical Eng., Burlington, VT 05405
**Arizona State University, Department of Chem. Bio. and Materials Eng., Tempe, AZ 85287

ABSTRACT

The growth of thin Si films by RF glow discharge undergoes a transition from amorphous to microcrystalline as power density is increased. This results in a substantial change in the film's electrical conductivity and activation energy for electrical conduction. The RF glow discharge has been characterized in terms of plasma density, plasma potential and electron temperature with emissive Langmuir Probe measurements. The structural transition has been observed with a transmission electron microscope.

INTRODUCTION

The possibility of obtaining microcrystalline Si with temperatures below $400^{\circ}C$ by RF glow discharge has generated considerable interest. This material could be used in a variety of thin film devices and should yield an improvement in carrier mobilities over that of a-Si:H. The doping efficiency of this material has been shown to be considerably higher than that of a-Si:H [1,2]. Y. Mishima, et.al. [3] have proposed a structural model in which microcrystallites of approximately 60 Å are imbedded in an amorphous medium. The amorphous medium is hydrogen rich with mono- and dihydride bonding. In the transition from amorphous to microcrystalline type behavior, it has been observed by x-ray diffraction that the volume fraction of microcrystallites increases from 36 to 65%. This appears to occur by an increase in the number of crystallites and not in the size of the crystallites. The size of the microcrystallites is believed to be limited by the total energy balance between microcrystalline and amorphous phases.

Electrical conductivity increases significantly with an increase in the microcrystalline volume fraction. This is attributed to a decrease in the amorphous phase which increases the number of conducting channels. The activation energy for electrical conduction decreases from .8 to .25 eV for the transition from an amorphous to microcrystalline structure. Although the activation energy is assumed to be associated with the potential energy barrier that exists between crystalline and amorphous phases, it is not clear what the precise mechanism is.

The microcrystalline phase has been found to grow when the plasma generates a high ratio of [H]/[SiH] species. High ratios of these species have been shown by optical emission spectroscopy to occur with high H_2/SiH_4 feed ratios or high RF power densities [3]. Another means would be to confine the plasma with a magnetic field, which increases the plasma density [4]. It is not clear how the increased [H]/[SiH] ratio increases the nucleation probability or the resulting higher microcrystalline volume fraction.

This investigation involves the growth of amorphous and microcrystalline thin films by RF glow discharge in a parallel plate reactor. The RF power density was increased and a transition from amorphous to microcrystalline phase was observed in the measured electrical properties. The plasma's

Mat. Res. Soc. Symp. Proc. Vol. 131. ©1989 Materials Research Society

density and potential were monitored with a Langmuir Probe. Finally, transmission electron microscopy was used to observe the microstructure of the deposited films.

EXPERIMENTAL

The samples were prepared in a custom designed reactor. The RF power at 13.56 MHz was capacitively coupled into the reactor chamber with two parallel 10 cm. diameter plates separated by a 3.8 cm. gap. The glass substrates were placed on the grounded holder which was heated with resistive cartridge heaters. All samples were deposited at 280°C as measured on the top surface of a glass substrate with a thin film thermocouple. The reactor pressure was maintained at 200 mTorr by a closed loop capacitance manometer/throttle value controller. The flow rates were 20 sccm SiH_4 and 20 sccm H_2. The input power was varied from 2 W to 35 W and was measured with a Bird Meter.

All electrical measurements were made with a four strip coplanar Al structure that was evaporated on the film immediately following deposition. The electrical conductivity activation energy was measured by varying the sample temperature for a series of I-V curves. The slope of the natural log of the conductivity verses 1/KT gives the activation energy. The photoconductivity was measured with an incandescent lamp whose intensity was set to AM1 with a NBS calibrated photocell.

The plasma's potential, density and electron temperature were monitored with an emissive Langmuir Probe. A sketch of the probe circuit and probe can be found in Figure 1a. and b. This probe circuit is similar to that used by J.R. Smith et. al. [4]. The d.c. power supply used to heat the filament was a 12 V car battery. It was necessary to filter high frequency interference with L_1C_1 and 60 cycle pick up with L_2C_2. The filament itself was a 1 cm. long, .005 cm. diameter W wire. The W wire filament was connected to a Cu support inside the glass sheath.

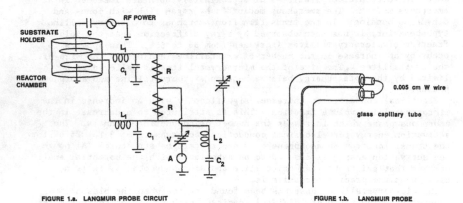

FIGURE 1.a. LANGMUIR PROBE CIRCUIT   FIGURE 1.b. LANGMUIR PROBE

The Langmuir Probe current trace was fitted to a third order polynomial expression. This expression was then differentiated, yielding a curve with two maxima. Smith, et. al. [4] have shown that the midpoint of these two maxima is an accurate determination of the plasma potential. The electron temperature can be obtained by taking the inverse of the slope of the natural log of the collection current corrected for primary electrons. This analysis yielded values for the electron temperature in the range of 4 eV.

An estimate of the plasma density has also been made in this investigation. It is assumed that at positive bias voltages, where the electron current saturates, the probe is behaving as a collecting probe and does not emit. Using the expression derived by Hershkowitz et. al. [5] for a partial wave rectified plasma potential;

$$n(cm^{-3}) \approx \frac{3.7 \times 10^8 <Ie(mA)>}{(1-B) \ S(cm^2) \ Te(eV)^{1/2}}$$

where

β — 1/2 for half wave rectified
$<Ie>$ — current
S — surface area of probe
Te — electron temperature (4eV)

DISCUSSION OF RESULTS

The conductivity of the deposited films as a function of RF power density can be found in Figure 2. As can be observed in this plot, dark conductivity increases exponentially with RF power density. Conductivities less than 10^{-7} (Ω cm)$^{-1}$ are typically assumed to be amorphous and those greater to be microcrystalline. Therefore in this investigation, power densities greater than .2 W/cm^2 produced microcrystalline material. This is also shown in a plot of the activation energy versus RF power density in Figure 3. The activation energy starts at ⁻.9 eV for low power densities and decreases rapidly to ⁻.25 eV as power density is increased to .5 W/cm^2. These results are similar to those obtained by Mishima, et.al. [3]. An interesting plot is that of dark conductivity verses activation energy in Figure 4. As the potential barrier is lowered, current increases exponentially.

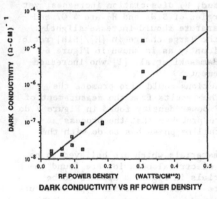

DARK CONDUCTIVITY VS RF POWER DENSITY

Figure 2.

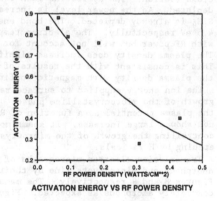

ACTIVATION ENERGY VS RF POWER DENSITY

Figure 3.

Also, measured was the photoconductivity of the films under AM1 irradiation. There was no clear trend observed with RF power, yet the photo-to-dark conductivity ratio plotted in Figure 5 showed a large variation with RF power density. A-Si:H is obviously a superior photovoltaic material over that of microcrystalline.

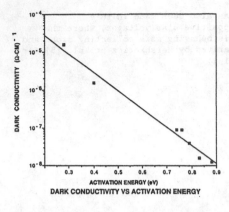

DARK CONDUCTIVITY VS ACTIVATION ENERGY

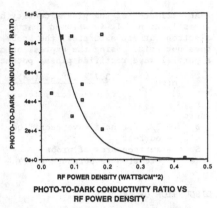

PHOTO-TO-DARK CONDUCTIVITY RATIO VS
RF POWER DENSITY

Figure 4. Figure 5.

In previous studies, high ratios of [H]/[SiH] species in the plasma have
been shown to be linked to the transition from amorphous phase growth to
microcrystalline phase growth [1, 2 and 3]. Although the exact mechanism
which favors microcrystalline nucleation is not known, it is believed to be
associated with the etching reaction of Si by H radicals. The increased H
concentration in the reactor pushes the etching reaction of weak Si-Si
bonds or Si-H components forward. Removal of these components leaves the
energetically favored crystalline material. The reason the H
concentration increases with increased power is most likely associated with
the increase in plasma density. At lower power densities the SiH_4 is
preferentially dissociated due to a large electron impact dissociation
cross section. The H_2 remains much less reacted where as the SiH_4 becomes
depleted. As the power level is increased, H_2 dissociation increases, yet
SiH_4 is already depleted. The bond energies of SiH_4 and H_2 are 4.07 and
4.5 eV respectfully. The electron temperature should increase slightly
with RF power but will not account for the large change in [H]/[SiH] ratio.
The plasma density does increase significantly as is shown in Figure 6.
This is consistent with the results of Hamasaki et.al. [1] who increased
the plasma density with magnetic confinement.

The ion energy supplied to surface reactions could also promote the
growth of the microcrystalline phase. The results from the measurement of
the plasma potential as a function of RF power density found in Figure 7 do
not show a large increase. It seems more probable that the process
controlling the growth of the microcrystalline phase has to do with the
etching by H radicals.

Transmission electron micrographs of materials which are believed to be
amorphous and microcrystalline by their electrical behavior are found in
Figures 8 and 9 respectively. The materials in Figure 8 appears to be
completely amorphous as expected. Figure 9 shows a sandwich type structure
with an amorphous region between two thin microcrystalline regions. This
was obtained for both samples which exhibited microcrystalline behavior.
It is not clear why the microcrystalline structure is only found at the
glass - Si interface and on the top surface of the film. Problems were
encountered with charging of the samples due to the insulating nature of
the glass substrates, but should not yield such a result. The only thing
that can be concluded about the microstructure of the film deposited at
high RF power density is that the microcrystalline phase is produced as
well as amorphous. It appears to indicate that built up stresses in center
of the Si film prevent crystallization.

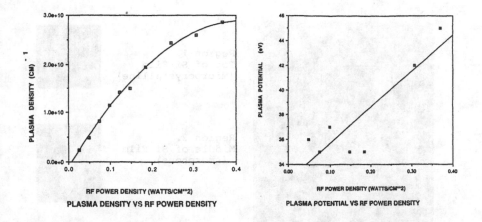

3.0e+10
2.0e+10
1.0e+10
0.0e+0
PLASMA DENSITY (CM) - 1
0.0
0.1
0.2
0.3
0.4
RF POWER DENSITY (WATTS/CM**2)
PLASMA DENSITY VS RF POWER DENSITY

Figure 6.

46
44
42
40
38
36
34
PLASMA POTENTIAL (eV)
0.00
0.10
0.20
0.30
0.40
RF POWER DENSITY (WATTS/CM**2)
PLASMA POTENTIAL VS RF POWER DENSITY

Figure 7.

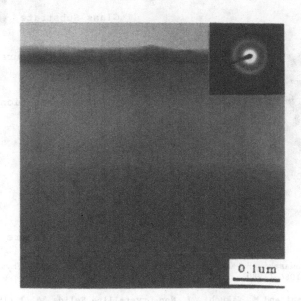

0.1um

Amorphous Si Thin Film on Glass Substrate

Figure 8.

Region 1.
Top of Si Film
(Microcrystalline)

Region 2.
Middle of Si Film
(Amorphous)

0.1um

Region 3.
Bottom of Si Film
(Microcrystalline)

Glass Substrate

Figure 9 a.

High Resolution TEM of
Region 3.

10nm

Figure 9 b.

REFERENCES

1.   T. Hamasaki, H. Kurata, M. Hirose and Y. Osaka, Appl. Phys. Lett.
      37, 1084 (1980).

2.   S. Usui and M. Kikuchi, J. Non-Crystalline Solids 34, 1 (1979).

3.   Y. Mishima, S. Mizazaki, M. Hirose and Y. Osaka, Phil. Mag. B 46,
      (1), 1 (1982).

4.   J.R. Smith, N. Hershkowitz and P. Coakley, Rev. Sci Instrum. 50,
      (2), 210 (1979).

5.   N. Hershkowitz, M.H. Cho, C.H. Nam and T. Intrator, Plasma Chemistry
      and Plasma Processing, 8, (1), 35 (1988).

IN SITU CHARACTERIZATION OF THIN FILM GROWTH BY FTIR IRRAS

J. E. BUTLER*, K. B. KOLLER*, AND W. A. SCHMIDT**
*Naval Research Laboratory, Chemistry Division, Washington, DC, 20375
**Naval Research Laboratory, Electronics Science and Technology Division,
Washington, DC, 20375.

ABSTRACT

In-situ analysis of low temperature Plasma Enhanced Chemical Vapor Deposition
(PECVD) SiO_2 films deposited on HgCdTe, Silicon, and Aluminum substrates was
performed by double modulation Fourier Transform Infrared Reflection Absorption
spectroscopy (FT-IRRAS). The sensitivity and selectivity of this technique are
sufficient for an in-situ assessment of the film quality and reaction conditions
at any stage of film growth. An oblique angle of incidence of ca. 55° was chosen
to yield maximum sensitivity for the 1260 cm^{-1} LO mode of SiO_2 on Si. The peak
frequency and shape of the LO mode absorption band varied with the quality of the
SiO_2 films. This diagnostic technique can be applied readily to in-situ analysis
of dielectric thin films formed under a variety of reaction conditions as long as
the gaseous ambient is partially transmissive to the IR radiation.

Introduction

The formation of dielectric thin films of SiO_2 at low substrate temperatures
by PECVD requires careful control of reaction conditions to yield high quality
dielectric/semiconductor interfaces. The ability to monitor the quality of the
thin dielectric film in-situ greatly facilitates the optimization process and
avoids the possibility of sample corruption by the process of removal and ex-situ
analysis. The authors have been interested in producing thin films of SiO_2 on the
compound semiconductor HgCdTe (MCT) and have found that the quality of the samples
is affected by multiple reaction variables (e.g., choice of reactants, flow rates,
pressure, carrier gas dilution, substrate temperature, reactor configuration) [1].

Normal incident IR spectra of SiO_2 is a proven method for the nondestructive
analysis of SiO_2 thin films [2-6]. The position and shape of the restrahlen band
(1080 cm^{-1}) of SiO_2 are sensitive to SiO_x stoichiometry, film density, and
impurities [2-6]. Impurity absorption bands for compounds such as Si-OH, Si-H, N-
H, and H-OH are also observable. The restrahlen band arises from interaction of
the IR photons with the transverse optical (TO) mode phonons and is most prominent
at a normal angle of incidence. In-situ analysis of SiO_2 thin films by
transmission spectroscopy on various substrates is limited by the severe
restrictions placed upon reaction chamber configuration and the difficulty of
transmission through the substrate. However, p-polarized light, at an appropriate
non-normal angle of incidence, interacts strongly with the longitudinal optical
(LO) mode of thin films of cubic crystals [7]. The LO absorption band is
observable in either non-normal angle of incidence transmission spectroscopy or by
external reflection spectroscopy when the thin films are on a metallic or
dielectric substrate. The frequency of longitudinal optical phonons, ω_l, is
related to that of transverse optical phonons, ω_t, by the Lyddane-Sachs-Teller
relation [8]:

$$\omega_l^2/\omega_t^2 = \epsilon_\infty/\epsilon_0 \tag{1}$$

where ϵ_∞ is the dielectric constant at the high frequency limit (where the ions
cannot follow the oscillation of the electric field and polarization is due
entirely to electron motion) and ϵ_0 is the dielectric constant at the low
frequency limit. The differences in both the peak frequency and intensity of the
LO mode from that of the TO mode are due to the sensitivity of the longitudinal

Mat. Res. Soc. Symp. Proc. Vol. 131. ©1989 Materials Research Society

components of infrared-active vibrations to long-range electric forces within the film in addition to the short-range interatomic forces that affect both modes [9]. Monitoring the LO mode of a thin film by p-polarized light at a non-normal angle of incidence has the benefit of being much more sensitive than normal angle of incidence transmission spectroscopy of the TO mode [10].

A double modulation technique [11-15] using a photoelastic modulator (PEM) in combination with a Fourier Transform Infrared Reflection Absorption Spectroscopy (FT-IRRAS) was used for this study. This method has the benefits of increased sensitivity and the cancellation of polarization insensitive absorptions. The latter attribute allowed for operation in a normal lab environment, i.e., a purged system was not required. A photoelastic modulator is used to modulate the phase delay between the s and p polarized components of the light at a high frequency.

Experimental

A modified version of a previously described PECVD reaction chamber [1] was used. The modification consisted of fitting the quartz reaction chamber with two KBr windows on the sides of the reactor tube, oriented so that the optical path was normal to the axis of the reactor tube and the angle between the incoming and reflected beams was 112 degrees at the sample. This modification is shown in more detail in Figure 1. The angle of incidence was chosen based on model calculations (based on published optical constants [16]) of the intensity of the spectral change, δS, for 100 Å of SiO_2 on silicon. It is noteworthy that the optimum angle of incidence varies with film thickness. Both Si and $Hg_{1-x}Cd_xTe$ (MCT)(x=.3) single crystal substrates were used for deposition of the dielectrics. The Si crystals were (100) oriented, n-type, 50mm dia. polished wafers. The MCT crystals were 15mm dia., unoriented, n type ($N_d < 5x10^{15} cm^{-3}$), wafers that had been polished and etched in Br/methanol solutions.

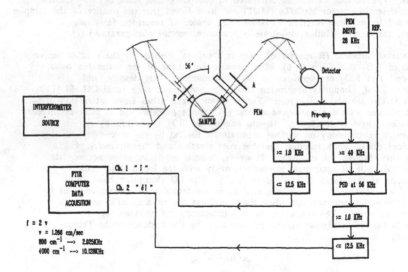

Figure 1: A schematic of the FTIR-IRRAS experimental apparatus is shown.

Silane (0.5% in argon) and nitrous oxide were used as reactants. The optimal reaction conditions were flow rates of 50 and 25 sccm for N_2O and SiH_4/Ar, respectively, at a pressure of 400 mtorr in the reactor. All layers were deposited on samples at 70° C and employed a rf power of 25 watts at 13.5 MHz. Extensive work had been performed previously to determine the conditions to give the best quality films [1]. Film thicknesses were measured by ellipsometric measurements at 632.8nm.

The spectra were obtained with a FTIR spectrometer (Mattson Instruments, Sirius 100) at 4 cm^{-1} resolution. The IR light was polarized by passing through an etched metallic grid on ZnSe (Science Services Inc.) and modulated at 28KHz using a photoelastic modulator (Hinds International). The IR signals, sensed by a liquid nitrogen cooled photoconductive MCT detector, consisted of interferograms at the various harmonics of the PEM carrier frequency. The normal (dc) interferogram was isolated by two low pass filters in series, while the interferogram carried on the second harmonic of the PEM frequency was demodulated by a lock-in amplifier (Brookdeal, model 9503) using the smallest possible time constant (0.1 milliseconds) in order to pass the necessary side band information. Additional analog filters isolated the frequencies of interest in the second harmonic (2ω) interferogram. 100 scans and 500 scans were averaged for the dc and 2ω interferograms respectively. The RF power was turned off and film growth stopped during measurements because of inadequate electrical shielding of the detector. The specific filter settings are given in figure 1.

A calibration spectrum was obtained from a reference aluminum mirror or from a bare substrate prior to CVD deposition by orienting both polarizers parallel to p (or s). The null interferogram was obtained when the second polarizer, A (see figure 1), was left in the p polarized position and the first, P, was adjusted to minimize the peak amplitude of the 2ω interferogram, i.e., setting "I_p equal to I_s". Phase information from the 2ω calibration interferogram was used to process the nulled signal 2ω interferogram.

Spectral change, δS, in the substrate was calculated between a modified surface and the original surface by:

$$\delta S = \left. \frac{(I_{2\omega}/I_{dc})null}{(I_{2\omega}/I_{dc})calib} \right|_{modified} - \left. \frac{(I_{2\omega}/I_{dc})null}{(I_{2\omega}/I_{dc})calib} \right|_{original} \tag{2}$$

where the $I_{2\omega}/I_{dc}$ obtained at the null position of the polarizers is ratioed to that obtained when both polarizers are parallel to either the s or p orientation of the surface.

Results and Discussion

Figure 2 shows the spectra for SiO_2 deposited by PECVD sequentially on an MCT substrate. The intensity of the spectral change in the LO peak of SiO_2 increased with the thickness of the film in a slightly nonlinear fashion up to 2500 Å thick and the values agreed with model calculations. No increase in LO mode peak intensity was observed for films over 2500 Å thick. However, the TO mode peak increased steadily with film thickness. On Si and MCT substrates, the LO peaks are negative and represent a loss in reflectivity whereas the TO peaks are positive and are caused by an increase in reflectivity. The TO peaks of SiO_2 were observed on Si and MCT substrates but were completely suppressed on an Al substrate. The peak frequency of the LO mode band varied with reaction conditions and was always less than that observed for thermal oxide grown on Si. IR spectra of three, equal thickness, SiO_2 films prepared in different ways, are shown in Figure 3. Two samples (Figure 3b and 3c) represent the observed extremes, due to reaction conditions [1], in PECVD growth quality while the third is a thermally grown oxide (Figure 3a) which indicates the optimum IR spectrum of an SiO_2 film.

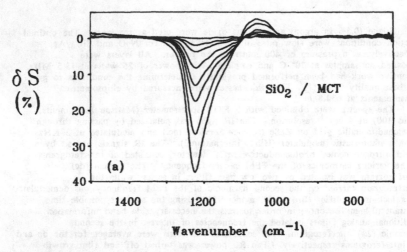

Figure 2: IRRAS spectra (δS, as defined by Eqn. 2) of PECVD SiO$_2$ on MCT was recorded after reaction times of 1, 3, 5, 10, 15, 20, 30, and 43 minutes. The thickness after deposition was 1795 Å, giving an average growth rate of 42 Å/min. The refractive index of the final film was 1.438. Spectra of PECVD SiO$_2$ grown on Si under the same conditions are identical to the spectra shown here.

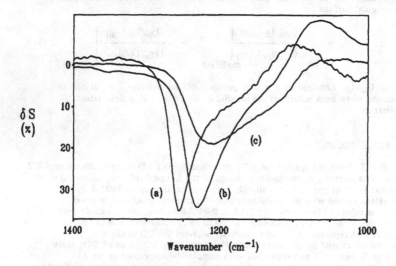

Figure 3: (a) IRRAS spectrum of thermally grown SiO$_2$ on Si, d (thickness) = 1161 Å, n.= 1.47, using a calibration spectrum from a different bare Si wafer. (b) typical spectrum of "good" PECVD SiO$_2$ on Si, grown at 70° C, d = 1079 Å, n = 1.422.(c) typical spectrum of "poor" PECVD SiO$_2$ on Si, grown at 70° C, d = 1127 Å, n = 1.33.

The results obtained for PECVD films deposited on Si, MCT, and Al substrates using similar reaction conditions were the same, with the exception of the suppression of the TO mode peak on Al. The peak frequency of the LO mode shifted to higher wavenumber by ca. 10 cm^{-1} as the film thickness was increased from ca. 50 Å to ca. 500 Å after which the peak frequency became independent of film thickness. SiO$_2$ films less than 50 Å thick exhibited absorption bands that were too broad for accurate determination of peak frequencies.

A shift in the optical properties of a PECVD sample was observed when annealed overnight in air at 450° C. The optical properties as determined by ellipsometry changed from an index of refraction and thickness of 1.420 and 2110 Å, respectively, to 1.425 and 1782 Å. The LO mode peak frequency moved from 1234 cm^{-1} to 1246 cm^{-1}.

The LO mode peak frequency and optical constants of several SiO$_2$ films, prepared by different methods, were determined: thermal oxide, 1257 to 1260 cm^{-1}; PECVD 70°C, 1205 to 1234 cm^{-1}; PECVD 400°C, 1242 cm^{-1}; PECVD annealed at 400°C after growth at 70°C, 1244 cm^{-1}. The frequency of the LO mode peak increases with the temperature of the preparation method.

Silicon wafers that were polished on both sides were placed on an aluminum mirror to evaluate the effect of increased reflectance from the mirror. It was observed that the sensitivity for the LO mode peak was not improved by the presence of the mirror as compared to samples that were polished on one side and rough on the back surface even though the intensity of the reflected light was significantly increased. An interstitial oxygen peak at 1106 cm^{-1} (25) was observed when this Si wafer (both sides polished) was placed on the aluminum mirror. This peak is not observed for samples with rough backs since the light that passes through the wafer is diffusely reflected and does not reach the detector. The presence of the native SiO$_2$ layer (ca. 15 to 20 Å thick) was observable when the spectrum from this Si wafer on Al was subtracted by that from the bare Al wafer. The signal from the native oxide on the interface of the Al is suppressed by the near normal angle of incidence (ca. 14°) at this interface.

These results have demonstrated the utility of the double modulation FT-IRRAS in-situ analysis technique for evaluating thin films of dielectric material. The characteristics of the LO mode band (frequency, intensity, and shape) are affected by many factors (stoichiometry, density, voids, and impurities) and are a sensitive indicator of film quality. While the LO mode absorption peak can not differentiate between these factors, their presence is indicated by comparison with the appearance of the absorption peak for the pure material. Impurities may also be detected by peaks in other regions of the IR spectrum.

Acknowledgements

This work was supported in part by a grant from the Office of Naval Research.

References

1. J. R. Waterman and W. A. Schmidt, Proceedings IRIS Detector Specialty Group Meeting (1986).

2. J. Wong, J. Electron. Mater. 5, 113 (1976).

3. P. G. Pai, S. S. Chao, Y. Takagi and G. Lucovsky, J. Vac. Sci. Technol. A 4, 689 (1986).

4. P. Pan, L. A. Nesbit, R. W. Douse and R. T. Gleason, J. Electrochem. Soc. 132, 2012 (1985).

5. A. L. Shabalov and M. S. Feldman, Thin Solid Films 151, 317 (1987).

6. G. Lucovsky, M. J. Manitini, J. K. Srivastava and E. A. Irene, J. Vac. Sci. Technol. B 5, 530 (1987).

7. D. W. Berreman, Phys. Rev. 130, 2193 (1963).

8. R. H. Lyddane, R. G. Sachs and E. Teller, Phys. Rev. 59, 673 (1941).

9. J. F. Scott and S. P. S. Porto, Phys. Rev. 161, 903 (1967).

10. R. G. Greenler, J. Chem. Phys. 44, 310 (1966).

11. W. G. Golden, D. S. Dunn, and J. Overend, J. Catal. 71, 395 (1981).

12. L. A. Nafie and D. W. Vidrine, in Fourier Transform Infrared Spectroscopy, Vol. 3, edited by J. R. Ferraro and L. J. Basile, (Academic Press, Inc., Orlando, 1982) p. 83.

13. S. N. Jasperson and S. E. Schnatterly, Rev. Sci. Instr. 40, 761 (1969).

14. A. E. Dowrey and C. Marcott, Appl. Spectro. 36, 414 (1982).

15. V. M. Bermudez and V. H. Ritz, Appl. Opt. 17, 542 (1978).

16. D. F. Edwards, and H. R. Philipp in Handbook of Optical Constants of Solids, edited by E. D. Palik (Academic Press, Inc., Orlando, 1985) pp. 547 and 749.

17. K. Krishnan and R. B. Mundhe, SPIE, Spectroscopic Characterization Techniques for Semiconductor Technology 452, 71 (1983).

CHEMICAL KINETICS OF SILICON-RICH OXIDE GROWTH IN AN LPCVD REACTOR

C.H. LAM[1] and K. ROSE
Center for Integrated Electronics
Rensselaer Polytechnic Institute
Troy, New York 12181

ABSTRACT

The chemical kinetics of Silicon Rich Oxide (SRO) growth in a N_2O-SiH_4 LPCVD reactor has been studied at deposition temperatures from 610 to 680°C and pressures from 0.4 to 0.5 torr. We can produce SRO films with a wide spectrum of input reactant ratios $\gamma = [N_2O]/[SiH_4] = 1$ to 40. The dependence of film composition on γ changes dramatically in a region around $\gamma = 2$.

Growth for $\gamma < 2$ is consistent with the chemical kinetics of SIPOS growth. Growth for $\gamma > 20$ can be explained by oxidation of silicon in the bulk of the growing SRO film. We can explain growth from $\gamma \approx 4$ to 20 by considering the chemical kinetics of possible binary surface reactions which may produce Si-Si or Si-O bonds. This allows us to accurately model the dependence of SRO growth rate in this region as a function of γ, pressure, and deposition temperature.

INTRODUCTION

By increasing γ, the ratio of $[N_2O]$ to $[SiH_4]$, in an LPCVD reactor one can produce SiO_x films with increasing oxygen content. At low γ these films approach polysilicon and are described as SIPOS, Semi-Insulating PolySilicon. SIPOS is commonly used as a surface passivant for high-voltage devices [1]. For high γ one approaches SiO_2, producing Off-Stoichiometry Oxides (OSO) which can be used as insulating layers in EAROM technology [2]. At intermediate γ one produces Silicon-Rich Oxides (SRO) which are useful as injecting layers in EAROM technology [3].

We report the growth kinetics of SRO films using silane (10% in nitrogen) and nitrous oxide as reactant gases. Deposition temperatures were varied from 610 to 680°C, deposition pressures from 0.4 to 0.5 torr. Growth and measurement techniques are discussed in the next section. This is followed by a comparison of our low γ results with the model of Hitchman and Kane [4] for SIPOS growth. An alternative model is developed to explain growth at higher values of γ. Anomalous deposition around $\gamma \approx 2$ is noted.

GROWTH AND MEASUREMENT TECHNIQUE

Two-inch p-type-silicon wafers were RCA cleaned within 24 hours prior to film deposition. This was followed by a 30-second dip in 10% aqueous HF to remove native oxide before the wafers were loaded into our reactor. The two heating zones toward the loading door were maintained at the same temperature, while the heating zone toward the pump and was maintained at a slightly lower temperature to avoid excessive deposition on the tube wall.

---

[1]Now at IBM, 1000 River Road, Essex Junction, Vermont 05452

Mat. Res. Soc. Symp. Proc. Vol. 131. ©1989 Materials Research Society

Wafers were loaded horizontally in the center zone of the reactor. Temperature variation in this zone was less than 2^0C. In our reactor, film thickness variation across horizontally loaded wafers was typically around 5% compared to as much as 50% from the center to the periphery for vertically loaded wafers. The loading configuration had little effect on the silicon content of our films.

Refractive index and thickness of the SRO films were measured with a microcomputer-controlled ellipsometer, the Rudolph Auto EL-II, at a wavelength of 632.8 nm. Measurements were performed on each SRO film immediately after deposition and after isothermal annealing in nitrogen. High temperature isothermal annealing converted amorphous silicon islands into microscopic crystallites [5]. We have used a common annealing temperature and time for all SRO films reported, 950^0C for 30 minutes. Ellipsometry measurements provided an accurate estimate of silicon content in annealed wafers over the full range of gas ratios examined [6,7].

SILICON RICH OXIDE GROWTH

Figure 1 shows the growth rate of SRO films at various temperatures over a wide range of reactant gas ratios, γ. The composition of materials shown in this figure covers the entire range of silicon-rich oxides (SRO), ranging from SIPOS at low values of γ to off-stoichiometric oxide (OSO) for $\gamma \geq 20$. For $\gamma < 2$, the growth rate variation can be explained using the kinetic model developed by Hitchman and Kane for SIPOS growth [1]. However at higher values of γ, this model does not provide a good fit to the data.

The deviation from SIPOS growth is greater than Fig. 1 suggests. Figure 2 shows the dependence of refractive index, γ, for the same films and growth conditions. There is clearly a region of anomalous behavior around $\gamma = 2$ that is not reflected in the monotonic behavior of Fig. 1. We have previously shown that the Bruggeman effective medium approximation allows us to monotonically relate silicon content to refractive index over the range of γ we have studied [6,7]. In the anomalous region, the increase in n with γ suggests a decrease of oxygen content in the films, even though the reactant ratio of nitrous oxide to silane is increasing. In fact, films grown in the anomalous region exhibit much higher etch rates, which indicates high porosity. Further, the dependence of growth rate on temperature is much stronger. All this suggests that a radical change in growth mechanisms occurs around $\gamma = 2$.

SIPOS GROWTH ($\gamma \leq 2$)

Hitchman and Kane [1] modeled the kinetics of SIPOS growth in an LPCVD reactor for $0 \leq \gamma \leq 0.6$ by extending a model of polysilicon growth previously developed by Hitchman and coworkers [8]. In particular, they explained the drop in growth rate with increasing γ by assuming that adsorbed N_2O blocked surface sites for SiH_4 pyrolysis. Indeed, N_2O inhibited growth to a greater extent than could be accounted for by dilution of silane. Thus, the polysilicon growth rate,

$$G = kn\theta, \tag{1}$$

became

$$G = kn\theta/(1 + x), \tag{2}$$

where k is the reaction constant for the dissociation of silane to silicon, n is the density per unit area of available adsorption sites, θ is the fractional coverage of the surface by adsorbed silane, and x is the oxygen-to-silicon ratio in SiO_x. If the surface of the film is covered by SiO_x, the number of silicon adsorption sites is reduced from n to $n/(1 + x)$.

The value of θ depends on γ in a complex way and depends on the products of the intermediate reaction steps. The simplest approach is to assume that θ is directly proportional to the fraction of silane in the gas flow. That is,

$$\theta = \frac{c[SiH_4]}{[SiH_4] + [N_2O]} = \frac{c}{1 + \gamma} \tag{3}$$

where c is a constant. Thus, assuming n in Eq. (2) is constant,

$$G^{-1} = K(1 + x)(1 + \gamma) \tag{4}$$

where $K = (k\,c\,n)^{-1}$. For $\gamma < 5$ we find that $x \approx 0.23 + 0.18\,\gamma$. Figure 3 shows the best fit of Eq. (4) to our data at 650^0C. $K^{-1} = 190$ Å/min. The fit is reasonable for $\gamma \leq 10$. For higher γ, the growth rate is somewhat faster than the model predicts. The value of G^{-1} was plotted to accentuate the region of lower growth rate at higher γ. Hitchman and Kane [4] developed a more elaborate model that considered reactant depletion, tube geometry, and intermediate reaction steps. Use of their more elaborate model did not change the fit to our data.

SRO GROWTH ($\gamma \geq 4$)

To analyze the growth of SRO beyond the anomalous region we consider the competition between the formation of Si-Si bonds (as Si adatoms are incorporated into a growing SRO film) and the formation of Si-O bonds (as O adatoms bond to available Si). Once a Si atom is adsorbed, it is unlikely to be desorbed at our moderate growth temperatures. Hence, it will either form an Si-Si bond with an available Si atom or an Si-O bond with an available O atom. The probabilities of forming these bonds must be complementary.

$$P^s(Si - Si) + P^s(Si - O) = 1 \tag{5}$$

An adsorbed O atom from an impinging N_2O molecule is much more likely to be desorbed. Its only chance to form a bond with Si is by direct impingement. Hence, we argue that the probability of its forming a bond with Si is comparable to that for an Si adatom to bond with Si,

$$P^n(Si - O) \approx P^s(Si - Si). \tag{6}$$

We have observed that SRO film growth is steady state, with the growth rate independent of film thickness. The SRO growth rate, G, is given by conversion of the flux of adsorbed silane, J, into the composite SRO film. $G = J\,V_o$ where V_o is the volume of a unit of SRO film.

In steady state growth, for a linear reaction, the flux of silane J is balanced by the reaction of adsorbed species.

$$J = k_1 P^s(Si - Si)n^a_{SiH_4} + k_2 P^s(Si - O)n^a_{SiH_4} \tag{7}$$

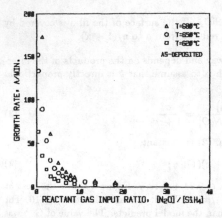

Figure 1. Growth rate of SRO films as a function of reactant gas ratio.

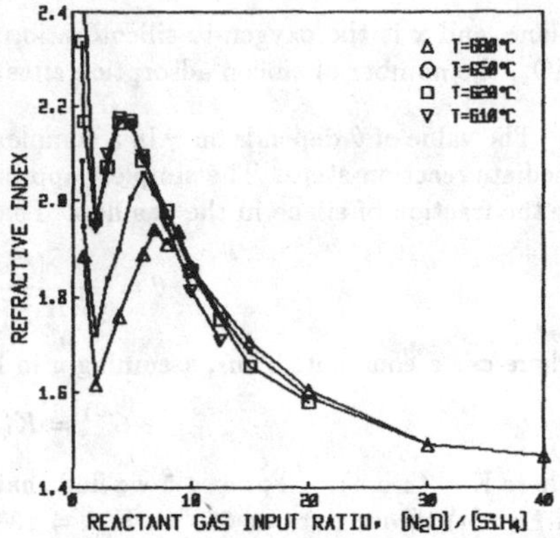

Figure 2. Variation of refractive index with growth conditions for SRO films.

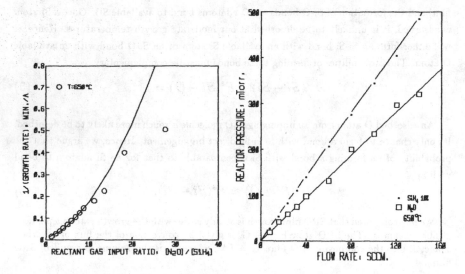

Figure 3. Hitchman-Kane model for SIPOS growth rate dependence on reactant gas ratio.

Figure 4. Linear dependence of reactor pressure on reactant flow rate.

where k_1 and k_2 are the surface-reaction-rate constants for silane to form Si-Si and Si-O bonds, respectively. The value of $n^a_{SiH_4}$ is the concentration of adsorbed silane molecules. Assuming diffusion of the reactant through a boundary layer,

$$J = h(n^g_{SiH_4} - n^a_{SiH_4}).$$ (8)

$n^g_{SiH_4}$ is the concentration of silane molecules in the gas stream. Since h is relatively large, $n^a_{SiH_4}$ can be replaced by $n^g_{SiH_4}$ in Eq. (7). Furthermore, the k_2 term can be neglected compared with the k_1 term in Eq. (7) because the growth of polysilicon is much faster than the growth of CVD oxide.

To find the unit volume of the resulting SRO film we assume a stoichiometric form for SiO_x which corresponds to Si crystallites in an SiO_2 matrix. The volume of a unit of this composite material is

$$V_o = \frac{1}{n_{Si}}\left(1 - \frac{x}{2}\right) + \frac{1}{n_{SiO_2}}\frac{x}{2},$$ (9)

where $n_{Si} = 5\text{x}10^{22}$ atoms cm^{-3} and $n_{SiO_2} = 2.22\text{x}10^{22}$ molecules cm^{-3} are the concentrations of crystalline Si and thermal SiO_2, respectively. Since we assume that all oxygen is bound as SiO_2, every oxygen atom requires an Si-O bond. Consequently, $x = P^s(Si - O) = 1 - P^s(Si - Si)$.

Combining Eq. (7) and Eq. (9) we obtain

$$G \approx \left[\frac{1 + P^s(Si - Si)}{2n_{Si}} + \frac{1 - P^s(Si - Si)}{2n_{SiO_2}}\right]k_1 P^s(Si - Si)n^g_{SiH_4}.$$ (10)

Assuming the silane in the gas stream behaves as an ideal gas at the deposition temperature T_d, we obtain

$$P_{SiH_4} = n^g_{SiH_4}k_B T_d.$$ (11)

Figure 4 shows the linear relationship between reactor pressure and the silane in nitrogen flow rate and nitrous oxide flow rate, respectively. Since we use 10 percent silane in nitrogen and the ratio of nitrous oxide to silane is γ, the partial pressure of silane is

$$P_{SiH_4} = P_{reactor}/(10 + \gamma)$$ (12)

For ideal gases, the reactor pressure should be proportional to total flow rate, f_T. Thus,

$$n^g_{SiH_4} = af_T/k_B T_d(10 + \gamma).$$ (13)

To complete the growth model, we need to relate $P^s(Si-Si)$ to γ. The number of Si-O bonds formed by the decomposition of silane must equal the number formed by the decomposition of nitrous oxide. This gives

$$P^s(Si - O)n^a_{SiH_4} = P^n(Si - O)n^a_{N_2O},$$ (14)

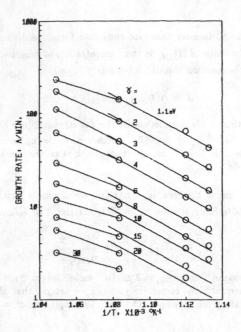

Figure 5. Fit of SRO growth rate to an Arrhenius temperature dependence.

Figure 6. Model for SRO
growth rate
dependence on
deposition
temperature and
reactant gas
ratio.

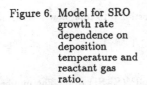

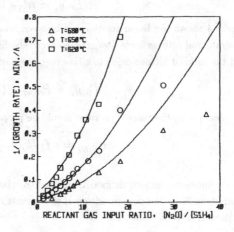

where $n^a_{SiH_4}$ and $n^a_{N_2O}$ are the respective concentrations of adsorbed silane and nitrous oxide per unit area. We expect $n^a_{N_2O}/n^a_{SiH_4} = \beta\gamma$.

Combining these equations gives

$$G \approx \frac{af_T k_1}{2k_B T_d}\left(\frac{2+\beta\gamma}{n_{Si}} + \frac{\beta\gamma}{n_{SiO_2}}\right)\left(\frac{1}{1+\beta\gamma}\right)^2\left(\frac{1}{10+\gamma}\right) \tag{15}$$

where we expect k_1 to have an Arrhenius form, $K_1 \exp(-E_a/k_B T_d)$. The exponential dependence of the reaction rate on temperature should dominate the temperature dependence of G. Figure 5 shows the extent to which G can be fit by a single activation energy. $E_a = 1.1$ eV provides a reasonable fit for $\gamma < 20$ and $T_d \leq 650^0$C.

The overall fit of our model to the observed growth rate is shown in Figure 6. Eq. (15) provides a better fit for intermediate values of γ from 6 to about 20 for 620 and 650^{0}C. We take $\beta = 0.4$, $E_a = 1.1$ eV, and $K_1 = 4\times10^{15}$ Å min^{-1}. For 680^{0}C, $E_a = 1.0$ eV provides a better fit. We take a = 3.45 mtorr/sccm and $f_T = 160$ sccm as fitting parameters in all cases. This theory fits our experimental results over a somewhat broader range of γ than the Hitchman-Kane model. In addition, it provides reasonable values for a number of parameters of kinetic interest.

REFERENCES

1. T. Matsushita et al., Jap. J. Appl. Phys. Supplement, 15, 35 (1976).

2. D.J. DiMaria et al., J. Appl. Phys., 54, 5801 (1983).

3. D.J. DiMaria, K.M. DeMeyer, and D.W. Dong, IEEE Elec. Dev. Lett. EDL-1, 179 (1980).

4. M.L. Hitchman and J. Kane, J. of Crystal Growth, 55, 485 (1981).

5. H. Hamasaki et al., J. Appl. Phys., 49, 3987 (1978).

6. K-T Chang, C. Lam, and K. Rose, Mat. Res. Soc. Symp. Proc., 105, 193 (1988).

7. K-T Chang, C. Lam, and K. Rose, Mat. Res. Soc. Symp. Proc., 106, 107 (1988).

8. M.L. Hitchman, J. Kane, and A.E. Widmer, Thin Solid Films 59, 231 (1979).

SILICON SUBOXIDES: THE "CO-DEPOSITION" OF a-Si:H AND SiO$_2$

D.V. Tsu, G. Lucovsky and M.W. Watkins, Departments of Physics and Materials Science and Engineering, North Carolina State University, Raleigh, NC 27695-8202

ABSTRACT

The deposition mechanism of silicon suboxides (SiO$_x$, x<2) prepared by remote plasma enhanced chemical vapor deposition (Remote PECVD) is investigated. These films were deposited in a Deposition/Analysis chamber designed to investigate the gas phase chemistry. In this technique, an O$_2$/He mixture is plasma excited, and the silane reactant is injected into the deposition chamber *down-stream* from the plasma tube. We show that if the plasma *after-glow* is prevented from extending into the deposition region by an electrical grid placed between the plasma tube and the deposition region, silicon dioxide is then deposited for *all* O$_2$/He mixtures investigated (0.1 to 1.0 %). In contrast, hydrogenated suboxides of silicon are deposited when the plasma after-glow is allowed to extend past the grid into the deposition region.

INTRODUCTION

In this paper, we discuss the deposition of *hydrogenated* suboxides of silicon (SiO$_x$:H x<2) by Remote PECVD. In this technique, a flowing mixture of O$_2$ and He is rf plasma excited (13.56 MHz) by inductive coupling in a vitreous silica tube remote from the deposition chamber. Silane (SiH$_4$), the source of the silicon atoms in the oxides, is introduced into the deposition chamber at a point between the end of the chamber, where the excited O$_2$/He mixture enters, and the heated substrate, where deposition occurs. Deposition therefore occurs "down-stream" from the plasma-*excitation* region. In an earlier publication [1], we demonstrated that the stoichiometry of the silicon oxides could be controlled by varying the relative concentration of O$_2$ in He: suboxides were produced for O$_2$ concentrations below about 1 %; oxides and/or superoxides (containing SiOH bonding groups), for O$_2$ concentrations in excess of 1 %. In recent studies [2,3], we have investigated the gas phase chemistry of silicon dioxide and amorphous hydrogenated silicon (a-Si:H) depositions by Remote PECVD, with a Deposition/Analysis System which incorporates a combination of optical emission spectroscopy (OES) and mass spectrometry (MS), in addition to thin film deposition. In particular, we have shown that the conditions (rf power, pressure, etc.) for the deposition of a-Si:H are such that the plasma after-glow must extend into the deposition region where the silane is introduced. However, we find no plasma-generated gas-phase fragmentation of the silane, so that the effects of the extension of the after-glow into the chamber region are qualitatively different from those obtained when the silane reactant is directly injected into a region in which the plasma is generated by rf excitation. In the present paper, we show further that if plasma "glow" is contained up-stream from the deposition region by the use of a grid, then silicon dioxide is deposited for *all* oxygen concentrations, including the range of O$_2$ concentrations between 0.1 and 1.0 %, where our previous studies had indicated only suboxide deposition [1]. For these concentrations we now show that suboxides are produced only if the plasma after-glow is allowed to extend into the deposition region.

EXPERIMENTAL APPARATUS

All of the thin films produced in this study were deposited in the Deposition/Analysis System which is described in a previous publication [2]. Figure 1 shows a schematic of a section of the chamber containing the plasma tube and the first gas dispersal ring. The chamber is made of 2 in. O.D. stainless steel (SS) tubing and contains five analyzing stations and two gas dispersal rings. Each analyzing station consists of three ports: one port for the MS and the other two for optical studies. The first two stations and the first gas dispersal ring are shown in the figure. The silane mixture (10% SiH_4:90% Ar) is introduced into the chamber through this gas dispersal ring at a flow rate of 10 standard cm^3/minute (sccm). The O_2/He mixture, delivered at a flow rate of 100 sccm, flows through the plasma tube and past the plasma bias plates into the chamber region. The relative concentration of O_2 in He was varied between 0 and 1.0 % by varying the relative flow rates of a 1.0 % O_2 in He source gas, and a pure He source gas. The plasma tube consists of 32 mm O.D. vitreous silica approximately 10 in. long and the rf power for this study was set at 29±1 W. The substrate/sample-heater block assembly can be positioned anywhere along the length of the chamber. For this work, it was positioned at Station #2 which is 7 in. down-stream from the end of the plasma tube. We have not as yet calibrated the substrate temperature relative to the sample block temperature; all of the films were deposited at a block temperature of 225°C. In addition, the sample block was biased -125 V with respect to ground. Under this bias condition, the dihydride (SiH_2) component in the resulting a-Si:H [3] and suboxide films is minimized. We have observed that the substrate bias does not affect the deposition rate for the a-Si:H films [3]. Finally, the pressure is measured with a Baratron and maintained at 300 mTorr using a down-stream throttle controller. The process gases are pumped by a 55 CFM rotary vain pump.

The plasma bias plates consist of four rectangular Al plates arranged in a *box-like* configuration coaxial with the chamber, and a circular #10 mesh SS screen grid capping the end of this box. The plates and the grid are supported by teflon and can be biased independently. In this study, only two conditions were varied: (i) the relative O_2 concentration in He, between 0 and 1.0 %; and (ii) the bias *state* of the end grid, either *electrically floating or grounded*. The four Al plates were electrically floating during the entire study. All of the other deposition parameters, e.g., the SiH_4/Ar flow rate, pressure, rf power, etc. were not varied. When the end grid is *grounded*, the plasma after-glow, as can be observed from the view port of Station #1, extends down-stream from the grid. On the other hand, when the grid is allowed to *float*, the plasma glow is observed to stop at the grid and the after-glow does *not* extend down-stream into the chamber region.

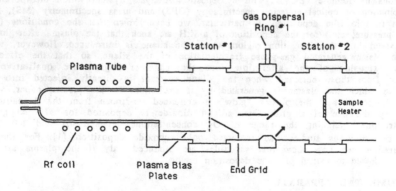

FIG. 1. The Deposition/Analysis chamber.

RESULTS

The thin films produced in the apparatus described above were deposited onto crystalline Si substrates (10 - 100 Ω cm) and examined with an infrared (ir) spectrophotometer, a Perkin-Elmer Model PE983. Figure 2 shows the ir spectra of two different films, each of which was deposited using a 0.30 % O_2 in He mixture. The top spectrum was obtained from a film deposited with the end grid *grounded* and the bottom spectrum with the end grid *floating*. The bottom spectrum is stoichiometric SiO_2 and contains the three characteristic Si-O vibrational absorption bands: the stretching band at about 1052 cm^{-1}; the bending band at 806 cm^{-1}; and the rocking band at 445 cm^{-1}. The peak position of the Si-O stretching band can vary between about 1080 and 1050 cm^{-1} in stoichiometric SiO_2 depending on the deposition and/or growth temperature [4]. For the temperatures used in the Remote PECVD of SiO_2 (<500°C), this frequency ranges between about 1050 cm^{-1} and 1065 cm^{-1}. The principle absorption feature in the top spectrum is also due to an Si-O stretching vibration [5], but in this case, the peak frequency has decreased to 1020 cm^{-1}. Lucovsky *et al.* [5], have shown that in suboxides, i.e., SiO_x for x < 2, there is an approximately linear relationship between the value of x and the peak frequency of the Si-O stretching band. For an oxygen atom imbedded into an amorphous silicon network, this frequency is about 960 cm^{-1}, whereas for SiO_2 it is significantly higher as indicated above. We then estimate that the value of x for the film of the top spectrum is between about 1.0 and 1.3. The other absorption features, in the top spectrum, below about 900 cm^{-1} are also consistent with a hydrogenated suboxide film, and are associated with SiH bond-bending vibrations [5].

The frequencies of the Si-O stretching band maximum, for films deposited with the end grid either *floating* or *grounded*, are plotted as a function of the relative O_2 concentration (in the O_2/He mixture) in Fig. 3. The changes in the *grid-floating* data (solid squares) are relatively small and can be accounted for by the incorporation of SiOH bonding groups, which increases with increasing O_2 [1]. The spectra of these films do not show any evidence for SiH bond-stretching or bond-bending absorption bands. In contrast, under certain conditions, both SiH and SiOH absorption features can be seen in the ir spectra of films deposited under the *grid-grounded* condition. The SiH features are seen for *all* of the films deposited under the *grid-grounded* condition; the SiOH bond-stretching absorption band (between 3700 and about 3200 cm^{-1}) is seen, however, only in films produced with O_2 concentrations greater than about 0.7 %. For the films prepared with O_2 concentrations above 0.5 %, the Si-O stretching frequencies have increased well-above 1050 cm^{-1} and is indicative of near SiO_2 stoichiometry. For these films, in addition to the three Si-O vibrational bands, the ir absorption spectra include the two SiH related features at about 2270 and 875 cm^{-1} associated respectively with localized SiH stretching and bending vibrations. These frequencies are higher than the corresponding values of the SiH vibrations in a-Si:H (1985 and 630 cm^{-1}) because of a chemical induction effect associated with the substitution of three oxygen nearest neighbors for silicon atoms in a-Si:H [6].

Next, we examine the deposition rates of the oxide and suboxide thin films. To make this determination, the film thickness was measured with a Dektak profilometer. In most cases, a sharp step could be achieved by protecting the thin film with Bee's wax and etching an edge region with concentrated HF acid. For the suboxides deposited using oxygen concentrations below 0.4 %, the etching technique did not work and we had to rely on physically masking the substrate during deposition to create the step. This data is plotted in Fig. 4; the solid lines represent linear least square fits to the data. When O_2 is removed from the deposition process, i.e., when pure He flows through the plasma tube, the grid-grounded condition results in a-Si:H thin films, whereas in the grid-floating condition, the pure He flow

292

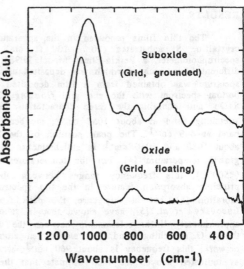

FIG. 2. Ir absorbance of two films deposited with an 0.3% mixture of O_2 in He. All of the other deposition parameters (rf power, flow rates, etc.) are the same. Only the condition of the end grid is changed.

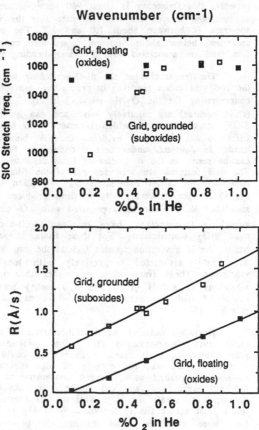

FIG. 3. Peak frequency of the SiO stretching band vs. the O_2 concentration for films deposited under the grid grounded and grid floating states.

FIG. 4. Deposition rate vs. the O_2 concentration for films deposited under the grid grounded and grid floating states.

(and no plasma after-glow in the region in which the silane is introduced) does not results in thin film deposition.

Finally, we have measured differences in the current to the sample probe between the floating and grounded states of the end grid for the -125 V sample bias condition. The applied DC voltage was measured with a Keithley Model 616 electrometer which has an input impedance of $> 2 \times 10^{14} \Omega$, and the current was measured with a Keithley Model 175 multimeter. The sample probe was positioned at Station #2 and pure He flowed through the plasma tube. The rf power was tuned to the same power used in the film depositions and the SiH_4/Ar flow was turned off. The multimeter was arranged so that negative current indicated electron flow and positive current indicated a positive ion flow. Under this condition, a positive current of about 0.4 μA was measured for the grid-floating condition, while an increase, by about a factor of four, in current was measured for the grounded condition. We have also measured the current as a function of the O_2 concentration for the two states of the end grid. We find for the grid-floating condition, a current of about 0.5 μA, which is independent of the O_2 concentration. For the grid-grounded condition, on the other hand, the current increases from about 1.5 μA to over 6 μA as the O_2 concentration increases to 1 %. This increase in current may be due to O_2^+ ions created in the plasma-glow region and identified by OES [2].

DISCUSSION and CONCLUSIONS

We have used the terms *glow* and *after-glow* as a convenient way of describing the physical differences in the nature of the plasma regions resulting from the electrical state of the grid screen, either floating or grounded. The bias condition of the end grid can have no effect on neutral excited species, e.g., He metastables, generated in the plasma tube, and transported through the grid down-stream into the deposition region. It can however, affect charged species as discussed above. The plasma after-glow extending down-stream from the grid in the grid-grounded condition can therefore be associated with an enhanced passage of electrons and ions from the plasma past the grid and into the chamber. In other words, the plasma is in effect allowed to "spill" into the deposition chamber. The physical mechanism is not clearly understood at this time, but preliminary studies show that this does not depend on whether the gas is flowing. The differences described above, due to the different electrical states of the grid, are also found under static, i.e., zero He flow, conditions for the pressures controllable in the our chamber (0.1 to 1.0 Torr). We have also found that the self-bias of the plasma bias plates decreases from about 15 VDC to about 3 VDC when the end-grid is grounded. We are currently investigating whether this change, and the corresponding changes in the after-glow, are caused by changes in the plasma potential [7].

We have shown that when the O_2/He plasma after-glow is allowed to extend down into the deposition region, silicon suboxides are produced. On the other hand, for the same O_2/He mixtures, eliminating a plasma after-glow by electrically floating the end grid results in stoichiometric silicon dioxide. In addition, we have shown that the deposition rate of the silicon dioxides is always less than the suboxides over the range of O_2 concentrations examined. When O_2 is eliminated for the grid-floating condition, there is no thin film deposition; i.e, a-Si:H thin films are not formed. Since neutral He metastables created in the plasma [2] can be transported into the chamber under the grid-floating condition, this means that the deposition of a-Si:H films can not involve excitation of the silane reactant by a neutral He metastable species. This observation, that noble gas metastables cannot account for a-Si:H film deposition, has also been obtained in the studies of Knights *et al* [8]. It does, however, support an SiO_2 deposition mechanism based on active neutral oxygen metastable species [2], and that in this process the neutral silane reactant is chemically consumed in the heterogeneous reaction whereby SiO_2 is formed. Fig. 4 shows that the increase in thin film deposition rate with O_2 concentration is effectively independent of the bias condition of the end grid, further supporting this mechanism of SiO_2 thin film deposition based on an interaction between neutral oxygen metastables and silane [2].

We therefore propose a model for the deposition of silicon suboxides by Remote PECVD based on the codeposition of a homogeneous alloy material with an hydrogenated amorphous silicon *bonding-component* and a silicon dioxide *bonding-component*. The a-Si:H deposition rate is controlled by the charged species (most likely the electrons) extracted into the plasma *after-glow* region, and the oxide deposition rate is controlled by the active neutral species (most likely O_2 metastables [2]). In this context, the silicon deposition can be switched on by grounding the end grid, and the oxide deposition can be switched on by introducing O_2 into the plasma. We have shown in previous work [3] that, even though it is necessary for the plasma to extend into the deposition region to obtain a film, the deposition rate of a-Si:H does not depend on the substrate bias condition. This codeposition of the a-Si:H and silicon oxide "phases" does not result in diphasic material, since the mixture of the associated local bonding groups occurs on an atomic scale. The linear shift of the center frequency of the Si-O bond-stretching band in the ir spectra of suboxides is consistent with a random incorporation of Si and O atoms into the films, i.e. with suboxide films are that are homogeneous alloys of Si and O, and in addition contain a small fraction, about 10 at.% of H in SiH bonding groups.

ACKNOWLEDGEMENTS

This research is supported by ONR, contract N00014-C-79-0133.

REFERENCES

1. D.V. Tsu, G.N. Parsons, G. Lucovsky and M.W. Watkins, Mat. Res. Soc. Proc. (Fall 1987, in press)
2. D.V. Tsu, G.N. Parsons, G. Lucovsky and M.W. Watkins, J. Vac. Sci. Technol. A, (1989, in press)
3. G.N. Parsons, D.V. Tsu, C. Wang and G. Lucovsky, J. Vac. Sci. Technol. A, (1989, in press)
4. G. Lucovsky, M.J. Mantini, J.K. Srivastava and E.A. Irene, J. Vac. Sci. Technol. B 5, 530 (1987)
5. G. Lucovsky, J. Yang, S.S. Chao, J.E. Tyler and W. Czubatyi, Phys. Rev. B 28, 3225 (1983)
6. G. Lucovsky, Solid State Commun. 29, 571 (1979)
7. B. Chapman, Glow Discharge Processes (John Wiley & Sons, New York, 1980)
8. J.C. Knights, R.A. Lujan, M.P. Rosenblum, R.A. Street, D.K. Bieglesen and J.A. Reimer, Appl. Phys. Lett. 38, 331 (1981).

DEPENDENCE OF PECVD SILICON OXYNITRIDE PROPERTIES ON DEPOSITION PARAMETERS

Aubrey L. Helms, Jr.[*] and Robert M. Havrilla[**]
[*]AT&T Bell Laboratories - Engineering Research Center, PO Box 900, Princeton, NJ 08540
[**]AT&T Microelectronics, 2525 N. 12th St., Reading, PA 19604

ABSTRACT

The properties of Plasma Enhanced Chemical Vapor Deposited (PECVD) silicon oxynitride thin films were determined for a variety of deposition conditions. The films were characterized with respect to stress, refractive index, deposition rate, hydrogen content, dielectric constant, and uniformity. The films were deposited in an Electrotech ND6200 parallel plate reactor using a silane - ammonia - nitrous oxide process gas chemistry. Deposition parameters which were investigated include process gas flow rate, power, and total pressure. The possible application of these films as both inter-layer and final passivation layers for use on GaAs ICs will be discussed.

INTRODUCTION

The deposition of high quality dielectric films at low temperatures is critical to the realization of GaAs ICs. The common use of Au-Ni-Ge ohmic contacts limits the processing temperatures of GaAs ICs to ~300 °C. This has driven the development of deposition techniques for dielectric materials to those based on plasma enhanced chemical vapor deposition (PECVD) processes. Traditionally, Si_3N_4 has been the material of choice because of its passivation properties, mechanical strength, barrier properties, and process compatibility [1-3]. One of the disadvantages of Si_3N_4 is its high dielectric constant which contributes to parasitic line capacitances that limit the device speed. The lower dielectric constant of SiO_2 makes this material appealing, however, the long term reliability of devices encapsulated in SiO_2 is not as good as those passivated with Si_3N_4 [4].

Recently, there has been much work on the development of silicon oxynitride (SiON) films [5-7]. The hopes are that these films will exhibit the best properties of both Si_3N_4 and SiO_2, namely the passivation and mechanical properties of Si_3N_4 and the low dielectric constant and low stress of SiO_2. The properties of these films are very sensitive to the composition and deposition parameters. The implementation of such a material into the manufacture of GaAs ICs requires that the properties of the thin films be well understood as a function of process parameters.

EXPERIMENTAL

The films were deposited in an Electrotech ET 6200ND plasma enhanced chemical vapor deposition (PECVD) system. This unit is a cassette-to-cassette system capable of handling as many as 14 wafers (3" diameter) in each deposition run. The reactor is a parallel plate design which operates at a plasma frequency of 380 kHz. The power, pressure, temperature, and process gas flows are microprocessor controlled. The process gases were SiH_4, N_2O, NH_3, and a carrier gas of N_2. In each deposition, a 3" Si monitor wafer was included to determine the stress of the film. It has been our experience that Si wafers tend to distort in a uniform manner to a given stress. Also included in each deposition was a piece of Si which had been polished on both sides. This piece was used for the FTIR studies to determine the hydrogen content of the films.

The deposition parameters which were selected as variables were power, pressure, SiH_4 flow rate, N_2O flow rate, and NH_3 flow rate. The variables and their levels are shown in Table I. A factorial experimental design was constructed by J. Eberhardt which resulted in a matrix of 28 experiments. Films of ~5000 A were deposited on 2" GaAs wafers which had been processed to include the bottom capacitor plates of the AT&T 1520B Laser Driver GaAs IC.

TABLE I

PROCESS PARAMETERS

Variable	Low Level	Medium Level	High Level
Power (W)	325	1750	3750
Pressure (mTorr)	500	800	1000
SiH$_4$ Flow (sccm)	110	130	150
N$_2$O Flow (sccm)	450	500	600
NH$_3$ Flow (sccm)	150	350	600

The thickness and refractive index of the films deposited on the Si monitor wafers were determined using a Rudolph Auto EL III ellipsometer. The wavelength was held constant at 623.8 nm and an incident angle of 70°. Five readings were taken on each wafer and used to determine the average thickness, refractive index, and uniformity. The stress of the film was determined by measuring the Si monitor wafers on a GCA Autosort Mark 11.150 wafer bow measuring system. The stress in the films was calculated from:

$$\sigma = (E_s T^2 8 \Delta h)/(6(1-\upsilon)tC^2)$$

where,

σ	Stress (dynes/cm^2)
$E_s/(1-\upsilon)$	1.81 x 10^{12} dynes/cm^2 for Si(100)
T	Wafer thickness
t	Film thickness
Δh	Change in wafer bow
C	Chord length over which wafer bow is measured

The hydrogen content of the films was determined by FTIR on a Nicolet 5DX FTIR [8]. The concentration of N-H bonds was determined by integrating the area under the peak centered at 3368 cm^{-1} from 3423 to 3167 cm^{-1}. This may also include a contribution from O-H bonds at 3380 cm^{-1}. The Si-H bond concentration was determined by integrating the area under the peak centered at 2232 cm^{-1} from 2294 to 2095 cm^{-1}. The Si-N bond absorption was centered at 1032 cm^{-1}. This area will also include contributions from any Si-O (1070 cm^{-1}) and Si-O bending (805 cm^{-1}) linkages which may be present. The processing of the 2" wafers was continued to complete the construction of the capacitor. The capacitance of the films was measured on the standard production test sets in the 2" GaAs manufacturing line at the AT&T Reading Works. The dielectric constant was calculated from the capacitance, capacitor area, and the measured film thickness [9].

Compositional information was obtained by Rutherford Backscattering Spectrometry (RBS) on an Ionex Thin Film Analysis system. The RBS spectra were obtained at a beam energy of 2 MeV. The atomic concentrations were obtained by measuring the step height of the RBS profile. Additional composition of the films was determined by Auger Electron Spectroscopy (AES) on a PHI-610 from Physical Electronics. The AES parameters were chosen to try and minimize the sample charging which was found to be quite severe in some cases.

RESULTS

The experimental matrix of 28 experiments allowed the investigation of the five variables over a wide range. The refractive index varied from a low value of 1.575 for a film deposited at high pressure and a high fraction of N$_2$O to a high value of 2.031 for a film deposited at low pressure and low fraction of N$_2$O in the process gas stream. The uniformity was very good, generally less than 5% with only a few exceptions. The refractive index was observed to generally be lower at higher powers. The refractive index was independent of the Si/N ratio in the process gas stream. One disconcerting trend which was apparent was that the refractive index did not correlate well with the dielectric constant as would be expected. The refractive index was observed to be independent of the absolute N$_2$O flow but decreased as a function of the fraction of N_2O fraction in the process gas stream (except for films deposited at low powers). The refractive index was expected to decrease with an increasing N_2O fraction since this in turn increases the relative amount of oxygen in the film and the film becomes more like SiO_2. The refractive index was also independent of the N_2O/NH_3 ratio in the process gas stream.

The hydrogen bond content of the films was determined by FTIR [8]. Many of the films showed no absorption in the N-H region of the spectrum (3423-3167 cm^{-1}). Those films which did indicate an absorption in this region correlated with the presence of an O-H shoulder on the envelope. It is assumed that the absorption in these cases is due to O-H and not N-H, although this cannot be stated unambiguously. It can be concluded that almost all of these films contain no N-H bonds within the accuracy of of IR. Additionally, those films which showed the presence of "O-H" bonds tended to have a Si-N lineshape whose peak shifted to higher wavenumbers implying an increased concentration of Si-O bonds. Most of the films had a total H-bond concentration between 0-1.0 x 10^{22} bonds/cm^3. Those films which had a very high H-bond concentration also showed the presence of the "O-H" absorption.

The stress of the films was generally low (<1.0 x 10^{-9} dynes/cm^2). The films which showed a very high stress were deposited at high power, low pressure, had low rates, indicated the presence of "O-H", had high Si-H concentrations, and had high H-bond concentrations. The higher stress films also showed Si-N IR lineshapes which peaked at higher wavenumbers, again, indicating the presence of Si-O bonds. The plot in Figure 1 shows that the stress decreases with the fraction of SiH_4 in the process gas stream. The opposite was found to be true for the case of the N_2O fraction in the process gas stream where the stress was observed to increase.

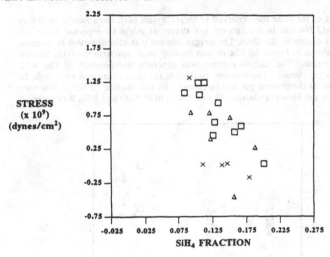

Figure 1. Stress vs SiH_4 fraction in the process gas stream. (Δ) - Power = 325 W; ($\square$) - Power = 1750 W; (x) - Power = 3750 W.

Many of the samples did not yield a satisfactory measurement of the dielectric constant. The measurements indicated that the capacitor plates were shorted. A number of possible explanations could account for this; metal flaps left from the lift-off process, pinholes in the film, over-etching the film during processing after deposition, and particle contamination prior to deposition. There was some data which indicated that the dielectric constant tended to decrease at high H-bond concentrations and also decreased with the N_2O flow but the data is not well behaved.

The deposition rate generally increased as a function of power as expected. The notable exception to this trend occurs at low pressures and high power. This is shown in Figure 2. The rate was generally independent of the N_2O or SiH_4 fractions in the process gas stream.

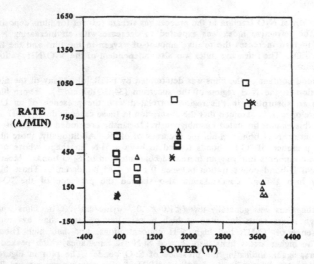

Figure 2. Rate vs Power. (Δ) - Pressure = 500 mTorr; (☐) - Pressure = 800 mTorr; (x) - Pressure = 950 mTorr.

The oxygen content of the films was observed to increase with both the absolute N_2O flow rate as well as the N_2O fraction in the process gas stream as might be expected since this serves as the source of oxygen in the films. The oxygen content was also observed to increase with higher power as shown in Figure 3. The oxygen content decreased with the NH_3 fraction in the process gas stream. The oxygen content was generally independent of the SiH_4 fraction in the process gas stream. The nitrogen content of the films decreased with both the N_2O and SiH_4 fractions of the process gas and increased with the fraction of NH_3. The ratio of oxygen to nitrogen in the films correlated well with the ratio of N_2O and NH_3 flow rates.

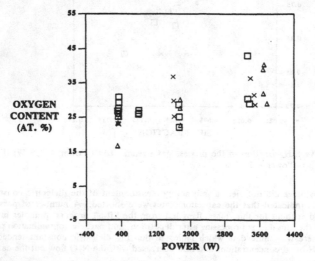

Figure 3. Oxygen content vs Power. (Δ) - Pressure = 500 mTorr; (☐) - Pressure = 800 mTorr; (x) - Pressure = 950 mTorr.

The refractive index was observed to decrease as a function of the oxygen content of the film. The oxygen content was observed to be high for films deposited under high power which generally led to high deposition rates. At these high deposition rates, the films generally contained a low number of hydrogen bonds.

DISCUSSION

The goal of this study was to determine the optimum process parameters for the deposition of PECVD SiON films with respect to the film properties of uniformity, low stress, low refractive index (low dielectric constant), low hydrogen content, and reasonable deposition rate (>300 A/min). The use of a factorial experimental design allowed the realization of this goal in a manageable number of experiments.

The uniformity of the films was very good in every case, indicating that all of the conditions would be suitable for a manufacturing process depending on the other film properties. The levels of stress were generally quite low, except for the films deposited under high power and low pressure. The high stress observed in these films is most likely due to the heavy ion and electron bombardment which the films would receive under these conditions. The inability to accurately measure the dielectric constant for many of the films was a disappointment. Using the refractive index as an indication of the dielectric properties of the films is an acceptable alternative.

The presence of hydrogen in PECVD SiON films is well documented [8,10-12]. The presence of hydrogen is a possible problem from the standpoint of long term reliability. The films deposited in the present study generally showed no N-H bonds. The peaks present in the N-H region of the IR spectra were assumed to be due to an O-H peak which occurs in the same region. The hydrogen which was detected was bonded to silicon. This is undesirable from a reliability point of view because Si-H bonds are weaker and less stable than N-H bonds. The lower hydrogen content observed for films which also contained the highest concentration of oxygen was encouraging because these films also tended to have the lowest refractive index.

The timely manufacture of GaAs ICs demands that the deposition processes be as efficient as possible. Films which are deposited at very low rates are susceptible to to the inclusion of particles either generated by gas phase nucleation or by the mechanical "flaking" of films deposited on the walls of the chamber during previous runs. A high deposition rate will also ensure that the throughput of the process remains high. Many of the conditions in this study had sufficiently high deposition rates to be used in manufacture. The parameters which appeared to have the most significant effect on the deposition rate were the power and the N_2O fraction in the process gas stream, although the model developed from the experimental design indicated that all five of the variables chosen as well as several of their cross-terms directly affected the deposition rate.

CONCLUSIONS

It has been shown that the properties of PECVD SiON films can be greatly affected by the deposition parameters. This is in agreement with similar studies by other authors [5-7]. The use of a factorial design to generate the experimental matrix of process conditions allowed the exploration of the process parameter space in a manageable fashion while retaining the data necessary to recognize and develop trends. For the system described above, the power and N_2O flow rates appear to have the greatest overall effect on the important film properties of stress, refractive index (dielectric constant), hydrogen content, film composition, and deposition rate. The knowledge of the trends and their effects on the film properties allow the development of a suitable dielectric film which will meet the needs of applications in the manufacture of GaAs ICs.

ACKNOWLEDGEMENTS

The authors gratefully acknowledge the assistance of the GaAs project team at the AT&T Reading Works. Specifically, J. Eberhardt for the factorial experimental design, K. Parks for assistance in using his product code for the dielectric constant measurements, P. Sakach for the AES, M. Berthinet for the FTIR measurements, and D. Condash for the deposition of the films. Additionally, we would like to acknowledge A-M. Lanzillotto and L. Hewitt of the SRI David Sarnoff Research Center for the RBS analysis.

REFERENCES

1. G.J. Valco, and V.J. Kapoor, J. Electrochem. Soc. 134(3), 685 (1987).

2. W.A.P. Claassen, W.G.J.N. Valkenburg, M.F.C. Willemsen, and W.M.v.d. Wijert, J. Electrochem. Soc. 132(4), 893 (1985).

3. V.S. Dharmadhikari, Thin Solid Films. 153, 459 (1987).

4. S.H. Wemple, W.C. Niehouse, H. Fukui, J.C. Irvin, H.M. Cox, J.C.M. Huang, J.V. Delorenzo, and W.O. Schlosser, IEEE Trans. Electron Devices. ED-28, 834 (1981).

5. J.E. Schoenholtz, and D.W. Hess, Thin Solid Films. 148, 285 (1987).

6. C.Y. Wu, R.S. Huang, and M.S. Lin, J. Electrochem. Soc. 134(5), 1200 (1987).

7. C.M.M. Denisse, K.Z. Troost, J.B. Oude Elferink, F.H.P.M. Habraken, W.F.v.d. Weg, and M. Hendriks, J. Appl. Phys. 60(7), 2536 (1986).

8. W.A. Lanford, and M.J. Rand, J. Appl. Phys. 49(4), 2473 (1978).

9. K. Parks, AT&T Microelectronics Product Engineer for the 1520B Laser Driver chip, private communication.

10. F.H.P.M. Habraken, Appl. Surf. Sci. 30, 186 (1987).

11. S.P. Speakman, P.M. Read, and A. Kiermasz, Vacuum. 38(3), 183 (1988).

12. J. Vuillod, J. Vac. Sci. Technol. A5(4), 1675 (1987).

CHEMISTRY OF NITROGEN-SILANE PLASMAS

DONALD L. SMITH, ANDREW S. ALIMONDA and FREDERICK J. VON PREISSIG
Xerox Palo Alto Research Center, 3333 Coyote Hill Rd., Palo Alto, CA 94304

ABSTRACT

The N_2-SiH_4 rf glow-discharge plasma has been analyzed by line-of-sight mass spectrometry of species impinging on the deposition electrode, including N atoms. Properties of $Si_xN_yH_z$ films deposited from this plasma have been examined. At high rf power and low SiH_4/N_2, almost all of the SiH_4 is consumed by reaction with N atoms at the film surface and becomes incorporated into the film. No Si-N precursor species are seen in the gas phase. This is in contrast to the NH_3-SiH_4 plasma, where the $Si(NH_2)_3$ radical is the key precursor. If power is insufficient or SiH_4 flow is excessive, Si_2H_m species are generated in the plasma. Under optimized conditions, films slightly N-rich with no Si-H bonding and only 7% H (as N-H) can be deposited at high rate. The film tensile stress of the NH_3 process is absent in the N_2 process due to the lesser amount of condensation that takes place during deposition. However, trench coverage is much better in the NH_3 process.

INTRODUCTION

"Silicon nitride" thin film ($Si_xN_yH_z$) deposited by plasma-enhanced chemical vapor deposition (PECVD) has been studied for over two decades and is widely used in the semiconductor industry. Yet until recently, little was known about the plasma chemistry of the deposition process. Earlier this year, we reported the first analysis of the plasma chemistry of silicon nitride PECVD from ammonia and silane, the most commonly used reactant mixture [1]. In the present work, the nitrogen-silane process is analyzed using the same reactor and techniques, and is compared to the results for ammonia. Others have found that the N_2 process results in films with much lower H content, with z = 5-10% vs. 20-30% for the NH_3 process. Low mobile H content is important when the film is used as a passivant for MOS devices. However, the NH_3 process is more widely used, because the N_2 process has been found more difficult to control and to give low deposition rates [2, 3]. It will be seen below that these problems can be avoided when the plasma chemistry is understood and properly controlled.

For comparison with the N_2 process, our results for NH_3-SiH_4 will be reviewed first [1, 4]. When the plasma rf power is sufficiently high, both NH_3 and SiH_4 become activated, and when there is sufficient excess NH_3, almost all of the SiH_4 reacts with it to form tetra-aminosilane, $Si(NH_2)_4$, and the triaminosilane radical, $Si(NH_2)_3$. The latter is the dominant precursor species for film growth. It decomposes on the surface and in a "condensation zone" beneath the surface in a process whereby an NH_2 and a H from neighboring precursor radicals combine to form an NH_3 molecule, which is evolved into the plasma. The resulting Si and N dangling bonds pull together to propagate the Si-N network, and this creates tensile stress. Films deposited under these conditions are N-rich (y/x > 4/3), and the ratio of total H to excess N is about 3/1, the same as in NH_3. The tensile stress and the degree of condensation towards Si_3N_4 both increase with deposition temperature. There is no Si-H bonding detectable by infrared absorption: all of the H is bonded to N, as in the precursor radical. Conversely, when either power is insufficient or SiH_4 flow is too high, there is not enough active NH_3 generated to complete the amination reaction, and silane radicals begin to react with each other to form disilane. Films deposited under such conditions exhibit Si-H bonding and lower y/x.

EXPERIMENTAL APPROACH

The apparatus has been described previously [1]. Briefly, with reference to Fig. 1,

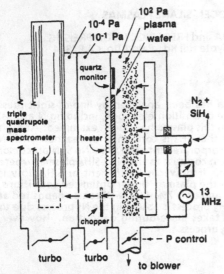

Fig. 1. PECVD reactor and mass spectrometric sampling system.

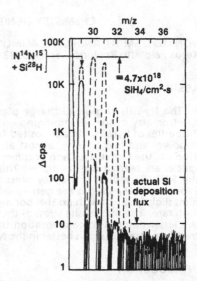

Fig. 2. Mass spectra of sampled beam from $N_2 + SiH_4$: gas (----) and 70 W plasma (—); 45 sccm N_2, 1.2 sccm SiH_4, and 60 Pa in reactor.

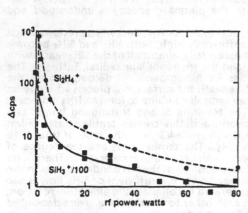

Fig. 3. Beam mass signals vs plasma power. Other reactor conditions as in Fig. 2.

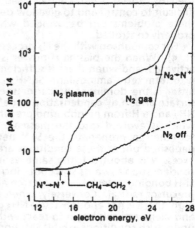

Fig. 4. DC m/z 14 signals from N_2 gas and 80 W plasma vs electron ionizing energy. Appearance potentials indicated are from the literature [8].

it consists of a conventional parallel plate, glow discharge plasma reactor with the added analytical tools of a quartz crystal mass deposition monitor and a mass spectrometer. The quartz crystal is situated in the wafer plane and operates at approximately the deposition temperature (for <360°C). An orifice array in the wafer plane samples impinging species into the mass spectrometer in a collisionless manner through a turbopumped chamber and a beam collimating orifice. A chopper and synchronous differential counting electronics have been added to help extract weak beam components from residual background in the mass spectrometer; the difference in counts per second between beam+background (chopper open) and background only (chopper closed) is reported as Δcps. Ionization was done at 40 eV except where indicated otherwise. The triple quadrupole mass spectrometer (TQMS; Extrel Corp.) dissociates mass-filtered "parent" ions by collision-assisted cracking in Ar and then sweeps the resulting "daughter ion" spectrum to determine the structure of the parent ion. In this way, for example, Si_2H_6 can be distinguished from $SiH_2(NH_2)_2$ at the same mass number.

$Si_xN_yH_z$ films were deposited on 3 in. diameter Si(100) wafers for measurement of various film properties as follows: index of refraction by ellipsometry at 633 nm, thickness by stylus profiling over HF-etched steps, density by direct weighing, mechanical stress by laser beam deflection determination of the change concavity of the substrate as a result of film deposition, infrared absorption (IR) in transmission using a fourier-transform instrument with 4 cm⁻¹ resolution, and N and Si content by Rutherford backscattering spectrometry (RBS) [5]. H content was measured in two ways: 1) by integration of the N-H absorption band at 3340 cm⁻¹ and use of a published calibration [6] and 2) by elastic recoil detection (ERD) of H [7] using the same 2.3 eV He⁺⁺ probe beam as for RBS.

MASS SPECTROMETRY

All measurements and deposition in this work were done at a fixed total pressure of 60 Pa and a N_2 flow of 45 sccm, to which a much smaller flow of pure SiH_4 was added. We will first look at the plasma chemistry under optimized deposition conditions. The mass spectrum of the gases at 1.2 sccm SiH_4 is shown by the dashed line in Fig. 2. The $N^{14}N^{15}$ isotope peak at m29 (= m/z or mass/charge 29 assuming z = 1) is 0.70% of the main N_2 peak at m28, which saturates the counter. Part of m29 and all of m30-m33 are due to cracking and Si isotopes of SiH_4. Upon application of 70 W of rf power, the SiH_4 peaks attenuate by more than x100, and a large H_2 peak (not shown) appears. No new peaks were seen for $Si_mN_nH_p$ products. The minimum size that a significant deposition precursor peak could be was determined as follows. The known impingement flux of SiH_4 gas gives a major peak height shown by the dashed line at m30 and thus calibrates the mass spectrometer. From the film deposition rate and Si content, the actual Si deposition flux is found to be 1.2×10^{15} Si/cm²-s, which corresponds to the level shown on Fig. 2, assuming the same ionization cross section for the depositing Si species as for the m30 component of SiH_4 gas. This is the minimum precursor impingement flux - the one for a unity value of the sticking coefficient, S; it would be higher for a given deposition rate if S were lower. It may be concluded, therefore, that there are no precursor Si-N species, and that deposition occurs instead by reaction of activated silane and nitrogen ⋅ ⋅ the wafer surface.

From the above Si deposition flux and supply flow of SiH_4, the fractional utilization of SiH_4 may be estimated by assuming uniform deposition over both of the 15 cm diameter electrodes, and it is about 86%, which is much higher than usual. A N_2 process has been reported recently in which high rates were achieved by constricting gas flow and power over the wafer as is done in Fig. 1 [3], but SiH_4 utilization was still only 9%. The N_2 flow was much higher there, so perhaps the residence time was too short for the reaction to go to completion.

The effects of decreasing power below the 70 W level are shown in Fig. 3. Peaks around m60 (exemplified by $Si_2H_4{}^+$) and silane ($SiH_3{}^+$) both begin to rise: more Si_2H_m are being generated, and silane utilization is less. The N_2 is beginning to behave as an inert diluent in which SiH_m react only with themselves. This is similar to the behavior of the NH_3 process, except that the critical power to activate NH_3 is only about 1/5 of that needed to activate N_2.

The degree of N_2 activation can be monitored directly by looking at N atoms at m14. Of course, N_2 also dissociates into N^+ in the ionization chamber, but it requires more electron energy to do so than does N ionization, by about the energy of the N_2 bond. Fig. 4 shows the DC signal at m14 vs. ionizing electron energy. Note that the appearance potential of N^+ from N_2 gas is about 24 eV, whereas that for N^+ from N atoms generated by the plasma is only about 14 eV. By using 23 eV ionization, the signal from N_2 gas dissociation is eliminated, although there may be some contribution from dissociative ionization of N_2 which has been raised into excited states in the plasma. The small signal shown below 24 eV from N_2 gas remained when the N_2 was turned off and is coming from residual CH_4 in the mass spectrometer; beam chopping eliminated this component. The chopped signal at 23 eV was found to rise monotonically with rf power, as expected.

Fig. 5 shows the effects of SiH_4 flow on the plasma chemistry and the film deposition in a 60 W plasma. The N^+ signal at 23 eV drops linearly and the deposition rate rises linearly with increasing flow up to about 1.4 sccm. Just above this "critical" flow rate, three things happen: 1) N^+ levels out at a residual value, 2) deposition rate slope tapers off, and 3) disilane begins to rise rapidly. These results may be explained as follows. As SiH_4 flow increases in the region below the critical flow, an increasing fraction of the active nitrogen arriving at the surface is consumed there by reaction with SiH_m, and almost all of the SiH_4 is thus incorporated into the deposit, as discussed above. Above the critical flow, all of the active nitrogen is being consumed, and additional SiH_4 has only itself to react with, resulting in disilane. In this latter regime, Si-H bonding and excess Si would be expected to develop in the film. (Part of the residual N^+ signal above the critical flow in Fig. 5 may be from some activated state of N_2 that dissociatively ionizes at 23 eV but that does not react with SiH_m.) Presumably, at higher power the critical SiH_4 flow for active nitrogen depletion would be higher, so that still higher deposition rates would be obtainable without resulting in Si-H or excess Si. We observed a similar SiH_4 critical flow for the NH_3 process.

FILM PROPERTIES

Fig. 6 shows IR spectra of films deposited at 450°C under the optimized plasma conditions determined above for the N_2 process (1.2 sccm SiH_4, 60 W) and previously [1] for the NH_3 process (1.8 sccm SiH_4, 18 W). Note that there is no Si-H bonding in either case, which is unusual for PECVD silicon nitride. The film from N_2 contains less N-H and no $N-H_2$, as others have observed. It also has somewhat more O contamination, which is common in the N_2 process and is believed to be absorbed upon air exposure after deposition [10]. The lack of Si-H is a consequence of there being enough active nitrogen impinging on the surface to displace all of the H from adsorbing SiH_m. N_2-process films having no Si-H were reported recently [3] using similar high power, low SiH_4 flow conditions. Si-H can be eliminated also by adding NF_3 to the plasma [11], but this increased film etch rate in buffered HF and also resulted in extensive hydrolysis upon air exposure if not carefully controlled. NF_3 addition also was reported to increase deposition rate by a factor of 2-4, but we saw no significant increase upon NF_3 addition to our optimized process due to the fact that we are already incorporating 86% of the supplied Si into the film.

Table I lists additional properties of N_2-process film along with those for NH_3-process film deposited in the same reactor at the same rate and temperature. Both are N-rich, but the N_2 film has much less excess N and much less H. This means that the film structure has developed more towards Si_3N_4. The high tensile stress of the NH_3 film is absent in the N_2 film, which is consistent with the absence of an NH_2-saturated precursor species such as the triaminosilane radical. The reactants at the surface in the N_2 process are probably N and SiH_{0-2}, which require much less condensation to develop the Si-N network and thus do not produce tensile stress. The N_2 process gives poor sidewall and bottom coverage of deep trenches, indicating

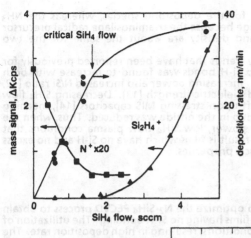

Fig. 5. Beam mass signals and film deposition rate *vs* SiH₄ flow in 60 W plasma. 45 sccm N₂ and 60 Pa in reactor.

Fig. 6. FTIR spectra of $Si_xN_yH_z$ films deposited under optimized plasma conditions (see text) at 450°C, using N₂-SiH₄ and NH₃-SiH₄. Peak assignments are from the literature [9].

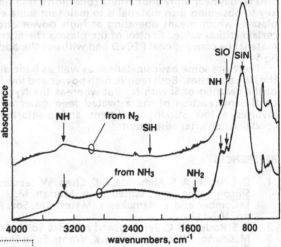

Property	N₂	NH₃
SiH₄ flow, sccm	1.2	1.8
rf power, W	60	18
dep. rate, nm/m	21	22
Si utilization, %	86 ± 20	62 ± 20
atomic % Si	37	31
N/Si ratio	1.39 ± .1	1.62 ± .1
at. % H by IR	7	24
at. % H by ERD	13	27
refractive index	1.84	1.81
density, g/cm³	2.68	2.66
stress, MPa	-142	+ 883
trench coverage	poor	excellent

Table I. Comparative properties of $Si_xN_yH_z$. Deposited at 450°C under optimized plasma conditions using 45 sccm of N₂ or NH₃ plus SiH₄.

a near-unity sticking coefficient, S, for the depositing species, whereas the NH_3 process gives excellent trench coverage because the triaminosilane radical precursor has $S<0.01$[4]. Refractive index and density are about the same for the two processes.

The above analysis explains many trends that have been reported previously for the N_2 process. The ratio of N-H to Si-H bonds was found to increase with both N_2/SiH_4 gas ratio [3] and power [12]. Increasing power also increased N/Si ratio [12, 13], as well as density, resistivity and dielectric strength [13]. Decreasing SiH_4 flow decreased the flat-band voltage shift upon stressing MIS capacitors [14], meaning that the rate of trapped charge buildup in the nitride was reduced. Thus, when the process is operated under the high power, low SiH_4 flow plasma conditions that maintain excess active nitrogen, the result is films which have no Si-H and no excess Si and which exhibit improved electrical properties.

CONCLUSIONS

Mass spectrometry can be used to optimize the N_2-SiH_4 PECVD process to obtain near-stoichiometric or slightly N-rich films having no Si-H bonding. The utilization of SiH_4 feed is near unity under these conditions, resulting in high deposition rate. The key to obtaining such material is to maintain some excess of active nitrogen in the plasma, which means operating at high power and keeping SiH_4 flow below a certain critical value. Control of the plasma chemistry allows the deposition of this material by conventional PECVD and without the addition of other reactants to the N_2-SiH_4.

There are some basic similarities as well as basic differences between the N_2 and the NH_3 processes. Both require high power and low SiH_4 flow to ensure completion of the reaction of Si with N. But whereas the N_2 process involves a high sticking-coefficient reaction of the activated feed gases at the surface, the NH_3 process involves a low sticking-coefficient amino-saturated precursor that undergoes condensation after adsorption.

REFERENCES

1. D. L. Smith, A. S. Alimonda, C-C. Chen, W. Jackson and B. Wacker in Amorphous Silicon Technology, edited by A. Madan, M. J. Thompson, P. C. Taylor, P. G. LeComber and Y. Hamakawa (Mater. Res. Soc. Proc. 118, Pittsburgh, PA 1988) pp. 107-112.
2. R. S. Rosler, W. C. Benzing, and J. Baldo, Solid State Technol., 19 (6), 45 (1976).
3. M. Chang, J. Wong, and D. N. K. Wang, Solid State Technol., May 1988, p. 193.
4. D. L. Smith, A. S. Alimonda, C-C. Chen, S. E. Ready, and B. Wacker, to be published (1989).
5. S. Baumann, Charles Evans & Associates, Redwood City, CA (1988).
6. W. A. Lanford and M. J. Rand, J. Appl. Phys. 49 (4), 2473 (1978).
7. B. L. Doyle and P. S. Peercy, Appl. Phys. Lett. 34 (11), 811 (1979).
8. R. D. Levin and S. G. Lias, Ionization Potential and Appearance Potential Measurements, 1971-1981 (National Bureau of Standards, Washington, DC, 1982).
9. D. V. Tsu, G. Lucovsky, and M. J. Mantini, Phys. Rev. B 33 (10), 7069 (1986).
10. S. Hasegawa, M. Matuura, H. Anbutu and Y. Kurata, Phil. Mag. B 56 (5), 633 (1987).
11. C-P. Chang, D. L. Flamm, D. E. Ibbotson, and J. A. Mucha, J. Appl. Phys. 62 (4), 1406 (1987).
12. H. Dun, P. Pan, F. R. White, and R. W. Douse, J. Electrochem. Soc. 128 (7), 1555 (1981).
13. M. Maeda and Y. Arita, J. Appl. Phys. 53 (10), 6852 (1982).
14. N-S. Zhou, S. Fujita, and A. Sasaki, J. Electronic Mat. 14 (1), 55 (1985).

CONTRIBUTIONS OF GAS PHASE REACTIONS TO CVD PROCESSES

S. B. Desu* and Surya R. Kalidindi**
* Department of Materials Engineering, Virginia Polytechnic Institute, Blacksburg, VA 24061;
**Department of Mechanical Engineering, MIT, Cambridge MA 02139.

Abstract

An analytical model to account for the effects of gas phase reactions in CVD processes is presented. In this model a system with only two reactions, gas phase production of an intermediate and reaction of the intermediate on the surface of the substrate to form the film is considered. The reaction kinetics, convective and diffusive transport mechanisms are coupled to analyse the thickness distribution over the length of the reactor.

Theoretical deposition rate profiles in the direction of gas flow are shown with varying deposition temperatures and gas flow velocities for both surface and gas phase reaction controlled mechanisms. It has been shown that by analyzing the nature of the deposition rate profiles, the rate limiting step in a CVD process can be identified.

Introduction

Chemical vapor deposition is a very versatile process for deposition of metals, semiconductors and dielectrics and is being used commercially in growing single crystalline and polycrystalline silicon, silicon carbide, II-VI and III-V semiconductors, binary metal oxide films for microelectronic device applications. In all these processes, besides the surface reactions, some sort of gas phase reactions may take place in the heated gas layer in contact with the deposition surface. In majority of these cases, the gas phase reactions produce activated species which then diffuse to the surface and decompose to produce the desired film. In many cases, the production of these intermediate species in the gas phase is the rate limiting step. Thus, understanding the gas phase reactions will enable us to obtain films with the desired quality.

The formation of SiH_2 in the gas phase for the CVD processes using silane was proposed in the literature [1-3]. This was further exploited in developing HOMOCVD concept for depositing films on substrates at low temperatures [4]. Similarly, gas phase formation of $SiCl_2$ in CVD reactions using silicon chlorides was also proposed [5]. Gas phase complex formation was shown in the deposition of III-V compounds by MOCVD [6-8]. Recently it was shown that the gas phase reactions control the deposition rate of SiO_2 in the pyrolysis of silicon alkoxides [9]. Despite all this evidence, most published models for CVD either neglect gas phase reactions or take them into consideration through a lumped overall rate equation [10]. Besides affecting the overall rate, gas phase reactions can significantly effect the step coverage and thickness nonuniformity. To understand these effects, the individual reactions (gas phase as well as surface reactions) should be explicitly considered in the analytical model.

Analytical Model

A typical CVD process comprising of a gas phase reaction and a surface reaction is considered for simplicity. In the gas phase reaction the reagent (R) decomposes to give an intermediate (I) and a gaseous byproduct (P_1), with an associated rate constant K_3 sec^{-1}.

$$R \xrightarrow{K_3} I + P_1 \tag{1}$$

On the surface of the substrate the intermediate decomposes to result in a film (F) and another gaseous byproduct (P_2), with an associated rate constant K_4 cm/sec.

$$I \xrightarrow{K_4} F + P_2 \tag{2}$$

Here the effects of adsorption, desorption and surface migration of the various species are neglected.

The governing mass transport equations with the corresponding boundary conditions for the reagent (R) and the intermediate (I) in a steady state condition in one dimension are

$$- D \, \partial^2 C_R / \partial x^2 + U \, \partial C_R / \partial x = -K_3 C_R \tag{3}$$

$$- D \, \partial^2 C_I / \partial x^2 + U \, \partial C_I / \partial x = -K_4 C_I A / V + K_3 C_R \tag{4}$$

Boundary Conditions:

At x = 0

$$U C_{Ro} = (U C_R)_{x=0} - D(\partial C_R / \partial x)_{x=0} \tag{5}$$

$$C_I = 0.0 \tag{6}$$

At x = L

$$\partial C_R / \partial x = 0.0 \tag{7}$$

$$\partial C_I / \partial x = 0.0 \tag{8}$$

where D, U are respectively the diffusion coefficient and the velocity (assumed to be same for both R and I), C_R and C_I are the concentrations of R and I respectively, L is the length of the reactor and A is the surface area and V is the volume of the reactor per unit length.

Introduce the following nondimensional Quantities

$$x^* = x/d \tag{9}$$

$$C_R^* = C_R / C_{Ro} \tag{10}$$

$$C_I^* = C_I / C_{Ro} \tag{11}$$

$$Pe = Ud/D; \quad \textbf{Peclet Number} \tag{12}$$

$$Da_R = K_3 d^2 / D; \quad \textbf{Gas Phase Damkohler Number} \tag{13}$$

$$Da_I = K_4 d/D; \quad \textbf{Surface Damkohler Number} \tag{14}$$

where d is chosen as the diameter of the reactor. The equations (3)-(8) then transform as

$$- \partial^2 C_R^* / \partial x^{*2} + Pe \, \partial C_R^* / \partial x^* = -Da_R C_R^* \tag{15}$$

$$- \partial^2 C_I^* / \partial x^{*2} + Pe \, \partial C_I^* / \partial x^* = -Da_I C_I^* (Ad/V) + Da_R C_R^* \tag{16}$$

Boundary Conditions:

At x* = 0

$$C_R^* = 1.0 - (1/Pe) \, (\partial C_R^* / \partial x^*)_{x^*=0} \tag{17}$$

$$C_I^* = 0.0 \tag{18}$$

At x* = L/d

$$\partial C_R^* / \partial x^* = 0.0 \tag{19}$$

$$\partial C_I^* / \partial x^* = 0.0. \tag{20}$$

It is therefore apparent from these equations that the deposition rate profiles depend on the nondimensional parameters: Peclet number (Pe), Gas phase Damkohler number (Da_R) and Surface Damkohler number (Da_I). Note that the nondimensional parameter Ad/V is characteristic of reactor geometry and is fixed for a given reactor configuration.

Finite element methods were used to solve the equations (15-20). Details of the mathematical model, numerical implementation and experimental confirmation were discussed elsewhere [9]. In order to calculate the actual deposition rates a fixed value for the initial reagent concentration C_{Ro} was used for the numerical results presented here. It can be easily seen from the equations that this value of C_{Ro} has no influence on the results discussed here.

Results and Discussion

The contribution of gas phase reactions to the film deposition can be clarified by analyzing the distribution of rate of deposition along the length of reactor. If the surface reaction is the limiting stage of deposition, the deposition rate decreases along the length of the reactor due to the depletion of the starting reagent. Under such conditions, with an increase in Pe (increasing gas flow velocity) the deposition rate increases all over the reactor zone due to a corresponding decrease in residence time. Figure 1 shows this effect clearly for $Da_R = 10.0$ and $Da_I = 5.0$. As the Pe is increased from 5.0 to 15.0 the deposition rate steadily increased. The rate of increase is higher at the entrance of the reactor compared to the end of the reactor which is specific to the reactor goemetry [9] used in this calculation.

In the case where the gas phase reaction is rate controlling, initially at the entrance of the reactor smaller amount of R decomposes. As R travels down the reactor the amount of its decomposition increases. Thus, the deposition rate increases along the length of the reactor. If Pe is increased the deposition rate falls, since at higher gas flow rates the residence time of the precursor is short. This case is illustrated in figure 2 for $Da_R = 2.0$ and $Da_I = 0.1$.

In figure 3, the deposition rate either increases or decreases depending on the location in the reactor with increasing Pe. The ascending segment of the profile is attributed to increased formation of the intermediate due to gas phase reaction, and the descending segment to the depletion of the reagent or entrainment of the intermediate by the gas stream. In this case both surface as well as gas phase reactions are important.

In the case where the gas phase reactions are rate controlling, thickness uniformity can be improved using a preheater. A preheater will heat the incoming gas such that a high concentration of intermediate is obtained at the entrance to the reactor. Figure 4 depicts the effect of preheater temperature in terms of Da_{PH} (Gas Phase Damkohler number in the preheater section) for a $Da_R = 2.0$ and $Da_I = 0.1$. For simplicity, in these calculations it is assumed that only Da_R changes in the preheater section. As the preheater temperature increases the deposition rate at the entrance of the reactor is increased, thus resulting in better uniformity. HOMOCVD concept is an extreme example of this case.

Figures 5 and 6 depict the influence of temperature on the deposition profile. For figure 5 Da_R is assumed to be insensitive to temperature, whereas Da_I is assumed to be constant in figure 6. The Damkohler number would be relatively insensitive to temperature if the associated activation energy is very low. For both surface control (figure 5) as well as gas phase control (figure 6) decreasing the Da (corresponds to decreasing temperature) improves the thickness uniformity.

Summary

With the aid of an analytical model for a reaction system it is shown that the rate limiting step in CVD can be identified simply by analyzing the nature of the deposition rate profile. If the deposition rate increases over the length of the reactor and decreases with increase of Pe, gas phase reaction is the rate controlling step. If the deposition rate decays along the length of the reactor and increases with increasing Pe, surface reaction is the rate controlling step. Guidelines are also provided for optimization of the process parameters to obtain uniform deposition rate profiles for both surface as well as gas phase reaction control regimes.

References

[1] K. J. Sladek, J. Electrochem. Soc., 118, 654 (1971).
[2] M. E. Coltrin, R. J. Kee and J. A. Miller, ibid., 131, 425 (1984).
[3] C. H. J. Van Den Brekel and L. J. M. Bollen, J. Cryst. Growth, 54, 310 (1981).
[4] B. A. Scott, in Semiconductors and Semimetals, edited by I. Pankove (Academic Press Inc., New York 1984) Vol. 21, Part A, pp. 123.
[5] P. Van Den Putte, L. J. Giling and J. Bloem, ibid., 31, 299 (1975).
[6] M. Tsuda, et. al., Jap. J. Appl. Phys., 26, L564 (1987).
[7] S. Ito, T. Chinohara and Y. Seki, J. Electrochem. Soc., 120, 1419 (1973).

310

[8] I. A. Frolov, et. al., Izv. Akad. Nauk SSSR, Neorg. Mater., **13**, 773 (1977).
[9] S. B. Desu, J. Amer. Ceram. Soc., to be published.
[10] K. F. Jensen and D. B. Graves, J. Electrochem. Soc., **130**, 1950 (1983).

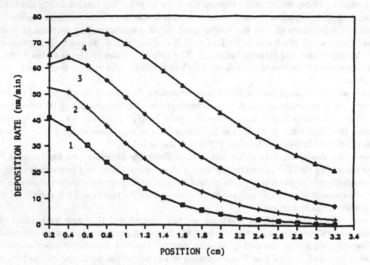

Fig. 1. Effect of Peclet number (Pe) variation on the deposition rate profile for $Da_R = 10.0$ and $Da_I = 5.0$ (surface control). The curves 1 to 4 are respectively for Peclet numbers 5, 7, 10 and 15.

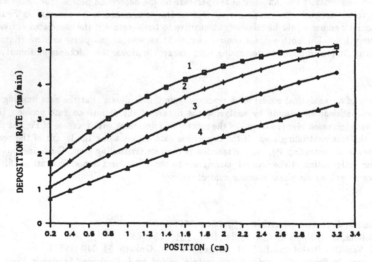

Fig. 2. Effect of Peclet number (Pe) variation on the deposition rate profile for $Da_R = 2.0$ and $Da_I = 0.1$ (gas phase control). The curves 1 to 5 are respectively for Peclet numbers 5, 7, 10, 15 and 20.

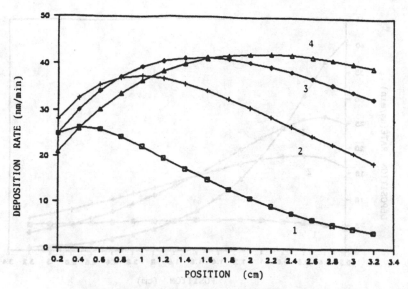

Fig. 3. Effect of Peclet number (Pe) variation on the deposition rate profile for $Da_R = 10.0$ and $Da_I = 1.0$ (gas phase control). The curves 1 to 4 are respectively for Peclet numbers 5, 10, 15 and 20.

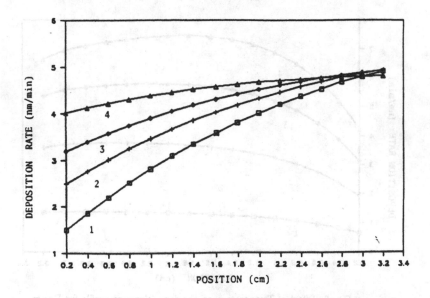

Fig. 4. Effect of Preheat setting (Ps) on the deposition rate profile for $Da_R = 2.0$, $Da_I = 0.1$ and $Pe = 8.0$. The curves 1 to 4 are respectively for Preheat settings 0, 5, 10 and 30.

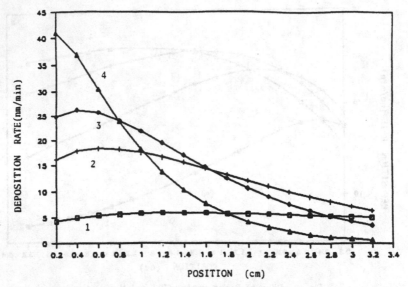

Fig. 5. Effect of Temperature on the deposition rate profile for $Da_R =$ 10.0 and $Pe = 5.0$, where Da_R is insensitive to temperature. The curves 1 to 4 are respectively for Da_I values of 0.1, 0.5, 1.0 and 5.0.

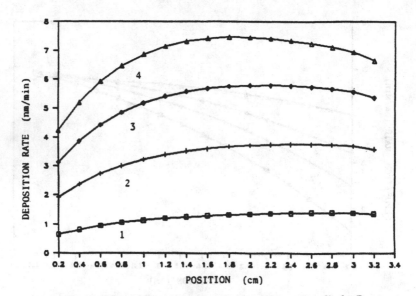

Fig. 6. Effect of Temperature on the deposition rate profile for $Da_I =$ 1.0 and $Pe = 5.0$, where Da_I is insensitive to temperature. The curves 1 to 4 are respectively for Da_R values of 0.1, 0.3, 0.5 and 0.7.

THE EFFECTS OF SURFACE TREATMENTS FOR LOW TEMPERATURE SILICON DIOXIDE DEPOSITION ON CADMIUM TELLURIDE.

Seong S. Choi, S. S. Kim, D.V. Tsu and G. Lucovsky, Departments of Physics and Materials Science and Engineering, North Carolina State University, Raleigh, N.C. 27695-8202

Abstract

We have successfully deposited thin films of SiO_2 on a cadmium telluride substrate at low temperature ($T_s =100°C-300°C$) by remote plasma enhanced chemical vapor deposition (Remote PECVD). The native oxide on the CdTe substrate has been removed, prior to deposition by either chemical etching in methanol and 1% bromine, or by dissolution in deionized water. After removal of the native oxide, the CdTe was inserted into a UHV-compatible deposition chamber and a He^+ plasma treatment was performed prior to deposition of an SiO_2 film. This treatment promotes strong adhesion between the deposited SiO_2 film and the CdTe surface. We find that the initial oxide removal process does not influence SiO_2 adhesion. The effect of the He^+ plasma treatment on the CdTe surface has been studied by Auger electron spectroscopy(AES), and Reflection high energy electron diffraction (RHEED).

Introduction

There is considerable interest in the formation of device structures that incorporate both CdTe and (Hg,Cd)Te layers. The majority of the proposed device structures also will require either surface passivation of junction boundaries, or the use of dielectric films as gate insulators in field effect transistor(FET) heterostructures. Many previous studies [1-3] have identified problems related to the adhesion of low temperature SiO_2 films onto CdTe surfaces. The maximum temperature to which CdTe surfaces can be exposed, without significant disproportionation is about 340°C. It has been suggested that this problem derives from a chemically inert CdTe surface state [4-6,8,10]. To circumvent this problem, alternative dielectric materials have been considered, e.g., the wide-band-gap II-VI material ZnS [2]; however the intrinsic resistivity of this material, and the quality of the thin films of ZnS generated by low temperature processes are not sufficient for many applications [2]. Therefore we have concentrated our effort on the CdTe surface processing prior to low temperature SiO_2 deposition.

Experimental Procedures

Substrates of cadmium telluride were obtained from Santa Barbara Research Corporation (SBRC). The substrates were prepared by an initial degreasing in trichloroethylene(TCE), acetone, and methanol. This was followed by removing the native oxide by either dissolution in deionized (D.I.) water {at least 30 minutes is required} [5], or by chemical etching in a 1% bromine/methanol solution [6,7]. After either of these processes, the sample was immediately inserted into a UHV multichamber surface process/surface analysis/dielectric deposition system [9] to minimize reoxidation and/or surface contamination. Upon introduction into the system, the substrates were first examined for chemical purity and stoichiometry by AES. The substrates were then annealed at either the system base pressure of $<10^{-8}$ torr, or in hydrogen, at a pressure of 300 mTorr, at temperatures of either 100°C or 300°C, for a period of 30 minutes. These processes tend to promote the formation of stoichiometric surfaces [6]. This thermal annealing was followed by a He plasma treatment for 1 minute and at a pressure of 100 mTorr. The He plasma treatment was performed just prior to the SiO_2 deposition by the Remote PECVD technique.

Mat. Res. Soc. Symp. Proc. Vol. 131. ©1989 Materials Research Society

The *in situ* characterizations, mentioned above, were performed in an integrated surface analysis chamber. The sample was transported, under UHV conditions, from the deposition or surface preparation chamber into the analysis chamber after the various surface treatments and/or oxide depositions. This surface analysis chamber is equipped with a AES that employs a single pass cylindrical mirror analyzer (CMA), and also Reflection high energy spectroscopy (RHEED). The primary energy for Auger electron spectroscopy (AES) was 3 keV, and the energy of RHEED was 10 keV. The characteristics of the SiO_2 deposited by Remote PECVD and analyzed in situ are described elsewhere [9].

Experimental Results

We have observed that the air-grown native oxide (see Fig. 1, note that the oxide lines occur at about the same position as the Te lines, and that this "500-eV feature" is different for the surfaces shown in Fig. 1 and those displayed in Figs. 2 and 3) on a CdTe substrate can be removed by either dissolution in deionized (DI) water, or by chemical etching in (1% bromine+methanol) (see Figs. 2 and 3). The derivative AES spectra taken from the sample processed only in DI water indicates a larger Cd/Te peak-height ratio than the corresponding spectra from a sample that has been subjected to the 1% bromine+methanol chemical etching. However, neither surface displays an AES spectrum indicative of CdTe stochiometry. In this regard, chemical etching in (1% bromine+methanol) has been shown to generate a surface which is Te-rich [6,7]. Our AES results support this observation, and also demonstrate that the DI water dissolution process results in a Te-rich surface possibly due to Te accumulation during the previous chemical etching. In addition, the RHEED pattern, obtained after the removal of the native oxide layers, indicates also a rough surface, probably due to some Te accumulation [6,7]. Figures 1-4 trace the evolution of the surface chemistry as determined by AES. Specifically, the DI water dissolution and bromine+methanol etch processes: remove native oxides, but leave the surfaces Te rich with respect to CdTe stoichiometry. Note that processing in DI water does not effectively remove the S-peak. Figures 3 and 4 are for the samples processed initially in DI water. The effect of thermal annealing at either 100°C or 300°C (see Fig. 3) drives the surface toward Cd/Te stoichiometry. Note here that the S peak is eliminated by heating at 300°C, but not 100°C. Finally, subjecting the CdTe surface to a He-plasma "etch": (1) maintains the surface stoichiometry; and in addition (2) remove the S peak from the sample initially annealed at 100°C.

The RHEED studies were performed in order to obtain a measure of the geometric perfection and/or reconstruction state of the (100) CdTe surface. In this sense these measurements complement the chemical characterization obtained by AES. After removal of the native oxide by either of the methods employed, the RHEED pattern, as shown in Fig. 5 for the bromine-etched substrate, indicated poor surface quality. Since both oxide-removal treatments leave the surface Te-rich, this result is anticipated [6,7]. Figs. 6a and 6b display the RHEED patterns obtained respectively for the DI water rinsed, and bromine+methanol etched surfaces after annealing at 300°C. Both patterns indicate an improved degree of surface perfection with respect to Fig. 5, [11,12]. The RHEED pattern for the DI water rinsed surface indicates a slightly higher degree of surface perfection compared with the pattern from the sample which was etched in 1% bromine+methanol. Figs. 7a and 7b indicate RHEED patterns after the He plasma treatments. This process results in a more nearly perfect surface.

Adhesion of SiO_2 Films

We have deposited SiO_2 thin films onto processed CdTe surfaces using the Remote PECVD process. These films were deposited with substrate temperatures ranging between 100°C and 300°C, and under conditions of gas flow which promoted SiO_2 stoichiometry and minimal SiOH contamination. We have observed correlations

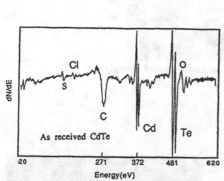

FIG.1. AES data from as-received CdTe.

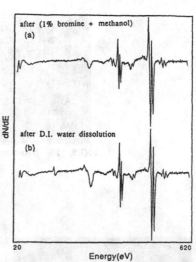

FIG.2. Removal of the oxide has been observed in both ways. S peak still remains after DI water dissolution.

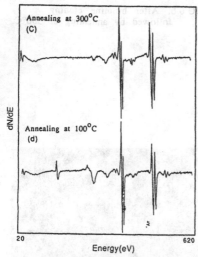

FIG.3 After DI water dissolution, Annealing at 300°C(c), 100°C(d) for 30min. was performed. C and S peaks have been observed after annealing at 100°C.

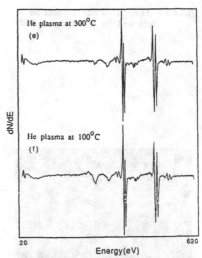

FIG.4. The substrate treated with 1 min., He plasma at 300°C, 100°C in addition to annealing,respectively. S peak has been removed after He plasma treatment at 100°C.

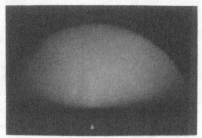

FIG.5. 1% Bromine/methanol etched sample.

FIG.6.a   After DI water dissolution
          followed by annealing at 300°C.

FIG.6.b. After bromine etching
          followed by annealing at 300°C.

FIG.7.a. After He plasma treatment
          in addition to (FIG.6.a.) procedure.

FIG.7.b. After He plasma treatment
          in addition to (FIG.6.b.) procedure.

between surface stoichiometry (determined by AES) and surface perfection (by RHEED), and the extent to which good adhesion can be obtained. We define the quality of surface adhesion by the *traditional scotch-tape test* [13].

It has been shown that annealing CdTe surfaces at temperatures between 300°C and 340°C is generally a necessary prerequisite for obtaining a surface which is: stoichiometric; and microscopically flat surface. This derives from the fact that Te evaporates faster than Cd for temperatures lower than 340°C, the *congruent evaporation* temperature of CdTe [11]. At temperatures in excess of 340°C, Te evaporates more readily. It has also been reported that after etching CdTe in bromine+ methanol, there is excess elemental Te on the CdTe surface [6,7]. This excess elemental Te tends to resist oxidation, and evaporates readily, e.g., the Te vapor pressure is reported to be $7x10^{-7}$ torr at a surface temperature of 225°C [16]. We believe that the adhesion of SiO_2 onto a CdTe surface will be adversely effected if the surface is Te-rich. This correlates with problems we have had in getting deposited SiO_2 films to adhere to surfaces that had not been subjected to either 300°C anneals and/or He plasma treatments.

We have observed that after annealing at either 300°C or 100°C with or without hydrogen gas present, there is a change in the AES spectra that indicates that the CdTe surface has ceased to be Te rich, and is either stoichiometric or Cd rich. The observation of a change in surface stoichiometry is accompanied by changes in the RHEED pattern indicating a higher degree of surface flatness. These changes are accompanied by somewhat better SiO_2 adhesion.

However, we find that the best adhesion is obtained only after both a UHV thermal anneal and a subsequent exposure to a He plasma. In this context, the combination of these two processes serves to: (1) drive the surface composition toward stoichiometric; and (2) promote active surface sites [14, 15] for oxide attachment.

Concluding Remarks

We believe that good adhesion of non-native dielectric materials to CdTe surfaces can only take place on surfaces that have Cd/Te compositions close to stoichiometry, and that in addition have surface sites that have been chemically activated for adhesion by charged particle bombardment. These observations are consistent with reports that elemental Te tends to resist oxidation, so that a Te rich surface is a poor receptor for attachment of a non-native oxide dielectric such as SiO_2. It is also well-known that excess Te from CdTe surfaces can be eliminated by thermal bakeout under UHV conditions. Excess Te is known to derive from etching in bromine-methanol solutions, and our studies have indicated that dissolution of native oxides in DI water also results in a Te rich surface condition.

We have previous reported that the oxidation of CdTe is very slow in *moist* laboratory air (about 50% room humidity) [5]. We have observed that for sputter-cleaned CdTe surfaces, it takes about 3 weeks to form detectable (Te-O) oxide peaks by X-ray excited photoelectron spectroscopy(XPS). This result is in agreement with the results of U. Solzbach and H.J. Richter [10]. We believe that there is a correlation between the rate of formation of native oxide on CdTe surfaces, and the formation of the local bonding environments required for non-native oxide attachment. This trend is supported by our studies.

We believe that it is necessary to activate a stoichiometric CdTe surface prior to achieving good mechanical adhesion between that surface and any non-native oxide. This can be achieved in at least two ways: (1) deposition of the oxide layer in the presence of UV radiation, and in the photox-type processes; or (2) activate the surface by exposure to a plasma treatment prior to film deposition. The first method has been reported [2,4] and yields films with good adhesion. The studies reported in his paper have shown that the second mechanism is also operative. The confinement of the He plasma inside the rf tube by increasing the pressure by 800 mtorr was made. At 100 mtorr, the pressure of a He plasma was low enough for the plasma to make contact with the sample substrate.

We have made some other observations regarding the He plasma activation process. We have biased the CdTe substrate, both negatively, (-100V and -25V) and positively (+25V) during the plasma treatment and find that this does not change the degree of adhesion. This suggests that: (1) either both He positive ions, and electrons can both provide surface activation; or (2) that surface activation involves a neutral species, e.g., the He metastable. We will investigate this point using biased-grid structures that can be placed between the end of the plasma tube and the substrate.

Acknowledgements

This work is supported under the Office of Naval Research (ONR) contracts (N00014-86-K-0760 and N00014-79-C-0133).

References

1. G. Lucovsky, J. Vac. Sci. Technol. A3, 346 (1985).
2. J. A. Wilson,V.A. Cotton, J.A. Silberman, D.Laser, W.E. Spicer, P. Morgan, J. Vac. Sci.Technol. A1(3), July/Sept. 1719, 1983.
3 W. E. Spicer, J.`A. Silberman, I. Lindau, A.B. Chen, A. Sher, J.A. Wilson, J. Vac. Sci. Technol. A1(3),.1735, 1983
4. B. K. Janousek, R. C. Carscallen and P. A. Bertrand, J. Vac. Sci Technol. A1(3),Jul/Sept.,1723, 1983.
5. S.S. Choi, G. Lucovsky, J. Vac. Sci. Technol.,B6(4), Jul/Aug., 1988
6. J. P. Haering, J. G. Werthen, R.H. Bube, L. Gulbrandsen, W. Jansen, and P. Luscher, J. Vac. Sci. Technol. A1(3), 1469, Jul/Sept. 1983.
7. A.J. Ricco, H.S. White, M.S. Wrighton, J. Vac. Sci. Technol. A2(2), 1984.
8. J.A. Silberman, D. Laser, I. Lindau, W.E. Spicer, and J.A.Wilson, J. Vac. Sci. Technol. A1(3), 1706, July/Sept. 1983
9. S.S. Kim and G. Lucovsky, presented at *International Topical Conference on Hydrogenated Amorphous Silicon Devices and Technology*, 21-23 Nov 1988, Yorktown Heights, New York, U.S.A.
10. U. Solzbach and H.J. richter, 191-205, Surf. Sci. 97(1980)
11. J. D. Benson, B. K. Wagner, A. Torabi, C.J. Summers, Appl..Phys.Lett. 49(16), 20 Oct. 1986
12. L.A. Kolodzieijsky, R.L. Guhshor, N. Otsuka, X.C. zhang, S.K. Chang, A.V. Nurmikko, Appl. Phys. Lett. 47(8), 15 Oct. 1985
13. D.M. Mattox, "Thin Film Adhesion and Adhesive Failure - A Perspective," in *"Adhesion Measurements of Thin film, Thick Films, and Bulk   Coatings. "* edited by K.L. Mittal, a symposium by American Society for Testing and Materials Philadelphia, Pa., 2-4, Nov. 1976.
14. Donald H. Buckley, Chapet 1, 5 (Adhesion) in *"Surface Effects in Adhesion, Friction, Wear, and Lubrication,"* Elsevier Scientific Publishing Company, 1981
15. V.F. Kiselev and O.V. krylov, Chapter 5, "Excited States in Adsorption and Catalysis," in *Adsoption Processes on Semiconductor and Dielectric Surfaces I*, Spinger-Verlag, Chemical Physics Series 32
16. R.G. Musket, 423-436, Surf. Sci. 74(1978)

CONTROLLED MASS FLOW
OF LOW VOLATILITY LIQUID SOURCE MATERIALS

ALAN D. NOLET, BRUCE C. RHINE, MARK A. LOGAN, LLOYD WRIGHT,
and JOSEPH R. MONKOWSKI, Monkowski-Rhine, Inc., 9250 Trade
Place, San Diego, California, 92126.

ABSTRACT

Chemical Vapor Deposition (CVD) of thin films for
microelectronic devices has historically used source
materials that are gases at room temperature [1]. The decision to use gases was largely a practical one based on the
relative ease with which the flow of gaseous materials can
be controlled. CVD of thin films plays a vital role in increased circuit density and performance of integrated circuits. Liquid sources offer alternative source composition,
reaction kinetics and reaction mechanisms to optimize a
given CVD process [2].

For example, CVD films of silicon dioxide (oxide)
and oxide films modified to lower the glass transition temperature such a borophosphosilicate glass (BPSG) have traditionally used gaseous source materials such as silane,
diborane and phosphine [3]. An all liquid system of
tetraethylorthosilicate (TEOS), triethylborate (TEB) and
triethylphosphine (TEPhine) has been found to offer superior
conformality and overall safety [4]. However, from a practical standpoint, the all liquid system has historically
suffered from reliable, reproducible mass flow control.

NEED FOR LIQUID MASS FLOW CONTROL

Traditionally microelectronic process steps such
as CVD, etching, diffusion, oxidation and ion implantation
have used source materials that are gases at room temperature. The use of liquids (and also solids) has been limited
to applications where mass flow control has not been critical. In the case of CVD, and LPCVD in particular, there is
a need for alternative chemistries to provide desired film
compositions and uniformities.

ALTERNATIVE METHODS

Five different liquid mass flow control methods
were evaluated. A summary description and brief outline of
advantages and disadvantages of each method is presented in
Table 1. The goal of the development program was to establish which of the five methods would be best for LPCVD applications. Sonic restriction [5] was eliminated because of
its complexity and the lack of readily available parts.
Traditional mass flow controllers were eliminated because of

the long flow stabilization time and the risk of condensation in the feed line. Direct vaporization was eliminated for the same reason. Carrier gas saturation was not considered because of the undesirable intrinsic requirement of adding a process gas diluent (i.e., the carrier gas) and the lack of process flexibility incurred.

The combination of metering the material in the liquid phase with subsequent and immediate vaporization was explored further [6]. This method virtually eliminates the need for large diameter plumbing to prevent condensation; provides for an exceptionally large flow rate range; and can accommodate premixed liquids.

By using premixed liquid precursors, only one liquid delivery system is required. This simplifies the design and improves system reliability. The liquid flow meter/vaporization concept is in effect a flash vaporizer. As such, a steady state mass balance on the system demonstrates that the number of moles of liquid equals the number of moles of gas flowing through the system. Accordingly, there is no accumulation of high boiling point materials and no change in premix composition.

EXPERIMENTAL

The method chosen was metering the liquid with subsequent vaporization. A number of commercially available liquid flow meters were evaluated. Because of instability and clogging problems, a prototype based on a proprietary meter design was developed and tested.

An analog signal (later digitized) from the meter is used to control either a stepper motor/needle valve combination or a solenoid valve. The stepper motor/needle valve combination was designed and a prototype built. The solenoid valve was adapted from a commercially available gas mass flow controller. Both were tested and worked well. Liquid flow rate was controlled using this combination of flow meter and feedback controlled stepper motor/needle valve. The system is controlled by a microprocessor.

Because the conductance of the system changed with flow rates and it was desirable to incorporate a wide flow rate range (2 to 50 grams per minute), a set of valve calibration tables was developed. Additionally, the unit is sensitive to the heat capacity of the liquid being processed. A set of heat capacity calibration constants was also developed. Both sets of calibration tables are made available to the microprocessor from on board ROM.

As presented in Figure I, a heated vaporizer is located immediately downstream of the control valve. Enough heat is provided to compensate for the heat of vaporization and to compensate for heat losses. The vaporizer is heated and of sufficient volume to allow enough residence time so that the material is completely vaporized.

RESULTS

In this set of experiments, TEOS, TEB and TEPhine were introduced in a new reactor configuration [7]. Process details have been provided elsewhere [8].

COMPARISON OF LIQUID MASS FLOW CONTROL ALTERNATIVES

Method	Description	Advantage	Disadvantage
Sonic Restriction	Material is vaporized, then passes through a small restriction where flow is in the sonic regime. Sonic flow is constant despite upstream pressure.	Flow is a function of orifice diameter	Expensive; orifice clogging problems; no feedback; may require large diameter plumbing and/or heat tracing.
Traditional Mass Flow Control	Large diameter, heated gas mass flow controller.	Good control / Available	May require large diameter plumbing and/or heat tracing; cannot use mixtures; long stabilization time.
Meter Liquid With Vaporization	Flow of liquid source materials is measured in the liquid phase, then injected into a vaporizer immediately prior to the reactor.	Good control; accommodates mixtures; fast response; no need for heat tracing or large diameter plumbing.	Not commercially available
Direct Vaporization	Source container plumbed to system. Vapor pressure from heated source provides driving force for mass flow. Controlled by valve restriction on source temperature.	Conceptually simple; inexpensive	May require large diameter plumbing and/or heat tracing; long to stabilize or by-pass to pump is required; not conducive to mixtures as volatile materials deplete & change composition; poor control; no feedback.
Carrier Gas Saturation	Carrier gas is metered into a flask through a down tube, saturated gas exits to process.	Conceptually simple; good control; relatively inexpensive.	Requires a diluent carrier gas that may be difficult to accommodate in the process; not conducive to mixtures as volatile materials deplete & change composition; no feedback; may require large diameter plumbing and/or heat tracing.

TABLE I

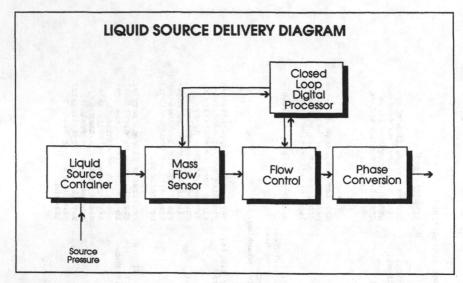

FIGURE I

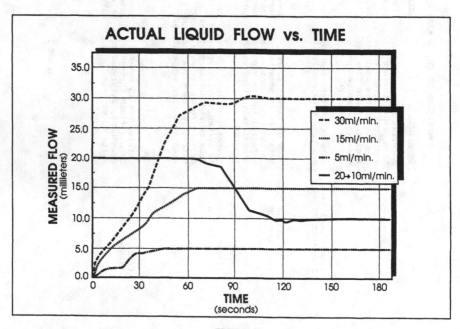

FIGURE II

BPSG FILMS DEPOSITED FROM TEOS, TRIETHYLBORATE
AND TRIETHYLPHOSPHINE

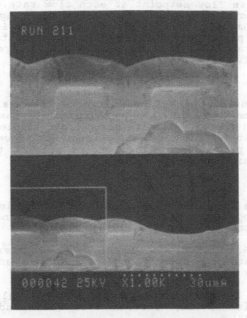

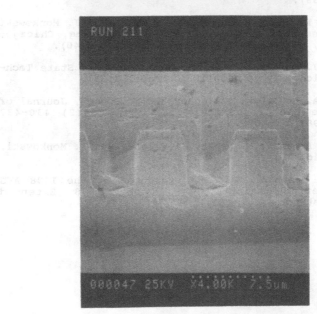

FIGURE III

The system stabilized within 90 seconds from start at 5, 15 and 30ml/min flow rates. System stability was +/- 3% of the mass flow setpoint at each flow rate. In addition, the system responded to a command to change the flow rate from 20ml/min to 10ml/min in less than 60 seconds. Results are presented in Figure II. Representative BPSG film results achieved in the novel CVD reactor are presented in Figure III.

CONCLUSIONS

A unique mass flow control system using liquid metering immediately followed by vaporization was designed, developed and tested. Results demonstrate stable mass flow at various rates. The mass flow concept was proven on mixed liquid BPSG precursors. The concept permits the use of premixed liquid materials without depletion.

REFERENCES

1.   See for example: W. Kern and V. Ban in Thin Film Processes, Edited by J.L. Vossen and W. Kern (Academic, New York, 1978), 257-331.

2.   F.S. Becker, D. Pawlik, H. Schafer, and G. Staudigl. Journal of Vacuum Science and Technology B 4, (3), 732-744 (May/June 1986).

3.   W. Kern presented at the 1988 ECS Fall Meeting, Chicago, IL, October 1988 (Extended Abstract #238).

4.   D. Freeman, M. Logan, L. Wright, and J. Monkowski presented at the 1988 ECS Fall Meeting, Chicago, IL, October 1988 (Extended Abstract #240).

5.   J.J. Sullivan, R.P. Jacobs, Jr. Solid State Technology, 29.10 1130118 (October, 1986).

6.   R.A. Lev, P.K. Gallagher and F. Schrey, Journal of the Electrochemical Socxiety, 134(2) 430-437 (1987).

7.   D. Freeman, M. Logan, L. Wright, and J. Monkowski, ibid.

8.   D. Freeman, W. Kern presented at the 1988 AVS Meeting, Atlanta, GA, October 1988 (Extended Abstract #).

PART VI

Aluminum Deposition Chemistry

ALUMINUM CHEMICAL VAPOR DEPOSITION USING TRIISOBUTYLALUMINUM: MECHANISM, KINETICS, AND DEPOSITION RATES AT STEADY STATE

BRIAN E. BENT*, LAWRENCE H. DUBOIS**, AND RALPH G. NUZZO**
*Current address: Department of Chemistry, Columbia University, New York, NY 10027.
**AT&T Bell Laboratories, Murray Hill, New Jersey 07974.

ABSTRACT

An important step in the chemical vapor deposition (CVD) of aluminum from triisobutylaluminum (TIBA) is the reaction between TIBA (adsorbed from the gas phase) and the growing aluminum surface. We have studied this chemistry by impinging TIBA under collisionless conditions in an ultra-high vacuum system onto single crystal Al(111) and Al(100) substrates. We find that when TIBA (340K) collides with an aluminum surface heated to between 500 and 600K, the aluminum atom is cleanly abstracted from this precursor with near unit reaction probability to deposit, epitaxially, carbon-free aluminum films. The gas phase products are isobutylene and hydrogen. From monolayer thermal desorption experiments, we have determined the kinetic parameters for the rate-determining step, a β-hydride elimination reaction by surface bound isobutyl ligands. Using these kinetic parameters and a Langmuir absorption model, we can predict the rate of aluminum deposition at pressures ranging from 10^{-6} to 1 Torr.

INTRODUCTION

In the late 1950's, Ziegler and coworkers found that aluminum alkyl compounds, particularly triisobutylaluminum (Al(C$_4$H$_9$)$_3$, TIBA) and diisobutylaluminum hydride (Al(C$_4$H$_9$)$_2$H, DIBAH) could be decomposed thermally to deposit high purity (>99 atom %) aluminum films [1]. The overall chemistry, as determined by analysis of the gas phase products, is as described in the following scheme:

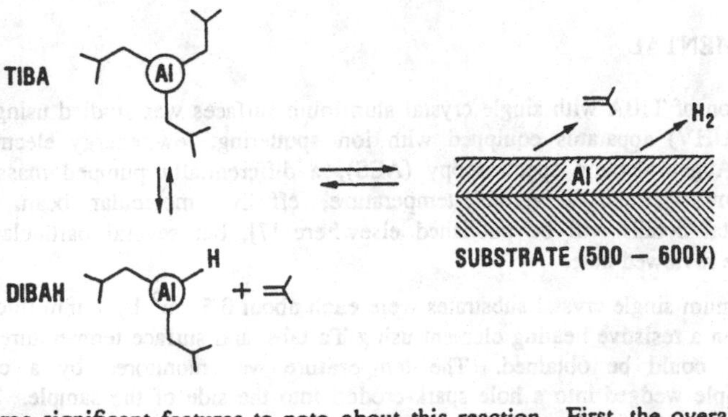

There are three significant features to note about this reaction. First, the overall process is reversible, which lead Ziegler et al. to suggest the potential of this system for refining aluminum. Second, the conversion of the isobutyl ligands into isobutylene can occur at least in part in the gas phase. Third, the reaction mechanism for conversion of the aluminum isobutyl moieties to isobutylene and hydrogen was judged to be a β-hydride elimination reaction; trimethylaluminum, which lacks β-hydrogens, decomposes thermally to produce aluminum carbide.

This aluminum deposition system was largely ignored [2] although interest was revived somewhat in 1982 when Cooke et al. demonstrated its potential to deposit conformal aluminum films for conductive contacts on silicon-based microelectronics devices [3]. Subsequent studies showed that the nucleation of these CVD aluminum films was strongly influenced by chemical pretreatments of the silicon substrates [4].

In order to examine the role of DIBAH and TIBA in this deposition, determine the relative importance of gas phase vs. surface chemistry, and characterize the surface reactions involved in the nucleation of film growth on silicon substrates and in the steady ·state deposition on aluminum, we have applied ultra-high vacuum techniques to study the surface chemistry of TIBA on single crystal substrates.

Our detailed studies of TIBA reactivity with silicon have been reported elsewhere [5]. This work showed that at room temperature TIBA has a reaction probability of less than 0.01 on silicon substrates heated to typical deposition temperatures of 550K. However, once aluminum is nucleated on the silicon surface, either by aluminum evaporation or by heating the silicon above 800K in the presence of the aluminum alkyl, TIBA readily decomposes to deposit aluminum at surface temperatures as low as 500K. The predominant gas phase products are isobutylene and hydrogen, but there are also alkyl silanes evolved as a result of a competing reaction with silicon diffusing to the metal surface (and reacting with the absorbed isobutyl groups and hydrogen). Details of this silicon etching reaction, which is potentially important both for aluminum deposition in silicon device processing and for the catalytic synthesis of alkyl silanes, will be described elsewhere [6].

In this paper we report on the steady state aluminum deposition chemistry which we have studied by impinging TIBA onto single crystal Al(111) and Al(100) substrates. We find that TIBA irreversibly adsorbs on clean aluminum surfaces with the isobutyl ligands remaining bound to the aluminum surface at temperatures below 500K. Above this temperature, a surface β-hydride elimination reaction occurs, resulting in the evolution of isobutylene and hydrogen. This surface reaction is the rate-determining step in aluminum CVD from TIBA, and a simple model which utilizes the kinetic parameters for this reaction allows us to accurately predict rates of aluminum deposition.

EXPERIMENTAL

The reaction of TIBA with single crystal aluminum surfaces was studied using an ultra-high vacuum (UHV) apparatus equipped with ion sputtering, low energy electron diffraction (LEED), Auger electron spectroscopy (AES), a differentially pumped mass spectrometer (with skimmer), and a variable-temperature, effusive molecular beam source. The experimental details will be published elsewhere [7], but several particularly significant aspects are reviewed here.

The aluminum single crystal substrates were each about 0.5 cm^2 by 2 mm thick. They were mounted on a resistive heating element using Ta tabs, and surface temperatures between 100 and 800K could be obtained. The temperature was monitored by a chromel-alumel thermocouple wedged into a hole spark-eroded into the side of the sample. The aluminum surfaces were cleaned *in situ* by cycles of sputtering at 600K with 1 kV Ar^+ ions and annealing in UHV at 700-750K. Despite extensive sputtering and a somewhat greyish to frosty appearance of the surface, the clean and annealed aluminum single crystal surfaces all showed sharp (1×1) LEED patterns.

The variable-temperature, effusive molecular beam source was a resistively heated, 3 mm OD stainless steel tube capped by a Ni disk bearing a 200μ pin hole. In the experiments reported here, both the doser and the gas handling lines were maintained at 330-350K to avoid TIBA condensation. The pressure behind the pinhole was varied between 10 and 200 mTorr. It is worth noting that the isobutylene and DIBAH, which are in equilibrium with TIBA, do not present a problem in these studies, since isobutylene does not react with aluminum surfaces under these conditions and DIBAH, being a trimer, has a negligible vapor pressure at 350K [7].

The mechanistic and kinetic results presented here were obtained from three types of experiments utilizing mass spectrometry. Thermal desorption studies (TDS) were performed by holding the adsorbate-covered surface within 1 mm of the skimmer to the mass spectrometer bearing a 3 mm diameter hole. A related technique called integrated desorption mass spectroscopy (IDMS), utilizes the same crystal/mass spectrometer geometry along with rapid scan mass spectrometry to obtain a complete mass spectrum of the thermally desorbing species [8]. Steady state scattering experiments were also performed using the same crystal/mass spectrometer geometry, but, in addition, TIBA molecules were impinged onto the heated surface at grazing incidence using the effusive beam source which was about 1 cm from the sample surface.

RESULTS AND DISCUSSION

We present and discuss our results in four subsections: (1) surface reaction mechanism, (2) surface reaction kinetics, (3) aluminum CVD model, and (4) predicted aluminum deposition rates.

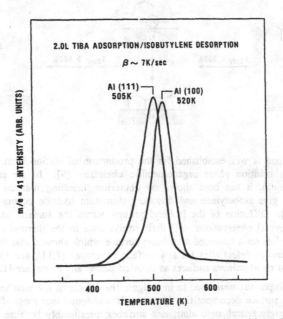

Fig. 1 Isobutylene (m/e=41) desorption observed after dosing either clean Al(111) (left trace) or Al(100) (right trace) with 2.0L of TIBA at 150K. The heating rate (β) was 7K/s in each case. A (1×1) LEED pattern is maintained during these experiments and Auger electron spectroscopy shows that the surfaces are carbon-free after each thermal desorption [7].

Surface Reaction Mechanism

Figure 1 shows the thermal desorption of isobutylene from Al(111) and Al(100) surfaces which have been dosed with 2.0L of TIBA at 150K. Only $m/e = 41$ desorption is shown here, but integrated desorption mass spectra [6], in which the entire cracking pattern of the desorption products are obtained, confirm that the desorbing species is isobutylene. It is noteworthy that isobutylene is evolved *at different temperatures from the Al(111) and Al(100) surfaces* and that, for each surface, *the isobutyl ligands are kinetically indistinguishable in TDS*. These points will be addressed below.

For submonolayer coverages of TIBA, the only other desorbing product besides isobutylene is hydrogen, which desorbs at virtually the same temperature as the olefin product [7]. No molecular TIBA desorption is detected until multilayer coverages are reached, at which point desorption of the multilayer is observed at about 220K. Consistent with the high purity aluminum films formed during CVD, no carbon is detectable on the surface by AES after the TDS experiment. The (1×1) LEED pattern for both the Al(111) and Al(100) surfaces is maintained, even after many TDS experiments, indicating that the aluminum left on the surface deposits epitaxially.

The desorption products from these single crystal aluminum substrates are suggestive of the surface β-hydride elimination pathway shown below:

Such a mechanism is well-established for the production of olefins from metal-coordinated alkyl groups in solution phase organometallic chemistry [9]. In the case of gas phase triisobutylaluminum, it has been shown by deuterium labelling that the elimination of 1 alkyl ligand to give isobutylene and diisobutylaluminum hydride occurs by a β-hydrogen abstraction [10]. Diffusion of the isobutyl groups across the surface, as shown above, is suggested by several observations: (1) their equivalence in the thermal desorption spectra, (2) previous studies on a sputtered aluminum surface which showed that the isobutyl groups preferentially bind at defect sites for low surface coverages [7,11], and (3) studies of alkyl iodides adsorbed on aluminum surfaces as detailed below and elsewhere [12].

Alkyl iodide compounds were used to investigate the effects of the attached aluminum atom in TIBA on the surface decomposition kinetics. It was found that most of these compounds readily and strongly adsorb onto aluminum surfaces, presumably because the weak carbon-iodine bond (50 kcal/mol) is broken to give a surface alkyl group and a coadsorbed iodine atom [12]. In the case of 1-iodo-2 methylpropane (an iodine atom attached to an isobutyl group), heating the resulting monolayer yields isobutylene and hydrogen at virtually the same temperature as that found for TIBA. This result suggests the reaction sequence shown below is analogous to the one given above for TIBA:

$\equiv$ + 1/2 H$_2$

$T_{SURF} < 500K$ $T_{SURF} > 500K$

Al   Al   Al

The most important point here is that both TIBA and the alkyl iodide yield identical organic products (isobutylene) with similar kinetic parameters, indicating that alkyl iodides can be used to model the surface chemistry of aluminum alkyls. This is a significant observation for several reasons, not the least of which is the fact that alkyl iodides are much more readily synthesized, isotopically labelled, and handled than the pyrophoric aluminum alkyls. In particular, we have utilized 1-iodopropane (ICH$_2$CH$_2$CH$_3$) and 1-iodopropane-2,2-d_2 (ICH$_2$CD$_2$CH$_3$) to determine whether or not the β-hydrogen atom is abstracted in the conversion of the alkyl to the olefin. The results of TDS and IDMS experiments with these compounds on an Al(100) surface are summarized schematically below [7,12]:

$$\begin{array}{c} CH_3 \\ | \\ CH_2 \\ | \\ CH_2 \end{array} \quad \xrightarrow{T_{max}=515K} \quad \left[\begin{array}{c} H \quad CH_3 \\ {}^H\!\!>\!C\!=\!C\!<_H \quad H \end{array} \right] \quad \longrightarrow \qquad \begin{array}{c} H \\ {}^H\!\!>\!C\!=\!C\!<^{CH_3}_H \end{array} + 1/2\,H_2$$

Al(100)   Al(100)   Al(100)

$$\begin{array}{c} CH_3 \\ | \\ CD_2 \\ | \\ CH_2 \end{array} \quad \xrightarrow{T_{max}=535K} \quad \left[\begin{array}{c} H \quad CH_3 \\ {}^H\!\!>\!C\!=\!C\!<_D \quad D \end{array} \right] \quad \longrightarrow \qquad \begin{array}{c} H \\ {}^H\!\!>\!C\!=\!C\!<^{CH_3}_D \end{array} + 1/2\,D_2$$

Al(100)   Al(100)   Al(100)

As shown, the perhydro compound produces propylene (m/e = 42) while its counterpart with two deuteriums on the β carbon produces exclusively propylene-d_1 (m/e = 43). These results substantiate the β-hydride elimination mechanism.

Surface Reaction Kinetics

Studies of isobutylene and hydrogen desorption from aluminum surfaces show that product desorption cannot be the rate-determining step in the surface β-hydride elimination. Both of these compounds desorb from aluminum over 100K below the temperature at which they are produced in TIBA decomposition [7]. High resolution electron energy loss spectroscopy (EELS) studies also show that isobutyl species are stable on the surface until near the isobutylene desorption temperature. The difference in the propylene TDS peak temperatures for the perhydro and 2,2-d_2 compounds above (515 vs 535K) implicates C-H (C-D) bond breaking is the rate-determining step. From these and other data [7], we conclude that the kinetic parameters for the TIBA TDS peaks are characteristic of the C-H bond breaking step in the surface β-hydride elimination.

It is evident from the peak temperatures in Fig. 1 that this C-H bond breaking rate is faster on Al(111) than on Al(100). We have determined the activation energies (E_a) and pre-exponential factors (A) for this reaction on these surfaces by measuring the shift in the thermal desorption peak temperatures (T_m) as a function of surface heating rate (β) and applying the analysis of Redhead [13]. Using a saturation exposure of TIBA and varying the surface heating rate from 1.5 to 16 K/s (peak temperature shift of 38K) on Al(100) and from 1 to 20 K/s (peak temperature shift of 40K) on Al(111), we obtain the results in Fig. 2. The slope of these plots is the activation energy divided by R (the gas constant). The pre-exponential factors are then determined from the kinetic expressions for a first order surface reaction [13]. This analysis makes the implicit assumption that both E_a and A are coverage and temperature independent. That this is reasonable is shown by the kinetic modeling discussed in the next section.

We find that for Al(111) the activation energy is 27.7 kcal/mol and the pre-exponential factor is 3.8×10^{11} s^{-1}. The kinetic parameters for Al(100) are dramatically different. The activation energy is 32.6 kcal/mol and the pre-exponential factor is 1.4×10^{13} s^{-1}. Were it not for a compensating effect between the kinetic parameters on these two surfaces, the differences in the TIBA decomposition rate would be much larger than the factor of 2-5 that is observed experimentally. It is interesting that the kinetic parameters on Al(111) are nearly identical to those reported for gas phase β-hydride elimination of one isobutyl group from TIBA (E_a = 26.6 kcal/mol and A = 1.6×10^{11} s^{-1} [10]). This result suggests that the β-hydride elimination on Al(111) might occur at individual aluminum atoms. It is not obvious, however, why the kinetic parameters are so different for TIBA on Al(100), and this point is under investigation.

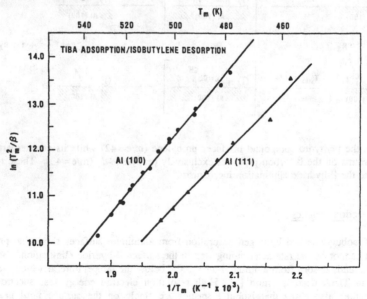

Fig. 2 A plot of $\ln(T_m^2 / \beta)$ vs $1/T_m$ is used to determine the activation energy for isobutylene (m/e=41) production on Al(100) (●) and Al(111) (▲) [13]. T_m is the temperature of the thermal desorption peak maximum (K) and β is the heating rate. The slope of these plots is E_a/R. Both surfaces were dosed with saturation amounts of TIBA at 300K.

Model For Aluminum CVD

Since we find that adsorbed TIBA decomposes on aluminum surfaces to deposit aluminum epitaxially at typical CVD temperatures, it seems probable that the steady state aluminum deposition rate is determined by the rate of this surface reaction. To test this hypothesis, we have developed a model which uses the kinetic parameters above to predict steady state aluminum deposition rates.

We assume that the rate of deposition is limited (for infinite flux of TIBA) by the rate of the surface decomposition reaction and that the kinetic parameters for this decomposition are independent of surface coverage. We also assume that only TIBA molecules incident on empty surface sites can be adsorbed. The rate of TIBA adsorption and the rate of isobutylene evolution are thus given by:

$$\text{Rate of TIBA Adsorption} = \sigma(1-\theta)s \quad (1)$$

$$\text{Rate of Isobutylene Evolution} = 3 A \theta n_s \exp(-E_a / RT) \quad (2)$$

where $\sigma = $ TIBA flux, $\theta = $ fractional surface coverage of TIBA, $s = $ sticking probably of TIBA on vacant surface sites, $A = $ Arrhenius pre-exponential factor, $n_s = $ number of adsorbed TIBA per unit area at saturation coverage, $E_a = $ activation energy, $R = $ gas constant, and $T = $ surface temperature. Note that the rate of isobutylene production includes a factor of three since there are three isobutyl groups per TIBA.

At steady state, the rate of TIBA adsorption must be 1/3 the rate of isobutylene evolution, allowing us to solve for θ:

$$\theta = \left[\frac{A n_s}{\sigma s} \exp(-E_a / RT) + 1 \right]^{-1} \quad (3)$$

Substituting for θ in equation (2) and dividing by three gives the rate of aluminum deposition (i.e. one third the rate of isobutylene production) in terms of parameters which can be determined experimentally:

$$\text{Rate of Aluminum Deposition} = \frac{A n_s \exp(-E_a / RT)}{\dfrac{A n_s}{\sigma s} \exp(-E_a / RT) + 1} \quad (4)$$

To convert this aluminum deposition rate from atoms/cm²/sec to a film growth rate (Å/sec), the number of aluminum atoms per cm² and the number of Å per layer of aluminum must be taken into consideration. While each of these factors individually is different for different crystal faces of aluminum, their combination must necessarily be the same, and this multiplicative factor is $\sim 1.66 \times 10^{-15}$ for an aluminum atom of radius 1.432 Å.

Aluminum Deposition Rates

We can use the model presented above to predict the rate of aluminum deposition for comparison with experiment. Two widely different experiments (the only two we know of which give Al CVD rates from TIBA as a function of surface temperature) will be modelled.

Scattering Experiments at 10^{-6} Torr. We have measured the rate of steady state aluminum CVD on our single crystal aluminum substrates by monitoring isobutylene evolution as a function of surface temperature while scattering TIBA from the surface. The results are shown in Fig. 3 for Al(111) and Al(100) surfaces. The points are the experimental results, which were obtained by ramping the crystal surface temperature while scattering TIBA into the mass spectrometer as described in the Experimental section. Throughout these experiments, the surface remained carbon free (as long as the surface temperature did not rise significantly above 600K [14]), and a (1×1) LEED pattern was maintained, indicating epitaxial aluminum deposition.

The curves in Fig. 3 are fits based on the model presented above. A sticking probability of 1 was assumed, and the number of TIBA molecules per cm^2 at saturation coverage was taken as $1.4×10^{14}$ on both surfaces [15]. To obtain the solid curves in Fig. 3, the kinetic parameters determined previously were applied, and the incident flux of TIBA was adjusted to give the best fit. This flux, the only adjusted parameter in the fit, could not be accurately measured since a TIBA/isobutylene mixture of unknown composition enters the ultra-high

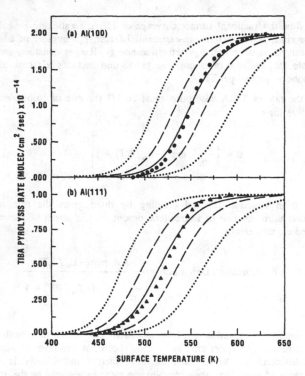

Fig. 3 Experimental (symbols) and calculated (solid line, equation (4)) TIBA pyrolysis rates are plotted as a function of surface temperature for aluminum film formation on (a) Al(100) and (b) Al(111). The dashed curves were calculated from equation (4) by varying the activation energy by ±1 kcal/mol with a fixed pre-exponential factor while the dotted curves were obtained by varying the pre-exponential factor by ±1 order of magnitude with a fixed activation energy.

vacuum chamber (see Experimental). It is noteworthy, however, that the fluxes used in the fits of Fig. 3 (2.0×10^{14} molecules/cm^2/sec in (a) and 1.0×10^{14} molecules/cm^2/sec in (b)) are within a factor of two of estimates based on the chamber background pressure of about 1×10^{-6} Torr during these experiments.

The dashed and dotted curves in Fig. 3 are the calculated results obtained by varying the activation energies and pre-exponential factors in the model by ± 1 kcal/mol and ± 1 order of magnitude, respectively. These curves show the sensitivity of the model to the kinetic parameters. It should be noted, however, that the sensitivity of the fit to these parameters does not necessarily imply that the experimentally determined values are extremely accurate. There can be compensating effects between pairs of parameters in the model which will offset individual inaccuracies. Still, we feel that the activation energies are accurate to better than 2 kcal/mol and the pre-exponential factors to within an order of magnitude.

Film Deposition at 1 Torr. The rate of aluminum deposition onto SiO$_2$ (from TIBA) as a function of reactor temperature at 1 Torr pressure has been reported by Cooke et al. [3]. Their results are plotted in Fig. 4 along with the predicted rates from our model assuming an *infinite flux* of TIBA to the surface. In this case the denominator of equation (4) goes to unity and we have

$$\text{Maximum Deposition Rate} = An_s \exp(-E_a / RT) \qquad (5)$$

Thus, substituting our kinetic parameters into equation (5) yields:

$$\text{Maximum CVD Rate (Al(111))} = 8.8 \times 10^{10} \exp(-1.39 \times 10^4 / T) \; (\text{Å} / \text{sec}) \quad \text{and} \qquad (6a)$$

$$\text{Maximum CVD Rate (Al(100))} = 3.25 \times 10^{12} \exp(-1.64 \times 10^4 / T) \; (\text{Å} / \text{sec}). \qquad (6b)$$

The agreement between our predictions derived from monolayer thermal desorption experiments in ultra-high vacuum and the experimental results at orders of magnitude higher pressure is quite remarkable. Since growth from flat, single-crystal aluminum surfaces was assumed in the predicted growth rates, part of the discrepancy in Fig. 4 could be due to the higher real surface areas which characterize the rough CVD aluminum films formed on SiO$_2$. We note also that, while the net macroscopic growth rate is actually an average of the different growth rates from different crystal faces, it is probable that the CVD film is strongly (111) textured [4a], consistent with the measured growth rate being closest to the predictions for Al(111).

Given our model for aluminum CVD, we can also determine aluminum deposition rates as a function of TIBA pressure by solving for σ in equation (4). Alternatively, we can calculate what fraction (η) of the maximum possible deposition rate is achieved for a given surface temperature and flux of TIBA (i.e. equation (4) + equation (5)).

$$\eta = \frac{\sigma}{\dfrac{An_s}{s} \exp(-E_a / RT) + \sigma} \qquad (7)$$

Note that when $\sigma = \dfrac{An_s}{s} \exp(-E_a / RT)$, $\eta = \frac{1}{2}$. This leads us to define

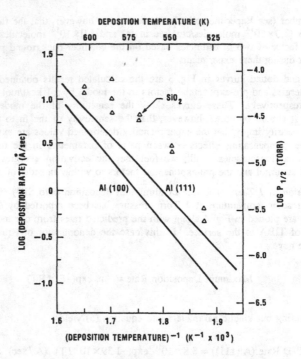

Fig. 4 The log of the aluminum deposition rate is plotted vs. 1/surface temperature on Al(100) and Al(111) single crystal substrates using equations (6a) and (6b) (solid lines) and this is compared to growth of aluminum films on SiO₂ (points) [3]. An infinite flux of TIBA to the surface is assumed for our model calculations. $P_{1/2}$, the pressure of TIBA (in Torr) required to give half of the maximum aluminum deposition rate, is defined in equations (9) and (10).

$\dfrac{An_s}{s} \exp(-E_a / RT)$ as $\sigma_{1/2}$, the flux necessary to achieve one half of the maximum deposition rate. Equation (7) can then be simplified to

$$\eta = \frac{\sigma}{\sigma_{1/2} + \sigma} \tag{8}$$

Converting TIBA fluxes to pressures yields

$$\eta = \frac{P}{P_{1/2} + P} \tag{9}$$

where for Al(111)

$$P_{1/2}(\text{Torr}) = 4.52 \times 10^5 \exp(-1.39 \times 10^4 / T) \tag{10a}$$

and for Al(100)

$$P_{1/2}(\text{Torr}) = 1.67 \times 10^7 \exp(-1.64 \times 10^4 / T) \tag{10b}$$

assuming s = 1.

To illustrate how these equations may be used, we present the following example. Let us suppose that one wants to achieve an aluminum deposition rate of 1 Å/sec. It can be seen from Fig. 4 (or eqn (5)) that a substrate temperature of at least 554K is required (this assumes a (111) orientation for the growing surface). To achieve 1 Å/sec deposition at 554K would require infinite pressure; however, we can see from equation (10a) that half this deposition rate is obtained with a pressure of only 5.3×10^{-6} Torr. Further, from equation (9), it is immediately evident that increasing the pressure by an order of magnitude to $10P_{1/2}$ will increase the deposition rate to 91% of the maximum, or 0.91 Å/sec.

A significant conclusion which can be drawn from equations (9) and (10) is that the pressures of TIBA required to achieve near maximal aluminum deposition rates in the temperature range of 500-600K are quite low. This observation suggests that TIBA would be a useful aluminum precursor in a low pressure chemical beam epitaxy process.

CONCLUSIONS

Our results show that aluminum surfaces readily effect the decomposition of TIBA at temperatures above 500K to deposit aluminum and evolve isobutylene and hydrogen. Below 600K, the deposited aluminum is carbon-free and grows epitaxially on the (111) or (100) substrates used in these studies. The rate-determining surface reaction, a β-hydride elimination, is 2-5 times faster on Al(111) than on Al(100) for surface temperatures of 470-570K. It has been demonstrated that the kinetic parameters determined from monolayer thermal desorption experiments for these surface reactions can be used to predict Al CVD rates for a wide range of TIBA pressures.

REFERENCES

[1] K. Ziegler, K. Nagel and W. Pfohl, Justus Liebigs Ann. Chem. 629, 210 (1960).

[2] Several studies of aluminum deposition from TIBA were reported. See, for example, H. O. Pierson, Thin Solid Films 45, 257 (1977).

[3] M. J. Cooke, R. A. Heinecke, R. C. Stern and J. W. C. Maes, Solid State Technol. 25, 62 (1982).

[4] (a) M. L. Green, R. A. Levy, R. G. Nuzzo and E. Coleman, Thin Solid Films 114, 367 (1984); (b) R. A. Levy, M. L. Green and P. K. Gallagher, J. Electrochem. Soc. 131, 2175 (1984); (c) C. G. Fleming, G. E. Blonder and G. S. Higashi, Mat. Res. Soc. Symp. Proc. 101, 183 (1988); (d) D. A. Mantell and T. E. Orlowski, Mat. Res. Soc. Symp. Proc. 101, 171 (1988).

[5]   B. E. Bent, R. G. Nuzzo and L. H. Dubois, Mat. Res. Soc. Symp. Proc. *101*, 177 (1988).

[6]   B. E. Bent, B. R. Zegarski, R. G. Nuzzo and L. H. Dubois, in preparation.

[7]   B. E. Bent, R. G. Nuzzo and L. H. Dubois, J. Am. Chem. Soc. *111*, 000 (1989).

[8]   L. H. Dubois, Rev. Sci. Instrum. *60*, 000 (1989).

[9]   J. P. Collman, L. S. Hegedus, J. R. Norton and R. G. Finke, *Principles and Applications of Organotransition Metal Chemistry, 2nd Edition*, University Science Books, Mill Valley, 386 (1987).

[10]   K. W. Egger, J. Amer. Chem. Soc. *91*, 2867 (1969); K. W. Egger, Int. J. Chem. Kin. *1*, 459 (1969).

[11]   B. E. Bent, R. G. Nuzzo and L. H. Dubois, J. Vac. Sci. Technol. *A6*, 1920 (1988).

[12]   B. E. Bent, B. R. Zegarski, R. G. Nuzzo and L. H. Dubois, in preparation.

[13]   P. A. Redhead, Vacuum *12*, 203 (1962).

[14]   At surface temperatures above 600K, carbon is incorporated into the growing aluminum film [7].

[15]   This value was approximated from the exposure necessary to achieve saturation coverage in the TIBA thermal desorption experiments. It is roughly consistent with the van der Waals radius of TIBA [7].

The Decomposition of Trimethylgallium and Trimethylaluminum on Si(100)

F. Lee, T. R. Gow, R. Lin, A. L. Backman, D. Lubben, and R. I. Masel[*], University of Illinois, 1209 W California St., Urbana Il., 61801.

ABSTRACT

The decomposition of trimethylgallium (TMG) and trimethylaluminum (TMA) on Si(100) is studied by TPD, XPS, and EELS. It is found that the decomposition of TMG is largely an intramolecular process. First one of the methyl groups in the TMG reacts with a hydrogen in another methyl group liberating methane. This leaves a CH_2 group bound to the gallium which is seen clearly in EELS. Subsequently, another hydrogen reacts with a methyl group producing additional methane. This leaves a gallium atom and a CH group on the surface. Hydrogen desorbs, while carbon is incorporated into the growing film. Careful calibration of the peak areas in the TPD data indicates that 2.1 ± 0.3 moles of methane, 0.5 ± 0.03 moles of hydrogen, 1 mole of gallium, and 0.95 ± 0.1 moles of surface carbon are produced for every mole of TMG which decomposes. Since, the decomposition ratio is independent of the state of the surface, the surface orientation, and the level of impurities on the surface, it appears that the surface is not a direct participant in the reaction.

TMA adsorbs as dimers. However, the dimers decompose upon heating to 400 K. Thereafter, the chemistry of TMA decomposition is similar to that for TMG. Nonetheless, the product ratio varies with the TMA coverage, the state of the surface, and the aluminum coverage. Also, silicon-carbon bonds are evident at 600 K in EELS. It is suggested, therefore, that the decomposition of TMA is similar to that of TMG, except that methyl groups in TMA have a tendency to migrate to the support. In both TMA and TMG, carbon incorporation is an intrinsic part of the decomposition process. This suggests that TMA and TMG would not be appropriate source gases when one needs to produce films with very low carbon levels.

INTRODUCTION

Trimethylgallium(TMG) and trimethylaluminum(TMA) are important source gases for MOCVD and MOMBE. Films grow easily. However, carbon incorporation during the film growth process has been a persistent problem. The objective of this paper is to examine the decomposition of TMA and TMG on a Si(100) substrate using standard surface spectroscopic techniques to try to get some insight on the carbon incorporation process.

EXPERIMENTAL

The experiments presented here were done using standard surface spectroscopic techniques. A 8 Ω-cm Si(100) sample was cut from a standard wafer. The sample was rinsed in HF, then mounted in a standard UHV system. The sample was then repeatedly sputtered and annealed until no impurities could be detected by AES. Next, the sample was then dosed with either TMA or TMG through a calibrated leak system. The sample was then examined with temperature programmed desorption (TPD), x-ray photoemission spectroscopy (XPS) or electron energy loss spectroscopy (EELS). All of the procedures were standard. One is referred to more complete descriptions of the work in references 1 and 2 for additional details.

---

*
Send Correspondence To This Author

Mat. Res. Soc. Symp. Proc. Vol. 131. ©1989 Materials Research Society

RESULTS

Trimethylgallium

Figure 1 shows a typical TPD spectrum of the decomposition of TMG on Si(100). One observes a rather broad methane peak between 400 and 800 K, a hydrogen peak at 780 K, and a gallium peak at 950 K. At higher exposures, a series of molecular TMG desorption features are also seen between 150 and 400 K. Calibration of the peak areas in the TPD data described elsewhere[1] indicates that 2.1 ± 0.3 moles of methane, 0.50 ± 0.03 moles of hydrogen, 1 mole of gallium, and 0.95 ± 0.10 moles of surface carbon are produced for every mole of TMG which decomposes. The decomposition ratio is independent of the TMG coverage as indicated in figure 2.

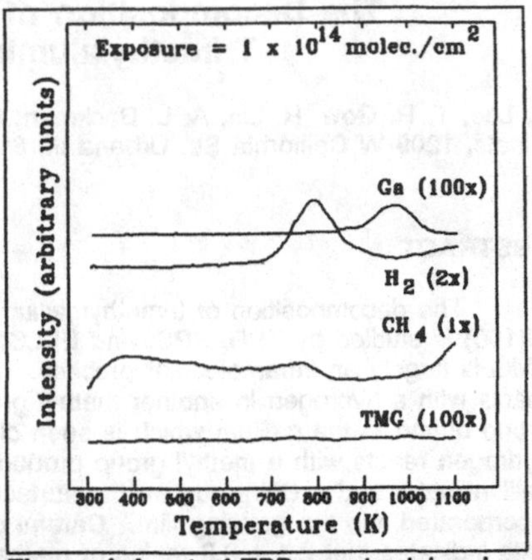

Figure 1 A set of TPD spectra taken by exposing a clean Si(100) sample to 1×10^{14} molecules/cm^2 of TMG then heating at 10 K/sec.

The effects of impurities and the state of the surface on the TMG decomposition process were also considered. It was found that the TPD spectra were essentially identical on Si(111), Si(100) and heavily sputtered silicon. There were some small peak shifts, when gallium or carbon was coadsorbed with the TMG. However, the relative sizes of the desorption peaks did not change. Runs were also made where deuterium atoms from a hot filament were adsorbed onto the silicon surface, then TMG was adsorbed. An attempt was then made to look for deuterium in the desorption products. Significant HD desorption was observed. However, no desorption of a CH$_3$D species was detected. The lack of influence of coverage, surface species, and surface orientation on the decomposition ratio, and the lack of incorporation of deuterium in the methane desorption product, lead us to suggest that the decomposition of TMG is largely an intramolecular process[1]. Two of the methyl groups from the TMG react with the hydrogens on the third methyl group to yield methane. There is not enough hydrogen in TMG to produce a third methane; surface hydrogen is not effective in hydrogenating the remaining CH group, and the CH group is too strongly bound on silicon to desorb as acetylene. As a result, the remaining carbon gets incorporated into the growing film.

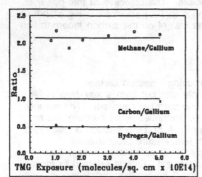

Figure 2 The number of moles of methane, hydrogen and surface carbon produced, per mole of TMG which decomposes as measured by TPD, XPS and AES.

X-ray Photoemission Spectroscopy (XPS) and Electron Energy Loss Spectroscopy were used to help identify some of the intermediates in the data above. Figure 3 shows an EELS spectrum taken by exposing a "clean" Si(100) sample to 1×10^{15} molecules/cm^2 of TMG then sequentially heating as indicated. The small peak in the "clean" spectrum is due to the presence of about 2×10^{13} molecules/cm^2 of residual carbon which was not removed during this particular cleaning cycle. At 300 K, the EELS spectrum looks just as expected for molecularly adsorbed TMG. There are broad peaks at 550, 740, 780, 1240, 1420, and 2950 cm^{-1}. By comparison, Kvisle and Ryttle[3] find major peaks at 540, 745, 790, 1168, 1425, 2955, and 2966 cm⁻¹ in the IR spectrum of solid TMG. Further, the same peak positions and relative peak intensities are seen in multilayer spectra at 100 K and monolayer spectra at 300 K. Thus, we conclude that the adsorption of TMG is molecular at 300K.

By comparison to the work of Kvisle and Ryttle, we assign the EELS peaks at 550, 750, 780, 1240, 1420, and 2950 cm⁻¹ to the Ga-C₃ stretch, the CH₃ A₂'' rock, the CH₃ E₂ rock, the CH₃ symmetric deformation, the CH₃ asymmetric deformation, and the CH₃ symmetric and asymmetric stretches respectively.

Figure 3 also shows the effects of heating on the EELS spectrum. The basic spectrum does not change significantly upon heating to 450 K. However, there is a substantial loss in the mode at 550 cm⁻¹, and a smaller decrease in the peak at 2950 cm⁻¹. In

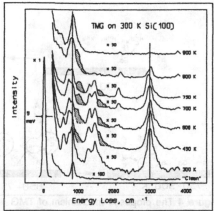

Figure 3 A series of EELS spectra taken by exposing a "clean" Si(100) sample to 1x10¹⁵ molecules/cm² of TMG then sequentially flashing to the temperatures indicated.

addition shoulders appear at about 900 and 1500 cm⁻¹. The shoulders are shaded in the figure. The shoulders grow upon heating to 600 K and a new peak appears at about 2070 cm⁻¹. Simultaneously, there is a general loss in intensity in the modes at 1240, 1420, and 2950 cm⁻¹. The modes at 1240, 1420, and 2950 cm⁻¹ lose further intensity upon heating to 750 K, while the 2070 cm⁻¹ peak grows. Simultaneously, the shoulder at about 900 cm⁻¹ shifts to 960 cm⁻¹ and becomes a distinct peak. The 550, 740, 1240, 1420, 1500, 2070, and 2950 cm⁻¹ peaks are all substantially attenuated at 800 K. However, the area of the 960 cm⁻¹ shoulder only decreases by 50%. The 2950 mode shifts to 3050 cm⁻¹. Only the 800 cm⁻¹ peak is evident at 900 K.

A series of reference spectra were measured to help clarify the peak assignments above. When hydrogen is adsorbed onto a clean Si(100) surface, a single loss peak is seen at 2070 cm⁻¹. Adsorbed carbon shows a single loss feature at 800 cm⁻¹. If TEG is decomposed on a surface, then hydrogen is readsorbed, a new loss feature is seen at 1860 cm⁻¹. We attribute the new loss feature to a gallium-hydrogen stretch.

There is not room here to discuss, in detail, the origin of the shoulders at 900 and 1500 cm⁻¹ seen in EELS. However, we attribute[2] these two modes to CH₂ deformation and bending modes. Youshinobu, et al [4] examined ethylene adsorption on Si(100) at low temperatures, and observed a CH₂ deformation mode at 960 cm⁻¹. The CH₂ bending modes in simple hydrocarbons usually lie between 1400 and 1500 cm⁻¹[5]. However, the CH₂ bending modes in methylene fluoride lie at 1507 cm⁻¹[5]. Thus, we tentatively assign the shoulders at 960 and about 1500 cm⁻¹ to CH₂ deformation and bending modes. More details are given in reference 2.

If these assignments are correct, then one can speculate about the nature of the surface intermediate produced between 450 and 600 K. The EELS data shows the presence of CH₂ groups. However, there is no evidence for gallium-hydrogen stretches, silicon-hydrogen stretches, or growth in the region expected for silicon-carbon stretches. Further, in data not shown, we have found that the C₁ₛ and Ga₂ₚ XPS of adsorbed TMG do not shift or broaden upon heating from 100 to 600 K. Gallium metal is known to form islands on silicon[7]. If such islands form, then one would expect that the gallium XPS peak position would be at the position for

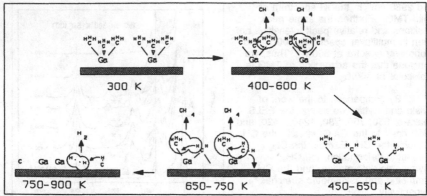

Figure 4 The proposed mechanism of TMG decomposition. TMG is a planar molecule. However, we have distorted the TMG in the figure to make the diagram easier to understand.

chemisorbed gallium, and not at the position for molecular TMG as is observed. As a result, we suggest that at 450-600K, CH_2 groups have formed but they are still attached to the gallium as indicated in figure 4.

Further heating causes major disruptions. The peaks associated with molecular TMG are strongly attenuated in the EELS spectrum. Simultaneously, a new peak appears at 2070 cm^{-1} which is characteristic of a silicon-hydrogen stretch[6]. Surprisingly, there is no evidence for a gallium-hydrogen stretch at 1860 cm^{-1}. Further, the Ga_{2p} XPS peak shifts toward that of chemisorbed gallium while the C_{1s} peak shifts toward that of chemisorbed carbon. Methane formation is seen in TPD. As a result, it appears that the bond between the CH_x species and the gallium atom has broken, and that we have mainly C, H, CH groups and gallium on the surface. From the measured stoichiometry, it appears that a total of two moles of methane, a half a mole of H_2, a mole of surface carbon and a mole of chemisorbed gallium are formed for each mole of TMG which decomposes.

As a result, we conclude that the decomposition of TMG on Si(100) follows the mechanism in figure 4. The TMG adsorbs molecularly at low temperature. Heating to 400-600, causes an intramolecular hydrogen shift to occur, liberating methane, and producing a CH_2 group. A second hydrogen is transferred upon further heating to leave, carbon, gallium, hydrogen and CH groups on the surface.

Trimethylaluminum

The TPD, XPS and EELS data for trimethylaluminum TMA decomposition on Si(100) were very similar to those for TMG decomposition above. For example figures 5 and 6 show a series of TPD and EELS spectra taken by adsorbing TMA at 100 K then heating as indicated. One observes TPD and EELS features much like those in figures 1 and 3. Of course, TMA dimerizes in the gas phase below 400 K. As a result, one observes an EELS peak at 340 cm^{-1} due to the dimers at low temperatures. However, the peak is substantially attenuated upon heating to 300 K, and disappears upon heating to 400 K. Thereafter, the TPD, XPS and EELS spectra from TMA look similar to those from TMG.

The one major difference between the results from TMA and TMG, however, is that while methane, hydrogen, aluminum and adsorbed carbon were the only products detected during both TMA and TMG decomposition, with TMA the ratio of the products varied with coverage surface structure, and surface cleanliness. No similar

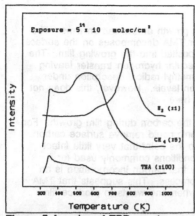

Figure 5 A series of TPD spectra taken by exposing a clean Si(111) sample to 5×10^{14} molecules/cm² of TMA then heating at 10 K/sec.

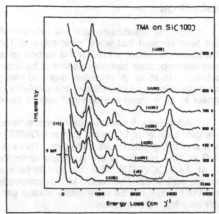

Figure 6 A series of EELS spectra taken by exposing a "clean" Si(100) sample to 1×10^{15} molecules/cm² of TMA then sequentially flashing to the temperatures indicated.

dependence is observed with TMG. Generally, when one adsorbed TMA on a clean Si(100) surface, one observed more hydrogen and surface carbon formation and less methane desorption than was observed with TMG. The amount of carbon deposition decreased with increasing TMA coverage. The carbon deposition also decreased when carbon or aluminum or other impurities were deposited on the surface. On a dirty surface, the decomposition ratio approached two methanes for every TMA which decomposed. The similarity of the spectra in figures 5 and 6 to those in figures 1 and 3 suggests that the TMA decomposition process is much like that in figure 4. However, the fact that the carbon deposition rate is higher with TMA than one could expect from the mechanism in figure 4, suggests that there is an additional pathway for carbon deposition during TMA decomposition which is not seen during TMG decomposition. That pathway is inhibited by surface carbon, aluminum etc. and appears to be affected by the surface structure. Therefore, it appears that the surface is directly involved with the process to deposit additional carbon. As a result, we suggest that the additional pathway involves migration of CH_x to the support prior to the desorption of methane.

There is some support for this mechanism in the EELS data in figure 6. Notice that in figure 6, a feature at about 800 cm⁻¹ grows into the spectrum at about 600 K. By comparison, with TMG we do not observe significant growth of the feature at 800 cm⁻¹ until the surface has been heated to 750 K, where most of the methane has already desorbed. The changes in the 800 cm⁻¹ peak are difficult to interpret since carbon-silicon bonds and CH_3 rocking modes both give peaks near 800 cm⁻¹. However, the changes in the EELS spectra are as expected if CH_x groups were migrating to the support. Thus, it appears that the EELS results are consistent with the notion that CH_x migrate to the support before they are hydrogenated, even if the EELS results, by themselves do not prove that migration of CH_x groups occurs.

In summary then, we conclude, that TMA decomposition also basically follows the mechanism in figure 4. The main reaction pathway involves intramolecular hydrogen transfer, liberating methane. However, in addition, there can be migration of CH_x groups to the support. As a result, the carbon deposition rate is higher with TMA than with TMG.

344

DISCUSSION

Still, the results here are not very promising growth of low carbon films. The data here suggest that when a molecule of TMG or TMA decomposes on the surface of a substrate, about a third of the carbon gets deposited into the growing film. The main decomposition pathway seems to be intramolecular hydrogen transfer leaving carbon. Lin et. al. [8-10] found that one can get methyl radical desorption under some conditions, which reduces the available carbon levels. However, this does not appear to be a major reaction pathway in the data here.

Admittedly, there are other ways to remove the carbon during film growth. For example, Luth et al[11] found that in MOMBE, arsenic could remove surface carbon from a growing gallium arsenide film. There is also the point that very little intact TMG or TMA make it to the substrate under the conditions commonly used for Ga$_x$Al$_{1-x}$As MOCVD[12]. Still, the results here show that carbon incorporation is an intrinsic part of the TMA and TMG decomposition process. This suggests that TMA and TMG would not be appropriate source gases when one wants to produce films with very low carbon levels.

CONCLUSIONS

In summary then, it was found that the decomposition of TMA and TMG on Si(100) is relatively simple. Both TMA and TMG adsorb molecularly. As the temperature is raised, intramolecular hydrogen transfer occurs, liberating methane. There is only enough hydrogen in each TMA or TMG molecule to remove 2/3 of the carbon in the molecule as methane, and so the remaining 1/3 remains behind on the substrate. Extra carbon gets deposited with TMA, due to CH$_x$ migration to the support. There are ways to remove some of the carbon under film growth conditions. However, the data here suggest that carbon incorporation is an intrinsic part of the TMA and TMG decomposition process. This suggests that TMA and TMG would not be appropriate source gases when one wants to grow films with very low carbon levels.

ACKNOWLEDGMENT

This work was supported by the National Science Foundation under Grant DMR 86-12860. Equipment was provided by NSF grants CPE 83-51648 and CBT 87-04667.

LITERATURE CITED

1. F. Lee, T. R. Gow, R. I. Masel, J. Electrochem Soc., to appear.
2. F. Lee, A. L. Backman, R. Lin, T. R. Gow, R. I. Masel, unpublished work.
3. S. Kvisle, E. Ryttle, Spectro. Acta, 40, 939 (1984).
4. J. Youshinobu, H. Tsuda, M. Onchi, M. Nishijima, J. Chem. Phys, 87, 7332 (1987).
5. L. J. Bellamy, The Infrared Spectra Of Complex Molecules, 3rd ed. Chapman and Hall, (1975).
6. J. A. Stroscio, S. R. Bare, W. Ho, Surface Sci. 154, 35 (1985).
7. B. Bourguignon, K. L. Carleton, S. R. Leone, Surface Sci. 204, 455 (1988).
8. D. W. Squire, C. S. Dulcey, M. C. Lin, Proc. Materials Research Soc. 54, 709 (1986).
9. D. W. Squire, C. S. Dulcey, M. C. Lin, Chem. Phys. Lett., 131, 112 (1986).
10. D. W. Squire, C. S. Dulcey, M. C. Lin, JVST B, 3, 1513 (1985).
11. N. Pütz, H. Heinecke, M. Weyers, H. Lüth, P. Balk, J. Crystal Growth, 74, 292 (1986).
12. S. P. DenBaars, B. Y. Maa, P. D. Dapkus, A. D. Danner, H. C. Lee, J. Crystal Growth, 77, 188 (1986).

ELECTRONIC STRUCTURE OF ADSORBED TRIMETHYLALUMINUM ON CLEAN Si(100) SURFACES

T. Motooka, P. Fons, and J.E. Greene

Department of Materials Science, University of Illinois, Urbana, IL 61801

ABSTRACT

The electronic structure of dimerized trimethylaluminum (TMA), $Al_2(CH_3)_6$, adsorbed on Si(100) surfaces has been investigated using molecular orbital (MO) calculations based on a cluster description of TMA/Si(100). The calculated results suggest that the interactions between TMA and the Si(100) surface are described by overlap of the TMA electron-deficient bond and Si surface dangling-bond orbitals. The electron-deficient bond orbital is the highest occupied MO of TMA and acts as an electron acceptor for charge transfer from a surface Si atom to TMA consistent with observed core-level and valence photoelectron spectra.

INTRODUCTION

Trimethylaluminum (TMA) is an organometallic molecule used as source material for metal and compound film growth by chemical vapor deposition. There has recently been a growing interest in UV-laser stimulated chemical reactions in surface-adsorbed TMA layers for direct writing of Al metallization on integrated circuits.[1,2] We have previously reported on the analysis of TMA adsorption on clean Si(100) using X-ray and ultraviolet photoelectron spectroscopy (XPS and UPS).[3] The valence electron spectra from condensed thick (≈10nm) TMA layers contained three peaks whose positions were in good agreement with the calculated molecular orbital (MO) energies of the isolated dimerized TMA, $Al_2(CH_3)_6$ molecule.[4] Core level spectra for C 1s, Al 2p, and Si 2p from single TMA layers adsorbed on Si(100) were acquired in both surface-normal and grazing directions and the Al 2p (Si 2p) peak from the interface region was found to shift toward lower (higher) binding-energies.[3] Based on these core level shifts, a structure model for TMA/Si(100) was proposed in which the dimerized TMA molecule adsorbs on Si(100) with its Al-Al axis perpendicular to the Si surface.

In this paper, we have investigated the electronic structure of dimerized TMA adsorbed on Si(100) using MO calculations based on a cluster description of the structure model for TMA/Si(100) described above. An essential feature in the adsorbed-TMA/Si interaction has been found to be the overlap between the Si surface dangling-bond and the electron-deficient bond orbitals in $Al_2(CH_3)_6$ resulting in charge transfer from the Si surface atom to the TMA molecule.

Mat. Res. Soc. Symp. Proc. Vol. 131. ©1989 Materials Research Society

EXPERIMENTAL RESULTS

The experiments were carried out in an ultra-high vacuum (< 10^{-10} Torr) system and the experimental details have been described previously.[3] UPS and XPS spectra were taken using a He resonance lamp and Mg K_α X-ray source, respectively. The substrates used in the experiments were optically flat 7-13 Ω-cm p-type Si(100) wafers and the surfaces were cleaned by in-situ sputter etching with 5keV Ne^+ ions without a subsequent anneal. The surfaces were free from C and O contamination, but amorphized.

Figure 1 shows typical UPS and XPS spectra from thick (≈10 nm) condensed TMA layers, single TMA adlayers absorbed on Si(100), and from the clean amorphized Si(100) surface.[3] The valence electron spectra obtained using He II 40.8eV photons are shown in Fig. 1a. The valence spectrum from the thick TMA layer exhibits three broad peaks labeled α, β, and γ. These peaks were assigned, based upon MO calculations, to correspond primarily to Al 3p + C 2p, C 2p + H 1s, and C 2s + H 1s bonding orbitals, respectively, in dimerized TMA.[4] The spectrum from the single TMA adlayer was similar to that of the thick TMA except for an overall shift of −0.5eV to the low-binding energy side.

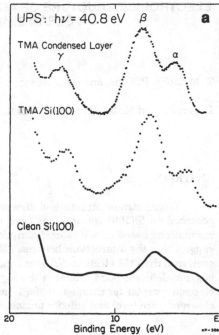

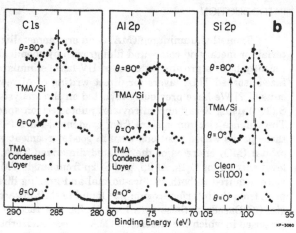

Fig. 1 Typical UPS (a) and XPS (b) spectra from condensed TMA, adsorbed TMA/Si(100), and sputter-cleaned Si(100). The substrate temperatures were -90, -80 and 20° C, respectively.

Figure 1b shows TMA Al 2p and C 1s and substrate Si 2p core spectra. The Al 2p spectrum from the single TMA adlayer in the surface normal direction (θ = 0°) included two peaks, but the low-energy peak was suppressed in the grazing-angle (θ = 80°) spectrum. The high-energy peak position was the same as that of the Al 2p signal from the thick TMA layer. The Si 2p peak

from the TMA-adsorbed-Si substrate shifted to higher-binding
energies compared with that of
the clean Si surface. The shift
was larger at $\theta = 80°$ than at $\theta = 0°$.

CLUSTER MODEL

The experimental results
described above indicate that the
TMA dimer, $Al_2(CH_3)_6$, molecularly adsorbs on Si(100) with the
Al-Al axis perpendicular to the
surface and charge transfers from
the Si surface to the $Al_2(CH_3)_6$
molecule.[3] Thus, for the following analysis we assume a structure for adsorbed TMA on Si(100)
as shown in Figure 2.

MO calculations were
employed to analyze the electronic structure of adsorbed TMA
using the self-consistent-field Xα
scattered-wave (SCF Xα SW)
method [4] based on a cluster description of the structure model shown in
Figure 2. Since the Si(100) surface was
amorphized, we did not account for surface reconstruction. Recent high-resolu-
tion electron energy loss spectroscopy
(HREELS) studies showed no differences
in the vibrational spectra of TMA adsorbed
on the Si(100) 2x1 and Si(111) 7x7 surfaces.[5] This suggests that surface effects on TMA adsorption can be essentially described by the interaction between the TMA molecule and a single
surface Si atom. The cluster used in the
calculations, therefore, included a Si atom
with two H atoms for the Si back-bond
termination.

CALCULATED RESULTS AND
DISCUSSION

Figure 3 shows the MO energy levels of the cluster as a function of h, the
height of the lower terminating-C atoms
above the Si surface (see Fig. 2). The

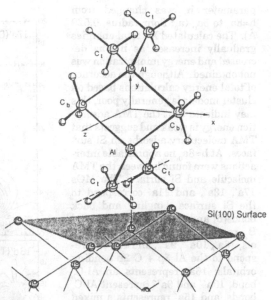

Fig. 2 A structural model for adsorbed TMA on Si(100).
$C_b(C_t)$ represent C atoms in bridging (terminating)
methyl groups. Unlabeled atoms in the TMA are H.

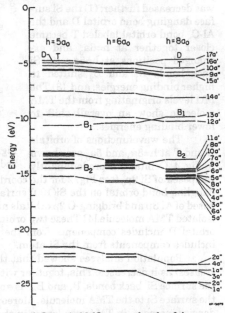

Fig. 3 Occupied MO energy levels of the
$Al_2(CH_3)_6SiH_2$ cluster at h=5, 6, and 8 a_0.

parameter h, was changed from h=$8a_0$ to $5a_0$ (a_0: Bohr radius, 0.529 Å). The calculated Xα total energies gradually increased as h was decreased and energy minimization was not obtained. Although the accuracy of total energy calculations based on cluster models is generally poor, this may indicate that the TMA adsorption energy is weak and suggests that TMA moleculary adsorbs on Si surfaces. At h=$8a_0$ no appreciable interactions were found between the TMA molecule and Si surface. The MOs 17a′, 13a′, and 11a′ correspond to the Si surface dangling and back bonds[6] and are labeled D, B_1, and B_2, respectively. The upper four levels, 16a′, 10a″, 9a″, and 15a′, correspond to the Al 3p + C 2p bonding orbitals; 16a′ represents an Al-C_b bond, 10a″ and 9a″ represent Al-C_t bonds, and 15a′ represents a mixed Al-(C_b+C_t) bond.[4] The MO levels started to change at h≈$7a_0$ and three prominent features were found as h was decreased further: (1) the Si surface dangling bond orbital D and the Al-C_b bond orbital labeled T became closer together at h=$6a_0$ and exchanged positions at h=$5a_0$; (2) Si back bonds B_1 and B_2 shifted to higher binding energies; and (3) The MO levels originating from the TMA molecule show an overall shift to lower binding energies.

The wavefunctions of orbitals D and T at h=$8a_0$ and $5a_0$ are shown in Fig. 4. Orbital D is predominantly

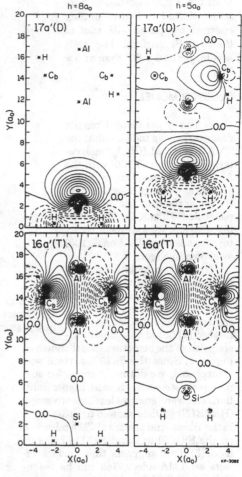

Fig. 4 Contour plots for wavefunctions of MOs 16a' and 17a' at h=5 and $8a_0$.

composed of Si $3p_y$ (see Fig. 2 for the coordinate system) corresponding to the surface dangling-bond orbital on the Si(100) surface.[6] On the other hand, orbital T is composed of Al 3p and bridging-C 2p orbitals and the highest occupied MO (HOMO) of the isolated TMA molecule.[4] These two orbitals are mixed as h decreases and at h = $5a_0$ orbital D includes components from the Al and C atoms in TMA while orbital T includes components from the Si atom.

Population analyses showed that the electron charge on Al(Si) increases (decreases) as h decreases. This, together with the increase of the ionization energies of the surface Si back bonds, B_1 and B_2, is an indication of electron charge transfer from the surface Si to the TMA molecule. Moreover, it was found that the amount of charge density increase in TMA was larger in the Al atom near the surface than in the one further from the surface. These calculated results are consistent with observed XPS and UPS spectra for adsorbed-TMA/ Si(100): (1) an increase of the binding energy for the 2p core level of the surface Si atom, (2) a splitting of the Al 2p XPS peak in which the lower binding-energy part is due to the Al atom near the Si surface, and (3) an overall shift in the UPS valence spectra of the thin TMA adlayer to the low-binding energy side.

Since the C_{2h} symmetry of the isolated TMA molecule is broken upon adsorption and is lowered , in the present model, to C_s symmetry, the four C atoms in TMA become non-equivalent suggesting a splitting in the C 1s XPS peak. However, the resolution of the present experiments was not sufficient to resolve these components and the C 1s signal was a broad asymmetric peak as shown in Fig. 1b.

Charge transfer from the Si surface to the TMA molecule can be attributed to the fact that orbital T is the HOMO and corresponds to the electron-deficient bond of TMA which can act as an electron acceptor. In addition, the Si(100) surface dangling bond D is predominantly composed of Si $3p_y$ and is known to be an active site for chemisorption on the Si(100) surface.[7] Thus, it is reasonable to consider that the geometry of adsorbed-TMA/Si(100) can be described

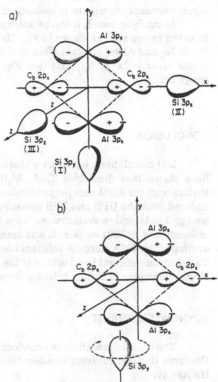

Fig. 5 Schematic diagram for the interaction between the surface Si 3p and TMA electron-deficient bond orbitals.

by the interaction between the orbitals T and D. The electron-deficient orbital T, primarily composed of Al $3p_x$ orbitals, is symmetric with respect to reflection across the x-y plane and antisymmetric with respect to C_2 or a 180° rotation around the z-axis. Among the three basic configurations of orbitals T and D illustrated schematically in Fig. 5a, configuration III gives rise to zero overlap between T and D due to the C_2 symmetry, while configurations I and II result in non-zero overlaps.

Given the above experimental and computational results, configuration I seems intuitively more correct. However, configuration II cannot be ruled out from simple symmetry considerations alone and total energy minimization is necessary for a more detailed theoretical analysis of the adsorption geometry. However, since the number of degrees of freedom is very large, it is extremely difficult to carry out a total energy minimization. Nevertheless, the angle-resolved XPS spectra for the Al 2p core levels shown in Fig. 1b support a model in which the Al-Al axis is perpendicular to the substrate surface. Moreover, in recent HREELS measurements of the vibrational spectra of the adsorbed-TMA on the Si(100)2x1 and Si(111)7x7 surfaces, the modes corresponding to the dipole moments parallel (normal) to the Al-Al axis were (were not) observed. This result is consistent with configuration I, since only surface-normal dipole moments contribute to inelastic dipole scattering.

In configuration I, it can be anticipated that the Al-Al axis will slightly deviate from the on-top site, as shown in Fig. 5b, in order to obtain a larger overlap between the Si $3p_y$ and Al $3p_x$ orbitals. This model suggests that the adsorbed TMA can freely rotate around the Si $3p_y$ axis (see Fig. 5b) and thus the adsorption geometry is irregular.

CONCLUSION

MO calculations based on a cluster description for adsorbed TMA on Si(100) have shown that dimerized TMA, $Al_2(CH_3)_6$, molecularly adsorbs on the Si(100) surface with the Al-Al axis perpendicular to the surface in accordance with the model deduced from the UPS and XPS measurements.[3] Based on the behavior of the MO energy levels and wavefunctions as a function of separation between the TMA molecule and Si(100) surface, it was found that the interactions are described by the overlap of the TMA electron-deficient bond and the Si surface dangling-bond orbitals. The electron-deficient bond orbital is the HOMO of the TMA molecule and acts as an electron acceptor in charge transfer from a surface Si atom to TMA.

ACKNOWLEDGEMENT

The authors gratefully acknowledge the financial support of the Office of Naval Research through contract number N00014-81-K-0568 administered by Dr. Krystl Hathaway.

REFERENCES

1. R.M. Osgood and H.H. Gilgen, Ann. Rev. Mater. Sci.15, 549 (1985).
2. T.E. Orlowski and D.A. Mantell in Laser and *Particle Beam Chemical Processing for Microelectronics*, edited by D.J. Ehrlich, G.S. Higashi, and M.M. Oprysko, Mater. Res. Symp. Proc. vol. **101** (North Holland, 1988), p. 165.
3. W.R. Salaneck, R. Bergman, J.-E. Sundgren, A. Rockett, T. Motooka, and J.E. Greene, Surf. Sci. **198**, 461 (1988).
4. T. Motooka, A. Rockett, P. Fons, J.E. Greene, W.R. Salaneck, R. Bergman, and J.-E. Sundgren, J. Vac. Sci. Technol. A **6**, 3115 (1988).
5. D. Lubben, T. Motooka, J.E. Greene, and J.F. Wendelken, Phys. Rev. B, in press.
6. M. Schmeits, A. Mazur, and J. Pollman, Phys. Rev. B 27, 5012 (1983).
7. F. Bozso and Ph. Avouris, Phys. Rev. Lett. **57**, 1185 (1986).

SURFACE REACTION MECHANISMS IN THE METALLISATION AND ETCHING OF SEMICONDUCTOR MATERIALS.

A WEE, AJ MURRELL, CL FRENCH, RJ PRICE, RB JACKMAN
AND JS FOORD[*]
University of Oxford, Inorganic Chemistry Laboratory, South Parks Road,
Oxford, OX1 3QR. UK. [*] to whom correspondence should be addressed.

ABSTRACT

Surface spectroscopic techniques have been used to investigate aluminium deposition form tri-methyl aluminium (TMA) on Si(100), and the etching of InP by chlorine. Thermal reactions and processes stimulated by UV lamps and ion beams are examined. The results are interpreted in the light of the adsorption states which are formed and the surface transformations of chemical states which are observed to occur.

1. INTRODUCTION

Chemical reactions at the semiconductor-vapour interface provide the basis for a complete processing scheme to fabricate opto- and micro-electronic devices from semiconductor materials. All of the dry processes employed depend crucially on the microscopic nature of the chemical reactions taking place at the interface involved. Thus for example etching utilises the reaction of a halogen containing species with the semiconductor surface to form absorbed halide phases which subsequently decompose to volatile products; in order to maintain a fast etch rate both steps must occur rapidly [1,2]. Similarly the chemical vapour deposition (CVD) of metallic films places tight constraints on the nature and rates of the differing processes which can occur in a CVD reactor. The situation is made more complicated by the desirability in many instances to drive the reactions using non-thermal excitation sources which stimulate specific processes at the interfaces present [3,4].

Dry chemical processing suggests great promise but nevertheless vast problems remain to be solved. Although progress can be made empirically an alternative and perhaps ultimately more profitable approach is to obtain a thorough understanding of the underlying science. In this paper we illustrate how an insight into the fundamental surface chemistry involved may be gained in two processes, namely the etching of chlorine by InP and the deposition of aluminium from TMA on Si(100).

2. EXPERIMENTAL

All experiments were carried out in two stainless-steel UHV systems equipped with an RFA for LEED and Auger analysis, a mass spectrometer for thermal desorption analysis, and an X-ray source with hemispherical electron energy analyser for XPS. A low energy (500ev) argon ion gun allowed ion beam irradiation of samples whilst a sapphire viewport permitted sample irradiation from an external deuterium lamp. TMA was dosed onto the sample via a leak valve and capillary tube directed at the front face of the crystal; chlorine dosing employed a solid state electrolytic cell [5].

Before experiments commenced, the Si(100) sample was cleaned by slow heating to 1500K after which treatment the crystal displayed the familiar 2x1 reconstructed surface and surface contaminants were below Auger detection levels [6]. The InP (100) sample was cleaned by Ar$^+$ ion bomabardment followed by annealing at 650K to produce the (4x2) reconstructed surface.

3. CHLORINE ETCHING OF INP

3.1 Adsorption-desorption processes.

Varying doses of Cl_2 were exposed to the InP surface at 300K, which was then subsequently analysed by Auger spectroscopy. A typical spectrum and uptake curve is presented in figure 1 (a) and the measured variations in In:P Auger signal intensities as a function of chlorine coverage are presented in figure 1 (b). It is apparent from the data that the Cl Auger signal shows a rapid increase initially as chlorine is admitted to the surface but then the rate of uptake slows substantially although the rate does not drop to zero, even for very high exposures. The change in adsorption rate coincides with the point where the detected P: In ratio begins to fall off quite markedly.

This behaviour is relatively common in halogen adsorption systems [8]; it arises from rapid adsorption into a halogen overlayer followed by much slower uptake into a multilayer corrosion phase. The implication from the data is that the corrosion phase is enriched in In in comparison with InP itself.

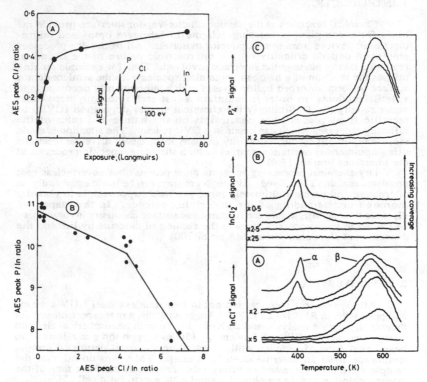

Figure 1 (a) AES uptake curve for Cl on InP. (b) In/P ratio as function of Cl coverage. All data derived from Auger intensity ratios.

Figure 2 Thermal desorption spectra for (a) $InCl$ (b) $InCl_2$ and (c) P_4 plotted as a function of increasing Cl exposure.

Thermal desorption spectra for varying chlorine coverages and desorbing masses are illustrated in figure 2. The only desorbing species detected were $InCl_x$ (x = 1-3) and P_x (x = 1-4); in particular no Cl_2 or PCl_x species were observed. Initially at low Cl coverages a single high temperature peak is displayed in the spectra (labelled β). The β peak grows to "saturation" as the Cl coverage rises to the point at which the break is seen in the Auger spectra and then a low temperature peak, labelled α, emerges. This second peak comes to dominate the spectrum at high Cl coverages. Cracking pattern analysis suggests that the species desorbing from the α state is exclusively $InCl_3$ while the β state products are P_4 and InCl.

Correlation of the thermal desorption data with the Auger results provides a clear picture of the thermal adsorption-desorption processes involved in the interaction of Cl_2 with InP. The clean InP surface reacts rapidly with Cl_2 to produce a halogen overlayer which is very strongly bound to the surface, desorbing only at high temperatures to yield InCl and P_4. This overlayer effectively passivates the interface and the reaction with chlorine slows considerably, resulting in the slow formation of a corrosion phase. This corrosion phase is only weakly bound and is highly enriched in In over P; it decomposes at relatively low temperatures evolving $InCl_3$. As expected repeated adsorption-desorption cycles were found to result in a significant depletion of In from the surface region of the InP sample. The data also clearly indicates why the steady state thermal reaction between chlorine and InP is rather slow; formation of the β state is rapid but desorption of etch products are slow from this state since it is strongly bound. Similarly although desorption of the α state is relatively rapid above 400K, its formation is slow since the surface is passivated by the presence of the β state. In no case is it possible to couple a fast adsorption step with a fast desorption step therefore the overall rate is slow.

3.2 Ion beam induced processes

In order to increase the rate of etching at low temperatures, some additional energy source is required and ion beams present one possible solution [9]; the influence of ion beams on the adsorbed phases formed in the InP/Cl_2 interaction is therefore of considerable interest.

Following exposure of InP to a high chlorine dose (to produce a large α state coverage) the surface was exposed to a 500eV Ar^+ ion beam for varying time intervals after which the surface was analysed by TDS. The results showed that the α state $InCl_3$ desorption peak very rapidly disappeared from the TDS spectra although the chlorine coverage remained almost unchanged (as monitored by AES) during the initial stages of ion bombardment. Simultaneously the InCl and P_4 desorption peaks increased in size. The implication of this is that the ion beam induces a chemical effect whereby phases evolving $InCl_3(g)$ are influenced by the ion beam such that they desorb as InCl at low temperatures. Clearly the effects of ion beams on adsorbed phases is not then simply linked to physical sputtering and indeed the chemical changes observed here appear to occur much more rapidly than the sputtering process. Studies of McNevin et al on the ion-beam induced steady-state etching of InP by Cl_2 observe a rapid increase in the etch rate at about 400K [9]. Interestingly this is the temperature at which α state desorption takes place (see above). A further implication of this work is that the presence of this state plays a significant role in controlling the etching reaction. A possible explanation of this is that the ion beam can disrupt very thin adsorbed phases and stimulates the overall etching reaction by promotion of a reaction between such phases and the underlying substrate (as implied by the data above). If the surface is covered by a thick corrosion phase such processes can no longer; the main promotion effect is lost and the reaction is limited by physical sputtering and the slow chemical reaction of the vapour phase with a passivated surface.

354

4. ALUMINIUM DEPOSITION FROM TMA ON Si

4.1 Thermal Processes

Experiments were performed by exposing the clean Si(100) sample at 77K to TMA and then analysing the phases formed using XPS and TDS techniques. Thick phases of condensed TMA (>XPS sampling depth) can readily be condensed on the sample at 77K and in figure 3 we plot out the changes in the XPS spectrum as these surface phases are heated to successively increasing temperatures. It is apparent that the multilayer phase desorbs at around 210K to leave approximately one physical adlayer behind which gradually decomposes further as the temperature is raised [10,11]. Thermal desorption studies provide a further insight into the processes occurring and representative spectra are illustrated in figure 4 for the desorption of TMA itself. The spectra show a single peak at ~220K demonstrating that the condensed phase desorbs as TMA (the high temperature peak should be ignored since it is associated with the supports). No other TMA peaks are seen in TDS showing that the chemisorbed state evolves as other products. TDS observations indicated that CH_4, C_2H_6, H_2 and a volatile aluminium alkyl species (not TMA) were evolved from the surface at 300-500K and we believe that these species represent the volatile products resulting from the decomposition of the chemisorbed form.

In summary, TMA can reversibly physisorb on Si(100) or it can become irreversibly trapped in a chemisorption well in which it decomposes to yield gas phase products (CH_4, C_2H_6, H_2, aluminium alkyls) above 300K and aluminium deposits.

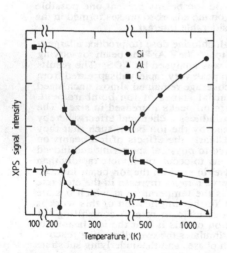

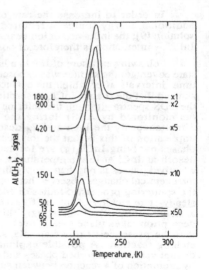

Figure 3 XPS signal intensitites vs surface temperature recorded during the heating of a condensed film of TMA on Si(100). TMA exposure 600L.

Figure 4 Thermal desorption spectra monitoring $Al(CH_3)_2{}^+$ of TMA adsorbed on Si at 77K, for temperatures up to 300K.

If TMA is beamed onto the surface at high temperatures, the rate of decomposition to Al can be followed by monitoring the intensity of the back-scattered TMA by mass spectrometry. Plots of signal intensity versus surface temperature are indicated in figure 5. The data presented above suggests that the decomposition of TMA is thermally activated and this is born out by the results in figure 5 since the detected TMA intensity falls as the surface temperature rises. However, the plots are at first sight puzzling since a secondary minimum appears at 1000K in the decomposition rate.

An explanation of the data is the fact that TMA decomposition is autocatalytic; deposited Al catalyses the further decomposition of absorbing TMA. This is shown is figure 5 where two experimental runs are illustrated for differing initial surface conditions, firstly for clean Si and secondly for Si with two monolayers of Al evaporated on to it. For the latter surface the TMA signal falls off at significantly lower temperature than in the former case demonstrating the autocatalytic effect. In order to explain the shape of the plots in figure 5, deposited Al films on Si were heated *in vacuo* and analysed by XPS. The results showed that Al dissolved from the surface at ~1000K, hence the autocatalytic effect is lost at this temperature, causing a reduction in the rate of TMA decomposition.

The results illustrate nicely how a complex decomposition rate can be explained by consideration of the surface microscopic effects involved. Some of the practical consequences of these effects are discussed elsewhere [12,13].

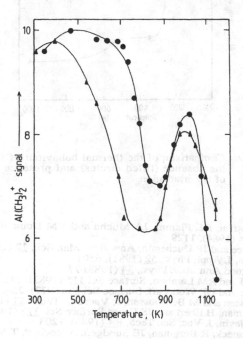

Figure 5   57 amu intensity vs surface temperature recorded during the steady state exposure of TMA towards Si. Circles: Si initially clean, triangles: Si with pre-adsorbed Al

4.2 UV induced reactions

Irradiation of the adsorbed TMA films formed at 77K was carried out using a deuterium lamp in order to investigate photon-induced processes. Following irradiation the surface was heated and XPS spectra were monitored as a function of temperature. Results are presented in figure 6. Interestingly while the condensed layer evaporates very readily from the surface, the irradiated films are found to be thermally stable. No changes in the composition are observed during irradiation but small chemical shifts were seen suggestive of a change in chemical state. Heating of the irradiated film above 350K caused copious amounts of CH_4 and C_2H_6 to be desorbed, leaving Al films on the surface. The influence of UV light in this system would therefore appear to centre on overcoming the thermal activation barrier for TMA dissociation resulting in the formation of Al species and the trapping of organic fragments on the surface; subsequently these organic fragments can desorb thermally yielding a stable Al film.

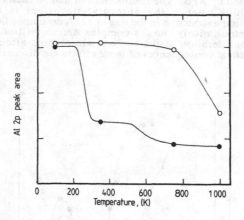

Figure 6   Comparison of the thermal behaviour of films heated in the absence (filled circles) and presence (open circles) of UV irradiation.

5 References

1   DE Ibbotson, DL Flamm, JA Mucha and VM Donnelly, Appl. Phys. Letts., 44 (1984), 1129
2   SD Hersee and JP Duchemin, Ann. Rev. Mat. Sci., 12 (1982), 65
3   SD Allen, J. Appl. Phys., 52 (1981), 6501
4   RM Osgood Ann. Rev. Phys., 34 (1983) 77
5   JS Foord and RM Lambert, Surface Sci 115 (1982), 141
6   SJ White and DP Woodruff, Surface Sci 63 (1977), 254.
7   JM Moissen and M Bensoussan, J. Vac. Sci. Tech., 21 (1982), 315
8   RB Jackman, H Ebert and JS Foord Surface Sci. 176 (1986), 183
9   SC McNevin, J. Vac. Sci. Tech. B4 (1986), 1203
10   WR Salaneck, R Bergman, JE Sundgen, A Rochett, T Motooka and JE Greene, Surface Sci., 198 (1988) 461
11   JF Lubben, T Motooka, JE Greene, JF Wenddken, JE Sundgen and WR Salaneck, Mat. Res. Soc. Symp. Proc., 1988
12   JE Bouree, J Flicstein and YL Nissim, Mat. Res. Soc. Symp. Proc., 1988
13   GS Higashi, GE Blonden and CG Fleming, Mat. Res. Soc. Symp. Proc., 1986.

NUCLEATION BARRIERS IN CHEMICAL VAPOR DEPOSITION OF TRIISOBUTYLALUMINUM ON SILICON

D.A. MANTELL
Xerox Webster ResearchCenter, Webster, NY 14580

ABSTRACT

The nucleation of chemical vapor deposition (CVD) using triisobutyl-aluminum (TIBA) on Si (100) surfaces is observed in situ with x-ray photoelectron spectroscopy (XPS). Oxygen from oxide on the silicon inhibits the rate of nucleation by reacting with adsorbed TIBA and forming a thin layer of oxidized organometallic. This layer blocks active adsorption sites and prevents further deposition. On a surface without oxide, the TIBA molecules decompose liberating aluminum that can migrate and nucleate into islands opening sites for further adsorption and film growth. By removing the oxide (native or thermal) in selected areas of the surface, the barrier to nucleation is removed and aluminum deposition can occur in a predetermined pattern.

INTRODUCTION

When the initial nucleation step of chemical vapor deposition (CVD) film growth is slow, slower than the subsequent film growth, poor film morphologies are usually the result. Differences in the time of the initial nucleation across the surface result in film thickness variations by the rapid film growth. Triisobutylaluminum (TIBA) chemical vapor deposition on silicon surfaces is known to exhibit this behavior[1,2]. In order to avoid poor quality films, the surface can be pretreated in various ways to lower this "nucleation barrier" before CVD[1,2]. In CVD deposition with TIBA on silicon it has been shown that if this pretreatment is done in an area selective manner, patterned CVD growth is possible[2-5]. This has been demonstrated with UV lasers which are used to initiate film growth by direct writing[3,4] or projecting[5] a pattern which grows by CVD (as long as the surface temperature is low enough to prevent nucleation in untreated parts of the surface).

The initial nucleation of CVD film growth for TIBA/silicon is examined in this paper. There is a vital role played by surface oxygen, from silicon's native oxide, in the adsorption of TIBA, the dissociation of TIBA, and the nucleation of CVD growth. As with WF_6/silicon[6], TIBA is found to selectively deposit (at low temperature 300°C) on silicon and not to deposit on silicon oxide. This is possible because film nucleation is suppressed by the surface chemical processes that occur at the organometallic silicon interface. In particular, the silicon (native) oxide is reduced by the aluminum[7], leaving an aluminum oxide (and carbon) layer that suppresses adsorption of more TIBA that is necessary for further aluminum deposition.

EXPERIMENT

The ultrahigh vacuum (UHV) system used for these experiments was pumped by a 500 l/sec ion pump and titanium sublimation pump (base pressure of 1×10^{-10} torr). Since it is impractical to passivate the walls of the UHV chamber with organometallics, a doser introduced the gas so that it hit the surface before the walls of the chamber. The effective pressure at the surface was 1×10^{-7} torr while the chamber pressure remained at 1×10^{-8} torr. Residual gas analysis with a quadrupole mass spectrometer in the UHV chamber revealed that introducing the gas directly from the liquid TIBA source introduced a large amount of hydrocarbon

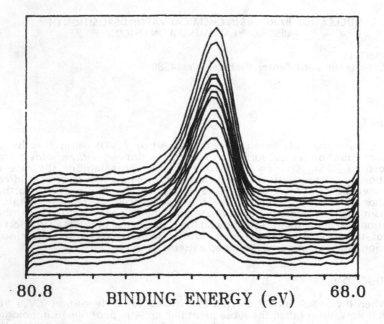

$$80.8 \quad \text{BINDING ENERGY (eV)} \quad 68.0$$

Figure 1:   Aluminum 2p spectra taken during CVD growth.

impurities, particularly isobutylene, a product of the spontaneous decomposition of TIBA. To remove these impurities, an auxiliary chamber was first dosed with TIBA vapor and then briefly pumped on. Most of the hydrocarbons, which are more volatile, were pumped away first. Then the surface was dosed with the vapor desorbing from the walls of the auxiliary chamber, though the actual fluxes of TIBA and isobutylene were not accurately determined.

RESULTS

A Si (100) wafer with its native oxide was transferred into the UHV system and then was prepared by heating to 400°C to remove water. When cooled the sample was sputtered lightly to remove surface carbon. An area of about a square centimeter was further sputtered to create a profile in the oxide thickness, with no native oxide in the center and increasing amounts of oxide away from the center. Thus XPS analysis of the surface with a small area XPS source (300 × 600 μm spot size) could be used during dosing and film growth. The effect of surface oxygen concentration at the surface was examined after film growth by examining various points across this profile.

With the x-rays focused at the center of the sputtered region, an aluminum film was grown on the sample. The aluminum 2p spectra taken during film growth are shown in figure 1. The time between spectra was four minutes. The entire sequence represents roughly 1½ hours of measurement and film growth with the surface maintained between 285° and 305°C[8]. Initially the peak binding energy was 74 eV but decreases as the film grows and takes on more metallic properties. Sputter profiling revealed that this film did not uniformly cover the surface, but instead growth proceeded through island formation.

The integrated aluminum and oxygen intensities after CVD growth versus position are shown in figure 2. Where there was the least amount of oxygen,

XPS Normalized Intensities

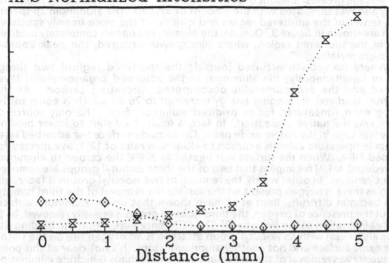

Figure 2:   Aluminum 2p and oxygen 1s corrected integrated intensities versus position across the surface. Zero corresponds to the center of the region in which the native oxide was sputtered away. Away from the center less of the native oxide was sputter removed which is demonstrated by the increasing amount of oxygen.

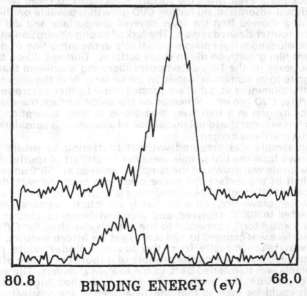

Figure 3:   Aluminum 2p spectra taken after CVD growth. Top curve is from inside the region the native oxide was removed by sputtering. Lower curve is from outside the sputtered region (with native oxide intact).

there was the greatest amount of deposited aluminum. The aluminum 2p binding energy also was correlated with the amount of oxygen. The aluminum 2p spectra at the center of the sputtered region and outside of the more heavily sputtered region are shown in figure 3. Outside, the aluminum appears completely oxidized while in the sputtered region, where film growth occurred, the peak appears much more metallic.

Where no growth occurred (outside the sputtered region) two things occurred simultaneously; the aluminum in the adsorbed organometallic layer oxidized and the organometallic decomposed, liberating carbon. As the aluminum oxidized, its binding energy increased to 76 eV which is equal to the binding energy measured for an oxidized aluminum film. The only source of oxygen was the native oxide itself. In fact, a small .2 eV shift to lower binding energy was seen in the native oxide peak. On an oxide surface the adsorbed layer at room temperature exhibits a carbon to aluminum ratio of 12:1, as expected for adsorbed TIBA. When the surface was heated to 300°C the carbon to aluminum ratio reduced 4:1. This implies that two of the three isobutyl groups are removed and one remains. In other words, the removal of two isobutyl groups is facile, but as long as there is oxygen present on the surface the removal of the third isobutyl group becomes difficult. Bent et al. have shown that on an aluminum surface, without the presence of oxygen, the third isobutyl group is as easily removed, by β-hydride elimination as the first one.[9] The difference here is that the binding site for the aluminum with an isobutyl group attached is an oxygen site on the silicon native oxide surface and not another aluminum atom. It's not clear at this point what suppresses removal of the remaining carbon. Perhaps β-hydride elimination is suppressed so the only channel remaining for carbon removal is decomposition of the remaining isobutyl group. Such a proposal has been shown with theoretical modeling (see G. Higashi - in this volume). When an oxide surface is further heated to 375°C, the carbon to aluminum ratio decreases further to 2:1, suggesting that the remaining isobutyl group decomposes rather than desorbs.

Once an aluminum-oxide (as well as carbon) layer was formed, no further adsorption took place even when the surface was cooled to room temperature. Without significant adsorption, no further CVD growth is possible on this surface. Sputter profililng showed that the layer covered the surface and did not form islands as in the sputter cleaned region. The lack of strong adsorption sites on top of the oxidized aluminum layer plays a crucial role in the inhibition of nucleation and subsequent film growth on silicon oxide surfaces. During CVD on the silicon surface without oxygen, the TIBA decomposes liberating aluminum that must be mobile enough to form aluminum islands on the surface. Both the silicon sites that open up and the aluminum sites that are formed allow further adsorption of TIBA and hence further CVD growth. Whereas on the oxide surface the aluminum is immobilized by oxygen in a thin layer that blocks further adsorption and CVD growth. So in this manner the initial nucleation of aluminum is considerably more facile on a surface without oxygen.

A second sample was prepared without sputtering to insure that the conclusions drawn from the first sample were not an artifact of sputter damaging the surface. An oxide was grown on the sample in an oven at 900°C under flowing oxygen. On half of the surface the entire oxide was removed with diluted HF while on the other half roughly 50Å of oxide was left. Then the sample was rinsed in deionized water, blown dry, and immediately put into the vacuum system. The sample was heated to 400°C, analyzed, and then transferred to another chamber (base pressure 5×10^{-8} torr), connected to the analysis chamber, for CVD growth. At 350°C, the surface was exposed to 10^{-5} torr of gas for fifteen minutes. The TIBA in this case was taken directly from the source cylinder while pumping continuously to keep down the partial pressure of isobutylene. Then after it had cooled, the sample was transferred back to the analysis chamber. The advantage of processing outside the analysis chamber was that higher fluxes of organometallic could be used so that the time needed for passivation of the dosing lines was much reduced and higher growth rates could be achieved. The major disadvantage was that due to the high reactivity of the deposited aluminum and the TIBA, the surface picked up oxygen (or water) that was liberated from

XPS Normalized Intensities

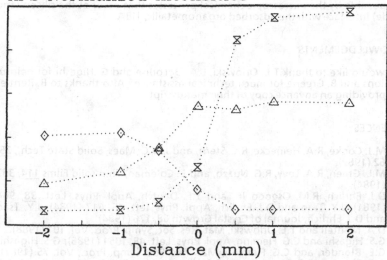

Distance (mm)

Figure 4: XPS corrected integrated intensities versus position across the surface starting with the region that was HF stripped to bare silicon and moving to the region with approximately 50Å of oxide. Silicon 2p-triangles, carbon 1s-squares, aluminum 2p-diamonds.

walls of the chamber during transfer.

The XPS peak intensities versus position across the surface is shown in Figure 4. The greatest amount of film growth was found on bare silicon. Very little nucleation was found on the oxide. Higashi et al. also observed some nucleation on oxide surfaces at these higher surface temperatures.[5] Even though there was no sputtering on the surface, the basic behavior is observed with this sample as with the previously discussed one; the nucleation of film growth is inhibited by the oxide but proceeds more easily on oxide free surfaces. This confirms that sputter damage does not substantially alter the basic mechanism that CVD of TIBA effectively nucleates where there is less oxygen present.

CONCLUSION

Selective deposition of TIBA on silicon has been demonstrated. Oxygen, in this case from the silicon oxide, plays a crucial role in inhibiting film nucleation of CVD with TIBA. At CVD growth temperatures on a surface with some oxygen, the aluminum reacts with oxygen (reducing the silicon oxide in both the numerical and the chemical sense) to form a thin aluminum oxide-carbon layer. Instead of the aluminum atom being liberated to migrate along the surface and nucleate aluminum islands (creating new sites for adsorption of and decomposition of the TIBA), the aluminum atoms are immobilized in a thin layer. This layer blocks the sites at which more TIBA could normally adsorb and remain on the surface for a period of time necessary for efficient decomposition and contribution to film growth. In this manner, patterned CVD is possible because the TIBA growth selectively nucleates on silicon but is inhibited from doing so on silicon oxides. Starting with a thermal oxide, it is possible to remove the oxide in a predetermined pattern using conventional lithographic techniques and to selectively grow aluminum only within that pattern (as is done with tungsten [6]) using TIBA in a chemical vapor deposition process. Thus this macroscopic process is made possible by the microscopic chemical processes that occur at the silicon (and its oxide) interface with the adsorbed organometallic, TIBA.

ACKNOWLEDGEMENTS

I would like to thank T.E. Orlowski, L.A. DeLouise and G. Higashi for helpful discussions and B. Greene for much technical assistance. Also thanks to B. Bent et al. for providing an advance copy of their manuscript.

REFERENCES

1. M.J. Cooke, R.A. Heinecke, R.C. Stern, and J.W.C. Maes, Solid State Tech., 25, 62 (1982).
2. M.L. Green, R.A. Levy, R.G. Nuzzo, and E. Coleman, Thin solid Films 114, 367 (1984).
3. D.J. Ehrlich, R.M. Osgood Jr., and T.F. Deutch, Appl. Phys. Lett. 38, 964 (1981); J.Y. Tsao and D.J. Ehrlich, Appl. Phys. lett. 45, 617 (1984); J.Y. Tsao and D.J. Ehrlich, Journal of Crystal Growth 68, 176 (1984).
4. D.A. Mantell and T.E. Orlowski, Mat. Res. Soc. Symp. Proc., Vol. 101 (1988).
5. G.S. Higashi and C.G. Fleming, Appl. Phys. Lett. 48, 1051 (1986); G.S. Higashi, G.E. Blonder, and C.G. Fleming, Mat. Res. Soc. Symp. Proc., Vol. 75 (1987); C.G. Fleming, G.E. Blonder, and G.S. Higashi, Mat. Res. Soc. Symp. Proc., Vol. 101 (1988), G.E. Blonder, G.S. Higashi, and C.G. Fleming, Appl. Phys. Lett. 50, 766 (1987).
6. N. Miller and R. Herring, 159th Meeting of Electrochem. Soc., 81, 712 (1981).
7. R.S. Bauer, R.Z. Bachrach, and L.J. Brillson, Appl. Phys. lett. 37, 1006 (1980).
8. Since the thermocouple was imbedded in the manipulator holder and was not directly on the silicon sample, this temperature may be an overestimate of the true surface temperature.
9. B.E. Bent, R.G. Nuzzo, and L.H. Dubois, in press; B.E. Bent, R.G. Nuzzo, and L.H. Dubois, Mat. Res. Symp. Proc., Vol. 101, 177 (1988).

EPITAXIAL GROWTH OF Al ON Si BY GAS-TEMPERATURE-CONTROLLED CVD

Tsukasa KOBAYASHI, Atsushi SEKIGUCHI, Naokichi HOSOKAWA, and
Tatsuo ASAMAKI
ANELVA Corporation, Yotsuya 5-8-1, Fuchu, Tokyo 183

ABSTRACT

Epitaxial Al(111) film was deposited on Si(111) by
low-pressure chemical vapor deposition with the use of
tri-isobutyl aluminum (TIBA) at the substrate temperature of
400 ℃ with the deposition rate of 0.9 μm/min. It was
necessary for epitaxy to preheat the TIBA gas just before the
deposition on the substrate. The preheat was made by
gas-temperature-controller provided in the chamber. The film
surface was so smooth that reflectance was higher than 90 %.
The film was recognized as a single crystal over the entire
surface of 4-in. wafer. The film contained about 0.1 % of Si and
20 ppm of O, C, and H. No hillock appeared on the film after 430
℃ annealing for 40 min. The interface of Al and Si was rather
stable so that no alloy penetration occured. The possibility of
epitaxial growth of Al(100) on Si(100) was also shown.

INTRODUCTION

Aluminum and aluminum alloy films are widely used for
metallization in VLSI processing. These films are currently
deposited by the magnetron sputtering technique. However,
this results in insufficient step coverage for a narrow and
deep hole of submicron size with increasing integration
density of IC's. The low pressure chemical vapor deposition
(LPCVD) of aluminum films have been recognized as a potential
technology.[1-7] However, it has been very difficult to obtain
a smooth surfaced film.
On the other hand, Holtzl tried to obtain fine-grained
silicon carbide film by controlled nucleation thermochemical
deposition (CNTD).[8] They preheated the reactant gases to make
appropriately activated or decomposed unstable molecules, which
arrived onto the heated substrate to form a smooth surfaced film
by the complete decomposition. Matsumoto et al. tried another
preheating activation process in CVD with tungsten filament to
obtain a diamond film from CH₄.[9] In Aluminum CVD, Solanki et al.
made an unstable intermediate product by laser from trimethyl
aluminum to obtain a smooth surfaced film.[10] Thus, preheating of
reactant gases has been sometimes effective to improve the
flattness of films in CVD.
In this work, the epitaxial growth of aluminum film is
attempted with the use of TIBA by a newly developed LPCVD
technique referred to as "gas-temperature-controlled CVD"
(GTC-CVD). A gas-temperature-controller was equiped in the
process chamber to preheat the reactant.[11] Though it has
been known that aluminum films were grown epitaxially on
alkalihalide and silicon by physical vapor deposition (PVD)
such as evaporation and ionized cluster beam deposition,[12-16]
epitaxial growth of Al on Si by CVD was found to be possible
for the first time.[17]

EXPERIMENTS

Two different reactors were used. These constructions and the film formation processes were almost same except the reactor size and a load lock chamber. The reactors were cold wall type and had a gas-temperature-controller to preheat the reactant gases. This controller consisted of a Cu cylinder and several Cu plates with many small holes and was usually maintained at 230 ℃. The TIBA was bubbled by Ar, and was introduced into the reactors. These gases were preheated just before their arrival to the Si substrare by the gas-temperature-controller. This preheating had an important role on the epitaxial growth. Substrate temperature was about 400 ℃. It was found that the substrate temperature of 370-400 ℃ was neccesary for epitaxy.

In Experiment A, [17] the TIBA was kept in a cylinder at 50℃. A load lock chamber combined with the reactor was used in order to reduce the oxgen contamination caused by the exchange of wafers. The background pressure before the deposition was lower than 3×10^{-6} Torr. Substrates were 4-in. Si wafers having (111), (100), and (511) orientation and SiO_2 layer.

Surfaces were precleaned with dilute HF. No other substrate treatment was employed. Pressure during the deposition was kept at 2 Torr. The typical growth rate was about 0.9 μm/min at an Ar flow rate of 150 cc/min.

Another reactor was used in Experiment B. [11] The reactor had no load lock chamber and was smaller in size than that of the Experiment A. The TIBA kept at 70 ℃ was bubbled by Ar and introduced to the reactor with Si_2H_6 gas to form Al-Si films.

The flow rate of Ar and Si_2H_6 were typically 40 and 5 sccm, respectively. Substrates were 4-in. Si(100) wafers. No other wafer having different orientation had been tried. The other conditions were same as the Experiment A.

RESULTS AND DISCUSSION

At first, results of the Experiment A are described. The 1 μm-thick Al film grown on Si(111) was characterized as follows. In Fig. 1, 30 keV RHEED pattern was shown. The incident angle of the electron beam was about 2° and the beam diameter was about 2 mm. Large surface area was subjected by the electron beam. Streaks were clearly observed even when the film was rotated about the axis normal to the film surface. This results showed that the film had a flat surface and at least its surface was a single crystal over the entire wafer. The X-ray diffraction patterns of the film was measured with CuKα radiation. Besides the reflection from the Si substrate, only (111) and (222) peaks were observed. The rocking curve of Al(111) diffraction was measured. The half width was smaller than 0.3°. This value was almost the same as that of Si wafers used. Transmission electron diffraction of the film was made as shown in Fig. 2. The diffraction patterns showed a sixfold symmetry. The pattern was not changed when the position of the electron beam was moved within a sample. From these results, it was concluded that this film was an epitaxial single crystal. The resistivity of the film was about 2.7 $\mu\Omega$cm, the same as that of the bulk value.

Fig. 1.   RHEED pattern of
Al(111) film grown on Si(111)
in the Experiment A.   Incident
electron beam was parallel to
Si[110].

Fig. 2.   Transmission
electron diffraction pattern
of Al(111) film grown on
Si(111) in the Experiment A.

Yamada et al. showed that the epitaxial Al(111) films had
been grown on Si(111) by ionized cluster beam (ICB)
deposition.[15,16]   In ICB deposition, the substrate was heated
up to 1000 ℃ in an ultrahigh vacuum before the deposition to get
a clean Si surface.   On the other hand, in our experiment, the
substrate was precleaned only with dilute HF.   The naturally
oxidized layer with a thickness of several tens of angstroms
might remain on the Si substrate. Nevertheless, the epitaxial
growth was possible.   A similar situation was reported in the
evaporation of Al film.[14]   Probably, the oxidized layer on Si
was deoxidized by deposited Al because of the high substrate
temperature of 400 ℃

The film of about 1 μm thickness had a mirror surface.
Reflectance was measured in the range from 200 to 900 nm, as shown
in Fig.3.   It was higher than 90 % in the range from 310 to 600
nm, which was comparable to a high-quality mirror of
evaporated Al much thinner film of about 0.1 μm thickness.   The
surface roughness R_{rms}, measured by WYKO TOPO-3D, was about 2.5 Å.

The impurity contents of O, C, and H in the epitaxial
Al(111) film were measured by SIMS.   The contents were as low
as 20 ppm, much lower compared with the conventional CVD.[4-6]
It will be possible to lower the impurity in GTC-CVD with
further reduction of the background pressure before the
deposition.   It was also found by SIMS measurement that the
film contained about 0.1 % of Si.   This result showed a very small
amount of Si diffused from the wafer into the film.   The interface
between Al and Si was observed after the Al film was etched off by
phosphoric acid.   No caves originated by alloy penetration was
observed on the Si substrate.   Only terraces of exact (111) plane
with ~0.3 μm width were observed.   The Al(111) epitaxial film was
annealed at 430 ℃ for 40 min in N_2 atomosphere.   No hillock
was observed on the film surface after the annealing.

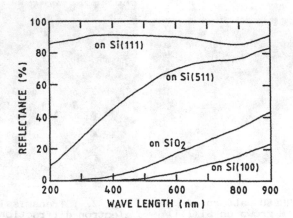

Fig. 3.   Wavelength dependence of reflectance of Al film grown on Si(111), Si(511), Si(100), and SiO₂ in the Experiment A.

In the Experiment B, 1 μm-thick Al film was grown on Si(100). Though the flatness of film surface was slightly degraded, it was much smoother than that of conventional Al-CVD. The reflectance of the film was about 50 % at the wavelength of 500 nm. It was possible to make in-situ Al-Si alloy film by addition of Si₂H₆. The Si content in the film, measured by SIMS, was 0.4 %. The impurity contents of O and C were 0.05 % and 0.02 %, respectively, which were slightly larger than those in the Experiment A because the reactor in the Experiment B had no load lock chamber. Resistivity of the film was about 3.1 μΩcm. X-ray diffraction of the film showed the film was strongly orientated in (100) direction. The half width of rocking curve of Al(200) peak was very narrow and comparable to that of Al(111) epitaxial film in the Experiment A. In Fig. 4, 80 keV RHEED pattern was shown. The spotty pattern assured that the film was also an epitaxial single crystal, i.e. Al(100)//Si(100). This pattern was not changed by shifting the position of the electron beam within a sample.

The preferred or epitaxial orientations of Al film grown on Si by several different methods were summarized in TableⅠ. We can see that the Al(111) epitaxial film could be obtained on Si(111) by all the methods listed. Epitaxial growth is considered to be easy because the nearest neighbour distances of Al and Si triangle lattices in (111) planes are 2.86 Å and 3.84 Å, respectively, whose ratio is almost the integer ratio 3 vs. 4 within an error lower than 1 %. The growth of Al(100) may be possible under some condition.[17] On the other hand, the orientations of Al on Si(100) are various, depending on the fabrication methods. The Al(110) film deposited by ICB was bicrystal. Ohmi et al. showed all the Al film formed on Si by specially arranged bias sputtering machine was epitaxial Al(111) in spite of different orientations of substrates.[18]

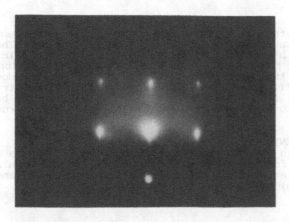

Fig . 4.   RHEED pattern of Al(100) film grown on Si(100) in the
Experiment B.   Incident electron beam was parallel to Si[100].

TABLE I.   Comparision of the preferred orientation of Al
films deposited on Si and SiO₂ by several different methods.
Epitaxial single crystals are underlined.

Methods	on Si(100)	on Si(111)	on Si(511)	on SiO₂
GTC–CVD	(100)[a], (110)*#	(111), (100)*c)	(110)*	poly*
ICB[15,16]	(110)[b]	(111)	---	(111)*
Bias-Sputter.[18]	(111)	(111)	---	(111)*
Evaporation[13,14]	(110)*	(111)*, (100)*#	---	---

*:measured by X-ray diffraction
#:accompanied by small peaks of other orientations
a):Single crystal film was grown only in the Experiment B by now.
     Others were the results of the Experiment A.
b):bicrystal
c):considered as the result of degradation of TIBA in the reactant
     tank after a long period experiment. [7]

In our GTC-CVD, the single phase epitaxial Al film on Si(100)
has been obtained only in the Experiment B but not in the
Experiment A.   It should be clarified how both experimental
conditions differed.   The effect of Si₂H₆ addition on epitaxy has
not been studied yet.   Further investigation is necessary.

SUMMARY

The epitaxial Al(111) film of about 1 μm thickness was grown on Si(111) substrate by GTC-CVD for the first time. The film had a very smooth surface. No hillock appeared on the film after 430 ℃ annealing for 40 min. No so-called alloy penetration was observed in the interface between Al and Si. Furthermore, the epitaxial Al(100) film was also grown on Si(100) substrate. The epitaxial Al film deposited by GTC-CVD is expected to be applicable to migration-free and hillock-free IC interconnects.

ACKNOWLEDGMENTS

We would like to thank T. Hashimoto and F. Kimura at Nikon corp. for measuring the reflectance, Dr. T. Mochizuki at Hoya laser lab. for measuring the surface roughness, and Dr. H. Okabayashi at NEC corp. for helpful discussions.

REFERENCES

1. M. J. Cooke, R. A. Heinecke, R. C. Stern, and J. W. C. Maes, Solid State Technol. **25**, No. 12 (Dec), 62 (1982).
2. D. R. Biswac, C. Ghosh, R. L. Layman, J. Electrochem. Soc. **130**, 234 (1983).
3. M. L. Green, R. A. Levy, R. G. Nuzzo, and E. Coleman, Thin Solid Films **114**, 367 (1984).
4. R. A. Levy, M. L. Green, and P. K. Gallagher, J. Electrochem. Soc. **131**, 2175 (1984).
5. R. A. Levy, P. K. Gallagher, R. Contolini, and F. Schrey, J. Electrochem. Soc. **132**, 457 (1985).
6. T. Kato, I. Ito, H. Ishikawa, and M. Maeda, Extended Abstracts 18th Int. Conf. Solid State Devices & Materials, Tokyo, 1986, (Business Center for Academic Societies Japan) p. 495.
7. T. Amazawa and H. Nakamura, Extended Abstracts 18th Int. Conf. Solid State Devices & Materials, Tokyo, 1986, (Business Center for Academic Societies Japan) p. 755.
8. R. A. Holtzl, Proc. 6th Int. Cof. Chemical Vapor Deposition (The Electrochem. Soc., Princeton, 1977) p. 107.
9. S. Matsumoto, Y. Sato, and M. Kamo, Jpn. J. Appl. Phys. **21**, L183 (1982).
10. R. Solanki, W. H. Ritchie, and G. J. Collins, Appl. Phys. Lett. **43**, 454 (1983).
11. A. Sekiguchi, T. Kobayashi, N. Hosokawa, and T. Asamaki, to be published in Jpn. J. Appl. Phys. **27**, No. 11, (1988).
12. S. Ino, D. Watanabe, and S. Ogawa, J. Phys. Soc. Jpn. **19**, 881 (1964).
13. J. J. Lander and J. Morrison, Surf. Sci. **2**, 553 (1964).
14. F. d'Heurle, L. Berenbaum, and R. Rosenberg, Trans. Metall. Soc. AIME. **242**, 502 (1968).
15. I. Yamada, H. Inokawa, and T. Takagi, J. Appl. Phys. **56**, 2746 (1984).
16. I. Yamada and T. Takagi, IEEE Trans. Electron Devices. **34**, 1018 (1987).
17. T. Kobayashi, A. Sekiguchi, N. Hosokawa, and T. Asamaki, Jpn. J. Appl. Phys. **27**, L1775 (1988).
18. T. Ohmi, H. Kuwabata, T. Shibata, N. Kowata, and K. Sugiyama, Proc. 5th Int. VLSI Multilevel Interconnection Conference, Santa Clara, 1988, p. 446.

EXCIMER LASER PHOTOFRAGMENTATION OF TMA ON ALUMINUM:
IDENTIFICATION OF PHOTOPRODUCT DESORPTION DYNAMICS

T.E. ORLOWSKI AND D.A. MANTELL
Xerox Webster Research Center, 800 Phillips Rd., Webster, NY 14580

ABSTRACT

New mechanistic details regarding aluminum deposition by ArF excimer laser photodecomposition of trimethylaluminum (TMA) adsorbed on aluminum covered SiO_2/Si substrates have been obtained using a time-of-flight quadrupole mass spectrometer. CH_3 radicals and Al-$(CH_3)_n$ (n = 1,2,3) species are efficiently photoejected from the surface with up to 0.22 eV of translational energy. Experiments at various TMA dosing levels reveal differences in desorbed fragment translational energy presumably associated with variations in surface site binding energy. No direct evidence is found for desorption of Al from the surface indicating that Al is more tightly bound than methyl-aluminum fragments. By carefully monitoring changes in fragment translational energy as an Al deposit forms on the clean SiO_2/Si substrate, we examine how the surface influences the onset of Al growth. No evidence of ethane or methane desorption from the sample surface is found implying that radical recombination and hydrogen abstraction are primarily secondary gas phase reactions which are not surface initiated.

INTRODUCTION

Recognition of the potential impact of laser based techniques upon the processing of semiconductor materials for microelectronic devices has been increasing over the past several years. Among the advantages laser processing offers are high spatial resolution for maskless patterning, selective photochemistry through choice of photon energy and low temperature alternatives to conventional thermal processing[1]. A process of great technological importance and one investigated in many laboratories recently is the patterned laser photodecomposition of organometallics to deposit metals on various surfaces (for a review see Ref. 2). Although it is generally accepted that high spatial resolution requires the photodissociation reactions to be confined to surface species[3], many fundamental mechanistic issues remain unresolved including the origin of carbon incorporation in the deposited metal film and the role of surface/adsorbate interactions in the photodecomposition process. In this study, we examine the photodecomposition of trimethylaluminum (TMA) adsorbed upon an aluminum covered SiO_2/Si substrate using a time-of-flight (TOF) mass spectroscopic technique. We identify products created on the surface during the photodecomposition process and obtain fragment desorption dynamics under various processing conditions. Some new aspects of the role surface/adsorbate interactions play in the photodissociation process are discovered.

EXPERIMENTS AND RESULTS

The apparatus used to obtain photofragment time-of flight data is described elsewhere.[4] Sample substrates [Si, p-type, 8-12 ohm-cm, (100)] were oxidized at 1000°C in dry O_2 to create a thin ($\sim$300Å) SiO_2 layer. After cooling, rinsing with deionized water and alcohol, they were mounted on a temperature-controlled stage in the sample chamber and heated to 200°C for 24 hours during system bakeout to remove all traces of water from the sample surface. To improve TMA gas purity, the TMA cylinder is cooled in a dry ice/acetone bath (195K) and pumped on with a liquid N_2 trapped mechanical pump. Under operating conditions, the sample is aligned with its surface normal chosen as the direction of observation. The surface (25°C) is dosed continuously with TMA ($\sim$1 x 10-6 Torr) by means of a stainless steel capillary doser. During dosing, a liquid nitrogen cryopump whose cold surfaces fill the entire open space above the sample effectively pump TMA molecules which do not directly adsorb on the substrate. A 1.5 mm spot on the sample is illuminated at an angle of 45° with pulses (17 nsec) from an ArF excimer laser (193 nm) at repetition rates of 1-30 Hz and per pulse fluences of 5-30 mJ/cm^2. Care is taken to insure that the laser pulse cleanly enters and exits the chamber windows without scattering off interior chamber surfaces. In addition, the laser beam is spatially filtered through two apertures before entering the chamber to insure intensity uniformity. Time-of-flight data is collected by setting the QMS to a desired mass and recording the number of counts per 10 μsec time interval on the multichannel scaler as a function of time following the laser pulse. The QMS is operated at an ionization energy of 70 eV and an emission current of 2 mA. Typically the data from 5000 laser pulses are averaged. The multichannel scaler and laser triggering pulses are carefully synchronized to insure that t = 0 corresponds to the laser pulse arriving at the sample surface.

Time-of-flight spectra at an incident laser pulse intensity of 15 mJ/cm^2 and a laser repetition rate of 5 Hz have been obtained at masses 72, 57, 42, 29, 27, and 15. Since the larger mass fragments can create smaller mass fragments in the QMS ionizer, care must be taken to ensure that the observed fragments at mass < 72 originate from the sample surface and not the QMS ionizer. This can be accomplished by checking for correlations in the transit times of the species.

Fig. 1 (top) displays a TOF transient corresponding to mass 72 [Al(CH$_3$)$_3$]. The data has been fit assuming a Maxwell-Boltzmann distribution in transit times using the expression[5]

$$N(t) = gt^{-4}\exp[-m(l^2/t^2)/2kT] \qquad (1)$$

where t is the flight time, l is the transit distance (l = 34 cm for our apparatus), m is the fragment mass, k is Boltzmann's constant, T is the effective temperature (K) and g is a scaling factor. Eq. 1 implies that fragments leave the surface with an average translational energy $E_{trans} = 2kT$. From the fit and Eq. 1, one obtains T = 388K (close to our calculated surface temperature of ~380 K) and a transit time of 804 μsec for the mass 72 data shown at the top of Fig. 1.

TMA is known to exist as a dimer Al$_2$(CH$_3$)$_6$ at room temperature. Observation of mass 72 indicates that either the laser is photodissociating the dimer and producing the monomer which then desorbs at nearly the surface temperature or the laser is photodesorbing the dimer (with an effective temperature of 776K) which then produces mass 72 in the QMS ionizer. It is hard to imagine that the dimer could leave the surface intact with a translational energy of .13 eV. At the present time, photodissociation of the dimer on the surface seems the more likely explanation for the mass 72 data. Other fragments are formed as well and discussed next.

Fig. 1 (middle) shows a TOF transient (taken under identical conditions) corresponding to mass 57 [Al(CH$_3$)$_2$]. The data is fit as a sum of two peaks with the later peak arriving at the transit time previously observed for mass 72 corresponding to mass 57 fragments created in the QMS ionizer when the mass 72 species arrive. In order to compare transit times for different masses, one must correct for differences in the ion transit time within the QMS. The ion transit time for the quadrupole mass filter used in these experiments is 3.40 $(m)^{\frac{1}{2}}$ μsec where m is the fragment mass in a.m.u. For the data in Fig. 1, and subsequent data, the fitting procedure corrects for this internal mass-dependent delay in the QMS and all transit time data shown corresponds to the time required to traverse the sample to QMS ionizer distance (34 cm). The first peak for the mass 57 data occurs at a transit time of 464 μsec with a Maxwell-Boltzmann temperature of 920 K corresponding to a fragment translational energy of 0.16 eV. These mass 57 fragments are leaving the surface "hot" due to photodissociation of TMA monomer:

$$Al(CH_3)_3 + h\upsilon \rightarrow Al(CH_3)_2 + CH_3 \qquad (2)$$

The mass 57 data in Fig. 1 (middle) could have been fit to three peaks to improve the quality of the fit for times >1200 μsec. An additional peak corresponding to a Maxwell-Boltzmann temperature of 77K would result. Since experimental conditions (slight intensity variations across the illuminated region of the sample and fragment residence times in the QMS ionizer) could also lead to deviations from an idealized Maxwell-Boltzmann distribution, we hesitate to fit any of our data to more than two peaks at the present time.

Methyl (CH₃) fragments are also observed leaving the surface. Fig. 1 (bottom) corresponds to a TOF transient at mass 15 (CH₃). The data is fit to two peaks with the first peak appearing at a transit time of 230 μsec when the second peak is fixed at the transit time observed for mass 57, the most intense TOF fragment observed. This procedure accounts for the CH₃ fragments created in the QMS ionizer when

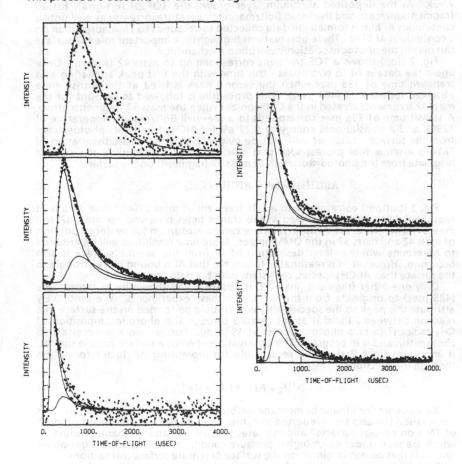

Fig. 1 TOF transients at a laser pulse intensity of 15 mJ/cm² and fits to Maxwell-Boltzmann distributions: Mass 72 (top); mass 57 (middle) and mass 15 (bottom).

Fig. 2 TOF transients at a laser pulse intensity of 15 mJ/cm² and fits to Maxwell-Boltzmann distributions: Mass 42 (top) and Mass 27 (bottom).

the mass 57 fragments arrive. A transit time of 230 μsec for mass 15 corresponds to a Maxwell-Boltzmann temperature of ~990K and a translational energy of 0.17 eV. Thus, CH_3 fragments leaving the surface are also "hot".

The results reported here differ markedly from other published time-of-flight data for laser photodissociation of adsorbed TMA. Higashi found[6] that CH_3 fragments photodesorb from hydroxylated oxide surfaces (Al_2O_3, SiO_2) with a subthermal velocity distribution characterized by a Maxwell-Boltzmann temperature of 150K. No other fragments were observed. TMA is strongly chemisorbed in this case because of OH groups on the surface which strongly bind Al (liberating CH_4) whereas in the experiment discussed here, TMA is physisorbed to an aluminum covered (laser-deposited) substrate. In fact, until an aluminum spot is visible on the sample surface, the TOF transients we observe are extremely weak. As the deposited aluminum layer grows, the TOF peak signals for all fragments increase and the fitted Boltzmann temperature increases as well until a continuous Al film is formed (all data collected corresponds to TMA adsorption on a continuous Al film). These observations highlight the important role the surface can play in the photodissociation/desorption mechanism.

Fig. 2 (top) shows a TOF transient corresponding to mass 42 ($AlCH_3$). Once again the data is fit to two peaks - this time with the first peak appearing at a transient time of 336 μsec when the second peak is fixed at the transit time observed for mass 57 (464 μsec). This procedure is followed to account for the mass 42 fragments created in the QMS ionizer when the mass 57 fragments arrive. A transit time of 336 μsec corresponds to a Maxwell-Boltzmann temperature of 1290K and a translational energy of 0.22 eV for $AlCH_3$ fragments photoejected from the surface. Later we will provide evidence that the photodissociation of TMA is a sequential process such that mass 42 fragments leaving the surface originate from the photodissociation of mass 57 fragments on the surface:

$$Al(CH_3)_2 + h\upsilon \longrightarrow AlCH_3 + CH_3 \tag{3}$$

Fig. 2 (bottom) corresponds to a TOF transient at mass 27(Al). The data is fit reasonably well with peaks fixed at the transit times observed for mass 42 and mass 57. Therefore, the Al observed here can be accounted for by decomposition of mass 42 and mass 57 in the QMS ionizer. More time resolution will be required to determine whether laser desorption of Al from the sample surface is also occurring. However, it is reasonable to assume that Al is bound more strongly to the surface than Al-CH_3 species, consistent with the data shown here.

Only one other fragment (mass 29) is observed and shown by its transit time (438 μsec) to originate from the surface in these experiments. We tentatively attribute this peak to the species AlH_2 which could be formed on the surface in a reaction between Al and H atoms created through the photodecomposition of CH_3 radicals[7] (a two photon process at 193 nm). Further work is necessary to confirm this and is in progress. The observation of AlH_2, a surface-mobile species, is important since it could be responsible for promoting Al cluster formation through the reaction:

$$AlH_2 + Al \longrightarrow Al\text{-}Al + H_2 \uparrow \tag{4}$$

No evidence for ethane or methane desorption from the surface is found. This observation has also been reported in earlier work on the thermal decomposition of TMA on various surfaces[8] and indicates that ethane and methane production which are seen under much higher pressure conditions[9] are secondary gas-phase reactions that do not originate on the surface or require surface interactions.

Experiments as a function of laser repetition rate reveal the role surface coverage plays in the laser desorption dynamics.[4] As the time between laser pulses is increased, the mass 57 intensity approaches a constant value. As stated earlier, the sample surface is dosed continuously during these experiments (chamber pressure ~1 x 10[-6] Torr). Thus, at any laser repetition rate, equilibrium is reached between dosing of the surface and laser photodesorption. If we attribute the saturation value in the mass 57 intensity to a surface coverage of one monolayer which previous studies have found the saturated TMA surface coverage to be,[10] then the effective dose rate (assuming a sticking coefficient of unity) is ~1 monolayer/sec for a laser repetition rate of 1Hz. Increased laser pulse rates deplete the amount of TMA adsorbed on the surface through photodissociation and desorption faster than it can be replenished by the doser. In this manner, the chosen laser repetition rate controls the extent of surface coverage at which the experiment is performed. We also find variations in the Maxwell-Boltzmann temperature corresponding to the TOF transit time of mass 57 fragments as the surface coverage is varied. At low surface coverage (high laser pulse rate) fragments leave the surface at a lower effective temperature indicating the desorption of more tightly bound fragments. At higher surface coverage (low laser pulse rate) most available binding sites are filled and the average fragment binding energy is less as indicated by a higher effective fragment temperature. To minimize the influence of collisions among desorbing species upon their translational energies, we have performed most of our experiments at a laser repetition rate of 5 Hz resulting in an effective surface coverage of ≈20%.

Experiments as a function of laser pulse intensity reveal the sequential nature of the photodecomposition process and that photodecomposition of TMA and photodesorption of products are distinct processes not occurring simultaneously. Shown in Fig. 3 is a plot of integrated intensity (5000 laser pulses) for the TOF transients for mass 57 and 42 versus laser pulse intensity. Both species exhibit a threshold behavior and more importantly, the appearance of mass 42 intensity lags the mass 57 intensity considerably. This situation could arise if mass 42 species were produced primarily from the photodecomposition of mass 57 species on the surface (Eq. 3) and mass 57 species were created by primarily the photodecomposition of mass 72 (Eq. 2). The mass 57 intensity does not decrease as the mass 42 intensity builds up (as one might expect for a sequential process) due to continuous dosing during the experiment. The "delay" then represents the difference in laser exposure required to build up the mass 42 intensity to a detectable level. To confirm this, it would be desirable to obtain data for mass 72

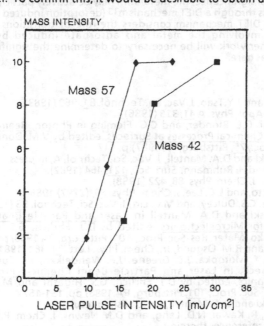

Fig. 3   Integrated intensity (5000 laser shots) for the TOF transients at mass 57 and mass 42 versus laser pulse intensity.

over this laser pulse intensity range. Unfortunately, signal levels are not large enough to perform a similar study on mass 72 directly because the parent ion contribution to the TMA mass spectrum is weak (~5% of total intensity). If we use the second peak in the mass 57 TOF data (which originates from mass 72 leaving the surface) we find that the mass 72 data nearly falls on top of the mass 57 data shown in Fig. 3. This suggests that both mass 72 and mass 57 are created during photodissociation of the TMA dimer:

$$Al_2(CH_3)_6 + h\upsilon \rightarrow Al(CH_3)_3 + Al(CH_3)_2 + CH_3 \qquad (5)$$

Much more evidence will be required to confirm this. However, a scenario consistent with our observations is that the TMA dimer adsorbs with its long axis perpendicular to the surface as recently proposed by Lubben, et al.[11] Upon laser irradiation, the monomer furthest from the surface photodissociates producing $Al(CH_3)_2$ and CH_3 leaving the other monomer intact on the surface. It then desorbs (possibly laser induced thermal desorption) or itself absorbs a photon and photodissociates.

Except for mass 72, all of the TMA photofragments observed here leave the surface with significantly more translational energy than the calculated surface temperature (<380 K) can provide and the TOF transient intensities are nearly linear in laser fluence consistent with a photon induced desorption process. However, fragment translational energy (mass 42, 57) varies linearly with laser fluence.[12] We hesitate to draw conclusions at this point regarding the TMA electronic states involved in the photodissociation mechanism and how the excess electronic energy provided by the 6.4 eV photons is accommodated both internally and through surface dissipation because of the role the surface may be playing. As stated earlier, we are observing photodesorption of TMA fragments from an aluminum covered (laser-deposited) SiO_2/Si substrate. Although it is important to determine what the TMA photoproducts are on this surface because of its technological significance, the metal film may be actively participating in the desorption process through a DIET mechanism[13] (desorption induced by electronic transitions). The DIET mechanism considers the role of transitions to repulsive electronic states involving the metal and adsorbate induced by electronic absorption. Further work will be necessary to determine the significance if any this mechanism has here.

References

1. D.J .Ehrlich and J.Y.Tsao, J. Vac. Sci. Technol. B1, 969 (1983).
2. F.A. Houle, Appl. Phys. A 41, 315 (1986).
3. G.S. Higashi, G.E. Blonder, and C.G. Fleming in Photon, Beam and Plasma Stimulated Chemical Processes at Surfaces, edited by V.M. Donnelly (Mater. Res. Soc. Proc., 75, Pittsburg, PA 1987) p. 117.
4. T.E. Orlowski and D.A. Mantell, J. Vac. Sci. Technol. A, in press.
5. G. Wedler and H. Ruhmann, Surf. Sci., 121, 464 (1982).
6. G.S. Higashi, J. Chem. Phys. 88, 422 (1988).
7. C. Ye, M. Suto, and L.C. Lee, J. Chem. Phys. 89, (2797) 1988.
8. D.W. Squire, C.S. Dulcey, and M.C. Lin, J. Vac. Sci. Technol. B3 (1513) 1985.
9. T.E. Orlowski and D.A. Mantell in Laser and Particle Beam Chemical Processing for Microelectronics, edited by D.J. Ehrlich, G.S. Higashi and M.M. Oprysko (Mater. Res. Soc. Proc., 101, Pittsburg, PA 1988) pp. 165-170.
10. D.J. Ehrlich and R.M. Osgood, Jr., Chem. Phys. Lett. 79, 381 (1981).
11. D. Lubben, T. Motooka, J.E. Greene, J.F. Wendelken, J.E. Sundgren, and W.R. Salaneck, in Laser and Particle Beam Chemical Processing for Microelectronics, edited by D.J. Ehrlich, G.S. Higashi and M.M. Oprysko (Mater. Res. Soc. Proc., 101, Pittsburg, PA 1988) pp. 151-157.
12. T.E. Orlowski and D.A. Mantell, unpublished results
13. Ph. Avouris, R. Kawai, N.D. Lang, and D.M Newns, J. Chem. Phys. 89, 2388 (1988), and references therein.

The Effect of Metal Atoms and Ligand Combinations in Organometallics on Excimer Laser Photoproducts and Yields

Y. Zhang and M. Stuke

Max-Planck-Institut für biophysikalische Chemie
Postfach 2841, D-3400 Göttingen, F.R. Germany

ABSTRACT

A systematic study to find the role of different metal atoms and ligand combinations on the yield of the photoproducts generated upon irradiation of gas phase organometallics (i.e. metal-alkyls) by uv excimer laser radiation was performed using laser ionization mass spectrometry.

INTRODUCTION

In laser induced chemical vapor deposition (laser-CVD), various volatile organometallics have been used as precursors of metals [1,2]. Different organometallics for a given metal have been studied to reduce carbon incorporation that results in poor conductors and to give highly conductive metal films. Recent experiments [3,4] demonstrated that replacing trimethylaluminum (TMA) [5] by dimethylaluminumhydride (DMAH) can improve the conductivity of deposited aluminum films. Triisobutylaluminum (TIBA) [1] was found to give highly conductive aluminum films, but TIBA has a low vapor pressure. Fig. 1(a) shows the resistivity of the deposited conductors made from TMA, DMAH and TIBA, respectively. The explanation of this relation is unknwon and carbon incorporation from organic ligands has been mentioned without giving any details. For this, a systematic study was performed to find the role that different metal atoms and ligand combinations have on the yield of photoproducts generated upon irradiation of gas phase organometallics by uv excimer laser radiation.

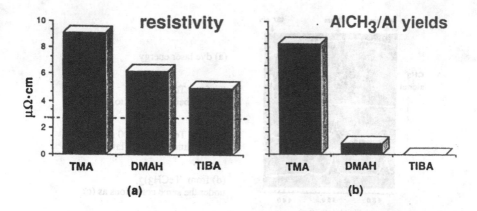

Fig. 1: (a) The resistivity of the aluminum films made from TMA [5], DMAH [3,4] and TIBA [1] with Laser-CVD processes, where the dashed line indicates that of bulk aluminum [4]; (b) The distribution of the AlCH3/Al yields from ArF laser dissociation of TMA, DMAH and TIBA, respectively.

EXPERIMENT

The technique used for analysis is laser photoionization mass spectrometry [6]. The main system studied is the isolated gas phase metal-alkyls [7] of the Al, Ga and Te series with different ligands (CH_3, C_2H_5 and $i-C_4H_9$) and ligand combinations (CH_3, H and Cl) under collision-free conditions (10^{-6}-10^{-5} mbar), since most of them have widely been used in Laser-CVD processes (see Ref.[1-6]). The laser system used for photolysis is a uv excimer laser (ArF at 193 nm, KrF at 248 nm and XeCl at 308 nm) at 0.1- 10 mJ/cm^2.

RESULTS

1. Detection of photoproducts:

In order to understand carbon incorporation in the metal film grown in laser-CVD processes, it is neccessary to find out the photofragments produced from uv laser dissociation of organometallics in detail. Using laser photoionization mass spectrometry, **neutral** photofragments can not only be detected directly by a single shot of the probe laser in the time-of-flight (TOF) mass spectrum [8,9], but also be identified by the multiphoton ionization (MPI) spectrum [10]. Table 1 shows all the detected photofragments under uv excimer laser photolysis, where, in addition to free metal atoms, which can easily be identified from the known spectral data [11], many **metal-containing species** were found by us **for the first time** (printed in bold). Carbon atoms [12] and methyl radicals [13] were also detected as photoproducts with small abundances. No alkenes were detected by this technique owing to further fragmentation by the probe laser.

The detection of molecular species is far more difficult than free atoms since polyatomic molecules have less defined spectral characteristics and are also easily fragmented by the probe laser. Some photoproducts with star (*) in Table 1 were identified with MPI spectra even though they could not directly be detected in the laser TOF spectrum, due to further fragmentation [10]. Table 2 gives some spectral characteristics used by us to identify the photofragments. For some molecular species which are not given in Table 2, however, their spectra are nearly structureless. Fig. 2 shows an example of how to identify the transient species **TeCH3** as uv excimer laser photoproduct through MPI spectra of CH_3^+ (m/e=15) ion signal. The spectrum of the CH_3^+ ions induced by the dye laser alone as shown in trace (b) of Fig. 2 has a resonance at 437 nm, corresponding to the two-photon resonance transition of the precursor molecules $Te(CH_3)_2$ [6].

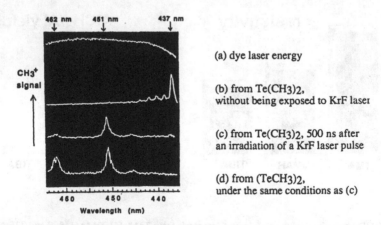

(a) dye laser energy

(b) from $Te(CH_3)_2$, without being exposed to KrF laser

(c) from $Te(CH_3)_2$, 500 ns after an irradiation of a KrF laser pulse

(d) from $(TeCH_3)_2$, under the same conditions as (c)

Fig. 2: The spectra of the CH_3^+ (m/e=15) signal induced by dye laser under different conditions.

Table 1: Detection of uv excimer laser photofragments by laser ionization mass spectroscopy

precursor	uv wavelength (nm)	detected-photofragments
Al(CH₃)₃#	248, 193	C, CH₃, Al, AlH, AlCH₃, Al(CH₃)₂
Al(C₂H₅)₃#	248, 193	Al, AlH, AlCH₃, Al(C₂H₅)₂
Al(i-C₄H₉)₃	248, 193	Al, AlH, Al(i-C₄H₉)₂
Al(i-C₄H₉)₂H#	248, 193	Al, AlH
Al(CH₃)₂H#	193	Al, AlH, AlCH₃
Al(CH₃)₂Cl#	193	Al, AlH, AlCH₃, AlCl
Al(C₂H₅)₂Cl#	193	Al, AlH, AlCl
Ga(CH₃)₃	248, 193	Ga, GaCH₃, Ga(CH₃)₂
Ga(C₂H₅)₃	248, 193	Ga, GaH*, Ga(C₂H₅)₂
Te(CH₃)₂	308, 248	C, CH₃, Te, TeCH₃*
Te(C₂H₅)₂	308, 248	Te
(TeCH₃)₂	308, 248	CH₃, Te, Te₂, TeCH₃*
(SeCH₃)₂	248, 193	Se, Se₂

Table 2: Spectral characteristics for identifying photofragments

neutral fragment	observed resonance of ionic signal	two-photon resonance transition	organometallic precursor
CH₃	333.5 nm (CH₃⁺)	$\tilde{A}\,^2A''_2 - \tilde{X}\,^2A''_2$	TMA
AlH	448.5 nm (AlH⁺)	$C^1\Sigma^+ - X^1\Sigma^+\,(0,0)$	TEA
AlCH₃	442.0 nm (AlCH₃⁺)	$\tilde{C}\,^1A_1 - \tilde{X}\,^1A_1$	TMA
AlCl	359.7 nm (AlCl⁺)	$(?) - X^1\Sigma^+$	DMACl
GaH*	444.5 nm (Ga⁺)	$C^1\Sigma^+ - X^1\Sigma^+$	TEG
TeCH₃*	462.2 nm (CH₃⁺)	$\tilde{B} - \tilde{X}$	Te(CH₃)₂

#: The organometallics may be partially dimers or even multimers, depending on the temperature and its pressure.

*: See the text.

After KrF laser irradiation of Te(CH3)2, this resonance peak disappears, which demonstrates that more than 95% of the precursor molecules can be dissociated by one shot of the 248 nm laser pulse [14], and at the same time, two new resonance peaks come out at 451 nm and 462 nm, respectively (see Fig. 2(c)). In order to confirm the small resonance at 462 nm, (TeCH3)2 molecules were introduced under the same conditions and the corresponding spectrum is given in Fig. 2(d). The resonance at 451 nm can result either from the three-photon transition of methyl radicals CH3 (γ_1(0,0) at 150.29 nm) [15] or from the two-photon transition of tellurium-monomethyl transient species TeCH3 (B-X at 225.3 nm) [16], whereas the resonance at 462 nm can only be due to the two-photon transition of TeCH3 (B-X at 231.1 nm) [16], thanks to a memory of the spectral signature of parent molecules TeCH3 in the CH3+ spectrum. CH3 as one photoproduct can be confirmed by a resonance at 333.5 nm [8]. Now it can be learnt from Fig. 2, that one KrF laser pulse can dissociate more than 95% of the molecules Te(CH3)2, with CH3 and the transient photofragments TeCH3 produced.

The diatomic species Se2 as well as Te2 [14] were found from the corresponding specific precursors (SeCH3)2 and (TeCH3)2 respectively. So far no SeCH3 was detected for lack of its known spectral data.

2. Distribution of yields:

In Table 1, main photoproducts are not only free metal atoms with **large** abundances but also metal-monomethyls and/or metal-hydride with **some** abundances. Some metal-dialkyls were detected with **small** abundances, which may be considered to be transient intermediates [9]. Detailed yield distributions of main photoproducts were studied with Al-alkyls and some main results are summarized as follows [17]:

Fig. 3 shows the yields of aluminum atoms generated upon ArF laser (at 193 nm) dissociation of six Al-alkyls with different ligands (trimethyl-, triethyl- and triisobutyl- aluminum, i.e. TMA, TEA and TIBA) and with different ligand combinations (dimethylaluminumhydride (DMAH), dimethylaluminumchloride (DMACl) and diethylaluminumhydride (DEAH)).

Fig.4 shows the distribution of relative yields of the main aluminum containing photoproducts with respect to Al-alkyls with different ligands, i.e. TMA, TEA and TIBA.

Fig. 5 shows the distribution of relative yields of the main aluminum containing photoproducts with respect to Al-alkyls with different ligand combinations, i.e. TMA, DMAH and DMACl.

Fig. 1(b) is a distribution of the AlCH3/Al yields from TMA, DMAH and TIBA, resulting from Figs. 4 and 5, in order to compare the relation of the resistivity of aluminum films made from the corresponding organometallic precursors in Fig. 1(a).

DISCUSSION

Since bonds between the central metal atom and different ligands have different energies, the uv laser dissociation of various organometallics may give remarkably different yields of the metal atom at a given photolysis wavelength (see Fig. 3), even if the precursors generally exhibit main dissociative continuum absorptions in about the 180-250 nm uv range. As shown in Table 1, KrF laser (at 248 nm) can efficiently dissociate TMA, TEA and TIBA molecules, but **not** DMAH, DMACl and DEACl and a dissociation of Te-alkyls can even be carried out by the XeCl laser at 308 nm.

If a comparison is made between the resistivity of aluminum films deposited and the yield distribution of AlCH3/Al from the coresponding aluminum-alkyls such as TMA, DMAH and TIBA (see Fig. 1 (a) and (b)), it is reasonable to consider the **metal-carbon** fragments, namely, AlCH3, (since metal-dialkyl was detected only with small abundances), as one source of carbon incorporation into the deposited aluminum film, in spite of the fact that the resistivity may have a more complex dependance on the actual deposition conditions. Since, as shown in Table 1, the

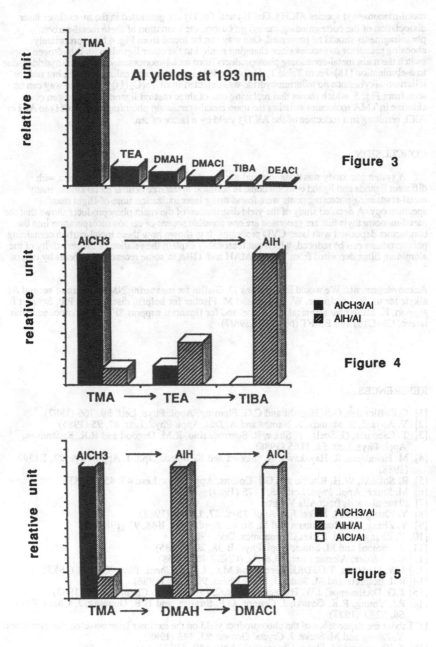

Al yields at 193 nm

TMA TEA DMAH DMACl TIBA DEACl

relative unit

Figure 3

AlCH3/Al AlH/Al

TMA → TEA → TIBA

Figure 4

AlCH3 → AlH → AlCl

AlCH3/Al AlH/Al AlCl/Al

TMA → DMAH → DMACl

Figure 5

metal-monomethyl species $AlCH_3$, $GaCH_3$ and $TeCH_3$ are generated in the uv excimer laser dissociation of the corresponding methyl precusors, the formation of these metal-carbon photofragments should be prevented. One way can be found from Fig. 4 by appropriately choosing precursor molecules since changing methyl to the other ligands with ß-hydrogen does switch the main metal-containing photoproducts from metal-monomethyls to metal-hydrides, due to ß-elimination [18]. From Table 1, this is true for aluminum- and gallium-alkyls, but not for tellurium-alkyls since no telluriumhydride was detected from $Te(C_2H_5)_2$. Another way can be seen from Fig. 5, which shows that replacing one of three methyl ligands by hydrogen or chlorine in TMA molecules switches the main metal-containing photofragment $AlCH_3$ to AlH or $AlCl$, resulting in a reduction of the $AlCH_3$ yield by a factor of ten.

CONCLUSION

A systematic study was done in metal-alkyl systems of the Al, Ga and Te series with different ligands and ligand combinations. In addition to the free central metal atoms, many metal-containing photofragments were found using laser ionization time-of-flight mass spectroscopy. A detailed study of the yield distribution of the main photoproducts shows that the metal-monomethyls that are generated are one possible source of carbon incorporation into the conductors deposited with laser-CVD processes. It is shown how these metal-carbon containing photoproducts can be reduced, which can reasonably explain the relation of the resistivity of the aluminum films deposited from TMA, DMAH and TIBA in some recent experiments by others.

Acknowlegements: We would like to thank D. Gudlin for measuring NMR spectra of several Al-alkyls for us, R. Larciprete, W. Richter and M. Fischer for helpful discussions, F.P. Schäfer for support, K. Müller for technical assistance, and for financial support SFB 93 (Photochemie mit lasern, C2+C15) and BMFT (Nr. 13N 5398/7).

REFERENCES

[1] G.E. Blonder, G.S. Higashi and C.G. Fleming, Appl. Phys. Lett. 50, 766 (1987)
[2] Y. Aoyagi, S. Masuda, S. Namba and A. Doi, Appl. Phys. Lett. 47, 95 (1985)
[3] T. Cacouris, G. Scelsi, P. Shaw, R. Scarmozzino, R.M. Osgood and R.R. Krchnavek, Appl. Phys. Lett. 52, 1865 (1988)
[4] M. Hanabusa, K. Hayakawa, A. Oikawa and K. Maeda, Jpn. J. Appl. Phys. 27, L1392 (1988)
[5] R. Solanki, W.H. Ritchie and G.J. Collins, Appl. Phys. Lett. 43, 454 (1983)
[6] M. Stuke, Appl. Phys. Lett. 45, 1175 (1984)
[7] all metal-alkyls from Alfa Ventron
[8] Y. Zhang and M. Stuke, Jpn. Appl. Phys. 27, L1349 (1988)
[9] Y. Zhang, Th. Beuermann and M. Stuke, Appl. Phys. B48, 97 (1989)
[10] Y. Zhang and M. Stuke, Chemtronics, Dec. 1988
[11] R. Fantoni and M. Stuke, Appl. Phys. B 38, 209 (1985)
[12] C.E. Moore, Atomic Energy Levels, (CNBS 467, 1958)
[13] J.W. Hudgens, T.G. DiGiuseppe, and M.C. Lin, J. Chem. Phys. 79, 571 (1983)
[14] R. Larciprete and M. Stuke, J. Phys. Chem. 90, 4568 (1986)
[15] T.G. DiGiuseppe, J.W. Hudgens and M.C. Lin, J. Phys. Chem. 86, 36 (1982)
[16] P.I. Young, R.K. Gosavi, J. Connor, O.P. Strausz, and H.E. Gunning, J. Chem. Phys. 58, 5280 (1973)
[17] about the dependence of the photoproduct yield on the excimer laser wavelengths please see: Y. Zhang and M. Stuke, J. Crystal Growth 93, 143 (1988)
[18] Y. Zhang and M. Stuke, Chem. Phys. Lett. 149, 310 (1988)

PART VII

II. Novel Precursors
Metals and Dielectrics

PURPOSEFUL CHEMICAL DESIGN OF MOCVD PRECURSORS
FOR SILICON-BASED SYSTEMS

BERNARD J. AYLETT
Department of Chemistry, Queen Mary College, London E1 4NS, U.K.

ABSTRACT

It is shown that volatile molecular compounds with silicon-metal bonds can act as effective MOCVD precursors to metal silicides, which are deposited as thin films under relatively mild conditions. Strategies for the design and synthesis of such "prevenient" precursors are explored, and possible extensions of this approach are considered.

INTRODUCTION

Chemical vapour deposition (CVD) techniques offer an extremely attractive route to microelectronics device structures. Thickness and composition of the deposited films can be readily altered, sharp horizontal interfaces between different phases can be achieved, adequate purity can usually be maintained, and epitaxial deposition is often possible. Most importantly, the experimental arrangements are far more conducive to routine large-scale production than alternative techniques such as molecular beam epitaxy. Variants such as photo-assisted or plasma-enhanced CVD can often provide significant advantages or new features.

This approach has been particularly successful in relation to III/V systems; organometallic precursors have been widely used, in reactions developed from the original work of Manasevit and Simpson [1]:

$$GaMe_3 + AsH_3 \rightarrow GaAs + 3CH_4 \quad (1)$$

to produce many binary, ternary, and quaternary phases such as $Ga_xAl_{1-x}As_yP_{1-y}$. This metallo-organic chemical vapour deposition (MOCVD) approach has led to a range of sophisticated device architectures, with wide electronic and optoelectronic applications.

It would clearly be of great significance if a similar variety of silicon-based device structures were available, especially if they could be produced relatively cheaply by MOCVD techniques. The huge amount of existing expertise in silicon-based systems could then be built on and enhanced.

PRESENT-DAY SILICON DEVICE MATERIALS

Existing CVD procedures in this area are chiefly directed towards deposition of polycrystalline or amorphous silicon and insulating/dielectric layers of silicon oxide, nitride, and oxynitride, e.g.:

$$Si_nH_{2n + 2} \rightarrow nSi + (n + 1)H_2 \quad (2a)$$
$$(n = 1,2....)$$

$$SiHCl_3 + H_2 \rightarrow Si + 3HCl \quad (2b)$$

$$Si(OEt)_4 \rightarrow SiO_2 + \ldots \qquad (3a)$$

$$SiH_4 + N_2O \rightarrow SiO_2 + N_2 + \ldots \qquad (3b)$$

$$3SiH_4 + 4NH_3 \rightarrow Si_3N_4 + 12H_2 \qquad (4)$$

Some work has also been reported [2] on the deposition of metal silicides by reactions such as:

$$WCl_5 + SiCl_4 \xrightarrow[900K]{H_2} WSi_2 \text{ etc.} \qquad (5)$$

$$WF_6 + SiH_4 \xrightarrow[570K]{plasma} WSi_2 \text{ etc.} \qquad (6)$$

It appears that silicide films made in this way usually contain halide as contaminant, although this need not always be the case [3].

Metal silicides for electronic devices are not made commercially in this way; normally a thin metal layer is deposited by sputtering or by evaporation on the silicon substrate, then heated. Mutual interdiffusion occurs, producing one or more silicide phases [4]. Annealing at temperatures up to 1150K may be required. These silicide layers are an essential part of modern VLSI technology [5]: they serve as diffusion barriers, interconnects, gates, components of Schottky devices, etc. In particular, their conductivity can range from near-metallic to that of a wide band-gap semiconductor.

PREVENIENT PRECURSORS FOR CVD OF SILICIDES

There are clear advantages in using a molecular precursor in which the strong silicon-metal bond is already present. If the ligands attached to silicon (L^1) and to metal (L^2) can be sufficiently easily and cleanly removed, then the silicon-metal skeleton that results will condense to form a solid silicide; moreover, the stoichiometry of the silicide should correspond to the silicon: metal ratio initially present in the precursor. For example:

$$
\begin{array}{ccccc}
L^1_x Si\text{-}ML^2_y & \rightarrow & Si\text{-}M & \rightarrow & MSi \qquad (7) \\
\text{molecular} & & \text{intermediate} & & \text{solid} \\
\text{precursor} & & & & \text{silicide}
\end{array}
$$

We refer to this as the prevenient approach, because the silicon and metal have already come together in the synthesis of the precursor. If a substrate is present, the silicide may, under the right conditions, be deposited as a coherent thin film.

The first molecular Si-metal compound, $Me_3SiFe(CO)_2$ (C_5H_5), [6] was made in 1956, but it was not until 1965 that the detailed study of such compounds began [7], with derivatives of the type $R_3SiCo(CO)_4$. Now a wide range is known, in which silicon is bonded to almost all of the transition metals [8]. The most volatile are those compounds in which H is bonded to silicon and CO to metal, the silyl metal carbonyls $H_3SiM(CO)_n$; for example, $H_3SiCo(CO)_4$ boils at 385K/101 kPa.

It soon became clear that pyrolysis or photolysis of these silyl compounds in a static system [9-11] led not only to removal of ligands from silicon and metal but also to disproportionation about silicon (e.g. reacn. (8)), abstraction of oxygen from CO by silicon (e.g. reacn. (9)), and possibly formation of metal carbido species [12]:

$$2H_3SiMn(CO)_5 \rightarrow SiH_4 + H_2Si[Mn(CO)_5]_2 \tag{8}$$

$$2H_3SiV(CO)_6 \rightarrow (H_3Si)_2O + V(CO)_6 + \frac{1}{n}[V(CO)_5C]_n \tag{9}$$

With a flow system, however, such as that shown diagrammatically in Fig.1, it proved possible to form metal silicide thin films on a variety of sub-strates, including doped Si, SiO_2, GaAs and InP [10, 13]. Using a substrate temperature of about 720K, a furnace temperature of about 670K, and a mole ratio of carrier gas (Ar or He) to precursor of about 100:1, total decomposition of the precursor occurred. Only CO and H_2 were found as volatile products; the solid phase was investigated by X-ray powder methods, electron-microprobe analysis and, in some cases, by Auger spectroscopy. Some typical results are:

$$H_3SiCo(CO)_4 \rightarrow CoSi \ (+ H_2 + CO) \tag{10}$$

$$(H_3Si)_2 Fe(CO)_4 \rightarrow \beta\text{-FeSi}_2 \ (+ H_2 + CO) \tag{11}$$

$$H_3SiMn(CO)_5 \rightarrow Mn_5Si_3 + MnSi_x \tag{12}$$
$$(x \quad 1.25)$$

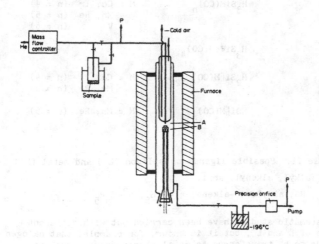

Fig. 1 The CVD apparatus

A, vapour exit nozzle from cooled inlet tube; B, substrate plate on holder with heater; P, pressure-monitoring devices.

It may be noted that while the cobalt and iron systems form a single stoichiometric silicide, with the same silicon: metal ratio as the precursor, the manganese system produces a metal-rich and a silicon-rich phase, although the overall Si:Mn ratio is 1:1. A recent report has described the use of $H_2Si[Mn(CO)_5]_2$ as a precursor [14], but with no carrier gas present; the pressure in the system was also higher than the partial pressure of precursor that we employed. The silicon: metal ratio was about 1:2, but significant amounts of C and O were incorporated in the film; we have earlier suggested [11, 13] that a large excess of carrier gas is beneficial in promoting the clean removal of ligands.

PRECURSOR AVAILABILITY AND DESIGN

Known hydridosilicon metal carbonyl species are shown in Table I. The larger molecules are naturally less volatile, and their decomposition temperatures are uncomfortably close to their sublimation temperatures. Notable absences include compounds containing Ti, Zr, Hf, Cr, Mo, W, Ni, Pd and Pt. If the range of precursors is to be extended to include other metals, what possibilities exist for altering the ligands on silicon or metal? Some potential examples are shown in Table II: many have in fact been incorporated into known compounds.

Table I   Hydridosilicon metal carbonyls

Si:M ratio	Compound		
2:1	$(H_3Si)_2 Fe(CO)_4$		
1:1	$H_3SiM(CO)_n$	M = Co, Ir	(n = 4)
		Mn, Re	(n = 5)
		V	(n = 6)
	$H_3SiFeH(CO)_4$		
1:2	$H_2Si[M(CO)_n]_2$	M = Co	(n = 4)
		Mn, Re	(n = 5)
1:3	$HSi[M(CO)_n]_3$	M = Mn, Re	(n = 5)

Table II.   Possible ligands on silicon (L^1) and metal (L^2)

L^1: halogen, alkyl, alkenyl, aryl.

L^2: PF_3, PR_3, NO, alkyl, η^2-alkene, η^5-C_5H_5, η^5-C_5R_5

Few systematic studies have been carried out with different combinations of L^1 and L^2, but it is known, for example, that halogen on silicon tends to be transferred to metal when compounds such as $F_3SiCo(CO)_4$ are pyrolysed. Also simple alkyl or aryl groups attached to silicon tend not to be lost cleanly on pyrolysis, but lead to incorporation of carbon in the product. It seems clear that L^1 = H is the ligand of choice, although alkyl groups that are both bulky and undergo β-elimination, such as i-Pr or s-Bu, might be suitable. At the metal centre, the choice for L^2 is perhaps wider: PF_3 seems a promising candidate, and PR_3 ligands would also be readily lost, although in the latter case the precursor itself would be of undesirably low volatility. Nitrosyl derivatives are nearly as volatile as the analogous carbonyls, but there is the danger of $Si \leftarrow ON-M$ interactions leading to incorporation of oxygen. Various π-bonded ligands could in principle be readily lost, and η^2-alkenes are obvious candidates. It has been shown that carbon incorporation occurs with precursors such as $H_3SiM(CO)_3(\eta^5-C_5H_5)$ or $H_3SiM(CO)_3(C_5Me_5)$ [15] (M = Cr, Mo, or W), although larger substituents on the ring may increase its readiness to leave [11]. Considerations of inadequate volatility and/or likelihood of carbon incorporation seem likely to rule out other classes of molecular Si-metal derivatives [8].

FUTURE OUTLOOK

Si-metal systems

The most pressing need in connection with silicide deposition is the synthesis of new precursor molecules, tailored to produce appropriate phases. As an illustration, the hydridosilicon metal carbonyls $H_3SiTa(CO)_6$ and $(H_3Si)_2W(CO)_5$, both presently unknown, are expected to provide commercially significant TaSi and WSi_2 deposits. At the same time, studies are needed to identify other "good" leaving groups L^1 and L^2 (eqn.7) that avoid problems of residual contaminants. Firmer correlations between composition, structure, and electrical properties of the deposited films will make the choice of appropriate phases for particular applications more rational, while theoretical and experimental studies of the deposition mechanism will provide guidance for the selection of reaction conditions.

Silicon oxides and nitrides

For oxide films, the prevenient precursor tetraalkoxysilane is already employed (eqn. 3a). Nevertheless, further development of Si-O precursors, perhaps containing also Si-H bonds, that decompose at lower temperatures and carry less risk of carbon contamination would be useful. Similar considerations apply to silicon nitride deposition: pyrolysis of volatile Si-N compounds offers an attractive route if precursors with cleanly-leaving groups on both Si and N can be synthesized. It may be noted that the plasma-enhanced CVD of a mixture of $(Me_2SiNH)_3$ and ammonia has been shown to give a carbon-containing product [16]; pyrolysis of $H_2Si(NMe_2)_2$ does the same [17].

REFERENCES

1. H.M. Manasevit and W.I. Simpson, J. Electrochem. Soc. 116, 1725 (1969).

2. H.M. Cooke, Vacuum 35, 67 (1985) and refs. therein; P.K. Tedrow, V. Ilderem and R. Reif, Appl. Physics Letters 46, 189 (1985).

3. A. Bouteville, A. Royer, and J.-C. Remy, Proceedings 6th EUROCVD Conference Jerusalem, Israel, March 1987, p.264.

4. B.J. Aylett, Brit. Polymer J. 18, 359 (1986).

5. S.P. Murarka, Silicides for VLSI Applications, Academic Press, New York, 1983.

6. T.S. Piper, D. Lemal, and G. Wilkinson, Naturwiss. 43, 129 (1956).

7. B.J. Aylett and J.M. Campbell, Chem. Comm. 1965, A.J. Chalk and
    J.F. Harrod, J. Am. Chem. Soc. 87, 1133 (1965).

8. B.J. Aylett, Adv. Inorg. Chem. Radiochem. 25, 1 (1982).

9. B.J. Aylett and A.R. Burne, unpublished observations.

10. B.J. Aylett and H.M. Colquhoun, J. Chem. Soc., Dalton Trans. 1977, 2058.

11. B.J. Aylett, in Transformation of Organometallics into Common and
    Exotic Materials (ed. R.M. Laine). NATO ASI Series E: Applied Sciences
    - No. 141, Martinus Nijhoff, Dordrecht, 1988.

12. see: B.J. Aylett and M.T. Taghipour, J. Organometal. Chem. 249, 55
    (1983).

13. B.J. Aylett and A.A. Tannahill, Vacuum, 35, 435 (1985).

14. G.T. Stauf, P.A. Dowben, N.M. Boag, L.M. de la Garza, and S.L. Dowben,
    Thin Solid Films 156, 327 (1988).

15. B.J. Aylett and M.J. Hampden-Smith, unpublished results.

16. T.A. Brooks and D.W. Hess, Thin Solid Films 153, 521 (1987).

17. B.J. Aylett and L.K. Peterson, unpublished results.

CHEMICAL VAPOR DEPOSITION OF COBALT SILICIDE

GARY A. WEST AND KARL W. BEESON
Allied-Signal, Inc., P. O. Box 1021R, Morristown, NJ 07960

ABSTRACT

Cobalt silicide films have been deposited by chemical vapor deposition using $Co_2(CO)_8$ or $HCo(CO)_4$ as the Co source and SiH_4 or Si_2H_6 as the Si source. The Co:Si ratio of the films increases with the deposition temperature, and $CoSi_2$ stoichiometry is obtained at 300° C using SiH_4 or at 225° C when Si_2H_6 is the Si precursor. Resistivities of films deposited in the range $CoSi_{2.0}$ to $CoSi_{3.0}$ are typically 200 microohm-cm and drop to 30 - 40 microohm-cm upon annealing at 900° C.

INTRODUCTION

Metal silicides can be used as gate, contact or interconnect metals for Si or GaAs discrete devices and integrated circuits. Silicides such as $CoSi_2$, WSi_2, $TaSi_2$, $MoSi_2$ and $TiSi_2$ are noted for their high temperature thermal stability and low resistivity [1]. In this group, polycrystalline $CoSi_2$ and $TiSi_2$ have the lowest measured resistivity (~15-20 microohm-cm) at room temperature [2].

With the notable exception of $CoSi_2$, there is a considerable body of literature covering chemical vapor deposition (CVD) processes for silicides. Metal halides such as WF_6 [3] $TaCl_5$ [4], $TiCl_4$ [5], MoF_6 [6] and $MoCl_5$ [7] are common gas sources for the metals. Cobalt halides, on the other hand, have relatively high melting points ($CoCl_2$, m.p. 724° C; CoF_2, m.p. ca. 1200° C) and low vapor pressures at moderate temperatures which renders them poorly suited for CVD. Aylett et al. [8] have demonstrated cobalt silicide deposition from the precursor $H_3SiCo(CO)_4$ which contains both Co and Si. However, the resulting films have a Co:Si ratio of 1:1.3 and this ratio cannot be easily varied. Metal depositions of W and Mo have been done by CVD from their respective carbonyls [9]. A common problem using metal carbonyls is incorporation of carbon and oxygen in the final films. A limited amount of CVD work has been published on Co deposition from $Co_2(CO)_8$ [10] which indicates that carbon and oxygen contamination is not a problem for this system. It is possible to indirectly form $CoSi_2$ using the Co CVD process by first depositing Co metal on a Si substrate and then thermally reacting the layered structure to form the silicide. In many cases, however, it may not be possible or desirable to use the substrate as the Si source. One example is the use of cobalt silicide Schottky gates on GaAs [11] where a Si source is not present. In this case, CVD of the silicide, not just the metal, is required.

We show that high quality stoichiometric $CoSi_2$ films can be produced by CVD using $Co_2(CO)_8$ or $HCo(CO)_4$ as the cobalt source and SiH_4 or Si_2H_6 as the precursor for Si. We characterize the deposition rates, film compositions, and film resistivities over a wide range of conditions. Carbon and oxygen contamination are low and the Co:Si film ratio is easily controlled by varying the deposition temperature.

Mat. Res. Soc. Symp. Proc. Vol. 131. ©1989 Materials Research Society

EXPERIMENTAL

The cold-walled CVD apparatus has been described elsewhere [5(b)]. Substrates are 1" square alumina to simplify measurements of resistivity and film composition which can be altered by diffusion from a silicon substrate. Silicon (111) single crystal substrates were used for several depositions to determine the feasibility of using CVD for epitaxial growth of $CoSi_2$ films. Substrate deposition temperatures range from 80 to 450° C. $Co_2(CO)_8$, from Strem Chemicals, Inc., is purified by vacuum sublimation and placed in a stainless steel reservoir cooled to -10 to 10° C. The $HCo(CO)_4$, which is a liquid at room temperature and has a much higher vapor pressure, is synthesized from $Co_2(CO)_8$ [12], and is used in a glass reservoir cooled to -77 to -63° C. The carrier gas for both Co sources is argon maintained at a flow rate of 10 sccm. The $Co_2(CO)_8$ and $HCo(CO)_4$ flow rates, which can be varied by changing the reservoir temperatures at fixed carrier gas flow, are estimated by a variety of methods, including mass spectrometry and weight change of the carbonyl reservoir, to be in the range 0.01 to 0.1 sccm. The SiH_4 or Si_2H_6 flow rates are varied from 0.1 to 2.0 sccm. The total deposition pressure is 0.1 to 1.0 Torr.

The deposited films are analyzed for composition by Auger electron spectroscopy (AES) which is calibrated using several films measured by Rutherford backscattering (RBS). Deposition rates are determined by measuring the film thickness using stylus profilometry and dividing by the deposition time. Film resistivities are calculated from four-point probe measurements and the film thickness. Other techniques used to analyze the films were scanning electron microscopy and x-ray diffraction spectroscopy. A quadrupole mass spectrometer attached to the CVD reactor is used to monitor reactant and product partial pressures prior to and during chemical vapor deposition.

RESULTS AND DISCUSSION

Film Composition

The cobalt silicide CVD film composition depends predominantly on the deposition temperature. Figure 1 shows a plot of the Si:Co film atomic ratio as a function of the deposition temperature for both the silane and disilane precursors using $Co_2(CO)_8$ as the cobalt source. As can be seen from the figure, the quantity of silicon incorporated in the film increases linearly with the deposition temperature. Stoichiometric $CoSi_2$ is achieved at 225° C with Si_2H_6 and at 300° C with SiH_4, whereas CoSi stoichiometry requires only 150° C with Si_2H_6 and 175° C with SiH_4. Below approximately 300° C, the film compositions are insensitive to the reactant gas compositions. For example, variations of the dicobalt octacarbonyl, silane, or disilane partial pressures by a factor of five do not measurably change the film compositions observed in Figure 1. Above 300° C the film compositions are more sensitive to the reactant gas conditions. Using a lower deposition pressure of 0.2 torr rather than 1 torr shown in Figure 1, causes the film composition to level off at a Si:Co ratio in the range 2.0 - 3.0 for both the silane and disilane precursors. Experiments done with $HCo(CO)_4$ as the cobalt

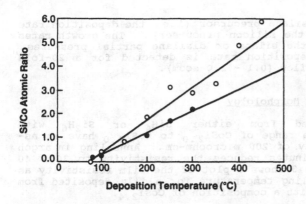

Figure 1

Film composition vs deposition temperature for silane (closed circles) and disilane (open circles).

precursor result in film compositions identical to the $Co_2(CO)_8$ precursor at a given deposition temperature.

Carbon and oxygen contamination are in the range of 15 - 30 atomic percent for films deposited below 100° C. These values decrease with increasing deposition temperature and drop to 3 - 6 atomic percent at 150° C. Above 200° C deposition temperatures both carbon and oxygen are below the Auger detection limit of 0.5 atomic percent.

Growth Rates

Cobalt silicide films begin to deposit at a substrate temperature of 60° C. The measured film deposition rate increases with increasing substrate temperature up to approximately 200° C, above which the deposition rate remains constant. The constancy of the deposition rate above 200° C most likely indicates a deposition rate limited by reactant diffusion to the substrate surface. A plot of the deposition rate vs the $Co_2(CO)_8$ partial pressure is shown in Figure 2. The linearity of this plot is consistent with a deposition rate limited by the carbonyl diffusion to the substrate surface. The data in Figure 2 is for a cobalt silicide film deposited from $Co_2(CO)_8$ and silane at 300° C, but a similar plot is

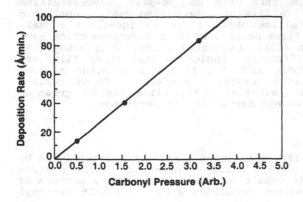

Figure 2

Film deposition rate vs cobalt carbonyl partial pressure.

observed for the disilane precursor (i.e. the deposition rate
does not depend on the silicon precursor). The growth rates
are independent of the silane or disilane partial pressures.
No change in the deposition rate is detected for a 20-fold
variation of silane flow (0.1 - 2.0 sccm).

Film Resistivity and Morphololgy

Films deposited from either SiH_4 or Si_2H_6 with
stoichiometry in the range of $CoSi_{2.0}$ to $CoSi_{3.0}$ have an as-
deposited resistivity of 200 microohm-cm. Annealing in argon
at 900° C for 30 minutes reduces the resistivity to 30 - 40
microohm-cm. Figure 3 shows a plot of the film resistivity as
a function of annealing temperature for a film deposited from
$Co_2(CO)_8$ and Si_2H_6 with a composition of $CoSi_{2.4}$.

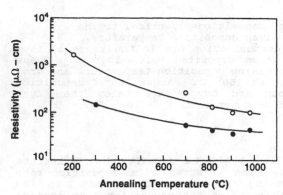

Figure 3

Film resistivity
vs annealing
temperature. The
upper curve is for
a film of $CoSi_{1.0}$
composition and
the lower for
$CoSi_{2.4}$.

Films with silicon content less than $CoSi_{2.0}$ or greater than
$CoSi_{3.0}$ have higher as-deposited and annealed resistivities.
An annealing curve for a film of $CoSi_{1.0}$ composition is
included in Figure 3 for reference.

The crystalline phases of the annealed cobalt silicide
films depend on the initial film composition. A film of
composition $CoSi_{1.5}$, after annealing at 800° C, shows x-ray
diffraction peaks due to both the CoSi (JCPDS file card No. 8-
362) and $CoSi_2$ (JCPDS file card No. 8-344) polycrystalline
phases. In contrast, a film of composition $CoSi_{2.0}$ contained
only the $CoSi_2$ crystalline phase after an identical anneal.

Cobalt silicide films several hundred angstroms thick have
been grown on silicon (111) substrates at 300° C from $Co_2(CO)_8$
and Si_2H_6. X-ray diffraction indicates that these films are
highly crystalline $CoSi_2$ with the $CoSi_2$ (111) plane aligned
parallel to the silicon substrate surface. These results
provide evidence that epitaxial $CoSi_2$ films can be grown on
silicon using CVD at modest deposition temperatures.

Reaction Chemistry

Cobalt silicide film deposition appears to be initiated by
decomposition of the cobalt carbonyl precursor. The minimum
temperature to deposit cobalt silicide films from a mixture of
cobalt carbonyl and silane coincides with the cobalt carbonyl

decomposition temperature. We have observed that cobalt metal films can be deposited from $Co_2(CO)_8$ at a temperature above $60°$ C. The linear dependence of the deposition rate on the cobalt carbonyl partial pressure shown in Figure 2 is also consistent with a reaction rate limited by the carbonyl decomposition at the substrate surface.

Mass spectroscopy of the gas stream near the substrate shows a progression of $Co(CO)_x$ (x = 1-4) mass peaks (using either $Co_2(CO)_8$ or $HCo(CO)_4$ as the cobalt precursor) at temperatures below $60°$ C. As chemical vapor deposition is initiated by raising the substrate temperature, the $Co(CO)_x$ mass peaks decrease with a simultaneous increase in the CO mass peak intensity. The ratio of the $Co(CO)_x$ mass peaks remains constant during deposition indicating that the production of $Co(CO)_x$ (x < 4) gaseous products is not a major process. Also no evidence is found in the mass spectrum during deposition of any adduct product containing both cobalt and silicon.

The above results are consistent with a 2-step deposition mechanism described by the reactions:

$$Co_2(CO)_8 \text{ (surface)} \longrightarrow 2Co(CO)_4 \text{ (surface)} \tag{1}$$

$$Co(CO)_4 + xSiH_4 \longrightarrow CoSi_x + 4CO + 2xH_2 \tag{2}$$

where cobalt carbonyl decomposes on the heated substrate surface and reacts on the surface with silane. Initial production of the $Co(CO)_4$ radical is expected from the relative bond energies of $Co_2(CO)_8$, since the metal-metal bond is weaker (E_{Co-Co} = 21 Kcal/Mole vs E_{Co-CO} = 32 Kcal/Mole) [13]. Consequently, $Co(CO)_4$ is shown as the silane coreactant in Equation 2. However, the participation of other cobalt radicals cannot be ruled out.

The direct reaction of cobalt carbonyl radicals with silane (or disilane), rather than the independent decomposition of silane on the substrate surface followed by reaction, is supported by the independence of the film composition and deposition rate on the silane reactor partial pressure. The modest cobalt silicide deposition temperature is also significantly below the reported decomposition temperature of $550°$ C for silane [14] or $450°$ C for disilane [15].

The higher silicon film content for Si_2H_6 compared to SiH_4 for a given deposition temperature (Figure 1) may be a result of the lower Si-Si bond strength (81 Kcal/Mole) [16] in disilane compared to the Si-H bond strength (92 Kcal/Mole) [17] in silane making the disilane reaction more facile. Apart from this kinetic consideration, the higher silicon film content may simply reflect the larger atomic percentage of silicon in disilane over silane.

SUMMARY

In summary, we have demonstrated a low-temperature CVD process for producing cobalt silicide films of high purity. The CVD reaction is initiated by decomposition of $Co_2(CO)_8$ on the substrate surface and the resulting Si:Co ratios in the films can be controlled over a wide range by the deposition temperature.

Acknowledgments We wish to thank Dr. R. Chin for AES, Dr. F. Reidinger for x-ray analysis and Dr. S. Baumann of Charles Evans and Associates for RBS.

REFERENCES
1. S.P. Murarka, "Silicides for VLSI Applications," Academic Press, New York (1983).
2. a) S.P. Murarka, J. Vac. Sci. Technol. 17, 775 (1980). b) F. Nava, K.N. Tu, E. Mazzega, M. Michelini, and G. Queirolo, J. Appl. Phys. 61, 1085 (1987).
3. D.L. Brors, J.A. Fair, K.A. Monnig and K.C. Saraswat, Solid State Technol. 26, 183 (1983).
4. D.S. Williams, E. Coleman, and J.M. Brown, J. Electrochem. Soc. 133, 2637 (1986).
5. a) G.A. West, A. Gupta, and K.W. Beeson, Appl. Phys. Lett. 47, 476 (1985). b) A. Gupta, G.A. West, and K.W. Beeson, J. Appl. Phys. 58, 3573 (1985). c) G.A. West, K.W. Beeson, and A. Gupta, J. Vac. Sci. Technol. A3, 2278 (1985). d) P.K. Tedrow, V. Ilderem, and R. Reif, Appl. Phys. Lett. 46, 189 (1985). e) R.S. Rosler and G.M. Engle, J. Vac. Sci. Technol. B2, 733 (1984).
6. a) P.J. Gaczi, in Proceedings of the Symposium on Multilevel Metallization, Vol. 87-4 (The Electrochemical Society, Pennington, NJ, 1987), p. 32. b) G.A. West and K.W. Beeson, in Proceedings of the Tenth International Conference on Chemical Vapor Deposition, Vol. 87-8 (The Electrochemical Society, Pennington, NJ, 1987), p. 720.
7. a) S. Inoue, N. Toyokura, T. Nakamura, M. Maeda and M. Takagi, J. Electrochem. Soc. 130, 1603 (1983). b) D.E.R. Kehr, in Proceedings of the Sixth International Conference on Chemical Vapor Deposition (The Electrochemical Society, Pennington, NJ, 1987), p. 511.
8. a) B.J. Aylett and H.M. Colquhoun, J. Chem. Soc., Dalton Trans., 2058 (1977). b) B.J. Aylett and A.A. Tannahill, Vacuum 35, 435 (1985).
9. a) J.J. Lander and L.H. Germer, Metals Technol. 14 (1947). b) L.H. Kaplan and F.M. d'Heurle, J. Electrochem. Soc. 117, 693 (1970). c) G.J. Vogt, J. Vac. Sci. Technol. 20, 1336 (1982).
10. a) V.F. Syrkin, V.N. Prokhorov and L.N. Romanov, Zh. Prikl. Khim. (Lenigrad) 49, 1301 (1976). b) M.E. Gross and K.J. Schnoes, in Proceedings of the Tenth International Conference on Chemical Vapor Deposition, Vol. 87-8 (The Electrochemical Society, Pennington, NJ, 1987), p. 759.
11. S.S. Lau, W.X. Chen, E.D. Marshall, C.S. Pai, W.F. Tseng and T.F. Kuech, Appl. Phys. Lett. 47, 1298 (1985).
12. H. W. Sternberg, I. Wender, R. A. Friedel, and M. Orchin, JACS 75, 2717 (1975)
13. P. J. Gardner, A Cartner, R. G. Cunninghane, and B. H. Robinson, J. Chem. Soc. Dalton Trans., p. 2582 (1975).
14. R. F. C. Farrow, J. Electrochem. Soc.: Solid State Science and Technology 121, 899 (1981).
15. T. L. Chu, S. S. Chu, and S. T. Ang, J. Appl. Phys. 59, 1319 (1986).
16. P. Ho, M. E. Coltrin, J. S. Binkley, and C. F. Melius, J. Phys. Chem. 89, 4647 (1985).
17. T. N. Bell, K. A. Perkins, and P. G. Perkins, J. Chem. Soc. Faraday Trans. 77, 1779 (1981).

LOW-TEMPERATURE ORGANOMETALLIC CHEMICAL VAPOR DEPOSITION OF TRANSITION METALS

Herbert D. Kaesz,[*] R. Stanley Williams,[*] Robert F. Hicks,[†] Yea-Jer Arthur Chen,[*] Ziling Xue,[*] Daqiang Xu,[*] David K. Shuh[*] and Hareesh Thridandam[†]

[*]Department of Chemistry & Biochemistry - UCLA - Los Angeles, CA 90024-1569
[†]Department of Chemical Engineering - UCLA - Los Angeles, CA 90024-1592

ABSTRACT

A variety of transition-metal films have been grown by organometallic chemical vapor deposition (OMCVD) at low temperatures using hydrocarbon or hydrido-carbonyl metal complexes as precursors. The vapors of the metal complexes are transported with argon as the carrier gas, adding H_2 to the stream shortly before contact with a heated substrate.

High-purity platinum films have been grown using $(\eta^5-C_5H_5)PtMe_3$ [1] or $(\eta^5-CH_3C_5H_4)PtMe_3$ [2] at substrate temperatures of 180°C or 120°C, respectively. The incorporation of a methyl substituent on the cyclopentadienyl ligand decreases the melting point of the organoplatinum complex from 106°C [1] to 30°C [2] and increases the vapor pressure substantially. Film deposition also occurs at a lower substrate temperature. Analyses by X-ray diffraction (XRD), Auger electron spectroscopy (AES) and X-ray photoelectron spectroscopy (XPS) indicate that the films are well crystallized and do not contain any observable impurities after sputter cleaning.

The substrate temperatures for the first appearance of other transition-metal films from organometallic precursors are as follows (°C): $Rh(\eta^3-C_3H_5)_3$ (120/Si), $Ir(\eta^3-C_3H_5)_3$ (100/Si), $HRe(CO)_5$ (130/Si) and $Ni(\eta^5-CH_3C_5H_4)_2$ (190/glass, 280/Si). These films are essentially amorphous and contain trace oxygen impurities (< 2%), except for the Re film, which was 10% oxygen and 20% carbon.

INTRODUCTION

Low-temperature deposition processes are desired for very large-scale integrated (VLSI) microelectronics to reduce wafer warpage, generation of defects, and redistribution of dopants(s). Organometallic chemical vapor deposition (OMCVD) often provides routes to desired materials at lower temperature than possible with corresponding inorganic precursors. The difficulties of deposition of most transition metals using CVD are, (a) the non-availability of volatile precursors and, (b) the high temperatures required to decompose the precursors. However, the possibilities of large throughput and good step coverage are so attractive that attempts continue to be made to find ways to deposit transition metals using OMCVD at low temperature.

SELECTION OF SOURCE MATERIALS

Much of the previous work involved transition metal complexes of *acac* (acetylacetonate), carbon monoxide, halogens and/or PF_3 [1-5]. These precursors frequently led to incorporation of heteroatoms into the films, and otherwise unsuitable results. Based on earlier observations in this laboratory [6] and elsewhere [7,8] of the decomposition of metal carbonyl and/or hydrocarbon complexes under an atmosphere of hydrogen, we were prompted to examine such derivatives for OMCVD in the presence of H_2. The complexes tested successfully and those that produced films with the smallest amounts of incorporated contaminants are presented in Table 1.

TABLE 1. Organometallic Precursors for OMCVD

COMPOUND	SOURCE Temp (°C)	DECOMPOSITION Temp (°C)	COMMENTS
$CpPtMe_3$	25	180	bright, crystalline film
$(MeCp)PtMe_3$	25	120	bright, crystalline film
$HRe(CO)_5$	-5	130	bright, amorphous film
$Rh(allyl)_3$	25	120	bright, amorphous film
$Ir (allyl)_3$	25	100	bright, amorphous film
$Ni(MeCp)_2$	35	280	bright, amorphous film

Cp: η^5-cyclopentadienyl ; MeCp: η^5-methylcyclopentadienyl ; allyl: η^3-C_3H_5

EXPERIMENTAL PROCEDURE

Synthesis of organometallic precursors

The organometallic precursors were synthesized according to literature methods with some modifications. All syntheses were carried out under an inert atmosphere with Schlenck techniques. $CpPtMe_3$ [9] and $(MeCp)PtMe_3$ [10] were made by the reaction of Me_3PtI [11] with sodium cyclopentadienyl [Aldrich] and sodium methylcyclopentadienyl in diethyl ether. $M(allyl)_3$ (M=Rh, [12] Ir, [13])were synthesized from MCl_3 (M=Rh, Ir) [Aldrich] and allyl magnesium chloride [Aldrich] in diethyl ether. $HRe(CO)_5$ [14] was obtained through the reaction of $BrRe(CO)_5$ [15] and Zn/H_3PO_4. The synthesis of $Ni(MeCp)_2$ was carried out parallel to that of $Ni(Cp)_2$ [16,17] by using a solution of Na(MeCp) in THF.

Deposition Studies

The product of the decomposition was deposited on glass slides or 13.6 ~ 18 ohm cm, p-type, (100) oriented Si substrates. The glass slides were cleaned with acetone prior to use. The silicon substrates were degreased in trichloroethylene, acetone and iso-propanol, dipped in hydrofluoric acid, rinsed in DI water and dried under argon gas prior to being loaded into the apparatus. All growth was undertaken in a horizontal OMCVD system operated at atmospheric pressure [6]. Argon was used as the carrier gas at 25 ml/min through a resistively heated zone. Hydrogen gas was introduced through a second port at 25 ml/min to assist the decomposition of the organometallic compound.

RESULTS

The structure and composition of transition-metal films were analyzed by X-ray diffraction, scanning electron microscopy (SEM), and XPS and AES depth profiling.

The XRD patterns of the platinum films from $CpPtMe_3$ and $(MeCp)PtMe_3$ precursors are shown in Fig. 1. The diffraction patterns indicated that the films were highly textured. The relative intensities of the X-ray peaks of the films, along with a Pt powder sample, are summarized in Table 2. The fact that some of the peaks from the films were significantly broader than those of the powder sample indicated a small grain size. All the remaining films appeared to be amorphous, since there were no peaks in their XRD patterns.

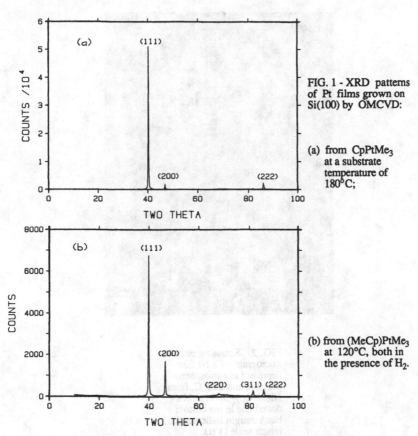

FIG. 1 - XRD patterns of Pt films grown on Si(100) by OMCVD:

(a) from CpPtMe₃ at a substrate temperature of 180°C;

(b) from (MeCp)PtMe₃ at 120°C, both in the presence of H₂.

TABLE 2. Comparison of Relative Intensities

Pt Source	200	220	311	222
Pt. ref. powder	53	31	33	12
CpPtMe₃/pyrex	40	14	11	11
CpPtMe₃/Si(100)	4	–	–	7
(MeCp)PtMe₃/pyrex	15	6	7	6
(MeCp)PtMe₃/Si(100)	60	–	12	7

* Relative to 111 reflection = 100%

The scanning electron micrograph of a typical Ni film from Ni(MeCp)$_2$, presented in Fig.2, shows that the deposit consists of particles approximately 1000Å in the lateral dimension. This morphology is typical of all the deposited amorphous films.

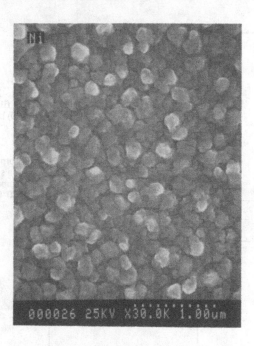

FIG. 2   Scanning electron micrograph of a Ni film deposited at atmospheric pressure and 280°C, from Ni(MeCp)$_2$ in H$_2$. The dotted line in the bottom black margin indicates the length scale (1 μ).

X-ray photoelectron spectroscopy and Auger electron spectroscopy indicate that the surfaces of the transition-metal films were highly contaminated with carbon and oxygen after several days of exposure to ambient conditions. The surfaces were ion-bombarded to remove several hundred Ångstroms, and subsequent analysis showed no detectable contamination in the platinum and less than 2% oxygen contamination in the rhodium, iridium and nickel films.

However, the rhenium film contained 10% oxygen and 20% carbon. An XPS spectrum of a lightly sputtered iridium film is shown in Fig. 3, and the AES spectrum of a nickel film in Fig. 4 shows representative results for the analysis of the films. Preferential sputtering rates were not taken into account in determining the level of contaminatioin in the films, but both the Pt and Ni films had much smaller contamination levels than that of sputter-cleaned foils, which were supposedly 99.999% pure.

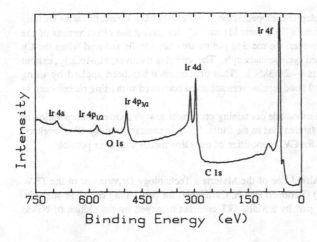

FIG. 3   XPS spectrum of an Ir film deposited from Ir (η^3–C_3H_5) at 100°C in H_2, showing the main Ir peaks and a small peak caused by O contamination.

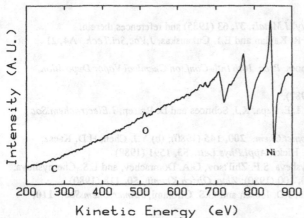

FIG. 4   AES spectrum of a Ni film deposited from Ni(MeCp)$_2$ at 280°C in H_2, showing a small peak caused by O contamination. This deposited film was significantly cleaner than the standard used: a carefully cleaned and ion-bombarded foil of nominally 99.999% pure Ni.

DISCUSSION

The transition-metal complexes containing cyclopentadienyl and allyl ligands are good candidates as OMCVD source materials for thin-film deposition. The metal-ligand bonds are much weaker than the bonds within the ligand, so that the decomposition should involve primarily the breaking of metal-to-ligand bonds, with the organic residue remaining in the vapor phase or desorbing from the surface, to be removed subsequently from the deposition system.

The hydrogen is essential for reducing carbon contamination. In the absence of H_2, nonreflecting black deposits were produced, and AES studies confirmed that there was a much higher concentration of carbon inside the films. Trapped hydrogen in the films deposited at low temperatures may also have been responsible for the amorphous nature of several of the metals. Thus, the hydrogen apparently inserts into the metal-carbon bonds of the precursors to form a stable hydrocarbon species that is easily removed from the system.

A great variety of homoleptic cyclopentadienyl complexes have almost the same melting points (173°C). Thus, the metals "lose their identities," illustrating the effectiveness of the shielding of Cp ligands. However, the melting points are dramatically reduced when the Cp ligand is replaced by substituted cyclopentadienyls. The methyl derivatives, $M(MeCp)_2$, exhibit extremely low melting points (~ 26-38°C). This phenomenon has been applied by using $Ni(MeCp)_2$ as a precursor, and good results were achieved compared with using nickelocene as a precursor [18].

The results from carbon monoxide containing precursors always showed incorporation of large amounts of C and O contaminations in the films. Thus, in contrast to many other workers, we avoid the use of carbonyls for CVD deposition of transition metals whenever possible.

ACKNOWLEDGMENTS

We wish to thank Dr. Alfred Lee of the Materials Technology Department of the TRW Space and Technology Group (Redondo Beach, California) for performing the SEM analysis. This work was supported in part by a SDIO/IST contract managed by the Office of Naval Research.

REFERENCES

1. M.L. Green and R.A. Levy, *J.Metals*, **37**, 63 (1985) and references therein.
2. A.D. Berry, D.J. Brown, R. Kaplan and E.J. Cukauskas, *J.Vac.Sci.Tech.* **A4**, 21 (1986).
3. M.E. Gross and K.J. Schnoes, *Proc. 10th Intl. Conf. on Chemical Vapor Deposition*, 759 (1987).
4. Y. Pauleau, *ibid.*, 685 (1987).
5. M.L. Green, M.E. Gross, L.E. Papa, K.J. Schnoes and D. Brasen, *J.Electrochem.Soc.* **132**, 2677 (1985).
6. (a) H.D. Kaesz, *J.Organomet.Chem.* **200**, 145 (1980); (b) Y.J. Chen, H.D. Kaesz, H. Thridandam, and R.F. Hicks, *Appl.Phys.Lett.* **53**, 1591 (1988).
7. Yu. A. Kaplin, G.V. Belysheva, S.F. Zhil'tsov, G.A. Domrachev, and L.S. Chernyshova, *J.Gen.Chem.* (USSR), **50**, 100 (1980); *Zhur.Obsch.Khimii* **50**, 118 (1980).
8. J.E. Gozum, D.M. Pollina, J.D. Jensen and G.S. Girolami, *J.Amer.Chem.Soc.* **110**, 2688 (1988).
9. S.D. Robinson, B.L. Shaw, *J.Chem.Soc.* **277**, 1529 (1965).
10. H.P. Fritz, K. Schwarzhans, *J.Organomet.Chem.* **5**, 181 (1966).
11. J.C. Baldwin and W.C. Kaska, *Inorg.Chem.* **14**, 2020 (1975).
12. J. Powell and B.L. Shaw, *J.Chem.Soc.* A, 583 (1968).
13. P. Chini and S. Martinengo, *Inorg.Chem.* **6**, 837 (1967).
14. M.A. Urbancic and J.R. Shapley, *Inorg.Synth.* **25**, in press.
15. S.P. Schmidt, W.C. Trogler and F. Basolo, *Inorg.Synth.* **23**, 40 (1985).
16. J.F. Cardes, *Chem.Ber.* **95**, 3084 (1962).
17. F.H. Kohler, *J.Organomet.Chem.* **110**, 235 (1976).
18. G.T. Stauf, D.C. Driscoll, P.A. Dowben, S. Barfuss and M. Grade, *Thin Solid Films*, **153**, 421 (1987).

A COMPARISON BETWEEN ENERGETICS OF DECOMPOSITION AND PHOTO-DEPOSITION OF Pd AND Pt FROM Pd(C$_5$H$_5$)(C$_3$H$_5$) AND Pt(C$_5$H$_5$)(C$_3$H$_5$)

KARL-HEINZ EMRICH**, G.T. STAUF*, W. HIRSCHWALD**, S. BARFUSS**, P.A. DOWBEN*, R.R. BIRGE*, AND N.M. BOAG***
*Laboratory for Solid State Science and Technology, Syracuse University, Syracuse, New York 13244-1130
**Institut für Physikalische Chemie, Freie Universität Berlin, Takustrasse 3, 1000 Berlin 33, Federal Republic of Germany
***Department of Chemistry and Applied Chemistry, Salford University, Salford, England

ABSTRACT

The energetics of decomposition of a variety of organometallic compounds have been determined from photoionization mass spectroscopy and electron impact mass spectroscopy. In particular, Pd(C$_5$H$_5$)(C$_3$H$_5$) and Pt(C$_5$H$_5$)(C$_3$H$_5$) have been studied in this fashion, and the information used to make patterned Pd/Ni/Si heterostructures by laser induced photolysis of the palladium compound and nickelocene in vacuum (MOCVD). Contamination in the thin films was determined by Auger electron spectroscopy, and compared with that found by other researchers for photo-deposition of Pt from the cyclopentadiene allyl [1]. Thermodynamic data is used to explain differing contamination levels in the Pd and the Pt coatings.

INTRODUCTION

The deposition of metal thin films by laser decomposition of metal alkyls and metal carbonyls has attracted considerable attention [2-4]. This method offers an attractive alternative to conventional photolithography processes in several applications, including photomask repair, real-time patterning of specialized circuitry, and in situ vacuum processing compatible with ion beam implantation. The focused nature of the laser beam prevents repair and fabrication processes from damaging nearby satisfactory areas, a problem with photolithography, and is ideal for selected area processing, in which it is used to promote CVD reactions [2].

While there is great interest in studying organometallic chemical vapor deposition (MOCVD or OMVPE) processes, only for a few source compounds have the energetics of decomposition been examined in detail. The η-C$_5$H$_5$ (cyclopentadiene or Cp) and η-C$_3$H$_5$ (allyl) ligands, in particular, are thought to make organometallic complexes which are well suited to use in MOCVD. Obviously, then, it is important to study the decomposition thermodynamics of some compounds of this type, to determine whether they are suitable for formation of pure metal thin films. Such studies have been done for ferrocene, nickelocene, and cobaltocene [5], as well as allylcyclopentadienyl palladium, or Pd(C$_5$H$_5$)(C$_3$H$_5$) [6]. These previous results for the palladium complex are herein compared to our recent studies of the similar compound, Pt(C$_5$H$_5$)(C$_3$H$_5$). We will also discuss contamination results for the Pd coatings we have made from Pd(C$_5$H$_5$)(C$_3$H$_5$), demonstrating that it is a suitable organometallic source for the photolytic deposition of Pd in Pd/Ni/Si heterostructures.

EXPERIMENT

The deposition chamber has been described previously [5]. Briefly, an all glass vacuum system with a base pressure of $\approx 10^{-5}$ Torr was used. Sample vapor from pure organometallic crystals at room temperature was continuously flowed through the system, with sample pressure remaining under 1 mTorr. A quartz optically flat window was used to admit the 3.68 eV (337 nm) radiation from a N$_2$ laser, also described previously [7]. A quartz lens outside the system focused the laser to a 1.26 mm^2 spot on a cleaned, but unannealed and unpolished Si(111) wafer section. The laser operated at a 10 pulse/sec rate, with a peak power of 0.4 MW/pulse, giving an energy density at the sample surface of not more than 0.434 W/cm^2. Deposition times ranged from 3 to 6 hours. A nickel mesh with 20 × 20 μm holes was used to make the pattern on the substrate by selectively blocking the laser light. The mesh was held against the substrate by two clamps which also held down the silicon wafer.

The nickel was deposited from $Ni(C_5H_5)_2$ (nickelocene) purchased from Strem and resublimated before use, while the palladium was deposited from $Pd(C_5H_5)(C_3H_5)$. This compound was synthesized by previously established methods [8]. The platinum compound whose thermodynamics we studied was synthesized similarly.

The electron and photon impact mass spectroscopy experiments were undertaken by using experimental apparatus described previously [6]. Briefly, a molecular beam of sample vapor generated in an alumina Knudsen cell was directed into the ionization region of either a Varian MAT single-sector magnetic field mass spectrometer (electron impact) or a Balzers QMG 511 quadrupole mass spectrometer (photon impact). In the case of photoionization, the synchrotron radiation source was the electron storage ring BESSY (Berliner Elektronenspeicherring Gesellschaft fuer Synchrotronstrahlung mbH) in Berlin, BRD.

In both cases, the energy of the ionizing electrons or photons was varied to obtain an ionization efficiency curve (IEC), i.e. a plot of ion intensity versus impact energy. These could be used to create the relative abundance plots seen later, which show amounts of the parent and major daughter fragments present with increasing energies. Calibration, data reduction, evaluation procedure, and analysis of the fine structure of the IEC's were undertaken using procedures outlined elsewhere [9, 10]. The palladium complex was studied both by electron impact and photoionization, in order to increase confidence in the bond energies, while the platinum complex was examined only by photoionization.

The Auger electron spectroscopy results were taken with a Perkin Elmer CMA following Ar^+ ion bombardment to remove surface contamination. Approximately 100 Å of the thin film was removed prior to acquiring representative AES results. In order to collect the XES spectra we used a KEVEX 5500 spectrometer fitted onto an ISI Super II SEM, with 25 keV incident electron energy.

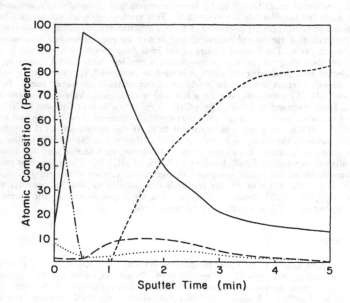

Figure 1. Depth profile of Pd/Ni heterostructure on silicon, made by AES and Ar^+ ion milling.

——————— Pd   — ·· — ·· — C   — — — — Si   — — — — Ni   ········· O

RESULTS

We were able to deposit large arrays of nickel and palladium squares on this silicon substrate. X-ray energy spectroscopy (XES) confirmed the presence of the metals, and showed that no appreciable deposition took place between the squares. XES was also used to estimate coating thicknesses and deposition rates, based on the amount of the silicon substrate signal seen through the thin film overlayer.

We found palladium to deposit at a substantially higher rate than nickel under these conditions, around around 8000 Å/hour, with Ni going down at only about 300 Å/hour (±20% in each case). AES showed the bulk of the Pd coatings to be quite pure, with no more than 5-10% carbon and less than 2% oxygen (below the detectable limit with this instrument). An elemental depth profile in Figure 1 (previous page) shows that oxygen contamination is limited to the surface, while carbon is at a low level all the way through both the nickel and palladium coatings. In spite of the very broad interfaces caused by ion beam mixing during the sputter depth profile, we see that MOCVD can be used to make a MOS (metal-oxide-semiconductor) structure. We have fabricated a silicon, SiO_2, Ni, Pd heterostructure 20×20 μm in size (Figure 1). The lack of contamination will be explained in the case of Pd in the discussion section of this paper.

We also collected ionic abundance information on $Pd(C_5H_5)(C_3H_5)$ and $Pt(C_5H_5)(C_3H_5)$. These data are presented in relative abundance plots in Figures 2 and 3, respectively. The captions in these figures show the major fragments we saw from each compound. Figure 2 was made with the electron impact mass spectrometer data, so fragments differing only by hydrogens could be distinguished. The $Pt(C_5H_5)(C_3H_5)$ plot, on the other hand, was made using photoionization at the synchrotron, as discussed above. Since $Pd(C_5H_5)(C_3H_5)$ was studied both with electron and photon impact, we can be certain that the difference in method does not make a significant difference in relative ionic abundances. The relationship of these plots to coating composition will be explained in the discussion section.

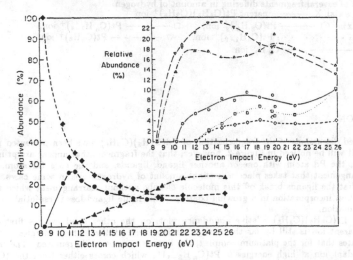

Figure 2. The breakdown diagrams for $Pd(C_5H_5)(C_3H_5)$ derived from electron impact IEC's. Relative intensities are plotted as a function of electron impact energy, with an intensity of 100 % implying that this is the only observed fragment. **Main plot:** ♦ — — — parent $Pd(C_5H_5)(C_3H_5)^+$ ion, ● ———— $C_3 H_5^+$ ion, ▲ —·—·—·— $C_3 H_3^+$ ion. **Corner inset plot:** ◇ — — — — Pd$^+$ ion, ○ ———— $PdC_5 H_5^+$ ion, □ ······· $PdC_3 H_5^+$ ion, △ —·—·— $C_5 H_5^+$ ion, ▽ — — — $C_5 H_6^+$ ion.

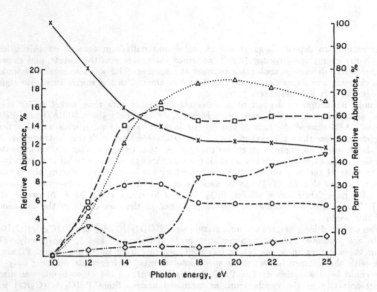

Figure 3. Relative abundance diagram for Pt(C₅H₅)(C₃H₅), based on photoioniza-tion. No other fragments were seen in greater than 1-2% abundance. Due to the lack of mass resolution of the QMS mass spectrometer, some of the ions listed below may consist of several fragments differing in amount of hydrogen.

Right axis: X ——————— parent Pt(C₅H₅)(C₃H₅)⁺ ion

Left axis: ◯ — — — Pt(C₃H₅)₂⁺ ion,   □ ——— Pt(C₆H₂₋₄)⁺ ion,
　　　　　△ · · · · · · · · · · Pt(C₅H₁₋₃)⁺ ion,   ▽ — · — · — Pt(C₅H₆)⁺ ion,
　　　　　◇ — ·· — ·· — C₃H₃⁺ ion.

DISCUSSION

Detailed discussions of the fragmentation of Pd(C₅H₅)(C₃H₅) have been presented previously [6]. It will be noticed in Figure 2, however, that the fragments that appear are primarily the parent, the Pd atom with one or another ligand, ligands, and the bare Pd atom. The only rearrangement that takes place is a slight amount of hydrogen loss in some cases. This indicates that the ligands break off this molecule cleanly, without rearrangement which might lead to carbon incorporation in a growing coating. This facile ligand loss is essential to good coating formation.

In the Pt(C₅H₅)(C₃H₅) relative abundance plot, on the other hand, we see first of all that the parent ion is still by far the most abundant even at the highest energies (≈50%). This indicates that for the platinum complex, the ligands are not easily removed. The second most abundant ion at high energies is Pt(C₆H₂₋₄)⁺, which comes either from the (C₅H₅) ligand picking up a carbon or, more likely, losing two carbons from a (C₅H₅), as well as many hydrogens. The Pt(C₃H₅)₂⁺ ion is seen to be a favorable ionic decomposition product as well. This presumably comes from two carbon loss again, though the Pt(C₃H₅)⁺ may be picking up another allyl. At no photon energy in this range, no matter how high, was the bare Pt⁺ ion seen in more than *half a percent* abundance. Nothing resembling the (C₅H₅)⁺ ion is observed in any great amount, and even the allyl ion is only present at 2% at most. In fact, ions with masses appropriate to (C₅H₅)₂⁺ and (C₅H₅)(C₃H₅)⁺ appear in amounts of 1-2%. These results indicate that ionic decomposition routes for the production of Pt⁺ are far less energetically favorable than is the case for Pd. Using the appearance potentials for fragment ions as well as the ionization potentials of $Pd(C_5H_5)(C_3H_5)$ (7.7 eV) and $Pt(C_5H_5)(C_3H_5)$ (6.9 eV) allows us to estimate some bond strengths for the significant Pd and Pt species. The results are summarized in Table 1.

TABLE I. Selected Ionic Bond Strengths

Ionic Bond	Appearance Potential (eV)	Bond Strength
$Pd(C_5H_5)^+ - C_3H_5$	$AP(PdC_5H_5)^+ = 10.8$	$10.8 - 7.7 = 3.1$ eV
$Pd(C_3H_5)^+ - C_5H_5$	$AP(PdC_3H_5)^+ = 12.0$	$12.0 - 7.7 = 4.3$ eV
$Pd^+ - (C_5H_5)(C_3H_5)$	$AP(Pd)^+ = 13.8 \pm 0.1$	$13.8 - 7.7 = 6.1 \pm 0.1$ eV
$Pt(C_6H_{2-4})^+ - C_2H_6$	$AP(PtC_6H_{2-4})^+ = 7.6$	$7.6 - 6.9 = 0.7$ eV
$Pt^+ - (C_5H_5)(C_3H_5)$	$AP(Pt)^+ = 23 \pm 2$	$23 - 6.9 = 16 \pm 2$ eV

Clearly there is a much stronger bond between the Pt atom and its ligands than exists between the Pd atom and its ligands. This fact does not bode well for coating attempts, since carbon containing ligands which are not cleaved from Pt atoms will become incorporated into the growing metal coating.

The thin film deposition experiments are consistent with this information. When we photodeposited palladium, as described previously [7], we got very low contamination levels. Figure 1 shows that, beyond a surface layer, there is very little carbon or oxygen. Quantitative AES gives well under 10% carbon and 2% oxygen present in the bulk of the palladium film. Considering the vacuum levels present in the system (10^{-5} Torr), some of this might well have come from background gases present.

In contrast, when coatings have been made from the photolysis of the $Pt(C_5H_5)(C_3H_5)$ compound, carbon levels were around 24% *after* annealing in air. Before this annealing process, carbon levels were as high as 90% [11]. The lasers used were both providing radiation in the UV (308 nm *vs.* 337 nm), and the differences can not explain such enormous carbon contamination differences. We believe that the relative abundance information discussed above does provide an explanation.

The nickel thin film in the heterostructure, also deposited by photolysis of nickelocene, was also relatively free of contamination, as shown in Figure 1. We believe this to be due to energetic considerations, based on its thermodynamic decomposition cycle as discussed previously in [5, 7]. Briefly, the bond energies of the (C_5H_5) ligands are low enough that they can be cleaved from the metal atom by the laser we used. The second (C_5H_5) in each case, in particular, does not require nearly as much energy to clip off as the first one. Thus, if the first (C_5H_5) is removed, very little additional energy will suffice to break the bond to the second. If the decomposition process is started at all, it is likely to go to completion.

CONCLUSION

We have previously demonstrated the practicality of making patterned thin metal films from palladium and nickel. We have now discussed how the contamination levels in the palladium coating can be related to thermodynamic decomposition considerations. A comparison between $Pd(C_5H_5)(C_3H_5)$ and $Pt(C_5H_5)(C_3H_5)$ shows the usefulness of the relative abundance information to predict coating contamination outcomes. The palladium coating from photodeposition of this compound had less than 10% carbon and 2% oxygen, while the platinum coatings had 24-90% carbon present, depending on annealing procedures. The differing ionic fragment abundances can be used to explain this, as indicated above. The platinum compound cannot be cleanly removed from its ligands, which will undergo considerable rearrangement before cleavage. The palladium complex, in contrast, loses its ligands cleanly in mass spectrometer experiments.

ACKNOWLEDGMENTS

This work was funded by the U.S. DOE through grant # DE-FG-02-87-ER-45319, the Deutsche Forschungsgemeinschaft/Sonderforschung Bereicht 6 (DFG/SFB 6), the Syracuse University Senate, and International Business Machines, through research agreement # 8074. We would like to thank G.O. Ramsayer for his assistance in obtaining the AES results.

REFERENCES

(1) G.T. Stauf and P.A. Dowben, Thin Solid Films 156, L31-L36 (1988).

(2) P.A. Dowben, J.T. Spencer and G.T. Stauf, Mater. Sci. & Eng. (1989) (in press).

(3) D.J. Ehrlich, R.M. Osgood and T.F. Deutsch, J. Vac. Sci. Technol. 21, 23 (1982).

(4) T.R. Jervis and L.R. Newkirk, J. Mater. Res. 1, 420 (1986).

(5) G.T. Stauf, D.C. Driscoll, P.A. Dowben, S. Barfuss, and M. Grade, Thin Solid Films 153, 421 (1987).

(6) G.T. Stauf, P.A. Dowben, K. Emrich, S. Barfuss, W. Hirschwald, and N.M. Boag, J. Phys. Chem. 92, 1988 (in press).

(7) G.T. Stauf and P.A. Dowben, Thin Solid Films 156, L31-L36 (1988).

(8) Duward F. Shriver, Inorganic Syntheses vol. 19, (John Wiley and Sons, New York, 1979), p. 220.

(9) M. Grade, J. Wienecke, W. Rosinger and W. Hirschwald, Ber. Bunsenges. Phys. Chem. 87, 355 (1983).

(10) W. Rosinger, M. Grade and W. Hirschwald, Ber. Bunsenges. Phys. Chem. 87, 536 (1983).

THE DEPOSITION OF METALLIC AND NON-METALLIC THIN FILMS THROUGH THE USE OF BORON CLUSTERS.

ZHONGJU ZHANG, YOON-GI KIM, P. A. DOWBEN, JAMES T. SPENCER*,
THE CENTER FOR MOLECULAR ELECTRONICS AND THE DEPARTMENTS OF CHEMISTRY
AND PHYSICS, SYRACUSE UNIVERSITY, SYRACUSE, NEW YORK 13244-1200

ABSTRACT

New borane clusters and their corresponding transition and rare earth metal complexes are currently being investigated in our laboratories for their utility as unique source materials for the formation of both metallic and non-metallic thin films. These borane cluster complexes exhibit highly favorable properties for use in OMVPE processes, such as; (1) relatively high volatility, (2) anticipated high stability of the ligand itself to provide clean ligand-metal dissociations, (3) high temperature stabilities of the complexes, (4) readily preparable in significant quantities, and (5) availability of theoretical and spectroscopic probes of structure-reactivity relationships. In this work, we have prepared both non-metallic thin films, including materials such as boron nitride, and metallic thin films (both the transition and rare earth metals) through the use of these unique cluster materials.

Boron nitride has been investigated as a potential hard coating for use as an insulating electrical layer and protective coating. We have investigated plasma enhanced chemical vapor deposition and pyrolytic deposition of boron nitride from readily available and easily handled borane clusters. Auger electron spectroscopy was used to show that the film was high purity boron nitride of uniform composition.

The deposition of transition and rare earth metal thin-film materials of controlled stoichiometry has recently received considerable interest. We have discovered the borane cluster-assisted deposition (CAD) of metallic thin-films involving both transition and rare earth metal materials. Through the use of this unprecedented borane cluster chemical transport process, films ranging in thickness from 100 nm to several microns have been straightforwardly and systematically prepared for numerous metal and mixed-metal boron-containing systems with controlled composition at relatively low temperatures. These new materials have been characterized by SEM and other techniques.

INTRODUCTION

The deposition of metallic (such as the transition metals) and main-group materials (such as GaAs and AlGaAs) using metal-organic epitaxial source compounds has recently received a great deal of attention. This research has dealt primarily with the development of new epitaxial techniques and the purification of existent organometallic source materials [1]. While these source compounds have proven adequate in many instances, further developments and improvements in epitaxial processes in the future must rely upon the cognizant and systematic design and synthesis of new source materials developed to exhibit enhanced chemical properties for depositional processes. Thus far, relatively little

* author to whom correspondence should be addressed

emphasis has been placed on this vitally important aspect of epitaxial science. The use of metallaborane cluster complexes offers many of the same advantages as other metal-organic source materials with, however, several other important benefits.

There are a number of important problems associated with the epitaxial source materials currently in use. These problems include; (1) high toxicity (and the related environmental and occupational concerns), (2) great difficulties in handling and manipulations, (3) undesired co-deposition of carbon and other elements, and (4) substrate reactivity problems. The use of borane clusters containing depositionally important elements is expected to circumvent many of the problems currently encountered with other source materials.

In order to achieve the goal of the development of improved CVD source materials, we are exploring the synthetic and deposition chemistry of boron-containing cluster materials. These materials exhibit highly favorable properties for use in MOCVD processes, such as; (1) relatively high volatility, (2) anticipated high stability of the ligand itself to provide clean dissociations, (3) high temperature stability of the source materials, (4) readily preparable in significant quantities, (5) availability of mechanistic pathways for clean depositions to occur, and (6) low toxicity and easily handleable materials. In main-group and metalloid film preparation, the use of boron-cluster main-group compounds for the preparation of III-V semiconductors has a vitally important advantage in that only group III and V elements are involved in the source materials and, therefore, do not introduce inherent impurities such as carbon.

An important factor governing the suitability of a particular ligand and its corresponding complexes for MOCVD and MBE is its stability, both uncomplexed and complexed. Borane and carborane cluster ligands, such as $B_5H_8^{-1}$ and $C_2B_4H_4^{-2}$, have been shown to form highly stable transition metal complexes and typically <u>exceed</u> the stability of the analogous η^5-cyclopentadienyl ligand complexes (well known in organometallic chemistry as highly stable species [2]). The extraordinary stability provided by these boron cluster ligands is apparent from the fact that metallaborane cluster complexes can often be prepared even when the corresponding η^5-cyclopentadienyl complexes are not known, such as in the cases of $\{(C_2B_9H_{11})_2Cu]^{-1}$ and $[(C_2B_9H_{11})_2Ni]$ [1,3-5]. Theoretical studies [6,7] have indicated that one of the major factors contributing to the enhanced stabilizing effects of these ligands arises from a rehybridization of the cage bonding orbitals, shown schematically in Figure 1. The orbitals used for bonding to the

Figure 1. Hybrid orbitals on the open face of typical *nido*-carborane systems (such as $R_2C_2B_9H_9$ and $R_2C_2B_4H_4$), available for metal-ligand bonding [6,7].

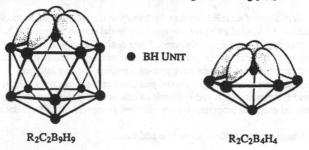

● BH UNIT

$R_2C_2B_9H_9$ $R_2C_2B_4H_4$

metal center on the open face of the cluster are oriented approximately 26° from the orthogonal configuration (as in the η^5-cyclopentadienyl ligand) toward the metal center. This preferred orbital orientation results in significantly better metal-ligand orbital overlaps with a concomitant molecular orbital energy level stabilization of approximately 2 eV. Other factors which contribute to the inherent stability of metallaboron-containing clusters include the relative chemical inertness of the cage itself to a wide variety of chemical reagents (due primarily to the delocalized, covalent cluster bonding) and its stability with respect to thermal degradation.

BORON NITRIDE DEPOSITION FROM DECABORANE(14)

The facile deposition of cubic boron nitride is of interest due to its potential as a hard coating. The hardness of BN is believed to be primarily due to the large cohesive forces of BN [8]. Additionally, the preparation of thin-films of boron nitride is of interest in semiconductor electronics because of their high resistivity, insulating and dielectric properties. Current technology for the deposition of boron nitride requires diborane(6), boric acid, boron trichloride, trimethylboron or borazine as the boron source [1,8]. We have explored the use of decaborane(14) as the boron source for the formation of boron nitride and pure boron thin films.

Decaborane(14), $B_{10}H_{14}$, is an air-stable colorless solid [mp = 99.7°C, decomposition only above 170°C (in vac) and vapor pressure (100°C) = 19 torr] [9]. The physical and chemical properties of this borane cluster make it an ideal source material for the deposition of boron nitride. In a standard R.F. plasma reactor chamber [R.F. voltage of 13.56 KHz and 20 watts], a mixture of 300 microns of N_2 and 20 microns of decaborane(14) yielded opaque, whitish thin films of boron nitride. These films were obtained on several sample substrates including GaAs and Si. The thickness of these coatings were readily varied in the range of at least 50 nm to 25 microns. The resulting boron nitride films had hydrogen contents no greater than 6.6%, which is far better than 13 to 16% obtained by other deposition methods [8]. Similar results were obtained when ammonia was used as the nitrogen source gas. In this case, the hydrogen content of the films was no more than 6.6%. Boron nitride deposition was found to occur readily in this R.F.-assisted process at substrate temperatures below 242°C with the reactor operating at 20 Watts. In these plasma-assisted processes, the higher thermal energy required for deposition is overcome by the use of greater electron kinetic energies. This allows the substrate temperature to be considerably lower, as described above. An Auger electron spectrum of a typical film is shown in Figure 2. Films with variable nitrogen contents (0 to approx. 20%) are formed on GaAs substrates with no impurities. These films are readily oxidized upon exposure to the air. Decaborane(14) itself contains only boron and hydrogen so that inherent impurities, such as carbon and oxygen, can be introduced into the films only by residual impurities in either the decaborane(14) or the deposition system. Similar work with other boranes is in progress. This method appears to be a convenient and efficient method of preparing films of boron nitride with controlled stoichiometry.

BORANE CLUSTER ASSISTED DEPOSITION

The deposition of thin films of metals and metal/non-metal materials has received a great deal of attention. In our studies to develop new source materials for epitaxial processes, we have discovered an unusual transport phenomenon involving borane

Figure 2. Auger electron spectrum of boron nitride film from the plasma deposition of decaborane(14) with nitrogen gas.

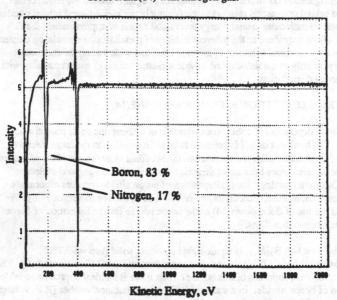

clusters in the formation of thin films of transition and lanthanide metal boron-containing thin film materials. In this reaction, a gas phase, high temperature reaction between borane clusters and an appropriate metal source occurs to form a transient, weakly bound borane-metal source complex in the gas phase. This weak complex then decomposes on a suitable substrate, such as pyrex, to yield thin films of the metal source material which contains a variable (and controllable) amount of boron incorporated into the film . This reaction has been shown to proceed for a variety of metal-source materials, as summarized in Figure 3. In an attempt to understand the mechanism of this chemistry, the reaction was attempted with one of the two reaction components absent (borane and metal source). No deposition has been observed in the absence of either the borane cluster or the metal source at significantly elevated temperatures (800°C). Preliminary studies involving the use of other potential transport agents at 300°C (such as CO, phosphines, phosphites, HCl and others) instead of the borane cluster have also resulted in no formation of thin film materials within the heated region of the reactor. Investigations are currently in progress to delineate the scope and to clarify the mechanism of this unusual transport reactions in thin film preparation.

Acknowledgements This work was funded by the grants from the Donors of the Petroleum Research Fund administered by the American Chemical Society, The Research Corporation, The Office of Naval Research [Grant No. N-00014-87K-0673] and the Syracuse University Senate Research Committee.

Figure 3. Cluster Assisted Deposition (CAD)[†]

$$MCl_3 + Borane\ Cluster \xrightarrow{300°C} M_{3-n}Cl_n\ Thin\ Film$$

$$M° + Borane\ Cluster \xrightarrow{300°C} M°\ Thin\ Film$$

$$\left\{\begin{matrix} MCl_3 \\ or \\ M° \end{matrix}\right\} + \left\{\begin{matrix} 800°C \\ or\ HCl\ (at\ 400°C) \\ or\ KH\ (at\ 400°C) \\ or\ H_2\ (at\ 400°C) \\ etc... \end{matrix}\right\} \longrightarrow No\ Thin\ Film$$

$$Borane\ Cluster \xrightarrow{500°C} No\ Thin\ Film$$

Representative Systems

Lanthanide Metals

MCl₃

$M = Gd*, Tb, Er, Pr, La, Sm*, Ho$

M⁰

$M = Sm*, Yb, Er, Dy$

Transition Metals

M⁰

$M = Cu*, Ni, Fe$

Mixed Metallic Systems

Cu/Sn*

[†] The metal thin-films derived from these reactions contain a controllable amount of incorporated boron.

[*] Preliminary characterization was by SEM. SEM and Auger studies are currently in progress for the other metal systems shown.

REFERENCES

1. P. A. Dowben, J. T. Spencer and G. T. Stauf *Mat. Sci. Eng. B,* in press (**1988**).

2. J. P. Collman and L. S. Hegedus, <u>Principles and Applications of Organotransition Metal Chemistry,</u> 2nd ed. (University Science, Mill Valley, California, **1987**).

3. L. F. Warren and M. F. Hawthorne *J. Am. Chem. Soc.* <u>90</u>, 4823-4828 (**1968**).

4. L. F. Warren and M. F. Hawthorne *J. Am. Chem. Soc.* <u>92</u>, 1157-1173 (**1970**).

412

5. M. F. Hawthorne, D. C. Young, T. D. Andrews, D. V. Howe, R. L. Pilling, A. D. Pitts, M. Reintjes, L. F. Warren, L. F. and P. A. Wegner *J. Am. Chem. Soc.* **90**, 879 **(1968)**.

6. M. J. Calhorda, D. M. P. Mingos and A. J. Welch *J. Organometal. Chem.* **228**, 309-320 **(1982)**.

7. D. M. P. Mingos *J. Chem. Soc., Dalton Trans.* 602-610 **(1977)**.

8. J.-G. Kim, P. A. Dowben, J. T. Spencer and G. O. Ramseyer *J. Vac. Sci. Technol.* **A7 (1989)**, in press.

9. Callery Chemical Company, Technical Bulletin CM-070 (1971).

ALLOY THIN FILMS FROM DISCRETE METALLABORANE CLUSTERS

T. P. FEHLNER*, M. M. AMINI*, M. V. ZELLER**†, W. F. STICKLE***, O. A. PRINGLE****, G. J. LONG****, F. P. FEHLNER*****

*Chemistry Department and **College of Engineering, University of Notre Dame, Notre Dame, IN 46556

***Physical Electronic Laboratories, Perkin-Elmer, Mountain View, CA 94043

****Departments of Physics and Chemistry, University of Missouri-Rolla, Rolla, MO 65401

*****R.D.&E. Division, Corning Glass Works, Corning, NY 14831

†Presently at NASA Lewis Research Center, Cleveland, OH 44135

ABSTRACT

The ferraborane $HFe_4(CO)_{12}BH_2$ sublimes at 50 $^{\circ}$C and decomposes by loss of H_2 and CO on various substrate surfaces at 160-180 $^{\circ}$C to yield films having metallic appearances with typical thicknesses of 4000Å. These films are amorphous by X-ray, conduct electricity and adhere well to glass, metal and silicon substrates. The composition was determined by ESCA and Auger spectroscopies. The films exhibit highly uniform composition and the Fe/B ratio is 4 showing that film stoichiometry is defined by the cluster core. The ESCA chemical shifts show the presence of boride and metallic iron. At present the bulk film contains a high level of oxygen in the form of iron and boron oxides. No α-iron is present by X-ray or Mössbauer spectroscopies. Some carbon and hydrogen (SIMS) are present. Mössbauer spectra show the presence of magnetic iron similar to that exhibited by authentic $Fe_{80}B_{20}$ samples but the films also contain substantial paramagnetic iron.

INTRODUCTION

Although the Metal Organic Chemical Vapor Deposition techniques have been utilized for a long period of time [1], inorganic and organometallic chemists only recently have become involved in

Mat. Res. Soc. Symp. Proc. Vol. 131. ©1989 Materials Research Society

synthesizing new molecular precursors for known materials [2-4]. Ease in handling and predefined, precise stoichiometry has made the preparation of advanced materials from organometallic precursors by MOCVD a fast growing area of research. Among new advanced materials, metallic glasses have a special position because of their properties such as hardness, strength, ease of magnetization [5] corrosion resistance [6] catalytic activity [7] and conductivity [8]. Metal borides were prepared for the first time by electrochemical deposition [9] and because of applications in the ceramic and electronic industries they are now prepared on a large scale by rapid quenching and melt-spining methods. Recently titanium, zirconium, and hafnium boride have been prepared from metal borohydrides [10,11] by CVD methods.

We have explored the utility of ferraboranes in the preparation of iron-boride thin films and we report below the utilization of $HFe_4(CO)_{12}BH_2$ to prepare metal boron films of the composition $Fe_{80}B_{20}$.

EXPERIMENTAL

The $HFe_4(CO)_{12}BH_2$ was prepared according to our procedures with purification by column chromatography followed by sublimation [12]. To prepare a film, the ferraborane was sublimed at 50 ^oC in a low pressure CVD reactor. The substrates (glass or silicon wafer, copper foil or aluminum foil) were resistively heated with typical deposition temperatures of 175 ^oC.

The films were analyzed by X-ray diffraction (XRD), X-ray photoelectron spectroscopy (XPS), Auger electron spectroscopy (AES), Mössbauer spectroscopy and secondary ion mass spectroscopy (SIMS). The XPS profiles were done using 4kV Ar^+ at a sputter rate of 30Å/min relative to Ta_2O_5. The multiplex windows which make up the profile were taken with a spectrometer resolution of 0.7 eV relative to Ag $3d_{5/2}$ using monochromatic Al X-rays.

RESULTS AND DISCUSSION

The exposure of a substrate to 4 X 10^{-6} - 4 X 10^{-5} torr pressure of $HFe_4(CO)_{12}BH_2$ at 175 ^oC in the pyrolysis chamber results in the deposition of a film with a metallic appearance. X-ray diffraction studies of a thin film deposited on glass ($\approx$4000 Å) reveal only a broad peak in the 5-65 (2θ) region characteristic of an amorphous solid. There is no indication of any α-Fe. Auger electron spectroscopy of a typical film shows approximately 59 at. % Fe, 14 at. % B with the remainder being carbon and oxygen. The amount of the latter two elements depends on deposition conditions. The AES depth profile analysis shows the carbon and oxygen content to be constant in the interior of the film. Consistent with the presence of oxygen, two peaks are observed for boron (169 and 179 eV) corresponding to oxidized and unoxidized boron respectively. These values compare well with those observed in a known metallic glass $Fe_{80}B_{20}$ [13].

The concentration and oxidation states of B and Fe are obtained from XPS data. An XPS survey scan of a CVD film as deposited on glass is shown in Figure 1. The XPS depth profile recorded in Figure 2 indicates that the bulk film, obtained after removing approximately 200Å of surface oxidation, has a uniform composition. The atomic concentration values, which are calculated using peak areas for each elemental region from the depth profile data after a twenty minute sputter time, are the following: 57 at. % Fe; 15 at. % B; 8 at. % C; and 20 at. % O.

In Figure 3, the high resolution scans for the B 1s and the Fe 2p regions reveal two distinct oxidation states for each element in the bulk film. The $2p_{3/2}$ peak shows that the major Fe species is metallic, with a strong peak at a binding energy of 706.8 eV. At 711.0 eV, there is a shoulder attributed to oxidized Fe. The B peak at 187.3 eV is representative of a boride and the peak at higher binding energy, 191.7 eV, is due to oxidized boron.

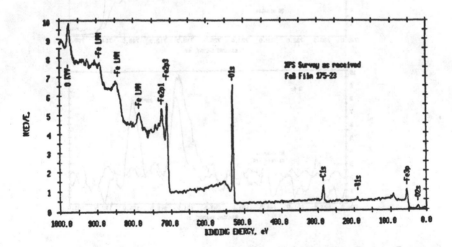

Fig.1. The XPS survey spectrum of a film deposited on glass by pyrolysis of $HFe_4(CO)_{12}BH_2$.

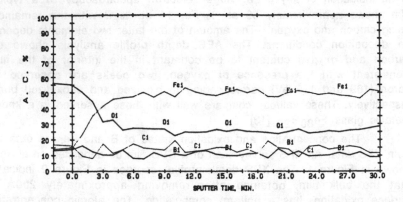

Fig. 2. The XPS depth profile for CVD metallic glass at a sputter
rate of 60 Å/min.

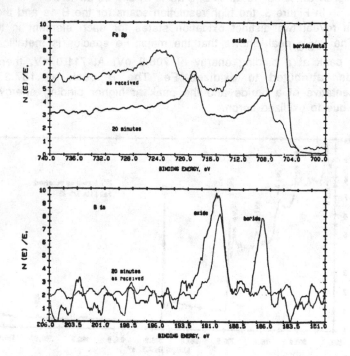

Fig. 3. Detailed XPS spectrum of a film deposited on glass
a) iron 2p spectra   b) boron 1s spectra (energy scale
uncorrected)

The Fe/B ratio of 4 determined by XPS and AES indicates that the film stoichiometry (neglecting the oxygen impurity) is defined by the cluster core. The detailed XPS spectra show that the oxygen combines with boron and iron to form oxides. The peaks at 191.7 and 711.0 eV are comparable to literature values for BO and FeO species from oxidized $Fe_{80}B_{20}$ [13-16]. Peaks corresponding to a metal boride and appearing at 706.8 eV (Fe $2p_{3/2}$) and 187.3 eV (B 1s) are enhanced below the surface. The boron 1s peak for three films with different oxygen contents are compared in Fig. 4 which shows that the boride peak grows with reduction of oxygen.

The constant oxygen content as a function of depth even during a continuous AES profile suggests oxidation of the ferraborane core during deposition. At the base pressure of the pyrolysis chamber (10^{-6} torr) it takes less than a minute to cover the surface with a monolayer of oxygen. In turn, this suggests that better deposition conditions will reduce the contamination.

Typical films have been examined by Mössbauer spectroscopy which reveals no α-iron. However, consistent with the XPS observations, two types of iron are observed. A broad magnetic sextet suggests the presence of a distribution of hyperfine fields typical of amorphous alloys such as $Fe_{80}B_{20}$[17]. In addition there is a broad quadrupole doublet which, no doubt, can be associated with the oxidized material known to be present. In spectra obtained at 85K the relative intensity of this doublet is greatly reduced, with a corresponding increase in the magnetic component. This indicates that the films may contain a substantial portion of their iron in a superparamagnetic form at room temperature.

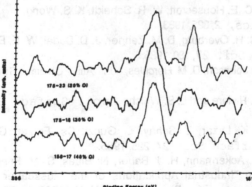

Fig. 4. Boron 1s spectra for films with different oxygen content.
Film 150-17, 40%, 175-16, 30% and 175-23, 20% oxygen.

ACKNOWLEDGEMENTS

The support of the Army Research Office (contract DAAL03-86-K-0136, MMA, TPF) and The Petroleum Research Fund of the American Chemical Society (18202-AC3, GJL, OAP) is gratefully acknowledged.

REFERENCES

1. C. F. Powell, J. H. Oxley, J. M. Blocher, Vapor Deposition, Wiley: New York: 1966.
2. G.T. Stauf, P.A. Dowben, N.M. Boag, L. Morales, De La Garza, S. L. Dowben, Thin Solid Films 156, 327, 1988.
3. C. L. Czekaj, G. L. Geoffroy, Inorg. Chem. 27, 10, 1988.
4. A. Kaloyeros, M. Hoffman, W.S. Williams, Thin Solid Films 141, 237, 1986.
5. V. Guntzel, K. Westerholt, J. Magn. and Magnetic Mater. 69, 124, 1987.
6. G. Savva, Y. Wasedaz, K.T. Ausy, Mat. Sci. Eng. 85, 157, 1987.
7. C. Yoon, D. L. Cocke, J. Non-Cryst. Solids 79, 271, 1986.
8. Boron and Refractory Borides, V. I. Matkovich, Ed., Spring-Verlag: New York, 1977.
9. A. Brenner, Electrodeposition of Alloys, Acad. Press N. Y. 1963.
10. J. A. Jensen, J. E. Gozum, D. M. Pollina, G. S. Girolami, J. Am. Chem. Soc. 110, 1643, 1988.
11. A. L. Wayda, L. F. Schneemeyer, R. L. Opila, App. Phys. Lett. 53, 361, 1988
12. T. P. Fehlner, C. E. Housecroft, W. R. Scheidt, K. S. Wong, Organometallics, 2, 825, 1983.
13. D. R. Huntly, S. H. Overburg, D. M. Zehner, J. D. Budai, W. E. Brower, Jr.,- Appl. Surface Sci. 27, 180, 1985.
14. D. J. Joyner, O. Johnson, D. M. Hercules, J. Am. Chem. Soc. 102, 1980, 1910.
15. S. Myhra; J. C. Riviere and L. S. Welch; Appl. Surface Sci. 32, 156, 1988.
16. G. Kisfaludi, K. Lazar, Z. Schay, L. Guczi, Cs. Fetzer, G. Konczos, A. Lovas, Appl. Surface Sci. 24, 225, 1985.
17. U. Gonser, M. Ackermann, H. J. Bauer, N. Blaes, S. M. Fries, R. Gaa, and H. G. Wagner, in "Industrial Applications of the Mössbauer Effect", G. J. Long and J. G. Stevens, Eds., Plenum Press, N. Y., 1986, p. 25.

MIXED METAL ORGANOMETALLIC CLUSTERS: PRECURSORS FOR
INTERMETALLIC POWDERS

K.E. GONSALVES AND K.T. KEMBAIYAN
Department of Chemistry and Chemical Engineering, Stevens Institute of
Technology, Hoboken, New Jersey 07030

ABSTRACT

Iron-cobalt organometallic clusters were pyrolyzed and yielded Fe-Co
intermetallic powders. These materials were characterized by SEM-EDAX.

INTRODUCTION

Transition-metal cluster compounds have been under scrutiny for their
potential catalytic applications [1]. Also, in non-catalytic areas, there
is a move to build alloy powders from large metal clusters [2]. Therefore,
we have initiated a program to develop intermetallic powders from mixed-
metal organometallic precursors [3]. Here we report the synthesis and
characterization of iron-cobalt intermetallic powders obtained from hetero-
organometallic clusters [4]. Our objective therefore is to (1) provide a
simple and effective means of synthesizing iron-cobalt intermetallics
(2) vary the ratio of iron to cobalt in such intermetallics, and (3) provide
a method to produce homogeneous products from   iron-cobalt metalorganic precursors specifically synthesized for this purpose. The resulting intermetallics may have unique magnetic characteristics.

EXPERIMENTAL

Materials and Equipment

All reactions were conducted in argon, using Schlenk techniques. In
all the synthesis the apparatus was freshly assembled immediately after
removal from the hot oven and subjected to repeated evacuation and purging
by argon. Prepurified argon gas (Matheson) was dried over concentrated
sulfuric acid and phosphorus pentoxide. Residual oxygen was removed by a
BASF deoxygenation catalyst. Reagent grade acetone was also further purified by refluxing with successive small quantities of potassium permanga-
nate until the violet color persisted. It was then dried with anhydrous
calcium sulfate, filtered from the dessicant and fractionated under argon.
The center cut was collected for subsequent use. Distilled water was de-
oxygenated by boiling for 10 hours, followed by cooling under an argon
stream.

Infra-red spectra were recorded on a Perkin Elmer 983 spectrometer
interfaced with a data station. ^1H NMR spectra were recorded on a 200
MHz Bruker spectrometer. Pyrolyses were conducted in a quartz combustion
tube in a tube furnace (Lindberg Model 54253) interfaced with a Eurotherm
model 810 microprocessor controller. All X-ray powder diffraction measurements were carried out by means of a General Electric Diffractometer [3].

The characterization of powders was carried out primarily by means of
a high resolution scanning electron microscope (JEOL-JSM 840) which enables
morphological observations of microstructures and elemental analysis. The
particle size, distribution and morphology were studied by secondary elec-

tron mode. Standardless semi-quantitative analysis was performed by means of a computer-controlled EDAX system. Several measurements were made at low magnification (<100X) to obtain the average composition. Specimen homogeneity was assessed by analysing different regions at high magnification

Synthesis and Reactions

Iron-tricobalt-hydrido-dodecacarbonyl

Iron-tricobalt-hydrido-dodecacarbonyl, $HFeCo_3(CO)_{12}$, was synthesized according to the method of Chini et al. [5]. Yd. 70%.

Thermolysis of a $HFeCo_3(CO)_{12}$ and $Fe(CO)_5$ Mixtures

In a typical experiment [3], $HFeCo_3(CO)_{12}$ (0.11g, 0.002 mole) and 1.2 ml (1.68 g, 0.09086 mol) of $Fe(CO)_5$ were placed in a Schlenk flask connected to a mercury overpressure bubbler. The mixture was stirred magnetically and heated on an oil bath for 45 minutes at 80°C. Evolution of gas was observed, possibly carbon monoxide. The flask was cooled to room temperature and the unreacted $Fe(CO)_5$ removed under vacuum. The residue was utilized for further characterization and subsequent high temperature pyrolysis experiments. Modifications of the above procedure involved thermolysis under pressure.

Pyrolysis

Pyrolysis of $HFeCo_3(CO)_{12}$ [3]

$HFeCo_3(CO)_{12}$ was transferred quickly to an alumina boat and the latter to a quartz tube which had been flushed with argon gas for 30 minutes. The quartz tube was heated in an electrical tube furnace from ambient to different final temperatures. In a typical run, pyrolysis was conducted in an argon flow at 300°C for 45 min. when maximum gas evolution occurred. Variations of this procedure included: changing the gas from argon to nitrogen; heating at ca 300°C for 45 minutes followed by ramping the temperature up to 1000°C. The weight loss was generally 65-70%. Similar methods were utilized for the products of thermolysis experiments of $Fe(CO)_5$ and $HFeCo_3(CO)_{12}$ described above. In the latter pyrolysis, weight loss was around 10%.

RESULTS AND DISCUSSION

Iron-tricobalt-hydrido-dodecacarbonyl, $HFeCo_3(CO)_{12}$, was synthesized according to the procedure of Chini [5]. An X-ray diffraction pattern closely matched previously published values for d spacings and relative intensities for $HFeCo_3(CO)_{12}$. The infra-red spectrum was also similar to that reported earlier [5].

To start with, the compound $HFeCo_3(CO)_{12}$ was pyrolysed in an argon atmosphere at 300°C for 45 minutes. Another portion of $HFeCo_3(CO)_{12}$ was pyrolysed at 330°C in a nitrogen atmosphere for 45 minutes. The char yield obtained in this experiment was 31.45%. The X-ray diffraction pattern of these pyrolysed powders were different from that of $HFeCo_3(CO)_{12}$. It was not readily amenable for structure determination owing to the presence of only a few diffraction peaks. The percentage of cobalt in the powders was 2:1 for Fe:Co.

Fig. 1. SEM(2000X) of homogeneous and fine powders

Fig. 2. Morphology of particles pyrolyzed at high
temperatures

To increase the iron-content, it was anticipated that thermolysis of a mixture of $HFeCo_3(CO)_{12}$ and $Fe(CO)_5$ would achieve the objective. Such a procedure would be dependent upon the solubility of $HFeCo_3(CO)_{12}$ in $Fe(CO)_5$. The solubility was ascertained to be 60 g/l at 20°C.

$Fe(CO)_5$ is known to decompose thermally at temperatures around 130°C to Fe. Therefore $HFeCo_3(CO)_{12}$ (0.11g) was mixed with $Fe(CO)_5$ (1.2 ml) and heated in an argon atmosphere for 45 min. at 80°C. The unreacted $Fe(CO)_5$ was removed under vacuum leaving a solid black residue. The residual mixture was a homogeneous powder of average Fe:Co ratio approximately equal to 1:3.

In order to enhance the iron content, the initial reaction temperature was raised to 138°C [$HFeCo_3(CO)_{12}$ 0.13g; $Fe(CO)_5$ 2 ml]. The resultant powder sample was was found to be finer and richer in iron content-Fe:Co ratio equal to 2:1. The sample homogeneity was confirmed by repeated EDAX measurements at several locations. This product was then further pyrolyzed in argon at 350°C and yielded very fine homogeneous particles of 0.2 - 2 μm in size (Fig. 1). The composition remained the same after pyrolysis.

In an attempt to study the effect of pressure in these reactions, $HFeCo_3(CO)_{12}$ was heated in a pressure tube at 138°C. The resultant powder exhibited pyrophoric properties. Upon contact with air, the powder caught fire immediately. This sample was not amenable for routine characterization. We will attempt to analyze these powders in inert atmospheres. However, a similar treatment of a mixture of $HFeCo_3(CO)_{12}$ and $Fe(CO)_5$ heated in a pressure tube at 138°C for 2 hours rendered the product non-pyrophoric. An interesting feature here is that under pressure, there was a decrease in the iron content (Fe:Co = 1:3) compared to a similar treatment conducted at atmospheric pressure described above. The latter produced powders where the ratio of Fe:Co was 2:1.

On the rapid pyrolysis of $HFeCo_3(CO)_{12}$ from ambient to 1000°C, the end product was found to contain an inhomogeneous mixture of grey and blackish particles. Many particles appeared as sintered agglomerations in the form of fibers, as revealed by the SEM micrograph in Fig. 2. The bulk average composition of Fe:Co was found to be in the ratio of 3:7.

CONCLUSION

This initial work has indicated that iron-cobalt intermetallic powders of a specific composition can be obtained via the pyrolysis of mixed-metal organometallic clusters. We are also attempting to isolate and identify possible iron-cobalt organometallic clusters obtained via the thermolysis of $Fe(CO)_5$ and $HFeCo_3(CO)_{12}$. The composition of the powders is being further evaluated by atomic absorption, XPS and Auger spectroscopy to determine any residual carbon and/or presence of oxides.

ACKNOWLEDGMENT

Acknowledgement is made to the GAF Chemicals Corp. for partial support of this research program.

REFERENCES

[1] "Mixed-Metal Clusters" by W.L. Gladfelter and G.L. Geoffrey in "Advances in Organometallic Chemistry", F.G.A. Stone and R. West Eds Vol. 18, p. 207, Academic Press, New York 1980.

[2] J. Haggin, Chem. & Engg. News, P. 31, Oct. 5, 1987.

[3]   K. Gonsalves, U.S. Patent pending.

[4]   K. Gonsalves and K.T.Kembaiyan, J. Mat. Sci. Lett. (in press); Solid State Ionics: 13th ISRS Proceedings (in press).

[5]   P. Chini, P. Corradini and S. Cassata, Gazzetta Chimica Italiana 90, 1005 (1960); U.S. Patent 3,332,749 (1967).

POLYALKYLSILYNES: SYNTHESIS AND PROPERTIES OF "TWO-DIMENSIONAL" SILICON-SILICON BONDED NETWORK POLYMERS.

Patricia A. Bianconi,*† Timothy W. Weidman,† and Frederic C. Schilling†
†AT&T Bell Laboratories, 600 Mountain Ave., Murray Hill, NJ 07974.
*AT&T Bell Laboratories Postdoctoral Fellow. Current address: Department of Chemistry, The Pennsylvania State University, University Park, PA 16802

ABSTRACT

The synthesis of the first soluble silicon-silicon bonded network polymers, the polyalkylsilynes [RSi]$_n$ was accomplished using high-intensity ultrasound to effect the reductive condensation of alkyltrichlorosilanes with liquid sodium-potassium alloy. The resulting polymers ($\bar{M}_w$ = 25,000 to 100,000) remain hydrocarbon-soluble and may be cast into transparent films. Spectroscopic data indicate a structure consisting of tetrahedral alkylsilicon units assembled via silicon-silicon bonds into amorphous networks. Films or solutions of the polysilynes exhibit an intense UV absorption which tails into the visible, blue-shifted but similar in shape and intensity to that of amorphous silicon. The photoreactivity and pyrolysis properties of these materials will be described.

INTRODUCTION

Polysilanes, linear polymers possessing an all silicon-silicon bonded backbone, have recently become the focus of intense research[1], and a number of potential applications for these materials (as SiC precursors[2], photoinitiators[3], photoresists[4], and photoconductors[5]) have already emerged. However, research has been limited to the study of linear polysilanes of formula [R$_1$R$_2$Si]$_n$, the silicon analogues of polyolefins. Although linear polysilanes bearing an impressive diversity of alkyl, aryl, and even trimethylsilyl substituents have now been prepared and investigated, there has been little progress towards the preparation and characterization of monoalkyl silicon polymers [RSi]$_n$[6]. By direct analogy to carbon-based polymers, materials with a 1:1 alkyl to silicon ratio could adopt either discrete aromatic structures or take the form of linear conjugated polymers analogous to polyacetylenes. Alternatively, and more consistent with the decreased tendency of silicon towards π-bonding, a network structure with no carbon analogue may be adopted. In either case, such materials, viewed as the halfway point between polysilanes and silicon, should exhibit an intriguing array of physical properties.

In a recent communication[7] we described a new sonochemical procedure for the synthesis of the first such polymer, [n-C$_6$H$_{13}$Si]$_n$, for which we proposed the name "poly(n-hexylsilyne)". We here describe in more detail the synthesis of this and related poly(alkylsilynes) and the photoreactivity and pyrolysis properties of these materials.

RESULTS AND DISCUSSION

Synthesis of polyalkylsilynes

The standard synthesis of polysilanes involves the reductive condensation of dichlorosilanes with metallic sodium (Eq. 1)[1]. An intrinsic difficulty associated

$$R^1R^2SiCl_2 + 2Na \longrightarrow \begin{bmatrix} R^1 \\ Si \\ R^2 \end{bmatrix} + 2NaCl \qquad (1)$$

with applying this procedure to the reductive condensation of alkyltrichlorosilanes stems from the heterogeneous nature of the reaction. Precipitation of incompletely reduced polymer, passivation of the reducing agent, and (under more forcing conditions) reductive depolymerization, appear to prevent the isolation of $[RSi]_n$ product. The only explicit mention of a homogeneous reductive condensation involved reduction of methyltrichlorosilane in THF with sodium naphthalide to give an intractable solid product[6]. In order to eliminate the problems associated with conventional heterogeneous reductive condensation syntheses, we began exploring the synthetic utility of sonochemically generated liquid Na/K alloy-hydrocarbon emulsions. In this technique, the two immiscible liquids, NaK alloy and an organic solvent, are emulsified by ultrasonic irradiation with a high-intensity immersion horn. Unlike sonicated solid sodium dispersions, the sonochemically emulsified NaK alloy exhibits reactivity more closely approaching that of a homogeneous reductant[8].

The use of sonochemically generated NaK emulsions in hydrocarbon solvents greatly facilitated the synthesis of the first alkyl silicon network polymers, the "polyalkylsilynes", $[SiR]_n$ (Eq. 2).

$$RSiCl_3 + 3.00eqNaK \xrightarrow[\text{pentane, 5min; THF, 5min}]{\text{375W, 20KH}_z \text{ SONICATION}} [RSi]_n \qquad (2)$$

High-intensity ultrasound allows reductive condensations of alkyltrichlorosilanes to be performed under more homogeneous conditions in inert alkane solvents, eliminating monomer-solvent side reactions and other complications associated with the need for ethereal solvents and electron-transfer reagents[6]. While THF was found to be essential to effect complete reduction and dissolution of polymeric intermediates, it was added only after sufficient sonication of the alkane with NaK alloy to ensure that all of the trichlorosilane had reacted. Using three full equivalents of NaK alloy in the reductive condensation does give high polymer, but the solubility of the polymers so obtained decreases with time, possibly due to hydrolysis of residual silicon-chlorine bonds or crosslinking of Si-OH and/or Si-H functionalities. This difficulty is circumvented by using only 95% of the required three equivalents of NaK alloy to effect the reductive condensation, and then "capping" the remaining silicon-chloride bonds with an appropriate alkyl group by use of a Grignard or lithium reagent. The reaction mixture is titrated with the appropriate alkylating agent (i.e., n-hexylmagnesium bromide for poly(n-hexylsilyne), etc.,) until a hydrolyzed aliquot of the mixture exhibits neutral pH, indicating that all silicon-chlorine bonds have been alkylated (Eq. 3). This procedure presumably introduces dialkylsilicon "edge groups" and trialkylsilicon end groups, but NMR and analytical data on the purified high molecular weight polymers suggest that these represent <1% of the total $[SiR]_x$ units.

$$RSiCl_3 + 2.85eqNaK \xrightarrow[\text{pentane, 5min; THF, 5min}]{\text{375W, 20KH}_z \text{ SONICATION RMgX,PH7}} [RSi]_n \qquad (3)$$

Polysilyne purification is accomplished by dissolution of the polymers in THF and sequential precipitation with water, methanol, and ethanol. The polymers are all isolated as yellow, moderately air- and light-sensitive powders that remain freely soluble in nonpolar organic solvents, from which they can be cast or spun into transparent films. IR spectra of the polymers prepared using the optimized pentane-95% NaK-THF-RMgCl procedure show no significant Si-H or Si-O-Si bands, and with the exception of the less soluble poly(n-propylsilyne), chemical analyses correct for the empirical formula [RSi]$_n$ were obtained. The yields of purified poly(n-butylsilyne) and poly(n-hexylsilyne) ranged from 11 to 35%. Molecular weights versus polystyrene, as determined by gel permeation chromatography, are between 17,000 and 24,000 daltons for poly(n-butylsilyne) and poly(n-hexylsilyne); the hexane-soluble fraction of poly(n-propylsilyne) (vide infra) had mean GPC molecular weights between 3000 and 8000. Actual molecular weights, as determined by light-scattering, appear to be approximately four times greater than those determined by GPC[9].

The choice of alkyl substituent has a major effect on the properties of the poly(alkylsilynes). As was the case with linear polysilanes, the solubility of the polymers decreases with decreasing chain length of the n-alkyl substituents. While poly(n-hexylsilyne) is extremely soluble in organic solvents, only the low molecular weight($\bar{M}_w$ < 8000) fraction of poly(n-propylsilyne) is readily soluble, and material obtained from the reduction of methyltrichlorosilane is completely intractable. Changing the steric requirements of the alkyl substituent also appears to affect the degree of polymerization attainable. For example, the reductive condensation of t-butyltrichlorosilane gave only oligomeric material($\bar{M}_w$ = 350)[10], suggesting that bulkier groups may inhibit formation of an extensive polysilyne network. Conversely, the hydrocarbon-insoluble fraction of poly(n-propylsilyne) had $\bar{M}_w$ = 1.5 x 10^5 and a high polydispersity. Such high molecular weight fractions are not observed in the polysilynes with larger alkyl substituents; possibly the smaller n-propyl substituent allows more extensive network formation.

Polysilyne Structure

X-ray powder patterns of the polyalkylsilynes were featureless, which suggests that the polymers adopt amorphous, glass-like structures. Solution and solid state ^{29}Si and ^{13}C NMR studies indicate that the major component of the polysilyne backbone consists of alkylsilicon units in which each silicon atom is sigma-bonded to three other silicons. No spectral evidence for the presence of silicon-silicon double bonds was seen[11], and thus a linear polyacetylene-like configuration was eliminated as a primary structural feature of the polyalkylsilynes. The NMR spectra were consistent with a very rigid, randomly constructed network structure of sheets or cages of fused rings, structurally intermediate between that of linear polysilanes and amorphous silicon. Thus, although the stoichiometry [RSi]$_n$ may suggest an analogy to polyacetylenes, polysilynes actually adopt a sigma-bonded network structure unprecedented in carbon-based polymers.

Properties and Potential Applications of Polyalkylsilynes

Polyalkylsilynes are moderately sensitive to oxygen in the presence of light, but are thermally stable to 300°C. Pyrolysis mass spectra at this temperature show only the respective 1-alkene and smaller alkyl fragments, and no major silicon-containing fragments are volatilized. In contrast, mass spectra of poly(n-hexylmethylsilane) and poly(dimethylsilane)(phenylmethylsilane) 50:50 copolymer[2] showed fragmentation into a variety of silylene and higher oligomeric components. This difference in pyrolysis behavior can be directly related to the difference in structure of the two classes of polymers: that is, to the greater propensity of the network structure over the linear to enforce retention of silicon atoms. When films of the polyalkylsilynes are pyrolyzed they are converted (without the pretreatment required for linear polysilanes) to mixtures of silicon and silicon carbide. At moderate temperatures (400° C to 600° C) red-brown amorphous material is formed, which on heating at 1400° C becomes black and glassy. The weight of retained silicon in the ceramic products is as high as 95% of the weight of silicon in the starting polymer[12], confirming the mass spectral results and suggesting that these polymers may be useful as high-yield precursors to silicon-based ceramic films.

While the linear diorganopolysilanes exhibit strong $\sigma\text{-}\sigma^*$ transitions (λ_{max} = 300 to 350 nm, ε = 2800 to 12000 per Si)[1]in the near UV, solutions or films of the polyalkylsilynes exhibit an even more intense broad absorption from $\lambda \leq 200$ nm tailing down into the visible. This absorption is blue-shifted but similiar in shape and intensity to that of amorphous silicon. The high extinction coefficient per silicon fragment (ε = 29,000 to 35,000 at 200 nm) of the polyalkylsilynes may be attributed to an extension of Si-Si $\sigma\text{-}$ "conjugation" from one dimension along a linear polymer backbone, as in the polysilanes, into three dimensions across the polysilyne networks. On broad band UV irradiation in solution or as films, both the polyalkylsilynes and linear polysilanes undergo bleaching, but that of the polysilynes proceeds far less rapidly. This may be attributed to the delocalized nature of the excitation or to the increased propensity of the network polymer over the linear to enforce recombination of photogenerated silicon radicals. While photooxidation of polysilanes results in degradation to low molecular weight cyclic siloxanes[1], irradiation of polyalkylsilynes in air leads to bleaching and crosslinking to give an insoluble siloxane network. This behavior suggests that films of the polysilynes could be used as negative photoresists for microlithography, or other applications where photopatterning is used.

CONCLUSIONS

The synthesis of the first [RSi]$_n$ network polymers, or poly(alkylsilynes), has been acheived by careful selection of monomer, reductant, and solvent in a sonochemically mediated reductive condensation procedure. The new polymers (with R = n-propyl, n-butyl, and n-hexyl) remain soluble in nonpolar organic solvents and are readily cast into amorphous, transparent yellow films. All chemical and spectroscopic data indicate that the polysilynes are constructed primarily of sp^3-hybridized alkylsilicon units assembled via Si-Si bonds into irregular networks. Thus the polysilynes provide the first examples of soluble σ-delocalized materials with a composition intermediate between that of linear polysilanes and elemental silicon. Investigations into the properties and potential applications of these and related new materials are in progress.

ACKNOWLEDGMENTS

We thank M. E. Galvin, M. Y. Hellman, A. M. Mujsce, and L. E. Stillwagon for experimental advice and assistance.

REFERENCES

1. (a) R. West, J. Organomet. Chem.300, 327 (1986). (b) J. M. Zeigler and L. A. Harrah, Macromolecules 20, 601 (1987). (c) R. D. Miller, B. L. Farmer, W. Fleming, R. Sooriyakumaran, and J. Rabolt, J. Am. Chem. Soc. 109, 2509 (1987). (d) K. A.

Klingensmith, J. W. Downing, R. D. Miller, and J. Michl, J. Am. Chem. Soc. 108, 7438 (1986). (e) P. Trefonas, R. West, and R. D. Miller, J. Am. Chem. Soc. 107, 2737 (1985) (f) F. C. Schilling, F. A. Bovey, and J. M. Zeigler, Macromolecules 19, 2309 (1986).

2. (a) R. West, L. D. David,, P. I. Djurovich, H. Yu, and R. Sinclair, Am. Ceram. Soc. Bull. 62, 899 (1983). (b) S. Yajima, J. Hayashi, M. Omori, andK. Okimura, Nature 261, 683 (1976).

3. R. West, A. R. Wolff, and D. J. Peterson, J. Radiation Curing 13, 35 (1986).

4. (a) J. M. Zeigler, L. A. Harrah, and A. W. Johnson, SPIE Adv. Resist Technol. Proc. II 539, 166 (1985). (b) D. C. Hofer, R. D. Miller, C. G. Willson, and A. R. Neureuther, SPIE Adv. Resist Technol. Proc. 469, 16 (1984).

5. R. G. Kepler, J. M. Zeigler, L. A. Harrah, and S. R. Kurtz, Bull. Am. Phys. Soc. 28, 362 (1983).

6. (a) R. West and A. Indricksons, J. Am Chem. Soc. 94, 6110 (1972). (b) R. West, Ann. N. Y. Acad. Sci. 31, 262 (1973). (c) D. Seyferth and Y. F. Yu, in Design of New Materials, edited by D. L. Cocke and A. Clearfield (Plenum Publishing, New York, 1987), pp.79-94.

7. P. A. Bianconi and T. W. Weidman, J. Am. Chem. Soc. 110, 2342 (1988).

8. T. W. Weidman, presented at the 192nd ACS National Meeting, Anaheim, CA, 1986 (unpublished).

9. P. A. Bianconi, L. E. Stillwagon, and T. W. Weidman, unpublished results.

10. P. A. Bianconi, E. W. Kwock, and T. W. Weidman, to be submitted for publication.

11. (a) M. J. Michalczyk and R. West, J. Michl, J. Am. Chem. Soc. 106, 82111 (1984). (b) R. West, M. J. Fink, and J. Michl, Science 214, 1981, 1343. (c) H. B. Yokelson, J. Maxka, D. A. Siegel, and R. West, J. Am. Chem. Soc. 108, 4239 (1986).

12. P. A. Bianconi, F. C. Schilling, and T. W. Weidman, unpublished results.

CHEMICAL VAPOR DEPOSITION OF SILICON CARBIDE USING A NOVEL ORGANOMETALLIC PRECURSOR

WEI LEE[*], LEONARD V. INTERRANTE[*], CORRINA CZEKAJ[*], JOHN HUDSON[**], KLAUS LENZ[**] AND BING-XI SUN[**]

[*]Departments of Chemistry and [**]Materials Engineering, Rensselaer Polytechnic Institute, Troy, NY 12180-3590.

ABSTRACT

Dense silicon carbide films have been prepared by low pressure chemical vapor deposition (LPCVD) using a volatile, heterocyclic, carbosilane precursor, $MeHSiCH_2SiCH_2Me(CH_2SiMeH_2)$. At deposition temperatures between 700 and 800°C, polycrystalline, stoichiometric SiC films have been deposited on single crystal silicon and fused silica substrates. Optical microscopy and SEM analyses indicated formation of a transparent yellow film with a uniform, featureless surface and good adherence to the Si(111) substrate. The results of preliminary studies of the nature of the gaseous by-products of the CVD processes and ultrahigh vacuum physisorption and decomposition of the precursor on Si(100) substrates are discussed.

INTRODUCTION

The chemical vapor deposition (CVD) of silicon carbide has a long history of development and successful application [1]. However, the full potential of this material, and the CVD method for its generation, has not been realized, in part due to the extreme temperatures (>1000°C) and exacting conditions required by the existing approaches. In particular, its large band gap, high-temperature stability, high thermal conductivity, high breakdown electric field, and high electron saturation velocity make it an attractive candidate for use as a high temperature, radiation-resistant semiconductor [2-5]. Similarily, its hardness, oxidation and corrosion resistance suggest a wide range of potential applications for protective, abrasion and corrosion resistant coatings. One of the problems associated with the use of SiC for such applications is the fact that it can exist in a variety of crystalline modifications and is difficult to obtain as a single-phase material in high compositional purity. Another major problem is the high temperature which is generally required to obtain high-quality SiC by the existing CVD methods.

Silicon carbide thin films of widely varying composition and morphology have been prepared by a range of chemical vapor deposition techniques [1]. These are usually based on the pyrolysis of mixtures of silicon and carbon containing compounds, such as $SiCl_4$ with CCl_4, $HSiCl_3$ with C_6H_{14}, or SiH_4 with C_3H_8. Single-component SiC precursors, such as CH_3SiCl_3, have also been employed. These processes are generally carried out at atmospheric pressure. A carrier gas, such as H_2, He, Ar, or N_2 is generally used, with H_2 often needed for the complete removal of chlorine as HCl. Deposition temperatures range from 800 to 1800°C, with temperatures greater than 1200°C being optimal.

The high deposition temperatures associated with these CVD processes often promote the decomposition and/or reaction of the substrate. This can lead to the deposition of films that are neither phase nor compositionally homogeneous and pure. In addition, film microstructure, composition, and adherence are adversely effected by the corrosive gaseous by-products of these processes [6]. Finally, control of the microstructure, thickness, and purity of the films is limited in lower temperature CVD processes. Carrying out these CVD processes at lower total pressures, 1-100 torr, can improve control over these properties; however, the associated deposition temperatures of greater than $1200°C$ limit the choice of potential substrates [7].

To circumvent these difficulties, we have investigated the synthesis and low pressure chemical vapor deposition (LPCVD) of volatile, heterocyclic, organometallic precursors. Previously, we reported the CVD of SiC using a cyclic carbosilane precursor, $[MeHSiCH_2]_3[8]$. More detailed studies have shown that the compound employed was actually a four-membered ring carbosilane, $MeHSiCH_2SiCH_2Me(CH_2SiMeH_2)$, a structural isomer of the expected six-membered ring compound. The research described herein provides additional information on the preparation and characterization of SiC films obtained from LPCVD studies employing this precursor at substrate temperatures between 600 and $900°C$.

EXPERIMENTAL

Synthesis and Characterization of Me(H)SiCH$_2$SiCH$_2$Me(CH$_2$SiMeH$_2$).

The preparation of the cyclic carbosilane was based on the Grignard coupling reactions discussed by Kriner, as summarized in equation 1 [9]. Spinning band distillation of the pale yellow liquid product of the reduction

$$MeSi(Cl_2)CH_2Cl + Mg \longrightarrow [MeSi(Cl)CH_2]_n \xrightarrow{LiAlH_4} [MeSi(H)CH_2]_n \quad (1)$$
$$\text{and other prods.} \qquad \text{and other prods.}$$

with LiAlH$_4$ yielded a colorless liquid, b.p.=$73-75°C$ at 33 torr, in ca. 15% overall yield. The gas chromatogram of this liquid showed two peaks indicating a mixture of 85% of the cyclic carbosilane, $MeHSiCH_2SiCH_2Me(CH_2SiMeH_2)$, and 15% of an impurity which was tentatively identified as $Me_2Si(CH_2SiH_2Me)_2$. Samples for analytical studies were collected directly from the GC; whereas the mixture was used in the CVD experiments. Subsequent analysis of the major component of this liquid fraction by ^1H and ^{29}Si NMR indicated that it was comprised of an approximately 50/50 mixture of the expected cis and trans geometric isomers of the cyclic carbosilane.

Mass spectrometric analysis: m/e 174 (M^{+1}), 173, 159, 129, 115, 99, 85, 73, and 59. IR: 2960 (s), 2900 (m), 2870 (w), 2120 (vs), 1400 (w), 1340 (m), 1245 (s), 1040 (vs), 940 (vs), 900 (vs), 800 (vs), 750 (m) and 700 (m) cm^{-1}. ^1H NMR: -0.15 (two triplets), -0.01 (multiplet), 0.07 (two triplets), 0.14 (multiplet), 0.25 (2 doublets, 2 singlets), 0.4 (multiplet), 4.1 (2 sextets), and 4.9 (multiplet) ppm. $^{29}Si\{^1H$ coupled): -39.1 (triplet), -15 (doublet), 7.5 (singlet), and 8.5 (singlet) ppm.

Apparatus and Procedure for the CVD of SiC from MeHSiCH$_2$SiCH$_2$Me(CH$_2$SiCH$_2$Me)

Details of the substrate preparation, horizontal, hot-wall CVD reactor, and analytical instrumentation are presented elsewhere [10]. In typical experiments, ca. 0.5-1.0 g of the carbosilane precursor was loaded into the precursor container fitted with a high vacuum greaseless stopcock and o-ring joint, in an N$_2$ filled glove box. After connecting the precursor container to the reactor and evacuating the system to 10^{-5} torr, the substrates (cleaned Si(111) or (100) wafer pieces and silica plates) were heated to 800°C for several hours. The furnace was set to the deposition temperature, between 600 and 900°C, the precursor frozen with liquid nitrogen, and the precursor container opened to the reactor and vacuum system. Subsequently, the precursor was warmed to 0-20°C and vaporized into the reactor using a mechanical/diffusion pump system. The typical steady-state pressure of the CVD reactor was 0.5-0.9 torr. After deposition, the reactor was cooled to room temperature and opened to air.

Gas phase pyrolysis products of the precursor were analyzed by gas chromatography, FTIR, and NMR. The thin films were examined by SEM, Auger spectroscopy, electronic absorption and FTIR spectroscopy, ellipsometry, and XRD measurements.

The adsorption behavior and decomposition of the carbosilane precursor on a clean Si(100) surface were studied in an ultra- high vacuum apparatus, containing a sample holder/positioner, a quadrupole mass spectrometer and a cylindrical mirror based system for Auger electron spectroscopy [11]. The precursor (equilibrium vapor pressure ca. 2 torr at room temperature) was admitted through a solenoid operated pulsed valve.

RESULTS AND DISCUSSION

Chemical Vapor Deposition of SiC Films in the Hot-Wall Reactor

Pure, dense, polycrystalline SiC films were deposited on silicon and fused silica substrates by the thermal decomposition of MeHSiCH$_2$SiCH$_2$Me(CH$_2$SiMeH$_2$) at temperatures of 700-900°C and pressures of 0.5-0.9 torr. Film thicknesses ranged from a few hundred Angstroms to one micron, depending on the time, quantity of precursor employed, and the substrate and precursor temperatures. The deposition rates were typically 100-150 A/min.

The surface morphology of the SiC films was studied by SEM and optical microscopy. Optical micrographs of the pale yellow, transparent films indicated a smooth, uniform, and featureless surface. Scanning electron micrographs of these SiC films showed a uniform, non-porous, and fine-grained surface. The scanning electron micrographs of SiC film cross-sections illustrate that the films are adherent to the substrate, with no indication of crystalline orientation or porosity. A representative scanning electron micrograph of an SiC film deposited on a Si(111) surface at 800°C is shown in Figure 1. X-ray powder diffraction studies indicate that these films are noncrystalline or nanocrystalline. TEM and selected area diffraction analyses to obtain additional information on film microstructure are in progress.

Compositional analyses of the SiC films prepared at 700-800°C by Auger electron spectroscopy indicated that the films were essentially stoichiometric, % Si=49 and % C=51, with oxygen levels in the bulk of less than 0.8%. These results were derived by comparison to measurements made on a single crystal of 6H-SiC of known 1:1 stoichiometry. A representative

434

Figure 1. SEM of a SiC-coated Si wafer piece broken to reveal the SiC layer.

Auger depth profile of an SiC film deposited on Si(111) substrate at 800°C is shown in Figure 2. The FTIR transmission spectrum of this film showed a single strong band at *ca.* 800 cm^{-1}, as expected for SiC. Optical absorption studies carried out on a *ca.* 0.5 μm, pale-yellow, transparent film deposited on SiO$_2$ at 800 °C showed no absorbance from 900 to 550 nm with a gradually accelerating increase in absorbance at lower wavelengths to a cut-off at around 300 nm.

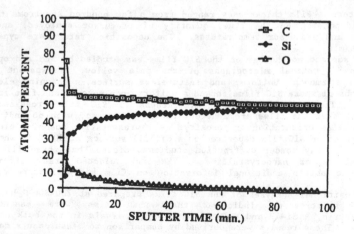

Figure 2. Auger depth profile of an 800 °C SiC film on Si.

At a deposition temperature of 600°C, no SiC was deposited, with most of the carbosilane precursor passing through the reactor and collecting in the liquid nitrogen-cooled trap. Films deposited at 900°C contained higher proportions of carbon than 1:1 and were black and shiny in appearance. Auger depth profiles on two such films showed Si:C ratios of 1:1.46 and 1:1.75. The oxygen content of these films was less than 0.8%.

Measurements of the refractive index, using an ellipsometer with a 70° incident light angle and a wavelength of 6328 A, gave a value (2.7±0.1) which was very close to that of single crystal SiC [12] for the films deposited at 700-800 °C. The refractive index values obtained for the carbon-rich SiC films deposited at 900 °C were considerably higher (3.0±0.1), possibly due to the non-stoichiometry and multiple scattering from SiC_{1+x} or C particles.

Preliminary investigations of the gas phase pyrolysis products of the CVD processes were carried out by collecting the gases in the liquid nitrogen trap and separating them by gas chromatography. In a separate experiment the precursor was swept with nitrogen past a SiO_2 probe, heated to ca. 800 °C with a nichrome heating element, and the gas mixture was analyzed downstream by GC. The latter analyses indicated the presence of H_2 and methane, with a small amount of other hydrocarbons. The gaseous products obtained on warming the contents of the liquid nitrogen trap to room temperature were mainly ethylene and acetylene, with small amounts of methane and ethane. In addition, a small amount of a liquid byproduct was obtained which ^{1}H NMR studies suggest contain one or more silane compounds in addition to unreacted precursor. More detailed analyses of the gas and liquid by-products of the CVD processes are in progress; however, it is apparent that the pyrolysis chemistry is quite complex, probably involving homolysis of the CH_3, H, and $CH_2SiH_2CH_3$ groups on the cyclic carbosilane as radical species, followed by subsequent radical coupling and H-abstraction to give the various hydrocarbon and silane byproducts observed.

Precursor Interactions with Si(100) in the Ultrahigh Vacuum System

Measurements were first made at low temperature to characterize the adsorption behavior of the intact precursor. While cooling the sample to 130K the pulse valve was operated for 100 to 5000 pulses to deposit the precursor on the Si(100) surface. The sample was then heated at a linear rate of 4.6K/sec and one of the precursor mass peaks (mass 73) monitored with the mass spectrometer. Two approximately equal area desorption peaks were observed in all cases, presumably corresponding to the two geometrical isomers of the cyclic carbosilane. In a second set of measurements the pulse valve was operated while the substrate temperature was increased in increments above room temperature, and the resulting waveform of signal vs time on the detector mass spectrometer, tuned to mass 73, was collected. For temperatures above about 400K, all waveforms had essentially the same shape, approximating that of the pulse valve, convoluted with the pumping time constant of the main chamber. As the substrate was heated beyond about 800K the amplitude of the waveform was attenuated signaling the onset of decomposition of the precursor on the surface. Attempts to detect waveforms of the expected gas-phase product species, such as CH_3, CH_4, C_2H_6, and H_2 were made; however, no signal coherent with the pulsing process was observed at any of these masses.

Auger spectra taken before exposure to the precursor species showed only peaks attributable to silicon. After exposure to precursor at temperatures above 800K, a carbon peak was also visible with a carbon-to-silicon ratio

436

consistent with the formation of a SiC layer on the surface.

At the conclusion of these studies, the samples were further investigated by scanning electron microscopy and transmission electron microscopy. Small etch features (pits) were observed by SEM in the area of the sample that had been exposed to the precursor beam. Overlying part of these etched areas and the intervening region between them was a smooth, continuous film. This observation, coupled with the failure to observe any hydrocarbon by-products during the CVD reaction, suggests that the Si(100) substrate is reacting, presumably with the carbon-containing byproducts, to form SiC in this low coverage regime. The absence of such etch pits in the case of the samples obtained from the hot-wall reactor suggests further that the surface and/or gas phase chemistry may be fundamentally different in these two systems.

One sample from the ultra-high vacuum studies was thinned electrochemically and examined in the transmission electron microscope. An image of the film is shown in Figure 3. Selected area diffraction measurements taken on this same area, shown in the inset to Figure 3, gave a pattern consistent with a fine-grained, polycrystalline silicon carbide. Because of the small differences among the various SiC polytypes, unequivocal identification of the phase(s) present was not possible; however, it was apparent that there was some preferred orientation in the SiC, with <111> SiC || <111> Si. While there is not complete epitaxial growth of SiC, it is significant that as-deposited some degree of lattice matching is occurring at the interfaces.

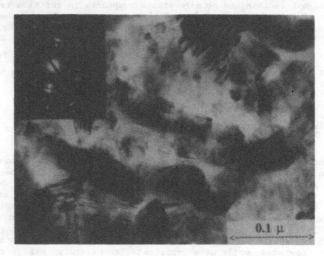

0.1 μ

Figure 3. Transmission electron micrograph of a SiC film deposited on Si(100). The diffraction pattern of a selected area of this film is shown in the inset.

ACKNOWLEDGEMENTS

We thank Mr. Chris Whitmarsh for synthesizing the carbosilane precursor and Mr. Ray Dove and Dr. Krisna Rajan for assistance with the SEM and TEM analyses. This work was supported by the Office of Naval Research under Contract No. N0001485K0632.

REFERENCES

1. Schlichting, J. *Powder Met. Intl.* 12, 141 (1980).
2. Lely, A. *Ber. Dtsch. Keram. Ges.* 32, 229 (1955).
3. O'Connor, J.R.; Smiltens, J., eds. <u>Silicon Carbide. A High Temperature Semiconductor</u>. New York: Permagon Press (1960).
4. Marshall, R.; Faust, J.; Ryan, C., eds. <u>Silicon Carbide-1973</u>. Columbia, S.C.: University of South Carolina Press (1974).
5. Campbell, R.B.; *IEEE Trans. Indus. Electr.*, IE-29, 124 (1980).
6. Bessman, T.M.; Stinton, D.P; Lowden, R.A.; *MRS Bull.* 45 (1988).
7. Fujiwara, Y.; Sakuma, E.; Misawa, S.; Endo, K.; Yoshida, S. *Appl. Phys. Lett.* 49, 388 (1986).
8. Interrante, L.V.; Czekaj, C.L.; Lee, W. *NATO Meeting Proc.-The Mechanisms of Reaction of Organometallic Compounds with Surfaces*, Lomdon: Plenum Press (1988), in press.
9. Kriner, W. *J. Org. Chem.* 29, 1601 (1961).
10. Interrante, L.V.; Lee, W.; McConnell, M; Lewis, N.; Hall, E. *J. Electrochem Soc.* in press.
11. Kurtz, E.A.; Hudson, J. *Surface Sci.*, 195, 15 (1988).
12. Palik, E.D., ed. <u>Handbook of Optical Constants of Solids</u>. New York: Academic Press (1985), p 587.

CHEMICAL VAPOR DEPOSITION OF CHROMIUM CARBIDE
FROM ORGANOMETALLIC PRECURSORS

N.M. RUTHERFORD, C.E. LARSON, AND R.L. JACKSON
IBM Almaden Research Center, 650 Harry Road, San Jose, CA 95120

ABSTRACT

Chemical vapor deposition from $CpCr(CO)_3H$, $Cr(CH_2CMe_3)_4$, and $Cr(NPr^i{}_2)_3$ at low pressures yields chromium carbide films at temperatures as low as 330 °C. The films are mirror-bright and extremely smooth with no significant features observable by SEM. In general, films prepared at higher temperatures had lower resistivities, and were more likely to be crystalline than those prepared at lower temperatures. However, amorphous films deposited at 330 °C were much harder than the crystalline films. Chemical vapor deposition from $Cr(NPr^i{}_2)_3$ resulted in carbide films with the most desirable properties.

INTRODUCTION

Interest in metal-organic chemical vapor deposition (MOCVD) of thin films has seen a resurgence during recent years, in areas from micoelectronics to tribology [1–3]. Much of the reason behind the renewed interest in the area of microelectronics has been in response to the need for lower processing temperatures.

Much success has been found with the use of organometallic compounds for deposition of semiconductors such as GaAs from $GaMe_3$ and AsH_3 [4], InP from $InEt_3$ and PH_3 [5], and ZnSe from $ZnMe_2$ and MeSeH or Me_2Se [6]. In addition, many main group and late transition metals have been deposited in pure form from metal alkyls, carbonyl complexes, and arene complexes [7–9]. The early transition metals have not yet been deposited in pure form, undoubtably due to their extreme reactivity. Nevertheless, there have been many cases where useful compounds such as early transition metal silicides [10], carbides [11], borides [12] and oxides [13] have been deposited at low temperatures where previously they had been deposited only at temperatures in excess of 1000 °C. In the present paper, we explore organometallic precursors to chromium carbide, Cr_3C_2, with an interest in observing how the metal ligands (arene, carbonyl, amide, or alkyl) affect the properties of the final solid state product.

EXPERIMENTAL

The low pressure MOCVD system is described below. The stainless steel chamber was pumped by a Balzers 50 l/s turbomolecular pump (TMP). The base pressure of the system was approximately 10^{-6} torr. During deposition, the TMP was isolated from the system by a vacuum gate valve, and the chamber was pumped by a liquid N_2-trapped mechanical pump. Helium, argon, and hydrogen carrier gases passed through O_2 and water filters before being introduced into the precursor chamber using mass flow controllers. Typical flow rates for the inert carrier gas and hydrogen gas were 17-20 and 3 sccm, respectively. The total pressure in the chamber, usually maintained at 300 mtorr, was controlled by throttling the mechanical pump. The silicon substrates were placed on a copper block (2" dia. x 1/4") heated by a tungsten filament situated directly beneath it. The temperature of the block was monitored using a thermocouple attached to the surface, while a proportional temperature controller maintained a constant temperature during deposition. The temperature of the silicon substrate was measured with an Omega OS-1000 infrared pyrometer. Deposition studies were carried out at 330 °C and 473 °C with typical film growth periods were on the order of 60 minutes.

Composition of the deposits was determined by Auger electron spectroscopy (AES) and X-ray photoelectron spectroscopy (XPS) performed by Surface Science Laboratories (Mountain View, CA). Microhardness data were recorded on a commercially available nanoindenter by Dr. Richard White of the IBM General Products Division. Film thicknesses were determined by profilometry (Alpha-step). Resistivities were determined by the four point probe method. CpCr(CO)$_3$H, Cr(CH$_2$CMe$_3$)$_4$, and Cr(NPr$^i_2$)$_3$ were synthesized according to literature procedures [14-16], and were purified by vacuum sublimation. Condensable by-products of the CVD reaction were analyzed by mass spectrometry.

RESULTS

Deposition from CpCr(CO)$_3$H

Sublimation of CpCr(CO)$_3$H occurs readily at room temperature and the measured vapor pressure of the compound is 155 mtorr at 25 °C. Low pressure chemical vapor deposition from CpCr(CO)$_3$H at a substrate temperature of 330 °C gave faintly yellow, mirror-like films on the surface of silicon substrates. Scanning electron micrographs of the films show them to be extremely smooth and uniform, with no significant features above the nanometer level. X-ray diffraction (XRD) scans show only broad, diffuse scattering in the 10-90° (2θ) region, suggesting a thin film structure which is amorphous. Nevertheless, the films adhere very well to the silicon substrates and are quite difficult to scratch. Microhardness measurements gave values between 8 and 11 GPa. Polycrystalline Cr$_3$C$_2$ is reported to have a microhardness of 12.75 to 17.65 GPa [17]. An average resistivity of 360 $\mu\Omega$-cm was found for the coatings, compared with a value of 75 $\mu\Omega$-cm expected for stoichiometric Cr$_3$C$_2$ [17]. Auger electron spectroscopy revealed a Cr to C ratio of 1.5:1. AES depth profile studies indicate a high oxygen content near the surface of the films which falls off to 17% in the bulk of the film. The oxide overlayer most likely results from surface oxidation of the films upon exposure to air, whereas oxygen in the bulk may be due to incomplete decomposition of the precursor, diffusion of oxygen from the surface into the film, or less than optimum vacuum conditions.

As with many CVD reactions, the thermal decomposition pathway from the precursor to the final product has not yet been established. In an effort to understand the chemistry behind the reaction, the condensable, volatile products were isolated from the CVD reaction. Not surprisingly, the products were identified as cyclopentadiene, dicyclopentadiene, and cyclopentene, suggesting that protonation or π-M-C$_5$H$_5$ bond homolysis had occurred. A recent labeling study of the vapor decomposition of C$_7$H$_8$Cr(CO)$_3$ demonstrated that less than 13% of the carbon in the deposited carbide could be attributed to the carbonyl ligands [18]. Thus, it seems likely in the present case that most of the carbon originates from the cyclopentadiene ring. The M-CO bonds are weaker than the Cp-M bond and are likely to be broken first during thermolysis [19], thus leaving a Cp-Cr$^+$ fragment. Bond homolysis may occur, leaving the Cp radical to decompose on a forming Cr surface, or perhaps a π-Cp to σ-Cp bond shift, followed by intramolecular decomposition to smaller metal bound hydrocarbons and ultimately carbide [20]. More detailed studies are obviously necessary before the actual mechanism can be elucidated.

Under identical pressure and temperature conditions, addition of 15% pure hydrogen (as a reductant) to the carrier gas stream prior to introduction into the precursor chamber did not significantly alter the composition or properties of the films.

At a substrate temperature of 473 °C, the appearance of the as-deposited films are again mirror-bright. Results of electrical measurements on the films gave an average resistivity of 265 $\mu\Omega$-cm, and no discernible XRD patterns were observed. AES results reveal a Cr to C ratio of 1.5:1, and depth profile studies indicate that the oxygen content in the bulk of the films has dropped to 2.5%. X-ray photoelectron spectroscopy results (Fig. 1) indicate that the chromium peak at 574.2 eV corresponds to chromium in a reduced form, while the peak at 575.8 eV (which decreases to negligible amounts with increasing argon ion sputtering times) can be attributed to Cr_2O_3. The carbon (1s) spectrum exhibits a peak at 284.6 eV, a value associated with hydrocarbon or C-C bonds, and another peak at 282.8 eV attributed to carbon in chromium carbide. The carbon peak at 284.6 eV disappears at a sputtering depth of 300 A, and at this depth one low intensity peak remains in the O(1s) spectrum at 530.6 eV corresponding to Cr_2O_3.

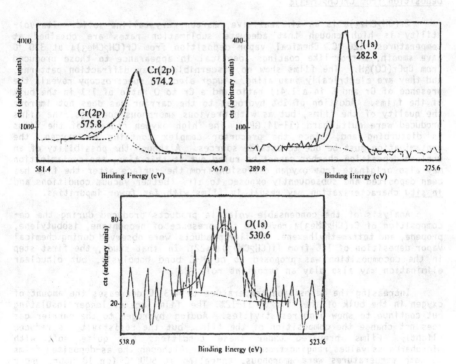

Figure 1: XPS spectra of chromium carbide film from CpCr(CO)₃H

Addition of 15% hydrogen to the carrier gas did not result in a significant change in the AES determined film composition. The appearance of sharp lines in the x-ray diffraction pattern, however, indicates that a crystalline carbide has formed. The pattern is most closely correlated with that of Cr_3C_2, although some shifting of the peaks from those found for Cr_3C_2

is evident. The slightly irregular pattern can be attributed to lattice strain caused by non-stoichiometry. Under these deposition conditions (473 °C, 15% H_2) the films appear somewhat cloudy due to surface roughness. Scanning electron micrographs show the surface to have a blastular texture. Formations of this type may be due to homogeneous gas phase decomposition causing the film to be deposited as fine particulates. High deposition temperatures have previously been observed to result in sooty powders in the decomposition of $CpFeCo(CO)_6$ [8] and $C_7H_8Cr(CO)_3$ [18]. Microhardness measurements show that the crystalline films are "softer" (2 to 8 GPa) than the amorphous low temperature films. This result is surprising in view of the observation that hydrocarbon inclusions and defect sites significantly decrease the hardness of a carbide, while often increasing the mechanical strength of the coating [21]. However, as would be expected, the resistivities of the crystalline coatings (110 $\mu\Omega$-cm) are lower than those found for amorphous films deposited at the same temperature.

Deposition from $Cr(CH_2CMe_3)_4$

$Cr(CH_2CMe_3)_4$ is an air-sensitive, maroon, crystalline solid. Its volatility is high enough that adequate sublimation rates are obtained at temperatures >120 °C. Chemical vapor deposition from $Cr(CH_2CMe_3)_4$ at 330 °C gave smooth, mirror-like coatings identical in appearance to those produced from $CpCr(CO)_3H$. The films show no discernible X-ray diffraction patterns, and they are electrically insulating. Auger electron spectroscopy reveals the presence of Cr and C in a 1.4:1 ratio and a Cr to O ratio of 1:1 in the bulk of the films. Addition of 15% hydrogen to the carrier gas does not improve the quality of the films, but as with previous amorphous coatings, the films produced were quite hard (11-14 GPa). The high oxygen content of the films is disturbing and, since the precursor complex contains no oxygen, the contamination must be due to outside sources. Although the possibility of an unclean deposition chamber cannot be ruled out at this time, the contamination may also originate from oxygen diffusion from the surface after the film has been deposited and subsequently exposed to air. Better vacuum conditions and in situ characterization may result in films with far fewer impurities.

Analysis of the condensable volatile products produced during the decomposition of $Cr(CH_2CMe_3)_4$ revealed the presence of neopentane, isobutylene, propene, and tetramethylhexene. Similar products were observed during chemical vapor deposition of TiC from $Ti(CH_2CMe_3)_4$ [22]. In that study, the first step in the decomposition was proposed to be M-C bond homolysis, but binuclear elimination may also play an important role [23].

Increasing the deposition temperature to 473 °C decreases the amount of oxygen in the bulk of the film to 11-12%. The films are no longer insulating but continue to show high resistivities. Adding hydrogen to the carrier gas does not change the composition of the films, but the resistivity is reduced slightly. Films prepared under these conditions are quite soft with microhardness values ranging from 2 to 3 GPa. Although the as-deposited films at both temperatures were amorphous, annealing at 600 °C for 18 hours under vacuum (10^{-3} torr) resulted in crystallization of the high temperature films, while the low temperature films remained amorphous.

Deposition from $Cr(NPr^i_2)_3$

$Cr(NPr^i_2)_3$ is an air-sensitive, deep red, crystalline solid. The magnetic susceptibility of the compound indicates that a monomeric, 3-coordinate species is present in the solid state [16]. Undoubtably the bulky isopropyl ligands prevent chromium from achieving a higher coordination. The volatility of

$Cr(NPr^i{}_2)_3$ is high enough that adequate sublimation rates are obtained at temperatures >110 °C. Films deposited at a substrate temperature of 330 °C by chemical vapor deposition from $Cr(NPr^i{}_2)_3$ were mirror bright and very difficult to scratch. AES studies gave signals for Cr, C and N with an oxygen signal near the AES detection limit. The corresponding atomic composition, 56% Cr, 29% C, 13% N and ca. 2% O gives a Cr to C ratio of 1.9 : 1. X-ray diffraction scans show that the films are amorphous as deposited, but when annealed for 18 hours at 600 °C under vacuum (10^{-3} torr), crystallization occurs. Again, the observed XRD pattern most closely correlates with the pattern found for Cr_3C_2.

Addition of 15% hydrogen to the carrier gas results in a slight increase in the carbon content of the films. XPS results on these films, sampled after argon ion etch to 300 Å, indicate that carbon is present in two forms: carbide (283.0 eV), and inorganic/organic carbon, C-R (R = C, H) (284.6 eV). Chromium was present in the reduced form (574.3 eV), attributed to chromium carbide, and in an oxidized form (576.0 eV) associated with Cr_2O_3. In addition, the films contained nitrogen in a reduced form which can be attributed to chromium nitride (397.3 eV). No organic fragments containing nitrogen were detected at this etch depth. Microhardness measurements indicate that the films deposited from $Cr(NPr^i{}_2)_3$ are the hardest films deposited thus far, with values ranging from 17.5 to 19 GPa. All of the deposited films had similar resistivities (265 to 291 $\mu\Omega$-cm) regardless of carrier gas composition.

Condensable products of the decomposition reaction contained diisopropylamine, propene and isopropylamine. The thermal decomposition of a similar precursor, $Ti(NMe_2)_4$, was previously reported to yield films of TiN [24]. Other studies have found, however, that titanium carbide is is deposited from the metal amide precursor [22,25]. The first step in the thermolysis of early transition metal amides has been shown to be intermolecular metallation with resulting M-C bond formation. However, further heating of the metalacyclic compound yields an imido metal complex, $RN=M(NR_2)_3$, with loss of hydrocarbon, R [26]. If this were the case in the present decomposition reaction, one would expect to see both diisopropylamine and propene in the reaction products. It is as yet unclear by what mechanism the carbide, rather than the nitride, is formed. For titanium, TiC is favored thermodynamically. In the case of chromium, Cr_3C_2 is not favored over Cr_2N or CrN, but formation of $Cr_{23}C_6$ and Cr_7C_3 are favored over either nitride [26].

At a substrate temperature of 473 °C, films deposited from $Cr(NPr^i{}_2)_3$ were black or brown in color, or were mirror-bright. XRD scans show that the films are partially crystalline, and have relatively low resistivities (110 $\mu\Omega$-cm). AES results show a Cr to C ratio of 1.4 : 1 and a slight increase in the amount of oxygen found in the bulk of the films formed at the higher deposition temperature. In contrast, the nitrogen content of the films decreased slightly.

Addition of 15% hydrogen to the carrier gas results in films which are quite crystalline and which have very low resistivities, although they still contain 8 to 9% nitrogen. Interestingly, the Cr to C ratio has dropped to 1.3 : 1, with the oxygen level below the AES detection limits in these films. The films are slightly cloudy in appearance and do not adhere to the silicon substrate as well as the amorphous or partially crystalline films. XPS data taken at an etch depth of 300 Å (Fig. 2) indicate that the films deposited at 473 °C contain carbon in the carbide form plus a smaller amount in organic fragments. The Cr(2p) spectrum shows a peak at 576 eV associated with Cr_2O_3 and a peak at lower binding energy which has been assigned to chromium carbide. The observation that chromium oxide is detected by XPS but not by AES at this depth may be due to reoxidation of the surface between the time the etching process stops and aquisition of the data. However, a low intensity peak at

444

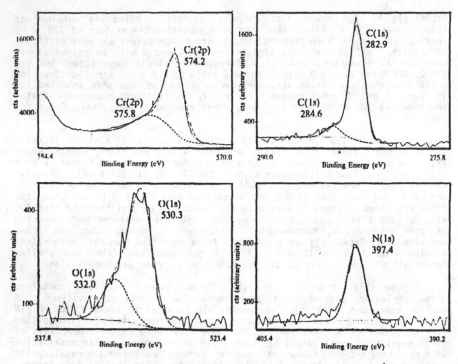

Figure 2: XPS spectra of chromium carbide film from Cr(NPr^i$_2$)$_3$

532 eV in the O(1s) spectrum indicative of C-O bonding was also observed. As was the case with the low temperature films, no peaks corresponding to organic nitrogen were observed. Films prepared from Cr(NPr^i$_2$)$_3$ under these conditions were some of the softest films deposited during this study, with microhardness values of 1 to 3 GPa.

CONCLUSIONS

Each of the precursor compounds studied gave films consisting predominantly of carbon-rich chromium carbide. Only films deposited from Cr(CH$_2$CMe$_3$)$_4$ at low temperatures contained appreciable amounts of oxygen, and this appeared to be due to deposition conditions, rather than to the nature of the precursor. Low levels of nitrides found in the films deposited from Cr(NPr^i$_2$)$_3$ did not adversely affect the film properties, and in fact, these films were the most crystalline, and had the lowest resistivities of all the films deposited. In contrast, high levels of oxygen in the bulk of the films correlated with higher resistivities. For all three of the precursors, two clear trends prevailed. First, the films deposited at 473 °C were more likely to be crystalline, and they had lower resistivities. Second, films deposited at 330 °C were amorphous and very hard. Finally, carrier gas compositions of 15% hydrogen and 85% argon or helium, versus 100% argon or helium, had no clear effect on the composition or properties of the films.

REFERENCES

1. J.O. Williams, Spec. Publ. R. Soc. Chem. <u>60</u> 1 (1986).
2. H.M. Manasevit, J. Cryst. Growth <u>13-14</u>, 306 (1972).
3. H.E. Hintermann, A.J. Perry, E. Horvath, Wear <u>47</u>, 407 (1978).
4. R.D. Dupis, Science <u>226</u>, 623 (1984).
5. K. Uwaik, H. Nakagome, K. Takahei, Appl. Phys. Lett. <u>50</u>, 977 (1987).
6. S. Fujita, T. Sakamoto, M. Isemura, S. Fujita, J. Cryst. Growth <u>87</u>, 581 (1988).
7. M.L. Green, R.A. Levy, R.G. Nuzzo, E. Coleman, Thin Solid Films <u>114</u>, 367 (1984).
8. L. Czekaj, G.L. Geoffroy, Inorg. Chem. <u>27</u>, 8 (1988).
9. D.E. Trent, B. Paris, H.H. Krause, Inorg. Chem. <u>3</u> 1057 (1964).
10. B.J. Aylett, H.M. Colquhoun, J. Chem. Soc., Dalton, <u>1977</u>, 2058.
11. G.S. Girolami, A.E. Kaloyeros, C.M. Allocca, J. Am. Chem. Soc. <u>109</u>, 1579 (1987).
12. J.S. Jensen, J.E. Gozum, D.M. Pollina, G.S. Girolami, J. Am. Chem. Soc. <u>110</u> 1643 (1988).
13. L.A. Ryabova, YA.S. Savitskaya, Thin Solid Films <u>2</u>, 141 (1968).
14. E.O. Fischer, Inorg. Synth. <u>7</u>, 136 (1963).
15. E. Mowat, N.J. Hill, A.J. Shortland, G. Wilkinson, J. Chem. Soc., Dalton, <u>1973</u>, 770.
16. E.C. Alyea, J.S. Basi, D.C. Bradley, M.H. Chisholm, Chem. Comm., <u>1968</u>, 495.
17. G.V. Samsonov, I.M. Vinitskii, <u>Handbook of Refractory Metal Compounds</u>, (Plenum, New York, 1980), Chapter III, V.
18. T.J. Truex, R.B. Saillant, F.M. Monroe, J. Electrochem. Soc. <u>122</u>, 1396 (1975).
19. J.A. Conner, Topics in Current Chemistry, <u>71</u>, 71 (1977).
20. J.E. Bercaw, R.H. Marvich, L.G. Bell, H.H. Brintzinger, J. Am. Chem. Soc. <u>94</u>, 1219 (1972).
21. L.E. Toth, <u>Transition Metal Carbides and Nitrides</u>, edited by J.L. Margrave (Refractory Materials, <u>7</u>, Academic Press, New York, 1971), Chapter V.
22. G.S. Girolami, personal communication, 1987.
23. P.J. Davidson, M.F. Lappert, R. Pearch, Chem. Rev. <u>76</u>, 219 (1976).
24. K. Sugiyama, S. Pac, Y. Takahashi, S. Motojima, J. Electrochem. Soc. <u>122</u>, 1545 (1975).
25. M.R. Hilton, G.J. Vandentop, M. Salmeron, G.A. Somorjai, Thin Solid Films <u>154</u>, 377 (1987).
26. T. Takahashi, N. Onoyama, Y. Ishikawa, S. Motojima, K. Sugiyama, Chem. Lett., 525 (1978).
27. M.W. Chase, Jr., C.A. Davis, J.R. Downey, Jr., D.J. Frurip, R.A. McDonald, A.N. Syverud, <u>JANAF Thermochemical Tables</u>, 3rd ed. (American Chemical Society <u>14</u>, suppl. 1, 1985).

NEW PRECURSORS FOR THE ORGANOMETALLIC
CHEMICAL VAPOR DEPOSITION OF ALUMINUM NITRIDE

WAYNE L. GLADFELTER*, DAVID C. BOYD*, JEN-WEI HWANG*, RICHARD T. HAASCH*, JOHN F. EVANS*, KWOK-LUN HO#, AND KLAVS F. JENSEN#
Departments of #Chemical Engineering and Materials Science, and *Chemistry, University of Minnesota, Minneapolis, Minnesota   55455

ABSTRACT

Organometallic aluminum azides have been found to be effective precursors for the low temperature chemical vapor deposition of thin films of aluminum nitride. Quantitative analysis of the gas phase products of the reaction are used to develop an understanding of the reaction. Rate studies of the deposition were performed in the temperature range from 400 to 800°C. Below 525°C, an activation barrier of 26.4 kcal/mol was found, while above 525°C, a value of 5.23 kcal/mol was obtained. The effects of the presence of N-C bonds and the type of Al-N interaction within the precursor are evaluated.

INTRODUCTION

Aluminum nitride has the wurtzite crystal structure and several physical properties that make it especially interesting, both as a bulk material and as a thin film. As a large band gap (6.2 eV) III-V compound[1], which is very hard, resistant to chemical attack and high melting (2400°C)[2], it has useful properties for coatings. Its piezoelectric nature makes it potentially important for surface acoustic wave devices, and its high thermal conductivity may be exploited in applications for packaging electronic microcircuits[3].

The nitrogen sources in these previous studies have included ammonia[4,5] (the most common source), hydrazine[6] and alkylaluminum amido compounds[7,8]. We describe here studies on the use of azide[9,10] group as the reactive nitrogen source and compare these results to precursors which contain direct N-C bonds.

THIN ALUMINUM NITRIDE FILMS FROM [Et$_2$AlN$_3$]$_3$

Two reactors were used to study the CVD of AlN. The "survey" reactor had a hot-wall quartz tube which was usually operated without carrier gas and had a base pressure of 5 x 10^{-5} torr. A removable liquid N$_2$ trap placed between the quartz tube and the pump allowed the trapping and quantitation of all volatile products except

N_2, H_2, and CH_4. A schematic of the second reactor in which all rate measurements were performed is shown in Figure 1.

Figure 1

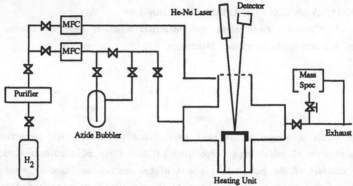

Pure hydrogen, produced by a Matheson Pd-alloy purifier, was used as the carrier gas for the $[Et_2AlN_3]_3$, and was also used to purge the viewport on the reactor chamber. The reactor chamber was a six-way stainless steel cross. The susceptor was a Mo plate heated radiatively by a Mo wire carrying a current of a few amperes. The growth rates were measured *in situ* with a 6328 Å He-Ne laser interferometry assembly[11]. Exhaust gases and unreacted precursor were cracked by passing them through a high-temperature furnace and filtering before venting to a hood.

Si(100) substrates were prepared by degreasing with trichloroethylene and methanol, oxidizing with nitric acid, and etching with HF. Typical growth conditions for this system were as follows: substrate temp., 400 - 800°C; $[Et_2AlN_3]_3$ temp., 40 - 60°C; gas manifold pres., 3 - 5 torr; reactor pres., 3 - 5 torr; azide/H_2 flow rate, 30 - 60 sccm; makeup H_2 flow rate, 5 - 20 sccm. The analytical data for the films, obtained using X-ray photoelectron spectroscopy (XPS), are summarized in Table 1, and they emphasize that the films produced in the two different reactors have similar compositions.

Gas Phase Products. The analysis of the gases produced from the reaction of $[Et_2AlN_3]_3$ ($T_{(precursor)}$ = 40°C, $T_{(reactor)}$ = 500°C) was obtained using gas chromatography and reproducibly showed 70% C_2H_4, 30% C_2H_6, and a very small amount of butane and butenes (C_4/C_2 = 0.03). Based on the mass of the precursor consumed during the deposition and the total pressure of the C_2H_4, C_2H_6, and C_4 products trapped, 89% of the precursor ethyl groups were accounted for in this analysis. Similar results were obtained when the furnace temperature was increased to 650°C (C_2H_4 = 76%, C_2H_6 = 24%, C_4/C_2 = 0.02, 95% recovery); or when the deposition was conducted under 1 torr of H_2 carrier gas (C_2H_4 = 65%, C_2H_6 = 35%, C_4/C_2 = 0.01). For comparison, analysis of the gaseous products from CVD of

$[Et_2AlNH_2]_3[12]$ under similar conditions, indicates the formation of C_2H_6 was significantly favored over C_2H_4 ($C_2H_4 = 25\%$, $C_2H_6 = 75\%$, $C_4/C_2 = 0.02$, 81% recovery).

Table 1. Atomic Compositions of Thin Films Using XPS after Ar^+ sputtering.

Precursor[a]	Reactor	Temp. (°C)	%Al	%N	%C	%O
$[Et_2AlN_3]_3$	quartz	500	43	45	10	2
$[Et_2AlN_3]_3$	metal	480	44	45	10	1
$[Me_2AlN_3]_3$	quartz	500	43	44	11	2
$[Et_2Al(NH_2)]_3$	quartz	550	44	41	7	8
$[Me_2Al(NMe_2)]_2$	quartz	700	42	19	38	1
$[Me_2Al(Azir)]_3$	quartz	500	34	18	33	15
$[Et_2Al(NH-t-Bu)]_2$	quartz	700	44	36	15	4
$Et_3AlNH_2(t-Bu)$	quartz	500	30	2	5	63

a) azir = NCH_2CH_2; Me = CH_3; Et = CH_2CH_3; t-Bu = $C(CH_3)_3$

These results are consistent with a β-hydrogen elimination mechanism operative in cleaving the Al-Et bond. An alternative mechanism, homolytic Al-Et bond cleavage, would form ethyl radicals which would couple or disproportionate to yield C_4H_{10} and C_2H_4/C_2H_6, respectively. The gas phase reactivity of ethyl radicals has been studied thoroughly by Lalonde and Price[13], from which they concluded that radical coupling is favored over disproportionation by a factor of 10. The analogous ratio under the experimental conditions employed in the current CVD studies is 0.03 indicating that liberation of gaseous ethyl radicals is not the predominant mechanism for Al-Et bond cleavage. The dramatic change in the C_2H_4 to C_2H_6 ratio that occurs upon changing the N-source to an NH_2 group is consistent with a change in the Al-Et bond cleavage path. Presumably, the availability of the weakly acidic hydrogens of the NH_2 group allows the facile protonolysis of the Al-C bond[12].

Rates and Mechanism of Film Growth Using $[Et_2AlN_3]_3$. Figure 2 shows the dependence of the growth rate on substrate temperature. In the low temperature regime (< 525°C), the growth rate increases with temperature with an apparent activation energy of 26.4 kcal/mol. At higher temperatures, the growth rate shows a weaker dependence on temperature (apparent activation barrier of 5.2 kcal/mol). This activation barrier is greater than that (~2 kcal/mol) corresponding only to mass transfer limited growth. Therefore, the change in the activation energy at high temperatures most likely represents a change in the chemical mechanism combined with mass transfer effects. Behavior such as this has been observed in other CVD studies[14]. Additional data is being collected to evaluate the relative importance of surface reactions versus homgeneous (gas phase) reactions.

Figure 2

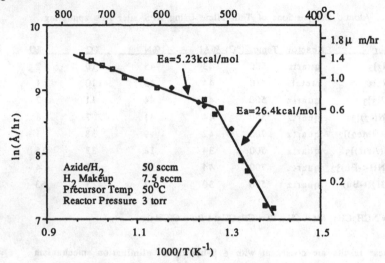

These results suggest elementary steps for several phases of the deposition process. Scheme 1 shows one of the possible scenarios for the mechanism of the reaction.

Scheme 1.

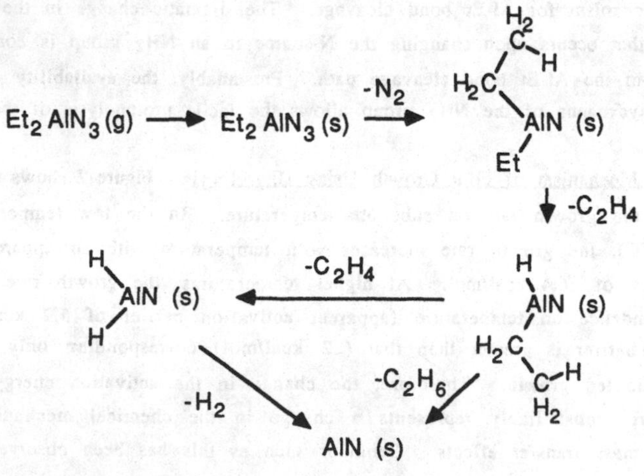

It must be emphasized that the relative timing of these events remains unknown. Further studies using temperature programmed reaction spectroscopy are planned to more completely elucidate the mechanistic details.

NITROGEN SOURCES CONTAINING N-C BONDS

Both azide (N_3) and amide (NH_2) are comparably effective as sources for nitrogen in the deposition of AlN. It was of interest to explore how much the variation in structure of the N-source would influence the AlN deposition, and we were particularly interested in determining the effect of incorporating a N-C bond into the precursor. The precursors studied included [Me$_2$Al(NMe$_2$)]$_2$[15], [Me$_2$Al(Azir)]$_3$ (Azir = aziridine)[16], Et$_3$AlNH$_2$(t-Bu)[17], and {Et$_2$Al[NH(t-Bu)]}$_2$[17]. All of these structures contain direct Al-N bonds, and with the exception of Et$_3$AlNH$_2$(t-Bu), all exist as cyclic four- or six-membered rings. The structures differ from the azide and amide precursors primarily due to the presence of the N-C bond. The substantial impact made by the presence of the N-C bond is highlighted in Table 1. All of these films contained large concentrations of carbon, and one contained only small amounts of nitrogen. In the case where the nitrogen source is the t–butylamine-triethylaluminum donor acceptor complex, the small concentrations of N indicate that a facile Al-N cleavage is operative, e. g. simple dissociation of the donor-acceptor bond. It should be noted that [Me$_2$Al(NMe$_2$)]$_2$, which also gave films with low N values, was the most stable of all the precursors examined. Even at oven temperatures of 450°C, this precursor would pass through the hot tube unchanged. The lower N content of the film may be the result of an Al-N bond cleavage process involving β-hydrogen elimination of the NMe$_2$ ligand. The intermediate case of [Et$_2$Al(NH-t-Bu)]$_2$ is interesting because it contains no β-hydrogens on the amido ligand. It is also related to [Me$_2$Ga(As-t-Bu$_2$)]$_2$ which was recently reported to give GaAs films[18]. The results show that most of the nitrogen is incorporated into the film; unfortunately, the carbon content is also high. The compound containing the 3-member, highly strained aziridine ring, [Me$_2$Al(Azir)]$_3$, gave results similar to those found for [Me$_2$Al(NMe$_2$)]$_2$.

ACKNOWLEDGEMENTS

This research was supported by a grant from the Materials Chemistry Initiative of the National Science Foundation (CHE-8711821).

REFERENCES

1. W. M. Yim, E. J. Stofko, P. J. Zanzucchi, J. I. Pankove, M. Ettenberg, and S. L. Gilbert, J. Appl. Phys. 44, 292 (1973).

2. R. C. Weast (ed) "Handbook of Chemistry and Physics, 51st Ed.", The Chemical Rubber Company.: Cleveland (1970).

3. R. T. Baker, J. D. Bolt, U. Chowdhry, U. Klabunde, G. S. Reddy, D. C. Roe, R. H. Staley, F. N. Tebbe, and A. J. Vega, Mat. Res. Soc. Symp. Proc. 121, 471 (1988).

4. H. M. Manesevit, F. M. Erdmann, and W. I. Simpson, J. Electrochem. Soc. 118, 1864 (1971).

5. K. Sayyah, B. -C. Chung, and M. Gershenzon, J. Crystal Growth 77, 424 (1986).

6. D. K. Gaskill, N. Bottka, and M. C. Lin, J. Crystal Growth, 77, 418 (1986).

7. L. V. Interrante, L. Carpenter, Jr., C. Whitmarsh, W. Lee, M. Garbaukas, and G. A. Slack, Mat. Res. Soc. Symp. Proc. 73, 359 (1986).

8. L. V. Interrante, W. Lee, M. McConnell, N. Lewis, and E. Hall, J. Electrochem. Soc., in press.

9. R. K. Schulze, D. R. Mantell, W. L. Gladfelter, and J. F. Evans, J. Vac. Sci. Technol. A6, 2162 (1988).

10. D. C. Boyd, R. T. Haasch, D. R. Mantell, R. K. Schulze, J. F. Evans, and W. L. Gladfelter, Chemistry of Materials, 1, in press (1989).

11. K. Sugawara, T. Yoshimi, H. Okuyama, and T. Shirasu, J. Electrochem. Soc. 121, 1233 (1974).

12. F. C. Sauls and L. V. Interrante, private communication.

13. A. C. Lalonde and S. J. W. Price, Can. J. Chem. 49, 3367 (1971).

14. J. M. Jasinski, B. S. Meyerson, and B. A. Scott, Ann. Rev. Phys. Chem., 38, 109 (1987).

15. N. Davidson and H. C. Brown, J. Am. Chem. Soc., 64, 316 (1942).

16. J. L. Atwood and G. D. Stucky, J. Am. Chem. Soc., 92, 285 (1970).

17. K. Gosling, J. D. Smith, and D. H. W. Wharmby, J. Chem. Soc. (A), 1738 (1969).

18. A. H. Cowley, B. L. Benac, J. G. Ekerdt, R. A. Jones, K. B. Kidd, J. Y. Lee, and J. E. Miller, J. Am. Chem. Soc. 110, 6248 (1988).

ADHESION OF METALS TO MIXED OXIDE COATINGS (Al & Cr, Mo, OR W) PREPARED BY SPRAY PYROLYSIS OF ORGANOMETALLICS.

JAMES M. BURLITCH*, GERARD J. DeMOTT**, AND DAVID L. KOHLSTEDT**

* Dept. of Chemistry, Cornell University, Ithaca, NY 14853-1301.
**Dept. of Materials Science & Engineering, Cornell University, Ithaca, NY 14853.

ABSTRACT

Spray pyrolysis of the mixed metal organometallics, $Al[M(CO)_3C_5H_5]_3$ (M is Cr, Mo or W), prepared in tetrahydrofuran from the mercury analogues and aluminum metal, gave mixed oxide gels on heated alumina substrates. Air was used as the propellant for M = Cr, whereas oxygen was used for M = Mo or W. Mixtures of $Al[Cr(CO)_3C_5H_5]_3$ and $Al(O-i-C_3H_8)_3$ having a Cr/Al ratio of 0.02 to 0.93 were used to prepare gels with intermediate dopant concentrations. When calcined at 500 °C, the gels produced amorphous coatings of two-component oxides that were characterized by SEM and EDS methods. The strength of adhesion of a E-beam evaporated chromium metal overlayer to the coating, as measured by a continuous microindentation method, increased in the order Cr < Mo $\cong$ W; adhesion was found to be directly proportional to the effective free energy of formation of the oxide mixture.

INTRODUCTION

Control of the strength of a metal/ceramic interface is important in many recent technological developments. For example, the use of ceramic substrates for integrated circuits [1] requires a metal conductor film bonded directly to the ceramic. Two mechanisms have been proposed to describe the bonding of metals to oxides. A 'chemical' interaction may occur when the metal-to-ceramic interface consists of a continuous, thin layer of an oxide of the metal[2]. Alternatively, a mechanical 'interlocking' of the metal and the oxide may be responsible for bonding[3]. Chromium metal has been employed as an intermediate layer between copper thin films and aluminum oxide substrates. Recent studies of the strength of adhesion of chromium to gel-derived oxide coatings by means of pull tests [4] and by a continuous microindentation method [5] have suggested that both of these mechanisms may be operative. The amount and type of a dopant metal which could be introduced into the coatings was limited by segregation in the gel process.

To learn more about the factors which may affect the adhesion of chromium to metal oxides and, in particular, to learn what might be done to improve the strength of the bond between these two dissimilar materials, we prepared a series of amorphous, oxide coatings on alumina substrates, in which a Group VI metal was introduced by means of the mixed metal complex, $Al[M(CO)_3C_5H_5]_3$ (M is Cr, Mo or W) [6], and we measured the chromium metal-to-oxide interface bond strength by the microindentation method[7].

EXPERIMENTAL

Preparation of Coatings

Tetrahydrofuran (dry, distilled from potassium benzophenone) solutions of the complexes $Al[M(CO)_3C_5H_5]_3$ (M is Cr, Mo or W) were prepared from the analogous mercury derivatives, $Hg[M(CO)_3C_5H_5]_2$ (1.0 mmole in 30 mL) and excess aluminum metal turnings, according to the method of Petersen [8], and were handled under an argon atmosphere. Aliquots of the filtered (Celite™ filter aid for M = Cr & Mo), clear solutions were transferred from the Schlenk reaction vessel [9] to serum stoppered vials by means of a gas-tight syringe, a drying control agent (DCA), bis-2-(2'-methoxyethoxy)ethanol, was added and a portion of the mixture was placed in a second gas tight syringe for spray pyrolysis (SP). The SP apparatus, set up in a fume hood, consisted of a 1 mL gas-tight syringe equipped with a 22 gage tapered needle and driven by Sage model 341A syringe pump. A 16 gage, blunt, syringe needle, connected via 1/4 in. plastic tubing to a compressed gas cylinder was placed at 90 ° to the syringe needle at a distance of ca. 2-3 mm and was directed at a 2 x 5 cm alumina substrate (IBM) held on an aluminum foil covered hotplate at a distance of ca. 5 - 9 cm. The hotplate surface temperature was ca. 165 ° as measured by a thermocouple gage.

In a typical experiment, a 0.5 mL portion of a mixture of 1.0 mL of the 0.5 μ filtered $Al[W(CO)_3C_5H_5]_3$ solution, and 50 μl of DCA was sprayed onto the heated (ca. substrate during a 4 min period with oxygen (at 10 psi) as the propellant. During the application, the hotplate was moved to evenly distribute the pale, tan coating on the substrate; the needle to substrate distance was ca. 5 cm.

Lower concentrations of chromium were achieved by admixture of the solution of $Al[Cr(CO)_3C_5H_5]_3$ with appropriate quantities of a 0.1 M solution of aluminum isopropoxide in tetrahydrofuran. In a typical experiment, 3.0 mL of $Al[Cr(CO)_3C_5H_5]_3$ solution was mixed with 1.5 mL of $Al(OC_3H_8)_3$ solution and 35 μl of DCA. A 1.0 mL portion of this solution was sprayed with air (10 psi) onto the heated substrate at a distance of ca. 9 cm at a rate of 0.25 mL/min.

All of the coatings were heated in high purity alumina crucibles in air to 500 °C, held at this temperature for one hour and then were cooled to room temperature at 400 °C per hour. Scanning electron micrographs of the fired specimens revealed that the surface roughness was similar within this series and was similar to that achieved by sol-derived coatings[4]. Several specimens, prepared by slightly different protocols had cracked. Curiously though, after a too thick coating had been applied, most of it cracked and flaked off but the remainder formed a thin, relatively smooth, crack-free layer, strongly attached to the alumina substrate. The surface roughness of this residual layer was similar to that of the thinner, as sprayed, specimens used in this study. The thickness of the coatings was ca. 0.1 μm for all specimens. Energy dispersive X-ray analysis was used to confirm the presence of the transition element. Chromium metal films (0.5 μm) were deposited on the coatings by electron beam evaporation at 10 Å/sec., the final substrate temperature being ca. 100 °C.

Microhardness Measurements

The apparatus and procedures used for the continuous measurement of microhardness by a diamond Vickers indentor has been described previously[5]. Seven indentations were made on each specimen at each of six maximum loads (0.02 - 0.2 N). The averaged load vs. plastic depth curve was converted to indentation pressure vs. plastic depth, based on the projected area of the indentor at each depth.

Analysis of Microhardness Data

As described in detail in previous reports[5], the hardness measured for a layered composite material is related to the volume of plastically deformed material in each layer; for a film on a substrate, this behavior is described by [10]:

$$H_c = \frac{V_f}{V} H_f + \frac{V_s}{V} H_s$$

where H_c, H_f, H_s and V, V_f and V_s are the hardnesses and volumes of plastic deformation for the composite, film and substrate, respectively. If the mechanical properties of the material are known, the plastic deformation volume, and its radius, b, beneath a Vickers indentor may be calculated[11]. In this adaptation of the volume law of mixtures[5b], the expansion of the zone of plastic deformation is truncated at the metal film/ceramic substrate interface. The transfer of stress across this interface is described by an interface parameter, χ, defined by the equation: $b_{si} = \chi \cdot b_{fi}$, which relates the radii of the zones of plastic deformation (at the interface) in the substrate and film respectively. Thus defined, $\chi = 1$ for an interface with perfect adhesion; with no adhesion, χ will be near zero. After the depth dependent hardness values (H_f and H_s) and the mechanical properties of the chromium and oxide coating were determined on bulk specimens, the interface parameter was determined from the best fit of the calculated H_c vs plastic depth curve to the measured H_c vs plastic depth curve. For the purposes of this study, the coated alumina substrate was considered as one 'layer' since the interface between the chromium metal film and the modified oxide layer is the only area of interest. The interface parameter for a Cr film evaporated directly onto an uncoated alumina substrate was found to be 0.515 ±0.002.

RESULTS AND DISCUSSION

The mixed metal organometallic complexes used in this study had been previously studied for their unusual mode of binding of a carbonyl oxygen to the aluminum center [6], viz. Al[OCM(CO)$_2$C$_5$H$_5$]$_3$ (M is Cr, Mo or W). They were used here because they were known to be extremely reactive toward air and moisture[8], and thus seemed likely to decompose cleanly in a spray pyrolysis procedure. IR spectroscopic analysis of greenish grey test coatings from spray pyrolysis of the chromium complex on heated sodium chloride plates showed no absorptions due to carbonyl groups when air was used as the propellant. With the molybdenum derivative, however, red coatings made with air as propellant had v(CO) at 2010(m), 1945(vs), 1918(vs,br) 1906(vs) and 1894(s) cm^{-1}, suggestive of the presence of [Mo(CO)$_3$(C$_5$H$_5$)]$_2$; this dimer has v(CO) in KBr: 1957, 1926, and

1904 cm^{-1}[12]). With oxygen as propellant, neither the deep blue molybdenum coatings nor the tan colored tungsten oxide coatings had $\nu(CO)$ absorptions.

As shown in Table 1, the interface parameter and the strength of the metal-to-ceramic interface increased when transition metal oxide components were added to the coating. Similar, though considerably smaller effects, had been observed for low concentrations of dopants, added to alumina sols[4,7]. To place these findings on a more quantitative basis, we have attempted to relate the strength of the metal-to-ceramic coating interface to the free energy of oxidation of the coating. Pepper showed that the shear strength of the interface between various metals and sapphire increased as the free energy of oxide formation became more negative, i.e. as ΔG°_f for oxidation of the metal film approached that of aluminum[13]. To accommodate the variable amount of transition metal oxide in the present coatings, we define an effective free energy of formation as the weighted average of ΔG°_f for each metal in the coating [14] as follows

$$\Delta G_f^{eff} = \sum_{i=1}^{n} x_i \, \Delta G_f^{\circ}$$

where x_i is the mole fraction of each metal in the coating.

The data in Table 1 and the plot of χ vs ΔG_f^{eff} shown in Figure I indicate that the adhesion strength of the Cr films increases linearly with the effective free energy of oxidation of the coating.

Table 1. Interface parameters and effective free energies of formation for chromium films on mixed metal oxides.

M	M/Al ratio	χ	ΔG_f^{eff} (Kcal/mole)
-	0.00	0.60 ±0.04	-377.6
Cr	0.02	0.60 ±0.02	-375.1
Cr	0.13	0.62 ±0.02	-362.8
Cr	0.93	0.70 ±0.03	-315.7
Cr	3.00	0.75 ±0.02	-281.3
Mo	3.00	0.85 ±0.03	-190.1
W	3.00	0.89 ±0.03	-189.1

Reaction of chromium metal with the surface oxide ions results in the formation of M(oxide)-O-Cr bonds, perhaps, at the partial expense of the M-O bonds; if the oxide is forced to take on extra electron density from chromium (which is assumed to be oxidized in the process of bond formation), this might result in partial population of the empty band of the oxide, an unfavorable process. The more readily this electron transfer occurs, however, the stronger the Cr-O bond will be. Thus, availability of a low-lying, empty band will facilitate the Cr-O bond formation.

An oxide that has a very negative ΔG_f° is likely to have a large band gap. Oxides like aluminum oxide will form bonds with chromium only reluctantly,

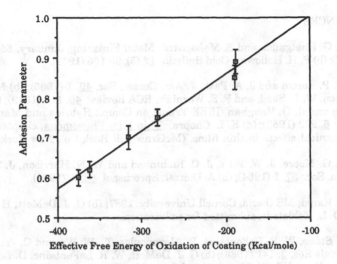

Figure I. A plot of χ vs ΔG_f^{eff} of mixed oxide coatings.

whereas an oxides of a metal, such as molybdenum or tungsten, which has a much more positive free energy of formation will have smaller band gap through which the charge transfer may occur. A small band gap might well have a deleterious effect on other useful properties of the ceramic, such as its dielectric constant, so that the use of very thin coatings to promote adhesion without affecting other bulk properties may provide a practical compromise.

This work has demonstrated the importance of interface chemistry in controlling the adhesion strength of chromium films on alumina substrates. The adhesion, as measured by the microindentation method, is dramatically increased by alteration of the interface chemistry so that the metal film bonds more tightly to the ceramic surface. Keeping the surface as smooth as possible while promoting strong metal-to-ceramic bonds may be particularly useful in meeting the increasingly stringent demands of microelectronic circuitry.

ACKNOWLEDGEMENTS

This work was supported by Corning Glass Works through its support of GJD, by IBM through the Cornell Ceramics Program, and by the U. S. Army Research Office, contract # DAAL03-03-87-K-0104. The electron microscopy was carried out in a central facility of the Cornell University Materials Center (NSF-MRL). Metal films were prepared in the National Nano-Fabrication Facility (NSF grant ECS-8619049). The microindentor was built and is maintained by the research group of Prof. C. -Y. Li.

REFERENCES

1. (a) L. G. Bhatgadde and S. Mahapatra, Metal Finishing, January, 55-57 (1987); (b) P. H. Holloway, Gold Bulletin 12 (3), 99-106 (1979).

2. (a) M. P. Borom and J. A. Pask, J. Am. Ceram. Soc. 49, 1 (1966); (b) M. Caulton, W. L. Sked, and F. S. Wozniak, RCA Review 40, 115 (1979); (c) R. W. Pierce and J. G. Vaughan, IEEE Trans. on Comp., Hybrids and Manuf. Tech. 6, 202 (1983); (d) K. L. Chopra. Thin Film Phenomena, chapter: "Mechanical effects in thin films."(McGraw-Hill Book Co., New York 1969).

3. (a) D. G. Moore, J. W. Pitts, J. C. Richmond and W. N. Harrison, J. Am. Ceram. Soc. 37, 1 (1954); (b) A. Dietzel, Sprechsaal 68, 3, (1935).

4. (a) H. Kanai, MS thesis, Cornell University, 1987; (b) G. J. DeMott, H. Kanai and D. L. Kohlstedt, submitted for publication.

5. (a) D. Stone, W. R. LaFontaine, P. Alexopoulos, T. -W. Wu and C. -Y. Li, J. Materials Res. 3, 141 (1988); (b) G. J. DeMott, W. R. LaFontaine, D. L. Kohlstedt, and C. -Y. Li, submitted for publication.

6. R. B. Petersen, J. J. Stezowski, C. Wan, J. M. Burlitch and R. E. Hughes, J. Am. Chem. Soc. 93, 3532 (1971).

7. G. J. DeMott, PhD thesis, Cornell University, 1989.

8. R. B. Petersen, PhD thesis, Cornell University, 1973.

9   J. M. Burlitch, How to Use Ace No-Air Glassware (Ace Glass Co. Vineland, N.J., Bulletin No. 3841).

10. P. J. Burnett and T. F. Page, J. Materials Sci. 19, 845 (1984).

11. B. R. Lawn, B. J. Hockey and S. M. Wiederhorn, J. Materials Res. 15, 1207 (1980).

12. R. B. King, in Organometallic Syntheses, Vol. 1, edited by J. J. Eisch and R. B. King (Academic Press, New York, 1965), p. 111.

13. S. V. Pepper, J. Appl. Physics 47, 801 (1976).

14. O. Kubaschewski and C. B. Alcock, Metallurgical Thermochemistry, (Pergamon Press, Oxford, 1983).

PART VIII

Laser-Induced Chemistry

SURFACE PROCESSES IN CVD: LASER- AND LOW ENERGY ELECTRON-INDUCED DECOMPOSITION OF W(CO)$_6$ ON Si(111)-(7x7)

CYNTHIA M. FRIEND, J.R. SWANSON, AND F.A. FLITSCH
Harvard University, Department of Chemistry, 12 Oxford St., Cambridge, MA 02138

ABSTRACT

The decomposition of W(CO)$_6$ adsorbed on Si(111)-(7x7) using low energy electrons and ultraviolet photons has been investigated under ultrahigh vacuum conditions. This work is motivated by a desire to understand the mechanism for laser- and electron-assisted chemical vapor deposition (CVD) of tungsten using volatile coordination complexes and to specifically understand the role of the surface in these processes. Both electron stimulated and photo-assisted decomposition of the adsorbed W(CO)$_6$ are observed. No thermal decomposition of the W(CO)$_6$ occurs under the conditions of these experiments, based on independent temperature programmed reaction experiments, ruling out the possibility of laser- or electron-induced heating as the cause of decomposition. Furthermore, the interaction of the W(CO)$_6$ with the Si(111)-(7x7) surface is shown to be exceedingly weak based on the fact that the desorption energy is 9.46 ± 0.77 kcal/mol. Desorption of CO is induced during both ultraviolet photolysis and electron bombardment. Carbon monoxide is exclusively evolved during ultraviolet photolysis: no W-containing fragments are desorbed. During electron bombardment, a small amount of the W(CO)$_6$ is desorbed, accounting for ~10% of the desorption. In both cases, CO-containing W fragments remain on the surface after decomposition at low surface temperature. The remaining surface fragments do not undergo further photolysis at 308 nm but do react thermally. Competing desorption and dissociation of CO are thermally induced resulting in carbide and oxide impurities in the deposited material. The fact that strongly bound W(CO)$_x$ fragments are trapped on the surface is proposed as a limiting factor in the purity of tungsten deposits using the decomposition of W(CO)$_6$.

INTRODUCTION

The decomposition of volatile refractory metal complexes, such as W(CO)$_6$, is being explored as a possible means of depositing refractory metal films using maskless technology.[1] Ideally, high conductivity films with good adhesion properties would be deposited on a silicon substrate using heat, photons or electrons. High conductivity films have not, however, been deposited in recent work. Instead, laser-assisted chemical vapor deposition (LCVD) of tungsten using W(CO)$_6$ as a precursor has lead to impure, porous films.[2,3] The goal of this work is to better understand the mechanism of W(CO)$_6$ decomposition on a silicon surface and to specifically investigate processes that occur on the surface rather than in the gas phase.

The ultraviolet photodecomposition of metal carbonyls on silicon under ultrahigh vacuum conditions has been recently investigated extensively. Photolysis of Fe(CO)$_5$ on Si(111)-(7x7)[4,5] and W(CO)$_6$ and Mo(CO)$_6$ on Si(111)-(7x7)[5] and Si(100)[6] involves primarily electronic excitation of the adsorbed molecule. During photolysis, only CO evolution is observed. The adsorbed products of photolysis of W(CO)$_6$ and Mo(CO)$_6$[5,6] are proposed to be partially decarbonylated fragments.

Mat. Res. Soc. Symp. Proc. Vol. 131. ©1989 Materials Research Society

In this study we compare the effect of low energy electrons and photons on $W(CO)_6$ adsorbed on Si(111)-(7x7). The study of electron stimulated decomposition of adsorbed $W(CO)_6$ was undertaken because of recent interest in electron-assisted CVD and their possible role in photo-assisted decomposition processes on the surface. Low energy electrons generated by ultraviolet light have been proposed to result in $Fe(CO)_5$ decomposition on glass and silver films.[7] Furthermore, high energy (2.5-3 keV) electron induced decomposition of $Fe(CO)_5$ on Si(100) produced a carbon contaminated Fe layer and a relatively large amount of electron stimulated desorption of the parent molecule.[8,9] Electron stimulated decomposition is anticipated since dissociation of gas phase $Fe(CO)_5$[10,11] and $W(CO)_6$[11] by low energy (< 200 eV) electrons is known. Low energy electron induced decomposition is of added interest because many surface sensitive analysis methods use or generate them and may, therefore, result in inadvertant decomposition.

EXPERIMENTAL

All experiments were carried out in two separate ultrahigh vacuum chambers described previously with base pressures of $\sim 1 \times 10^{-10}$ Torr.[4] Briefly, several *in situ* spectroscopic probes, including multiple internal reflection Fourier transform infrared, laser induced desorption, temperature programmed desorption/reaction, Auger electron spectroscopies and low energy electron diffraction, were used in our studies. In one of the chambers, line-of-sight detection was used to monitor desorption during electron bombardment, ultraviolet photolysis and temperature programmed reaction using a quadrupole mass spectrometer (UTI 100C). The ionizer of the quadrupole mass spectrometer is employed as the low energy electron source. The mass spectrometer is shielded with a small stainless steel cap with a 6 mm diameter aperture. The $W(CO)_6$ is adsorbed onto the Si(111)-(7x7) crystal at 145 K. In all desorption experiments, except those specifically studying the electron-induced chemistry, the crystal is biased at -60 V to prevent electrons from the ionizer from impinging on the sample. This was done to prevent inadvertent electron-stimulated chemistry during adsorption, temperature programmed reaction and photolysis experiments. The sample was not biased during collection of the infrared data, obtained in a separate chamber, but the overlayer was never exposed to the mass spectrometer ionizer during infrared data collection. Electron bombardment (55 V, 80-100 μA) is performed by grounding the crystal while it is in front of the aperture. Ultraviolet photolysis was affected using a Lumonics Excimer laser. ^{13}C-enriched $W(CO)_6$ is used in all desorption experiments because the lower chamber background pressure at mass 29 (^{13}CO) as compared to that at mass 28 (^{12}CO), allows more sensitive detection of desorbing species. Methods for sample preparation and adsorption are described in previous papers.[4,15]

RESULTS AND DISCUSSION

Thermal Chemistry

Temperature programmed reaction spectroscopy was used as an independent method of investigating the thermal chemistry of the $W(CO)_6$. No thermal reaction of the adsorbed $W(CO)_6$ was observed in a temperature programmed reaction experiment with the initial adsorption temperature of 145 K. Molecular desorption of the $W(CO)_6$ is exclusively observed as shown in Figure 1.

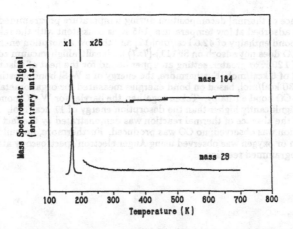

Figure 1. Temperature programmed desorption of ^{13}C-enriched W(CO)$_6$. Data is shown for two fragments formed in the mass spectrometer ionizer. The top trace is the signal for W, m/e 184, and the bottom trace is for ^{13}CO. The ratio of the two peaks is the same as for authentic samples of gaseous W(CO)$_6$ and the peak temperatures and lineshapes are the same for the two fragments. The heating rate was constant and 5K/s.

The W(CO)$_6$ desorption is zero-order based on the observation that the leading edge of the desorption peak is independent of coverage. The desorption energy of the W(CO)$_6$ is calculated to be 9.46 $\pm$0.77 kcal/mol by plotting the log of the desorption intensity as a function of 1/T, demonstrating that there is a very weak interaction between the W(CO)$_6$ and the Si(111)-(7x7) surface. The weak interaction is confirmed by infrared studies which exhibit a single peak in the CO stretch region at 1942 cm^{-1}, the same as the most intense band in the solid state spectrum assigned as the $\nu_6(T_{1u})$ mode.(Figure 2.)

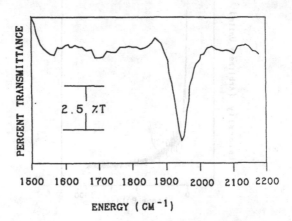

Figure 2. Infrared spectrum of W(CO)$_6$ adsorbed on Si(111)-(7x7). The W(CO)$_6$ exposure was 0.5L and the surface temperature 145K. Data was acquired for 1024 scans at 8 cm^{-1} resolution requiring a total of 16 min. of data acquisition time.

The absence of thermal decomposition during temperature programmed reaction when $W(CO)_6$ is adsorbed at low temperature (145 K) is consistent with the relatively large average W-CO bond enthalpy of 42.6 kcal/mol[12] and the low adsorption energy of CO on Si(111)-(7x7). CO does <u>not</u> adsorb on Si(111)-(7x7) under ultrahigh vacuum conditions at temperatures of 120 K or greater, setting an upper bound for the heat of adsorption of CO on Si(111)-(7x7) of 6 kcal/mol. Furthermore, the energy of a W-Si bond is estimated to be on the order of 30 kcal/mol, based on bond energies measured for organometallic complexes, less than the W-CO bond enthalpy. Taken together, the barrier for W-CO bond breaking is expected to be significantly higher than the desorption energy of 11 kcal/mol. Experimentally, the absence of thermal reaction was demonstrated by the fact that only $W(CO)_6$ desorption was observed; no CO was produced. Furthermore, no buildup of tungsten, carbon or oxygen was observed using Auger electron spectroscopy after temperature programmed reaction.

Photochemistry

Decomposition of adsorbed $W(CO)_6$, yielding gaseous CO and adsorbed CO-containing W species, is induced at low surface temperature (145 K) by ultraviolet laser light. Importantly, no decomposition of $W(CO)_6$ is indued by visible light (λ=720 nm). Light with wavelength of 720 nm is strongly absorbed by the Si(111)-(7x7) substrate but not by the $W(CO)_6$. This suggests that the ultraviolet decomposition of the adsorbed $W(CO)_6$ is due to electronic excitation of the adsorbed $W(CO)_6$, not due to electron-hole pairs created in the substrate. CO is the only gaseous product detected during laser photolysis. The fact that no W-containing molecules are desorbed during photolysis demonstrates that thermal processes are not contributing to the observed chemistry since laser heating should result in a significant amount of $W(CO)_6$ desorption. The amount of CO desorbed during a single laser pulse decreases rapidly as a function of the number of laser pulses as shown in Figure 3. This decrease in desorption intensity is attributed to depletion of intact $W(CO)_6$ on the surface.

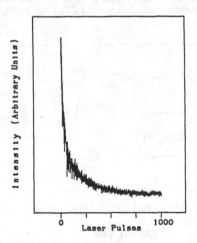

Figure 3. Desorption of ^{13}CO as a function of the number of laser pulses during photolysis of ^{13}C-enriched $W(CO)_6$ adsorbed on Si(111)-(7x7). The wavelength of light was 308 nm and the fluence during a single pulse 25 mJ/cm^2. The exposure of $W(CO)_6$ was ~0.5 L and the substrate temperature 145 K.

Not all of the CO ligands bound to $W(CO)_6$ are ejected during photolysis. After photolysis, one or more tungsten species containing CO remain on the surface. The remaining strongly bound tungsten species must have lost one or more CO since carbon monoxide is evolved as the only gas phase product during photlysis. The presence of these $W(CO)_x$ fragments, x<6, is clearly demonstrated by temperature programmed reaction data obtained subsequent to laser photolysis as shown in Figure 4b. The formation of gaseous CO at temperatures above 200 K is indicative of the kinetics for W-CO bond breaking in the adsorbed fragments. Since CO does not adsorb on Si(111)-(7x7) under the conditions of this experiment, the CO formed in temperature programmed reaction must be initially bound to tungsten. Notably, no new desorption features for W-containing molecules are observed. $W(CO)_6$ sublimation below 200 K is detected with diminished intensity compared to unphotolyzed $W(CO)_6$ at the same coverage. Infrared spectra obtained after ultraviolet photolysis are also consistent with the formation of $W(CO)_x$ fragments. A broadening of the infrared band on the low frequency edge is consistent with loss of one or more CO from the adsorbed $W(CO)_6$ during photolysis. (Data not shown.) A shift of ~ 25 cm^{-1} to lower frequency is anticipated for loss of each CO based on matrix isolation[13] and gas phase studies.[14] Unfortunately, a specific peak attributable to $W(CO)_x$ is not resolved after photolysis due to the intrinsically broad line width with the FWHM measured to be 33 cm^{-1} prior to photolysis. After photolysis with 50 laser pulses of 249 nm light with a fluence of 1mJ/cm^2, the FWHM increases to 66 cm^{-1}. We note that the fragments remaining on the surface may also have coalesced to form tungsten clusters. Neither the infrared or temperature programmed reaction data identifies a specific surface fragment.

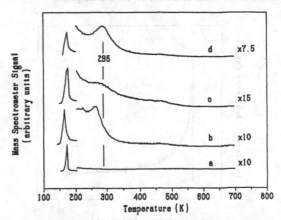

Figure 4. The ^{13}CO temperature programmed desorption/reaction spectrum is shown after: a) no treatment; b) photolysis with 1000 pulses of 308 nm laser light at a fluence of 25 mJ/cm^2 and the crystal floated negative 60 V; c) low energy (55 V, 80 μA) electron bombardment for 19 min, and; d) photolysis with 1000 pulses of 308 nm laser light with simultaneous electron bombardment. In all cases, the approximately linear heating rate is 5 K/s.

The $W(CO)_x$ fragments undergo thermal decomposition in competing CO formation, described above, and CO dissociation. The CO dissociation leads to strongly bound carbide and oxide impurities detected by Auger electron spectroscopy after photolysis and subsequent temperature programmed reaction. Thus, the presence of surface $W(CO)_x$ fragments is associated with impurities in laser deposited tungsten.

The $W(CO)_x$ fragment products that remain on the surface after ultraviolet photolysis do not undergo further photochemical decomposition at 308 nm. The photochemistry of the strongly adsorbed tungsten carbonyl fragments was investigated by photolysis of $W(CO)_6$ at low temperature followed by heating to 200 K to desorb residual parent $W(CO)_6$. Thus, the fragment(s) formed during photolysis are isolated on the surface. Subsequent irradiation with ultraviolet laser light ($\lambda=308$ nm) did not result in detectable desorption of any species. Furthermore, temperature programmed reaction data obtained following the second photolysis step were qualitatively the same as that obtained after only one photolysis step without annealing, the only difference being that the $W(CO)_6$ sublimation peak at 185 K was absent after annealing to 200 K and performing a second photolysis, as expected. The lack of efficient photochemistry of the adsorbed fragment(s) indicates that either the dissociation dynamics or rates of energy transfer to the solid are substantially different for the strongly bound $W(CO)_x$, x<6, fragment(s) than for the weakly adsorbed parent $W(CO)_6$. Furthermore, the absence of photochemical decomposition of the adsorbed $W(CO)_x$ fragments suggests an intrinsic limitation in forming pure deposits from pure photochemical activation of $W(CO)_6$.

Electron-Induced Chemistry

Low energy electrons also result in the decomposition of $W(CO)_6$ to produce gaseous CO and one or more CO-containing W fragments.[15] CO is evolved during bombardment with 55 eV electrons as shown in Figure 5. The crystal temperature rises ~ 2 K during the bombardment, which is too small to induce $W(CO)_6$ desorption.

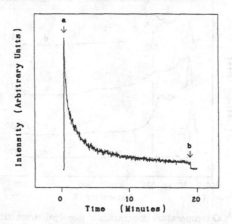

Figure 5. The ^{13}CO evolution signal as a function of electron bombardment time is shown in the figure. The bombardment was started at time (a) and stopped at time (b). The electron energy was 55 V and the current to the crystal was 100 μA. Approximately 10 % of the observed ^{13}CO signal is due to electron stimulated desorption of $W(CO)_6$.

Temperature programmed reaction spectra obtained after bombardment with low energy electrons results in the evolution of gaseous CO as is evident in Figure 4c. No new tungsten containing peaks are observed in temperature programmed reaction spectra obtained after electron bombardment. The new ^{13}CO peak is clearly attributable to decomposition of tungsten carbonyl fragments on the surface. Furthermore, a comparison of the ^{13}CO temperature programmed reaction spectra (Figure 4c) obtained after electron bombardment (55 V, 80 μA, 19 min) to that (Figure 4b) obtained after UV laser photolysis ($\lambda=308$nm, 25 mJ/cm², 1000 pulses) with the crystal floated negative 60 volts clearly demonstrates a difference in the kinetics of gas phase CO formation in the decomposition of the tungsten species formed from electron bombardment versus photolysis. The difference in the decomposition kinetics is clear evidence that the surface products formed from electron bombardment and ultraviolet photolysis are different. Auger electron spectroscopy experiments conducted after electron bombardment and temperature programmed reaction detect significantly larger amounts of carbon, oxygen and tungsten on the crystal as compared to after photolysis. This is further evidence for the proposed differences in surface products formed from photolysis and electron bombardment. Our future work will make a more detailed comparison of electron and photon assisted surface reactions of metal carbonyls.

Our observations also demonstrate the importance of protecting the sample from unwanted electron bombardment when investigating surface reactions of these molecules. Figure 4d depicts a ^{13}CO temperature programmed reaction spectrum after a photolysis experiment performed without floating the crystal but otherwise identical to that of Figure 4b. The sample is therefore simultaneously exposed to photons and electrons. The clear difference in ^{13}CO desorption temperature and peak shape shown in Figures 4b and 4d demonstrates that electrons alter the photolysis results. Just as these low energy electrons perturb a photolysis experiment so could they affect other experiments which generate low energy electrons directly or indirectly.

CONCLUSIONS

The decomposition of $W(CO)_6$ adsorbed on Si(111)-(7x7) is induced by both ultraviolet photons and low energy electrons. The fact that only light absorbed by the adsorbed $W(CO)_6$ leads to decomposition suggests that the photolysis is due to electronic excitation of a dissociative excited state of the adsorbate rather than electron hole pair production. The interaction of the $W(CO)_6$ with the Si(111)-(7x7) substrate is very weak, indicating a minimal perturbation of the adsorbed molecule.

Gas phase CO and strongly bound CO-containing W fragments are formed from both the ultraviolet photolysis and electron bombardment of the adsorbed $W(CO)_6$. The nature of the $W(CO)_x$ fragments formed by ultraviolet photolysis are distinctly different than those formed from electron bombardment. The exact nature of the surface fragments is not defined by our experiments, however.

The surface bound $W(CO)_x$ fragments do not efficiently undergo further laser photolysis at 308 nm. Therefore, there is an intrinsic limitation to the purity of tungsten deposited from $W(CO)_6$ using purely photochemical mechanisms.

ACKNOWLEDGEMENTS

We thank and acknowledge the National Science Foundation (CHE-83-09455 and CHE-84-51307) and the Harvard University Materials Research Laboratory (NSF DMR-83-169790) for support of this work, and Professor D.J. Darensbourg for the ^{13}C-enriched $W(CO)_6$.

REFERENCES

1. See, for example, D.J. Ehrlich, R.M. Osgood Jr., T.F. Deutsch, IEEE J. Quantum Elec. 16, 1233,(1980) and references therein.

2. S.D. Allen, J. Appl. Phys. 52, 6501 (1981).

3. R. Solanki, P.K. Boyer, and G.J. Collins, Appl Phys. Lett. 41, 1048 (1982).

4. J.R. Swanson, C.M. Friend, and Y.J. Chabal, J. Chem. Phys. 87, 5028 (1987).

5. N.S. Gluck, Z.Ying, C.E. Bartosch, and W. Ho, J. Chem. Phys. 86, 4957 (1987).

6. J.R. Creighton, J. Appl. Phys. 59, 410 (1986).

7. P.M. George and J.L. Beauchamp, Thin Solid Films 67, L25 (1980).

8. J.S. Foord and R.B. Jackman, Chem. Phys. Lett. 112, 190 (1984).

9. J.S. Foord and R.B. Jackman, Surf. Sci. 171, 197 (1986).

10. B.C. Hale and J.S. Winn, J. Chem. Phys. 81, 1050 (1984).

11. L. Sallans, K.R. Lane, R.R. Squires, and B. S. Freiser, J. Am. Chem. Soc. 107, 4379 (1985).

12. J.A. Connors, Topic in Current Chemistry, 71, 72-107, (1977).

13. See, for example, M. Poliakoff, J.J. Turner, J. Chem. Soc.-Dalton Trans., 2276, (1974).

14. W. Weitz, J. Phys. Chem. 91, 3945 (1987).

15. J.R. Swanson, F.A. Flitsch, C.M. Friend, Surf. Sci., (1988), Submitted.

SURFACE REACTIONS LEADING TO CONTAMINATION OF METAL FILMS PHOTOCHEMICALLY DEPOSITED FROM THE HEXACARBONYLS

K. A. SINGMASTER, F. A. HOULE AND R. J. WILSON
IBM Research Division, Almaden Research Center, 650 Harry Road, San Jose, CA 95120

ABSTRACT

A systematic study of the origin of contaminants in metal films photochemically deposited from the group VI hexacarbonyls is described. Background gas present in the cell during deposition, exposure to air and incomplete removal of CO groups from the surface of the growing film all affect C and O incorporation. The data are compared to results of recent experiments examining surface photoproducts of the metal carbonyls.

INTRODUCTION

The group VI metal hexacarbonyls present several important advantages as precursors to film deposition by ultraviolet (UV) photolysis: they are reasonably volatile (100 mTorr at room temperature), have a high absorption cross section in the region of 250 nm (10^{-17} cm^2) and dissociate in the gas phase with near unit quantum yield to form partly carbonylated species such as $M(CO)_4$, M = Cr, Mo and W [1-4]. Their chief disadvantage as precursors for photochemical deposition is that the films are of notably poor quality, being heavily contaminated by carbon and oxygen [3,4]. Although UV laser induced deposition can lead to good quality films if laser power densities are high enough to cause appreciable heating of the films [5], elevated temperatures are not always desirable. In this report we describe results of a systematic study of sources of contamination in films deposited from Cr and Mo hexacarbonyls by low power photochemical deposition [6]. The goals of this work are twofold: to investigate the possibility of controlling aspects of the deposition process to eliminate contaminants altogether, and to learn as much as possible about surface photochemical and thermal reactions for comparison to studies of photolysis of condensed metal hexacarbonyls on crystalline surfaces.

EXPERIMENTAL SECTION

All films were deposited on Si(111), doped n- or p-type (0.1 Ω-cm), cleaned in an acid oxidant bath and left covered by its native oxide. The light source was a frequency doubled Ar ion laser, providing up to 4 mW at 257 nm. After separation from the 515 nm beam, the UV beam was expanded and focussed to a 5 μm spot at the substrate surface. Films were grown during exposures ranging from 5s to 15 min using incident power densities from 40-2300 W/cm^2. Two deposition cells were used: a stand-alone cell which could be evacuated to the mid 10^{-4} Torr range, and an ultrahigh vacuum (UHV) cell with a base pressure in the low 10^{-9} range. The latter could be coupled to a vacuum suitcase for transportation of samples to a remote laboratory for analysis without air exposure. For all depositions the cells were backfilled with vapor of crystalline $M(CO)_6$ which had been subjected to several freeze-pump-thaw cycles. No buffer gases were used.

Three sets of measurements were carried out for each metal: growth in low vacuum and exposure to air prior to analysis (LV/air), growth in high vacuum and exposure to air (HV/air) and growth in high vacuum with transfer in vacuum (< 10^{-6} Torr for 30 min) for analysis (HV/vac). All films were analyzed by scanning Auger microscopy (SAM) and Ar ion depth profiling. Air-exposed films were characterized in a PHI 600 instrument using a 10 keV electron beam at 4 nA emission current, 200 nm spot. Films not exposed to air were transferred under vacuum to a VG Escalab (10 keV, 30 nA, 1 μm spot). In both instruments 2 kV Ar ion beams were used for sputtering. Scanning electron micrographs (SEM) were taken of several of the LV/air films.

Special care was taken to assess the importance of beam damage during analysis. The compositions of films deposited from $Cr(CO)_6$ were highly resistant to electron beam damage, although the thicker deposits ($> 10^{-5}$) tended to crack and peel. No evidence was found for preferential loss of Cr, C or O during sputtering. Films grown from $Mo(CO)_6$, on the other hand, were very unstable to electron beam exposure at emission currents above 30 nA, exhibiting both compositional changes and cracking. Preferential loss of O during sputtering was also observed. Although calibration of this loss was not possible using Mo oxycarbides, data for sputtering of MoO_3 powder pressed in In suggested the preferential loss of oxygen to be profound (see below) [7].

RESULTS

A typical film is shown in Fig 1. It exhibits deep ripples (410 nm wide) characteristic of deposition with a linearly polarized laser beam under conditions where photo-induced surface reactions are rate limiting [8]. This permits comparison of surface and gas phase photochemical and photophysical processes.

Auger data for films deposited from $Cr(CO)_6$ and $Mo(CO)_6$ are presented in Tables I and II. Each entry is an average of compositions for 15-50 films. Peak-to-peak height ratios of O (KLL, 503 eV) to Cr (LMM, 529 eV), C (KLL, 272 eV) to Cr, O to Mo (MNN, 221 eV) and C to Mo are reported, together with the spread in measurements. Excepting the O content of sputtered Mo films, the spread can be equated with error limits. Data

FIG. 1. SEM micrograph of a Cr film deposited by 257 nm photolysis of $Cr(CO)_6$ (300 W/cm², 10 s exposure, LV/air).

TABLE I. Auger data for Cr films deposited by UV photolysis of $Cr(CO)_6$

Films	Auger data		% composition		
	O/Cr	C/Cr	O	C	Cr
LV/air					
unsputtered	2.6±0.6	--	68	--	32
sputtered	2.1±0.3	--	65	--	35
HV/air					
unsputtered	2.5±0.4	0.3±0.1	50	25	25
sputtered	2.3±0.5	0.1±0.1	60	10	30
HV/vac					
unsputtered	2.3±0.3	0.2±0.05	54	20	26
sputtered	1.5±0.3	0.21±0.02	43	24	33
HV/vac					
off spot	2.2±0.2	0.4±0.01	43	33	24
Cr_2O_3 film					
unsputtered	2.4	--	66	--	34
sputtered	2.0	--	62	--	38
Predicted					
Cr_2O_3	1.9	--	60	--	40
$Cr(CO)$	1.2	0.28	33	33	33
$Cr(CO)_2$	2.5	0.57	40	40	20

TABLE II. Auger data for Mo films deposited by UV photolysis of $Mo(CO)_6$

Films	Auger data		% composition		
	O/Mo	C/Mo	O	C	Mo
LV/air					
unsputtered	4.9±0.8	0.6±0.2	50	30	20
sputtered	2.2±0.4	0.1±0.1	52	< 9	39
HV/air					
unsputtered	3.9±0.9	0.9±0.2	41	43	16
sputtered	1.5±0.2	0.3±0.2	35	30	35
HV/vac					
unsputtered	1.5±0.5	0.63±0.1	26	48	26
sputtered	0.75±0.15	0.61±0.1	15	55	30
HV/vac					
off spot	1.5±0.1	0.74±0.1	24	52	24
MoO_3 powder					
unsputtered	4.5	--	75	--	25
sputtered	1.5	--	50	--	50
Predicted					
MoO_3	4.5	--	75	--	25
$Mo(CO)$	1.5	0.34	33	33	33
$Mo(CO)_2$	3.0	0.68	40	40	20

obtained for authentic oxide and carbide samples and predicted peak height ratios [9] for selected stoichiometries are also included at the end of each table. Comparison of the predicted ratios with the experimental data permits assessment of approximate stoichiometries for the films as a function of deposition conditions. This type of information is valuable for learning about surface reactions, as discussed in the next section. Elemental compositions (presented in percentage units without error limits) have been derived from the Auger data and are included for purposes of comparison to previously published studies [1,3,4].

Depth profiling showed all films to be homogeneous in composition after removal of the top 1-2 nm, so data for sputtered films were averaged over the profile. Data for unsputtered films reflect compositions of the near-surface region. No variations in composition were found over the laser power density range used, so data for all intensities are combined.

Because of extensive gas phase photolysis during deposition a thin (5 nm) metal containing film covers the substrate. Analyses of this material (HV/vac, off spot) are listed in Tables I and II for comparison to films grown by surface illumination.

All carbon lineshapes for the Mo containing films were carbidic and those of Cr were mainly amorphous or graphitic [10]. This indicates that any CO groups remaining in the films are fully dissociated.

The Auger data show that film compositions depend strongly on deposition conditions. The LV/air films have as much oxygen in them as the thermodynamically most stable oxide, with carbon present only in the Mo films. Comparison of LV/air and HV/air films shows that oxygen-containing background gases react with the films during deposition, decreasing the amount of carbon in them. This is so efficient in the case of Cr that no detectable amounts of carbon are left. In fact, even the brief exposure to 10^{-6} Torr background gas during transfer is sufficient to oxidize the surface region of the Cr HV/vac films, as can be seen by comparing compositions of sputtered and unsputtered films (Table I). The effect of air exposure is quantified by comparison of HV/air and HV/vac films. Air exposure results in a much higher oxygen content, unsurprisingly, and an inhomogeneous distribution of carbon. Indeed the carbon concentration at the surface and in the bulk is observed to be time-dependent, increasing exposure time to air resulting first in bulk depletion of carbon, and then in net removal of carbon from the surface. Comparison of HV/vac and HV/vac off spot films probes differences in surface reactions of $M(CO)_x$ with and without incident UV photons. No differences are found in the Mo system, but the Cr off spot films show roughly double the amount of carbon and oxygen.

DISCUSSION

The Auger analyses provide clear evidence that three independent sources of contamination affect the compositions of films deposited photochemically from $Cr(CO)_6$ and $Mo(CO)_6$. Drawing on previous studies of the photodeposition process, it is possible to learn something about important surface and solid state reactions leading to C and O incorporation from the present data even though chemical information on reactive species could not be obtained because of the small size of the films.

It has been suggested that film growth from the metal hexacarbonyls is mediated by gas phase photolysis, that is, that surface reactions of mainly $M(CO)_4$ are responsible for film growth at 257 nm. Under the present conditions, gas reactions are saturated, and surface photochemical reactions are rate limiting. (This is supported by the existence of surface ripples, Fig 1.) Examination of HV/vac film compositions suggests that the film stoichiometries are consistent with CrCO and MoC_2O. Away from the laser beam stoichiometries are consistent with $Cr(CO)_2$ and MoC_2O. Thus, it appears that condensation of $M(CO)_4$ leads to loss of about 2 CO (or CO and CO_2 in the case of Mo) via a thermal route. The remaining C and O cannot be removed photochemically in the Mo system, as judged by film compositions in illuminated areas. In contrast, there exists a well-defined photochemical route for loss of one additional CO from Cr. It should be noted that even though the thermal and photochemical decomposition pathways are not chemically distinct in the Mo system, the photochemical reaction rate is much faster.

It is interesting to compare these results to studies of surface photolysis of $Mo(CO)_6$ condensed on Si, Mo or Rh crystal surfaces at low temperature [11]. Although details of the experimental procedures varied, the data consistently showed that decarbonylation under UV light was incomplete (2-3 COs removed) and that the CO groups remaining on the surface were not dissociated. Raising the surface temperature to 300 K resulted in loss of additional CO. In the present study, carried out under film growth conditions rather than monolayer adsorption, the initial adsorbate is already decarbonylated, the surface temperature is near ambient, and the surface is an amorphous metal oxycarbide. Despite the marked differences in reaction conditions, it appears that at least qualitatively the increased extent of CO loss and the complete dissociation of remaining CO groups may be accounted for by the higher surface temperature. The surface composition may also affect the chemistry since preadsorbed C and/or O is known to inhibit CO dissociation on Mo [12] and Cr [13], and may thus facilitate CO desorption during deposition.

Exposure of these (M,C,O) films to air results in their oxidation and in removal of carbon. Oxidation is not confined to the near surface region, but appears to occur through the entire thickness of the film. This suggests that the films are quite permeable, which property is no doubt enhanced by the ripple structure. The carbon profiles in the air exposed films indicates that carbon is immiscible in the oxide, segregating to the surface at an appreciable rate. The surface carbon concentration drops over time, suggesting that it is efficiently converted to volatile products (possibly CO or CO_2) even under ambient conditions. Oxidation by background gas during deposition also leads to removal of carbon. It is not possible to say whether the removal reaction involves C atom reactions at the surface as observed during air exposure, or simple displacement of CO groups from the surface prior to their dissociation.

Contamination of photochemically deposited Cr and Mo films by incomplete removal of CO, reaction with background gas during deposition and reaction with air is so efficient that we were unable to find reaction conditions under which pure metal films were formed. As noted above, heating during deposition with UV light can lead to formation of high quality films [5]. It would be of interest to understand why CO can be thermally but not photochemically labile on the surface of the growing film.

ACKNOWLEDGEMENTS

We thank IBM East Fishkill for providing postdoctoral support to KAS, to Vaughn Deline and Shirley Chiang for their cooperation in obtaining some of the SAM data, and to Jorge Goitia for technical assistance.

REFERENCES

1. R. Solanki, P. K. Boyer and G. J. Collins, Appl. Phys. Lett. *41*, 1048 (1982).

2. D. K. Flynn, J. I. Steinfeld and D. S. Sethi, J. Appl. Phys. *59*, 3914 (1986).

3. N. S. Gluck, G. J. Wolga, C. E. Bartosch, W. Ho and Z. Ying, J. Appl. Phys. *61*, 998 (1987).

4. R. L. Jackson and G. W. Tyndall, J. Appl. Phys. *64*, 2092 (1988).

5. H. H. Gilgen, T. Cacouris, P. S. Shaw, R. R. Krchnavek and R. M. Osgood, Appl. Phys. B*42*, 55 (1987).

6. A part of this study has been reported by K. A. Singmaster, F. A. Houle and R. J. Wilson, Appl. Phys. Lett. *53*, 1048 (1988).

7. T. T. Lin and D. Lichtman, J. Vac. Sci. Technol. *15*, 1689 (1978).

8. R. M. Osgood, Jr. and D. J. Ehrlich, Opt. Lett. *7*, 385 (1982).

9. L. E. Davis, N. C. MacDonald, P. W. Palmberg, G. E. Riach and R. E. Weber, "Handbook of Auger Electron Spectroscopy", Perkin Elmer Corporation, Eden Prairie MN (1978).

10. B. Lesiak, P. Mrozek, A. Jablonski and A. Jozwik, Surf. Interface Anal. *8*, 121 (1986).

11. J. R. Creighton, J. Appl. Phys. *59*, 410 (1986); N. S. Gluck, Z. Ying, C. E. Bartosch and W. Ho, J. Chem. Phys. *86*, 4957 (1987); C. C. Cho and S. L. Bernasek, J. Vac. Sci. Technol. A*5*, 1088 (1987); T. A. Germer and W. Ho, J. Chem. Phys. *89*, 562 (1988).

12. E. I. Ko and R. J. Madix, Surf. Sci. *109*, 221 (1981).

13. N. D. Shinn and T. E. Madey, J. Vac. Sci. Technol. A*3*, 1673 (1985).

MOLECULAR BEAM STUDY OF THE KrF* LASER
PHOTODISSOCIATION OF Cr(CO)$_6$ AND Mo(CO)$_6$

GEORGE W. TYNDALL AND ROBERT L. JACKSON
IBM Almaden Research Center, San Jose, California 95120-6099

ABSTRACT

The one-photon and sequential two-photon KrF* laser (248 nm) photodissociation of Cr(CO)$_6$ and Mo(CO)$_6$ were studied in a molecular beam using mass spectrometry to detect the photoproduct molecular ions. For Cr(CO)$_6$, the major one-photon product, Cr(CO)$_4$, undergoes secondary photodissociation to give predominantly Cr(CO)$_2$. The photodissociation cross section of Cr(CO)$_4$ is found to be 1.6 × higher than that of Cr(CO)$_6$ at 248 nm. For Mo(CO)$_6$, the major one-photon products, Mo(CO)$_5$ and Mo(CO)$_4$, undergo secondary photodissociation to give predominantly Mo(CO)$_3$.

INTRODUCTION

Ultraviolet multiphoton dissociation (UV MPD) of metallic compounds provides a mild technique for producing metal atoms and highly reactive unsaturated metallic species in the gas phase [1]. UV MPD has been used successfully to produce high quality thin films at relatively low substrate temperatures for microelectronic applications [2]. Metal atom formation via UV MPD appears to occur predominantly by a sequential mechanism, where a series of absorption/fragmentation processes eventually results in the loss of all ligands from the metal compound [3]. Unfortunately, the identity of the molecular intermediates involved in the UV MPD of most metallic compounds has not been determined. This is largely because the techniques typically used to study UV MPD, including multiphoton ionization, emission spectroscopy, and laser-induced fluorescence, are not amenable to the characterization of unsaturated metallic species. These species typically do not fluoresce, and thus far, they have eluded observation by standard multiphoton ionization techniques.

In this paper, we examine the products of KrF* laser MPD of Cr(CO)$_6$ and Mo(CO)$_6$ with emphasis on identifying the products formed in the sequential two-photon dissociation process. Photodissociation of Cr(CO)$_6$ and Mo(CO)$_6$ are examined in a molecular beam, using mass spectrometry to detect the molecular ions of the photoproducts. One reason for choosing Cr(CO)$_6$ and Mo(CO)$_6$ in this study is that each molecule readily yields metal atoms upon MPD at 248 nm [1, 4, 5]. In addition, the products formed upon single-photon KrF* laser dissociation of Cr(CO)$_6$ have been well-characterized [6]. We show that Cr(CO)$_2$ is the major product formed in the second step of the sequential two-photon dissociation of Cr(CO)$_6$. Cr(CO)$_2$ is produced primarily by photodissociation of the major single-photon product Cr(CO)$_4$. We also show that Mo(CO)$_5$ and Mo(CO)$_4$ are significant products formed in the one-photon dissociation of Mo(CO)$_6$, and that these species undergo secondary photodissociation to give predominantly Mo(CO)$_3$.

EXPERIMENTAL

The experimental techniques employed in this work have been described in detail elsewhere [7], so only a brief summary will be given here. Photodissociation is carried out in a differentially pumped, dual-chamber vacuum system. One chamber houses a pulsed molecular beam valve and the other chamber houses a quadrupole mass spectrometer (Extrel model C-50). The chambers were separated by a skimmer with a 0.1 mm diameter orifice. Cr(CO)$_6$ vapor was delivered to the molecular beam valve as a very dilute mixture (< 0.02%) in argon or helium (total pressure 500 torr). The high dilution combined with the small diameter of the valve nozzle (0.05 mm) limited the formation of Cr(CO)$_6$ clusters. The mass spectrometer was mounted on the molecular beam axis to permit observation of both reactant dissociation and product formation. Ionization of the reactants and products was accomplished by electron impact at an electron energy of 15 eV for Cr(CO)$_6$ and 17 eV for Mo(CO)$_6$. The laser beam intersected the molecular beam at right angles just in front of the aperture leading into the ionization region of the mass spectrometer. The laser beam was found experimentally to irradiate 48% of the metal hexacarbonyl molecules in the molecular beam that are detected by the mass spectrometer.

The signals for the reactant and product molecular ions obtained during each pulse of the molecular beam valve are sampled by a gated boxcar integrator and are stored separately in a PC. The gate is set at the temporal maximum of the ion signals, which occurs 150 μs after the laser pulse, corresponding to the time required for the molecules to travel from the laser interaction region into the mass spectrometer. Typically, data from 2000-6000 laser pulses were collected and averaged for the reactant and each product

Mat. Res. Soc. Symp. Proc. Vol. 131. ©1989 Materials Research Society

ion under one set of experimental conditions. The ion signals are corrected for the relative mass sensitivity of the quadrupole mass spectrometer. The product ion signals are also corrected for the background due to fragmentation of the parent metal hexacarbonyl molecular ion upon electron impact. After normalization of the product ion signals to reflect the extent of photodissociation of the parent metal hexacarbonyl, we obtain quantitative yields for each product ion.

RESULTS AND DISCUSSION

Our goal in this work is to determine the identity of the photoproducts formed in the sequential MPD of $Cr(CO)_6$ and $Mo(CO)_6$ and to determine the number of photons involved in the formation of each product. We thus performed detailed measurements of the yield vs. laser fluence for each product ion. In the simplest case (no saturation), the yield of a photoproduct is proportional to I^n, where I is the light intensity and n is an integer representing the number of photons involved in formation of the product. In no case, however, did we observe such simple behavior. The fluence dependence data are thus treated by a simple kinetic model of the sequential MPD process that explicitly accounts for saturation. For a sequential two-photon dissociation process, given in general form by

$$A \xrightarrow{h\nu} B \xrightarrow{h\nu} C, \tag{1}$$

the yield of the reactant A and the yields of the one- and two-photon dissociation products B and C are given by

$$A = A(0)\{1 + \kappa[\exp(-\sigma_A \Phi / h\nu) - 1]\}, \tag{2}$$

$$B = A(0)\kappa \frac{\sigma_A}{\sigma_B - \sigma_A}[\exp(-\sigma_A \Phi / h\nu) - \exp(-\sigma_B \Phi / h\nu)], \tag{3}$$

$$C = A(0)\kappa \left\{ \frac{\sigma_B}{\sigma_B - \sigma_A}[1 - \exp(-\sigma_A \Phi / h\nu)] - \frac{\sigma_A}{\sigma_B - \sigma_A}[1 - \exp(-\sigma_B \Phi / h\nu)] \right\}. \tag{4}$$

where σ_A and σ_B represent the photodissociation cross sections for A and B, respectively, at the radiation wavelength, ν, κ is the laser beam/molecular beam overlap factor (0.48, see above), and Φ is the laser fluence.

The behavior of the product ion yields as a function of laser fluence may be directly fit by these equations. The model does not give actual photoproduct yields, however, since we have not accounted for the different velocity profiles and the different ionization probabilities of the various products. We also do not account explicitly for fragmentation of the photoproduct molecular ions in the mass spectrometer. Photoproduct ion fragmentation may introduce uncertainty into the relative product yields determined in our experiments. For example, the apparent yield of a photoproduct $M(CO)_x$ will be reduced if $M(CO)_x^+$ fragments significantly upon electron impact, but the apparent yield may increase if the photoproduct ion $M(CO)_y^+$ ($x < y$) fragments to form $M(CO)_x^+$. For some photoproducts, however, ion fragmentation does not affect our ability to measure the photoproduct yield as a function of laser fluence. It is known, for example, that $Cr(CO)_4$ is the predominant photoproduct formed upon single-photon KrF* laser photodissociation $Cr(CO)_6$, whereas $Cr(CO)_5$ is a very minor product [6,8]. We observed the formation of $Cr(CO)_5^+$ as a photoproduct ion in our experiments, but the yield was too low to quantify and was much less than the yield of $Cr(CO)_4^+$. As a result, the apparent yield of $Cr(CO)_4^+$ will not be affected significantly by fragmentation of $Cr(CO)_5^+$, and the yield of $Cr(CO)_4^+$ vs. laser fluence will accurately reflect the yield of the $Cr(CO)_4$ photoproduct vs. laser fluence.

MPD of $Cr(CO)_6$

The behavior of the $Cr(CO)_4^+$ yield vs. laser fluence offers insight into the sequential two-photon dissociation of $Cr(CO)_6$. The yield of $Cr(CO)_4^+$ is linear in laser fluence at low fluences, but falls off substantially at higher laser fluences. The falloff is due to secondary photodissociation of the $Cr(CO)_4$ photoproduct. The data shown in Fig. 1 for $Cr(CO)_4^+$ may be fit using Eq. 3, where B corresponds to the yield of $Cr(CO)_4^+$. The photodissociation cross section for $Cr(CO)_4$, corresponding to σ_B in Eq. (3), is treated as an adjustable parameter. A good fit is obtained for $\sigma_B = 9.0 \times 10^{-17}$ cm^{-2}, which is 1.6 times the cross section of $Cr(CO)_6$.

The yield of the $Cr(CO)_3$ photoproduct vs. laser fluence is not as clearly established from the fluence

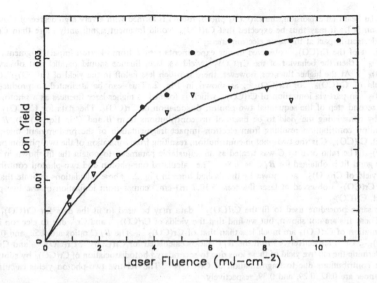

Fig. 1. Yield of $Cr(CO)_4^+$ (•) and $Cr(CO)_3^+$ (∇) as a function of laser fluence. The data are indicated by the points, while a fit to the data using the kinetic model outlined in Eqs. (1)-(4) is given by the lines.

dependence data for $Cr(CO)_3^+$. This ion may be formed by ionization of the $Cr(CO)_3$ photoproduct or by fragmentation of $Cr(CO)_4^+$. We note, however, that the behavior of the $Cr(CO)_3^+$ yield vs. laser fluence parallels that observed for $Cr(CO)_4^+$, as shown in Fig. 1. We fit the data for $Cr(CO)_3^+$ using Eq. (3) with the same value of σ_B as that used to fit the $Cr(CO)_4^+$ data. The quality of the fit suggests that $Cr(CO)_3^+$ results almost entirely from fragmentation of $Cr(CO)_4^+$, indicating that $Cr(CO)_3$ is not a significant product of either the first step or the second step of the sequential two-photon dissociation of $Cr(CO)_6$. This is consistent with previous work showing that $Cr(CO)_3$ is not formed to a significant extent in the single-photon KrF* laser photodissociation of $Cr(CO)_6$ [6].

Our conclusions for $Cr(CO)_3^+$ suggest that the the $Cr(CO)_4$ photoproduct fragments extensively upon electron impact ionization. This is confirmed by examining the relative yields of the photoproduct ions obtained at a fluence of 1.8 mJ-cm^{-2} (see Table I), where single-photon dissociation of $Cr(CO)_6$ is the only important photoprocess. The relative yield of $Cr(CO)_4^+$ is not only much smaller than unity, but is only slightly larger than the relative yield of $Cr(CO)_3^+$. The relative yields of $Cr(CO)_2^+$, $Cr(CO)_1^+$, and Cr^+ are significant, suggesting that these ions are also formed by electron impact fragmentation of the $Cr(CO)_4$ photoproduct. Such extensive fragmentation is not expected at the low electron energies used in this work (15 eV). We also note that electron impact fragmentation of $Cr(CO)_6$ under the same ionization conditions is much less extensive than that observed for $Cr(CO)_4$. The $Cr(CO)_6^+$ molecular ion makes up 78% of the ions produced upon electron impact ionization/fragmentation of $Cr(CO)_6$ at 15 eV in our mass spectrometer. The difference in the extent of fragmentation observed for $Cr(CO)_6$ and $Cr(CO)_4$ may be explained by noting the degree of vibrational excitation of each species prior to ionization. While the vibrational temperature of $Cr(CO)_6$ is well below room temerature, due to cooling in the

Table I. Relative ion yields for the one-photon products formed in the photodissociation of $Cr(CO)_6$ at 1.8 mJ-cm^{-2}. The ionizing electron energy in the mass spectrometer was 15 eV.

Ion	Relative Yield
$Cr(CO)_4^+$	0.25
$Cr(CO)_3^+$	0.21
$Cr(CO)_2^+$	0.31
$Cr(CO)_1^+$	0.15
Cr^+	0.08

molecular beam jet expansion, the $Cr(CO)_4$ photoproduct is formed with a very high degree of vibrational excitation [6]. It may thus be expected that $Cr(CO)_4$ would fragment significantly more than $Cr(CO)_6$ upon electron impact at the same electron energy.

If most of the $Cr(CO)_2^+$ observed in our experiments results from electron impact fragmentation of $Cr(CO)_4^+$, then the behavior of the $Cr(CO)_2^+$ yield vs. laser fluence should parallel that observed for $Cr(CO)_4^+$. At the higher fluences, however, there is much less falloff in the yield of $Cr(CO)_2^+$ than in the yield of $Cr(CO)_4^+$ or $Cr(CO)_3^+$, as shown in Fig. 2. This can be attributed to production of $Cr(CO)_2^+$ in part via ionization of $Cr(CO)_2$, which is formed at higher laser fluences as a photoproduct in the second step of the sequential two-photon dissociation of $Cr(CO)_6$. The data for $Cr(CO)_2^+$ may be fit by considering the yield to be made of up contributions from B and C in Eq. (3)-(4). B is the one-photon contribution resulting from electron impact fragmentation of the predominant one-photon product, $Cr(CO)_4$. C is the two-photon contribution, resulting from ionization of the two-photon product, $Cr(CO)_2$. The ratio of B to C was treated as an adjustable parameter to obtain the fit shown in Fig. 2. A very good fit is obtained for $B/C = 0.86$. The calculated one-photon and two-photon contributions to the yield of $Cr(CO)_2^+$ are shown by the dashed lines in Fig. 2. These calculations indicate that most of the $Cr(CO)_2^+$ observed at laser fluences > 10.7 mJ-cm^{-2} comes from ionization of the two-photon product, $Cr(CO)_2$.

The same procedure used to fit the $Cr(CO)_2^+$ data may be used to fit the data for $Cr(CO)_1^+$ and Cr^+. These fits are not shown, but we find that the yields of $Cr(CO)_1^+$ and Cr^+ due to electron impact fragmentation of $Cr(CO)_4$ are much less than that of $Cr(CO)_2^+$. The B/C ratios are 0.52, and 0.18 for $Cr(CO)_1^+$, Cr^+, respectively. From the B/C ratios calculated for $Cr(CO)_2^+$, $Cr(CO)_1^+$, and Cr^+, we can determine the relative yield of these ions due to secondary photodissociation of $Cr(CO)_4$ by subtracting out the contributions due to fragmentation of $Cr(CO)_4^+$. The relative two-photon yields calculated in this manner are 0.32, 0.29, and 0.39, respectively.

The relative two-photon yields given above indicate that Cr^+ is the major product *ion* formed upon ionization/fragmentation of the two-photon dissociation products. We note, however, that significant fragmentation of the two-photon products is expected upon electron impact, as was observed for the one-photon products. We thus conclude that $Cr(CO)_2$ is the major product formed in the second step of the sequential two-photon dissociation of $Cr(CO)_6$, and that $Cr(CO)_2$ fragments upon electron impact to give $Cr(CO)_1^+$ and Cr^+. $Cr(CO)_1$ and Cr may also be significant two-photon products. We showed in earlier emission spectroscopy experiments that Cr is indeed formed via two-photon sequential dissociation of $Cr(CO)_6$, although the yield was not determined [5].

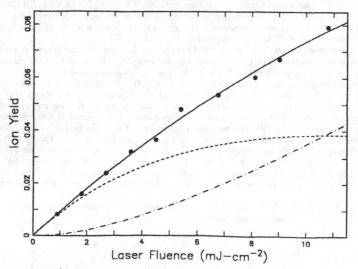

Fig. 2. Yield of $Cr(CO)_2^+$ as a function of laser fluence. The data are indicated by the points, while a fit to the data using the kinetic model outlined in Eqs. (1)-(4) is given by the solid line. The dashed lines represent the calculated one-photon (----) and two-photon (-.-.-) contributions to the $Cr(CO)_2^+$ ion yield.

MPD of Mo(CO)$_6$

Our data for Mo(CO)$_6$ are more limited than those for Cr(CO)$_6$, but we are able to reach some important conclusions regarding the single-photon and sequential two-photon dissociation of Mo(CO)$_6$ by studying the Mo(CO)$_x^+$ ion yields vs. laser fluence. We find that M(CO)$_5^+$ is a significant product ion formed at low laser fluences, where single-photon dissociation of Mo(CO)$_6$ is the only important photoprocess. The yield of Mo(CO)$_5^+$ as a function of laser fluence is shown in Fig. 3. The yield is not a linear function of laser fluence, but falls off significantly at higher laser fluence as was observed for Cr(CO)$_4^+$. We conclude that secondary photodissociation of Mo(CO)$_5$ occurs at laser fluences < 10 mJ-cm^{-2}. The fit to the data shown in Fig. 3 was obtained using Eq. (3) with $\sigma_B = 1.1 \times 10^{-16}$ cm^{-2}, corresponding to the photodissociation cross section for Mo(CO)$_5$. The behavior of the Mo(CO)$_4^+$ ion yield vs. laser fluence is similar to that observed for Mo(CO)$_5^+$, but the data (not shown) cannot be fit by Eq. (3) using the same value of σ_B as that used to fit the Mo(CO)$_5^+$ data. This leads us to conclude that Mo(CO)$_4^+$ is not formed exclusively by electron impact fragmentation of Mo(CO)$_5^+$, but also by direct ionization of the Mo(CO)$_4$ photoproduct.

The KrF* laser photodissociation of Mo(CO)$_6$ has not been reported, but our results may be compared to those obtained for KrF* laser photodissociation of Cr(CO)$_6$ and W(CO)$_6$. The predominant product observed in the single photon KrF* laser dissociation of Cr(CO)$_6$ and W(CO)$_6$ is the tetracarbonyl species, while the pentacarbonyl species is a very minor product in each case [6, 8]. We find, however, that Mo(CO)$_5$ is a significant product formed via single-photon dissociation of Mo(CO)$_6$. The first metal-CO bond dissociation energies for Cr(CO)$_6$, Mo(CO)$_6$, and W(CO)$_6$ are comparable (37, 41, and 46 kcal/mole, respectively [9]), so our results suggest that the second metal-CO bond dissociation energy for Mo(CO)$_6$ may be substantially higher than that of either Cr(CO)$_6$ and W(CO)$_6$. These energies are not known for Mo(CO)$_6$ and W(CO)$_6$ at the present time, so this hypothesis cannot be verified.

The identity of the predominant product formed in the second step of the sequential two-photon dissociation of Mo(CO)$_6$ is revealed by examining the yield of Mo(CO)$_3^+$ vs. laser fluence, shown in Fig. 4. Much less falloff in the yield of Mo(CO)$_3^+$ is observed at the higher laser fluences than is found for either Mo(CO)$_5^+$ and Mo(CO)$_4^+$. This behavior is similar to that observed for Cr(CO)$_2^+$ vs. Cr(CO)$_4^+$. We conclude, therefore, that Mo(CO)$_3^+$ is produced in part via ionization of Mo(CO)$_3$, which is formed at the higher laser fluences as a secondary photoproduct via dissociation of Mo(CO)$_5$ and Mo(CO)$_4$. From our data, we cannot determine whether photodissociation of Mo(CO)$_5$ or Mo(CO)$_4$ makes a more important contribution to the formation of Mo(CO)$_3$, so we did not make an attempt to fit the data for Mo(CO)$_3^+$.

The yields of Mo(CO)$_2^+$, Mo(CO)$_1^+$, and Mo$^+$ are very small at low laser fluences but increase

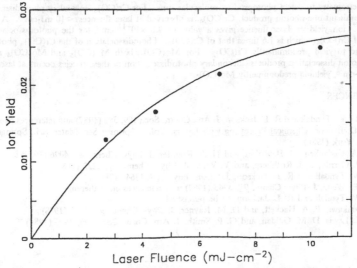

Fig. 3. Yield of Mo(CO)$_5^+$ as a function of laser fluence. The data are indicated by the points, while a fit to the data using the kinetic model outlined in Eqs. (1)-(4) is given by the line.

480

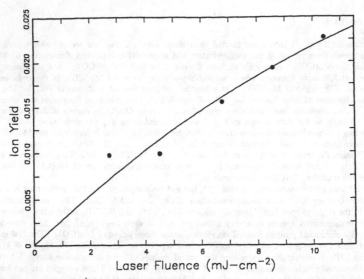

Fig. 4. Yield of Mo(CO)$_3^+$ as a function of laser fluence. The line represents an aid to the eye, rather that an actual fit to the data.

steadily with increasing laser fluence. This behavior indicates that these ions are formed predominantly via ionization/fragmentation of the products formed in the second step of the sequential two-photon dissociation of Mo(CO)$_6$. The yields of these ions is substantially below that of Mo(CO)$_3^+$ at all laser fluences, however, leading us to conclude that Mo(CO)$_3^+$ is the major two-photon product.

CONCLUSION

The KrF* laser dissociation of Cr(CO)$_6$ and Mo(CO)$_6$ has been examined in a molecular beam using mass spectrometry to detect photoproduct molecular ions. For Cr(CO)$_6$, secondary photodissociation of the predominant one-photon product, Cr(CO)$_4$, is observed at laser fluences < 10 mJ-cm^{-2}. A fit of the Cr(CO)$_4^+$ ion yield vs. laser fluence gives a value of 9.0 ×10^{-17} cm^{-2} for the photodissociation cross section of Cr(CO)$_4$, which is 1.6 times that of Cr(CO)$_6$. Photodissociation of the Cr(CO)$_4$ photoproduct was found to yield predominantly Cr(CO)$_2$. For Mo(CO)$_6$, both Mo(CO)$_5$ and Mo(CO)$_4$ are major single-photon dissociation products. Secondary photodissociation of these species occurs at laser fluences < 10 mJ-cm^{-2}, yielding predominantly Mo(CO)$_3$.

REFERENCES

1. See G. W. Tyndall and R. L. Jackson, J. Am. Chem. Soc., 109, 582 (1987) and references cited therein.
2. See D. Bauerle, Chemical Processing with Lasers, Vol. 1, Springer Ser. Mater. Sci., Springer-Verlag, New York (1986).
3. See A. Gedanken, M. B. Robin, and N. A. Kuebler, J. Phys. Chem., 86, 4096 (1982).
4. D. P. Gerrity, L. J. Rothberg, and V. Vaida, J. Phys. Chem., 87, 2222 (1983).
5. G. W. Tyndall and R. L. Jackson, J. Chem. Phys., 89, 1364 (1988).
6. See E. Weitz, J. Phys. Chem., 91, 3945 (1987) and references cited therein.
7. G. W. Tyndall and R. L. Jackson, to be published.
8. Y. Ishikawa, P. A. Hackett, and D. M. Rayner, J. Phys. Chem., 92, 3863 (1988).
9. K. E. Lewis, D. M. Golden, and G. P. Smith, J. Am. Chem. Soc., 106, 3905 (1984).

DYNAMICS OF THE KrF* LASER MULTIPHOTON DISSOCIATION OF A SERIES OF ARENE CHROMIUM TRICARBONYLS

GEORGE W. TYNDALL AND ROBERT L. JACKSON, IBM Almaden Research Center, San Jose, CA 95120

ABSTRACT

The KrF* (248 nm) laser multiphoton dissociation (MPD) of a series of (arene)chromium tricarbonyls has been investigated in the gas-phase using emission spectroscopy to detect the excited state photoproducts. In the MPD of all compounds studied, chromium atoms are formed in a variety of electronically excited states via a two-channel dissociation mechanism. The predominant pathway for formation of the ground electronic state and the lowest excited states is by a sequential absorption/fragmentation process, where the product of the one-photon dissociation of the parent molecule absorbs an additional photon and dissociates to Cr(I). The higher energy Cr(I) states are formed exclusively by a direct dissociation process, where the parent absorbs multiple photons prior to dissociation. The distribution of excited chromium atoms formed in the direct channel is statistical for all compounds studied and is independent of the nature of the arene ligand. In contrast, the distribution of Cr(I) states formed via the sequential dissociation channel is strongly dependent on the vibrational density of states in the arene ligand.

INTRODUCTION

The gas-phase photodissociation of organometallic compounds plays a key role in the deposition of metals via laser chemical vapor deposition. In the present work we have studied the KrF* (248 nm) laser MPD of: a) $Cr(CO)_3(\eta^3 C_6H_6)$, b) $Cr(CO)_3(\eta^3 CH_3C_6H_5)$ c) $Cr(CO)_3[\eta^3(n-C_3H_7)C_6H_5]$, and d) $Cr(CO)_3[\eta^3(t-C_4H_9)C_6H_5]$. We present evidence that the formation of Cr(I) occurs via two fragmentation channels: 1) a sequential absorption/fragmentation process involving an intermediate formed in the single photon dissociation of the parent (arene)chromium tricarbonyl, and 2) a direct dissociation process which involves a multiphoton absorption occuring prior to any dissociation step.

EXPERIMENTAL

The experimental apparatus has been described in detail in ref. [1]. Formation of Cr(I) is monitored in these experiments by measuring the fluorescence obtained following irradiation of compounds (a) - (d) at 248 nm. The focused output of an excimer laser (Lambda Physik model 101E) makes a single pass through a photolysis cell containing the vapor of one of the chromium compounds. A buffer gas, either Ar, He or CO, was passed through a reservoir containing the solid organochromium compounds, and the mixture was introduced to the cell through a molecular leak valve. The buffer gas was also passed over the laser entrance window, to minimize the deposition of chromium. Under normal operating conditions, the partial pressure of the (arene)chromium tricarbonyls was of the order of 1 mtorr, and the total cell pressure was < 1 torr. Emission normal to the laser beam is wavelength resolved with a 1 meter scanning monochromator and detected with a photomultiplier tube. Signal averaging is performed with a variable gate boxcar integrator and the output plotted on an x-y recorder. The data was collected with a spectral resolution of between 1-2 Å.

RESULTS

Throughout this paper we take the bond dissociation energy to be that energy required to form the ground state of both Cr(I) and the intact ligands from the parent chromium compound at 298K. The energy required to form a particular excited state of Cr(I), denoted ΔE_{298}, is then the bond dissociation energy plus the difference in energy between that excited state and the ground state of Cr(I). The bond dissociation energies for benzeneCr(CO)$_3$ and tolueneCr(CO)$_3$ are 5.08 eV and 5.04 eV respectively [2]. The bond dissociation energies for $t-$ butylbenzeneCr(CO)$_3$ and propylbenzeneCr(CO)$_3$ have not to our knowledge been reported, but are not expected to deviate by more than 2% from benzeneCr(CO)$_3$ [2].

Mat. Res. Soc. Symp. Proc. Vol. 131. ©1989 Materials Research Society

Following the MPD of each of the chromium compounds studied in this work, the spectral region from 300 nm to 550 nm was scanned. In all cases, well-resolved Cr(I) fluorescence was the only emission was detected. Emission from the $t^5F_2^o$ (ΔE_{298} = 11.06 eV) $\rightarrow b^5G_3$ and $v^5D_3^o$ (ΔE_{298} = 11.01 eV) $\rightarrow a^5P_{2,3}$ transitions, located at 361.0 nm and 384.9 nm respectively, are the most intense in the emission spectrum. The strong emission intensities from these states arises because the states are formed via resonant transitions at 248 nm from the a^5S_2 (ΔE_{298}) and a^5D_2 (ΔE_{298}) low-lying states, which are formed in the MPD of benzeneCr(CO)$_3$. The strength of the fluorescence intensity from the $v^5D_3^o$ and $t^5F_2^o$ states, coupled with the relatively low transition probabilities for their formation from a^5S_2 and a^5D_2, indicates that the populations of the a^5S_2 and a^5D_2 states are very much higher than those found for the other excited states. In addition to the emission from the a^5S_2 and a^5D_2 states, flourescence is also observed from the $z^7P_j^o$ (ΔE_{298} = 7.97 - 7.99 eV) $z^5P_j^o$ (ΔE_{298} = 8.40 eV), $y^7P_j^o$ (ΔE_{298} = 8.52 - 8.54 eV) $y^5P_j^o$ (ΔE_{298} = 8.73 - 8.78 eV) and $z^5F_j^o$ (ΔE_{298} = 8.90 - 8.96 eV) electronic states of Cr(I).

The laser fluence dependence of the emission intensity from the $z^7P_j^o$, $y^7P_j^o$ and $v^5D_3^o$ states was measured. Emission from the $z^7P_j^o$, $y^7P_j^o$ and $v^5D_3^o$ states varies as the cube of the laser fluence over most of the range of laser fluences used in this work. At the highest fluences used, however, a slight curvature in the data for emission from the $z^7P_j^o$ and $y^7P_j^o$ states suggests the onset of saturation. In addition emission from the $y^7P_j^o$ state varies as the square of the laser fluence the lowest fluences studied. The three-photon fluence dependence found for the $v^5D_3^o$ emission intensity is consistent with that expected based on the bond dissociation energy of benzeneCr(CO)$_3$ (5.08 eV). Formation of a^5S_2 (ΔE_{298} = 6.02 eV) requires absorption of two photons by benzeneCr(CO)$_3$. The third 248 nm photon then pumps the system to the fluorescent $v^5D_3^o$ state. Formation of the $z^7P_j^o$ and the $y^7P_j^o$ states requires a minimum of two photons which is also in agreement with that found experimentally.

The dependence of the Cr(I) emission spectra on the buffer gas pressure was investigated using Ar, He and CO. We find that the emission intensity from the $t^5F_2^o$ and $v^5D_3^o$ states decreases as the buffer gas pressure increases over the range of <1 - 25 torr. A representative example of the quenching of the $t^5F_2^o \rightarrow b^5G_3$ transition by Ar is shown in figure 1. The observed quenching at a given buffer gas pressure decreases as the incident laser fluence increases. The extent to which the emission is quenched is comparable for Ar and He at a given pressure, but is greater when CO is used as the buffer gas. The quenching we observe from the $t^5F_2^o$ and $v^5D_3^o$ states following the MPD of benzeneCr(CO)$_3$ cannot be fit by a simple Stern-Volmer analysis. The remaining five Cr(I) states observed in these emission studies are not buffer gas pressure dependent. The relative populations of these states were calculated using the analysis outlined in ref. [1], and are adequately described by a statistical distribution with an electronic temperature of 8500 ± 1000K.

The emission spectra observed in the MPD of compounds (b) - (d) are very similar to the Cr(I) spectrum obtained in the MPD of benzeneCr(CO)$_3$. The populations found for the $z^7P_j^o$, $z^5P_j^o$, $y^7P_j^o$, $y^5P_j^o$ and $z^5F_j^o$ states indicates that a statistical distribution of these states is formed in the 248 nm MPD of each compound. The electronic temperatures describing the Cr(I) distributions formed in the MPD of the (arene)chromium tricarbonyls, are all the same, within the experimental error, as that obtained in the MPD of benzeneCr(CO)$_3$, i.e. 8500 ± 1000K. While the formation of the $z^7P_j^o$, $z^5P_j^o$, $y^7P_j^o$, $y^5P_j^o$ and $z^5F_j^o$ states is independent of the arene ligand, the relative emission intensity from the $v^5D_3^o$ and $t^5F_2^o$ states varies significantly as the arene ligand is changed (see below). No quantitative quenching studies were carried out on the Cr(I) emission spectra obtained from the laser MPD of these compounds, but limited results suggest that emission from the $t^5F_2^o$ and $v^5D_3^o$ states is quenched by added buffer gas in analogy with that found in the MPD of benzeneCr(CO)$_3$.

DISCUSSION

Before discussing the MPD of the (arene)chromium tricarbonyls, it is instructive to review the main results of our previous study on the excimer laser MPD of Cr(CO)$_6$.[1,3] We have previously reported on the quenching of emission from the $t^5F_2^o$ and $v^5D_3^o$ states of Cr(I) formed in the 248 nm MPD of Cr(CO)$_6$. Since emission from these states was quenched, while emission from none of the other numerous Cr(I) states observed was quenched, we proposed that the MPD of Cr(CO)$_6$ to form Cr(I) occurs via two channels. Formation of the low-lying states of Cr(I), i.e. the ground state and the first two excited states (a^5S_2 and a^5D_2), was proposed to occur predominantly via a sequential mechanism, where successive absorption/fragmentation steps result in the complete loss of all ligands. Quenching of the a^5S_2 and a^5D_2 states of Cr(I) upon addition of a buffer gas was shown to be consistent with a competition between up-pumping of the vibrationally hot intermediate and either

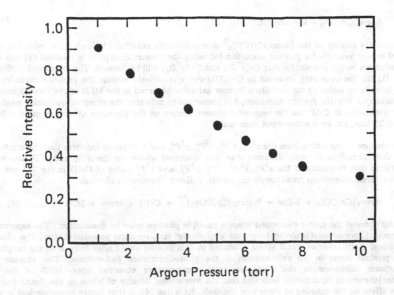

Figure 1. Dependence of the emission intensity from the $t^5F_2^o$ excited state of Cr(I) on argon pressure. The laser fluence was 20 mJ-cm^{-2}. All data were collected using a 3 ns boxcar gate placed 12-15 ns after the temporal maximum in the laser intensity.

vibrational cooling of this intermediate by collisions with the buffer gas, or reaction of the intermediate in the case of added CO. The lack of any significant quenching from the high-lying Cr(I) states (e.g. $z^7P_J^o$, $z^5P_J^o$, $y^7P_J^o$, etc.), was then taken to be evidence of a second dissociation channel. This channel was proposed to involve the direct multiphoton excitation of Cr(CO)$_6$ to a dissociative continuum.

Based on the similar quenching results obtained in the 248 nm MPD of *benzeneCr(CO)*$_3$, we propose that the dissociation dynamics can be described by an analogous two-channel dissociation mechanism. The dominant pathway for formation of Cr(I) in the low-lying a^5S_2 and a^5D_2 excited states is a sequential absorption/fragmentation process,

$$(arene)Cr(CO)_3 + 1h\nu \rightarrow (arene)Cr(CO)_x^\dagger + (3-x)CO \qquad (1)$$

$$(arene)Cr(CO)_x^\dagger + 1h\nu \rightarrow Cr(a^7S_3, a^5S_2, a^5D_2) + arene + xCO \; [4]. \qquad (2)$$

The dissociation of the parent organochromium compound in this channel is characterized by the absorption of a single 248 nm photon followed by the rapid photoejection of a CO ligand [5]. A considerable fraction of the initially absorbed energy should be retained by the (arene)Cr(CO)$_x$ fragment following the inital dissociation event. Because this energy may be sufficient to induce further thermal loss of CO, the exact identity of the intermediate is not known. We have therefore written the intermediate in the general form (arene)Cr(CO)$_x$, where x=(2,1,0). This intermediate then absorbs an additional 248 nm photon and subsequently fragments to Cr(I). The observed quenching in the formation of the a^5S_2 and a^5D_2 states arises from a competition between up-pumping of the (arene)Cr(CO)$_x^\dagger$ intermediate to these states via reaction (1), and collisional energy transfer via equation (3).

$$(arene)Cr(CO)_x^\dagger + M \rightarrow (arene)Cr(CO)_x + M^* \qquad (3)$$

Vibrational cooling of the $(arene)Cr(CO)_x^\dagger$ intermdiate via reaction (3) leads to a reduction in the total energy available for product excitation following the photon absorption in reaction (2) and hence a reduction in the probability that $Cr(a^5S_2)$ and $Cr(a^5D_2)$ will be formed. The treatment outlined in ref. [1] for the quenching observed in $Cr(CO)_6$ can qualitatively explain the pressure dependence of the quenching, including the non-Stern-Volmer behavior observed in the MPD of the (arene)chromium tricarbonyls. For the present discussion, it is important to note that the observed quenching establishes that formation of Cr(I) via the sequential process occurs on the timescale of one or more collisions at < 25 torr, i.e. on a nanosecond time scale.

The absence of quenching from the $z^7P_J^o$, $z^5P_J^o$, $y^7P_J^o$, $y^5P_J^o$, and $z^5F_J^o$ states indicates that the mechanism for their formation is vastly different than that discussed above for the a^5S_2 and a^5D_2 states. We postulate that formation of the $z^7P_J^o$, $z^5P_J^o$, $y^7P_J^o$, $y^5P_J^o$, and $z^5F_J^o$ states of Cr(I) in the 248 nm MPD of the (arene)chromium tricarbonyls occurs via a direct dissociation channel:

$$(arene)Cr(CO)_3 + 2\text{-}3h\nu \rightarrow \{(arene)Cr(CO)_3\}^{**} \rightarrow Cr(I) + arene + 3CO. \qquad (4)$$

In this channel the parent compound absorbs multiple photons prior to dissociation. The superexcited compound formed most probably fragments 'explosively' via a repulsive potential surface. The observed statistical distribution of metal atoms produced in this process could arise from a strong coupling of the product states in the exit channel of the photodissociation half-collision. The absence of a significant difference in the Cr(I) electronic temperatures observed upon MPD of the four (arene)chromium tricarbonyls indicates that the vibrational density of states in the ligand has very little effect on the coupling of these exit channels. Reaction (4) is thus highly non-statistical in terms of the vibrational degrees of freedom. This suggests that the fragmentation represented by reaction (4) occurs on a picosecond or faster timescale, which is consistent with the lack of quenching found for Cr(I) states formed by the direct mechanism. A pictorial diagram illustrating the formation of Cr(I) via this two-channel dissociation mechanism is shown in figure 2.

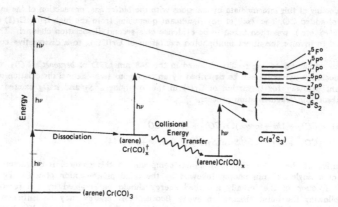

Figure 2. Schematic representation of the two-channel dissociation mechanism for the formation of Cr(I) in the 248 nm MPD of the (arene)chromium tricarbonyls.

We now turn our attention to the effect of changing the arene ligand on the relative branching between formation of the a^5S_2 and a^5D_2 states via the sequential mechanism and the higher energy Cr(I) states formed via the direct mechanism. Since the distribution of Cr(I) states formed in the direct dissociation process can be described by a single temperature independent of the arene ligand (see above), we take the emission intensity of the $y^7P_j^o \rightarrow a^7S_3$ transitions to be proportional to the total amount of Cr(I) formed via the direct channel. The relative branching between formation of the a^5S_2 and a^5D_2 states via the sequential dissociation channel and those states formed via the direct channel is then easily obtained by comparing the relative emission intensities from the $v^5D_3^o$ and $t^5F_2^o$ states with the emission intensity from the $y^7P_j^o$ state. This has been done in figure 3 for the case of the $t^5F_2^o$ state, formed via resonant up-pumping of the a^5D_2 state. Similar results were obtained in the case of the $v^5D_3^o$ state, formed via resonant up-pumping of the a^5S_2 state. It is immediately apparent from figure 3 that the relative branching into the a^5D_2 state decreases as the complexity of the arene ligand increases.

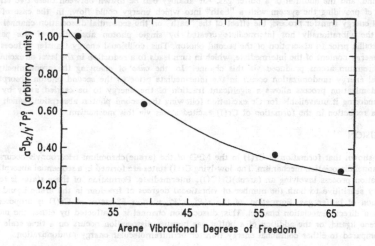

Figure 3. Plot illustrating the effect of the vibrational degrees of freedom in the arene ligand on the relative branching between formation of the a^5D_2 state via the sequential mechanism, and the $y^7P_j^o$ state formed via the direct mechanism.

These results clearly indicate that the vibrational degrees of freedom in the arene ligand strongly influence the partitioning of energy in the sequential dissociation channel. The most likely mechanism consistent with our data is that prior to the up-pumping of reaction (2), complete intramolecular vibrational energy randomization occurs in the (arene)Cr(CO)$_x$ intermediate formed in reaction (1). Because of the large number of vibrations in the intermediate, as well as its relatively high level of internal excitation, the vibrational density of states should be sufficiently large that complete energy randomization will occur on a subnanosecond time scale [7]. The energy randomization process will thus be complete prior to reaction (2), since our quenching results establish that the average time between formation of the intermediate and further up-pumping is on the order of nanoseconds. Increasing the vibrational density of states in the arene ligand by lengthening the alkyl side chain can affect the probability of forming the relatively high energy a^{5}D and a^{5}S products in two ways. If reaction (2) is a quasi-statistical dissociation process in terms of the vibrational degrees of freedom, the amount of energy carried away by the arene ligand will be roughly proportional to the vibrational density of states in the arene. As more energy is carried away by the arene, less energy will be available for excitation of the Cr atom, favoring formation of lower energy products such as Cr(a^{7}S$_3$) and the various molecular products. If reaction (2) is non-statistical in terms of the vibrational degrees of freedom, similar to the 'explosive' process described for the direct dissociation mechanism, then a more important factor will be the relative amount of energy within the intermediate that resides in the arene internal vibrations prior to reaction (2). This is because very little energy flow from the ligand vibrations to the chromium atom electronic degrees of freedom is expected to occur in a direct dissociation process. Because vibrational energy randomization is complete in the intermediate prior to reaction (2), more energy will reside in the arene internal vibrations as the alkyl side chain is lengthened, i.e. as the arene vibrations make up a larger fraction of the total number of vibrations in the intermediate. Again, lengthening the arene side chain causes more energy to be carried away by the ligand in reaction (2), favoring formation of lower energy products. Either explanation is consistent with our observation that the probability of forming $Cr(a^5S)$ and $Cr(a^5D)$ decreases as the number of arene internal vibrations increases.

As has been discussed above, those states that are formed via the sequential absorption/fragmentation process are found to be very sensitive to both the number of vibrational degrees of freedom in the arene ligand, and the addition of a buffer gas. An analogy can be drawn between these two effects in terms of providing the system with a "bath" into which energy could flow. In the case of the collisional energy transfer process, the effect of the "bath" on the sequential dissociation channel was to cool the vibrationally hot intermediate created by single photon absorption of the parent organometallic prior to absorption of the second photon. This collisional energy transfer reduces the internal energy content of the intermediate, which in turn leads to a reduction in the level of excitation in the chromium atoms produced via this channel. In the case of changing the arene moeity, a vibrational energy randomization occurs in the intermediate prior to the second photon absorption. This randomization process allows a significant fraction of the energy to be carried away by the ligand, rendering it unavailable for Cr excitation following the second photon absorption, which also leads to a reduction in the formation of Cr(I) excited states via this mechanism.

CONCLUSION

We have shown that formation of Cr(I) in the MPD of the (arene)chromium tricarbonyls occurs via a two channel dissociation mechanism. The low-lying Cr(I) states are formed in a sequential absorption/fragmentation process involving an $(arene)Cr(CO)_x$ intermediate. Formation of these states is found to be very sensititive to both the number of vibrational degrees of freedom in the arene ligand and the addition of buffer gas. Formation of the $z^7P^0_j$, $z^5P^0_j$, $y^7P^0_j$, $z^5F^0_j$ states of Cr(I) is proposed to occur via a direct dissociation channel. This dissociation channel is unaffected by either the nature of the arene ligand, or the addition of a buffer gas because dissociation occurs on a time scale that is fast compared to either collisional energy transfer or intramolecular energy randomization.

REFERENCES

1. G.W. Tyndall and R.L. Jackson, J. Chem. Phys., 89, 1364, (1988).

2. J.A. Conner, J.A. Martinho-Simoes, H.A. Skinner, and M.T. Zafarani-Moattar, J. Organometal. Chem., 179, 331, (1979). F.A. Adedeji, D.L.S. Brown, J.A. Connor, M.L. Leung, I.M. Paz-Andrade and H.A. Skinner, ibid., 97, 221, (1975).

3. G.W. Tyndall and R.L. Jackson, J. Am. Chem. Soc., 109, 582, (1987).

4. The products of reaction (2) imply that an intact arene moeity is formed. This has not been established, however. Reference 6 reports that $C_6H_6^+$ was not observed in the MPI of benzene chromium tricarbonyl.

5. D. Rooney, J. Chaiken, and D. Driscoll Inorg. Chem., 26, 3939, (1987).

6. G.J. Fisanick, A. Gedanken, T.S. Eichelberger, N.A. Keubler, and M.B. Robin, J. Chem. Phys., 75, 5215, (1981).

7. J.B. Hopkins, D.E. Powers, and R.E.Smalley, J. Chem. Phys., 73, 683, (1980).

EXCIMER LASER PHOTODISSOCIATION STUDIES OF DISILANE AT 193 NM

J. M. JASINSKI*, J. O. CHU*, AND M. H. BEGEMANN**
*IBM RESEARCH DIVISION, T. J. WATSON RESEARCH CENTER, YORKTOWN HEIGHTS, NY, 10598
** DEPARTMENT OF CHEMISTRY, VASSAR COLLEGE, POUGHKEEPSIE, NY, 12601

ABSTRACT

Optical emission, laser spectroscopic and mass spectrometric techniques have been used to study the ArF excimer laser induced photochemistry of disilane at 193 nm. Evidence is found for the formation of a number of photofragments from single and multiphoton dissociation. Effects due to secondary photolysis are observed at high excimer laser repetition rates.

INTRODUCTION

Disilane, Si_2H_6, is a convenient silicon source for photchemical CVD of silicon containing thin films because it is the simplest silicon hydride which can readily be excited by standard photolysis sources such as 185 nm Hg lamps and 193 nm ArF excimer lasers [1]. In order to understand the mechanism of thin film growth in photo-CVD processes it is necessary to understand the gas phase photochemistry of disilane. In particular it is important to understand what the primary photoproducts are and whether thay are kinetically stable enough to diffuse to the film growth surface. The only previous study of disilane photochemistry in this spectral region was performed at 147 nm using light from a Xe resonance lamp [2]. The available information on the electronic spectroscopy and possible photodissociation pathways for disilane has recently been reviewed [3].

In this paper we summarize the results of optical emission, laser spectroscopic and mass spectrometric studies of the 193 nm photodissociation of disilane. We report absolute primary quantum yields for disilane loss and for the formation of one stable product, silane. We present direct spectroscopic evidence for the formation of silicon atoms, silylene and silylidyne from both single and multiphoton dissociation processes. We present indirect evidence that the single photon photochemistry is dominated by the formation of silicon monoradical and/or closed shell unsaturated silicon species and we provide evidence that as the time between laser pulses is decreased relative to the average residence time of the gas in our photolysis cell secondary photolysis processes dramatically change the photochemistry.

EXPERIMENTAL

All experiments were conducted in a stainless steel flow cell under conditions of constant total pressure and constant total gas flow. In optical experiments, typically 40 - 200 mTorr of disilane diluted in 2 - 10 Torr of a buffer gas such as helium at total total flow rates of 50 - 300 sccm was photolyzed by the unfocused output of an ArF excimer laser at repetition rates of 2 - 40 Hz. A generalized schematic of the experimental arrangement is shown in Figure 1. Ultraviolet and visible emission was monitored at right angles to the excimer laser beam through a suprasil window with a monochrometer and photomultiplier. Infrared emission was monitored at right angles through a KCl window with an InSb detector. For laser absorption experiments, the output of a visible dye laser or infrared diode laser was directed down the length of the cell and onto a PIN photodiode or an InSb detector, respectively. For laser induced fluorescence studies, the output of a pulsed dye laser was directed down the length of the cell and fluorescence was collected at right angles with a monochrometer and photomultiplier. Full experimental details for the flow cell operation and the laser spectroscopic techniques can be found elsewhere [4 - 6]. For mass spectrometric studies, the gas in the flow cell was sampled by inserting an orifice into the cell at right angles to the excimer laser beam. The orifice generated a molecular beam which was directed to the ionizer region of a differentially pumped quadrupole mass spectrometer.

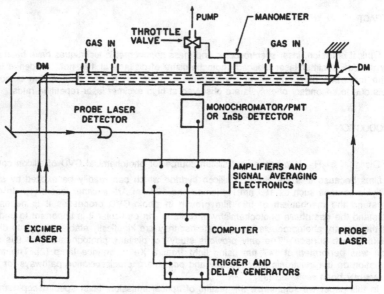

Figure 1. Generalized schematic of the experimental setup.

QUANTUM YIELDS AND DETECTION OF STABLE PRODUCTS

Using time-resolved infrared diode laser flash kinetic spectroscopy we have measured the primary quantum yields for loss of disilane and formation of silane under conditions where the excimer laser repetition rate is comparable to the residence time of gases in the photolysis cell [5]. The quantum yields were determined from changes in the absoprtion of an IR diode laser probe beam by Si-H stretching modes of silane and disilane. Typical transient infrared absorption curves are shown in Figure 2. The measured quantum yields and relevant absorption cross sections are given in Table I.

In complementary molecular beam sampling mass spectrometry studies, we find that in addition to silane, trisilane is an easily observed stable silicon containing photoproduct. No tetrasilane is observed. More importantly, as the time between excimer laser pulses is decreased relative to the gas residence time, the quantum yield for disilane loss increases to at least 4, at which point depletion of disilane from the cell becomes a limiting factor.

489

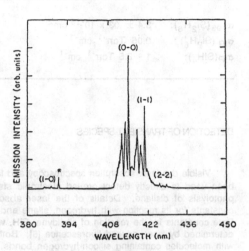

Figure 3. SiH $A^2\Delta \longrightarrow X^2\Pi$ emission. The numbers in parentheses represent various vibronic bands. The line with the asterisk is the Si(I) $4s_{(1)}{}^1P^0 \longrightarrow 3p_{(0)}{}^1S$ transition.

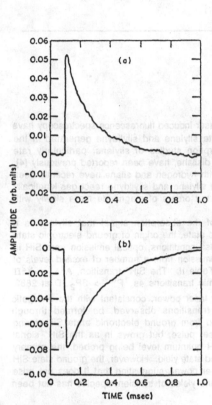

Figure 2. Infrared transient absorption signals showing (a) loss of disilane and (b) formation of silane.

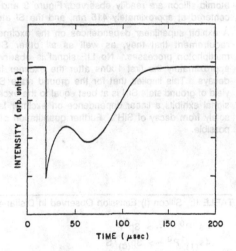

Figure 4. Time resolved infrared emission at $2.5\mu m \leq \lambda \leq 5.5\ \mu m$.

TABLE I: Absorption Cross Sections and Primary Quantum Yields

$\sigma_{193}(Si_2H_6)$:	$3.4 \pm .03 \times 10^{-18}\ cm^2$		$\Phi_{diss}(Si_2H_6)$: 0.7 ± 0.1
$\sigma_{IR}\ (Si_2H_6)$:	$0.05\ Torr^{-1}\ cm^{-1}$		$\Phi_{form}(SiH_4)$: 0.1 ± 0.03
$\sigma_{IR}(SiH_4)$:	$1 - 2.5\ Torr^{-1}\ cm^{-1}$		

DETECTION OF TRANSIENT SPECIES

Visible dye laser absorption spectroscopy and laser induced fluorescence spectroscopy have been used to directly detect ground electronic state silylene and silylidyne generated in the photolysis of disilane. Details of the laser absorption studies of silylene, particularly rate constants for its reaction with hydrogen, silane and disilane, have been reported previously [4]. Rate constants for the reaction of silylidyne, SiH, with hydrogen and silane have recently been determined by laser induced fluorescence [6]. Both silylene and silylidyne react gas kinetically with molecules containing silicon-hydrogen bonds, but orders of magnitude more slowly with molecular hydrogen.

While it is clear from the prompt risetime of the absorption signal that silylene is a primary photoproduct formed in its ground electronic state, the origin of ground electronic state SiH is less certain. Under single excimer laser pulse conditions, optical emission from SiH in its first excited electronic state, SiH*, as well as emission from a number of excited levels of atomic silicon are readily observed, Figure 3 and Table II. The SiH* transition, $A^2\Delta \longrightarrow X^2\Pi$ centered at approximately 415 nm, and the Si atomic transitions $4s\ ^1P^0 \longrightarrow 3P_2\ ^1D$ at 2882 Å exhibit superlinear dependences on the excimer laser power, consistent with the energetic requirement that they, as well as all other Si transitions observed, be formed through multiphoton processes. No LIF signal is observed from ground electronic state SiH during approximately the first 100ns after the excimer laser pulse, but grows in as the SiH* signal decays. This implies that for the ground state SiH quantum level being probed, the primary yield of ground state SiH is at best equal to the excited state yield. However, the ground state SiH signal exhibits a linear dependence on excimer laser power, suggesting that it does not arise solely from decay of SiH*. Further quantitation of the yields of transient species has not been possible.

TABLE II. Silicon (I) Emission Observed in Disilane Photodissociation at 193 nm

$4s^3P^0 \longrightarrow 3p^3P$	$3d_{(1)}\ ^1P^0 \longrightarrow 3p_{(2)}\ ^1D$
$4s_{(1)}\ ^1P^0 \longrightarrow 3p_{(0)}\ ^1S$	$3d_{(2)}\ ^1D^0 \longrightarrow 3p_{(2)}\ ^1D$
$4s_{(1)}\ ^1P^0 \longrightarrow 3p_{(2)}\ ^1D$	$3d_{(3)}\ ^1F^0 \longrightarrow 3p_{(2)}\ ^1D$
$3d_{(1)}\ ^1P^0 \longrightarrow 3p_{(0)}\ ^1S$	

INFRARED EMISSION

When the time between excimer laser pulses is short compared to the gas residence time in the flow cell, the apparent disilane quantum yield increases, as reported above. Under these conditions an intense infrared emission signal with complicated temporal and spectral properties is observed (Figure 4). At lower excimer laser repetitions rates, the sharp, sort time feature dominates the emission, while at higher excimer laser repetition rates the long time feature appears to become faster and more intense until it dominates the emission. Filter studies in the 2 - 6 μm region show that the fast componenet of the signal is predominantly emission in the region around 5 μm, suggestive of Si-H emission, while the slow component is spectrally broad and is observed at all wavelengths which we are presently capable of detecting (~1 - 6 μm), suggesting that it may be black body emission. The infrared signal can be quenched by addition of propylene, an efficient silicon hydride radical and silicon atom scavenger, to the gas mixture.

DISCUSSION

From the summary of experimental results presented above, it is clear that excimer laser induced photodissociation of disilane at 193 nm is a complicated process which must be discussed in terms of single and multiphoton processes occurring under single photolysis pulse conditions and in terms of additional processes which occur when the gas mixture is photolyzed by multiple laser pulses.

Since the disilane loss and silane formation quantum yields are independent of laser intensity and since our measured ultraviolet absorption cross section for disilane at 193 nm agrees quantitatively with previous measurements under lower intensity conditions [5,7], we conclude that under our experimental conditions disilane photochemistry is dominated by single photon absorption. While disilane spectroscopy is not well understood, the most recent results suggest that single photon absorption at 193 nm excites the 4s <--- $2a_{1g}$ Rydberg transition [7]. Since the state is not purely dissociative and since the transition does show weak vibronic structure, we believe that radiative decay may explain our observation of less than unity quantum yield for disilane loss.

Based purely on thermodynamic considerations, single photon photodissociation of disilane at 193 nm can generate up to twelve distinct chemical species through twenty energetically allowed reaction pathways [3]:

Products	ΔH_R kcal/mol	Products	ΔH_R kcal/mol
$H_2SiSiH_2 + H_2$	38.0	$SiH_4 + SiH_2(A^1B_1)$	102.1
$SiH_4 + SiH_2$	57.8	$SiH_4 + Si(3p^1D_2) + H_2$	114.8
$H_3SiSiH + H_2$	61.3	$2SiH_2 + H_2$	118.3
$Si(H_2)Si + 2H_2$	70.4	$SiH_3 + SiH + H_2$	120.7
$H_2SiSiH_2(^3B) + H_2$	72.9	$Si_2 + 3H_2$	121.9
$2SiH_3$	76.5	$SiH_4 + SiH + H$	133.2
$SiH_4 + SiH_2(a^3B_1)$	77.4	$SiH_2(a^3B_1) + SiH_2 + H_2$	137.9
$H_2SiSi + 2H_2$	85.3	$H_2SiSiH + H_2 + H$	138.8
$H_3SiSiH_2 + H$	88.7	$SiH_4 + Si(3p^1S_0) + H_2$	140.7
$SiH_4 + Si + H_2$	96.8	$H_2SiSiH_2 + 2H$	142.2

As disscussed in detail elsewhere [5], the diode laser quantum yield studies, combined with known or estimated rate constants for the reactivity of the possible radical species involved and the results of scavenging experiments, lead to the conclusion that less than 20% of the primary photoproducts are silylenes and silylidynes and that the quantum yilds of 0.7 and 0.1 are the primary quantum yields for loss of disilane and formation of silane. This further leads to the conclusion that the majority of the transient photodissociation products are relatively kinetically stable silicon monoraicals and/or unsaturated silicon hydride species.

In addition to the twenty single photon channels, there are thirty-eight channels for two photon dissociation below the disilane ionization limit. Our optical emission results provide evidence for the following pathways:

Products	ΔH_R kcal/mol	Products	ΔH_R kcal/mol
$SiH_3 + SiH(A^2\Delta) + H_2$	190.4	$SiH_4 + Si(4s\ ^3P^0)$	210.0
$SiH_4 + SiH(A^2\Delta) + H$	202.9	$SiH_4 + Si(4s\ ^1P_1^{\ 0})$	213.7

We see no optical evidence for the formation of ions or for dielectric breakdown of the gas. In particular we do not observe uv-visible emission from excited states of hydrogen atoms, atomic silicon ions or SiH^+.

Under multiple photolysis conditions, which may more closely approximate the conditions typically used in excimer laser photo-CVD from disilane, the most striking result is the observation of strong infrared emission. We suggest that this emission correlates with the onset of gas phase nucleation processes which ultimately lead to the well known formation of macroscopic particles of brownish amorphous hydrogenated silicon powder. Nucleation may result from the buildup of species such as unsaturated silicon hydrides or simply from excessive production of silicon atoms by secondary photolysis of higher silicon hydrides produced from the reactions of the primary photodissociation products. Further study is required to test this hypothesis. Nonetheless, it is clear that under multiple photolysis conditions, disilane photochemistry is markedly different and even more complicated than under single laser pulse conditions.

CONCLUSION

Single photon photodissociation of disilane under single laser pulse conditions is clearly a complicated process which can and does lead to a plethora of photoproducts. We have presented evidence that the majority of these products are silicon hydride monoradicals and/or closed shell unsaturated species rather than silylenes and silylidynes. While evidence for multiphoton processes is easily observed in the form of uv-visible emission, these processes are probably minor compared to single photon processes under conditions where the excimer laser is unfocused. When multiple photolysis of the gas mixture occurs, unidentified products build up which lead to an apparent increase in the disilane loss quantum yield and to strong infrared emission. We suggest that the infrared emission signal is related to the onset of gas phase nucleation processes.

REFERENCES

1. See, for example: A. Yoshikawa and S. Yamaga, Jpn. J. Appl. Phys., Pt.2, 23, L91 (1984); K. Kumata, U. Itoh, Y. Toyoshima, N. Tanaka, H. Anzai and A. Matsuda, Appl. Phys. Lett., 48,

1380 (1986); Y. Mishima, M. Hirose, Y. Osaka and Y. Ashida, J. Appl. Phys, $\underline{55}$, 1234 (1984); T. Sugii, T. Ito and H. Ishikawa, Appl. Phys. $\underline{A46}$, 249 (1988).

2. G. G. A Perkins and F. W. Lampe, J. Am. Chem. Soc., $\underline{102}$, 3764 (1980).

3. H. Stafast, Appl. Phys. $\underline{A45}$, 93 (1988).

4. J. M. Jasinski, J. Phys. Chem., $\underline{90}$, 555 (1986); J. M. Jasinski and J. O. Chu, J. Chem Phys., $\underline{88}$, 1678 (1988).

5. J. O. Chu, M. H. Begemann, J. S. McKillop and J. M. Jasinski, Chem. Phys. Lett. (in press).

6. M. H. Begemann, R. W. Dreyfus and J. M. Jasinski, Chem. Phys. Lett. (in press.)

7. U. Itoh, Y. Toyoshima, H. Onuki, N. Washida and T. Ibuki, J. Chem. Phys., $\underline{85}$, 4867 (1986); M. A. Dillon, D. Spence, L. Boesten and H. Tanaka, ibid., $\underline{88}$, 4320 (1988).

8. R. Robertson, D. Hils, H. Chatham and A. Gallagher, Appl. Phys. Lett., $\underline{43}$, 544 (1983).

LASER CVD OF a-Si:H: FILM PROPERTIES AND MECHANISM

K. HESCH AND P. HESS
Institute of Physical Chemistry, University of Heidelberg, Im Neuenheimer Feld 253,
D-6900 Heidelberg, F.R.G.
H. OETZMANN AND C. SCHMIDT
Asea Brown Boveri AG, Corporate Research
Eppelheimer Strasse 82, D-6900 Heidelberg, F.R.G.

ABSTRACT

Films of a - Si: H were deposited from SiH_4 in a parallel configuration employing a cw CO_2 laser. Infrared spectra, photoconductivities and dark conductivities of the films were studied as a function of surface temperature between 250°C and 400°C. The results are compared with data obtained from deposition experiments using Si_2H_6 and an ArF laser. Implications for the mechanism of LICVD are discussed.

INTRODUCTION

The laser-induced chemical vapor deposition (LICVD) of amorphous, hydrogenated silicon (a-Si:H) has found increasing interest recently. Especially the processes using a cw CO_2 laser and SiH_4 [1-4] and a pulsed ArF laser and Si_2H_6 [5-7] in a parallel surface - beam configuration have been studied in more detail.

From a thermodynamic point of view a-Si:H is a metastable material, and therefore its properties and relative stability strongly depend on the conditions of preparation. It is generally believed that the laser methods are soft deposition methods, which allow an accurate control of the conditions during film growth. Therefore, the use of laser photons to initiate the deposition process may not only lead to a-Si:H films with optimized properties, but also to a better understanding of the mechanism and the chemical processes involved in the gas phase and at the surface.

In this paper results are reported for the chemical composition and bonding configuration obtained by IR spectroscopy and the dependence of the photoconductivity and dark conductivity of the a-Si:H films on substrate temperature. These film properties measured for a - Si:H deposited from SiH_4 with a cw CO_2 laser are compared with the corresponding results observed for films grown from Si_2H_6 employing a pulsed ArF laser.

EXPERIMENTAL

Deposition of a-Si: H films was achieved by decomposing SiH_4 with a cw CO_2 laser beam in a configuration with the laser beam parallel to the substrate surface. More details of the experimental setup and the deposition conditions such as gas flow, pressure, laser power, etc. can be found elsewhere [3].

In the parallel configuration the laser radiation directly heats the gas above the surface and the surface temperature can be controlled more or less independently. Thus, the processes occurring in the gas phase, namely laser-induced chemistry, and the processes taking place at the surface determining the film structure and composition may be synergistic. The substrate surface temperature is of crucial importance for the film properties and often was not measured with sufficient accuracy in previous experiments. Therefore, the surface temperature was measured by an evaporated Ni sensor with an accuracy of ± 1K. This device is described in more detail in [4].

The films prepared for FTIR analysis were deposited on 0.3 mm thick single-crystalline silicon wafers. The spectra were recorded with a Nicolet Fourier transform spectrometer type MX 1. For the measurements of the electrical conductivity two Al electrodes (10 mm long, separated by 0.5 mm) were evaporated onto the film and contacted by thin silver wires with a conducting glue. The photoconductivities were measured under AM1 illumination. The SiH_4 sample used was supplied by Messer Griesheim with the following impurity specifications: H_2: < 100 ppm, Ar: < 1ppm, N_2: < 5ppm, CH_4: < 1ppm, CO:< 2ppm, O_2:< 1ppm and H_2O: < 1ppm.

Mat. Res. Soc. Symp. Proc. Vol. 131. ©1989 Materials Research Society

RESULTS

Figure 1 shows the infrared transmission spectra obtained for films grown at 250°C, 300°C, 350°C and 400°C. The film grown at 250°C with a hydrogen content of about 22 % has absorptions at 2090 cm^{-1}, 890 cm^{-1}, 840 cm^{-1} and 635 cm^{-1}. Following the assignments given in [8-10], peaks at 845 and 890 cm^{-1} together indicate a structure of polymeric chains of $(SiH_2)_n$, which should be accompanied by an absorption at 2100 cm^{-1} rather than 2090 cm^{-1}. The 2090 - 2100 cm^{-1} peaks are assigned to the SiH_2 and $(SiH_2)_n$ stretch modes, the 890 cm^{-1} peak to SiH_2 and $(SiH_2)_n$ bend scissors and the 845 cm^{-1} peak to polymeric $(SiH_2)_n$ wagging. The ratio of the absorption coefficients of α (845 cm^{-1}) / α (890 cm^{-1}) has been considered as a measure of $(SiH_2)_n$ chain formation. According to this interpretation an appreciable part of the hydrogen in the SiH_2 bonding configuration belongs to a polymeric structure $(SiH_2)_n$. The film deposited at 300°C possesses absorption bands at 2080 cm^{-1}, 890 cm^{-1}, (845 cm^{-1}) and 635 cm^{-1} in good agreement with previously reported data for LICVD near 300°C [11, 12]. The small peak at 845 cm^{-1} due to polymeric chains may be observable or not at this substrate temperature depending on deposition conditions. The data suggest that there is a transition from polymeric to isolated SiH_2 groups with rising surface temperature and decreasing growth rate. In the spectrum of the film grown at 350°C substrate temperature the absorption patterns below 1000 cm^{-1} remain similar to those observed for the 300°C film. The absorption at 2080 cm^{-1}, however, broadens to values below 2000 cm^{-1}. This indicates absorption due to monohydride, because the 2000 cm^{-1} peak is associated with the SiH stretch mode. In the spectrum recorded for the film deposited at 400°C mainly the 2000 cm^{-1} peak is left, indicating predominantly SiH with little SiH_2 bonding in this film with about 3 % hydrogen content.

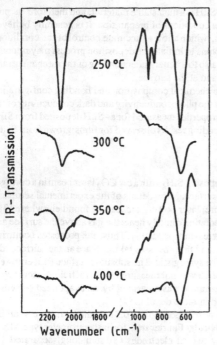

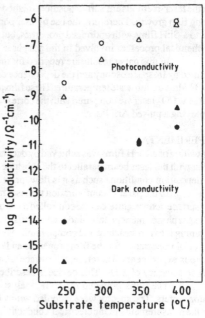

Fig. 1   FTIR transmission spectra of films grown at 250°C, 300°C, 350°C and 400°C substrate temperature.

Fig. 2   Photoconductivities (o, Δ), and dark conductivities (●, ▲) at 298 K as a function of substrate temperature for two different samples at each temperature.

According to these results the hydrogen bonding configuration changes from mainly SiH_2 bonding with a substantial proportion of SiH_2 groups involved in polymeric structures at 250°C to SiH bonding at 400°C. This type of bonding behavior has been observed for all laser-deposited a-Si:H materials. Within experimental error the spectra obtained for a-Si:H grown with a CO_2 laser from SiH_4 agree with those measured for films deposited from Si_2H_6 with an ArF laser [5]. Therefore, we conclude from the IR spectra that the different laser-induced deposition processes lead to films with similar chemical composition and bonding configuration. This may also point to a similar chemistry occurring in the gas phase and at the surface in the case of IR and UV laser-induced deposition in the parallel substrate-beam configuration.

Figure 2 shows the photoconductivity and dark conductivity values at 298 K as a function of the substrate temperature during film deposition. The dark conductivity increases from about 10^{-16} $(\Omega cm)^{-1}$ at 250°C to about $10^{-10}(\Omega cm)^{-1}$ at 400°C. The lowest dark conductivity values measured directly were in the $10^{-11}(\Omega cm)^{-1}$ range. The lower conductivity values were obtained by a linear extrapolation of the high temperature measurements. Figure 3 gives an example of such an extrapolation to a value as low as $7.8 \times 10^{-15}(\Omega cm)^{-1}$. The photoconductivity exhibits a similar increase from about 10^{-9} $(\Omega cm)^{-1}$ to about $10^{-6}(\Omega cm)^{-1}$ between 250°C and 400°C. The ratios of the corresponding photoconductivities to dark conductivities ranged from about 10^4 to 10^6. These results are in good agreement with the conductivity data reported in [2] for deposition from SiH_4 with a CO_2 laser. However, LICVD of a-Si:H from Si_2H_6 with an ArF laser yielded somewhat different results. For a deposition rate of 6nm/min the dark conductivity varied roughly between 10^{-11} and $10^{-9}(\Omega cm)^{-1}$ and the photoconductivity between 10^{-8} and $10^{-4}(\Omega cm)^{-1}$ in the considered temperature interval [5]. With increasing deposition rate, the conductivities decreased and approached the present values near the highest temperatures at a deposition rate of 24nm/min, but were still considerably higher than the present data around 250°C. The decrease of the conductivity by 4 orders of magnitude from 400°C to 250°C should at least partly be due to an increase of recombination centers near midgap. The activation

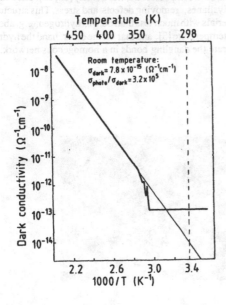

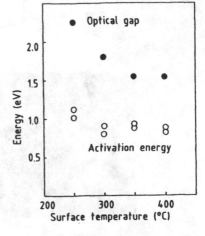

Fig. 3   Arrhenius plot of the dark conductivity with extrapolation to the room temperature value.

Fig. 4   Dependence of the optical gap and the activation energy of the dark conductivity on the substrate temperature.

energy of the dark conductivity and the optical gap increase with increasing hydrogen concentration and these effects may contribute to the decrease of the dark conductivity and photoconductivity, respectively.

The activation energy for the dark conductivity was determined from the high temperature range of the Arrhenius plot, because of a small curvature in the low temperature part of some films. The activation energy increases from about 0.8 eV at 400°C to about 1.1 eV as the surface temperature decreases to 250°C. On the other hand, the optical gap increases from about 1.5 eV to about 2.2 eV in this temperature range [3], as reported in [2] for films deposited by the present method. This result confirms a suggestion made in [2], namely that the activation energy likely represents the energy difference between the Fermi level and the nearest mobility edge, presumably that of the conduction band. As can be seen in Fig. 4, the value of the activation energy is approximately half the value of the optical gap. This indicates a position of the Fermi level near the middle of the optical gap as observed for LICVD films before [2].

The deposition of a-Si:H below a characteristic temperature that depends on the method employed leads to films with high hydrogen content and low spin density. The IR spectra show bonds which are characteristic for $(SiH_2)_n$ chain segments in the amorphous network. In fact, it was shown several years ago that films obtained by plasma deposition from silane under suitable conditions do not consist of a random continuous network but possess a columnar morphology [13]. The columnar or rod-like structures with dimensions in the range of 10 nm resulted from the growth of a finite number of nuclei into islands that failed to coalesce. Figure 5 shows a transmission electron micrograph of a LICVD film grown at a substrate temperature of 250°C. It was taken after dissolving the NaCl substrate. The island-like structures are about 30-50 nm in diameter, which is between a third and half of the film thickness. A scanning electron micrograph of the same film did not reveal any corresponding surface structure. This may be interpreted in terms of a columnar morphology for this material, too. It has been suggested for plasma deposited material, and it could also be true in our case, that the islands consist of a coherent network, while the regions in between possess a lower density [14, 15]. They may contain high concentrations of hydrogen and polysilanes, removing defects and stress. This structure explains how $(SiH_2)_n$ chains can be found in materials with much less than 66.6 % hydrogen, e.g. about 20 %-25 % in LICVD films of 250°C substrate temperature [3], and that on the other hand the hydrogen content is much higher than needed to saturate the dangling bonds in a homogenous network.

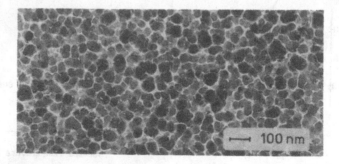

Fig. 5   Transmission electron micrograph of a 100 nm thick film grown at 250°C on a NaCl substrate at a deposition rate of 3 nm / min.

DISCUSSION

The a-Si:H films prepared by the CO_2 laser - SiH_4 and the ArF laser - Si_2H_6 deposition method show similar IR spectra. The general features are the existence of polymeric $(SiH_2)_n$ structures at the lowest surface temperatures studied, the predominance of SiH_2 groups at medium temperatures and the prevalence of the SiH bonding configuration at the highest surface temperatures investigated. However, the spectroscopic data are not accurate enough to allow a more detailed interpretation. Not only must the surface temperature be measured with a higher accuracy than performed in [5] but it is also necessary to take the dependence on the deposition rate into account. Preliminary results indicate a change in the hydrogen bonding and structure with the growth rate. This implies that distinct differences in the IR spectra may be observed for films deposited by different laser techniques but also for films grown at different rates. In fact, the conductivity measurements prove that these effects exist. The photoconductivities and dark conductivities of films deposited by the ArF laser - Si_2H_6 method at a deposition rate of 24 nm / min agree much better with the corresponding conductivities of films grown by the CO_2 laser - SiH_4 method at about 2-5 nm / min than those obtained at a deposition rate of 6 nm / min by the UV method. This can be attributed to small changes in the hydrogen content and configuration with the deposition method and rate leading to somewhat different film structures.

The precise measurements of the surface temperature during deposition not only allow a better comparison of results obtained by different techniques but may also be used in order to come to a better mechanistic understanding. As stated earlier on the basis of the precise surface temperature measurements with a Ni sensor [4], a rate-limiting surface reaction step can be excluded due to the large growth rate variation observed at almost constant surface temperature. The rate-limiting step in the gas phase is generally believed to be SiH_4 decomposition leading to the diradical SiH_2. The decomposition of Si_2H_6 may also involve the generation of diradicals such as SiH_2 and SiH_3SiH, however, with a higher rate, explaining the higher growth rates observed in the latter case. The diradicals are highly reactive. As shown recently, the rate constant for SiH_2 insertion into SiH_4 and Si_2H_6 is near the gas kinetic limit [16, 17]. Therefore, insertion reactions will occur during the diffusion of diradicals to the surface forming higher silanes, as already proved by mass spectrometry [3, 11]. Whether under LICVD conditions, especially at the relatively high silane pressure of several mbar, enough diradicals reach the surface to explain the deposition rate is not clear. Alternatively, we may assume that only the comparatively more stable higher silanes reach the surface, and, after adsorption, will be incorporated into the growing network. That deposition with a distribution of higher silanes is possible under comparable thermal conditions has been shown [18].

The surface chemistry responsible for the composition and structure of the film is not understood presently. The three-step mechanism including SiH_2 insertion at the surface, H_2 elimination and cross linking, first suggested for glow-discharge deposition [19], has also been applied to LICVD [2]. As discussed before, it is highly questionable whether a sufficient number of diradicals really reaches the surface under LICVD conditions operating much nearer thermodynamic equilibrium than glow discharge. It would not be surprising if the dynamics of surface bond creation, hydrogen elimination and bonding as well as cross linking to a three-dimensional network cannot be understood on the basis of a simple model developed from gas phase chemistry.

ACKNOWLEDGEMENTS

We are grateful to Dr. Schröder and coworkers from the University of Kaiserslautern, Department of Physics, for recording and discussing the FTIR spectra. Financial support of this work by the Bundesministerium für Forschung und Technologie (BMFT) under contract No. 13N5363 8 and by the Fonds der Chemischen Industrie is gratefully acknowledged.

REFERENCES

[1]   M. Meunier, J.H. Flint, J.S. Haggerty and D. Adler, J. Appl. Phys. 62, 2812 (1987)

[2]   M. Meunier, J.H. Flint, J.S. Haggerty and D. Adler, J. Appl. Phys. 62, 2822 (1987)

[3]   D. Metzger, K. Hesch and P. Hess, Appl. Phys. A **45**, 345 (1988)

[4]   K. Hesch, P. Hess, H. Oetzmann and C. Schmidt, Appl. Surf. Sci., to be published

[5]   A. Yamada, M. Konogai and K. Takahashi, Jap. J. Appl. Phys. **24**, 1586 (1985)

[6]   T. Taguchi, M. Morikawa, Y. Hiratsuka and K. Toyoda, Appl. Phys. Lett. **49**, 971 (1986)

[7]   D. Eres, D.H. Lowndes, D.B. Goehegan and D.N. Mashburn, MRS. Symp. Proc. **101**, 355 (1988)

[8]   G. Lucovsky, R.J. Nemanich and J.C. Knights, Phys. Rev. B **19**, 2064 (1979)

[9]   G. Lucovsky, J. Yang, S.S. Chao, J.E. Tyler and W. Czubatyi, Phys. Rev. B **28**, 3225 (1983); **28**, 3234 (1983)

[10]   G. Lucovsky and W.B. Pollard in: Topics in Appl. Phys., Vol **56** ed. by J. D. Joannopoulos and G. Lucovsky (Springer; Berlin, Heidelberg, New York, Tokyo 1984), p. 301

[11]   R. Bilenchi, I. Gianinoni and M. Musci, J. Appl. Phys. **53**, 6479 (1982)

[12]   H.M. Branz, M. Meunier, S. Fan, J.H. Flint, J.S. Haggerty and J. Adler, in: Proc. 18th IEEE Photovoltaic Spec. Conf. , 513 (1985)

[13]   J. C. Knights and R.A. Lujan, Appl. Phys. Lett. **35**, 244 (1979)

[14]   J. C. Knights, J. Non-Cryst. Sol. **35 & 36**, 159 (1980)

[15]   R. J. Nemanich, D. K. Biegelsen and M.P. Rosenblum, J. Phys. Soc. Japan **49**, 1189 (1980)

[16]   G. Inoue and M. Suzuki, Chem. Phys. Lett. **122**, 361 (1985)

[17]   J. M. Jasinski and J.O. Chu, J. Chem. Phys. **88**, 1678 (1988)

[18]   S. C. Gau, B.R. Weinberger, M. Akhtar, Z. Kiss and A.G. MacDiarmid, Appl. Phys. Lett. **39**, 436 (1981)

[19]   F.J. Kampas and R.W. Griffith, Appl. Phys. Lett. **39**, 407 (1981)

EXCIMER LASER INDUCED PHOTOCHEMISTRY OF
SILANE-AMMONIA MIXTURES

J. M. JASINSKI, D. B. BEACH and R. D. ESTES

IBM Research Division, Thomas J. Watson Research Center, Yorktown Heights, NY 10598

ABSTRACT

Steady-state and time resolved molecular beam sampling mass spectrometry have been used to study the ArF excimer laser induced photochemistry of silane-ammonia mixtures at 193 nm. Under both steady state and single laser shot conditions, products as complicated as tetraaminosilane are formed promptly. A mechanism which accounts for the formation of all observed products is proposed and evaluated.

INTRODUCTION

Argon fluoride excimer laser photolysis of silane-ammonia mixtures at 193 nm has been demonstrated as a possible low temperature large area photo-CVD technique for the deposition of amorphous hydrogenated silicon nitride thin films [1]. Since silane is for all practical purposes transparent at 193 nm [2], the photodeposition process is initiated by the well known photodissociation of ammonia [3]. The major photoproducts are hydrogen atoms and amidogen radicals, eq. 1.

$$NH_3 \rightarrow NH_2 + H \qquad (1)$$

Hydrogen atoms are known to react with silane to produce silyl radicals [3], eq 2., while NH_2 is less reactive with silane [5].

$$SiH_4 + H \rightarrow SiH_3 + H_2 \qquad (2)$$

Eqs. 1 and 2 result in the formation of reactive gas phase species containing both silicon and nitrogen, and it is possible that film growth proceeds directly from the impingement of these species on surfaces. It is also possible that these radicals lead to gas phase chemistry which produces molecules containing silicon-nitrogen bonds, and that those species are the immediate film growth precursors.

In this paper we report preliminary results of a mass spectrometric study of the 193 nm photochemistry of silane-ammonia mixtures. We find that the gas phase radical chemistry following photolysis is complicated and leads, unexpectedly, to the formation of aminosilane molecules as large as $Si(NH_2)_4$. Mass spectrometric studies of the glow discharge chemistry [6] and of the mercury sensitized photochemistry [7] of silane-ammonia mixtures have recently been reported. Our results for direct photolysis are strikingly similar to the glow discharge results and markedly different form the mercury sensitized photochemical results.

Mat. Res. Soc. Symp. Proc. Vol. 131. ©1989 Materials Research Society

EXPERIMENTAL

A schematic of the excimer laser photolysis/molecular beam sampling mass spectrometer is shown in Figure 1. In typical experiments, silane and ammonia, 1-25 mTorr each, dilute in helium buffer gas at total pressures of 0.5-1.0 Torr and ambient temperature are photolyzed in the flow cell with the unfocused output of an ArF excimer laser operating at repetition rates of 1-30 s^{-1}. The flow cell is operated at constant total gas flow and constant total pressure. Gas from the flow cell is formed into a molecular beam by a sampling orifice, and the beam is directed into the ionizer region of a differentially pumped quadrupole mass spectrometer. For steady state studies, the laser is operated at a repetition rate which is fast compared to the average residence time of gas in the flow cell and the mass spectrum is scanned. In order to detect laser induced chemistry, the spectrum is digitized and a background spectrum, acquired in the absence of laser irradiation, is subtracted. For time resolved studies, the mass spectrometer is tuned to the mass of interest and the time dependence of the mass signal is observed and digitized following the firing of the laser. The laser is operated at a low enough repetition rate that all photoproducts are flushed from the sampling region before the next laser pulse.

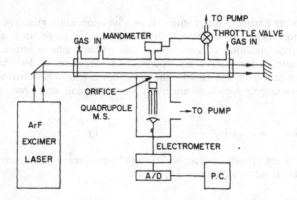

Figure 1. Schematic of the experimental apparatus.

RESULTS AND DISCUSSION

A steady state mass spectrum over the mass range of the observed photoproducts is shown in Figure 2. Products were identified from known mass spectral cracking patterns and from changes in cracking patterns upon substitution of deuterated ammonia or deuterated silane for ammonia or silane in the photolysis mixture. Hydrazine was also observed as a product when deuterated ammonia was used. It cannot be observed in the undeuterated case because of mass spectrometric interference from silane. Triaminosilane and tetraaminosilane cracking patterns are consistent with those reported by Smith et al. [6], once the trisilane spectrum is subtracted from the region of the tetraaminosilane peak at mass 92.

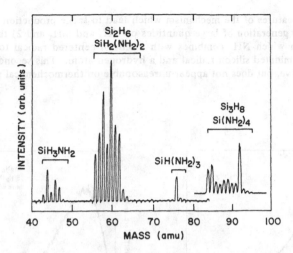

Figure 2. Steady state mass specturm of photolysis products.

Because of the complexity of the products and the apparent simplicity of the photoinitiation steps discussed in the introduction, it seemed likely that the higher aminosilanes were formed from secondary photolysis of species such as disilane and silylamine, which can readily be formed from radical reactions of SiH_3 and NH_2. That this is *not* the case is demonstrated by the time resolved data shown in Figure 3. All product signals rise and decay following a single excimer laser pulse. The risetime reflects a convolution of the apparatus sampling time and the kinetics of the reactions which form the products, while the decay of the signals is due to gas being swept out of the sampling volume.

A plausible chemical mechanism which reproduces all of the observed photochemistry is presented in Scheme I. This mechanism has been quantitatively evaluated using a stochastic mechanism simulator [8] which runs on a personal computer and reaction rate constants taken from the literature or estimated when literature values were not available [9]. A detailed discussion of the mechanism and the choice of rate constants and simulation parameters is beyond the scope of this paper. Typical simulation results are shown in Figure 4. The mechanism and the simulation reproduce the experimental findings semi-quantitatively. All observed species are accounted for and the relative yields of products are reproduced qualitatively. Further refinement of the rate constants, the simulation conditions and the sampling time of the apparatus are required to determine the origin of the quantitative difference in risetime of the experimental and simulated signals.

The key features of the mechanism which lead to facile production of aminosilanes are: 1) generation of large quantities of NH_2 and SiH_3 and 2) the proposed chain steps in which NH_2 combines with a silicon centered radical to produce a more highly aminated silicon radical and a hydrogen atom. This second step is the most speculative, but does not appear unreasonable on thermochemical grounds.

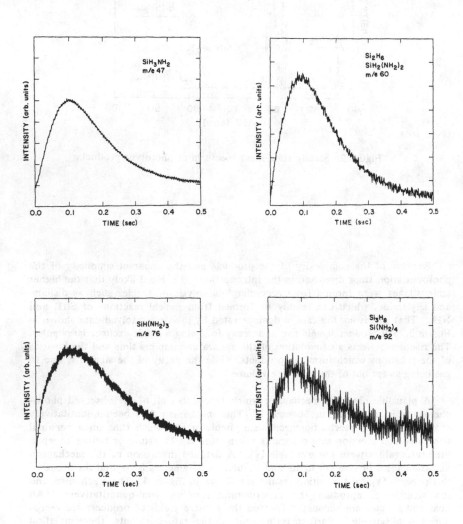

Figure 3. Time resolved mass signals for various photoproducts.

SCHEME I

Initiation

$$NH_3 \rightarrow H + NH_2 \quad (1)$$
$$H + SiH_4 \rightarrow SiH_3 + H_2 \quad (2)$$
$$NH_2 + SiH_4 \rightarrow SiH_3 + NH_3 \quad (3)$$

Higher Silanes

$$2SiH_3 \rightarrow Si_2H_6 \quad (4)$$
$$\rightarrow SiH_2 + SiH_4 \quad (5)$$
$$\rightarrow SiH_3SiH + H_2 \quad (6)$$
$$SiH_2 + SiH_4 \rightarrow Si_2H_6 \quad (7)$$
$$SiH_3SiH + SiH_4 \rightarrow Si_3H_8 \quad (8)$$

Amino Silanes

$$SiH_2 + NH_3 \rightarrow SiH_3NH_2 \quad (9)$$
$$SiH_3 + NH_2 \rightarrow SiH_2NH_2 + H \quad (10)$$
$$SiH_2NH_2 + H \rightarrow SiH_3NH_2 \quad (11)$$
$$SiH_n(NH_2)_m + NH_2 \rightarrow SiH_{n-1}(NH_2)_{m+1} + H \quad (12)$$
$$SiH_{n-1}(NH_2)_{m+1} + H \rightarrow SiH_n(NH_2)_{m+1} \quad (13)$$

Additional Termination Steps

$$2H \rightarrow H_2 \quad (14)$$
$$2NH_2 \rightarrow N_2H_4 \quad (15)$$

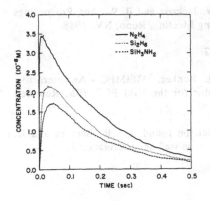 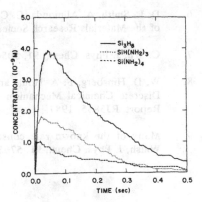

Figure 4. Simulated time dependences of photoproducts.

CONCLUSIONS

We have demonstrated that even under single laser pulse conditions, the excimer laser induced photochemistry of silane-ammonia mixtures results in formation of a range of products, including all possible aminosilanes. These observations parallel those made under plasma CVD conditions [6], where it has been suggested that decomposition of the aminosilanes on heated substrates is a significant film growth pathway. It is possible that the aminosilanes also contribute significantly to film growth under photo-CVD conditions since substrate temperatures of 200-600 °C are typically employed. The products from direct photolysis of silane and ammonia mixtures are remarkably different from those observed under mercury sensitized conditions. In the mercury sensitized photolysis, no aminosilanes higher than silylamine are observed but disilazane, $(SiH_3)_2NH$, and disilaneamine $Si_2H_5NH_2$ are observed [7].

REFERENCES

1. J. M. Jasinski and B. S. Meyerson, J. Appl. Phys., **61**, 431 (1987), and references therein.

2. J. H. Clark and R. G. Anderson, Appl. Phys. Lett., **32**, 46 (1978).

3. H. Okabe, *Photochemistry of Small Molecules*, (Wiley, New York, 1978), p. 269.

4. D. Mihelcic, V. Schubert, R. N. Schindler and P. Potzinger, J. Phys. Chem., **81**, 1543 (1977).

5. J. O. Chu and J. M. Jasinski, unpublished.

6. D. L. Smith, A. Alimonda, C. Chen, W. Jackson and B. Wacker, Proceedings of the Materials Research Society Spring Meeting, Reno, NV, 1988.

7. C. Wu, J. Phys. Chem., **91**, 5054, 1987.

8. W. D. Hinsberg, F. A. Houle and D. L. Bunker, "MSIMPC - An Interactive Discrete Chemical Mechanism Simulator for the IBM PC", IBM Research Report, RJ5855, 1987.

9. Many of the known rate constants can be found in: R. Becerra and R. Walsh, J. Phys. Chem., **91**, 5765 (1987), as well as in reference 3.

DEFECTS AND THEIR CONTROL IN SiO_2 FILMS PREPARED BY D_2-LAMP PHOTOCHEMICAL VAPOR DEPOSITION

HIDEHIKO NONAKA, KAZUO ARAI, AND SHINGO ICHIMURA
Electrotechnical Laboratory, 1-1-4 Umezono, Tsukuba, Ibaraki 305 Japan

ABSTRACT

Amorphous silica films deposited from the mixture of gases (Si_2H_6 and Si_2F_6) by deutrium-lamp CVD were studied by IR, vacuum UV, EPR and Auger electron (AE) spectrometries. The F-doping enhanced the film growth and removed defects in the film such as -H, -OH, and E' centers. A model on deposition and defect formation mechanisms was proposed based on the thermodynamic stabilities of resultant HF in the reactions. The AES study showed that the film surface modified by activated oxygen had an increased hardness against electron beams.

INTRODUCTION

Photochemical vapor deposition (photo-CVD) is one of the promising methods in low-temperature processes for preparing amorphous silica (a-SiO_2). The photo-CVD has its inherent problems, such as residual -H and -OH in the films and having an unrelaxed network due to a low-temperature process. In the present view of investigating the growth mechanisms of the films, the photo-CVD has an advantage due to its rather simple decomposing processes compared to, for example, plasma-CVD; the reaction paths can be limited as radicals are selectively produced by photodissociation of reactant gases according to photoexcitation energy.

We have shown that when a-SiO_2 films are deposited from disilane (Si_2H_6), oxygen, and perfluorodisilane (Si_2F_6) by deuterium-lamp (D_2) photo-CVD, the characteristics including defects of the films strongly depend on the preparation conditions [1]. Fluorine (F)-related reactions should play an important role in the film growth and defect behavior. In addition, F-doping itself is interesting, because it has an effect on the radiation hardness of the MOS devices [2] and fibers [3].

To investigate the nature of defects in the films, comparison with knowledge obtained from experiments in bulk SiO_2 glass will be helpful. In bulk SiO_2 glass, an absorption edge in the vacuum ultraviolet (VUV) region has been found to be sensitive to the existence of diamagnetic defects such as -OH, -Si-Si- (oxygen-deficient center, ODC), and -O-O- (peroxy-linkage) [4]. In the case of -OH, for instance, the absorption arises from around 7.5 eV and an apparent band edge, that is almost an exponential feature, proportionally shifts to the lower energy side as OH-content is increased. In the case of -Si-Si-, a clear absorption peak appears at 7.6 eV.

Paramagnetic defects such as E' centers (hole trap in oxygen-deficient center, ODC) were induced by intense UV light such as excimer lasers [5-7].

In this paper, we measured the VUV absorption and electron paramagnetic resonance (EPR) of the films prepared by D_2-lamp photo-CVD under systematically changed conditions.

The ODC in a-SiO_2 thin films is said to be detectable by Auger electron spectroscopy (AES) [8,9]. However, there is the possibility of causing damage to the sample surface by use of high-energy electron beams for the excitation itself, which makes the characterization of the film difficult even in a semi-

Table I. Characteristics of the films.

No.	Si_2H_6 /sccm	Si_2F_6 /sccm	thick-ness /μm	growth rate /Åmin^{-1}	[OH] mol.%	[H] at.%	[F] at.%	880 cm^{-1} /cm^{-1}	[E'] /cm^{-3}
1	0.5	0	1.0	50	0.7	0.5	---	8	7.7×10^{17}
2	0.5	1.9	1.3	100	<0.1	~0	2.2	~0	$<1 \times 10^{16}$
3	2.4	0	1.3	342	1.1	0.8	---	54	3.9×10^{17}
4	2.4	2.4	1.1	629	<0.1	0.5	2.3	33	$<1 \times 10^{16}$

quantitative way. We attempted to use AES to investigate the surface modification effect brought about by supplying atomic or low-energy ionic oxygen onto the surface of the film, after determining the damage-free measuring conditions of AES.

EXPERIMENTAL AND RESULTS

Amorphous SiO_2 films were deposited from Si_2H_6 (and Si_2F_6) and O_2 using the D_2-lamp photo-CVD method at 200°C on n-type Si wafer. The VUV light from the D_2-lamp was irradiated perpendicular to the substrate. In order to measure their VUV absorptions, the films were deposited on single crystals of lithium fluoride (LiF).

The films for EPR measurement were deposited on Si wafers of high resistivity (7000-20000 Ω cm). The EPR spectra of the films were measured at room temperature using an EPR spectrometer, Bruker ESP 300. The film on the substrate was cut into pieces (5 × 10 mm) and placed in a silica sample tube (11 mm ϕ). The sample tube was set in the microwave cavity so as to make the magnetic field perpendicular to the film surface. Non-saturating microwave power of 0.2 mW was applied for the evaluation of the spin density of the resonance. We observed the well-known E' centers occasionally along with P_b centers (see Fig. 1). As seen in Table I, the films deposited without F-doping have E' centers of the order of 10^{17} per cm^3, while the films with F-doping have that of less than one order of magnitude.

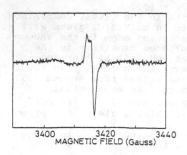

Fig. 1. EPR spectrum of film no.3 with E' centers. The broad band in the lower magnetic field is known as P_b centers.

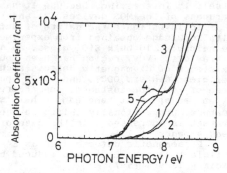

Fig. 2. VUV absorption of the films. The numbers correspond to those in Table I. Film no.5 was deposited under the same conditions as film no.4, having a thickness of 3 times as much as that of no.4.

The VUV absorption measurement was done using synchrotron. radiation in the same way as previously reported [4]. In brief, the absorption at 6.2 - 11.3 eV was measured through a 1 m monochromator (Nikon/McPherson 225) using a GaAsP-type Shottky photodiode as a detector and the spectral resolution was 2.5 nm. Figure 2 shows the results. The characteristics of the film for each spectrum in Fig. 2 may correspond to those given in Table I because the samples were prepared under the same conditions except for the substrate.

As seen in Fig. 2, the absorption edges of the spectra shift to the higher energy side of photon energy as the OH-content is decreased. Therefore, the effect of -OH on the VUV absorption of the a-SiO$_2$ films agrees with that of the bulk glass. For the film no.4, which was deposited at high growth rate with F-doping, a peak at 7.6 eV appears in the spectra which is attributable to -Si-Si- in case of the bulk a-SiO$_2$. It was confirmed that the peak arose from the bulk of the film and not from the interface between the film and LiF substrate, because there is no difference in the height of the peak when the film is made three times thicker (see spectra 4 and 5 in Fig. 2). The density of ODC in film no.4 was estimated to be about 5×10^{19} per cm^3 according to the absorption coefficient obtained in the bulk SiO$_2$ glass.

REACTION MECHANISM

The effects of admixing Si$_2$F$_6$ to the reactant gases are summarized as follows :
i) the growth rate of the film is considerably enhanced (no film deposition occurs without Si$_2$H$_6$).
ii) the number of impurity defects such as -OH and -H and paramagnetic defects (E') are effectively decreased.
iii) fluorine is incorporated into the film at the concentration of about 3 % at maximum, independently of the mixing ratio of Si$_2$F$_6$ or growth rate.
iv) too much enhancement in the growth rate by admixing Si$_2$F$_6$ causes oxygen-deficient defects (-Si-Si-) in the films.
Since no such effects were observed when SiF$_4$, which cannot be decomposed by D$_2$ lamp, was used instead of Si$_2$F$_6$, we have concluded that photodissociated radicals of Si$_2$F$_6$ play an important role in the reactions which eventually result in those effects.

We have proposed a model where the photodissociated radicals of Si$_2$F$_6$, possibly SiF$_3$, pull H off from Si$_2$H$_6$ in the gas phase and on the growing surface of the film to form HF. According to the model, the observed enhancement in the film growth is explained by excitation of both Si$_2$H$_6$ and the growing surface by this pull-H reaction. Since HF is thermodynamically so stable, fluorine may also react with H in the film, for instance, as schematically shown in Fig. 3

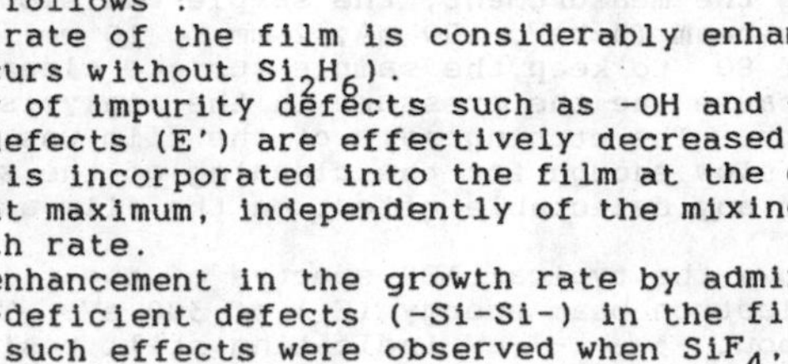

$$\equiv Si \overset{H}{\underset{F}{\diagup}} Si \equiv \xrightarrow{-HF} \equiv Si - Si \equiv$$

$$\equiv Si \overset{OH}{\underset{F}{\diagup}} Si \equiv \xrightarrow{-HF} \equiv Si - O - Si \equiv$$

Fig. 3. Schematic description of fluorine reactons.

In IR, a peak at 880 cm^{-1} occasionally appears for H-rich a-SiO$_2$ films. A good correlation between the intensity of 880 cm^{-1} absorption and that of -OH or -H in IR spectra indicates that the band at 880 cm^{-1} is due to an H-related defect such as a pair of SiH's or that of SiH and SiOH. We believe that the defect must have a definite structure judging from the sharpness of the absorption.

AES OF THE FILMS

The ODC, -Si-Si- is said to be observed as a peak at 89 eV for LVV transitions in the first derivative Auger spectra, dN(E)/dE curves. The assignment is based on the result that the peak appeared only in the oxygen-deficient films, and on the calculation of the energy level of the defect [9]. However, another report suggested that a peak centered at about 86 eV may have been associated with paramagnetic defects such as E' centers and peroxy radicals based on the results of EPR measurements [10].

Two kinds of films were prepared for AES measurement. One was the slowly deposited film without F-doping corresponding to the film no.1 in Table I (A) and the other was the fast deposited film with F-doping corresponding to the film no.4 (B). As noted before, the former contains E' centers of about 8×10^{17} per cm^3, while the latter contains the ODC of about 5×10^{19} per cm^3. The AES measurement was carried out in situ. The thickness of each film was about 1000 Å in order to prevent the film from a strong charging. During the measurement, the sample was slowly etched by rastering Ar$^+$ ion beam (3 keV, 100 nA, 2 mm $\times$ 10 mm) at an incident angle of 80 to keep the sample surface clean from contamination because the the pressure in the analysis chamber was about 8×10^{-9} Torr. The etching rate of the film was about 0.4 Å/min, which was slow enough for the cleaning of the surface without damage or any detectable effect on the film as described below.

Figure 4 shows the typical AES spectra of the films measured at the primary electron beam energy (E$_p$) of 300 eV. The spectra are shifted by about -3 $\sim$ -7 eV indicating still positive charging of the surface probably by ejection of more secondary electrons than primary beams. The spectra show only well-known Si peaks in a-SiO$_2$ and the 89 eV peak was not originally observed at the sample current of less than 1.50 μA for film A and 0.85 μA for film B, respectively. Therefore, the films measured had no detectable defects, attributable to the 89 eV peak irrespective of the film characteristics. By keeping the E$_p$ and sample current as low as 300 eV and around 1 μA, respectively, the AES measurement can be carried out without damaging the sample surface.

The existence of thresholds for the generation of the 89 eV peak is clearly demonstrated in Fig. 5 showing the intensity of the peak normalized by that of the 76 eV peak against the total sample current. During the measurement, the sample current was operated to increase after a time interval of more than 15 minutes. The films A and B show different onsets for the appearance of the 89 eV peak, indicating the difference in the qualities of the films; the film containing more ODC's seems less durable against electron beams. The separate onsets in the graphs A-1 and A-2 in Fig. 5 indicate that the threshold for the peak is determined by the sample current used and not by the accumulated sample current. The same slopes for the growing intensity of the 89 eV peak of the film with different amounts of electron doses supports the above assumption. The saturating tendency in the intensity of the peak against the sample current above the threshold implies that there is a certain kind of equilibrium between the defect formation and reconstruction of the surface.

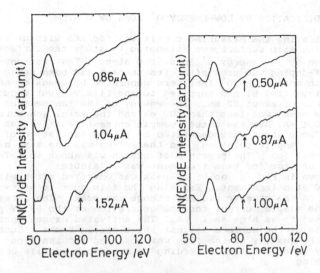

Fig. 4. AES spectra (Si,LVV) of the a-SiO₂ films.
Left: Film A, slowly deposited without F-doping.
Right: Film B, fast deposited with F-doping.
Sample current is indicated for each spectrum.
Arrows indicate 89 eV peaks. Electron energy is
nominal due to charging-up of the sample.

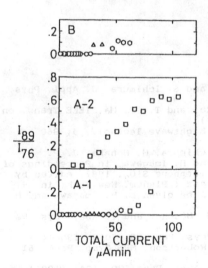

Fig. 5. Relation between the
intensity of 89 eV peak normalized
by 76 eV peak and the total sample
current for films A and B. Sample
current were (o) ≤0.7, (△) 0.85,
(o) 1.0, (□) 1.5 μA.

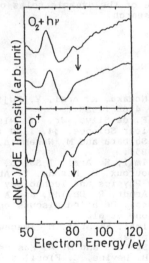

Fig. 6. AES spectra of film B
showing the surface modification by either O atoms (upper)
or O ions (lower).

SURFACE MODIFICATION BY LOW-ENERGY O^+ IONS OR O ATOMS

By employing the measuring conditions for AES without causing damage to the film surface, we attempted to study the surface modification by low-energy O^+ ions or O atoms. The fast deposited film with F-doping was used after its surface had been damaged by the electron beam. The O^+ ions were generated by means of an ion gun of RF type (100 W) and supplied to the film without focusing at a distance of about 80 mm. The energy of the ion beam of 10 mm ϕ at the anode was less than 100 eV and the total ion current was about 0.1 mA, but the actual sample current density was unknown. The O atoms, were generated by photodissociation of oxygen by D_2 lamp irradiation, and the set-up was the same as that for the deposition. The pressure of oxygen was about 10^{-4} Torr. The period of time for each treatment was 30 minutes.

As shown in Fig. 6, no 89 eV peak was observed after either O^+ ion or O atom treatment. Exposing the film just to an oxygen atmosphere in the chamber did not change the peak. The threshold value of the sample current for the generation of the 89 eV peak was increased to as high as 1 μA. The activated oxygen increased the hardness of the surface against the electron beams. Such change in the quality of the film was observed for the first several layers of the film according to the depth profile obtained by Ar^+ etching.

The generation of E' centers observed may be due to the D_2 lamp irradiation. The F-doping suppressed the E' center generation in the film as noted before. The effect may have relation to the resultant decrease of mobile defects of -H and -OH [11]. The result of AES also suggests the dynamic nature of defect creation by electron beam. Investigating the microscopic picture of the dynamic roles of -H and -OH will be the next target for research.

REFERENCES

1. H. Nonaka, K. Arai, Y. Fujino, and S. Ichimura, J. Appl. Phys. 64, 4168 (1988).
2. E. F. da Silva, Jr., Y. Nishioka, and T.- P. Ma, IEEE Trans. on Nuclear Science, 34, 1190 (1987).
3. S. Shibata and M. Nakamura, J. Lightwave Technol., 3, 860 (1985).
4. H. Imai, K. Arai, T Saito, S. Ichimura, H. Nonaka, J. P. Vigouroux, H. Hosono, Y. Abe, and H. Imagawa, in Proceedings of the Physics and Technology of Amorphous SiO_2, 1987, edited by J. Arndt, R. Devine, and A. Revesz (Plenum, New York, in press). Detailed discussion will be given by H. Imagawa and H. Hosono et al.
5. H. Imai, K. Arai, H. Imagawa, H. Hosono, and Y. Abe, Phys. Rev. in press.
6. H. Stathis and M. A. Kastmer, Phys. Rev. 89, 7079 (1984).
7. A. B. Devine, C. Fiori, and J. Robertson, MRS Symp. Proc. 61, 177 (1986).
8. K. Maki and Y. Shigeta, Jpn. J. Appl. Phys. 20, 1047 (1981).
9. S. S. Chao, J. E. Tyler, Y. Takagi, P. G. Pai, G. Lucovsky, S. Y. Lin, C. K. Wong, and M. J. Mantini, J. Vac. Sci. Technol., A3, 1574 (1986).
10. C. Fiori and R. A. B. Devine, Phys. Rev. Lett., 52, 2081 (1984).
11. H. Nonaka, K. Arai, and J. Isoya, unpublished.

THE DIRECT PHOTO-CVD OF SILICON DIOXIDE FROM Si_2H_6 AND N_2O_3

YASHRAJ K. BHATNAGAR AND W. I. MILNE
Department of Engineering, University of Cambridge, Trumpington Street, Cambridge CB2 1PZ, England.

ABSTRACT

The direct photo-CVD of silicon dioxide on c-Si substrates has been achieved using a new combination of Si_2H_6 and N_2O_3 gases, with an external deuterium lamp as the VUV source. The variation of the deposition rate and the refractive index with process parameters is reported. FTIR studies on these oxides show that the oxide is very nearly stoichiometric with no Si-N bonds and only trace amounts of Si-H bonds. C-V, I-V and breakdown measurements made on these films are reported. a-Si:H thin film transistors have also been fabricated using this oxide as the gate dielectric.

INTRODUCTION

The direct photo-chemical vapour deposition (photo-CVD) of SiO_2 has been reported by other workers from SiH_4 or Si_2H_6 diluted with an inert gas and mixed with oxidants such as O_2, N_2O or NO_2 [1-6]. In experiments using pure disilane, we found that even at N_2O/Si_2H_6 ratios as high as 150 to 200, the oxide films had very high refractive indices indicating oxygen deficiency in the films. Infrared absorption spectroscopy of these films revealed absorption bands associated with Si-N, Si-H and O-H bonds, in addition to the Si-O peaks. When pure disilane was mixed with O_2, we found that gas phase reactions predominate even at low total pressures, resulting in profuse powder formation. When NO_2 was used as the oxidant, there was little powder formation, but the surface reactions were spontaneous and we observed high deposition rates without photo-excitation even at room temperature. These films were visibly porous. Therefore, an oxidant had to be found which would react only upon photo-excitation and have enough oxygen to oxidise the disilane more completely than N_2O. This paper reports on the direct photo-CVD of SiO_2 from Si_2H_6 and N_2O_3 mixtures. At room temperature, N_2O_3 is largely dissociated into NO and NO_2 and in the light of further experiments, we believe that NO and its photo-dissociated fragments play the dominant role in the deposition process.

EXPERIMENTAL

The deposition chamber used in the direct photo-CVD of SiO_2 has been described elsewhere [7]. No deposition was observed without irradiation and it was found that successive depositions could be done without the need for cleaning the lamp window. The film thickness and refractive index were measured using a Gaertner model L117 ellipsometer. IR absorption spectra of the films were measured on a Perkin-Elmer 983 FTIR spectrometer to a resolution of 2 cm^{-1}. Laser induced ion mass spectroscopy (LIMA) was carried out at the department of Materials Science and Metallurgy of the University of Cambridge. Although it is not a quantitative technique, LIMA has a sensitivity of the order of 20 ppm for most elements. Aluminium dots 1mm in diameter were evaporated onto the samples for electrical measurements. The capacitance-voltage characteristics of these oxides were measured using the standard 1MHz high frequency technique on MOS capacitors. For breakdown measurements, a high dc voltage was applied to the aluminium dot and breakdown was assumed to have occurred at the electric field value where the current density through the oxide exceeded 1.3 A m^{-2}.

RESULTS

Figures 1a-d show the variation with deposition conditions of the deposition rate (▲) and the refractive index (□). Fig. 1a shows that the deposition rate is clearly photon-limited. Fig. 1b indicates that the optimum flow ratio of N_2O_3/Si_2H_6 is between 30 and 50. The decrease in both deposition rate and refractive index, in fig. 1c, with increasing substrate temperature has also

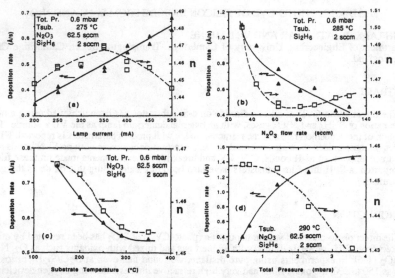

Fig. 1a-d : The variation of the deposition rate (▲) and the refractive index (□) with deposition conditions. Tsub = substrate temperature; Tot. Pr. = total pressure.

been observed by Okuyama et al [8] using Si_3H_8 and O_2 and they attributed this behaviour to a decrease in the adsorption of radical species. Fig. 1d shows that the operating pressure for good quality films should be about 0.6 mbar.

Fig. 2 shows the IR absorption spectra of a photo-CVD film before and after annealing at 320°C in wet N_2 flowing at 0.5 l/min. From the peak position of the Si-O stretching mode peak [9], we estimate that x > 1.9 in the annealed SiO_x film. The near absence of nitrogen from the

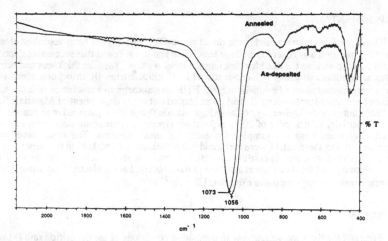

Fig. 2 : IR absorption of a film as deposited and annealed at 320 °C in wet N_2.

films has been comfirmed by LIMA. Fig. 3 shows the thickness dependence [10] of the full width at half maximum (FWHM) and the peak position of this mode, both in the as-deposited (∎) and the annealed (×) states. Increasing the annealing time shifts the peak position to higher wavenumbers (●), which shows that the film stoichiometry improves with anneal time.

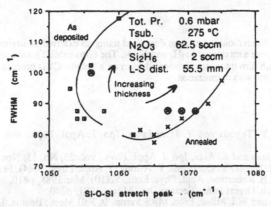

Fig. 3 : Dependence of the FWHM and the Si-O-Si stretch peak position on film thickness.
∎ As deposited; × Annealed; ● Effect of changing anneal time.

The C-V curves showed very little hysteresis with an oxide fixed charge of about 10^{11} cm^{-2} and an interface trap density of about 10^{11} cm^{-2}eV^{-1}. Typically the breakdown strength was found to be 7MV cm^{-1}. The lamp acts as a point source which leads to poor film uniformity and this makes accurate measurement of film thickness very difficult. We estimate the etch rate to be about 12 Å s^{-1}.

We have made some preliminary devices using the photo-CVD process. Fig. 4 shows the characteristics of an inverted staggered thin film transistor fabricated by photolithoghaphic

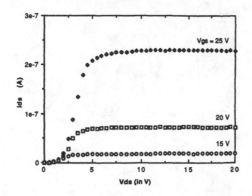

Fig. 4 : On-state characteristics of thin film transistors
fabricated from photo-CVD deposited materials.

techniques on Corning 7059 glass discs, using a-Si:H and SiO_2 deposited by photo-CVD. The device shows a turn on ratio of about 4 orders of magnitude. However, the effective mobility of carriers in the a-Si:H is very low and the threshold voltage is in the range of 10 volts. Leakage through the gate electrode is about 1 pA, which confirms the quality of the oxide.

CONCLUSIONS

Good quality silicon dioxide has been deposited using an external deuterium lamp to photo-chemically decompose a mixture of Si_2H_6 and N_2O_3. The films exhibit good optical and electrical characteristics. Thin film transistors fabricated from an all photo-CVD process show promise of excellent performance with optimization.

REFERENCES

1. M. Okuyama, Y. Toyoda and Y. Hamakawa, Jpn. J. Appl. Phys., vol. 23, No. 2, Feb. 1984, pL97.
2. Y. Tarui, J. Hidaka and K. Aota, Jpn. J. Appl. Phys., vol. 23, No. 11, Nov. 1984, pL827.
3. Y. Mishima, M. Hirose, Y. Osaka and Y. Ashida, J. Appl. Phys., 55(4), Feb. 1984, p1234.
4. J. Marks and R.E. Robertson, Appl. Phys. Lett., 52(10), Mar 1988, p810.
5. J.W. Peters, Tech. Digest Intnl. Electron Dev. Meet., 1981, p240.
6. P.A. Robertson and W.I. Milne, Proc. MRS Symp. B, Fall Meet., Boston, Dec. 1986, p257.
7. Y.K. Bhatnagar and W.I. Milne, to be published in Thin Solid Films.
8. M. Okuyama, N. Fujiki, K. Inoue and Y. Hamakawa, Jpn. J. Appl. Phys., vol. 26, No. 6, June 1987, pL908.
9. P.G. Pai, S.S. Chao and Y. Takagi, J. Vac. Sci. Technol. A, vol. 4(3), May/June 1986, p689.
10. I.W. Boyd, Electron. Lett., vol. 24, No. 17, Aug. 1988, p1062.

PYROLYTIC AND LASER PHOTOLYTIC GROWTH OF CRYSTALLINE AND AMORPHOUS GERMANIUM FILMS FROM DIGERMANE (Ge2H6)

DJULA ERES, D. H. LOWNDES, J. Z. TISCHLER, J. W. SHARP, D. B. GEOHEGAN, and S. J. PENNYCOOK; Solid State Division, Oak Ridge National Laboratory, Oak Ridge, TN 37831-6056

ABSTRACT

High-purity digermane (Ge_2H_6, 5% in He) has been used to grow epitaxially oriented crystalline Ge films by pyrolysis. Amorphous Ge:H films also have been deposited by pyrolysis and ArF (193 nm) laser-induced photolysis. The amorphous-to-crystalline transition and the film's morphology was studied as a function of deposition conditions. The film's microstructure, strain and epitaxial quality were assessed using x-ray diffraction curves and scanning and transmission electron microscopy. It was found that commensurate, coherently strained epitaxial Ge films could be grown pyrolytically on (100) GaAs at low (0.05–40 m Torr) Ge_2H_6 partial pressures, for substrate temperatures above 380°C.

INTRODUCTION

The epitaxial Ge/GaAs interface is a subject of considerable research interest because it is almost perfectly lattice-matched (the relative difference in room temperature lattice constants is less than 8×10^{-4}) and it is relatively simple to grow [1]. Heterostructures involving Ge or Ge alloy buffer layers on other semiconductor materials (e.g., Si) are technologically interesting since epitaxial Ge-based systems seem to be the most suitable for lattice matching to GaAs and other III–V compounds. Ge epitaxial film growth methods have included evaporation, sputtering [2], pyrolysis [3], and more recently MBE [1], ion beam deposition [4] and laser-induced chemical vapor deposition [5]. Epitaxial growth has been claimed in a rather wide temperature range, from room temperature to 700°C. However, substrate temperatures between 400 and 500°C seem to be necessary for high quality epitaxy.

In this paper we report what are believed to be the first published results of pyrolytic and ArF laser photolytic chemical vapor deposition (CVD) experiments using pure digermane. Comparisons with germane (GeH_4) CVD also are reported. Systematic investigations of deposition conditions and of GaAs substrate preparation procedures were carried out in order to define a variable space for high quality epitaxial Ge film growth.

EXPERIMENTAL

The deposition chamber [6] was a six-way turbo-pumped stainless steel cross, configured as a flowing-gas reactor. It was evacuated typically to the low 10^{-6} Torr range prior to deposition. The source gases entered the chamber through a rectangular slit with a 30×0.2 mm^2 cross-section, positioned about 5 mm above and 25 mm away from the substrate. The source gases were mixtures of either GeH_4 (10%) or Ge_2H_6 (5%) in helium. The chamber was equipped with 25 mm diameter Suprasil windows that permitted photolysis of source gas molecules with the laser beam parallel to the substrate surface. We eliminated the problem of film deposition on the windows through which the laser beam traversed the chamber by internally flushing each window with an independently controlled He flow, in the range of 50–150 sccm/window.

The pulsed excimer laser (Questek, Model 2640) was operated with ArF (193 nm). The laser beam was not focused, but was collimated by a rectangular slit (6×20 mm^2). The typical transmitted power density was 600 mW/cm^2 at 30 Hz. The laser beam intersected the source gas flow perpendicularly, with the bottom edge of the beam about 1 mm above the substrate surface.

The film surface temperatures were monitored by an IR radiation thermometer (IRCON, Type W). In the temperature range where GaAs is partially transmitting, appropriate corrections were made using a crystalline silicon sample of similar thickness. The film deposition rates were monitored in situ, by an optical reflectivity technique that has been described elsewhere [6,7]. A brief description of this technique is as follows: A high stability HeNe-laser beam (632.8 nm) is reflected from the top and bottom interfaces of the growing film, giving rise to a series of interference oscillations in the reflected intensity. The characteristics of the oscillations are determined by the optical constants of the substrate and the overlayer, as well as the surface morphology of the growing film. Non-specular reflectivity caused by a rough surface can be modeled by a simple decaying Gaussian envelope, whose parameters include the height and lateral size of the surface features. This function is convoluted with the calculated reflectivity signal for a smooth surface, using appropriate values of the optical constants of the substrate and film [8].

The GaAs substrate was first degreased by rinsing in trichloroethylene, acetone, methanol and deionized water and then chemically etched with a solution based on a $H_2SO_4/H_2O_2/H_2O$ mixture. Before loading into the deposition chamber it was rinsed with deionized water and blown dry with nitrogen. Prior to deposition the GaAs substrate was thermally cycled to 550–600°C for 5 min, in order to desorb the surface oxide, and then was taken to the growth temperature.

RESULTS AND DISCUSSION

Figure 1 shows the dependence of the Ge pyrolytic deposition rate on the substrate temperature for three incidence rates (partial pressures) of Ge_2H_6, and for one GeH_4 incidence rate. Several interesting features of Fig. 1 should be noted. For each incidence rate there is a transition temperature above which single crystalline (epitaxial) Ge film growth was observed. In qualitative agreement with other thin-film growth experience, the deposition rates were found to depend strongly on both the substrate temperature and the incidence rate [9]. However, the general features of the four growth curves are very similar, indicating that the same growth mechanism operates. In agreement with the theory of nucleation and thin-film growth, the growth rates initially increase with increasing substrate

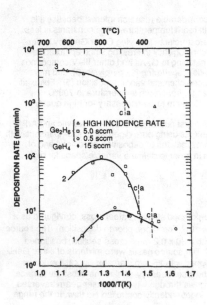

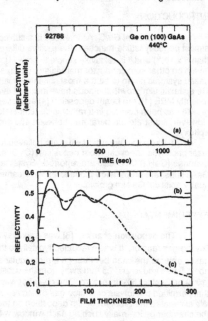

Fig. 1. Logarithm of deposition rate vs. reciprocal substrate temperature. The total pressure during deposition was 40 mTorr and 1 mTorr for 5.0 and 0.5 sccm digermane (5% in helium), respectively, and 100 mTorr for 15 sccm digermane (10% in helium). The a–c transition temperature for a particular incidence rate is given by the dashed vertical line.

Fig. 2. (a) Experimental reflectivity curve for Ge-film growth on (100) GaAs at 440°C. Reflectivity curve (b) is a model calculation for a specularly reflective Ge film on GaAs. Curve (c) is the specularly reflected component for a Ge film with surface features illustrated qualitatively in the inset.

temperature, reach a maximum, and then decrease as the substrate temperature is increased further and desorption becomes dominant. For fixed substrate temperature, the growth rates increase with increasing incidence rate [10]. The amorphous-to-crystalline (a–c) transition temperature shifts toward lower substrate temperatures with decreasing incidence rate. This is in agreement with results obtained from epitaxial Ge film growth by evaporation where a similar shift was observed [11].

Epitaxial Ge films were obtained on (100) GaAs for all substrate temperatures between 380 and 600°C. However, the films surface morphology was a strong function of both the substrate temperature and the incidence rate of Ge–bearing molecules. The surfaces of films grown at 400°C and low incidence rate were very rough. The films became smoother with increasing substrate temperature. Based on these observations, we infer that epitaxial Ge film growth proceeds by formation and growth of three-dimensional nuclei. This is in qualitative agreement with classical nucleation and growth theory, which predicts that for certain substrate-overlayer combinations, two-dimensional nucleation (layered growth) becomes favored over three-dimensional nucleation (island growth), with increasing saturation of the growth surface by the precursors to film growth [12]. In our case, increasing saturation amounts to increasing the incidence rate of the Ge-bearing molecules and/or to increasing the substrate temperature, since the latter increases the thermal generation rate of growth precursors. The data for growth curve 3 was derived from films grown under conditions of high saturation. These films were specularly reflective large-area epitaxial films.

In light of the above discussion, and the fact that the photodissociation cross-section of Ge_2H_6 is much larger [13] than that of GeH_4, laser enhancement of the Ge epitaxial growth rate might be expected. We conducted experiments in which the Ge_2H_6 flowing over (100) GaAs substrates was ArF laser-irradiated above the downstream half of the substrate. The film thickness and morphology were compared in the irradiated and unirradiated halves. Below the a–c transition temperature, laser-enhanced deposition was observed. The laser enhancement of deposition rate increased with decreasing substrate temperature, but the films were always amorphous. Qualitative inspection suggests that their hydrogen content also increased with decreasing temperature. Above the transition temperature, the thermal decomposition rate always dominated photo-decomposition, due primarily to the low duty cycle of the pulsed excimer laser. Consequently, no laser enhancement of Ge epitaxial growth rates was observed in the parallel-beam geometry. (In some current experiments, there are preliminary indications of epitaxial growth rate enhancement at low substrate temperatures, by very low-fluence direct surface irradiation.) Our results appear to disagree with a previously published report of ArF laser-induced growth of Ge films on GaAs at substrate temperatures as low as 285°C, as a result of ArF laser photolysis of GeH_4 in the parallel-beam geometry [5]. Those authors hypothesized that the c–Ge grew from low concentrations of digermane that were formed from GeH_4 photolysis products. In their experiments, initial film growth of 40 to 70 nm thickness also proceeds by island growth (judging by the rough surface morphology of the epitaxial deposit). However, past this thickness the films were amorphous. This switch from crystalline to amorphous Ge deposition could be associated with a change in growth mode from island growth to a particular case of layered growth, for which fine-grained polycrystalline or amorphous film growth is more likely [14]. This change in the growth mode, which is precursor incidence-rate and substrate-temperature dependent, could be caused by the laser-induced build-up of film precursors, which translates into an increased incidence rate. A similar dependence of the growth mode on the deposition conditions—but with opposite results—was observed in our experiments. We succeeded in growing very uniform, specularly reflective, large-area epitaxial Ge films by utilizing high supersaturation to alter the crystalline growth mode from island to layered growth. The experimental conditions and procedures used to control the growth mode will be reported elsewhere [15].

Information about the growth mode and the a–c transition temperature was obtained from the reflectivity signal which was used for in situ thickness monitoring, and from post-growth SEM, TEM, and x-ray diffraction observations. As shown in Fig. 2, a typical reflectivity signal for a Ge film grown on (100) GaAs consists of interference oscillations damped by absorption. The damping of the interference oscillations is ordinarily much stronger for a–Ge than for c–Ge. However roughness of the growing c–Ge surface, due to island growth, manifests itself in a non-specular component of reflectivity. Non-specular reflection first becomes noticeable when the film reaches a thickness corresponding to the first peak in the reflectivity oscillations (~28 nm). As Fig. 2 illustrates, the successive oscillations characteristic of a perfectly layered film are lost in the rapid decay of the specular reflectivity signal, as the surface feature size increases with the film thickness. For such rough films, the rate and extent of the reflectivity decay depends on the scale of the surface features. Plan view SEM (Fig. 3) and TEM (Fig. 4) micrographs were used to determine the final size of the surface features after film growth was completed under slow growth (low incidence rate) conditions. The roughness parameter used in the model calculations then could be compared with the experimentally determined feature sizes [8], and the experimental and calculated reflectivity decays compared (Fig. 2). The transition temperatures for epitaxial growth were determined by the appearance of the characteristic reflectivity signal decay associated with a rough crystalline surface, and were verified by both x-ray diffraction and TEM data.

One significant result of this work is that epitaxial Ge films that are coherently strained and completely commensurate with a (100) GaAs substrate were grown by digermane pyrolysis, for all substrate temperatures in the range 380 < Ts < 600°C. Because the bulk lattice constant for Ge is larger than that for GaAs (a[GaAs] =0.56535 nm [16], a[Ge] = 0.5677 nm [17], a completely commensurate epitaxial Ge film is under compression in the plane of the Ge–GaAs interface, and so it is in tension perpendicular to a (100) GaAs substrate. It is this perpendicular lattice constant that is measured by the (400) x-ray diffraction peak. For a completely commensurate, coherently

Fig. 4. TEM micrograph showing partial relaxation of strain in the Ge–GaAs interface for a thick (~2 μm) Ge film.

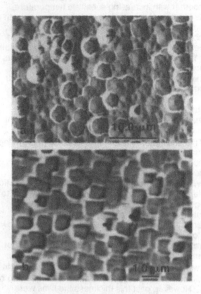

Fig. 3. SEM micrographs showing surface features of Ge films grown on (100) Ge (top) and (100) GaAs (bottom), 110 and 260 nm thick, respectively.

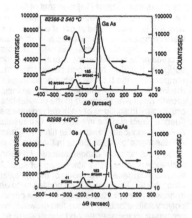

Fig. 5. X-ray diffraction curves for Ge films on (100) GaAs. The vertical dashed line marks the expected peak position for a c–Ge film with the bulk (relaxed) Ge lattice constant.

strained Ge film, the Ge(400) peak is expected to occur at -173 arc sec from the GaAs (400) peak, while if the Ge film has its bulk (relaxed) lattice constant the Ge (400) peak should be found at -100 arc sec, using the lattice constants given above [18]. As Fig. 5 and Table 1 show, all of our Ge films had the (400) peak at -173 (±10) arc sec, indicating that they were completely commensurate with the GaAs substrate. A much thicker (~2 μm) Ge film was found to be partially relaxed ($\Delta\theta$ = -129 arc sec, and cross-section TEM revealed what may be an array of edge dislocations, spaced at intervals of ~0.8 μm, in the plane of the Ge–GaAs interface (Fig. 4.). However, even in this case there were no dislocations or extended defects of any kind extending into either the Ge film or GaAs substrate.

The full width at half-maximum (FWHM) intensity of the Ge and GaAs (400) x-ray peaks also was measured. The experimental Ge FWHM was compared with the minimum width possible for diffraction by the finite number of lattice planes in each c–Ge film [17]. As shown in Table 1, the experimental Ge (400) peak widths were the minimum possible for T_S> 500°C, and increased to only about 75% more than the minimum for films grown near 380°C. Thus, even the films grown at relatively low temperatures are of high crystalline quality. The slight increase in Ge FWHM may be due to voids or vacancies that are not annealed out during low-temperature growth. The GaAs substrate (400) peak widths were found to decrease monotonically with increasing substrate temperature (and with increasing Ge film thickness, see Table 1), apparently because the much shorter times required for film growth at the higher T_S result in much less exchange of Ge and As by interdiffusion during growth. The Ge–As exchange reaction, which was observed in x-ray photoemission studies of MBE-deposited Ge on (100) GaAs, leads to changes in the atomic arrangement near the interface [18].

From the low-temperature part of growth curve 2, an activation energy of 1.4 ± 0.1 eV is derived. This value agrees with that obtained for epitaxial growth of Ge films by thermal

Table I. Analysis of Ge and GaAs (400) X-ray diffraction peaks.

SAMPLE	T_{sub} (°C)	THICKNESS (nm)	Germanium SHIFT (arc sec)	FWHM (arc sec) Expt	Calc	GaAs FWHM (arc sec)
90188-1	380	710	−183	47	25	35.5
82988	440	750	−183	41	24	23.5
83088-1	480	400	−180	63	45	~25
83088-2	515	370	−167	52	48	18.5
82388-2	540	420	−165	40	42	18.5
83088-3	560	310	−172	57	57	12.5

decomposition of GeH_4 [3]. However, actual film growth is a result of a complex set of simultaneous, and to a certain degree synergystic, processes that include thermal decomposition of the Ge-bearing molecules, surface adsorption, desorption and diffusion, and formation of the critical nuclei. Therefore, the apparent activation energy is very sensitive to a change in growth conditions and should be understood accordingly. For example, growth curve 1 in Fig. 1 has less steep slope than curve 2. Curve 3 does not show the decrease of growth rates with increasing substrate temperature that is characteristic of island growth, because the large incidence rate prevents desorption from being the growth rate-limiting step.

In order to establish a basis for comparing the film growth process for two different Ge-bearing molecules, GeH_4 and Ge_2H_6, we grew a few films by thermal decomposition of GeH_4. Figure 1 shows that the general features of the growth curves are similar and growth curve 1 shows that the Ge film deposition process is not affected significantly by differences in the chemical pathways that lead to film growth. However, film growth rates from Ge_2H_6 are much faster than from GeH_4, for a given partial pressure and substrate temperature (Fig. 1).

CONCLUSIONS

High quality epitaxial Ge films were grown on (100) GaAs by pyrolytic CVD from GeH_4 and Ge_2H_6. Under appropriate conditions the pyrolytic growth rates from Ge_2H_6 were up to two orders of magnitude higher than from GeH_4. The microstructure, strain and epitaxial quality of films grown from Ge_2H_6 were assessed using x-ray diffraction and high resolution scanning and transmission electron microscopy. It was found that commensurate, coherently strained epitaxial Ge films could be grown pyrolytically on (100) GaAs at low Ge_2H_6 partial pressures, for substrate temperatures > 380°C. Photochemical enhancement of deposition rates by ArF laser-induced dissociation of Ge_2H_6 also was investigated. Laser-enhancement of deposition rates was observed only at low substrate temperatures and the films were always amorphous.

ACHNOWLEDGMENTS

We would like to thank P. H. Fleming, J. T. Luck, and C. W. Boggs for skilled technical assistance. This work was sponsored by the Division of Materials Science, U. S. Department of Energy, under contract De-AC-5-84OR21400 with Martin Marietta Energy Systems, Inc.

REFERENCES

1. The Technology and Physics of Molecular Beam Epitaxy, edited by E. H. C. Parker (Plenum Press New York, 1985).
2. Handbook of Thin Film Technology, edited by L. I. Maissel and R. Glang (McGraw-Hill New York,1983), p. 10.22.
3. H. A. Krautle, P. Roentgen, and H. Beneking, J. Cryst. Growth 65, 439 (1983).
4. T. E. Haynes, R. A. Zuhr, S. J. Pennycook, B. C. Larson, and B. R. Appleton, J. Vac. Sci. Technol. (accepted for publication)

522

5. V. Tavitian, C J. Kiely, D. B. Geohegan, and J. G. Eden, J. Appl. Phys **52**, 1710 (1988).

6. D. Eres, D. H. Lowndes, D. B. Geohegan, and D. N. Mashburn, Mater. Res. Soc. Symp. Proc.**101**, 355 (1988).

7. D. H. Lowndes, D. B. Geohegan, D. Eres, S. J. Pennycook, D. N. Mashburn, and G. E. Jellison, Jr., Applied Surface Science (in press).

8. J. W. Sharp, D. Eres, and D. H. Lowndes (in preparation).

9. R. W. Vook, Intern. Metals Rev. **27**, 209 (1982).

10. D. Walton, J. Chem. Phys. **37**, 2182 (1962).

11· B. W. Sloope and C. O. Tiller, J. Appl. Phys. **36**, 3174 (1965).

12. I. Markov, and R. Kaischew, Thin Solid Films **32**, 163 (1976).

13. The dissociation cross section for Ge_2H_6 apparently has not been measured at 193 nm.

However, it can be inferred from our photolytic deposition rate measurements, and from the analogy with SiH_4 and Si_2H_6, that it is much larger than the 193 nm cross section of 3×10^{-20} cm^2 for GeH_4.

14. D. Kashchiev, Thin Solid Films **55**, 399 (1978).

15. D. Eres et al. (to be published)

16. J. F. C. Baker and M. Hart, Acta Cryst. **A31**, 364 (1975).

17. J. F. C. Baker, M. Hart, M. A. G. Halliwell, and R. Heckinbottom, Sol. St. Electronics **19**, 331 (1976).

18. A. Segmuller and M. Murakami, "X-ray Analysis of Strains and Stresses in Thin Films," in Analytical Techniques for Thin Films, edited by K. N. Tu and R. Rosenberg (Academic Press, San Diego, 1988), p. 153.

19. B. E. Warren, X-ray Diffraction (Addison-Wesley New York, 1969), p. 252.

20. R. S. Bauer and W. Sang, Jr., Surface Sci. **132**, 479 (1983).

VALENCE HOLE LOCALIZATION IN MOLECULAR AUGER DECAY

D.A. Lapiano-Smith, C.I. Ma, K.T. Wu,* and D.M. Hanson Department of Chemistry, State University of New York,Stony Brook, N.Y. 11794-3400
*On leave from Department of Chemistry, State University of New York, College at Old Westbury, Old Westbury, New York 11568

ABSTRACT

Studies of the unimolecular decay, following the excitation of core electrons of the carbon and fluorine atoms in carbon tetrafluoride and silicon and fluorine in silicon tetrafluoride by monochromatic, synchrotron radiation, provided evidence for a "valence bond depopulation" fragmentation mechanism. The fragmentation processes were examined using time-of flight mass spectroscopy. The mass spectra show the distribution of ions collected in coincidence with low and high energy electrons. Distinct changes in the mass spectra with atomic site of excitation and photon energy are observed. The observation of F^{2+} ions in the time-of-flight mass spectra following excitation of a fluorine 1s electron in SiF_4 is significant because it provides direct evidence for the formation of a localized, two-hole, final valence state that persists on the time scale of fragmentation. In contrast, the lack of F^{2+} ions from CF_4, indicates that the fragmentation occurs through a delocalized two-hole state.

I. INTRODUCTION.

Evidence for a "valence bond depopulation" mechanism is reported in this paper. The term "valence bond depopulation" refers to an Auger final state in which the two valence holes produced by Auger decay are highly correlated and localized on the atomic site of the core hole. The concept of localization has been proposed to explain the Auger spectra of some metals,[1,2] semiconductors,[3] and gas phase molecules[4,5,6] and electron and photon stimulated desorption of ions from solid surfaces.[7] Two valence holes created by Auger decay of the core hole become confined or localized on the same atomic site because of the large energy that must be dissipated if they separate. The rate of this dissipation may be slow compared to that for other processes. The presence of such localized final states is dependent upon the magnitude of ionicity and polarity in the bonds, bond lengths and the Coulomb repulsion between the holes. An interesting possibility, stemming from the explanation of the spectra, is that the electronic relaxation and subsequent bond dissociation will be localized around the atomic site of excitation. This is possible if the two valence holes created by Auger decay of the core hole, can be localized around a single atom for a time sufficient to allow bond rupture to occur.

Carbon and Silicon tetrafluoride were specifically chosen in order to observe whether or not two hole correlation and localization occurs in the valence shell of fluorine. For an Auger transition, involving a particular bonding orbital, the process will occur through that region of the valence electron density immediately around the atomic site of excitation.[6] However, due to the large magnitude of ionicity of the Si-F bonding orbitals, relative to the covalent C-F bonding orbitals, one would expect that the resulting two hole density should appear around the fluorine atomic site in SiF_4 rather than in CF_4. If the localized two hole state persists on the time scale of fragmentation, a high yield of the doubly charged ion, F^{2+} is expected. The observation described below of F^{2+} ions following excitation of a fluorine 1s electron in SiF_4 provides direct evidence for the formation of a localized two hole state in Auger decay and for its persistence on the time scale of dissociation. In contrast, the mass spectra obtained for CF_4 indicate, due to the high yield of F^+ and no observation of F^{2+}, that fragmentation is characteristic of a delocalized hole state.

II. EXPERIMENTAL TECHNIQUES.

The experiments were conducted at the National Synchrotron Light Source at Brookhaven National Laboratory on the U15 beam line of the VUV storage ring. This beam line utilizes a toroidal grating monochromator (TGM) that can be used in the spectral range of interest, from 280 to 900 eV, and provides a resolution, $E/\Delta E$, of about 300 and a photon flux of about 10^{11}/sec.

The gases, CF_4 (Matheson, 99.998%) and SiF_4 (Pfaltz & Bauer, Inc., 99.99%) were used as received. The experimental techniques have been discussed in detail previously.[8,9] In short, the apparatus consists of a skimmed molecular beam source and an ionization chamber. The ionization chamber is equipped with a low resolution retarding potential electron energy analyzer and a time of flight (TOF) mass spectrometer.

III. EXPERIMENTAL RESULTS.

Total ion yield spectra of carbon tetrafluoride in the regions of the carbon and fluorine K edges and of silicon tetrafluoride in the region of the fluorine K edge were obtained but are not presented here.

Mass spectra of ions produced in coincidence with electrons of different energies can be obtained by setting the retarding potential of the analyzer at different values. This feature is important because electrons with different kinetic energies result from various relaxation channels that lead to different fragmentation patterns. For the data reported here, the interaction region (ionization volume) was biased at 1390 V, and the retarding potential was set at 1400 V to pass both low energy electrons and Auger electrons. The largest contribution comes from low energy electrons, which are collected about 20 times more efficiently than the Auger electrons, because the low energy electrons are accelerated toward the electron energy analyzer. The Auger electrons, created with high kinetic energies, are not sensitive to this focusing effect. The processes that produce low energy electrons are shake-off in photoionization, double Auger decay, autoionization following Auger decay, and, depending upon the photon energy, also direct photoionization.

Mass spectra for CF_4 and SiF_4 are shown in Figs. 1 and 2, respectively. Mass spectra for CF_4 were obtained with excitation at the prominent resonances at 298 and 693 eV and in the carbon and fluorine K ionization continua at 316 and 743 eV. The peaks seen in the spectra are attributed to C^+ (m/q = 12), CF^{2+} (15.5), F^+ (19), CF_2^{3+} (25), CF^+ (31), CF_2^+ (50), and CF_3^+ (69). Mass spectra for SiF_4 were taken with excitation below the fluorine K edge (669 eV), at the two resonances (690 and 719 eV), and in the continuum (788 eV). The peaks are attributed to Si^{4+} (7), F^{2+} (9.5), Si^{2+} (14), F^+ (19), SiF^{2+} (23.5), Si^+ (28), SiF_2^{2+} (33). The relative yields of the ionic fragments for each of the excitation energies are given in Table I for CF_4 and Table II for SiF_4. The values in these tables were obtained from the peak areas and averaged for two and four data sets, respectively. The average deviation in these values varies for different peaks, but the uncertainty can be estimated conservatively at 10%.

IV. DISCUSSION.

A. Carbon Tetrafluoride.

TOF mass spectra, obtained with four photon excitation energies are shown in Fig. 1. Only small differences are apparent between resonance and continuum excitation. The most significant features in these spectra are the ratio of F^+ to C^+ ions, the absence of F^{2+} ions and the presence of a very sharp CF_2^{2+} line. Although several relaxation processes contribute to these spectra, some definite conclusions can be reached.

The similarity of the fragmentation patterns obtained with resonance and continuum excitation is somewhat unusual. Clearly single electron autoionization of the core hole resonance in the neutral molecule leads to different electronic configurations than Auger decay of the core hole in the +1 ion. The greater depletion of the valence electrons in the latter case can result in an increased yield of smaller ionic fragments. The lack of a difference here may indicate the importance of multiple electron ionization in the decay of the core hole resonance.

The most prominent feature of the fragmentation patterns is the variation in the ratio of F^+ to C^+ ions. This ratio is largest for excitation of a fluorine core electron and decreases

525

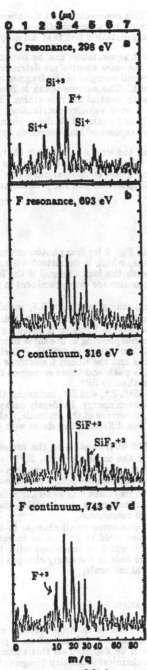

FIG. 1. CF₄ time-of-flight mass spectra

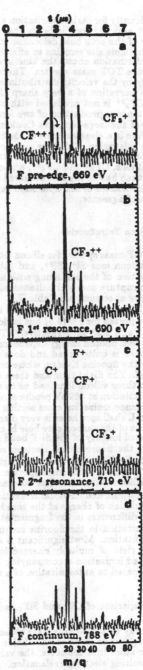

FIG. 2. SiF₄ time-of-flight mass spectra.

significantly for carbon excitation as shown in Table III. This observation and the absence of C^{2+} and F^{2+} ions indicate that the valence holes created by autoionization or Auger decay of the core hole delocalize on the time scale of fragmentation but that memory of the core hole site remains to affect the ratio of F^+ to C^+.

Information about the kinetic energy released in the fragmentation can be obtained from the TOF mass spectra. The width of the peaks in the mass spectra are determined mainly by the velocity distributions associated with the recoil energies of the fragments.[10] The observation of a very sharp CF_2^{2+} peak is significant. The narrow width indicates that CF_2^{2+} is not produced with other ions but rather with neutral fluorine atoms. The simultaneous production of two ionic fragments leads to a wide velocity distribution, because of the large repulsive Coulomb energy, and hence broad peaks. This decay channel therefore is a clear example of the failure of the "Coulomb explosion" mechanism to apply to the relaxation of shallow core holes in molecules.

The mass spectra reveal the complete absence of both the singly and doubly charged parent ions, CF_4^+ and CF_4^{2+}, and only small amounts of CF_3^+ and CF_2^+. F^+ accounts for more than half of the ions observed in each mass spectrum. These considerations underline the importance of multiple bond rupture in the chemical relaxation and the production of neutral fragments.

B. Silicon Tetrafluoride.

TOF mass spectra for silicon tetrafluoride are shown in Fig. 2 for four photon energies. The silicon ions Si^+, Si^{2+}, and Si^{4+} all have narrow lines, which is consistent with the production of the ion along with neutral fragments or with the lack of recoil if the four bonds rupture nearly simultaneously. The three silicon ions also are very prominent in the pre-edge spectrum, see Table II.

The peak observed at a flight time of 2.41 μs, corresponding to $m/q = 9.5$, matches that expected for $F^{2+}(9.5)$ but is close to $Si^{3+}(9.3)$. These ions have calculated flight times of 2.41 and 2.38 μs, respectively. The assignment of this peak to F^{2+} also is based on its width and photon energy dependence. Unlike the peaks corresponding to the three Si ions, this peak is quite broad and does not appear at all with excitation at 669 eV, which is below the fluorine K-edge, where only the valence electrons and the silicon L electrons are ionized. The flight time and these differences between this peak and those corresponding to the three silicon ions lead us to assign it to F^{2+} rather than to Si^{3+}.

Excitation at 669 eV produces only atomic ions: Si^{4+}, Si^{3+}, F^+, and Si^+, indicating that the valence ionization cross section is negligible at this photon energy. Evidently decay of silicon L shell core holes is very disruptive of the bonding structure of the molecule, unlike the case of the carbon core hole in carbon tetrafluoride. This difference correlates with the increased ionicity of the Si-F bond relative to C-F.

At the fluorine core electron resonances (690 and 719 eV), relative to the pre-edge region, F^+/Si^{2+} ratio increases significantly and F^{2+} and the molecular ions, SiF^{2+} and SiF_2^{2+}, appear in the mass spectra. These changes reflect the transfer of the core hole site from silicon to fluorine. The information concerning the site of the core hole is preserved in the electronic and chemical relaxation processes in that the fragment ions generally reflect the depletion of charge at the core hole site. Within the experimental uncertainties, there are no differences in the fragmentation patterns at the two resonances.

Excitation in the fluorine continuum at 788 eV produces some small changes in the fragmentation. Most significant is a decrease in the relative yield of F^+ and an increase in the yield of multiply charged ions Si^{4+}, F^{2+}, and Si^{2+} which is consistent with the increased ionization accompanying Auger decay of the core hole in the singly charged ion as compared to autoionization of a resonance in the neutral molecule.

C. Comparison of CF_4 and SiF_4 and Valence Hole Localization.

Decay of a core hole state either by resonant autoionization or an Auger process can produce two or more valence holes. Since the core hole is essentially localized on a particular atom in a molecule, the valence charge density can be depleted on this atom in the resulting electronic relaxation. The differences in the chemical relaxation (fragmentation) associated with the core hole states in CF_4 and SiF_4 are consistent with a greater localization of the valence holes for the case of SiF_4.

The concept of valence hole localization has been used previously [8,11] to explain the Auger spectra of CF_4 and SiF_4. These Auger spectra were analyzed in terms of contributions from localized two hole states and delocalized two hole states. The contributions of the localized states was larger for SiF_4 relative to CF_4. This difference was attributed to the larger size, greater F-F distance, and higher bond ionicity of SiF_4, all of which decrease the interactions that lead to valence hole delocalization. [6,11]

The fragmentation patterns that we observed are consistent with this analysis. In particular, the localized two hole state F^{2+} is detected directly for SiF_4. In a time-dependent picture, the present study shows that this localized valence hole state persists for a time after Auger decay at least comparable to the time scale for bond rupture to occur. The observation of F^{2+} for SiF_4 and not for CF_4 implies that in some molecules a localized picture of bonding (the valence bond model) may be more appropriate for describing the electronic and chemical relaxation accompanying decay of a core hole than a molecular orbital model, which appears to be relevant in other cases.[4] To emphasize this point, we identify the former as a valence bond depopulation mechanism and the latter as a molecular orbital depopulation mechanism.

V. SUMMARY.

Electronic and chemical relaxation following core electron excitation in CF_4 and SiF_4 has been investigated by electron energy analysis and time-of-flight mass spectroscopy. In CF_4, the yield of low energy electrons and the similarity of the fragmentation patterns for resonance excitation and continuum ionization provide evidence for the high efficiency of multiple electron ionization processes. The presence of CF_2^{2+}, produced with a low recoil kinetic energy, provides a clear illustration of the failure of the "Coulomb explosion" mechanism to apply to this compound. The small change in the F^+/C^+ ratio as the site of the core hole is varied from carbon to fluorine indicates that the valence holes produced in the Auger decay are delocalized but that memory of the initial core hole site is retained.

In SiF_4, no molecular ions were detected below the fluorine K-edge where only valence and silicon core electrons can be ionized, which indicates that the valence ionization cross section must be negligible at this photon energy. The presence of F^{2+} and the significant variation of the fragmentation pattern with the atomic site of the core hole indicate that the valence holes resulting from Auger decay are localized to a greater extent than in CF_4. This enhanced localization correlates with the fluorine-fluorine distances and the bond ionicity, which favor valence hole localization for the silicon compound. Such localization had been proposed to explain the differences in the Auger spectra of these two compounds as well.

Support for this research from the National Science Foundation under Grant CHE-8703340 is gratefully acknowledged. Part of the work was conducted at the National Synchrotron Light Source, which is supported by the Department of Energy, Division of Material Science and Division of Chemical Science.

References

1. (a) E. Antonides, E.C. Janse, and G.A. Sawatsky, Phys. Rev. 8 1 5, 1669 (1977);
   (b) G.A. Sawatzky, Phys. Rev. Lett. 39, 504 (1977);
   (c) G.A. Sawatzky and A. Lenselink, Phys. Rev. B 21, 1790 (1980).
2. M. Cini, Surf. Sci. 87, 483 (1979); Solid State Commun.
   20, 605 (1976); 24, 681 (1977); Phys. Rev. B 17, 2788 (1978).
3. (a) P.J. Bassett, T.E. Gallon, M. Prutton, and J.A.D. Matthew,
   Surf. Sci. 33, 213 (1972); (b) D.R. Jennison, J. Vac. Sci. Technol. 20, 548 (1982).

4. (a) T.D. Thomas and P. Weightman, Chem Phys. Lett. 81, 325 (1981);
   (b) P. Weightman, T.D. Thomas, and D.R. Jennison, J. Chem. Phys. 78, 1652 (1983).
5. R.R Rye, D.R. Jennison, and J.E. Houston, J. Chem. Phys.73, 4867 (1980).
6. R.R. Rye, and J.E. Houston, J. Chem. Phys. 78, 4321 (1983).
7. (a) D.R. Jennison, J.A. Kelber, and R.R. Rye, Phys. Rev. B 25,
   1384 (1982); (b) P.J. Feibelman, Surf. Sci. 102, L51 (1981); (c) D.E. Ramaker,
   C.T. White, and J.S. Muday, J. Vac. Sci. Technol. 18, 748 (1981).
8. J. Murakami, M.C. Nelson, S.L. Anderson, D.M. Hanson,
   J. Chem. Phys., 85. 5755 (1986)
9. D.A. Lapiano-Smith, C.I. Ma, K.T. Wu and D.M. Hanson, J. Chem. Phys., in press.
10. N. Saito and I.H. Suzuki, Chem. Phys. Lett. 129, 419 (1986).
11. R.R. Rye and J.E. Houston, Acc. Chem. Res. 17, 41 (1984).

TABLE I: CF_4 Relative Ion Yields (% of total yield)

EXCITATION	C^+	CF^{+2}	F^+	CF_2^{+2}	CF^+	CF_2^+	CF_3^+
C RESONANCE	19	3	55	3.5	14.5	3	2
C CONTINUUM	17.5	4.5	55	3.5	17	1	1.5
F RESONANCE	15.5	4	56	5	14	4	1.5
F CONTINUUM	12	3	54	6	19	4	2

TABLE II: SiF_4 Relative Ion Yields (% of total yield)

EXCITATION	Si^{+4}	F^{+2}	Si^{+2}	F^+	SiF^{+2}	Si^+	SiF_2^{+2}
PRE-EDGE	12.0	—	31.4	49.0	—	7.6	—
1st RESON.	3.8	5.7	17.7	46.8	7.6	9.8	8.6
2nd RESON.	2.7	8.6	19.8	45.9	11.9	4.6	6.5
CONTINUUM	5.2	14.3	22.5	38.4	7.6	7.8	4.2

TABLE III: Relative Ion Yields

EXCITATION	F^+ / C^+
C RESONANCE	2.9
C CONTINUUM	3.1
F RESONANCE	3.6
F CONTINUUM	4.5

PART IX

Ion Beam Chemistry

ION BEAM ASSISTED DEPOSITION OF METAL FILMS

Kenji Gamo and Susumu Namba
Faculty of Engineering Science, and Research Center for Extreme
Materials, Osaka University, Toyonaka, Osaka 560, Japan

ABSTRACT

The characteristics of ion beam assisted deposition are discussed and compared with those of photon beam assisted deposition. Effects of various deposition parameters including ion species, beam energy and substrate temperature are discussed. Deposited films usually include impurities such as C and O. Inclusion of oxygen takes place by enhanced oxidation by background oxygen and may be reduced by depositing in a clean vacuum. Promising applications of maskless ion beam assisted deposition are also discussed.

INTRODUCTION

Maskless processing using directed beams is increasingly important for electronic device processing. With the increasing scale of integration, complexity of device processing increases. Maskless processing is important to reduce the complexity by direct fabrication of device patterns without using lithography steps.

Focused ion beam technology is a new fabrication technology for future device fabrication processing. It has many promising capabilities; First, it has very high resolution. Delineation of 30nm patterns has been demonstrated[1]. Second, it has a very high current density of >1A/cm^2 with a submicron beam spot size. The current density is 10^3-10^6 times larger than for conventional ion beam systems. Third, it has a capability of maskless fabrication for doping, etching and deposition.

The current intensity of the present focused ion beam (FIB) systems is only a few nA at the most, though an increase may be possible by improving the beam optics using various techniques including variable shaped beam techniques[2]. The low current intensity limits the practical applications of focused ion beams. Nevertheless, high current density and high resolution of FIB provides unique applications as, for example, mask repair and circuit diagnosis tools[3-5]. X-ray masks, for example, are fabricated using very fragile thin(several micronmeter thick) membrane. For such membrane mask, it is not easy to repair transparent or opaque defects using conventional techniques. For FIB, transparent defects are repaired by depositing heavy metals such as W, Ta or Au using the ion beam assisted deposition technique and opaque defects can be repaired by sputter etching without using lithography process. Circuit trimming can be done similarly, by cutting metal lines by sputter etching or connecting lines by metal deposition.

In the present paper, characteristics of metal film deposition using focused ion beams and some applications are discussed.

EXPERIMENTAL PROCEDURES

Energetic ion beam irradiation induces various kinds of chemical effects in target materials. Adsorbed molecules and substrate atoms may be excited or decomposed by electronic and nuclear collisions. This phenomena has useful applications to maskless ion beam assisted etching techniques. For example, it was found that 10-100 times etching rate enhancement over physical sputter etching has been observed for various materials including GaAs[6], Si[7], Al[8] and SiO_2[9].

Effective decomposition induced by ion beam irradiation has been observed for both organic and inorganic materials. By decomposition, both volatile and nonvolatile species are produced. Volatile species are subject to release from the material and film deposition occurs for nonvolatile species.

This can be applied for maskless deposition. Figure 1 shows an experimental setup for ion beam assisted deposition. The film deposition is achieved by bombarding focused or flood ion beams in a gas atmosphere of molecules which contains elements to be deposited. Samples are set in an inner subchamber or gas jet set in close contact to the point of ion incidence. These arrangements allow to obtain high gas pressure only near the sample surface, while keeping high vacuum pressure in the system enough to obtain stable operation for a liquid metal ion source and to sustain high voltages.

The gas molecules can be introduced up to the pressure of several tens mTorr without suffering significant scattering because the total scattering cross section of a tens of keV ion beam is around $10^{-15}/cm^2$ and the mean free path is tens of mm in a gas pressure of tens of mtorr. At this gas pressure, the surface reaction is dominant over the gas phase reaction because the gas density is about $10^{14}/cm^3$ in the gas phase while the density of adsorbed molecules is $10^{15}/cm^2$ or larger for multiadlayers.

The decomposition reaction is induced as a result of

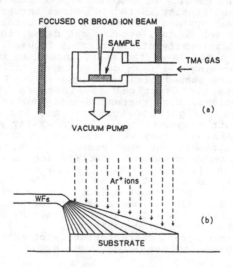

Fig. 1 Experimental setup for ion beam assisted deposition.

kinetic process in collision cascades, any ions reactive or inert are used.

METAL DEPOSITION

Up to now, various kinds of metal films have been deposited using focused or flood ion beams. Some of the results are summarized in Tab. 1. Film deposition using laser or UV photon beam induced reaction is also included for comparison.

Metal film deposition has been done using organometalic compounds, metal halide or carbonyl gas which have a vapor pressure of a few mTorr up to a few Torr at room or deposition temperature. The typical deposition yield is 10-30 atoms/ion. Because the deposition is determined by the balance between sputtering and deposition, this yield depends on the adlayer thickness and the ratio of the molecular flux to the ion flux as well as various irradiation parameters. The gas impinging rate of trimethyl aluminum, for example, is 5×10^{18}/cm^2/sec at 20mTorr, which corresponds to a beam current density of 0.8 A/cm^2. Therefore, at very high current density, the gas supply may be insufficient and the deposition rate may be limited by gas supply. This limit, however, may be removed by scanning the beam and reducing the effective current density.

Figure 2 shows Al patterns deposited by irradiating 35 keV focused Ga$^+$ beam in trimethyl aluminum ambient at the pressure of 30 mTorr[10]. The cross sectional profile of the deposited film reflects the Gaussian distribution of the focused beam.

Films deposited both by ion beams and photon beams generally include impurities such as C and O. For deposition of Al using trimethyl aluminum, the carbon concentration ranges from 2.8% for unfocused Ar$^+$ irradiation to more than 50% for focused Au$^+$ irradiation[11]. The detailed mechanism of the inclusion is not clear but it may be the result of insufficient decomposition. The inclusion of oxygen may be induced by enhanced oxidation from residual background gas. For photon induced deposition, it is reported that the impurity inclusion depends on the reaction path, i.e., pyrolytic or photochemical[12].

Figure 3 shows the depth dependence of the atomic composition of W films deposited using WF$_6$ by unfocused 50 keV Xe$^+$. The deposition was performed at WF$_6$ pressure of 100mTorr. The atomic composition was measured by Auger electron spectroscopy. Carbon may be due to the contamination which occurred by the exposure to the laboratory atmosphere but no fluorine inclusions were observed, which suggests effective decomposition by the ion bombardment.

Ion and photon beam assisted deposition give similar resistivity, more or less, which is a few times larger, at best, than the bulk resistivity. The lowest resistivity obtained so far by ion beam assisted deposition is 2×10^{-5} ohm·cm, which is obtained for the deposition of Au films and is about 10 times larger than the bulk value.

EFFECT OF DEPOSITION PARAMETER

The deposition yield and film properties are of primary concern from practical point of view. These are affected by various deposition parameters. These include ion species, ion

Tab. 1 Summary of deposition

Films/source	Deposition procedure	Composition of films	Yield (atoms/ion)	Resistivity (ohm·cm)	Remarks	Ref.
Al/TMA	50keV Au	Al(<42%), C(>33%), O(<24%)	10-20	10^{-1}-10^{0}		11
Al/TMA	50keV Ar 35keV Ga	Al(58%), C(2.8%), O(39%)				10,16
Pd/Pd-AC	20keV Ga	C(≤30%) O(≤30%)		3×10^{-4}		14
Ta/ Ta(OC$_2$H$_5$)$_5$	50keV He, Ar,Ne,Xe	Ta(56%) C(17%) O(27%)	10-50			15
W/WF$_6$	50keV He, Ar,Ne,Xe	W(75%) O(25%)	10-50	10^{-1}-10^{0}		15
W/W(CO)$_6$	Ga$^+$	W(50%) Ga(18%) C(30%),O(2%)	1	2×10^{-5}-4×10^{-4}		5
Au/DMGFA	15keV Ga$^+$	Au(75%) Ga(20%) O(<5%),C(<5%)	4-5	5-13×10^{-4}		13
	0.75keV Ar$^+$ 50keV Si$^+$	Au(95%) Au(75%) O(10%),C(10%)	1 8-16	2×10^{-5}		
Al/TIBA	KrF laser			5×10^{-6}		17
Cr/Cr(CO)$_6$	UV	C, O		5×10^{-3}		18
Mo/Mo(CO)$_6$	UV	Mo(60%) C(20%) O(20%)	>10nm/min	high r		19
Pd/Pd-Ac	Ar laser	Slight amount of C and O		4-24×10^{-5}		20
W/WF$_6$	UV		45nm/min	10^{-5}		19
Au/DMPNG	UV laser	Au>95% Au 25-70%, C		4-20Xbulk high r	Pyrolytic Photochemical	21

TMA; Trimethyl aluminum, DMGFAC; Dimetyl gold hexafluoro acetylacetonate, DMPNG; Dimethyl-(2,4-pentanedionate) gold, TIBA; Triisobutyl aluminum, Pd-Ac; Palladium acetate

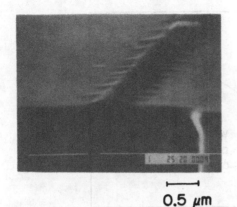

Fig. 2 Al line pattern deposited by 35 keV Ga$^+$ and trimethyl aluminum.

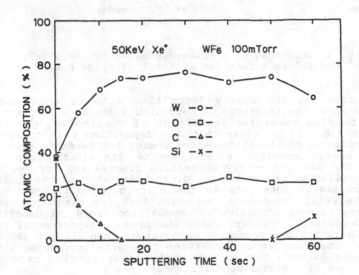

Fig. 3 Depth distribution of composition of W film deposited using 50keV Xe$^+$ and WF$_6$.

beam energy, beam scanning speed, substrate temperature and effect of the background gas.

Ion species

The deposition process is determined by energy deposited by collision. Different ions have different sputtering yield and total stopping power with different contributions between electronic and nuclear collisions. Therefore, ion species is one of the important parameters.

It can be expected that the deposition rate should increase

536

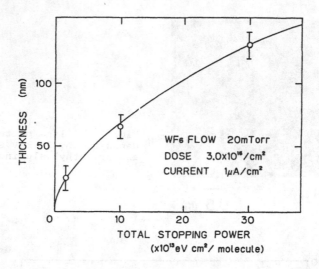

Fig. 4 Thickness of W films deposited by 50kev He⁺, Ar⁺ and Xe⁺ as a function of total stopping power.

with increasing the energy deposition rate by collisions. Figure 4 shows the thickness of deposited films as a function of total stopping power for deposition of W using WF_6 gas[15]. From the plot, it is clear that the deposition rate increases with increasing the total deposited energy but tends to saturate at high energy deposition rates. For He, the electronic collision is the dominant energy deposition process and for Xe, the nuclear collision is dominant. For Ar, the ratio of the nuclear collision energy loss rate to the electronic collision energy loss rate is 3. Therefore, we think that for the heavy ions, sputtering becomes effective and deposition yield per deposited energy decreases. However, under the present experimental condition where the gas supply is much larger than the ion imping-ing rate, the increase of deposition rate with increasing stop-ping power exceeds the increase of the sputtering rate and heavier ions gave higher deposition rate per ion.

Ion energy

The collision process also depends on the beam energy. The contribution of electronic collisions increases with increasing the beam energy. Figure 5 shows the thickness of films deposited by He, Ar⁺ and Xe bombardment in WF_6 gas ambient, and the total, nuclear and electronic stopping powers as a function of the beam energy. For Ar⁺, nuclear collision is dominant. The total stopping power is almost constant at an energy >10keV and decreases at an energy <10keV. For deposition using Ar⁺, the deposition rate increases with increasing the beam energy but the change of the observed deposition rate does not follow the change of the total stopping power. This may be partially due to the decrease of relative contribution of electronic col

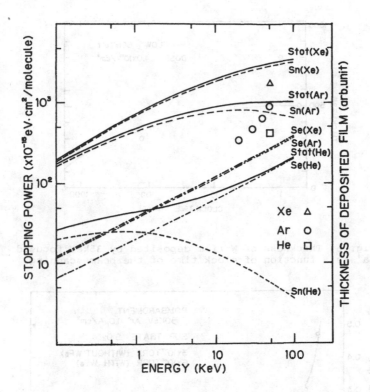

Fig. 5 Thickness of W films deposited by He$^+$, Ar$^+$ and Xe$^+$ bombardment in WF$_6$ ambient, and total, nuclear and electronic stopping powers as a function of the beam energy. The gas was introduced by a gas jet.

lision and the relative increase of an sputtering effect over the deposition process.

Beam scan speed

As the deposition is determined by a balance between deposition and sputter etching, beam scan speed or current density are important parameters to the determine deposition rate. If ion impinging rate is too high compared to the gas supply rate, sputtering becomes larger and the deposition rate decreases. Figure 6 shows thickness of W films deposited using 35keV focused Ga$^+$ and WF$_6$ ambient as a function of the clock time of the beam scanning. In the present experiment, the ion impinging rate and the WF$_6$ gas supply rate is 3×10^{18} and 6×10^{18}/cm^2/s, respectively. It is observed that the deposition rate decreases with increasing the clock time. At a clock time of 1msec, roughly a monolayer may be removed by the sputtering if we assume sputtering yield of 1 atom/ion and the decrease is ascribed to an increase of the sputtering effect.

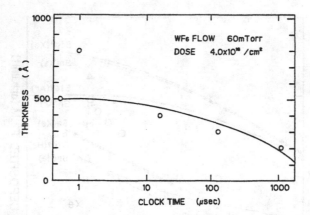

Fig. 6 Thickness of W film deposited by 35keV focused Ga⁺ as a function of clock time of the beam scanning.

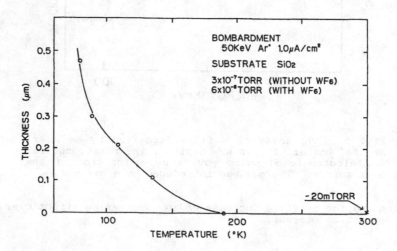

Fig. 7 Deposited film thickness as a function of the substrate temperature. The deposition was done at a dose of $7.5 \times 10^{14}/cm^2$ using WF_6 and 50 keV Ar⁺. See text for details.

The limitation of the deposition rate at high current density can be removed by rapidly scanning the beam and reducing the effective current density.

Substrate temperature

Substrate temperature affects the deposition process in two

different ways. At low temperature, a thick condensed layer is formed and the deposition rate may increase. This was confirmed as shown in Figure 7[15]. This figure shows the temperature dependence of the thickness of films deposited using WF_6 gas. For the deposition, the gas jet was used except for the deposition at 300K, where the subchamber arrangement was used. The pressure of the target chamber was 6×10^{-6} Torr and 3×10^{-7} Torr with and without WF_6 gas jet, respectively. Note that this pressure is only a relative value and the effective pressure at the sample surface is much larger. At the temperature of 80K, the deposition rate is about 5000 times larger than the deposition rate obtained at room temperature.

At high temperature, migration of deposited atoms occurs and film microstructure or impurity inclusion may be different from deposition at low temperature. For deposition of Au, for example, Ro and coworkers[22] found that films deposited at room temperature are made up of discontinuous columnar structure and have poor conductivity, while films deposited at 160°C have continuous and similar microstructure to conventionally evaporated films. They also observed high carbon incorporation(35-50%) for the deposition at room temperature.

Background gas pressure

The deposited films contain high density of oxygen. This may not cause severe problems for some applications such as mask repair, but for many applications, film purity is important. The oxygen included in the deposited films comes, probably, from residual background oxygen gas and the inclusion may be enhanced under beam irradiation as often observed as enhanced oxidation of metals or semiconductors [23]. Therefore, it can be expected that oxygen concentration may be decreased if relative concentration of oxygen is reduced. Figure 8 shows the relative oxygen concentration in the deposited W films as a function of relative oxygen concentration in the source gas for the deposition using WF_6 gas and 50 keV Ar^+ ions. In the figure, the relative oxygen concentration in the deposited film and in the source gas is taken as the peak height ratio of W to O of the Auger spectra and of the observed quadruple mass spectra, respectively. In the present experiment, the WF_6 gas was varied from 20 mTorr to 90 mTorr to vary the oxygen relative concentration while the background oxygen gas was kept constant. As expected, it was observed that the oxygen concentration in the film decreased if oxygen partial pressure is decreased. This result suggests that it is useful to reduce oxygen partial pressure to decrease inclusion of oxygen in the deposited films.

Recently, we have been repeating the W film deposition and observed that for the deposition in an oil-free vacuum ambient at a background pressure of 2×10^{-7} Torr, the deposited films contains only <2.5% of oxygen and shows no W-O bond peaks in XPS spectra.

REDUCTION OF DEFECTS

As shown in Fig. 5, the deposition rate may be higher for deposition using high energy beams. However, high energy beams induce a lot of damage in the substrate, compared to photon beam assisted deposition. To reduce damage, use of low energy beam

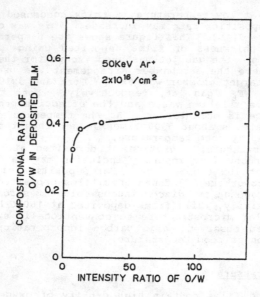

Fig. 8   Composition ratio of oxygen to tungsten in the
deposited film as a function of the relative concentration of oxygen to tungsten in the source gas.

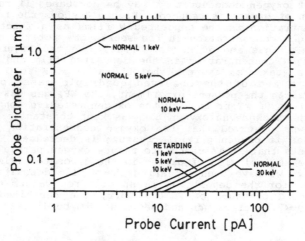

Fig. 9 Probe diameter as a function of probe current
for normal and retarding field FIB systems.   For the
retarding field system, the beam is first accelerated
to 30keV and retarded to the final energy as indicated.   The ion source is assumed to have a beam intensity of 20uA/sr and an energy spread of 10eV, typical for Ga sources.

is desirable[24]. Low energy focused ion beams are difficult to form because chromatic aberration increases with decreasing a beam energy but recently, submicron focused beams have been formed using retarding field techniques[25,26]. Figure 9 shows the calculated beam spot as a function of beam current. Using the retarding field technique, we can expect to obtain more than two orders of magnitude increase in beam intensity. Experimentally, 1keV Ga^+ focused beam with a diameter of 0.2-0.3 um has been obtained and reduction of damage has been confirmed using the low energy FIB for bombardment of GaAs[24].

From Fig. 5, we can speculate that deposition rate may decrease an order of magnitude or more at a beam energy around 1 keV. In fact, Shedd et al[13] deposited relatively pure and low resistive Au with a deposition yield of 1 atoms/ion for a deposition of Au using 0.75 keV Ar^+ and dimethyl gold hexafluoro acetylacetonate ambient. This value is roughly one tenth of the deposition rate obtained for 50keV Si but is still high enough for some applications like mask or circuit modification and repair.

APPLICATION

One of the promising applications of focused ion beam assisted deposition techniques is mask or circuit repair and modification. Laser assisted deposition also can be applied for selective deposition but it is difficult to deposit submicron patterns. Recent focused ion beam systems form a fine focused beam with a diameter smaller than 100nm. Using such fine focused beam, a submicron patterns with sharp side walls can be deposited. Moreover, maskless etching can also be performed by sputter or ion beam assisted etching. For circuit repair or diagnosis, therefore, focused ion beams are useful to make lateral or multi-layer interconnects.

Figure 10 shows an example of W lines deposited between two 1um thick aluminum lines and contact pads to test buried Al lines[5]. W patterns were deposited by using $W(CO)_6$ and Ga^+

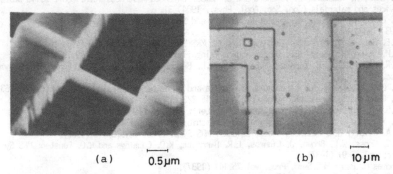

(a)   0.5μm   (b)   10μm

Fig. 10 (a) W interconnect and (b) contact pad. The interconnect was formed between 1um thick aluminum lines and the contact pad was formed to test buried aluminum lines by removing the passivation layer by Ga^+ FIB sputter etching followed by the ion beam assisted etching using $W(CO)_6$ jet with a flow rate of 10^{17} molecules/cm^2s.

focused ion beams. In Fig. 10(a), good step coverage is obtained by the ion beam assisted deposition and in Fig. 10(b), a contact pad is formed to test buried aluminum lines by opening contact holes in the passivation layer by maskless sputter etching followed by the deposition process. It takes several minutes to deposit the contact pads. It is desirable to reduce the process time but still, this is very promising because no other techniques are available to form contacts with buried layers.

SUMMARY

Characteristics of ion beam assisted deposition techniques are discussed. Combining focused ion beam techniques provide unique applications such as mask and circuit repair or modification. This application may be increasingly important with increasing complexity and integration of VLSI devices. Improvement of film characteristics such as purity and electrical resistance is necessary. To reduce oxygen inclusion, it may be helpful to deposit in ultra high vacuum background.

REFERENCES

1. T. Shiokawa, Y. Aoyagi, P. Kim, K. Toyoda and S. Namba: Jpn. J. Appl. Phys. **23** L232 (1984).
2. H.N. Slingerland, J.H. Bohlander, K.D.v.d. Mast and E. Koets:Microcircuit Eng. **5** 155 (1986).
3. L.R. Harriott, A. Wagner and F. Fritz: J. Vac. Sci. Technol. **B4** 181 (1986).
4. K. Saitoh, H. Onoda, H. Morimoto, T. Katayama, Y. Watakabe and T. Kato: J. Vac. Sci. Technol. **B6** 1032 (1988).
5. Y. Mashiko, H. Morimoto, H. Koyama, S. Kawazu, T. Kaito and T. Adachi: 25th Annual Proc. 1987 Intern. Reliability Phys. Symp. (IEEE, 1987) 111.
6. Y. Ochiai, K. Gamo and S. Namba: J. Vac. Sci. Technol. **B3** 67 (1985).
7. Y. Ochiai, K. Shihoyama, T. Shiokawa, K. Toyoda, A. Masuyama, K. Gamo and S. namba: J. Vac. Sci. Technol. **B4** 333 (1986).
8. Y. Ochiai, K. Shihoyama, T. Shiokawa, K. Toyoda, A. Masuyama, K. Gamo and S. Namba: Jpn. J. Appl. Phys. **25** L526 (1986).
9. Z. Xu, K. Gamo, T. Shiokawa and S. Namba: Nucl. Inst. Methods (to be published).
10. K. Gamo, N. Takakura, D. Takehara and S. Namba: Extended Abstracts 16th. Conf. Solid State Devices and materials (Jpn. Soc. Appl. Phys., 1984) 31.
11. K. Gamo, N. Takakura, N. Samoto, R. Shimizu and S. Namba: Jpn. J. Appl. Phys. **23** L293 (1984).
12. J.E. Bouree, J. Fliscstein and Y.I. Nissim: MRS Symp. Proc. vol.75 129 (1987).
13. G.M. Shedd, H. Lezec, A.D. Dubner and J. Melngailis: Appl. Phys. Lett. **49** 1584 (1986).
14. L.R. Harriott, K.D. Cummings, M.E. Gross, W.L. Brown, J. Linnros and H.O. Funsten: MRS Symp. Proc. Vol.75 99 (1987).
15. K. Gamo, D. Takahara, Y. Hamamura, M. Tomita and S. Namba: Micelectronic Eng. **5** (1986) 163.
16. K. Gamo and S. Namba: MRS Symp. Proc. Vol. 45 223 (1985).
17. G.S. Higashi and C.G. Fleming: Appl. Phys. Lett. **48** 1051 (1986).
18. H. Yokoyama, F. Uesugi, S. Kishida and K. Washio: Appl. Phys. **A37** 25 (1985).
19. G.A. Kivall, J.C. Matthews and R. Solanki: MRS Symp. Proc. vol.75 145 (1987).
20. M.E. Gross, W.L. Brown, J. Linnros, L.R. Harriott, K.D. Cummings and H.O. Funsten: MRS Symp. Proc. vol.75 91 (1987).
21. Thomas H. Baum: MRS Symp. Proc. vol.75 141 (1987).
22. J.S. Ro, A.D. Dubner, C.V. Thompson and J. Melngailis: J. Vac. Sci. Technol. **B6** 1043 (1988).
23. W. Reuter and K. Wittmaack: Appl. Surf. Sci. **5** 221 (1980).
24. H. Miyake, K. Gamo, Y. Yuba and S. Namba: Jpn. J. Appl. Phys. (to be published).
25. D.H. Narum and R.F.W. Pease: J. Vac. Sci. Technol. **B6** 966 (1988).
26. H. Sawaragi, H. Kasahara, R. Aihara, K. Gamo, S. Namba and M. Hassel Shearer: J. Vac. Sci. Technol. B6 974 (1988).

FOCUSED ION BEAM-INDUCED CARBON DEPOSITION

L.R. HARRIOTT AND M.J. VASILE
AT&T Bell Laboratories Murray Hill, New Jersey 07974

ABSTRACT

A process for repair of micron and submicron sized transparent defects on photomasks is described. Opaque films are deposited at the intersection of the flux of organic monomers from a gas jet and a 20 keV Ga ion beam. Focused ion beam-induced deposition differs from other ion-induced, electron beam and laser processes due to the very high ion current density and the sputtering of the material as it is being deposited. We have explored the deposition-sputtering rate competition for several precursor materials as a function of monomer flux and ion beam dose rate. Our results suggest a model for deposition which requires polymerization of the precursor through carbon-carbon double bonds to favor deposition over sputtering by creating high molecular weight material at the target.

INTRODUCTION

Defect repair for lithographic masks is currently the largest commercial application for focused ion beams[1-6]. Masks used for photolithography are usually composed of a glass or quartz substrate with patterned Cr as a light absorber. Defects in the form of excess Cr (opaque defects) can be removed by physical sputtering with a finely focused ion beam as illustrated in Fig. 1a. The ion beam is usually Ga focused to a submicron spot at 10-30 keV beam energy. In mask repair the ion beam is first used to form an image of the defect area in order to locate it precisely (defect locations are only approximately known as a result of the optical inspection process). The ion beam is scanned over the defect based on information from the ion image. Typical ion doses required for removal of the (850 Å) Cr film are 5×10^{17} ions/cm^2. This results in repair times of the order of seconds per square micron for available ion beam systems.

Repair of clear defects or missing Cr features can also be accomplished by sputtering. Refracting structures such as prisms, gratings, or lenses are machined into the glass surface with the focused ion beam. When the mask is used in a projection printing system, light will be directed away from the entrance pupil of the imaging lens and the area of the mask will print dark. Clear repair by this procedure is effective, although slow (tens of seconds per square micron) and not reversible.

Figure 1b illustrates the alternative approach of ion induced deposition for clear defect repair. A precursor gas is introduced into the vacuum chamber through a needle-sized tube (~100 μm dia.) in close proximity to where the ion beam strikes the target. Under the right conditions, a deposit is formed at the substrate only within the area irradiated by the ion beam.

The primary difference between focused ion beam induced deposition and other (electron or laser beam) techniques is the sputtering caused by the ion beam. During the irradiation of the deposition precursor adsorbed at the substrate, the ion beam not only causes bond breaking and other chemical processes but is also sputtering the adsorbed layer (and perhaps the substrate) at a rate generally greater than unity. That is, several (2-10) precursor or target atoms are sputtered away for each incoming ion. The result is that the flux of precursor material must roughly match the arrival rate of the ions and that some chemical amplification must occur in the deposition process. Otherwise, the precursor will be sputtered away at a higher rate than the deposit is formed resulting in a net removal of material from the substrate. This deposition/sputtering rate balance is the central issue in focused ion beam induced deposition. For typical finely focused ion beams, the current density is about 1 A/cm^2 corresponding to an ion flux of 6x10^{18} ions/cm^2/sec. Therefore, if the precursor flux is to be of that order, the equivalent pressure in the inlet tube is roughly 1 Torr. The gas jet arrangement shown in Fig. 1 is thus used to maximize the precursor flux in the area (~100 μm dia.) scanned by the ion beam on the target while minimizing the impact on the ion gun vacuum system.

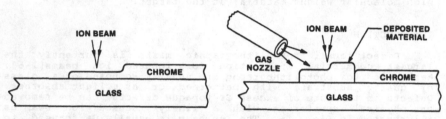

Figure 1 Focused ion beam mask repair by a) sputtering to remove opaque defects and b) deposition to repair clear defects.

Organic precursor materials have been used with finely focused ion beams to produce carbonaceous deposits for several reasons. A wide variety of relatively non-toxic materials are available with sufficient vapor pressure to develop the necessary flux at the substrate. In addition, carbon has a relatively low sputtering yield. The original inspiration though, probably drew form experience with contamination in scanning electron microscopes where opaque deposits are sometimes (unintentionally) formed when the electron beam irradiates a sample contaminated by pump oil from the vacuum system.

There are several approaches that might be taken to ensure that the rate balance favor deposition over sputtering for a practical process. Firstly, the ion beam may be scanned repetitively at a high rate to form the patterned deposit so that the time-averaged ion flux is reduced. Although this eases the requirements on precursor flux, it results in a process which may have a narrow practical latitude due to the ion dose rate sensitivity. It is preferable to use a process which is ion dose rate independent since it will be more robust. The approaches generally taken include using a high molecular weight precursor so that the ions may be outnumbered by carbon atoms. These materials tend to have low vapor pressures at room temperature, making it difficult to obtain sufficient gas flux at the target for a dose rate independent process. This may be overcome by heating the supply of precursor and the delivery tube to raise the vapor pressure. However this method has the drawback of producing deposits on all cool surfaces within the ion beam system in addition to the region irradiated by the ion beam resulting in contamination of ion optical and other system components. Another approach is to use a carrier gas to bring the high molecular weight material into the chamber. Generally this will result in prohibitive gas loading of the vacuum system before sufficient flux of the precursor is achieved at the substrate. Our approach [7] has been to use precursors of relatively low molecular weight that will polymerize readily and have vapor pressures of several Torr at ambient temperature. The idea is that the high vapor pressure will ensure an adequate molecular flux while the tendency for polymerization will result in a higher molecular weight material being formed at the substrate during the initial stages of deposition, giving the necessary gain to the process.

EXPERIMENTAL PROCEDURES

A 20 keV Ga ion beam with a spot size of 0.5 um and current of 1 nA was used in a series of experiments designed to evaluate several precursors for their ability to form opaque deposits at various ion beam dose rates. A 0.2 um beam with current of 160 pA was later used to produce fine features. The gas delivery system used a tube of 100 um inner diameter placed less than one diameter above the substrate. The pressure in the tube was regulated by a leak valve and measured with a capacitance manometer. Tube pressures of 0.5 to 3 Torr were used resulting in a flux at the surface of up to $1x10^{18}$ molecules/cm^2/sec and a chamber pressure rise from $5x10^{-8}$ Torr to $1x10^{-6}$ Torr during the deposition experiments.

Each material was evaluated with an inlet pressure of 2 Torr and the beam current, energy, and spot size fixed. The ion dose rate was varied by changing the number of scans over which a fixed dose of $5x10^{17}$ ions/cm^2 was delivered for rectangular features 2 X 10 um and 1,10,100, and 1000 scans of the beam. Deposition attempts were made on photomask substrates with the features covering both Cr and glass areas. Net deposition or removal was determined by transmission optical microscopy of the glass areas, rendered opaque in the case of net deposition, and the Cr areas, rendered clear in the case of net sputtering.

Initially four materials were tested: benzene, toluene, methyl methacrylate, and styrene. Benzene and toluene were chosen since they are known to form deposits in discharge tubes and methyl methacrylate and styrene are obvious polymer building blocks. Each of these materials is liquid at room Benzene and toluene yielded very thin translucent films at the lowest dose rate (highest scan rate) conditions. The higher dose rate conditions resulted in net removal of substrate material.

Methyl methacrylate did yield films over a wider range of scan rates and inlet pressures than benzene or toluene, but these films were also thin, translucent and contained pinholes, i.e. the films were not continuous.

Styrene resulted in net deposition of an opaque film over the range of scan rates tested. The deposition process was dose rate insensitive at inlet pressures above 1.4 Torr. At lower pressures, opaque deposits were only formed for the low dose rate (high scan rate) conditions. We calculate a molecular flux of 3×10^{19} molecules/cm^2/sec at the tube orifice based on pressures and calibrated pumping speeds. Although the flux is lower at the substrate, this may be used as an upper limit estimate. Film thickness as a function of ion dose was measured with a profilometer on larger features for a constant styrene inlet pressure of 2.2 Torr. The dependance was linear at a rate of 330 A/1×10^{17} ions/cm^2, which corresponds to 1 or 2 carbon atoms per incident Ga ion, indeed a close balance between deposition and sputtering. White light optical absorption was also measured for styrene films as a function of ion dose and followed an exponential dependance with an absorption coefficient of 2.2×10^{-3} per Angstrom. Features produced with a dose of 5×10^{17} ions/cm^2 (the same as the dose for Cr removal in opaque defect repair) are thus about 1650 A thick with an optical density of 3 (similar to that of the Cr film. The optical absorption of the carbon films is expected to be higher for the UV light used in photolithography.

Large area (0.2mm square) films were analyzed by AES. The Auger depth profile of the film showed that the Ga concentration does not exceed 10% in the first 200 A, indicating an efficient process in terms of the number of C atoms per implanted Ga ion. The atomic concentration of Ga rises sharply between 200 and 500 A to a steady concentration of 27 at% at larger depths, corresponding to the projected range of 20 keV Ga ions in this sort of polymer film. The presence of a high atomic concentration of Ga throughout most of the thickness of the films used for mask repair aids in the optical absorption (more so in the UV).

DISCUSSION

XPS analysis of films deposited from styrene showed the C(1S) signal characteristic of other ion-bombarded polymers and the pi-pi* satellite normally observed for aromatic rings was missing. The films are thus not polystyrene and the polymer formation mechanism probably includes ring opening induced by the ion bombardment. The normal polymerization route for styrene, through the vinyl bond, may be responsible for the strong chemisorption of styrene to radical sites at the growing polymer surface. Subsequent ion irradiation will open the aromatic ring and cause fragmentation, leading to a film which does not contain aromatic rings. This proposed mechanism of film growth restricts deposition to the area under ion bombardment and is consistent with the physical and chemical properties of the material.

To test the model of polymerization-aided deposition we evaluated, using the methods described above, several other materials which met the vapor pressure requirement but differed in whether a C-C double bond was present on a substituant to the aromatic ring. Phenylacetylene, indene, and benzofuran each produced opaque deposits over the range of dose rates for inlet pressures of 2 Torr. The properties of the resulting films did not qualitatively differ form those produced with styrene as a precursor. Thus any of the four materials would provide an adequate process. On the other hand, ethylbenzene which is very similar to styrene except for the C-C double bond, did not produce deposits at any of the dose rates tested for an inlet pressure of 2 Torr and always resulted in a net sputtering of the substrate. Indane, which differs from indene also by the

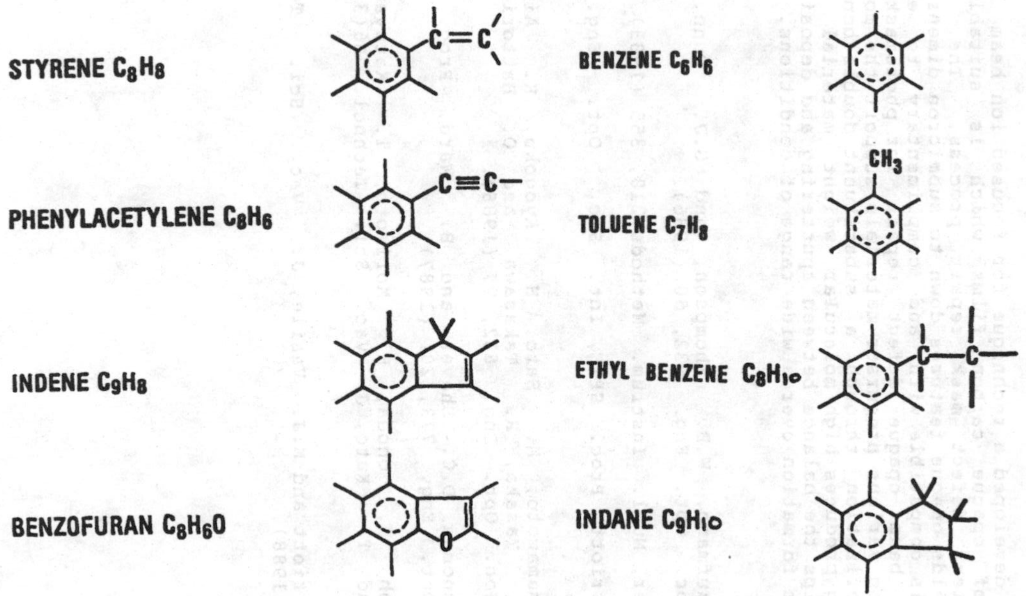

Figure 2 Precursors used for this work; those in the lefthand column produced opaque deposits while those in the righthand column did not.

lack of a double bond, produced the same negative results. For comparison the structural formulas for the various precursors are shown in Fig. 2.

CONCLUSION

We have developed a technique for focused ion beam induced deposition of opaque carbon films which is suitable for a practical clear defect mask repair process. The process described yields opaque features down to submicron dimensions in a way which is compatible with and complimentary to existing focused ion beam opaque defect repair for photomasks. Our experiments on various precursor materials support the model in which polymerization through a substituant double bond to an aromatic ring produces high molecular weight material at the target and tips the balance between sputtering and deposition in favor of film formation over a wide range of conditions.

REFERENCES

[1] H.C. Kaufmann, W.B. Thompson, and G.J. Dunn, Proc. SPIE,Int. Soc. Opt. Eng. 632, 60 (1986).

[2] A. Wagner, Nucl. Instrum. Methods 218, 355 (1983).

[3] L.R. Harriott, Proc. SPIE, Int. Soc. Opt. Eng. 773, 190 (1987).

[4] M. Yamamamoto, M. Sato, H. Kyogoku, K. Aita, Y. Nakagawa, A. Yasaka, R. Takasawa, and O. Hattori, Proc. SPIE, Int. Soc. Opt. Eng. 632, 97 (1986).

[5] N.P. Economou, D.C. Shaver, and B. Ward, Proc. SPIE, Int. Soc. Opt. Eng. 773, 201 (1987).

[6] K. Saitoh, H. Onoda, H. Morimoto, T. Katayama, Y. Watakabe, and T. Kato, J. Vac. Sci. Technol. B6(3), 1032 (1988).

[7] L.R. Harriott and M.J. Vasile, J. Vac. Sci. Technol. B6(3), 1035 (1988).

HYDROGEN ION BEAM SMOOTHENING OF
OXYGEN ROUGHENED Ge(001) SURFACES

K.M. Horn, E. Chason, J.Y. Tsao, and S.T. Picraux
Sandia National Laboratories, Albuquerque, NM 87185

ABSTRACT

A smooth Ge(001) surface can be severely roughened by
chemical etching with oxygen gas and then returned to its
original smoothness by exposure to hydrogen. The rate of
recovery of the surface depends strongly on temperature, as well
as on whether the hydrogen is introduced as a low energy ion
beam or simply as a gas. The roughness induced in the surface
by etching is "locked-in" after removal of the oxygen gas
through pinning of ledges by residual contamination. This
pinning prevents the surface from smoothening thermally. The
introduction of hydrogen to the surface promotes a chemical
reaction which frees the pinned sites and allows thermal
smoothening to proceed.

INTRODUCTION

The presence of hydrogen has long been thought beneficial
to molecular beam epitaxy (MBE) of GaAs, as well as to other
processes such as the plasma deposition of diamond-like films.
Calawa has reported that the dominant effect of the introduction
of hydrogen during MBE growth of GaAs is to reduce the
incorporation of oxygen and carbon in the epitaxial layer [1].
Bozack, et.al. have found that passivating Si(110) dangling
bonds with hydrogen before exposing to propylene prevents its
adsorption [2], while introducing hydrogen **after** the Si has been
exposed to propylene increases the silicon-propylene chemical
reaction [3].

It is therefore not unreasonable to expect that the
application of hydrogen ion beams to surfaces can induce
chemical reactions that result in a change of the surface
morphology. In this work we have studied an oxygen-hydrogen
surface chemical reaction which leads to smoothening of a
surface which has been severely roughened by oxygen etching.
The use of a low energy hydrogen beam, and even hydrogen gas, to
promote this reaction results in the creation of a surface
suitable for epitaxial growth, without the need of a high
temperature anneal and further growth.

EXPERIMENT

The measurements were performed in an MBE growth chamber
with a base pressure of 1×10^{-9} Torr. Controlled amounts of
oxygen can be admitted to the chamber through a leak valve and
monitored with both an ion gauge and a residual gas analyzer.
Hydrogen is admitted to the chamber via a Pd leak valve.
Through use of a standard electron-impact-ionization sputter
gun, hydrogen ion beams of 100 to 1000 eV can be directed onto

the sample surface. The Ge(001)±1° substrate used in these measurements was prepared by sequential pad polishes in Br:methanol and methanol. The initial in situ surface preparation consisted of two repetitions of a 750 C anneal for 20 minutes followed by growth of a 2000 angstrom buffer layer at 500 C.

Our principle diagnostic tool is reflection high-energy electron diffraction (RHEED). Using a 15 keV electron beam, incident at a glancing angle, diffraction patterns are observed on a phosphor screen. A video camera records the diffraction patterns which are digitized and stored in a computer. Live-time analysis of the diffraction patterns allows the display of the lineshape and spot intensity with a collection rate of better than two frames per second. The diffraction condition used is known as the out-of-phase condition [4] where electrons scattered from surfaces displaced by one atomic step interfere destructively. This diffraction condition provides the greatest sensitivity to surface roughness. Maximum intensity of the diffraction spot corresponds to a smooth surface, with lower intensities corresponding to progressively rougher surfaces. During all RHEED measurements the sample was positioned so that the electron beam was incident at approximately 6° off the <100> azimuth, to minimize the number of multiple scattering artifacts present in the background, such as Kikuchi bands.

RESULTS

The roughening caused by oxygen etching of the Ge(001) surface was measured in situ using the RHEED technique. Fig. 1 displays the roughening curves measured over a range of oxygen pressures at a sample temperature of 450 C. Roughening curves measured at higher sample temperatures exhibit slower roughening rates for a fixed pressure.

From this measurement we determined the temperature and pressure conditions required to create rough surfaces. While the oxygen exposed surfaces created in the preceeding roughening measurements would be considered severely rough for any MBE application, they do not exhibit the transmission-like "spotty" patterns characteristic of a three dimensional rough surface. Such surfaces are nearly always encountered during the initial stages of growth on new samples, freshly inserted into the MBE chamber. In order to create a surface which displays this "spotty" RHEED pattern, the initially smooth surface of a well characterized sample was roughened by exposure to $1x10^{-4}$ Torr oxygen for 3300 seconds. Hydrogen smoothening measurements were then carried out by evacuating the chamber to $8x10^{-7}$ Torr before back-filling with $5x10^{-5}$ Torr hydrogen gas and exposing the roughened surface to an hydrogen ion beam. The sample temperature was maintained at 500 C during both the oxygen roughening and hydrogen smoothening.

Presented in Fig. 2 is a plot of the peak intensity of the out-of-phase diffraction spot over the course of an oxygen roughening and hydrogen smoothening measurement. The lineshapes above the upper x-axis are sample diffraction spot profiles recorded at the indicated times. The initially smooth surface roughens rapidly when exposed to the oxygen atmosphere, as indicated by the decrease in the specular peak intensity. While still in the out-of-phase diffraction condition, two side peaks appear as the surface continues to roughen and the overall

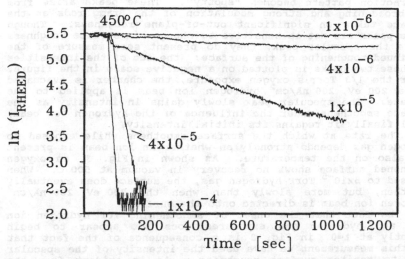

Fig. 1   Ln RHEED intensity during oxygen roughening at 450 C. Labeled pressures are in Torr.   The oxygen is admitted at t=0.

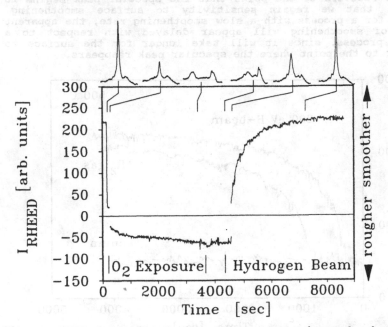

Fig. 2   RHEED intensity during oxygen roughening and subsequent hydrogen smoothening of Ge(001).   Intensity profiles of the diffraction pattern are shown above the upper x-axis. Sample temperature is 500 C.

diffraction pattern becomes "spotty". These peaks arise from the broadening and strong modulation of the Bragg rods as the surface takes on significant out-of-plane roughness. Though these peaks are not as sensitive to changes in surface roughness as is the specular peak, they do present some measure of the continued roughening of the surface; the sum of the intensities of these two peaks is plotted on a negative scale in the figure. After the 1/3 Torr·s oxygen exposure, the chamber is evacuated and a 200 eV, 200 nA/cm² hydrogen ion beam is applied to the surface. The specular peak slowly gains in intensity as the surface smoothens under the influence of the hydrogen ion beam, until finally it regains its initial intensity.

The rate at which the surface smoothens while exposed to hydrogen gas depends strongly on whether an ion beam is present and also on the temperature. As shown in Fig. 3, an oxygen roughened surface shows no recovery in vacuum at 500 C. When exposed to 5×10^{-5} Torr hydrogen gas, the surface does eventually smoothen, but more slowly than when the 200 eV, 200 nA/cm² hydrogen ion beam is directed onto the surface.

The fact that the onset of smoothening for hydrogen ion beam and hydrogen gas exposures does not appear to begin promptly at t=0 in Fig. 3 is a consequence of the fact that for this measurement we have used the intensity of the specular beam to monitor surface roughness. As is evident from the lineshapes shown in Fig. 2, it is not until the roughened surface recovers to the point that the specular peak begins to emerge, that we regain sensitivity to surface smoothening. Hence, for a process with a slow smoothening rate, the apparent onset of smoothening will appear delayed with respect to a faster process, since it will take longer for the surface to recover to the point where the specular peak reappears.

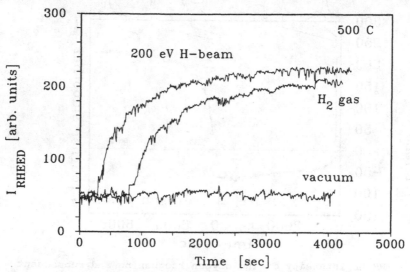

Fig. 3 Comparison of the smoothening effect of vacuum, hydrogen gas, and 200 eV, 200 nA/cm² hydrogen ion beam on an oxygen roughened Ge(001) surface at 500 C. The hydrogen gas and hydrogen ion beam are turned on at t=0.

The hydrogen smoothening effect also displays a pronounced temperature dependence. Below a temperature of 375 C, no smoothening of the surface is apparent, while at temperatures above 400 C the smoothening of the surface becomes quite pronounced. This fact is explained in terms of the smoothening model presented below.

DISCUSSION

The oxygen roughening curves of Fig. 1 exhibit an exponential decay with a first order dependence of the roughening rate on oxygen pressure. This result is consistent with thermal desorption measurements made by Surnev and Tikhov [5] which indicate first order kinetics for low oxygen coverages as well as the model of Madix et.al. [6] which proposes the migration of adsorbed oxygen atoms to the edges of pits where GeO is formed and desorbed.

The chemistry of the hydrogen-smoothening effect can be considered in terms of three possible mechanisms: (1) hydrogen-assisted detachment of Ge from ledges mediating island ripening, (2) the direct desorption of germane preferentially at steps, and (3) the unpinning of ledges by the release of residual contaminants from the surface which allows thermal smoothening to proceed. By thermal smoothening we refer to the tendency of a rough, yet uncontaminated surface to smoothen through the growth of large clusters as a result of the dissolution of smaller clusters [7].

The first two "hydrogen-assisted" smoothening mechanisms can be ruled out for the following reasons. Clean, smooth Ge surfaces which have been roughened by low temperature growth and then thermally smoothened, exhibit no enhancement of the smoothening when exposed to either hydrogen gas or hydrogen ion beams. Furthermore, smooth Ge surfaces which have been exposed to 200 eV, 200 nA/cm^2 hydrogen ion beams at temperatures below 350 C have been observed to actually roughen.

We propose that upon completion of the oxygen roughening, residual contaminants remain on the surface, immobilizing terrace ledges and impeding the dissolution of small islands. Exposure to hydrogen frees these pinned sites on the oxygen-etched surface, after which thermal smoothening can proceed. In this way, the hydrogen acts only to scrub the surface of contaminants (oxygen and possibly carbon) through chemical processes alone, when exposed just to hydrogen gas, and both chemically and physically (via sputtering) when a hydrogen ion beam is used. We include the possibility that carbon may contribute to the pinning of ledges since it is likely that while the oxygen is etching the Ge surface, it is also acting on the chamber walls, freeing hydrocarbons and CO which might then be adsorbed on the sample.

Preliminary measurements indicate that below a threshold temperature of about 375 C, the oxygen roughened surface does not noticeably smoothen, even when exposed to a hydrogen ion beam. This is to be expected since thermal smoothening proceeds very slowly at that temperature. However, if after completion of the hydrogen exposure the ion beam is turned off and the chamber evacuated, the surface will promptly smoothen when heated to 500 C. In contrast, an oxygen roughened surface does not smoothen even at 500 C if it is not exposed to hydrogen. This is strong additional evidence for the two step smoothening process we propose.

CONCLUSION

The roughening of Ge(001) surfaces by oxygen etching and subsequent smoothening of the surface by hydrogen ion beam exposure have been studied in real time using RHEED. In situ measurements of the oxygen roughening rate of Ge exhibit a first order dependence on pressure. Exposing the oxygen roughened surface to an hydrogen ion beam effectively smoothens the surface to a state suitable for epitaxial growth. Exposure to hydrogen gas also smoothens the surface, but not as quickly. The mechanism we propose for this smoothening is a two step process in which contaminants are removed from pinned sites by either chemical processes (in the case of exposure to H gas alone) or both physical and chemical processes (when H ion beams are used). Once the pinned sites have been freed, thermal smoothening can proceed, rapidly returning the surface to its original state of smoothness.

We would like to acknowledge the assistance of Dan Buller. This work was performed at Sandia National Laboratories supported by the U.S. Department of Energy under contract DE-ACO4-76DP00789.

REFERENCES

1. A.R. Calawa, Appl. Phys. Lett., 33, 12,1978.

2. M.J. Bozack, W.J. Choyke, L. Muehlhoff, J.T. Yates Jr., Surf. Sci., 176, 547, 1986.

3. M.J. Bozack, P.A. Taylor, W.J. Choyke and J.T. Yates Jr., Surf. Sci., 179, 132, 1987.

4. J.M. van Hove, P.R. Pukite and P.I. Cohen, J. Vac. Sci. Tech. B3 563 (1985).

5. L. Surnev and M. Tikhov, Surf. Sci., 123, 505, 1982.

6. R.J. Madix and A.A. Susu, Surf. Sci, 20, 377, 1970.

7. E. Chason, J.Y. Tsao, K.M. Horn and S.T. Picraux, J. Vac. Sci. Tech. B, in press.

MECHANISMS IN THE ION-ASSISTED ETCHING OF TiSi$_2$

W.L. O'BRIEN*, T.N. RHODIN* AND L.C. RATHBUN**
*School of Applied and Engineering Physics, Cornell University, Ithaca NY 14853
**National Nanofabrication Facility, Cornell University, Ithaca NY 14853

ABSTRACT

A detailed investigation into the mechanisms of the ion-assisted etching of TiSi$_2$ using 1000 eV argon ions and chlorine gas is reported. X-ray photoelectron spectroscopy (XPS) was used to investigate surface product distributions and modulated ion beam mass spectroscopy (MIBMS) was used to investigate gas phase product distributions. Identical experiments were performed on silicon and titanium single crystal substrates. Information on chemical mechanisms during ion-assisted etching was obtained by comparing the product distributions of the three substrates. For each substrate, the di and tetrachlorides were the major gas phase products. These products were emitted with high thermal velocities following a surface residence process. TiCl and TiCl$_2$ products were found on the titanium surface after simultaneous exposure to chlorine and argon ions, but not on the TiSi$_2$ surface. These observations are discussed in terms of specific chemical mechanisms which are critical during ion-assisted etching.

I. INTRODUCTION

The mechanisms of ion-assisted etching have been the subject of extensive research during the past decade. Typically, these mechanisms are interpreted in terms of physical processes associated with ion bombardment and/or ion-induced surface chemistry. Silicon, etched with chlorine and argon ions, is the most widely studied system[1,2,3]. This is due to the importance of silicon etching to the microprocessing industry. Other systems which have received detailed investigations are Si/XeF$_2$/Ar[4], GaAs/Cl$_2$/Ar[5,6], W/XeF$_2$/Ar[7] and Ti/Cl$_2$/Ar[8]. In this paper we present new experimental results on the ion-assisted etching of TiSi$_2$ using chlorine gas and argon ions. Comparison of these results is made to results obtained for the Si/Cl$_2$/Ar$^+$ and Ti/Cl$_2$/Ar$^+$ systems using identical experimental conditions. These results are interpreted in terms of specific chemical mechanisms involved in ion-assisted etching.

It is appropriate to discuss the present day knowledge of ion-assisted etching and then to discuss the specific systems Si/Cl$_2$/Ar$^+$ and Ti/Cl$_2$/Ar$^+$ to provide a background to understanding the new work on the TiSi$_2$/Cl$_2$/Ar$^+$ system. When an energetic ion strikes a surface it induces a collisional-cascade. The depth of this cascade is $\approx$20-30 Å for the ion energies and substrates discussed in this paper. During the duration of this cascade physical (collisional-cascade) sputtering takes place. If an ion strikes a surface in the presence of a reactive adsorbate, "ion-mixing" occurs. These ion-mixing processes have been discussed in detail by Dieleman et. al.[1] and have been investigated using Rutherford backscattering by Mayer et. al.[2]. Briefly, the incident ion implants adsorbed species by direct momentum transfer, or induces the adsorbate and substrate atoms to diffuse through the region of the cascade. These two processes result in the formation of an altered overlayer with a depth of $\approx$20-30Å (the depth of the cascade). This overlayer is in dynamic equilibrium with the unaltered substrate and emission to the gas phase. As etching proceeds, the outermost constituents of the overlayer are emitted into the gas phase, while the substrate below is altered through damage, implantation and diffusion.

Modulated ion beam mass spectroscopy (MIBMS) has been used by a number of research groups to investigate the mechanisms of ion-assisted etching. In MIBMS, a reactive gas doser, ion source and mass spectrometer are simultaneously focused onto the sample. While the reactive gas flux is held constant, the ion flux is periodically modulated in time. Mass filtered product waveforms are detected by the mass spectrometer. These periodic waveforms contain information on the specific products time-of-flight (kinetic energy distribution) and any surface residence distribution. By changing the substrate to detector distance, surface residence effects can be isolated. Typical parameters of interest in these experiments are ion energy and the reactive gas/ion surface flux ratio.

Dieleman et. al.[1] have used MIBMS to study the ion-assisted etching of silicon using chlorine gas and argon ions. They found the major gas phase products to be Si, SiCl and $SiCl_2$. These products were found to be emitted by collisional-cascade sputtering at high incident ion energies and low chlorine/argon ion surface flux ratios. A small percent of the products were thermally desorbed when the argon ion energy was decreased and/or the surface flux ratio was increased. These observations were interpreted[1] in terms of an assumed formation of SiCl and $SiCl_2$ trapped in surface "voids" during ion-mixing. Subsequent ions would release these "trapped" molecules. Etching was interpreted to proceed primarily through physical processes. The argon ion angle of incidence was 60° from the sample normal and products were detected at only one flight distance from the surface. No surface analysis was reported.

Rossen and Sawin[3], using MIBMS, determined that $SiCl_2$ and $SiCl_4$ were the major gas phase products during the ion-assisted etching of silicon with chlorine gas and argon ions. Furthermore, using detection at multiple flight distances, Rossen and Sawin[3] determined that emission proceeded by thermal desorption following a first order surface process. The values of the emission parameters for $SiCl_2$ were 2900 K and 150 μs for the translational temperature and relaxation time constant. For $SiCl_4$ emission the values of the emission parameters were 315 K and 180-300 μs. The relaxation time constant measured for $SiCl_4$ emission was depenedent upon the chlorine/argon ion surface flux ratio. A model was proposed[3] which involved the formation of an ion-induced surface reactant which would subsequently react with available surface chlorine to form the detected gas phase product. Chemical processes during ion bombardment were important in this interpretation. The argon ion angle of incidence was 23° from the surface normal.

The ion-assisted etching of titanium has been studied by O'Brien, Rhodin and Rathbun using argon ions and chlorine gas[8]. It was shown, by investigating the Ti-2p photoemission peak, that argon ion bombardment induced the formation of an overlayer containing TiCl and $TiCl_2$ on the titanium surface during simultaneous exposure to chlorine and argon ions. No overlayer was formed in the absence of ion bombardment. Thermodynamic calculations showed the first chlorination step, $Ti_{(s)} + Cl_{(s)} \rightarrow TiCl_{(s)}$, to be endothermic, while subsequent chlorinations are exothermic. The ion beam supplied the activation energy for this initial chlorination. The major gas phase products were found to be $TiCl_2$ and $TiCl_4$ using MIBMS. Detection of products at two flight distances suggested that emission was by thermal desorption following a first order surface process. The value of the emission parameters for $TiCl_2$ were 4000 K and 260 μs and 3500 K and 370 μs for $TiCl_4$ emission. Both physcial and chemical processes were found to be important.

II. EXPERIMENTAL PROCEDURES

The experimental procedures have previously been described in detail[5,8]. Briefly, an ultra high vacuum system with modulated ion beam mass spectroscopy, x-ray photoelectron spectroscopy and Auger electron spectroscopy (AES) capabilities was used. Chlorine gas (1×10^{-5} Torr) and 1000 eV argon ions were used throughout the study. The average chlorine/argon ion surface flux ratio was 16. Samples studied were Si(111), unorientated single crystal titanium and TiSi$_2$. The TiSi$_2$ was formed by sputter depositing 1 μm of titanium onto a Si(100) wafer and annealing to 1100 K for 2 hours. Formation of TiSi$_2$ was verified with a Guinier camera experiment. The results of the titanium investigation have been reported in detail[8]. Reference to these results will be made for comparison to the results of the new investigations using silicon and TiSi$_2$.

The gas phase product distributions were detected in real time (during simultaneous exposure to chlorine gas and argon ions) using MIBMS. Unfortunately, the surface product distributions could not be detected in real time. This was due to the geometrical constraints caused by the large size of the double pass cylindrical mirror energy analyzer used for both AES and XPS. However, useful information could be obtained by performing XPS and AES after various sample pretreatments. XPS is chemically sensitive and was used to determine the surface product distributions. While AES is also chemically sensitive, interpretation is more difficult due to the multiple transitions allowed. AES was used to check sample cleanliness. The samples were cleaned by etching 1/2 hour before collecting any data, this also guaranteed steady state etching conditions. The samples could potentially etch through to the silicon substrate. This possibility was periodically monitored using XPS and AES.

III. EXPERIMENTAL RESULTS

The major gas phase products detected during MIBMS were the di and tetrachloride species for each surface studied. Detection by mass spectroscopy requires electron impact ionization of the neutral products. This ionization also results in the fragmentation of product species. The detected signal intensities, obtained by integrating product waveforms and correcting for the velocity dependence of the ionization probability, are given in Table.I. (Only products consisting of the major isotopes are considered). These intensities are also compared to the mass fragmentation patterns of SiCl$_4$[9] and TiCl$_4$[10] (Table.I). It is noted that the exact details of mass fragmentation depend on the specific ionization conditions. It appears that the di and tetrachlorides are the major gas phase etchant products.

Product waveforms were collected at two distances from the surfaces, 14.0 and 21.5 cm. These product waveforms were interpreted in terms of the collisional-cascade sputtering model, the normal thermal desorption model and the thermal desorption model following a first order surface process model. In each case the best fits were obtained using the thermal desorption following a first order surface process model. The translational temperatures and reaction time constants obtained from the best fits are given in Table.II. The waveforms measured at m/e values corresponding to the trichloride ions were used for the tetrachloride fits. In each case these waveforms were identical to those obtained at the m/e value corresponding to the associated tetrachloride ion. The waveforms measured at m/e values corresponding to the dichloride ions were used to obtain the best fits for the dichloride product waveforms once the appropriate tetrachloride waveform was properly subtracted. Any errors in the mass fragmentation pattern will lead to some error in the dichloride product waveforms. However, since the waveforms for the tetra and trichloro ions were identical in each case, the product waveforms used for the tetrachloride species are accurate.

Representative product waveforms are shown in Fig.1 and Fig.2. These waveforms were obtained with a surface to ionization region distance of 21.5 cm. Equally precise fits were obtained at the shorter flight distance. Models not allowing for the occurrence of surface residence times could not fit the two sets of data obtained at the different flight distances. The estimated error in the values for the emission parameters are $\pm$ 500 K and

Table I. Abundance of mass fragments detected from various surfaces during ion-assisted etching compared to mass fragmentation pattern of tetra-chlorides. All results obtained using 1000 eV argon ions and a chlorine/argon ion flux ratio of 16.

	SiCl+	SiCl2+	SiCl3+	SiCl4+
from Si	235	55	100	50
SiCl4	43	10	100	48
from TiSi2	130	32	100	50

	Ti+	TiCl+	TiCl2+	TiCl3+	TiCl4+
from Ti	176	130	175	100	36
TiCl4	55	48	46	100	38
from TiSi2	73	75	106	100	44

Table II. Emission temperature and first order reaction time constant obtained from best fits to data for emission from silicon, titanium and TiSi2. All results obtained using 1000 eV argon ions with a chlorine/argon ion surface flux ratio of 16.

	From elemental	From silicide
SiCl2	Temp = 4300 K, τ = 400 µs	Temp = 5000 K, τ = 250 µs
SiCl4	Temp = 1900 K, τ = 1210 µs	Temp = 4000 K, τ = 980 µs
TiCl2	Temp = 4000 K, τ = 260 µs	Temp = 4500 K, τ = 280 µs
TiCl4	Temp = 3500 K, τ = 370 µs	Temp = 3800 K, τ = 780 µs

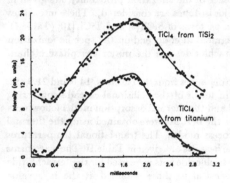

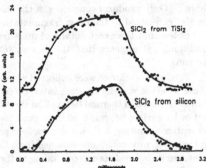

Fig. 1 TiCl$_4$ emission from TiSi$_2$ and titanium with the best fits to the data. Parameters obtained from best fits are presented in Table II.

Fig. 2 SiCl$_2$ emission from TiSi$_2$ and silicon with the best fits to the data. Parameters obtained from best fits are presented in Table II.

±50 μs. Since increasing the emission temperature causes a similar effect on the product waveform as decreasing the time constant, there was a finite area in parameter space which gave best fits to the data. The estimated error given for the emission parameters roughly defines an area in parameter space within which the root mean square deviation of the fit was within 95% of its minimum value.

As stated in the Introduction, an overlayer containing TiCl and $TiCl_2$ was observed on the titanium surface after simultaneous exposure to argon ions and chlorine gas. No such overlayer was observed on the $TiSi_2$ surface after identical treatment. No shifts in the Si-2p photoemission peak were detected for either the silicon or the $TiSi_2$ surface. The reported chemical shifts of SiCl and $SiCl_2$ are 0.93 and 1.8 eV respectively[11]. (A small peak, shifted this much, may be below the detection limit for the experimental technique used). The Si-2p counting rates are low due to the low cross section of the 1253 eV (Mg K_α) photons used. Nothing more will be said about the existence of the silicon chloride surface species. However, it has been shown that the presence of silicon either prevents the $TiCl_y$ overlayer from forming, or destabilizes it. It should be noted that the XPS experiments were performed 10 minutes after sample exposure.

IV. DISCUSSION

The first order surface process has been modeled previously for $AsCl_3$ emission from GaAs during MIMBS using argon ions and chlorine gas[5]. It is not appropriate to discuss this model in detail here, this will be done later. Briefly, the incident ions induce the formation of a surface reactant. This reactant has a lower activation energy for subsequent chlorination. If a subsequent chlorination is exothermic, and the reaction takes place on the physical surface, the resulting product can be emitted with high translational temperatures. The observed surface residence time is a combination of two effects then. The first is the lifetime of the ion-induced surface reactant, and the second is the reaction time constant for the chlorination(s) leading to the detected product. In this model the observed time constant for highly chlorinated products may be a combination of a number of such processes.

The data presented in Table.II are the best set of data which can be used to compare emission from the different surfaces, even though the results of Rossen and Sawin[3] and Dieleman et. al.[1] are more complete for silicon chloride emission from silicon (they used a variety of argon ion energies and chlorine/argon ion surface flux ratios). The data in Table.II was obtained under identical experimental conditions and interpreted in exactly the same manner. Thus any trend in the data should be accurate, and not a reflection of experimental conditions. It should be noted that these new results for silicon chloride emission from silicon are very similar to those obtained by Rossen and Sawin[3]. The significant similarities are the identity of the major products and the observation of surface residence times.

These new results for silicon chloride emission from silicon differ from those obtained by Dieleman et. al.[1] One possible explanation for this discrepancy is the choice of ion angle of incidence. In this work we used a 22° angle of incidence from the sample normal for ion bombardment. This is very similar to the ion angle of incidence used by Rossen and Sawin[3], 23°. The use of a larger angle of incidence by Dieleman, 60°, causes a much higher adsorbed chlorine sputtering yield than the lower angle of incidence[2]. Thus, for equivalent chlorine/argon ion surface flux ratios, the surface investigated by Dieleman et. al.[1]is characterized by a much lower chlorine surface concentration than the surface investigated in this and Rossen and Sawin's[3] study. This can explain the lower chlorination of the products detected by Dieleman et. al.[1]. Furthermore, if the lack of chlorine becomes

560

the rate limiting step in the formation of $SiCl_2$ and $SiCl_4$, physical sputtering of lower chlorinated species would dominate.

The most significant trend indicated in Table.II is that, while the emission temperatures for specific products from different surfaces are identical within experimental error (except $SiCl_4$), the time constants are not. In fact, the time constants for the titanium chlorides are greater for emission from the $TiSi_2$ surface while the time constants for silicon chlorides are greater for emission from the silicon surface. A possible explanation for this observed trend will be discussed in terms of the chemistry of the altered surface.

One possible explanation for this observed trend would be the occurrence of a surface reaction of the type

$$SiCl_x + TiCl_y \rightarrow SiCl_{x+1} + TiCl_{y-1}, \tag{1}$$

on the $TiSi_2$ surface. This type of reaction would increase the production rate of silicon chloride species at the expense of titanium chloride species.

X-ray photoemission results on the Ti-2p peak from titanium and $TiSi_2$ after simultaneous exposure are consistent with this proposed explanation. As stated earlier, the presence of silicon has the effect of either preventing $TiCl_y$ compounds from forming on the altered surface of $TiSi_2$ or causing them to decompose once the ion beam is turned off. Since similar titanium chloride gas phase products are emitted from the titanium and $TiSi_2$ surfaces, it is most probable that the altered surfaces from which the emission proceeds are similar. Therefore, the silicon probably has the effect of decomposing the $TiCl_y$ species in the altered surface by a reaction similar to that described in eqn(1).

V. CONCLUSIONS

MIBMS and XPS were used to investigate the ion-assisted etching of $TiSi_2$. Comparison was made to identical experiments using silicon and titanium substrates. The experimental results obtained for the ion-assisted etching of silicon were similar to those obtained by Rossen and Sawin. The differences in these results and those obtained by Dieleman et. al. were interpreted in terms of the argon ion angle of incidence. The differences in the observed product distributions and emission parameters for the three surfaces were interpreted in terms of additional chemical pathways available on the $TiSi_2$ surface. Photoemission studies of the Ti-2p peak are consistent with this explanation. These observations suggest that surface chemical effects during ion bombardment are important aspects of the ion-assisted etching of $TiSi_2$, silicon and titanium. This work was supported by NSF Grant DMR-87-05680 and the National Nanofabrication Facility (NSF-ECS-86-19049). W.L. O'Brien is an IBM Manufacturing Research Fellow.

1. J. Dieleman, F.H.M. Sanders, A.W. Kolfschoten, P.C. Zalm, A.E. deVries and A. Haring, J. Vac. Sci. Technol. **B3** 1384 (1985).
2. T. Mizutani, D.J. Dale, W.K. Chu and T.M. Mayer, Nuc. Instru. Meth. Phys. **B7** 825 (1985).
3. R.A. Rossen and H.H. Sawin, J. Vac. Sci. Technol. **B3** 1595 (1987).
4. H.F. Winters and F.A. Houle, J. Appl. Phys. **54** 1218 (1983).
5. W.L. O'Brien, T.N. Rhodin and L.C. Rathbun, J. Appl. Phys. (to be published).
6. S.C. McNevin and G.E. Becker, J. Appl. Phys. **58** 4670 (1985).
7. H.F. Winters, J. Vac. Sci. Technol. **A3** 700 (1985).
8. W.L. O'Brien, T.N. Rhodin and L.C. Rathbun, J. Chem. Phys. **89** 5264 (1988).
9. S.C. McNevin and G.E. Becker, J. Vac. Sci. Technol. **B3** 485 (1985).
10. I.L. Agufonov, M.U. Zuera, and V.G. Rachkov, Ah. Neorg. Khim. **15** 273 (1970).
11. K.C. Pandy, T. Sakuraiund and H.D. Hagstrum, Phys. Rev. B, **16** 3648 (1977).

PART X

Microelectronics Processing and Technology

SPIN-ON SILICON OXIDE (SOX): DEPOSITION AND PROPERTIES

GERALD SMOLINSKY and VIVIAN RYAN
AT&T Bell Laboratories, Murray Hill, NJ 07974-2070

ABSTRACT

High quality SiO_2 films are obtained by spin-coating wafers with a sol/gel of silicic acid in either a 2, 3, or 4-carbon linear-aliphatic alcohol. Some properties of the deposited film depend upon the solvent: such as density, tensile stress, and infrared spectrum. However, Rutherford-back-scattering analysis indicates the O:Si ratio (2.00 ±0.05) to be independent of the solvent. The infrared spectrum of the oxide exhibits Si–OSi absorption in the range 1070–1080 cm^{-1} depending on the curing temperature and solvent system. (The weaker Si–OSi band is found at 804–810 cm^{-1}) In addition, low-temperature-cured (<500 °C) films show Si–OH absorption. Films hot-plate baked at 150–350 °C are stable but not fully cured. Films from propanol baked at 400 °C have a refractive index of 1.41–1.42 and a wet-etching rate in 30:1 BOE of ~1250 Å/min. Films cured at 900 °C have a refractive index of 1.42–1.43, a wet-etching rate of ~430 Å/min, and are more dense by a factor of ~1.25. Dry-etching with CHF_3/O_2 occurs at rates comparable to those of CVD oxides. Multiple applications lead to crack-free films as thick as 0.6–0.8 µm. Deposition over aluminum-patterned topography results in a smoothing of the surface and suppression of hillock growth in the aluminum even after a 450 °C cure. SOX adheres to silicon, aluminum, and silicon dioxide. A boron-doped SOX is readily prepared.

INTRODUCTION

The search for a spin-on SiO_2 process has taken on some aspects of the legendary quest for the Holy Grail. Undoubtedly the discovery of a spin-on oxide possessing the properties of a high-temperature-deposited material would be a very useful addition to the integrated-circuit maker's art. There are several commercially available silicon-containing materials commonly referred to as "spin-on glasses" (SOG). (These materials are sold by several companies, such as Allied Chemical Co., Hitachi Chemical Co., and Ohka America Inc.) They can be grouped as either polysilicates or polysiloxanes: The former are composed primarily of the elements of silicon and oxygen (although some hydrogen may be present in the form of hydroxyl groups), while the later contain alkyl or aryl groups bonded to the silicon. The physical and electrical properties of a polysiloxane SOG are determined by molecular weight, polydispersity, and the amount and nature of the organic moiety (usually methyl or phenyl) bonded to the silicon atom. The physical properties of a polysilicate SOG are a function of how the material is formulated, while the electrical properties are more a function of the curing temperature. Another difference between these two classes of materials is the stability of their respective solutions: polysilicates are formulated as sols with a relatively short (days to weeks) shelf-life, while polysiloxanes are formulated as true solutions with a much longer (months) shelf-life. In this paper we will describe the synthesis and properties of a novel polysilicate spin-on oxide with the acronym of SOX.[1]

MATERIALS AND METHODS

Preparation of SOX solutions

SOX solutions were prepared by dissolution of silicon(IV) acetate in an appropriate alcohol: ethanol, 1-propanol, or 1-butanol. The acetate, a commercially available solid, was used as obtained from the manufacturer; the alcohols were distilled at atmospheric pressure under nitrogen in quartz equipment and stored in plastic bottles. The reagents had minimum contact with metals and glass. The SOX solution had no contact with metals and all reactions and filtering were implemented in plastic ware composed of any of the following materials: polycarbonate (PC), polyethylene (PE), polypropylene (PP), or polytetrafluoroethylene (PTFE).

Reactions were carried out on a scale of about 20 g of acetate; larger amounts were obtained by combining the requisite number of these preparations. A 4-weight-percent silicon-content SOX solution was prepared as follows: Alcohol (33.2 g) was added rapidly to 20.0 g of silicon acetate contained in a wide-mouth, 125-cc-PE bottle. The reactants were stirred with a PTFE coated magnetic stirring bar for several minutes and then filtered. Filtration was accomplished using low-pressure (<5 psi) nitrogen and a series of four filters: a prefilter composed of borosilicate glass followed by three PTFE filters in decreasing pore size of 10, 5 or 3, and 0.2 μm, respectively. Silicone O-rings were used as required. About 24 hours after preparation, DI water was added to the filtrate. The quantity of water was equal to a molar equivalent of silicon in the SOX solution. Brief swirling or shaking dissolved the water. For the case of 1-propanol, this solution was ready for spin-coating wafers 72 hours after dissolution of the silicon acetate in the alcohol. It was usable for an additional 48 hours (i.e., 72 to 120 hours after preparation) and was discarded 120 hours following preparation.

Boron-doped SOX was readily prepared by adding to the prefiltered propanol solution followed by stirring, either 0.05 mole of boron(III) oxide or 0.1 mole of triethyl borate, per mole of acetate.

Coating wafers

Wafers were coated with SOX in an automated, programmable spin-coater equipped with a three-hot-plate wafer track. The SOX solution was transferred through PTFE tubing from its reservoir to the point of application; the propellant was low-pressure helium. A disposable 0.5 μm (PTFE) filter was fitted in-line about 18 inches from the dispensing nozzle. A 0.5–1 mL puddle was dispensed onto the surface of a statically held 4-inch wafer, followed by abrupt acceleration (30000 rpm/min) to 3500 rpm for 25 seconds. Surfscan-defect counts of less than 100 were achieved.

Curing films

The freshly coated wafer was passed down a track of three hot-plates maintained at approximately 75, 100, and 150 °C, respectively. The residence time on each hot plate was 1 minute followed by 15 seconds on a room-temperature plate. All volatile components were expelled during this procedure. However in order to complete the curing process, the wafer was transferred to a 400 °C, oxygen-ambient furnace for 2 hours. As many as three successive SOX coatings were applied to unpatterned wafers using this curing cycle. Two coatings were applied to aluminum-patterned wafers without apparent deleterious effects such as cracking, delamination, or hillock growth.

SOX films deposited on patterned silicon or thermal oxide were successfully cured for 30 minutes at 900 °C and coated again; as many as four succesive layers were applied in this way. In the case of patterned wafers, each coating was fully cured through the 900 °C stage before application of a subsequent coating. However for unpatterned wafers, each coating was cured only through the 400 °C stage before the next coating; only after the final coating was the composite heated to 900 °C.

RESULTS AND DISCUSSION

Dissolution of silicon(IV) acetate in an alcohol leads to a chemical reaction:

$$Si(OAc)_4 + 4ROH \rightarrow 4ROAc + Si(OH)_4. \qquad (1)$$

The silicic acid thus formed spontaneously reacts with itself to extrude a molecule of water:

$$Si(OH)_4 + Si(OH)_4 \rightarrow H_2O + (HO)_3Si-O-Si(OH)_3 \rightarrow etc. \qquad (2)$$

The reaction continues in this vein to produce a three-dimensional polymeric silicon-oxygen bonded cluster (sol) having Si–OH groups on its surface. The rates of the alcoholysis step and subsequent polymerization depend upon the nature of the alcohol, that is, the number of carbon atoms and degree of substitution of the hydroxyl-bearing carbon atom. Thus the alcohol affects the shelf-life of the sol and the physical properties of the oxide film. No matter what the alcohol, all sols eventually gel. Some properties of SOX films formulated in different alcohols are given in TABLE I.

TABLE I

SOME PHYSICAL PROPERTIES OF CURED SOX FILMS FROM DIFFERENT ALCOHOLS:
3% ETHANOL (BORON-DOPED), 4% PROPANOL, & 4% BUTANOL

FILM PROPERTIES	Ethanol 400 °C-cured	Propanol 400 °C-cured	Propanol 900 °C-cured	Butanol 400 °C-cured
Thickness, 3500 rpm, Å: on Si	~1900	~2100	~1900	~2350
on Al	—	~1600	—	—
from 3% soln.	~1900	~1600	—	—
Index of Refraction	1.42	1.41–1.42	1.42–1.44	1.42
RBS Analysis: O/Si ratio, ±0.05	2.07	2.02	1.98	—
density (SOX:thermal oxide)	0.87	0.77	0.95	—
Wet-Etching Rate (30:1 BOE)	~550	~1250	~430	~1900
Tensile Stress, 10^9 dyn cm^{-2}	1.47	0.4	—	0.94

SOX from propanol

At 3500 rpm, a propanol solution (4 wt% Si) deposits a uniform visually unblemished coating over an unpatterned 4-inch wafer. Scanning electron microscopy (SEM) evinces a smooth featureless surface. The films adhere strongly to silicon, silicon dioxide, and aluminum. When the solution is dispensed on an aluminum-patterned wafer, the trenches between narrow (<2 μm) lines and spaces are coated to a greater depth than the top of the lines. This results in some degree of planarization of the topography. Indeed, two applications produce a quite smooth surface with SEM showing no cracks in the SOX. (This is true for 0.5 μm high aluminum features with as much as 0.4 μm thick SOX in the trenches.) A SOX-coated aluminum feature can be heated for one hour at 450 °C without the formation of hillocks in the aluminum.

Transmission electron microscopy (TEM) cross-sectional micrographs of 400 °C-cured material reveal some disparities: At a magnification of 1.27K, the film appears to be stratified into two layers of equal thickness, with the more dense layer at the substrate interface. In addition, "voids" with an approximate diameter of 50 Å occupy approximately 35% of the volume of the upper, less dense, layer. These supposed voids in fact may be regions of considerably lower density material. We believe this sponge-like microstructure can account for the high degree of water absorption found for SOX. (A SOX film absorbs 4–6 wt% water.)

The infrared spectrum (500–4000 cm^{-1}) of 400 °C-cured SOX(propanol) exhibits essentially four absorption bands: a broad, varying intensity band in the O–H region attributed to absorbed water; a strong and much weaker band at ~1070 and ~805 cm^{-1}, respectively, due to Si–OSi; and a weak band at ~936 cm^{-1} due to Si–OH.[2] After a 900 °C anneal in oxygen, the latter band vanishes and the strong Si–OSi absorption shifts to ~1080 cm,$^{-1}$ which is where this band is found in thermal oxide.

The wet-etching rate of 400 °C-cured oxide is 5–6 times that of thermal oxide, while that of 900 °C-cured material is only ~2 times faster. The decrease in the rate is a reflection of the increase in the density of the high-temperature-cured material. High-temperature-cured boron-doped SOX has a density that is almost that of thermal oxide and a wet-etching rate that is less than that of thermal oxide. Dry-etching rates of 400 and 900 °C-cured SOX, and thermal oxide are all about the same.

A mass spectrometric analysis was made of the products evolved on heating a SOX solution prepared in propanol. A liquid film of this solution was heated from room temperature to 500 °C. Propanol, propyl acetate, acetic acid, and water were detected. Propanol was expected; propyl acetate was one of the products from the reaction of silicon acetate and propanol (EQ. 1); water was the product of the polymerization of silicic acid (EQ. 2). Acetic acid resulted from the hydrolysis of the propyl acetate in the low (~3) pH solution.

SOX from butanol

SOX formulated in butanol is less stable toward gelling than that formulated in propanol, and thus the former solution has a shorter shelf-life. Furthermore, a SOX film deposited from butanol etches at almost twice the rate as one cast from propanol. However, the dry-etching properties are about the same. (In general, dry-etching is a far less discriminating procedure than wet-etching because of the preponderant physical rather than chemical nature of the process.) The infrared spectrum of SOX(butanol) is very similar to that of SOX(propanol). Although these films have no striations or bubbles, the shorter shelf-life and less dense oxide obtained from butanol make this solvent system less desirable than propanol.

SOX from ethanol

SOX formulated in ethanol (3–4 wt% Si) gels only after a period of many weeks and therefore has a very long shelf-life. Concurrently, ethanol solutions require exceedingly long periods to "ripen" and become usable for depositing oxide films. The addition of 5–10 mole-percent of hydrofluoric acid greatly accelerates the ripening process and dopes the film[3] with fluorine (F-SOX). Doping with boron (B-SOX; i.e., adding (EtO)$_3$B or B$_2$O$_3$) accelerates the ripening time as well as influences the properties of the resulting oxide film: such as density, wet-etching rate, stress, and infrared spectrum. The index of refraction of SOX films is essentially independent of the solvent system or dopant, but is sensitive to the curing temperature. However, the density of the films is a function of both dopant and solvent. Doping with boron makes a more pronounced difference than changing the alcohol from propanol to ethanol. (Interstingly, the density of boron-doped SOX prepared in propanol is essentially unchanged from that of the undoped SOX.)

Although F-SOX(ethanol) is only slightly more dense than SOX(propanol), its stress is twice, while its wet-etching rate is only ~75% that of the latter. Doping with boron results in an even greater increase in density and stress and a concomitant marked decrease in the wet-etching rates. Whereas the increased density and accompanying decreased etching rate of SOX from ethanol are desirable, the price paid for these salutary changes in properties is increased tensile stress, a greater tendency to crack over aluminum features, and the formation of cavities and bubbles in the film. A further point of interest is the effect of curing time on the wet-etching rate of B-SOX. After only a two hour cure, the rate is 720 Å/min but curing for ~18 hours causes the rate to drop to 550 Å/min. A thermal gravimetric analysis of boron-doped oxide films cast from ethanol sols, showed that ~92% of the weight loss occurred within the first 15 minutes at 400 °C. Approximately 1% additional loss occurred during the next 90 minutes; but 12 more hours of heating was required to drive-out the remaining volatile components. In contrast, undoped SOX from propanol undergoes 100% of its weight loss within ~30 minutes at 400 °C. For the boron-doped films, an explanation for the slow evolution of volatile components may be associated with the higher density of this oxide. This argument is consistent with the observed decrease in the wet-etching rate of B-SOX after prolonged heating.

A mass spectrometric analysis was made of the products evolved on heating a SOX solution prepared in ethanol with the addition of HF. A liquid film of this solution was heated from room temperature to 500 °C. Ethanol, ethyl acetate, acetic acid, water (see EQ. 1 & 2), and above 400 °C hydrogen fluoride were detected. (The slow evolution of volatile components is probably associated with the release of HF.) It is conjectured that fluorine is present in F-SOX as Si–F and that at elevated temperatures a reaction occurs with Si–OH groups:

$$Si–F + Si–OH \rightarrow Si–O–Si + HF. \qquad (3)$$

This conjecture is consistent with the absence of Si–OH absorbtion in the infrared spectrum of F-SOX. It is also consistent with a SIMS analysis which indicated that fluorine is expelled from the SOX on prolonged heating: Fluorine is uniformly detected (~10^{18} atoms/cc) throughout a film cured at 400 °C for two hours, but is not detected (<10^{16} atoms/cc) after curing for an additional 18 hours.

Besides absorption due to water, the infrared spectra of SOX films from ethanol exhibit absorption attributable[2] to Si–OSi bonding at ~1072 (vs) and ~808 cm^{-1} (w). The boron-doped film also exhibits a band at ~1390 cm^{-1} attributable to the B–O stretching mode as well as a band at ~916 cm^{-1} due to Si–OB bonding.[2] SOX deposited from propanol and cured at 400 °C always has an absorption band at ~936 cm^{-1} due to Si–OH; curing at 900 °C results in the absence of this band. Fluorine and/or boron doped SOX also lacks Si–OH absorption.[3]

SOX from ethanol-propanol

The rate of ripening of a SOX solution in ethanol-propanol is a function of the percent ethanol. In the two component alcohol solutions, the differential light scattering pattern indicates a much more highly polydisperse sol than in solutions of either neat alcohol.[4] SOX prepared in 100% ethanol ripens on the order of 50 days; SOX in 100% propanol ripens in ~5 days. In 50% ethanol-propanol, ripening occurs in about 10–12 days. In 25% ethanol, ripening requires only about a day; in 5% ethanol, in less than a day! Why the non-linear behavior?

The following model is proposed as an explanation: The conversion of (AcO)$_4$Si to SOX in a solution of alcohols is a two stage process. Keeping in mind that the alcoholysis of silicon(IV) acetate is a multi-step reaction, the over-all reaction for the first stage can be summed-up by EQ. 1. The second stage is the polymerization of the silicic acid, Si(OH)$_4$, and the growth of the sol, EQ. 2. The rate of the first stage is a function of the reactivity of the alcohol toward the acetate, while the rate of the second stage is a function of how good a solvent the alcohol is for the ripening sol. If ethanol solvolyzes the acetate more rapidly than propanol, and if ethanol is a far better solvent than propanol, then it follows that in ethanol the first stage is very much more rapid than the second stage. In propanol (and especially butanol) the opposite is likely to be true.

The time to ripening in the ethanol-propanol mixtures now can be understood: In 100% ethanol, the first stage is fast but the second stage, the ripening and shelf-life of the sol, is quite slow because ethanol is a very good solvent. In 100% propanol, the first stage is slower than in neat ethanol but the second stage is much faster because propanol is not as good a solvent. In 50% ethanol, the first stage, the alcoholysis of the acetate, is probably as fast as in neat ethanol and the second stage also is fast because the "goodness" of 1:1 ethanol-propanol as a solvent for the sol is compromised. In both 5 and 25% ethanol, there is sufficient ethanol present for a relatively fast first stage reaction but the quality of the solvents is so much more like that of neat propanol that the second stage of the process is accelerated.

The exact property that makes a solvent "good" is elusive. In the case of alcohols it seems to be associated with the length of the hydrocarbon chain: Ethanol is a better solvent than propanol which in turn is better than butanol. From this result it would appear that less polar (i.e., more hydrocarbon-like) solvents will not be as good as more polar (i.e., less hydrocarbon-like) solvents. Oxygenated compounds such as ketones and cyclic ethers may prove to be good solvents. In order to extend the shelf-life of the sol in propanol, an additive is required that improves the solvent quality of the propanol and yet does not appreciably effect the rate of alcoholysis of the acetate.

CONCLUSIONS

Dissolution of silicon(IV) acetate in a low-molecular-weight linear alcohol results in the formation of silicic acid. The latter condenses to produce a sol that gels to a film upon spin coating on wafers. Curing in oxygen at temperatures >400 °C, converts this film to a silicon oxide. The higher the curing temperature the more SiO_2-like the film. Best results are obtained with 1-propanol, although ethanol and 1-butanol can be used. Films from butanol are less dense and wet-etch faster than films from propanol. Films from ethanol contain cracks, cavities, and surface bubbles. Boron doping of the SOX lowers the wet-etching rate. Boron doped SOX cured at 900 °C exhibits properties similar to those of thermally grown oxide.

REFERENCES

[1] Some alternative methods of preparing silicate sol/gel solutions: a) R.B. Pettit and C.J. Brinker, *Solar Energy Mater.*, **14**, 269 (1986); b) B.E. Yoldas, *J. Polym. Sci., Polym. Chem.*, **24**, 3475 (1986).

[2] J. Wong, *J. Electronic Mater.*, **5**, 113 (1976).

[3] a) E.M. Rabinovich and D.L. Wood, in "Better Ceramics Through Chemistry II," edited by C.J. Brinker, D.E. Clark, and D.R. Ulrich (*Mater. Res. Soc. Proc.*, **60**, Pittsburgh, PA 1986) p. 251. b) D.L. Wood and E.M. Rabinovich, *J. Non-Cryst. Solids*, **82**, 171 (1986).

[4] V. Ryan and G. Smolinsky, in "Chemical Perspectives of Microelectronic Materials," edited by M.E. Gross, J.T. Yates, Jr., and J. Jasinski (*Mater. Res. Soc. Proc.*, **131**, Pittsburgh, PA 1989) this issue.

SPIN-ON SILICON OXIDE (SOX): PHYSICAL PROPERTIES OF THE SOL

VIVIAN RYAN and GERALD SMOLINSKY
AT&T Bell Laboratories, Murray Hill, NJ 07974-2070

ABSTRACT

This paper describes an analysis of the physical properties of the sol using several complementary light scattering techniques. Polymerization and aggregation kinetics were followed through time-dependent changes in the size, shape, and density of the sol particles. The sol growth rate was controlled by choice of solvent and silicon concentration. Changes in viscosity and pH were small during the reaction period. Three different particle-growth regimes exist in which either the particle density increased, decreased, or remained the same. The addition of boron, hydrofluoric acid, or water accelerated the reaction. The sol experimental data correlate with the density and wet-etching rate of the cured films. After curing, high-density films were obtained from sols with three common characteristics: an average particle diameter >450 Å, a relatively high polydispersity, and a low particle density. These criteria were generally satisfied by solutions one to two days old.

INTRODUCTION

A novel polysilicate spin-on oxide, SOX, has been developed for use as a dielectric layer in the fabrication of integrated circuits.[1] Stock solutions are prepared by dissolving silicon(IV) acetate in an alcohol. After a period of time, the solution forms a sol which can be spin-coated on a patterned wafer. With proper choice of solvent, sol ripening time, and cure treatment, the resulting film can have exemplary properties: a low cure temperature (150–200 °C) but still able to withstand heating to 900 °C, a low wet-etching rate, good mechanical properties, and the ability to smooth underlying topography. In this report, we relate the physical properties of sols to the characteristics of cured films. We employed light scattering methods to monitor details of polymerization and aggregation in sol solutions. These data enabled us to correlate the reaction rate of the silicic acid in solution and the size and shape of the particles in suspension with the wet-etching rates of the films.

LIGHT SCATTERING ANALYSIS

Light incident on a suspension is scattered in a complex angular pattern dependent on many factors. The attenuation in the transmitted light due to this scattering is called the absorbance, A; this is related to the sol turbidity, τ, through the relation, $A = \tau/L$, where L is the path length of the optical cell. For relatively dilute suspensions (i.e., individual particles do not multiply scatter and their positions are statistically independent of each other) of particles that are small compared to the wavelength of the incident light, absorbance measurements can be interpreted using Debye's equation:[2,3]

$$\tau = \frac{32\pi^3}{3\lambda^4} \mu_0^2 \frac{(\mu-\mu_o)^2}{\phi} V \qquad (1)$$

where V is the volume of a solute particle, $\phi = nV$ is the volume of solute per cc of solution, λ is the wavelength of incident light, μ_o and μ are the indexes of refraction of the solvent and solution, respectively, and n is the number of particles per cc. During polymerization and aggregation, the volume of particles increases as the number of particles decreases.

Debye's calculations include expressions which relate the angular intensity pattern of scattered light to the size and shape of the particles. A sphere will scatter light according to the relation:

$$I = \left[\frac{3}{X^3} \left[\sin X - X \cos X \right] \right]^2 \qquad (2)$$

where

$$X = \frac{2\pi D}{\lambda} \sin \frac{\theta}{2}, \qquad (3)$$

I is the intensity of scattered radiation, the angle θ is measured with respect to the forward direction, and D is the diameter of the particle.

In differential light scattering, dissymmetry, $I_\theta / I_{180-\theta}$, is measured. The angular pattern is compared to theoretical curves for scattering particles of different shapes and sizes. Once the shape of the particles is known, a single measurement is used to determine the characteristic dimension. When a distribution of particle sizes is present, the mass-averaged principal dimension is determined by this method.[4] In this study, it is reasonable to assume that Smoluchowski's theory of colloidal flocculation is an appropriate description of sol particle growth.[5] Light scattered by solutions undergoing such polymerization is predicted to vary with time, t, according to: $d\tau/dt \propto C^2$, where C is the weight concentration of the monomer.[6]

EXPERIMENTAL METHODS

Materials

Details of preparation are reported elsewhere.[1] One molar equivalent of water was added to selected stock solutions several hours after dissolution of the acetate because previous experiments had shown that films formed from sols prepared in aqueous alcohol were of poor quality. (Addition of water too late in sol formation made little difference in the quality of the cured films).

Light Scattering Equipment

All light scattering measurements were made on a Brice-Phoenix 2000 Series photometer[7] using a 365 nm Hg-line. Because unpolarized light was used and because the scattering volume is a function of angle, the observed intensities had to be corrected by the factor $\sin\theta/(1 + \cos^2\theta)$. An additional correction had to be made to the intensity of detected light to account for the component of incident light reflected back through the solution and scattered at the reverse angular dependence.[8] Refractive indexes for solutions were determined with a Reichert Mark II Abbe Refractometer having an accuracy of ±0.0001 units. The wet-etching rates for the films were measured in 30:1 BOE using laser interferometry.

RESULTS AND DISCUSSION

Change of Absorbance with Time and Solvent

Figure 1 shows the change of absorbance, A, with time of SOX/alcohol solutions (ethanol, propanol, and butanol) with, and without the addition of water at 24 hr. The reaction rate is given by the increase with time. Polymerization of silicic acid is most rapid in butanol with a rate of 0.11/day. The rates in propanol and

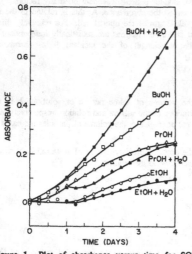
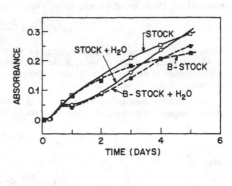
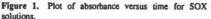

Figure 1. Plot of absorbance versus time for SOX solutions.

Figure 2. Plot of sol absorbance versus solution age for undoped and boron-doped SOX solutions in propanol.

ethanol are similar to each other measuring 0.07 and 0.06/day, respectively. However, ethanol solutions require a 30 hr incubation period before particles of a detectable size appear. The addition of water to the butanol stock results in an increase in rate to 0.20/day. Water added to propanol or ethanol results in slightly altered rates of 0.08 and 0.04/day, respectively, after an initial adjustment period of a few hours. The sol-to-gel transition, which takes place after ~72 hours in butanol stock, is not marked by a sharp change in light scattering. Indeed, the gelled solution continues to react further even after the material no longer flows on a macroscopic scale. Our experiments indicate that SOX sols evolve continuously at rates determined by the solvent. Therefore, the window in time during which the SOX suspension has an optimum size distribution for spin-coating high quality films is controllable by the choice of the proper solvent system. Reaction rates in butanol and ethanol are deemed unacceptably fast and slow, respectively. Hence, propanol is the solvent of choice. Figure 2 shows similar trends are observed for solutions prepared with and without boron, albeit the value of A is generally less with boron. The absorbance, A, decreases by 0.013 during the initial 8-hour period following addition of water to 15-hour-old stock, and then increases more rapidly than no-added-water stock. In all alcohols, doping solutions with fluorine greatly accelerates the reaction rates. For butanol and propanol, shelf life is reduced to ~3 hr. Fluorine-doped ethanol sols have a shelf life of 8 hr., however, the films obtained are of poor quality.[1]

Shape of the Sol Particles

Examples of dissymmetry measurements are shown in Figure 3 for stock solutions in propanol. The lines are the best-fit theoretical curves for a Debye-sphere scatter model. All the sols we investigated exhibit pre-gel dissymmetry values in close agreement with predictions based on the sphere model.

Change in Diameter of Particles with Time

Dissymmetry values (I_{45} / I_{135}) were measured for preparations as a function of solution age. Particle diameters (Figure 4 for propanol solvent) were calculated from the data using Debye's relation for spherical scattering particles (EQ. 2). Sol particles monotonically increase in size with time. Particles formed in the sols containing dissolved B_2O_3 are larger than those formed in the absence of B_2O_3 at a comparable solution age. Within one hour after the addition of water, the diameter grows an additional 20 Å. Henceforth, the mean rates of growth of particle size are greater in solutions to which water had been added: average polymer size after 5 days is 630 Å in the stock solution, 786 Å in the stock + water solution, 693 Å in boron-stock, and 872 Å in boron-stock + water.

Effect of Silicic Acid Concentration

If two solutions at different concentrations follow the same diffusion-limited polymerization process, $d\tau/dt \propto C^2$ predicts that the rates of change of turbidity with time should vary inversely with the square of the

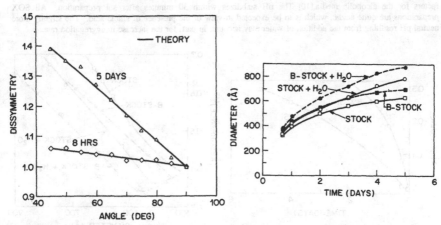

Figure 3. Plot of dissymmetry versus scattering angle for SOX solutions 8 hr. and 5 days after preparation in propanol. The smooth line is the best fit curve to a spherical-scatterer model in the Debye theory.

Figure 4. Plot of calculated particle diameter versus solution age for undoped and boron-doped SOX solutions in propanol.

concentrations. Figure 5 shows there is reasonable agreement between theory and experiment for 8 and 4% silicon solutions for the first few days. The growth rate at later times for the 8% solution probably slows because the reaction entered a chemical-limited region: the acetate consumed most of the available alcohol.

Relationship Between Absorbance and Diameter

Plots of absorbance, A, versus diameter, D, are shown in Figure 6 for propanol solvent. The value of A for the undoped stock solution from day one to three is roughly proportional to D^2, while from day three to five is proportional to D^3. For the doped stock solution, A is consistently proportional to D^3. When water is added to 15-hour-old doped or undoped stock solutions, there is a decrease in A over the next eight hours followed by a five-day period during which A is proportional to D^3.

In order to interpret the meaning of these relationships, it is necessary to measure the index of refraction, μ, of the solutions. Figure 7 shows the mean indexes from triplicate measurements. Little difference is found between doped and undoped solutions. (For propanol and propanol + B_2O_3, μ is the same, 1.3854). In stock solutions, μ decreases slightly during the first two days after preparation, then remains constant at 1.3883. When water is added to a 15-hour-old stock solution, μ increases over the next 5 hours to a plateau value of 1.3916.

From Figures 6 and 7, we can identify three modes of behavior in SOX sols. The first mode is characterized by $A \propto D^3$ and a constant value for μ, this is true for 3–5 day-old stock solutions, 0.5–5 day-old B-doped stock solutions, and 1–5 day-old stock + water solutions (either doped or undoped). In these cases, the particles are increasing in size with little change in ϕ, the volume of solute per cc of solution. These conditions describe simple aggregation in which the physical attraction of the particles cause no additional changes besides coalescence. The second mode occurs in 1–3 day-old stock solutions and is characterized by $A \propto D^2$ and a decreasing value for μ. According to EQ. 1, ϕ must be decreasing significantly such that $(\mu - \mu_o)^2/\phi \propto D^{-1}$. Similar observations have been reported for related sol formulations undergoing silicic acid polymerization[9], and are attributable to desolvation and the formation of water by the condensation of Si–OH groups. In the present work, the increase in particle size is accompanied by densification. The third mode occurs during the 8-hour period following addition of water to 15-hour-old stock solutions and is distinguished by a decreasing value of A and an increasing value of μ. A chemical reaction is strongly implicated because μ is observed to increase when water (which has a lower μ than propanol) is added to the system. From EQ. 1 it is seen that the increase in ϕ must be quite large to result in an overall decrease in A. Hence, the particle density decreases significantly as its size increases.

pH

The pH of solutions was measured to ascertain the effect of electrostatic charge repulsion on sol stability. Standard-pH-scale data for various propanol preparations are shown in Table I and include appropriate scaling factors for the alcoholic media.[10] The pH stabilizes within 30 minutes after sol preparation. All SOX preparations are quite acidic, which is to be expected in view of the presence of silicic acid. The slightly more neutral pH resulting from the addition of water may account, in part, for the increase in aggregation rate.

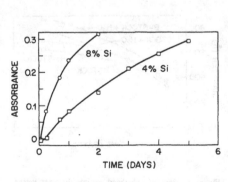

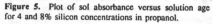

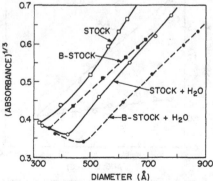

Figure 5. Plot of sol absorbance versus solution age for 4 and 8% silicon concentrations in propanol.

Figure 6. Plot of cube root of absorbance versus particle diameter for undoped and boron-doped SOX solutions in propanol.

TABLE I: pH OF SOL SOLUTIONS

SOLUTION	Stock	Stock + H$_2$O	B-Stock	B-Stock + H$_2$O
pH	2.96	3.24	2.68	3.11

Attempts were made to titrate the sols with KOH in propanol: The solution rapidly gelled when the pH increased by as little as one unit. We interpret this to mean that SOX solutions as formulated are quite self-stabilizing and that any attempt to further reduce the aggregation rate would require addition of a strong acid.

Relationship of Sol Parameters and Wet-Etching Rate

Wet-etching-rate data for films as a function of particle diameter in propanol sols is plotted in Figure 8. The data show that low etching rates correlate with sols containing particles greater than ~450 Å. Accordingly, these suspensions also are undergoing aggregation without densification. Since a low etching rate is a consequence of high film density, we conclude that sols with particles >450 Å in size produce the densest cured films. (The etching rate–film density relationship is bolstered by the observation that SOX films from young solutions shrink to a greater extent, 80% of the original thickness, than films from old solutions, 90%, when annealed at 900 °C). That relatively large-particle size produces higher-density films, can be explained as follows: Larger particles have not pre-densified in solution and thus are able to deform readily and cross link as they impinge on one another during solvent removal. This dependency has been found previously for other related systems.[11] Moreover, a state of high polydispersity accompanies advanced aggregation[5] and results in more efficient packing of the particles.

The improvements in etching rate for formulations with added water can be understood in terms of the absorbance–diameter curves shown in Figure 6. During the 8-hour period after the addition of water, the density of the particles in the sol decreases as their diameter increases; whereas, the density of the particles in the stock solution increases as their diameter increases. Hence 24 hours after preparation, the density of sol particles in solutions with added water is less than that of particles in stock solutions.

Absorbance–diameter data collected soon after preparation of the sol solution gives an estimate for the ratio of particle densities in undoped and boron-doped sols of 1.0:0.7. In addition, Figure 6 shows that boron-doped stock solutions commence particle growth without densification at an earlier age than undoped stock solutions. These data suggest that the lower density of boron-doped sol particles in comparison to undoped particles may account for some of the observed reduction in the etching rate of B-SOX.

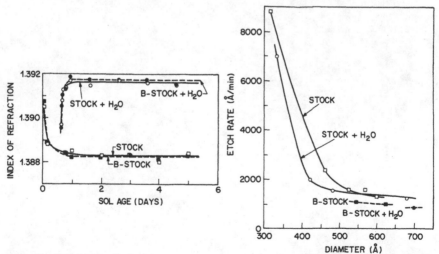

Figure 7. Plot of index of refraction versus solution age for undoped and boron-doped SOX solutions in propanol.

Figure 8. Plot of wet-etching rates for cured SOX films versus sol particle diameter at time of deposition.

574

CONCLUSION

The density of cured SOX films is a function of the physical nature of the sol solution from which it was deposited. Sol particle size, polydispersity, and density change with solution age; polymerization and aggregation are probably largely diffusion controlled. The densest cured films are spin-deposited from boron-doped, water-added sols that exhibit a particle size >500 Å and have a high polydispersity and a low particle density. Sols prepared in propanol retain this desirable state for at least three to four days. The light-scattering technique is a highly useful method to test for optimization of the sol formulation and yet is simple enough to monitor quality control in a manufacturing environment.

REFERENCES

[1]   G. Smolinsky and V. Ryan, in "Chemical Perspectives of Microelectronic Materials," edited by M.E. Gross, J.T. Yates, Jr., and J. Jasinski (*Mater. Res. Soc. Proc.*, **131**, Pittsburgh, PA 1989) this issue.

[2]   P. Debye, *J. Phys. Chem.*, **51**, 18 (1947).

[3]   M. Kerker, *The Scattering of Light and Other Electromagnetic Radiation*; Academic Press: New York, 1969; p. 427.

[4]   B.J. Berne and F. Pecora, *Dynamic Light Scattering*; John Wiley & Sons, Inc.: New York, 1976; p. 174.

[5]   S. Chandrasekhar, *Rev. Modern Phys.*, **15**, No. 1 (1943).

[6]   G. Oster, *J. Colloid Sci.*, **2**, 291 (1947).

[7]   B.A. Brice, M. Halwer, and R. Speiser, *J. Opt. Soc. Am.*, **40**, 768 (1950).

[8]   Y. Tomimatsu and K.J. Palmer, *J. Phys. Chem.*, **67**, 1720 (1963).

[9]   E. Heymann, *Trans. Faraday Soc.*, **32**, 462 (1936).

[10] R.G. Bates, *Determination of pH: Theory and Practice*; John Wiley & Sons, Inc.: New York, 1954; Chapter 8.

[11] C.J. Brinker, K.D. Keefer, D.W. Schaefer, and C.S. Asley, *J. Non-Crystalline Solids*, **48**, 47 (1982).

ELECTRICAL BEHAVIOR OF SPIN-ON SILICON DIOXIDE (SOX) IN A METAL-OXIDE-SEMICONDUCTOR STRUCTURE

N. Lifshitz, G. Smolinsky, and J.M. Andrews
AT&T Bell Laboratories, Murray Hill, NJ 07974-2070

ABSTRACT

We studied the behavior of mobile charges, which we believe are hydrogen ions, in a novel spin-on oxide, SOX. Our analysis is based on the Triangular Voltage Sweep (TVS) technique, which is normally used to detect the presence of mobile alkali ions in silicon dioxides. We found that the TVS traces of protons exhibit several unique features that allow us to distinguish a proton signal from that of other common contaminants, such as sodium. We suggest a model to explain these unique features. We show that the presence of hydrogen in the SOX material is due to the high water retention of this porous oxide.

INTRODUCTION

At the present stage of MOS VLSI integration, one avenue for future miniaturization is a three-dimensional integration, wherein conducting interconnects are formed in more than one level. Isolation of the different metallic interconnect levels is one of the most important problems in VLSI manufacturing. An important constraint on the intermediate dielectric is the limitation on the deposition temperature: the dielectric is deposited on the first layer of aluminum metallization. Therefore, the deposition temperature cannot exceed 450 °C. Another requirement, especially important for submicron design rules, is good step coverage. In this respect, spin-on silicon dioxides appear particularly attractive since they can be used in combination with better quality oxides, such as low-temperature plasma-deposited oxides from tetraethoxysilane (PE-TEOS).

In this report, we discuss the electrical behavior of a novel spin-on oxide SOX in the Metal-Oxide-Semiconductor (MOS) system. The main tool of our study is the Triangular Voltage Sweep (TVS) technique. This method was developed by Kuhn and Silversmith [1] to study ionic contamination in thermally grown SiO_2 of MOS structures. The technique is based on the measurement of the ionic displacement current in an MOS capacitor driven by a slow linear-voltage ramp. Usually the measurements are carried out at elevated temperatures (up to 300 °C), because the mobility of the ions in oxides is sufficiently high. The ionic motion is revealed as a current peak on the displacement-current-versus-voltage curve; the concentration of the mobile charge is calculated from the area under the peak. A typical TVS trace of thermal oxide contaminated with sodium is shown in Fig. 1. While the shape of the peaks obtained during the forward and reverse directions of the sweep and their displacement from zero bias may differ, the area under the peaks, which reflects the total mobile charge, should be the same for both sweep directions.

For the MOS structures used in the present study, the composite dielectric consisted of 500 Å of thermal oxide, 2000 Å of SOX, and 1500 Å of PE-PTEOS (phosphorus-doped PE-TEOS). The thermal oxide prevented ohmic leakage through the structure. The PE-PTEOS layer was added to imitate the actual structure of the intermediate dielectric.

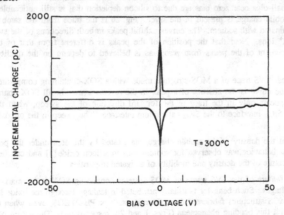

Fig. 1. TVS trace of thermal oxide contaminated with Na.

EXPERIMENTAL

Sample preparation

Four-inch-diameter, epitaxial-(p/p$^+$)-<100>-silicon wafers were cleaned and prepared for TVS measurements as follows: First, a 500 Å oxide layer was grown at 1000 °C in HCl and oxygen. Second, ~2000 Å of SOX was spin-deposited and cured at 400 °C in oxygen. Third, 1500 Å of PE-PTEOS was deposited at 400 °C in a plasma-enhanced-CVD reactor. (Prior to deposition the substrate was held in the evacuated reactor at 400 °C for 20 minutes. The oxide was deposited at a rate of ~200 Å/min.) Finally, 1 μm of aluminum was magnetron sputtered onto the surface of the wafer. The aluminum was patterned into 2.4×10^{-2} cm^2 area electrode dots. The fabricated MOS devices were annealed at 450 °C in hydrogen.

A detailed account of the chemistry of SOX preparation, deposition, and curing is published elsewhere. [2] It is important to note that the density of the cured oxide, as determined by the Rutherford Back Scattering technique, was found to be only 70–80% of that of thermal oxide.

Electrical measurements

The dielectrics were characterized using a digital system that approximated a triangular-ramp voltage with a stepping-ramp voltage. A Keithly Model 617 Programmable Electrometer was connected in the coulomb mode in conjunction with a built-in Programmable Voltage Source. The electrometer and the voltage source were controlled by a Hewlett-Packard Model 9836 Desktop Computer by means of an IEEE-488 Bus Interface. The voltage source was connected to the chuck that held the wafer. *Note that in the following discussion, all references to the sign of the applied bias are made relative to the silicon substrate, rather than to the top electrode.* The device wafer was maintained at a constant elevated temperature with a Thermochuck Model TP36 System manufactured by Temptronic, Inc.

The MOS capacitor was stressed at a negative holding voltage of chosen duration (usually 2 min) and magnitude so as to collect all positive mobile charge in the dielectric at the Si/SiO$_2$ interface. The voltage on the capacitor was then incremented by a selected value, ΔV, (usually 1 V) and maintained for a time, Δt, (usually 2 seconds) sufficient for equilibration of the mobile charge. The Coulomb meter was automatically zeroed prior to every voltage increment. The total displacement charge appearing in the system as a result of the ΔV was measured and the value stored in a two-dimensional array. Thus the data could be plotted later in the form of the dependence of the incremental charge on the applied voltage.

When the voltage reached the holding voltage of opposite sign, the capacitor again was stressed so as to collect the mobile charge at the SiO$_2$/aluminum interface. During this second part of the test, the bias increments had a sign opposite to that of the first part and therefore at the end of the cycle the bias returned to its original value.

RESULTS

Fig. 2 shows a comparison of several typical TVS traces obtained with different samples. Fig. 2a is from a MOS structure that consisted of 3500 Å of PE-PTEOS on 500 Å of thermal oxide. One can see that the trace is virtually flat (the small dips near zero bias are due to silicon depletion that is still noticeable at 300 °C). This indicates that no mobile charge is present in the layer. Fig. 2b is the trace from the same structure that was intentionally contaminated with sodium. The curves exhibit peaks in both directions of the sweep cycle that are characteristic of Na$^+$ ions. Note that the position of the peaks is different from that of thermal oxide (see Fig. 1). The displacement of the peaks from zero bias is believed to depend on the quality of oxide and its interfaces. [1]

Fig. 2c is a typical TVS trace of a MOS structure made with a SOX-containing composite oxide. The most unusual feature of the curve is the absence of a peak in the reverse trace. In all SOX-containing samples we measured, a peak appears only in the trace of the forward direction of the sweep, when the positive charge moves from the Si/SiO$_2$ interface to the SiO$_2$/aluminum interface. The sweep in the opposite direction shows no sign of a peak.

Fig. 3 shows that the density of the mobile charge, manifested by the area under the peak, increases with temperature. Similar behavior was observed for sodium ions in silicon oxide [1] and can be explained by the increase with temperature of the density and mobility of activated impurities.

A characteristic feature of the TVS traces of MOS systems containing SOX is the marked change in slope observed at high voltages. Such behavior is usually attributed to leakage through the oxide layer. [3] However, the TVS traces of MOS structures fabricated from thermal oxide or PE-PTEOS, even when contaminated with sodium, do not exhibit this bending phenomena (Figs. 1 and 2b, respectively). The nature of this current must

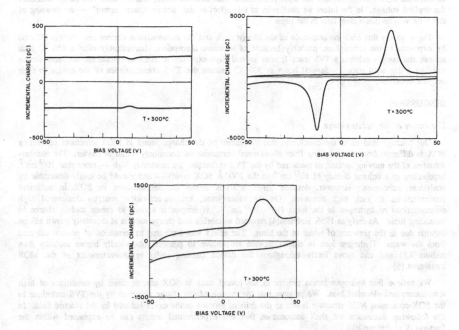

Fig. 2   TVS traces of different composite oxides: a) 3500 Å PE-PTEOS on top of 500 Å thermal SiO_2; b) 3500 Å PE-PTEOS on top of 500 Å thermal SiO_2, intentionally contaminated with Na; c) 1500 Å PE-PTEOS on top of 2000 Å SOX. The SOX layer was spun onto 500 Å of thermal oxide.

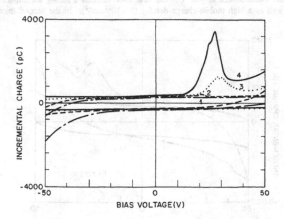

Fig. 3   TVS traces of the composite oxide containing SOX at different temperatures.

be different from leakage because the SOX is deposited over thermal oxide, and the latter can easily withstand the applied voltage. In the future we shall refer to this effect as the "accumulation current" — the meaning of this term will be made clear later in the paper.

Fig. 4 shows that both the amplitude of the charge peak and the accumulation current are strongly affected by exposure to the atmosphere, probably because of moisture absorption: Immediately after a 450 °C final anneal, the sample exhibits a TVS trace 1; after several days exposure to ambient conditions, the same device exhibits a degraded trace 2. Several hours at 300 °C restores the TVS characteristics of the device to their original condition.

DISCUSSION

The nature of the mobile charge

Our findings lead us to the conclusion that the nature of the charge found in MOS structures containing SOX is different from that resulting from alkali-metal contaminants commonly found in oxides. The number-densities of the moving species, as measured by the TVS technique, are extremely high — more than 10^{12} cm^{-2} (equivalent to a volume density of 10^{17} cm^{-3} in the 2000 Å SOX layer) — and should be easily detectable by analytical techniques. However, detailed SIMS analysis found no contaminants in SOX in sufficient concentrations at such high densities. On the other hand, neutron-activation analysis discovered high concentrations of hydrogen in the films (10^{21}–10^{22} cm^{-3}). Hydrogen is known to create mobile charge in insulating films. As early as 1966, Hofstein[4] reported instability and charge motion in thermally grown silicon dioxide due to the presence of water in the films. The effect was attributed to formation of protons released from the water. Hydrogen ions in silicon dioxide are known to possess substantially higher mobility than sodium[4,5] and can move freely throughout the device causing shifts in characteristics of the MOS transistors.[6]

We believe that hydrogen atoms present in the bound state in SOX can be freed by conditions of high temperature and electrical bias. We propose that the motion of the charge observed by the TVS technique in the SOX-containing MOS structure is due to the drift of protons in the oxide induced by the electric field. In the following discussion we shall demonstrate that our experimental results can be explained within the framework of this model.

Association of the asymmetric peaks with the accumulation current

In the course of our experiments we discovered that the two peculiar features of the SOX-TVS spectra — the asymmetric mobile charge signal and the accumulation current — are correlated in their intensity. For example, Fig. 5 shows several sequential TVS traces obtained on the same sample. Between measurements, the sample was left on the Thermochuck at 300 °C. The first trace exhibits a strong accumulation current (above 10^{-8} A at -50 V) as well as a high mobile-charge density (9×10^{12} cm^{-2}). In the second trace, made about 25

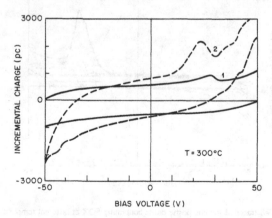

Fig. 4 Effect of moisture on the TVS traces. Curve 1 was taken immediately after the final 450°C anneal. Curve 2 was obtained on the same device several days later.

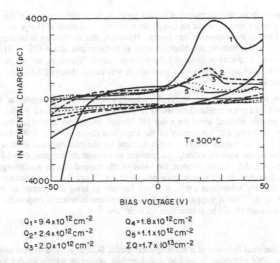

$Q_1 = 9.4 \times 10^{12} \, cm^{-2}$ $Q_4 = 1.8 \times 10^{12} \, cm^{-2}$
$Q_2 = 2.4 \times 10^{12} \, cm^{-2}$ $Q_5 = 1.1 \times 10^{12} \, cm^{-2}$
$Q_3 = 2.0 \times 10^{12} \, cm^{-2}$ $\Sigma Q = 1.7 \times 10^{13} \, cm^{-2}$

Fig. 5   Effect of 300 °C annealing on the TVS traces. Prior to the experiments the device was exposed to atmosphere for several weeks. The second trace was obtained immediately after the first, and the third 3 hrs after the second. The forth run was conducted next morning. Between the runs the wafer was kept on the chuck at 300 °C.

minutes after the first, both effects are considerably reduced. In the third trace, measured about 3 hours later, the changes are less dramatic. The last trace, obtained next morning, still is distinguishable from the preceding one.

We suggest the following mechanism to explain the correlation: Proton charge is formed and displaced not only during the application of the holding voltage, but throughout the entire cycle. The constant motion of this charge under the applied field is reflected in the slope of the curves. At high electric fields at or near the holding voltage, the rate of charge formation (rate of proton dissociation) is enhanced, causing the sharp downward sloping of the curve. This is in contrast to the case of sodium contamination, where all mobile ions present in the oxide are pushed to the interface by the holding voltage and no new charge is formed during the sweep, resulting in an absolutely horizontal trace. The large proton current flowing to the Si/SiO_2 interface leads to proton collection at the interface. This, in turn, produces a strong mobile-ion signal when the bias changes sign and protons move toward the opposite interface. At strong electric fields of opposite sign, the proton current flows to the SiO_2/aluminum interface, resulting in the sharp upward sloping of the curve.

Several other facts implicate the involvement of water in the charge generation: For example, the increase in the accumulation current and charge motion after prolonged storage at ambient conditions can be explained by moisture absorption and diffusion. Similarly, the strong decrease in the intensity of both these effects after a 300 °C anneal can be explained by dehydration of the device structure.

The large water-retention capacity of the SOX material is consistent with its low density, only about 70–80% that of thermal oxide. We believe that the peculiar behavior of SOX in the MOS system is mainly due to its porosity. Indeed, when SOX is densified at 900 °C all the phenomena which we ascribe to proton motion disappear. This observation emphasizes the connection between the electrical and structural properties of oxides.

The unidirectional motion of the charge

If one accepts the idea that the mobile charge introduced in the SOX-MOS structures is due to protons originating from absorbed water, still unexplained is the asymmetry of the current peaks with the direction of the voltage sweep. We suggest that this observed behavior is a result of the asymmetry of the MOS system itself, that is, the SiO_2/aluminum interface interacts with protons differently then the Si/SiO_2 interface.

580

One facile explanation for the asymmetry is the discharge of the arriving protons at the aluminum electrode. Since neutralized hydrogen atoms possess no charge, they will not contribute to charge motion and thus will not be detected in the trace of the reverse-voltage sweep. However, this mechanism is unlikely because aluminum as well as silicon electrodes become non-blocking only at temperatures above 400 °C. [1,4] Besides, like any electrochemical process, the discharge is usually thermally activated. Therefore, we would expect to see a peak in the reverse trace at lower temperatures — a peak which was never observed, however, even at temperatures as low as 200 °C (see Fig. 3).

Another possible mechanism for the observed asymmetry is trapping of protons at the SiO_2 / aluminum interface. In his work on the water-related instability in water-contaminated thermal oxides, Hofstein observed a strong trapping of protons released from the water. [4] Though he did not determine the exact nature of the trapping mechanism, he found that the density of the traps was above 10^{13} cm.$^{-2}$ We believe that a trapping phenomenon at the SiO_2 / aluminum interface might account for the unidirectional motion of the protons, and adopt this mechanism as our working model. All attempts to exhaust the trapping sites by repeated cycling of the voltage were unsuccessful. In one attempt, the value of the trapped charge was brought to 1.7×10^{13} cm^{-2} (see Fig. 5), and still no peak in the reverse trace was observed. Since the charge density, both displaced and trapped, decreases in every subsequent cycle, it is not feasible to bring the density of trapped protons to a substantially higher value. Thus if trapping at the SiO_2 / aluminum interface is responsible for the asymmetry, then the density of the traps should be greater then 1.7×10^{13} cm.$^{-2}$

SUMMARY

We studied the electrical behavior of a novel spin-on oxide, SOX in a MOS system using a modification of the TVS technique. This technique is used to detect mobile charge in silicon oxides by the appearance of characteristic peaks in the charge–voltage curves. We detected a massive motion of positive charges in the material. The density of the charge approaches 10^{17} cm.$^{-3}$ The TVS traces, however, exhibit a peculiar behavior: their shape indicates that the charge moves only in one direction, namely, from the Si / SiO_2 interface to the SiO_2 / aluminum interface. No charge motion in the opposite direction was ever detected. Another unusual feature of the MOS system containing SOX is the so-called accumulation current, which depends on the history of the device. For example, after long storage both the amplitude of the asymmetric peaks and the accumulation current are extremely high. On the other hand, when the same sample is subjected to prolonged annealing at 300 °C, the intensity of both effects decreases by several orders of magnitude.

We believe that the unique shape of the TVS traces is a signature of hydrogen ions that originate from absorbed water and move in the oxide under the influence of an electric field. We come to this conclusion on the basis of the following considerations: 1) Sensitive analytical methods found no metallic impurities present in the material in sufficient concentrations to produce such high quantities of the mobile charge. 2) A high concentration of hydrogen (10^{21}–10^{22} cm^{-3}) was detected in SOX by the neutron activation technique. 3) Many experimental facts are explained by our model. For example, the increase of the mobile-charge density after prolonged exposure to the atmosphere and, conversely, its decrease after a 300 °C anneal. 4) In the framework of our model, the unidirectional motion of the charge can be understood by invoking a proton-trapping mechanism at the SiO_2 / aluminum interface. We found that the density of the traps should exceed 1.7×10^{13} cm.$^{-2}$

We believe that all the unusual features of electrical behavior of SOX are due to high water retention that is facilitated by the porosity of the material. Since hydrogen may cause instabilities in fine-line MOS devices, it is important to have a technique for evaluating the hydrogen content in microelectronic materials. Unique features of the proton-TVS traces suggest this technique as an easy method of detecting hydrogen in electronic dielectrics.

REFERENCES

1. M. Kuhn and D. J. Silversmith, *J. Electrochem. Soc.*, **118**, 966 (1971).

2. G. Smolinsky and V. Ryan, in "Chemical Perspectives of Microelectronic Materials," edited by M.E. Gross, J.T. Yates, Jr., and J. Jasinski (*Mater. Res. Soc. Proc.* **131**, Pittsburgh, PA 1989) this issue.

3. E. H. Nicollian and J R. Brews, *MOS Physics and Technology*, (Wiley Interscience Publication, New York, 1982), p 597.

4. S. R. Hofstein *IEEE Trans. Electron Devices*, ED-13, 222 (1966).

5. G. Hetherington, K. H. Jack, and M. W. Ramsay, *J. Phys. Chem. Glasses*, **6**, 6 (1965).

6. R. C. Sun, J. T. Clemens and J. T. Nelson, 18th Annual Proceedings, Reliability Physics 1980, p. 244

UV EXCIMER LASER INDUCED PRENUCLEATION OF SURFACES FROM SPIN-ON FILMS FOLLOWED BY ELECTROLESS METALLIZATION

Hilmar Esrom, Georg Wahl and Michael Stuke*
Asea Brown Boveri, CRH, D-6900 Heidelberg, FRG
*Max-Planck-Institut f. biophys. Chemie, D-3400 Göttingen, FRG

ABSTRACT

UV excimer laser light-induced prenucleation of various surfaces from spin-on films in air is described. Good results were obtained using palladium acetate on substrate materials ranging from ceramics to polymers and even paper. Virtually any surface material can be prenucleated in this way. The Pd-prenucleated planar and curved surfaces can be metallized by a vast variety of metals using electroless plating solutions (Cu, Ni, Au, Pt, Co, Sn, and even alloys). Results are described in detail for copper from a standard electroless solution (Doduco or Shipley). Contact metallization through via holes using this process was achieved with excellent contact properties.

INTRODUCTION

In recent years, there has been considerable work on direct deposition of materials by laser driven processes. These methods usually involve laser-induced pyrolytic and photolytic deposition of films from gas phase, liquids and thin metallorganic films[1]. Characteristically of all these direct deposition methods is in most cases the low deposition rate, relative high impurities in the films and in individual cases the self-limiting of film thickness[2]. To overcome these problems, we developed a two-step hybrid process. The first step is the UV photo-induced deposition and patterning of a thin catalyst on substrate surfaces with excimer lasers followed by the second step, the usage of the activated surfaces for the selective electroless metal deposition. The prenucleation of substrate surfaces can be achieved by laser decomposition of thin metallorganic or inorganic films on non active surfaces[3]. The decomposition process in the solid phase has similarities to chemical vapor deposition (CVD) processes, however with the advantages of increased safety and ease of handling. The decomposition of the films can be performed in air, and no specialized vacuum equipments are needed.
Recently, R.C. Sausa et al.[4] reported the pyrolytic Ar+ laser decomposition of platinum metallorganic films and H.S. Cole et al.[5] reported the photolytic Ar+ laser decomposition of palladium acetate films for electroless copper plating.
In this paper, we report parts of our investigations[6] concerning the deposition of structured palladium catalysts by means of excimer laser photolytic decomposition of thin palladium acetate films on rough ceramic surfaces for electroless copper plating.

EXPERIMENT

A commercial excimer laser (Lambda Physik, model EMG 103) was used for the UV irradiation of the spin-on Pd acetate films. The intensity distribution across the rectangular beam profile (20×6 mm^2) was inhomogeneous and was therefore apertured with slits to minimize the spatial intensity inhomogenity. The laser pulse energy was measured with a joulemeter (Gen Tec PRJ-D, ED 500). We have studied the decomposition of the palladium acetate films by projection patterning and printing with contact masks. Without special cleaning or pretreatment the films were spin coated onto Al$_2$O$_3$ substrates. A 0.47×10^{-3} mole/l palladium acetate solution was used leading to ~0.2 µm thick Pd acetate films on the substrate surfaces. After laser exposure, the unexposed palladium acetate film was removed by washing in a chloroform solution. The optical transmittance of the palladium acetate film on a quartz substrate was measured with a Perkin Elmer UV Spectrophotometer 550. The film thickness and morphology were determined by scanning electron microscopy (JEOL JXA-733 ELECTRON PROBE). The adhesion was measured with a Sebastian Adherence Tester(QUAD-Group).

RESULTS

Fig. 1 gives the measured absorption spectrum of a 0.06 µm thin palladium acetate film showing a distinct absorption maximum at about 250 nm. The absorption coefficient for KrF laser light (248 nm) is ~45000 cm^{-1}. The photon penetration depth τ is estimated to be 0.15 µm for KrF and 1.5 µm for XeF light (351 nm) because of the lower absorption coefficient. To obtain good adhesion of the deposited films it is important that the light penetrate through the film to the film-substrate interface. Then the light can interact with the substrate surface to yield catalysts with good adhesion. Therefore, in the case of KrF light, the thickness of the palladium acetate film should not be greater than 0.2 µm.

The photolytic decomposition of a thin palladium acetate film on glass with excimer laser light leads to discontinous films of nuclei with average diameters of 0.1 µm. Since discontinuous clusters of palladium are more catalytic than continuous palladium deposits[7], the excimer laser-induced palladium films could be plated with copper easily with commercially available copper plating solutions (e.g. Shipley CP-78). The copper plating rate was determined to be 0.1 µm/min. For special applications many other metals like nickel, gold, platinum, palladium, cobalt, zinc, tin, aluminum, etc. can be plated from suitable electroless solutions. In the following, we give some results of the hybrid deposition method.

The edge quality of the photo-induced decomposition of the palladium acetate was tested with the line focus projection method. The line focus (16mm×150µm) of the KrF laser light (248 nm) was achieved using a cylindrical lens (f=156 mm). In this experiment 30 pulses were chosen at a pulse energy of 2.3 mJ (rate 1 Hz). After removing the unexposed Pd acetate in CCl$_4$, the palladium stripe was copper plated for about 1 min.

Fig. 2a shows a copper stripe on Al$_2$O$_3$. The Cu thickness is about 0.1 µm. Remarkable is the sharp edge without any palladium particles in the surrounding areas that are often present with lines written using an Ar$^+$ laser.

UV-spectrum of palladium acetate

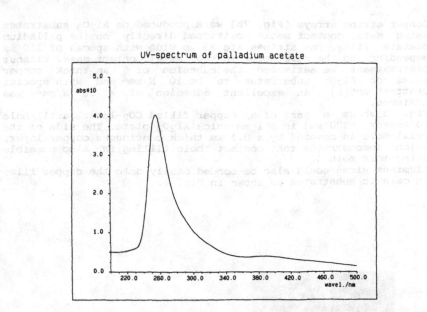

Fig. 1 Absorbance of a palladium acetate film in the
UV region

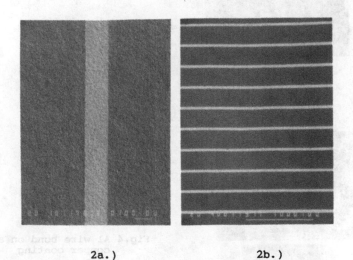

2a.)   2b.)

Fig. 2 Copper lines after electroless plating of excimer
laser-induced Pd stripes produced by using 2a)
the line focus and 2b) a contact mask
λ=248nm

Copper stripe arrays (Fig. 2b) were produced on Al_2O_3 substrates using metal contact masks positioned directly on the palladium acetate films. The stripes are 35 µm wide with spaces of 210 µm depending on the structures in the metal contact mask. Without pretreatment we estimated the adhesion of 6 µm thick copper films on Al_2O_3 substrates to be 10 N/mm^2 and with special pretreatment[6] an excellent adhesion of > 30 N/mm^2 was achieved.

Fig. 3 shows a part of a copper filled CO_2-laser drill-hole (diameter ≈300 µm) in a 1 mm thick Al_2O_3 plate. The side of the drill-hole is covered by a 0.1 µm thick continuous copper layer, which demonstrates that contact hole filling is also possible using this method.

Aluminum wires could also be bonded easily onto the copper films on ceramic substrates as shown in Fig. 4.

Fig. 3 Copper filled drill-hole
in ceramic substrates

Fig.4 Al wire bond on a
copper coating

Further experiments have shown, that various substrate surfaces like AlN, glass, polyimide, teflon, glaced ceramics and also paper can be structured by this method.

ACKNOWLEDGEMENT

The authors would like to thank Dr. J.Demny and Uta Feller for SEM and XPS investigations and Franz Schmaderer for helpful discussions. This work was partly financed by the Bundesministerium für Forschung und Technologie[Nr.13 N 5397/6]

REFERENCES

1. D.Bäuerle, Chemical Processing with Lasers, Vol.1, Springer Series in Materials Science (Springer, Berlin, 1986)

2. T.H.Baum, E.E.Marinero, and C.R.Jones, Appl.Phys.Lett.49(18), 1213(1986)

3. G.J.Fisanick, M.E.Gross, J.B.Hopkins, M.D.Fennell, K.J.Schnoes, and A.Katzir, J.Appl.Phys.57(4),1139(1985)

   G.J.Fisanick, J.B.Hopkins, M.E.Gross, M.D.Fennell, and K.J.Schnoes, Appl.Phys.Lett.46(12),1184(1985)

   M.E.Gross, G.J.Fisanick, P.K.Gallagher, K.J.Schnoes, and M.D.Fennell, Appl.Phys.Lett.47(9),923(1985)

   A.Auerbach, Appl.Phys.Lett.47(7),669(1985)

   M.E.Gross, A.Appelbaum, and K.J.Schnoes, J.Appl.Phys.60(2), 529(1986)

   M.E.Gross, A.Appelbaum, and P.K.Gallagher, J.Appl.Phys.61(4), 1628(1987)

   A.Gupta and R.Jagannathan, Appl.Phys.Lett.51(26),2254(1987)

4. R.C.Sausa, A.Gupta, and J.R.White, J.Electrochem.Soc.,Vol.134,No.11,2707(1987)

5. H.S.Cole, Y.S.Liu, J.W.Rose, R.Guida, L.M.Levinson, and H.R. Philipp, in Laser Processes For Microelectronic Applications, Vol.88-10 of the Proceedings of the Electrochemical Society, edited by J.J.Ritsko, D.J.Ehrlich, and M.Kashiwagi (Pennington,NJ,1988)

6. H.Esrom, G.Wahl, and M.Stuke, to be published

7. M.Paunovic, in Electroless Deposition of Metals and Alloys, Vol.88-12 of the Proceedings of the Electrochemical Society, edited by M.Paunovic and I.Ohno (Pennington,NJ,1988)

CATALYSIS OF ELECTROLESS COPPER PLATING USING Pd^{+2}/POLY(ACRYLIC ACID) THIN FILMS

ROBERT L. JACKSON
IBM Almaden Research Center, San Jose, California 95120-6099

ABSTRACT

A new method is described for initiating electroless copper deposition onto dielectric substrates. The method employs a thin poly(acrylic acid) film, applied to the substrate by dip-coating, to bind Pd^{+2} from PdSO$_4$ solution. Uptake of Pd^{+2} takes place by H$^+$/Pd^{+2} ion exchange at the carboxylic acid sites of the polymer. Upon immersion of a substrate treated with Pd^{+2}/poly(acrylic acid) into an electroless copper plating bath, reduction of Pd^{+2} to Pd^0 takes place, forming an active catalyst for the initiation of electroless copper plating.

INTRODUCTION

Electroless copper plating has been proposed as a method for producing circuit lines in printed circuit boards (PCB) (1). Fabrication of PCB's by electroless plating offers an important advantage over existing fabrication methods that employ chemical etching of a blanket copper layer to define the circuit lines. Because the chemical processes etch the copper isotropically, the sidewalls of the circuit lines are eroded as the desired thickness of copper is removed from the PCB substrate. Electroless deposition on the other hand is a totally additive process, so it is possible to form fine-pitch, high-aspect-ratio circuit lines with nearly rectangular cross sections.

Electroless copper plating cannot be initiated on PCB substrates (typically glass fiber/epoxy composites) without first depositing a catalyst onto the substrate surface. In a common method used to initiate electroless copper plating on dielectric substrates, the substrate is dipped into an aqueous colloidal suspension of palladium to deposit 1-2 nm diameter particles of palladium on the substrate surface (2-6). This process is fairly simple, but has some problems. Because the colloid is prepared by reduction of Pd^{+2} to Pd^0 by SnCl$_2$, the palladium particles are surrounded by a tin chloride shell (7,8). Tin chlorides are not active catalysts for electroless copper deposition, and thus the shell inhibits the activity of the palladium particles. The shell can be dissolved away by an "acceleration" step, which typically involves immersion of a catalyzed substrate in aqueous HCl or NaOH (2-6). Acceleration must be carried out soon after applying the colloid, however, since the shell oxidizes in air to form tin oxides that are very difficult to remove. The acceleration step also causes agglomeration of the palladium particles on the substrate surface (2-6).

This paper describes a new process for initiating electroless deposition of copper onto dielectric substrates. In this process, the substrate is coated with a very thin film of poly(acrylic acid) and is then immersed in an aqueous solution of PdSO$_4$. The carboxyl groups of the poly(acrylic acid) film complex with Pd^{+2} supplied by the PdSO$_4$ solution. Upon immersion in the electroless copper plating bath, the complexed Pd^{+2} is reduced to Pd^0, forming an active catalyst that initiates electroless copper plating.

EXPERIMENTAL

Plating experiments were carried out on standard, mechanically roughened FR-4 epoxy/glass fiber composite PCB substrates. The substrates were cleaned in acetone, aqua regia, and deionized water prior to use. In a typical experiment, the PCB substrate was immersed into an aqueous solution of poly(acrylic acid) (MW 250,000) immediately after it had been rinsed in deionized water as part of the cleaning process. After remaining in the poly(acrylic acid) solution for about 30 seconds, the substrate was removed from the solution and placed vertically, allowing the excess polymer solution to run off. After the substrate had dried under ambient conditions, it was placed in a stirred aqueous solution of PdSO$_4$. After removing the substrate from the PdSO$_4$ solution, the substrate was rinsed in deionized water for 30 seconds and blown dry in a stream of N$_2$.

In some experiments, the quartz crystal from a quartz crystal microbalance (QCM) was used as the substrate to allow the determination of film masses. The crystal frequency was monitored with a frequency counter linked to a PC. For measurements of the thickness of the poly(acrylic acid) film on the crystal surface, the frequency of the uncoated crystal was measured, after which the crystal was removed from its holder and coated with the polymer film as described above for the PCB substrates. After drying, the crystal was replaced in the holder and the new frequency recorded. The crystal frequencies before and after coating were measured three times; between each measurement, the crystal was removed from its holder and replaced to be certain that minimal error was introduced by changing the position of the crystal within its holder. In no case did the frequency variation due to repositioning the crystal in its holder exceed 20 Hz, which is less than 2% of the smallest frequency difference that was measured. The film mass per unit area was determined from the frequency change using the relation of Sauerbrey (9). This relation is accurate for the frequency changes observed in this work. For measurements of the rate of Pd^{+2} uptake in the poly(acrylic acid) film, one face of a polymer-coated crystal was exposed to aqueous $PdSO_4$ and the oscillator frequency was recorded every 0.3 seconds.

Electroless copper deposition was performed in all experiments using a bath composition similar to that reported by Bindra and Tweedie (10). A stock solution was prepared containing all components except formalin. The pH of the stock solution was adjusted to 11.7 at 25°C using NaOH. For each experiment, 25 mL of the stock solution was placed in a test tube. Formalin was added and the tube was stoppered and placed in a constant temperature bath operating at 73 ± 1°C. After the electroless deposition bath reached the desired temperature, the test tube was opened, the sample was placed inside, and the stopper was replaced. During copper deposition, the electroless bath was not stirred continuously but was occasionally agitated.

Palladium surface concentrations were measured for both the PCB and quartz substrates by Galbraith Laboratories (Knoxville, Tennessee) via extraction of the substrate with aqua regia followed by analysis of the extract via either plasma emission or atomic absorption spectroscopy X-ray photoelectron spectroscopy was performed by Surface Science Laboratories (Moutain View, California). IR spectra were obtained using an IBM IR44 FTIR spectrometer.

RESULTS AND DISCUSSION

Figures 1 and 2 show scanning electron micrographs of a circuit line and a plated hole produced using the Pd^{+2}/poly(acrylic acid) catalyst described here. These samples were prepared using an aqueous solution of 1% poly(acrylic acid) a 0.1 M aqueous solution of $PdSO_4$ as the coating solutions. The residence time in the $PdSO_4$ solution was 5 minutes. Complete coverage of the substrate is observed after a plating time of 2 hours, during which 6 μm of copper is deposited. This is sufficient to cover the surface roughness of the PCB substrates. For the patterned sample shown in Fig. 1, all resist processing was carried out after the poly(acrylic acid) coating step and before the $PdSO_4$ coating step.

To determine the poly(acrylic acid) film thickness, several crystals of a quartz crystal microbalance were dip-coated in the same manner as that described above for the PCB substrates. The film thickness on quartz crystals will not be exactly the same as that on a mechanically roughened PCB substrate, but the microbalance measurements give an estimate of the film thickness expected on a PCB substrate. The film mass per unit area for the poly(acrylic acid) coating prepared from a 1% solution is 13.1 ± 0.8 $\mu g\text{-cm}^{-2}$

Fig. 1. Scanning electron micrograph of a circuit line produced on a PCB substrate as described in the text. The white bar at the bottom left corresponds to 100 μm. Plating was stopped after 2 hrs. (~6 μm of copper deposited) to show uniform coverage of the plated copper. The photoresist used was Du Pont Riston T-168 spray developed in 1,1,1-trichloroethane.

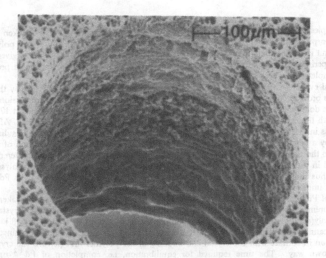

Fig. 2: Scanning electron micrograph of a plated hole in the PCB substrate shown in Fig. 1.

(11). Dividing this value by the density of poly(acrylic acid) (1.18 g-cm^{-3}, Ref. 12) gives a film thickness of 111 ± 7 nm. This shows that a very thin film is required to catalyze electroless plating using the Pd^{+2}/poly(acrylic acid) system. Film thicknesses obtained on quartz crystals using other solution concentrations of poly(acrylic acid) are shown in Table I.

The palladium concentration on the substrate surface obtained in the PdSO$_4$ immersion step was determined for both quartz and PCB substrates that had been dip-coated in 1% poly(acrylic acid) solution. The palladium surface concentrations were found to be 9.0 ± 0.5 μg-cm^{-2} for quartz substrates and 9.0 ± 0.7 μg-cm^{-2} for PCB substrates (11).

X-ray photoelectron spectroscopy (XPS) was performed on quartz crystal samples that had been dip-coated from 1% aqueous poly(acrylic acid) solution and then immersed for five minutes in 0.1 M PdSO$_4$. The only elements detected by XPS were carbon, oxygen, and palladium, both on the surface and after argon-ion etching to nominal depths of ~10 and 20 nm. No sulfur was detected (sensitivity ~0.5 at. %). High resolution spectra (±1 eV FWHM) were obtained for the C(1s) peak and the Pd(3d) peak. Anchoring the energy scale to the C(1s) peak at 284.6 eV (13), the Pd(3d$_{5/2}$) peak was found at 336.2 eV. This is close to the literature value of 336.1 eV given for Pd^{+2} in the predominantly covalently bonded oxide PdO (13). XPS was also performed on an identically prepared sample after immersion in the copper electroless plating bath for one minute followed by a 30 second rinse in deionized water. In this spectrum, copper and sodium are detected in addition to carbon, oxygen and palladium. The Pd(3d$_{5/2}$) peak is found at 335.3 eV, which is close to the literature values of 334.9 eV (13) and 335.3 eV (14) given for Pd^0. These data indicate that the Pd^{+2} deposited on the surface in the PdSO$_4$ immersion step is reduced to Pd^0 in the electroless copper plating bath. This is consistent with the distinct color change from light brown to black observed upon immersion of the sample into the electroless plating bath. Since Pd^0 is known to be a catalyst for electroless copper deposition, Pd^0 formed *in situ* is most likely the active catalyst that initiates copper plating in this system.

The data reported here suggest that Pd^{+2} is taken up by the poly(acrylic acid) film in aqueous PdSO$_4$ by H$^+$/Pd^{+2} ion exchange at the carboxylic acid sites of the polymer. Since no sulfur was detected by

Table I. Films thicknesses obtained on quartz crystals coated from aqueous poly(acrylic acid) as a function of the polymer solution concentration. Thicknesses were determined using a quartz crystal microbalance.

Concentration (%)	Thickness (nm)
0.5	45
1.0	111
2.0	269
3.0	444

XPS in samples coated with poly(acrylic acid) and immersed in $PdSO_4$, Pd^{+2} is not taken up in the film by simple sorption of $PdSO_4$ solution. Also, the number of Pd atoms found per poly(acrylic acid) monomer unit is 0.46, as determined from the Pd and poly(acrylic acid) concentrations given above. This is within experimental error of the value (0.50) expected for a H^+/Pd^{+2} ion-exchange process that has gone to completion.

Additional evidence in favor of an ion exchage mechanism for the uptake of Pd^{+2} by the poly(acrylic acid) films is provided by examining the IR spectrum of the film before and after immersion into aqueous $PdSO_4$. The dominant peak in a freshly coated sample is the carbonyl stretching peak at 1710 cm^{-1}. This coincides with the carbonyl stretching peak found for poly(acrylic acid) in a KBr pellet. After immersion for 5 minutes in 0.1 M aqueous $PdSO_4$, the 1710 cm^{-1} peak in the film IR spectrum has disappeared and is replaced by a new broad band centered at 1530 cm^{-1}. This is indicative of conversion of the carboxylic acid group in the poly(acrylic acid) film to the carboxylate salt form. A broad band centered at 1530 cm^{-1} is also found in the KBr pellet IR spectrum of poly(sodium acrylate) (obtained from Polysciences). The IR spectra thus indicate that complete conversion of the carboxylic acid groups to the Pd(II) salt form occurs upon immersion of a poly(acrylic acid) film into aqueous $PdSO_4$

A study of Pd^{+2} uptake by poly(acrylic acid) films from aqueous $PdSO_4$ was undertaken using *in situ* QCM measurements. The rate of Pd^{+2} was determined for films prepared on quartz crystals from 0.5% and 1.0% poly(acrylic acid) solutions. $PdSO_4$ solution concentrations of 0.03 M and 0.1 M were used. These data cannot be used to obtain the actual mass of Pd^{+2} taken up by the films, since the presence of the solution in contact with the crystal and swelling of the polymer film perturb the crystal frequency in an unknown way. The time required for equilibration, i.e. completion of Pd^{+2} uptake, can be determined from these measurements, however, by noting the time required for the crystal frequency to fall to a steady frequency. The data indicate that the equilibration time is proportional to both the poly(acrylic acid) film thickness and the $PdSO_4$ solution concentration. This is consistent with the kinetics observed upon exchange of H^+ from a weak acid ion exchange resin for a strongly bound cation from solution (15). For the solution concentrations used to prepare the samples shown in Figs. 1 and 2 (1.0% poly(acrylic acid), 0.1 M &pdso4), uptake of Pd^{+2} is nearly complete in three minutes.

CONCLUSION

A new method has been described for initiating electroless plating of copper onto dielectric substrates. The process employs a thin film of poly(acrylic acid), applied to the substrate by dip-coating, to bind Pd^{+2} upon immersion of the substrate in aqueous $PdSO_4$. This method is comparable in simplicity to the colloidal palladium method of initiating electroless copper deposition on dielectric substrates. No acceleration step is required, however, which has been found to cause agglomeration of colloidal palladium on the substrate surface. Uptake of Pd^{+2} by the poly(acrylic acid) film takes place by an ion exchange process. Upon immersion of the Pd^{+2}/poly(acrylic acid) film into an electroless copper bath, reduction of Pd^{+2} takes place, giving Pd^0 which serves as the active catalyst for initiation of electroless copper plating.

Acknowledgment.

The author would like to thank Dr. Roy Magnuson and Dr. Ronald McIlatton for numerous helpful discussions.

REFERENCES

1. See the discussion by D. P. Seraphim, IBM J. Res. Develop., 26, 37 (1982).
2. N. Feldstein, M. Schlesinger, N. E. Hedgecock, and S. L. Chow, J. Electrochem. Soc., 121, 738 (1974).
3. T. Osaka, H. Rakematsu, and K. Nikei, J. Electrochem. Soc., 127, 1021 (1980).
4. T. Osaka, H. Nagasaka, and F. Goto, J. Electrochem. Soc., 127, 2343 (1980).
5. J. Kim, S. H. Wen, D. Y. Jung, and R. W. Johnson, IBM J. Res. Develop., 28, 697 (1984).
6. J. Horkans, J. Kim, C. McGrath, and L. T. Romankiw, 134, 301 (1987).
7. R. L. Cohen and K. W. West, J. Electrochem. Soc., 120, 502 (1973).
8. R. L. Cohen and R. L. Meek, J. Colloid Interface Sci., 55, 156 (1976).
   G. Sauerbrey, Z. Physik, 155, 206 (1959).
   P. Bindra and J. Tweedie, J. Electrochem. Soc., 130, 1112 (1983).

11. Error bars represent one standard deviation.
12. Polymer Handbook, J. Bandrup and E. H. Immergut, ed. (Wiley, New York, 1975).
13. Handbook of X-ray Photoelectron Spectroscopy, G. E. Muilenberg, ed. (Perkin-Elmer Corp., Eden Prairie, Minn., 1979).
14. G. Kumar, J. K. Blackburn, R. G. Albridge, W. E. Moddeman, and M. M. Jones, Inorg. Chem., 11, 296 (1972).
15. F. Helfferich, J. Phys. Chem., 69, 1178 (1963).

LOW TEMPERATURE TRANSFORMATION OF THE SURFACE OF COPPER OXIDE FILLED PTFE VIA TREATMENT IN A HYDROGEN PLASMA

DOV B. GOLDMAN
Olin Corporation, 201 Roosevelt Place, Palisades Park, NJ 07650

ABSTRACT

The surface conductivity of PTFE-copper oxide filled composites containing between 10 and 25% of the oxides increases significantly after treatment in a low pressure hydrogen plasma at temperatures below $100^{\circ}C$. The plasma-reduction of copper oxides (CuO and CuO/Cu_2O mixture) was investigated and compared with thermal reduction at different hydrogen partial pressures.

INTRODUCTION

Preparation of conductive surfaces of polymers or polymer matrix composites can be carried out via reduction of a metal oxide mixed with a polymer matrix. Reduction of the oxide can be performed in a liquid chemical solution [1] or using a reducing gas, e.g. hydrogen. The gas-oxide reaction normally occurs at elevated temperatures. It is desirable to conduct the reaction at the lowest possible temperatures. Plasma treatment offers such an opportunity [2-4]. This work demonstrates the feasibility of a low temperature transformation (T $\leq 100^{\circ}C$) of copper oxide-filled PTFE surfaces from an electrically insulating to a conducting state by using an RF hydrogen plasma.

EXPERIMENTAL

Preparation of the composites consisted of mixing PTFE (Teflon® 8, E.I. du Pont de Nemours & Company) powder with copper oxide (Cu_2O, J.T. Baker, technical grade), followed by compaction and sintering into 50mm diameter and 4mm thick discs. The samples contained 10, 15, 20, and 25% by volume of the oxide. The sintering was conducted at 370 $^{\circ}C$ in air. Samples of Cu_2O or CuO ("Baker Analyzed" reagent) about 0.5mm thick layer on a PTFE substrate were prepared to study reduction of the oxides in plasma. The oxide powder was poured over the PTFE powder in the compacting mold. After compaction the two-layer samples were sintered under the same conditions as the composites. The samples had the same dimensions as the samples of the composites.

High purity grade hydrogen, and prepurified grade nitrogen (both from Matheson) were used in the experiments.

The temperature-controlled reduction of the oxide powders was conducted in a flow-through apparatus as described in [5] and a Perkin-Elmer model TSG-2 TG analyzer.

XRD analysis with CuK_a radiation identified the phases in the reactions.

The plasma treatment was conducted in a plasma desmear apparatus Model 400 manufactured by Advanced Plasma Systems [6]. Samples were suspended between the electrodes as shown in Fig. 1. A Plasmaloc 2 RF power supply (E.N.I.) operated at 42kHz was used. The power was varied between 1200 and 1400 W. The sample temperature was measured using an iron-constantan thermocouple attached to the sample surface. Sample heating was carried out

594

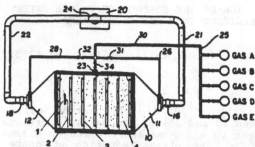

Fig 1. Schematic diagram of the plasma chamber. 1 - sample; 2 - themocouple; 3 - coupled electrodes; 4 - glow discharge; 11, 12 - pyramidal gas diffusion chamber; 16, 18 - exhaust valves; 20 - vacuum pump; 24 - exhaust tee; 25 - manifold; 26, 28 - gas inlets; 31, 32,34 - inlet valves.

in nitrogen. At the desired temperature nitrogen was replaced by hydrogen. The hydrogen flow rate of 1 l/min established a plasma chamber pressure of 2.2×10^{-1} torr. Adjustment of the RF power allowed us to maintain the desired temperature of the experiment within $\pm 5^{\circ}C$. Control samples were treated under identical temperature and hydrogen flow conditions but without RF power.

RESULTS AND DISCUSSION

Figure 2 shows the dependence of resistivity of the composite samples on the Cu_2O content before and after sintering. Densification of the samples and partial transformation of Cu_2O to more conductive CuO reduced the resistivity. The XRD pattern for the surface of a Cu_2O sample on PTFE taken after the sintering (Fig.3), confirms the transformation.

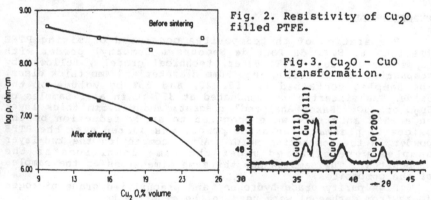

Fig. 2. Resistivity of Cu_2O filled PTFE.

Fig.3. Cu_2O - CuO transformation.

Metallic copper forms during the exposure of the oxides in the hydrogen plasma at temperatures between 50 and $120^{\circ}C$. Figure 4 presents the XRD pattern for CuO and mixed Cuo/Cu_2O samples taken before and after the exposure to the hydrogen plasma for different periods of time at $100^{\circ}C$. The relative intensity of the major peak for copper is plotted against the time of the exposure in Fig. 5. The resistivity of the copper oxide surface shown in the same figure decreases steeply after about six minutes of exposure to the plasma at $100^{\circ}C$. We did not detect formation of metallic copper on the control samples, and their resistivity did not change.

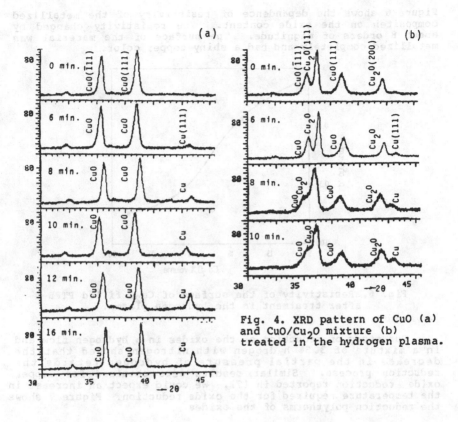

(a)   (b)

Fig. 4. XRD pattern of CuO (a)
and CuO/Cu₂O mixture (b)
treated in the hydrogen plasma.

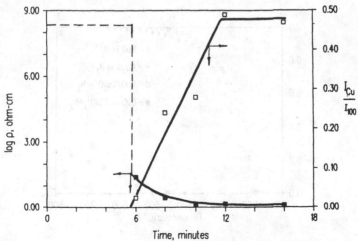

Fig. 5. Relative intensity of Cu (111) peak and resistivity
of CuO after treatment in the hydrogen plasma.

Figure 6 shows the dependence of resistivity of the metallized composites on the oxide content. The resistivity changed by about 8 orders of magnitude. The surface of the material was metallized completely and had a shiny copper color.

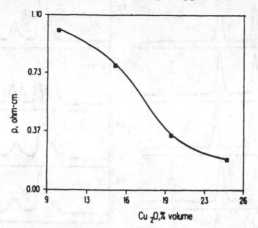

Fig. 6. Resistivity of the surface of Cu_2O filled PTFE after treatment in the hydrogen plasma.

Our study of reduction of the oxides in a hydrogen flow and in a mixture of 3.5% hydrogen with nitrogen showed that the decrease in the partial pressure of hydrogen retards the reduction process. Similar results were obtained for copper oxide reduction reported in [7]. We could expect an increase in the temperature required for the oxide reduction. Figure 7 shows the reduction polytherms of the oxides.

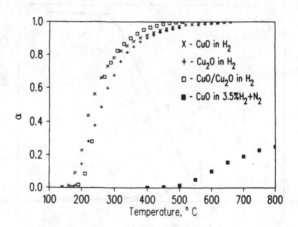

Fig. 7. Reduction polytherms for CuO, Cu_2O, and Cu_2O/CuO mixture in hydrogen and 3.5% hydrogen-nitrogen mixture.

The reduction rate decreases considerably with the decrease in the partial pressure of hydrogen. According to that observation we should expect a lower reduction rate in the plasma chamber without application of the RF power.

The observed reduction at lower temperatures in a hydrogen plasma should relate to the higher concentration of the ionized hydrogen atoms. According to Vol'kenstein's theory [8] the probability of a reaction between the gas molecules and the semiconductor oxide surface or a transition to the strong chemical absorption depends on the position of the Fermi level which is temperature dependent. The changes in temperature and the resulting changes in the position of the Fermi level will increase the probability of ionization of the gas molecules at the oxide surface. Plasma ionization of the gas assists this process and increases the probability of the reaction between the gas and the oxide surface. That can explain the observed oxide reduction at lower temperatures.

CONCLUSIONS

The RF plasma assists the conversion of copper oxides to metallic copper at hydrogen pressures and temperatures lower than that required for their thermal reduction.

Treatment of the PTFE-copper oxide filled composites in the hydrogen plasma results in metallization of their surface, enhancing significantly the surface conductivity.

REFERENCES

1. Potentially Conductive Substrates. (Rhone-Poulenc Co. publication, January 1986).
2. P. Capezzuto, F. Cramarossa, P. Maione, E. Molinari, Gaz. Chim. Ital. 104, 1109 (1974).
3. M.R. Wartheimer, J.-P. Bailon, J. Vac. Sci. Technol. 14, 699 (1977).
4. Y. Sakamoto, Y. Ishibe, Jap. J. Appl. Phys. 19 (5), 839 (1980).
5. D.B. Goldman, J. Therm. Anal. 30, 1071 (1985).
6. F. Fazlin, U.S. Patent No. 4 328 081 (4 May 1982).
7. Yu.B. Naumov, Yu.L. Pavlov, A.A. Vasilevich, A.M. Alekseev, E.A. Novikov, G.G. Shchibrya , Kinetika i Kataliz, 25 (4), 818 (1987).
8. F.F. Vol'kenstein, Electron Theory of Catalysis on Semiconductors, Nauka, Moscow, 1960.

DEGRADATION OF THE POLYIMIDE/COPPER INTERFACE

K. K. Chakravorty, V.A. Loebs, and S.A. Chambers
Boeing Electronics High Technology Center, P.O. Box 24969, M/S 9Z-80,
Seattle, Washington 98124-6269

ABSTRACT

A comparative investigation of polyimide/Cu interface degradation has been carried out for ultrathin photosensitive, non-photosensitive, and preimidized polyimide precursor films cured while in contact with a Cu substrate. The role of curing byproducts and environmental conditions on interface degradation has been elucidated by means of X-ray photoelectron spectroscopy. Immediately after curing, we observed some oxidation of the Cu in contact with non-photosensitive and photosensitive polyimide overlayers. On the other hand, only negligible oxidation was observed for the preimidized polyimide/Cu interface. Experiments in which samples were stored in vacuum, air and a humidity chamber show a dependence of the oxidation kinetics on air/moisture exposure. Preimidized and photosensitive polyimide/Cu interfaces, stored in air, became more extensively oxidized with time relative to identical samples stored in vacuum. Moreover, all three polyimide/Cu interfaces exhibited significantly more oxidation after 3 days in a humidity chamber than after 19 days of storage in air. Taken together, these data clearly demonstrate that absorbed water and its interaction with curing byproducts are key factors in the extent of Cu oxidation at the interface.

INTRODUCTION

Excellent thermal stability, adherence, and favorable dielectric properties have resulted in the widespread use of polyimide for microelectronics applications. Polyimide has been used as an interlevel dielectric [1] and a final passivant [2] for multilevel metal transistors. More recently, polyimide has been used as a dielectric in multilayer thin film interconnect structures such as multi-chip modules [3,4]. Most of these applications involve a substantial number of metal-polyimide interfaces. Therefore the nature of interfacial bonding between the two materials is a key parameter in the overall reliability of the structures. Consequently, a number of studies have been performed in which the chemistry of interface formation for metals evaporated onto cured polyimide films in an ultrahigh vacuum system has been investigated using X-ray photoelectron spectroscopy [5-7]. The first observation of Cu oxidation at the inverted polyimide on Cu interface was recently reported by Chambers and Chakravorty [8]. Most of these studies have focused on a nonphotosensitive, PMDA-ODA precursor type of polyimide. Similar information is lacking for the other types of polyimide used for microelectronics applications, such as the photosensitive precursor-based and preimidized versions. Due to their significantly different precursor composition and curing chemistry, different interfacial interactions are expected. Therefore, our goal in the present work has been to examine polyimide/Cu interfaces for these three different types of polyimide and to bring out any systematics that may exist between curing chemistry, environmental conditions and interface degradation.

EXPERIMENTAL

We selected Du Pont Pyralin PI-2525 as a representative of the nonphotosensitive PMDA-ODA precursor class of polyimides. Selectiplux HTR3 of EM Industries, Inc. and Pyralin- LTP PI-2590D of Du Pont company were selected as a photosensitive and a preimidized polyimide, respectively. Polyimide solutions were diluted and spun at 6000 rpm onto 1000 Å of evaporated Cu on SiO_2/Si wafers. The oxide present on the metal films was stripped by immersion in 1.2 N H_2SO_4 followed by prolonged rinsing in de-ionized water prior to spin coating. The thin polyimide films were then cured by heating to 380 °C over $2^1/4$ hours and holding at 380 °C for $3/4$ hour. Upon cooling to room temperature in the oven, the samples were divided into three groups, each containing the three kinds of polyimide. The first group was stored in vacuum, the second group was stored in air at room temperature and the third group was aged in a controlled humidity chamber at 85 °C and 85% relative humidity. Cu 2p X-ray photoelectron spectra were obtained every few days for a period of three weeks for each sample. The spectrometer employed was a Surface Science Instruments Model 301 with a focussed, monochromatic AlKa X-ray source. All spectra were obtained with an analyzer pass energy of 150 eV, an X-ray beam diameter of 1000μ, and a collection angle of 90° with respect to the surface.

RESULTS AND DISCUSSION

In order to insure that the polyimide films were continuous, scanning Auger microprobe and scanning electron microscopy analysis was done. Characterization of several regions of each sample showed that the films were indeed continuous and free of voids. Point spectra exhibited C,N and O KLL peaks yielding atomic concentrations in good agreement with those obtained by XPS. The Cu LMM peak was clearly present, but significantly attenuated, again indicating continuous coverage of the polymer overlayers. The continuity of the films prevents direct contact of oxygen with the Cu substrate so that any observed oxidation will be caused either directly or indirectly by the polyimide rather than by oxygen.

In figure 1 we show Cu $2p_{3/2}$ spectra for the three Cu surfaces with overlayers of preimidized, nonphotosensitive and photosensitive polyimide stored in air as a function of time. Also included at the top of each column is a Cu $2p_{3/2}$ reference spectrum obtained for a Cu surface, freshly stripped of its surface oxide and without any polyimide overlayer. The absence of a shoulder to higher binding energy establishes complete removal of the metal oxide in the starting metal surfaces. Spectra for the three polyimide/Cu interfaces stored in the humidity chamber are shown at the bottom. The first set of spectra (corresponding to day 1) were recorded within a few hours of the samples being removed from the oven. Based on attenuation of the total Cu $2p_{3/2}$ peak intensity relative to that for a clean Cu surface, we estimate that the film thicknesses were 30, 45, and 65 Å for preimidized, nonphotosensitive, and photosensitive polyimide samples, respectively.

Immediately after curing, there was evidence of partial oxidation of Cu at the interface by non-photosensitive and photosensitive polyimide overlayers. The amount of oxide at the preimidized polyimide/Cu interface was negligible. A shoulder on the Cu $2p_{3/2}$ peak, shifted aproximately 2 eV to higher binding energy, shows that some of the Cu at the interface had been oxidized to a probable combination of +1 and +2 oxidation states within one hour of being removed from the oven. Exposure to air increased the extent of oxidation of Cu at the interface for all three polyimides.

Although it appears that the extent of oxidation was highest at any given time for the Cu surface supporting the photosensitive polyimide film, this observation may be due to the higher polyimide film thickness for this sample (65 Å) compared to the other two (45 and 30 Å). In this case, the ratio of oxidized to elemental Cu peak intensity is artificially increased because of preferential attenuation (by the thicker polyimide film) of photoelectrons originating below the interfacial region. Exposure of samples to an increased level of humidity accelerated the rate of Cu oxidation for all three kinds of polyimide. After 3 days of aging at 85 °C and 85% relative humidity conditions, all three Cu interfaces were considerably more oxidized than after 19 days of air exposure.

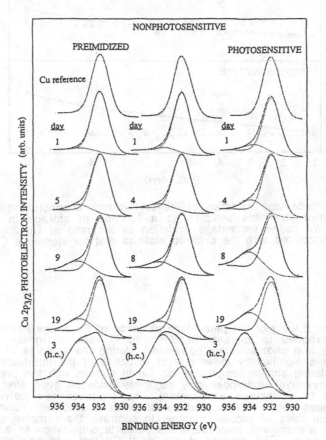

Figure 1. Cu 2p$_{3/2}$ spectra for three Cu surfaces · with overlayers of preimidized, nonphotosensitive and photosensitive polyimide stored in air as a function of time. Decomposition of the spectra into contributions from the elemental and the oxidized state is indicated with Gaussians. The last set of spectra labeled h.c. were recorded for samples aged in a humidity chamber at 85 °C and 85% relative humidity. Also included at the top of each column is a Cu 2p$_{3/2}$ reference spectrum for a bare Cu surface.

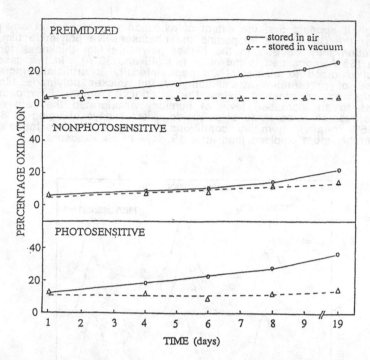

Figure 2. Percentage oxidation of the preimidized, nonphotosensitive and photosensitive polyimide samples as a function of storage in air and vacuum. We define percentage oxidation as the ratio of Cu $2p_{3/2}$ peak intensity associated with the oxidized state to that for elemental Cu.

In figure 2 we plot the percentage oxidation of Cu at the interface for samples exposed to air and those stored in vacuum. In the case of preimidized and photosensitive polyimides, storing the samples in vacuum resulted in a significantly lower rate of oxidation of the interfacial Cu than did storing similar samples in air. Cu in contact with the preimidized polyimide overlayer undergoes only negligible oxidation even after 9 days of storage in a vacuum environment. The photosensitive polyimide/Cu interface does not oxidize beyond the stage associated with curing, even after several days of vacuum storage. In contrast, the samples stored in air showed a monotonic increase in the percentage oxidation as a function of storage time. The amount of interfacial Cu that was oxidized in the air-stored samples increased by factors of 12 and 3 for the preimidized and photosensitive polyimides, respectively, over the 19 day period. In contrast, the nonphotosensitive polyimide/Cu interface showed nearly the same rate of oxidation in both air and vacuum.

The different oxidation behavior described above suggests that Cu/polyimide interface degradation depends on curing chemistry in the precursor overlayer. The three polyimide samples included in the present study differ significantly in their curing chemistry. The preimidized version is imidized at the precursor level. Therefore, a high-temperature curing step involves primarily desorption of the solvent [9]. The absence of any significant oxidation of Cu at the interface after curing this polyimide suggests that neither the polymer nor the solvent system (80/20 N-methyl-2-pyrrolidone/aromatic hydrocarbon [9]) interacts chemically with the Cu surface at elevated temperatures. The hygroscopic nature of polyimide films and consequent degradation of their dielectric properties in humid environments are well established [10]. It appears from the different rates of oxidation in air and in vacuum that the uptake of water in the hygroscopic polyimide film, brought about by air exposure is the driving force in the oxidation of the buried Cu surface.

The nonphotosensitive polyimide employed in the present study undergoes an imidization reaction of the polyamic acid precursor at elevated temperatures. The initial cure of the polyamic acid precursor (poly (4'4'-oxydiphenylpyromellitamic acid)) is suggested to proceed through transimidization rather than cyclization. Cyclization of some of the trans imide is achieved in the subsequent elevated-temperature annealing step [11]. The imidization reaction is accompanied by formation of water. Partial oxidation of the Cu surface in contact with the polyimide overlayers immediately after curing the polyimide film suggests a reactive interaction between the interfacial Cu atoms and water that is present from the imidization reactions and/or absorbed during air exposure. However, the near constancy in the rate of oxidation in air and vacuum strongly suggests that water produced in the imidization reaction is sufficient to drive the oxidation, and that absorbed water plays only a minor role.

Photosensitive polyimide precursors generally consist of polyamic acid esters. The ester group, such as oxyethylmethacrylate [12], forms an insoluble crosslinked chemical intermediate upon irradiation with UV light. High temperature annealing causes imidization and polyimide ring closure via cleavage of the ester groups. The crosslinked photosensitive groups are depolymerized and volatilized. In the case of the polyimide precursor which contains an oxyethylmethacrylate photoreactive group, the curing by-products are expected to be mono-and-poly hydroxyethylmethacrylate [12]. Significant oxidation of the Cu surface immediately after curing of this polyimide is suggestive of a chemical interaction involving curing byproducts and Cu. However, it appears that absorbed water is key to sustaining the oxidation process, inasmuch as the reaction comes to a halt when the sample is stored in vacuum.

Enhancement of Cu surface oxidation upon exposure to a humid environment for all three polyimide samples lends considerable support to a moisture-induced oxidation mechanism. This reaction may be brought about by an ionic-contaminant-catalyzed corrosive action [8]. Du Pont Kapton H type polyimide samples, metallized with evaporated Cu, show increased polymer-metal intermixing but apparently no oxide formation at the polyimide/Cu interface [13]. Our observation of substantial oxidation of Cu at the polyimide/Cu interface after a three-day exposure in a 85/85 humidity chamber is indicative of significant chemical activity. Moisture absorption by the cured polyimide film accelerates the interface reaction started during the curing of the precursor groups. The O1s spectra for these polyimide/Cu interfaces underwent significant changes as oxidation of the Cu surface occured and may yield further insight into the details of interfacial degradation. We are currently analyzing these spectra and will report on them at a later time.

CONCLUSION

The process of curing ultrathin films of photosensitive and nonphotosensitive polyimide over Cu causes significant oxidation of the Cu surface. In contrast, negligible oxidation is observed for a preimidized polyimide sample. The oxidation kinetics after curing show a strong dependence on moisture absorption from air and a highly humid environment. The observed oxidation behavior appears to support a mechanism whereby moisture absorbed by the cured polyimide film accelerates the interface reaction started during curing of the precursor groups.

ACKNOWLEDGMENT

We are indebted to Mr. Jay Cech for assistance in the humidity chamber related experiments.

REFERENCES

1. K. Sato, S. Harada, A. Saiki, T. Kimura, T. Okubo, K. Mukai, IEEE Trans. on Parts, Hybrids & Packaging, PHP-9, No. 3, 175 (1973).

2. S. Miller, Circuits Manufacturing, April 1977, 39.

3. R.J. Jensen, J.P. Cummings, and H. Vora, IEE Trans. CHMT, CHMT-7, No. 4, 384, (1984).

4. K.K. Chakravorty, J.M. Cech, C.P. Chien, K.P. Metteer, L.S. Lathrop B.W. Aker, M.H. Tanielian, and P.L. Young, ECS symposia Proc., Atlanta Georgia, May 1988, To be Published.

5. N. J. Chou and C.H. Tang, J. Vac. Sci. Technol. A 2, 751 (1984).

6. P. S. Ho, P.O. Hahn, J.W. Bartha, G. W. LeGoues, and B.D. Silverman, J. Vac. Sci. Technol. A 3, 739 (1985).

7. P.N. Sanda, J.W. Bartha, J.G. Clabes, J.L. Jordan, C. Feger, B.D. Silverman, and P.S. Ho, J. Vac. Sci Technol. A 4, 1035 (1986).

8. S.A. Chambers and K.K. Chakravorty, J. Vac. Sci. Technol., A 6, 3008 (1988).

9. Pyralin LTP- PI-2590-D, Technical Data Sheet, E-81791, 1986, Du Pont Company.

10. G. Samuelson and S. Lytle, Polyimide: Synthesis, Characterization and Applications, edited by K.L. Mittal (Plenum Press Publisher, New York, 1984), p. 751.

11. E. Sacher, J. Macromol. Sci.-Phys., B 25, 405 (1986).

12. R. Rubner, H. Ahne, E. Kuhn, and G. Kolodziej, Photograph. Sci. Eng., 23 (5), 303 (1979).

13. C. Chauvin, E. Sacher, A. Yelon, R. Groleau, and S. Gujrathi, Surface and Colloid Science in Computer Technology, Ed. K.L. Mittal, 1987, 267.

ELECTRICAL PROPERTIES OF HARD CARBON FILMS

WALTER VARHUE*, KIRIL PANDELISEV** and BRIAN SHINSEKI***
* Dept. of Electrical Eng., University of Vermont, Burlington, VT 05405
** Cominco Ltd./P.O. Box 3000, Trail, British Columbia, Canada V1R 455
***Intel Corporation, Chandler, AZ

ABSTRACT

The electrical resistivity, optical band gap and activation energy for electrical conduction have been determined as a function of preparation conditions. The operating conditions for the glow discharge reactor have been interpreted in terms of ion energy and reactive species production. The change in the electrical properties could not be explained as a percentage of $[SP_3]$ versus $[SP_2]$ bonding ratio. Rather, these two species are embedded in an amorphous medium which determines the materials electrical properties.

INTRODUCTION

Hard carbon or diamond-like films have generated considerable interest because of their unusual electrical, chemical and mechanical properties [1]. J. C. Angus, Et. Al. [2] have assembled an extensive bibliography on matters concerning the deposition process and physical properties of the films. The electronic properties of a-C:H have been carefully reviewed in references [3] and [4]. The later reference contains a large bibliography on the electronic nature of the films.

Theoretically, because of carbon's low atomic mass, a high thermal conductivity can be expected. This, coupled with a large band gap, makes diamond films a very attractive possibility for some specialized solid state devices. An example would be a semiconductor device operating in an environment with high temperatures or ionizing radiation. Carbon films are also extremely hard and chemically resistive. These properties make it an excellent encapsulation layer for high density integrated circuits. The film would efficiently act to conduct heat away to a sink while protecting the underlying circuitry.

Films have been prepared by a variety of techniques including sputtering from solid sources and by glow discharge from gaseous hydrocarbon sources. Both techniques have produced, under proper conditions, films which are classified as "dense carbon". The use of hydrocarbon sources is an attractive method and have yielded films with a large hydrogen content as high as 60% [5]. The structure, hydrogen content and electrical properties of the films are extremely sensitive to the energy with which carbon ions hit the surface. This energy is controlled by reactor configuration and operating parameters such as power and pressure. The ion energies can approach 1 KeV which could result in heating a small volume of the film to 10^7 °K for a brief instant. These high temperatures and short times are capable of forming metastable entities such as diamond.

In this investigation we have performed a systematic study of the effect of operating conditions on the electrical conductivity and chemical structure of hard carbon films. The operating conditions have been translated into effective ion energies with measurement of the plasma potential and the dc bias found on the powered electrode.

EXPERIMENTAL

The samples were prepared in a custom-designed reactor chamber with two parallel, 10 cm. diameter plates separated by a 3.8 cm. gap. The substrates were placed upside down on the top powered electrode and the temperature did not exceed 50°C. The substrates used were 3", n type, 1 Ω-cm resistivity Si wafers. The reactor pressure was maintained at 200 mtorr by a closed loop capacitance manometer/throttle valve controller. The flow rate of methane was 20 sccm. The input power was varied from 5 to 50 W and was measured with a Bird Meter. The d.c. self bias on the powered electrode was measured by filtering the line with a 7 mH RF choke. The average ion energy is assumed to be the sum of the dc self bias and the plasma potential. The plasma potential was monitored with an emissive Langmuir Probe. The probe circuit and technique used to measure the plasma potential can be found in another paper presented at this symposium [6].

The electrical resistivity measurement was made with an Al/a-C/Si wafer sandwich structure. The Al dots were evaporated through a perforated sheet and had a diameter .1 cm. The measurement was made at an applied electric field of 1×10^6 V/cm.

DISCUSSION OF RESULTS

The resistivity of the deposited carbon film as a function of RF power can be found in Figure 1. The resistivity is observed to decrease with increasing RF power. Next, resistivity is shown in Figure 2 to decrease with increasing pressure. The figures show lines connecting the data points, this was done for ease of viewing and does not assume the curve is continuous.

It is expected that ion energy striking the substrate surface will increase with RF power and decreasing pressure. The sheath potential which accelerates the ions toward the substrate is the sum of the negative dc self bias on the powered electrode and the average plasma potential. The average plasma potential was not insignificant as often assumed and was measured with an Emissive Langmuir Probe. A plot of measured plasma potential with RF power can be found in Figure 3. The total sheath voltage is shown in Figure 4. It is observed that the total sheath voltage increases almost linearly with RF power.

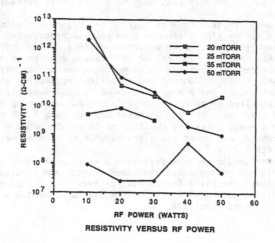

RF POWER (WATTS)

RESISTIVITY VERSUS RF POWER

FIGURE 1

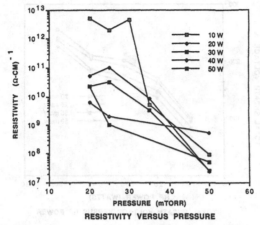

RESISTIVITY VERSUS PRESSURE

FIGURE 2

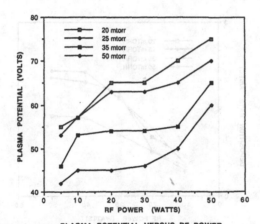

PLASMA POTENTIAL VERSUS RF POWER

FIGURE 3

The resistivity has been found to increase with decreasing RF power. The RF power supplied to the discharge also determines the rate of production of reactive chemical species which are the precursors to film growth. The deposition rate as shown in Figure 5 is a strong function of RF power. A plot of deposition rate as a function of reactor pressure showed little change with pressure. The lower deposition rates permit the adatoms more time to migrate across the surface seeking the lowest energy bonding site.

The results of FTIR spectroscopy can be used to give the concentration ratios of SP_3 versus SP_2 type bonding. The bonding configurations and theoretically predicted wave numbers have been tabulated by B. Dischler [7]. These peaks lie close together and there is significant overlap. A deconvolution analysis of the spectrum is required. We have made use of a commercial available code distributed by Nicolet entitled IRCON. As with all analysis of this type, the problem with false structure is always present. The data were used to obtain only a ratio $[SP_3]/[SP_2]$ bonding concentrations in a single sample. The results of this analysis are plotted in Figures 6 and

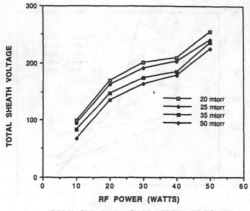

TOTAL SHEATH VOLTAGE VERSUS RF POWER

FIGURE 4

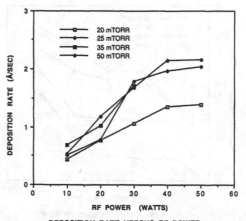

DEPOSITION RATE VERSUS RF POWER

FIGURE 5

7. The uncertainty in these plotted points represent error in the codes
ability to locate the peak and to properly determine its height. A
conservative estimate of the certainty of these points has been placed at ±
20%. Figure 6 shows the variation in $[SP_3]/[SP_2]$ aromatic (graphitic) bonding
as a function of power. Its variation with pressure showed no pattern and is
not shown. Figure 7 shows the variation of $[SP_3]/[SP_2]$ olefinic (polymeric)
bonding as a function of pressure. The variation of this ratio with power
showed no pattern. In both cases ion energy can be used to explain the
increase in $[SP_3]/[SP_2]$ bonding concentration. Ion energy increases with RF
power or decreasing pressure, resulting in more efficient production of the
metastable $[SP_3]$ allotrope. The increased $[SP_3]/[SP_2]$ aromatic ratio with RF
power is not consistent with the resistivity measurements. The resistivity
decreased with RF power, or as the diamond/graphite component increases.
These results indicate that the conductivity of the material cannot be
explained by a simple model of the relative concentration of the components.
These species are most likely embedded in an amorphous hydrogenated medium.

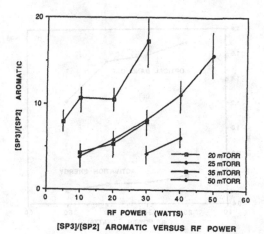

[SP3]/[SP2] AROMATIC VERSUS RF POWER

FIGURE 6

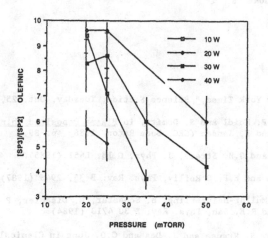

[SP3]/[SP2] OLEFINIC VERSUS RF POWER

FIGURE 7

The optical band gap, Eg, was measured using a Tauc Plot where the intensity varied over two orders of magnitude. The conductivity activation energy was obtained by measuring the conductivity as a function of temperature. A plot of the optical band gap and conductivity activation energy are shown in Figure 8 and are much less than that of diamond, (-5eV), which indicates a more complex conduction mechanism. The decrease in Eg with increasing power is consistent however with the resistivity measurements. From a structural point of view it is not clear why the band gap is decreasing with RF power while the $[SP_3]/[SP_2]$ aromatic ratio is increasing.

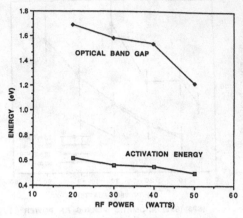

OPTICAL BANDGAP AND ACTIVATION
ENERGY VS RF POWER

FIGURE 8

REFERENCES

1. See "The New York Times," Science Section, Tuesday, Oct. 25, 1988.

2. J.C. Angus, P. Koidl and S. Domitz, in _Plasma Deposited Thin Films_, edited by J. Mort and F. Jansen (CRC, Boca Raton, 1986), p. 89.

3. J.L. Brédas and G.B. Street, J. Phys. C _18_, L651 (1985).

4. J. Robertson and E.P. O'Reilly, Phys. Rev. B _35_, 2946 (1987).

5. J. Fink, T. Müller-Heinzerling, B. Scheerer, B. Dischler, P. Koidl, A. Bubenzer, and R.E. Sah, Phys. Rev. B 30 4713 (1984).

6. W. J. Varhue, S. Krause and J. Dea and C.O. Jung in _Clemical Perspectives of Microelectronic Materials_, edited by M.E. Gross, J. Jasinski, and J.T. Yates, Jr. (Mater, Res. Soc. Proc. 131, Boston, MA 1988).

7. B. Dischler, A. Bubenzer and P. Koidl, Solid State Communications Vol. 48, No. 2, 105 (1983).

INTERACTION OF AN AROMATIC MOLECULE WITH A SURFACE

G. VIDALI[a], M.KARIMI[b]
[a]Physics Department, Syracuse University, Syracuse, N.Y.13244-1130
[b]Physics Department, Utica College, Utica, N.Y. 13502

ABSTRACT

The interaction between an aromatic molecule and the graphite basal plane is constructed using a Lennard-Jones 6-12 potential. The corrugation of the potential across the surface, the binding energies and vibrational frequencies of the aromatic molecules on the basal plane have been calculated and compared with experimental data. We have also calculated the interaction between a rare-gas atom and an aromatic molecule using two independent models, one based on the pairwise sum of the Lennard-Jones potential and the other on the Effective Medium Theory (EMT) for the repulsive part and a Van der Waals dispersion term for the attractive part. Our results are then compared with available experimental data.

Introduction

Considerable attention has been devoted lately to the study of the interaction of an aromatic molecule and a graphite surface [1-4]. Different scientific communities are interested in the study of this problem, since the interaction is relevant in processes occurring in coal gasification as well as in the interstellar medium (polyclyclic aromatic hydrocarbon molecules interacting with graphitic dust grains) [1].

Many experimental studies have been done on the interaction (adsorption/desorption, decomposition etc.) of aromatic complexes on carbon or graphitic surfaces. However, most of them were done under poorly characterized surface conditions. Until recently, not much serious theoretical effort has been devoted to the understanding of the interaction at a microscopic level.

There are some common features in the interaction of aromatic molecules with surfaces of metals and semimetals: (1) The adsorption of a monolayer of benzene on graphite yields two distinct commensurate phases which differ from one another by a rigid rotation with respect to the graphite lattice [4]; (2) Benzene lies flat on the surface; (3) The overlayer is "smooth"; and (4) vibrational modes (within the benzene molecule) are not greatly influenced by adsorption [5].

In this paper, we would like to report preliminary calculations for benzene and coronene adsorbed on graphite; we also report on the interaction of Van der Waals complexes formed by the adsorption of rare gases on a free standing benzene or coronene molecule. This work is part of an on-going theoretical and experimental effort to understand the interaction of aromatic molecules with surfaces. We chose to calculate the interaction of benzene and coronene with the basal plane of graphite for the following reasons: (1) it is a system of interest to the astrochemical and physico-chemical communities, as explained above; (2) these molecules don't react with graphite; and (3) there are available pair potentials for C–C and C–H interactions [3, 6, 7].

A more technical reason for doing this calculation was the desire to explore the possibility to extend our new model (based essentially on the effective medium theory" (EMT) [8] and van der Waals dispersion terms) to these systems. Previously we and others have applied this model quite successfully to the interaction of rare gases and molecular hydrogen with surfaces of insulators (oxides and alkali halides) [9–11]. We are also interested in comparing the results obtained using this method with the ones obtained employing Lennard–Jones type potentials.

The Model

A. Rare–gas atoms on benzene and coronene

We have evaluated the interaction potential of a rare–gas atom adsorbed on a single benzene and coronene molecule using a Lennard–Jones potential (hereafter this will be called Model 1), and EMT (Model 2).

For Model 1 we have used the following parameters. For the He–C potential we have used the parameters given in Ref. [12] which fitted remarkably well the atom scattering data for He–graphite [13]; the H–C parameters are taken from Ref [7]. For the parameters of the other systems, please see Table 1.

In Model 2, the repulsive part V_R is built using EMT which assumes a proportionality between the electron charge density at the surface and

$$V_R(\vec{r}) = \alpha \rho(\vec{r}) \qquad (1)$$

where ρ is taken from Ref. [14]. The attractive part of the potential is a superposition of dipole–dipole Van der Waals terms:

$$V_A(r) = \Sigma_i \, C_6/r_i^6 \qquad (2)$$

The electron charge density $\rho_i(\vec{r}) = A \exp(-\gamma r_i)/r_i$, where $A = 10.83 \, \text{Å}^{-2}$, $\gamma = 3.67 \, \text{Å}^{-1}$ [15] and C_6 are taken from Ref. [11].

Table 1.
Lennard–Jones Parameters from Ref. [22].

System	σ(A)	ϵ(meV)
Ne–C	2.80	2.65
Ar–C	3.10	6.17
Kr–C	3.21	7.38
Xe–C	3.36	8.93
Ne–H	2.78	2.82
Ar–H	3.15	4.73
Kr–H	3.27	5.99
Xe–H	3.40	6.65

B. Benzene and coronene/graphite

For these systems we haven't employed Model 2 since values of α as in Eq.(1) above, are not available for C–C and C–H interactions. Therefore we have used Model 1 with the following parameters: σ_{CC} = 2.4 Å, ϵ_{CC} = 2.41 meV, σ_{HC} = 2.98 Å, and ϵ_{HC} = 1.46 meV [3, 16, 17].

Here we report the calculations for the laterally averaged interaction potential. This was accomplished by summing the interaction of each H and C atom in the aromatic molecule with each atom of the graphite surface, i.e.:

$$V_{00}(z) = \Sigma_i \left[V_i^c(z) + V_i^H(z) \right] .$$

This potential was inserted in the Schrodinger equation and we obtained the eigenvalues E_n and the expectation value $<z>$ of the molecule above the graphite plane. We also calculated the vibrational frequencies for benzene and coronene over graphite. We did so by approximating the bottom of the potential well with a harmonic potential:

$$V_{00}(z) = -D + \frac{1}{2} k (z-z_{min})^2$$

where k is the spring constant.

Results and discussion

We checked the parameters of the He–C potential by calculating the He–graphite interaction as a sum of He–C interactions. We obtained the following eigenvalues (in parenthesis are the experimental values from Ref.[13]): 12.33 (12.42), 6.65 (6.59), 3.17 (3.05), 1.3 (1.24), 0.11 (0.16) meV.

When Model 2 was used, in order to fit the experimental data as above, we had to decrease the value of C_6 by 5%. Such change shouldn't be considered unexpected, since C_6 values are not very well known for these systems. We also found $<z>$ = 3.15 Å for He–graphite. This should be compared with $<z>$ = 3.1 Å [11]. Model 1 gives $<z>$ = 2.9 Å, while neutron scattering data give $<z>$ = 2.85 Å [18]. See also Ref. [19] for a comparison with other calculations. In general it is found that in Model 1 $<z>$ increases with the size of the atom, while it decreases when Model 2 is used [9–11]. The latter results seem more reliable, since similar trends are obtained with different semi ab–initio approaches.

There are some general features that emerge from the calculations reported in Table 2: 1) the interaction between a rare–gas atom and an aromatic molecule gets stronger as the size of the molecule gets larger; 2) in the limit of a very large aromatic molecule, the well depth D compares with D for the rare–gas/graphite case (as we would expect); 3) the expectation value $<z>$ = 3.1 Å for He/benzene is in good agreement with the

Table 2. Key parameters of rare-gas/benzene or coronene and benzene, coronene/graphite interactions. D, <z>, E_0 and E_i are the well depth, expectation value of adatom position and first two bound states. V_A, V_B and V_C are the minima of the potential at different adsorption sites: A, center of hexagon; B, on top of a C atom; and C, between two C atoms. Energies in meV and distances in Å.

System	D	<z>	E_0	E_1	V_A	V_B	V_C
He–benz[a]	11.5	2.9	7.7	2.7	11.5	7.8	8.5
	12.0	3.1	8.3	3.1	12.0	8.7	9.3
Ne–benz[a]	25.9	2.8	23.0	17.7	25.9	19.2	20.4
	29.5	2.8	26.5	21.0	29.5	22.1	23.5
Ar–benz[a]	57.5	3.1	54.3	48.2	57.5	44.2	46.7
	104.8	2.7					
Kr–benz	71.4	3.3	68.6	63.1	71.4	55.7	58.8
He–benz	84.3	3.4	81.6	76.3	84.3	67.6	70.9
He–Cor	15.8	2.8	11.5	5.4	–	–	–
Benz–gr	277.6	3.38	274.1	260.4	–	–	–
Cor–gr	1103.0						

[a] The top line is obtained with Model 1 (Lennard–Jones potential, bottom line is with Model 2 (EMT).

experimental value of 3.19 ± 0.37 Å [17]. For benzene, coronene/graphite we find that the binding energy per carbon atom is about 45 meV; this should be compared with 23 meV [20], 100 meV [21], or 68 meV [3]. These differences should be ascribed mainly to different choices of the Lennard–Jones parameters.

Finally, we discuss the effect of anisotropy of the He–C interaction on the results presented here [4]. The effect of the anisotropy is to modify the Fourier components of the potential which describe the variation of the potential across the unit cell. Since, for the systems considered here, these Fourier components are believed to be very small then we conclude that the role of anisotropy should be indeed small.

We may summarize our findings as follows: the study of the interaction of aromatic molecules with surfaces is still at the beginning, both from an experimental and theoretical viewpoint. EMT results are in general more reliable and closer to the experimental data. Unfortunately, some of the calculations required for the systems under consideration are very hard to carry out.

ACKNOWLEDGEMENTS

One of us (G.V.) would like to thank the Alfred P. Sloan Foundation for supporting this work through a fellowship.

REFERENCES

1. See articles in <u>Polycyclic Aromatic Hydrocarbons and Astrophysics,</u> Eds. A. Leger, L. d'Hendecourt, and N. Boccara (D. Riedel Publishing Co., Boston, 1987).
2. P. Jakob and D. Menzel, Surf. Sci. <u>201</u>, 503 (1988).
3. L. Battezzati, C. Pisani, and F. Ricca, Trans. Faraday Soc., <u>71</u>, 1629 (1975).
4. C. Bondi and G. Taddei, Surf. Sci. <u>203</u>, 587 (1988).
5. G. Vidali and M. Karimi (submitted).
6. W. A Steele, <u>The Interaction of Gases with Solid Surfaces</u> (Pergamon, Oxford, 1974); H. Hoinkes, Rev. Mod. Phys. <u>52</u>, 933 (1980).
7. J. M. Phillips and M. D. Hammerbacher, Phys. Rev. <u>B29</u>, 5860 (1984) and references cited therein.
8. J. K. Norskov, Phys. Rev. <u>B26</u>, 2875 (1982); N. D. Lang and J. K. Norskov, Phys.Rev., <u>B27</u>, 4612 (1983). M. W. Cole and F. Toigo, Phys. Rev. <u>B31</u>, 727 (1985).
9. M. Karimi and G. Vidali, Phys. Rev. <u>B38</u>, 7759; in <u>Diffusion at Interfaces: Microscopic Concepts,</u> Eds. H. Kreuzer, J. J. Weimer and M. Grunze (Springer Series in Surface Science), v.12, <u>43</u> (1988).
10. M. Karimi and G. Vidali, Phys. Rev. B 1988 (in press); Surf. Sci. (in press).
11. F. Toigo and M. W. Cole, Phys. Rev. <u>B32</u>, 6989 (1985).
12. W. E. Carlos and M. W. Cole, Phys. Rev. Lett. <u>43</u>, 697 (1979); Surf. Sci. <u>91</u>, 339 (1980).
13. Atom beam scattering and thermodynamic data are summarized in: M. W. Cole, D. R. Frankl, and D. L. Goodstein, Rev. Mod. Phys. <u>53</u>, 199 (1981).
14. M. W. Cole and F. Toigo, Phys. Rev. <u>B31</u>, 727 (1985).
15. M. W. Cole, private communication.
16. W. A. Steele, J. Phys. Chem. <u>82</u>, 817 (1978).
17. S. M. Beck, M. G. Liverman, D. L.Monts, and R. J.Smalley, J. Chem. Phys. <u>70</u>, 232 (1979).
18. K. Carneiro, L. Passell, W. Thomlison, and H. Taub, Phys. Rev. <u>B24</u>, 1170 (1981).
19. S. Leutwyler and J. Jortner, J. Phys. Chem. <u>91</u>, 5588 (1987).
20. L. A. Girifalco and R. A. Lad, J. Chem. Phys. <u>25</u>, 693 (1956).
21. D. P. DiVincenzo, E. J. Mele, and N. A.W. Holzwarth. Phys. Rev. <u>B27</u>, 2958 (1983).
22. G. Vidali, M. W. Cole, and J. R. Klein, Phys. Rev. <u>B28</u>, 3064 (1983).

DEPOSITION OF THIN ALUMINA FILMS FROM SUPERCRITICAL WATER JETS

J. I. Brand and D. R. Miller
Dept. of Applied Mechanics and Engineering Sciences, B-010
University of California, San Diego , La Jolla, Ca. 92093

ABSTRACT

Thin alumina films are grown on a silicon substrate, in vacuum, at a rate of about 10 Å per second, and at substrate temperatures below 150° C. Alumina is dissolved in supercritical water and the resulting solution expanded in a supersonic free jet, which is directed at the substrate. The films are characterized by ESCA and FTIR .

INTRODUCTION

We recently reported on the use of supercritical water to dissolve alumina, followed by a free jet expansion of this solution to deposit thin films[1]. This is a technique developed by Smith and colleagues [2], who grew particles and fibers of polymers and silicas. Water above its critical point will dissolve sparingly soluble materials such as quartz and alumina. The subsequent free jet expansion lowers the solution pressure rapidly, causing precipitation of the solute. Free jet expansions are well understood and commonly used for molecular beam studies [3]. At the pressures used to date for these supercritical expansions, the free jet forms a strong normal shock wave through which the clustered species must pass before striking a surface or equilibrating with the ambient background. The kinetics of the precipitation or clustering is not understood, and certainly involves a non-equilibrium process. We completed a supersonic gas-dynamic model of the nozzle flow and free jet expansion using equilibrium solubilities as a first approximation. This model showed that, under proper conditions, such precipitation is thermodynamically prohibited from occurring before the exit of the nozzle and must occur within the free jet expansion or beyond [1]. Further, our calculations showed that either a short capillary tube or a simple orifice in a plate could serve equally well as the nozzle.

We have consistently grown thin films of alumina which appear to be in the anhydrous, hexagonal alpha phase. In this report, we discuss the vacuum facility we have recently completed to investigate the supercritical expansions. We also report on the characterization of an alumina film grown in vacuum, instead of atmospheric conditions. This film is the first we have grown using an orifice nozzle rather than a capillary tube nozzle. We also present operating characteristics of the jet source calculated using our gas-dynamic model.

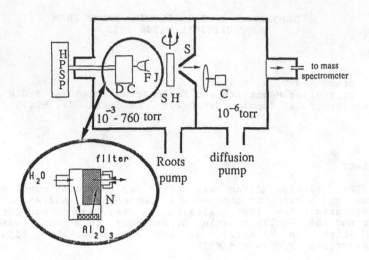

1. Schematic of Ceramic Beam and Film Growth Apparatus. HPSP = high pressure solvent pump; DC = dissolving cell; N = nozzle; FJ = free jet; SH = substrate holder; S = skimmer; C = chopper.

EQUIPMENT AND OPERATING CHARACTERISTICS

Figure 1 shows the vacuum facility used for the present study. The first chamber is pumped by a 330 CFM Roots pump and can be operated at pressures between one atmosphere and a few microns. The free jet expansion occurs in this chamber. The high pressure solvent pump is a Varian 8500 HPLC pump and provides continuous flow at water pressures up to 40 MPa . The dissolving cell is a Swagelok, stainless steel, high pressure, in-line, 7 micron filter, with an internal volume of 1.5 cm³ ; previously we had used 60 and 90 micron filters. The dissolving cell is wrapped with heating tapes, monitored with thermocouples, and can be heated above 600° C. The nozzle orifice is specially machined out of a standard Swagelok, 1/8th-inch, high pressure cap. A 0.05 cm diameter circular area is milled in the end of the cap to a wall thickness of 0.012 cm. A 0.0075 cm hole is then spark eroded with a copper electrode (EDM) through the thinned section. This hole is the nozzle orifice through which the high pressure solution passes before expanding in the supersonic free jet. This nozzle has operated as high as 40 MPa and 600° C. Previous nozzles were capillary tubes [1]. Advantages of the new geometry are a more accurately measurable diameter and more nearly isentropic flow, since friction losses are smaller. We are in the process of making smaller diameter nozzles in order to match the flow rates to our pumping speeds more effectively.

The substrate holder for the film reported here was a glass slide onto which a silicon crystal was clamped. A thermocouple clamped between the slide and the silicon crystal, 1 cm from the

center of the jet, indicated a maximum temperature of 150° C
during film growth. This method of growing alumina films is a
comparatively low temperature process and is a potential
technique to grow thin films without thermal destruction of the
substrate.

If the jet is directed toward a skimmer instead of a
substrate, a small fraction of the beam will pass into a second,
high vacuum chamber. In this second chamber, the beam is
mechanically chopped into pulses, and then it continues on to a
quadrupole mass spectrometer. The time of flight from the chopper
to the mass spectrometer is measured and used to determine the
velocity and temperature of the beam. The source temperature can
then be calculated from an energy balance. More importantly, the
mass spectrometer also monitors the beam composition, including
the nature of alumina species present. Although we have measured
the energy of the primary beam at various source conditions, we
have not yet been able to make accurate composition
determinations. Two sources of difficulties are the high
pressure in the jet in front of the skimmer, and the low
concentration of alumina in the beam. We are currently
attempting to overcome these difficulties by redesigning the
skimmer and adding cryogenic pumping to the mass spectrometer.

Figure 1 also shows the outline of shock structure expected
near the nozzle exit [3]. As the background pressure drops, this
shock structure lengthens and the shock effects weaken. The film
reported here was grown at a background pressure of about 133 Pa.

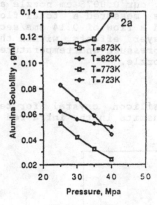

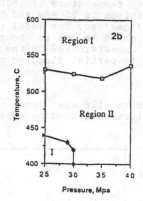

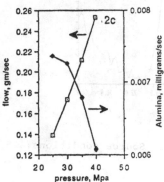

2. Operating Characteristics.
a: solubility data of ref [4] ;
b: regions where precipitation
will occur outside of the nozzle (II)
and may ,thermodynamically, occur
inside the nozzle (I);
c: mass flow rates, total and alumina,
calculated for an isentropic nozzle,
0.0075cm, T=500° C.

Figure 2 shows operating characteristics of the 0.0075cm orifice nozzle predicted by our previously described model [1]. Figure 2a shows the interpolated solubility data of Ganeyev and Rumyantsev [4] used in the calculations. Although the data are sparse and have substantial scatter, they are the only experimental data available in our pressure and temperature range. Since the solubility is quite sensitive to both pressure and temperature, and since the gas dynamics of the supercritical flow are non-ideal, determining the operating conditions can involve subtle choices. Figure 2b shows the regions of source pressure and temperature in which precipitation may (I) or may not (II) occur before the flow exits the nozzle orifice. The location of these calculated regions is sensitive to the solubility data used. Since the actual clustering is limited by kinetics, such models can only serve as guidelines to operating such a source. Operating in region II avoids clogging the nozzle since the alumina remains in solution until the rapid free jet decompression. One serious operating difficulty is that the lower branch of region I must be traversed during the start-up and shut-down periods. This may cause clogging of the nozzle. Operating near the upper boundary between regions I and II, at high pressures, maximizes the initial solubility of the alumina at some temperatures. At other temperatures, operating near this boundary may actually decrease alumina deposition rates. Figure 2c shows the effect of pressure on both total mass flow rate and the flow rate of alumina through a 0.0075cm nozzle orifice, at a fixed source temperature of 500° C. For our 0.0075 cm nozzle at source conditions of 25 MPa and 500° C, we measured a total flow rate of 0.13 gms/sec. The model predicted a flow of 0.14 gms/sec. Differences could be due to boundary layer effects within the nozzle, to uncertainties in the source pressure and temperature, or possibly to partial clogging of the nozzle.

RESULTS

An alumina film was grown on a silicon crystal for ten minutes. Silicon was used because it transmits in the infra-red

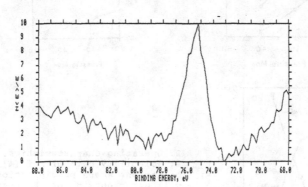

3. ESCA Spectrum of Alumina/Si Film. Source conditions: 25 MPa and 550° C; Al 2p peak is at 75.2 eV.

region of the expected Al-O stretch, and it is a suitable substrate for X-ray photoelectron spectroscopy (ESCA). This film was circular in shape, about 1 cm in diameter, and estimated to be about 5000 Å thick. It adhered to the Si very well and could not be scraped off easily.

Figure 3 shows the detailed ESCA spectrum in the region of the Al 2p peak at 75.2 eV binding energy. Literature values [5] vary from 74 to 75.3 eV for oxidized Al; literature values for the unoxidized form are between 72 and 73 eV. Peaks from the Si substrate showed no shifts compared to spectra of blank Si crystals. Our previously reported results [1] showed significant peaks at 685.6 and 599.8 eV, corresponding to fluorine. The present film contains no fluorine. Other recent films have contained fluorine and the source of this impurity is still

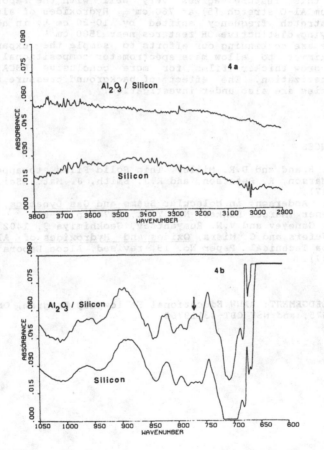

4. FTIR Spectra for Alumina Film. a: region near OH-stretch; b: region near Al-O stretch. Spectra for the blank Si substrate and the thin film on Si are shown for each region. Arrow indicates 764 cm⁻¹ feature.

uncertain. Contamination of the sample after film growth and contamination of the solvent are two possibilities. Mass spectrometer investigation of the beam composition will distinguish between them.

Figure 4 is a composite of FTIR spectra taken with a Nicolet 60 SXB. Figure 4a shows the region around 3500 cm^{-1} where the OH groups of water or hydroxides should appear. Both the blank Si substrate and the alumina film spectra are shown. There is no strong feature in this region. This indicates that this film, like our previous films [1], may be anhydrous. Figure 4b shows the region near 750 cm^{-1} where the Al-O stretch should appear. Unfortunately, silicon dominates the spectrum and alumina features, if any, are very small. The only feature possibly due to the film is the small peak at 764 cm^{-1}. While the signal-to-noise does not make this result convincing, it is encouraging since this feature agrees very well with the reported alpha aluminum Al-O stretch [5] at 760 cm^{-1}. Hydroxides of alumina show this stretch frequency shifted by 10-20 cm^{-1}, in addition to displaying distinctive OH features near 3500 cm^{-1}.

We are continuing our efforts to sample the expansion with the skimmer to allow mass spectrometer compositional analysis, and to grow thicker films for more conclusive ESCA and FTIR characterization. The effect of background pressure and source impurities are also under investigation.

REFERENCES

1. J.I. Brand and D.R. Miller, Thin Solid Films, to appear 1988
2. D. Matson, R. Peterson, and R.J. Smith, J. Mat. Sci. 22 ,1919 (1987).
3. J.B. Anderson, in Molecular Beams and Gas Dynamics, ed. by P. Wegener (Marcel Dekker, New York, 1974) p. 1-67.
4. I.G. Ganeyev and V.N. Rumyantsev, Geokhimiya 9, 1402 (1974)
5. K. Wefers and C. Misra, Oxides and Hydroxides of Aluminum, Alcoa Technical Paper No. 19, revised, Alcoa Laboratories (1987).

ACKNOWLEDGEMENT: AAUW Educational Fellowship Program, ONR N00014-87-K-0675, and NSF CBT-83119762

Proximity Effect in the Low Pressure Chemical Vapor Deposition of Tungsten.

N. Lifshitz

AT&T Bell Laboratories
Murray Hill, New Jersey 07974

ABSTRACT

The self-limiting effect during Low Pressure Chemical Vapor Deposition of tungsten is manifested by a sudden interruption of the reduction of tungsten hexafluoride WF_6 by silicon, so that only very thin (self-limited) films of silicon-reduced tungsten can be grown. It has been shown that the self-limiting effect is caused by formation of the non-volatile subfluoride WF_4. The temperature dependence of the self-limiting thickness of the tungsten films grown in a hot-wall reactor exhibits a characteristic maximum at temperatures near 350°C, which indicates that at this temperature the rate of formation of WF_4 is lower than at temperatures above and below. In the present paper we attempt to explain this peculiar dependence. We suggest a mechanism responsible for formation of the blocking agent WF_4. We demonstrate that the presence of a hot tungsten surface is essential for the strong self-limiting effect. This leads us to the discussion of the proper selection of the reactor type (hot-wall vs. cold-wall) for different process requirements.

INTRODUCTION

The self-limiting effect during Low Pressure Chemical Vapor Deposition (LPCVD) of tungsten has puzzled the scientific community for several years. Deposition of tungsten via reduction of tungsten hexafluoride WF_6 by hydrogen is described by the reaction:

$$WF_6 + 3H_2 \rightarrow W + 6HF \qquad (1)$$

When a silicon surface is exposed, reduction of WF_6 by silicon can also occur:

$$2WF_6 + 3Si \rightarrow 2W + 3SiF_4 \qquad (2)$$

In a neutral argon atmosphere only reaction (2) takes place. The self-limiting effect is manifested by a sudden interruption of reaction (2) at the very early stages of the process. In the present work we will be mainly concerned with the process of WF_6 reduction by Si. Later in the paper we will relate our findings with the process of tungsten deposition that involve both reactions (1) and (2).

Recently a model was proposed to explain the self-limiting effect by formation of a non-volatile subfluoride of tungsten, most likely tungsten tetrafluoride WF_4 [1]. This subfluoride is found in large amounts in silicon-reduced tungsten films. It is solid in the deposition temperature range, with a three-dimensional polymeric structure [2], and may block (chemically or physically) further interaction between WF_6 and silicon, thus causing the self-limiting effect. The exact nature of the blocking action still needs further investigation.

Interestingly, analogous chemistry employing the reduction of molybdenum hexafluoride, MoF_6, by silicon does not exhibit the self-limiting effect, so that silicon-reduced Mo films grow linearly with time [3]. This is expected because all subfluorides of molybdenum are volatile [4].

The extent of Si reduction before the onset of the self-limiting mechanism is an important parameter of the deposition process because it determines the extent of damage inflicted on the Si device by exposure to WF_6 [5]. It is not *a priory* obvious that the extent of silicon consumption during hydrogen reduction of WF_6 should be the same as in the case of WF_6 reduction by Si in an argon atmosphere. However, several experimental facts support this hypothesis. First, in the case of the Mo chemistry, where there is no self-limiting effect and reduction by Si proceeds indefinitely, it has been shown that both reactions — reduction by Si and by hydrogen — contribute in the process and the resulting film is the superposition of the two reactions [3]. It has also been demonstrated that the reduction chemistries involving hexafluorides of both refractory metals, tungsten and molybdenum, would be similar but for the self-limiting effect which appears only in the case of tungsten [6]. This implies that *before the onset of the self-limiting mechanism* the tungsten and molybdenum processes take a similar course and, therefore, the reaction of Si reduction of WF_6 proceeds to the same extent in both hydrogen and argon atmospheres.

It was noted by several authors that the self-limited thickness obtained by silicon reduction of WF_6 in an argon atmosphere strongly depends on temperature. In Fig. 1 two examples of the temperature dependence of

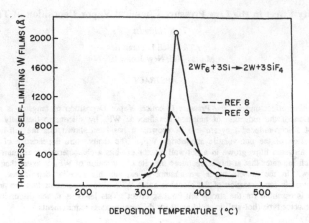

Fig. 1 Dependence of the self-limited thickness on the deposition temperature in a hot-wall reactor.

the self-limited thickness are shown [7,8]. The films were deposited in hot-wall reactors and the thickness of the films was determined by the Rutherford Backscattering (RBS) technique. In spite of the difference in the absolute values of the film thickness, the similarity is obvious: both dependences exhibit a characteristic maximum at some temperature T_M that varies for different experimental set-ups from 350 to 450°C (see Ref. 7 and references therein). This sharp increase of the self-limited thickness near T_M suggests that the self-limiting mechanism is curtailed at these temperatures. We shall assume that the self-limiting mechanism is associated with formation of WF_4 and try to correlate the temperature dependence of the self-limited thickness with that of the kinetics of formation of WF_4.

The shape of the curves in Fig. 1 suggests that there are two competing mechanisms (one being responsible for the formation of WF_4, and the second with the opposite effect) that determine the dependence of the self-limited thickness on temperature. The rate of formation of WF_4 would be enhanced in the temperature ranges sufficiently above and below T_M. At temperatures close to T_M the effect of both competing mechanisms leads to low WF_4 formation rates resulting in thicker self-limited films.

DISCUSSION

We believe that the main mechanism of formation of WF_4 is the process of reduction of WF_6 by tungsten available in the reactor. This process follows two steps [9,10]. The first step is a formation of a precursor, gaseous tungsten pentafluoride:

$$W + 5WF_6 \rightarrow 6WF_5 \tag{3a}$$

which is unstable, and readily disproportionates into WF_6 and WF_4:

$$2WF_5 \rightarrow WF_4 + WF_6 \tag{3b}$$

This process is used as a standard method of preparation of WF_4 [11]. Detailed analysis of tungsten removal rates via reaction (3) is given in Ref. 12. The reported results indicate that the reaction does not take place at temperatures below 230°C, and at higher temperatures the reaction rates depend on both temperature and partial pressure of WF_6. From the curves in Ref. 12, at 400°C and WF_6 partial pressure of 2.5 torr tungsten removal rate is ~3000 Å/min. We corroborated these results for lower pressures by observing a complete disappearance of 1000 Å W film subjected to a WF_6 / Ar atmosphere for 1 hr. at 400°C. The partial pressure in the reactor was 0.02 torr. This corresponds to tungsten etch rate of ≥16 Å/min. Our result supports the previous observation [12] that W etch rate is roughly proportional to WF_6 partial pressure.

The non-zero rate of W removal at temperatures above 200°C indicates that even at such low temperatures the process of formation of WF_4 via reaction (3) takes place. This reaction is expected to produce WF_4 in a hot-wall reactor because of the large area of hot tungsten on the walls of the reactor.

However, at temperatures above 300°C another process curtailing the rate of formation of WF$_4$ becomes important. This process is the disproportionation of WF$_4$, with WF$_6$ and metallic tungsten as the reaction products:

$$3WF_4 \rightarrow W + 2WF_6 \qquad (4)$$

The effect of temperature on this reaction is illustrated by the thermogravimetric curve for WF$_4$ [13] plotted in Fig. 2. (left-hand scale). The curve was obtained by heating a certain amount of WF$_4$ in a closed environment at various temperatures and recording the weights of the remaining WF$_4$ and formed metallic tungsten, together with WF$_6$ pressure. Importantly, the sharp onset of the disproportionation process occurs between 300 and 330°C, i.e. precisely where the self-limited thickness exhibits the sharp rise (Fig. 1). At 350°C the process reaches a new equilibrium. We hypothesize that the increase in the self-limited thickness at temperatures approaching T$_M$ from below is associated with the disproportionation of the blocking agent WF$_4$. We also argue that at temperatures above T$_M$ the increase in the WF$_4$ formation rate via mechanism (3) is responsible for the new onset of the self-limiting effect. In order to qualitatively explain our model in Fig. 2 we superimposed on the gravithermal curve from Ref.13 the temperature dependence of tungsten removal rates (which are proportional to the rate of formation of WF$_4$). Of course, this comparison is only illustrative: the disproportionation curve is determined by the thermodynamics of the process (4), whereas tungsten etch rate in WF$_6$ is determined by the kinetics of the process (3) under constant WF$_6$ flow and pressure conditions.

It is important to understand that the two reactions — of formation of WF$_4$ (3) and its disproportionation (4) — occur simultaneously, so that the observed rate of tungsten removal is a net result of both reactions (3) and (4).

SELECTION OF THE DEPOSITION REACTOR

The above discussion leads us to important predictions. It is clear that the presence of an exposed hot metallic tungsten surface is vital for the formation of WF$_4$, and therefore, for the self-limiting effect. It implies that in a cold-wall reactor, when the walls are not covered with tungsten, the self-limiting effect should be suppressed, and the silicon consumption via reaction (2) should be more extensive. Indeed, recently reported results support this hypothesis. It has been reported[14] that in a GENUS cold-wall reactor the self-limiting effect was observed only at temperatures below 300 or above 500°c. In the range 300-500°C, the thickness of the silicon-reduced films grew linearly with time. In another instance, the thickness of the films increased with temperature and reached several microns at 500°C[15]. These results indicate that in a cold-wall reactor the self-limiting effect is indeed suppressed, and silicon consumption is unimpeded. Hence, tungsten tetrafluoride plays a very important and positive role: it reduces silicon consumption during the deposition process.

This result is extremely important for the VLSI applications of tungsten LPCVD. Thus, the described model of WF$_4$ controlling Si consumption can explain a puzzling pattern of the distribution of the leakage current of shallow (~1000 Å) diodes across the silicon wafer. The wafers were subjected to tungsten deposition

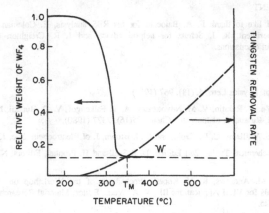

Fig. 2   Gravithermal curve for disproportionation of WF$_4$ from Ref.12 (left-hand scale). For illustration, the rate of tungten removal in a WF$_6$ / Ar atmosphere is also shown in the figure (right-hand scale).

626

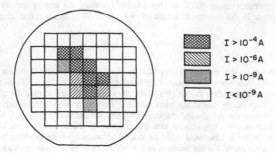

$I > 10^{-4} A$

$I > 10^{-6} A$

$I > 10^{-9} A$

$I < 10^{-9} A$

Fig. 3   Distribution across the Si wafer of the leakage current of shallow p-n junctions subjected to tungsten deposition in a hot-wall reactor.

at 325°C in a hot-wall reactor. After the deposition the diodes exhibited some increase in the leakage current, which indicates that some silicon was consumed. A typical distribution of the leakage after the deposition is shown in Fig. 3. As can be seen in the figure, the devices situated near the center of the wafer (and, therefore, farthest removed from the hot tungsten walls) exhibited much worse degradation than the devices at the edge of the wafer (the space between the wafer and the reactor wall was 1.5 cm).

However, the presence of WF_4 in a deposition chamber may have yet another effect, the loss of selectivity. In a recently reported experiment [16], WF_4 formed via reaction (3) contaminates the surface of silicon dioxide. Then at the deposition temperature WF_4 disproportionates, forming small tungsten particles that act as nucleation centers on the SiO_2 surface. Hence, it is not surprising that the loss of selectivity is especially severe in a hot-wall reactor.

As a reminder, the loss of selectivity was never, under any circumstances, observed for the analogous Mo process [3]. This fact becomes understandable in the model for selectivity loss suggested in Ref. 16: since all molybdenum subfluorides are volatile, they can not form nucleation centers on the SiO_2 surface.

From the above discussion, it becomes clear that the choice of the deposition equipment is determined by the specifics of the underlying structure: if sensitive silicon devices are exposed the silicon consumption should be minimized. Then it is advantageous to use a hot-wall reactor in the blanket deposition mode. On the other hand, if the specifics of the fabrication process calls for selective deposition in small areas (for example, filling the contact windows for the second metallization level), then the cold-wall reactor is the best choice.

ACKNOWLEDGEMENT

The author would like to thank F. A. Baiocchi for his RBS analysis, E. Coleman for conducting the tungsten etching experiment, R. J. Schutz for helpful advice and J. R. Creighton of Sandia National Laboratories for fruitful discussions.

REFERENCES

1. N. Lifshitz, Appl. Physics Lett., 51(13), 967 (1987).

2. V. S. Pervov, Yu. V. Rakitin, V. M. Novotortsev, A. T. Falkengof, V. D. Butskii, N. N. Savvateev and B. E. Dzevitskii, Russian J. of Inorganic Chem., 25(16), 2327 (1980).

3. N. Lifshitz, D. S. Williams, C. D. Capio, and J. M. Brown, J. of Electrochem. Soc, 134, 2061 (1987).

4. JANAF Thermochemical Tables, 2nd Edition, D. R Stull and H. Prophet, Editors, NBS, Washington DC (1971).

5. N. Lifshitz, J. M. Andrews, R. V. Knoell, Proceedings of the Workshop on Tungsten and Other Refractory Metals for VLSI Applications III , V. A. Wells. Editor, Material Research Society, Pittsburgh, PA (1988), p. 225.

6. N. Lifshitz, M. L. Green, J. of Electrochem. Soc., 135 1832 (1987).

7. M. L. Green, Y. S. Ali, T. Boone, B. A. Davidson, L. C. Feldman and S. Nakahara, J. of Electrochem. Soc., *135* 2285 (1987).

8. N. Lifshitz, unpublished.

9. J. Schroeder and F. J. Grewe, Chem. Ber., *103*, 1536 (1970).

10. A. V. Gusarov, V. S. Pervov, I. S. Gotkis, L. I. Klyuev, V. D. Butskii, Dokl. Akad. Nauk SSSR, *216*, 1296 (1974)

11. V. D. Butskii and V. S. Pervov, Russian J. of Inorgaic Chem., *22*(1), 6 (1977).

12. G. Dittmer, A. Klopfer and J. Schroeder, Philips Res. Repts., *32*, 341 (1977).

13. V. D. Butskii, V. S. Pervov, and V. G. Sevastyanov, Russian J. of Inorganic Chemistry, 22(5), 771 (1977).

14. R. V. Joshi, "Non-selflimiting nature of WF_6 and Si reaction in a LPCVD cold-wall system", to be published.

15. During his talk given at Bell Laboratories, Murray Hill, NJ, Dr. I. Nakayama of ULVAC Corporation, Kanagawa, Japan, reported that an unlimited growth of silicon reduced films had been observed in a cold-wall reactor. The film thickness increased with temperatures and reached several microns at 500°C.

16. J. R. Creigton, Proceedings of the Workshop on Tungsten and Other Refractory Metals for VLSI Applications, E. K. Broadbent, Editor, Material Research Society, Pittsburgh, PA (1987), p.43.

AN ANTIMONY PLANAR DIFFUSION SOURCE

JACK WILSON*, ROBERT GUSTAFERRO*, ALAN BONNY**, AND ED MADDOX*
*BP America, Research Center Warrensville, 4440 Warrensville Center Road,
Cleveland, OH 44128
**Now with Marathon Venture Partners, San Francisco, CA 94105

ABSTRACT

Antimony is a useful n-type dopant for buried layers, due to its low rate of lateral diffusion in subsequent processing steps. To date, no solid planar diffusion source has been available. Reported here is the development of such a source.

The source is made by high temperature synthesis in a hot press. A mixture of antimony trioxide and silicon powders is placed in the press, and heated under pressure. The reaction

$$2Sb_2O_3 + 3Si \rightarrow 3SiO_2 + 4Sb$$

results in a solidified product. Wafers are sliced from this product to form the actual planar diffusion sources.

The sources are used in a two-step doping process: a predeposition followed by a drive-in. Oxygen is required in both steps, due to the chemistry involved. Typical properties of the doped silicon are: sheet resistance, 10 Ω/square, junction depth, 4-6 microns, and antimony surface concentration, 3-8 x 10^{19}/cm^3.

INTRODUCTION

In order to diffuse antimony into silicon in a reasonable time, diffusion temperatures must be around 1200°C. Antimony melts at 630°C, and antimony oxide melts at 656°C, so these materials alone could not be suitable solid sources. However, if trapped in a solid matrix of some kind, such that the antimony evolves from it relatively slowly, antimony or antimony oxide could be useful sources of antimony. Since the matrix has to be non-contaminating, the most obvious choice for a matrix material is silicon itself, with silicon dioxide as a second choice. This suggests the possibility of using the high temperature synthesis reaction between antimony trioxide and silicon, i.e.,

$$2Sb_2O_3 + 3Si \rightarrow 3SiO_2 + 4Sb$$

This reaction was first reported by Kahlenberg and Trautmann [1], although these authors stated that the reaction did not occur on heating with a Bunsen burner, and only reacted with difficulty on igniting with a spark. The product was a black molten mass containing a few globules of antimony. Ugai et al. [2] state that silicon and antimony trioxide react, starting at 560°C, but that only 6% of the silicon is oxidized. Most of the antimony trioxide is converted to antimony tetraoxide, Sb_2O_4, which is inert to silicon.

SOURCE PREPARATION

In exploratory experiments using microwave heating, it was found that a mixture of silicon and antimony oxide powders produced a satisfactory source after heating to a temperature around 1400°C. The powders were mixed together, cold pressed to form a disc, and then heated rapidly (i.e., in around 2 minutes) in the microwave unit to the temperature desired, when the microwave was switched off. The disc was transformed from a green compact to a solid, ceramic-like body.

Repeated experimentation showed that it was only necessary to heat to 600°C for the process to occur.

Scanning electron microscopy combined with x-ray energy dispersive spectrometry showed that the product consisted of spheres of elemental antimony, and particles of excess silicon in a matrix of silica containing some antimony; see figure 1. There was no antimony oxide remaining.

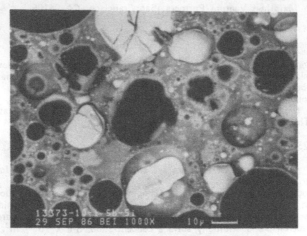

Figure 1   Scanning Electron Microscope Photograph of Mixture of Silicon and Antimony Trioxide after Reaction. The White Regions are Antimony, the Black Regions are Silicon, and the Grey Regions are Antimony-containing Silica.

The appearance of the product depended on composition. A stoichiometric mixture yielded a black, friable, spongy mass, similar to that described by Kahlenberg. Introduction of excess silicon led to a strong, grey solid. Differential thermal analysis on a mixture of 68% antimony oxide, 32% silicon showed that oxygen can have a significant effect on the reaction. Analyses done in air and nitrogen were totally different (fig. 2). In nitrogen, a strong exothermic reaction starts at 480°C, and takes off rapidly at 570°C. In air, a comparatively weak

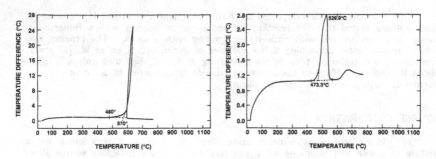

Figure 2   Differential Thermal Analysis Trace of a Mixture of 68% Sb_2O_3/32% Si heated (a) in Nitrogen, (b) in Air. Note Factor of Ten Difference in Ordinate between (a) and (b).

exothermic reaction starts at 525°C, peaking at 530°C, followed by a second reaction starting at 612°C. The difference is that in air, antimony trioxide oxidizes to form the tetraoxide, which does not react with silicon [2]. Thus, if the reaction is attempted in air, the reaction appears not to proceed (Kahlenberg and Trautmann's Bunsen burner experiment), or to proceed with low yield (Ugai et al.). However, if the reaction is performed either in an inert atmosphere as in the differential thermal analysis, or very rapidly, as in Kahlenberg and Trautmann's spark experiment, or the present microwave experiments, the reaction will go to completion.

Following the microwave experiments, the reaction was performed in a nitrogen purged hot press. The mixed powder was loaded into the press, pressure applied, and then heated until the reaction triggered. This proved to work very well. Moreover, cylindrical slugs of material could be made this way, from which the discs comprising the planar diffusion sources could be sliced. The technique is scaleable; to date the largest billet produced is five inches in diameter, but there appears to be no reason that larger diameter billets could not be made.

DIFFUSION TESTING

Method

The sources were loaded into a tube furnace along with 2 silicon wafers for each dopant source. The silicon wafers were (100), p-type, 10-30 ohm cm, and 8-23 mils thick. The diffusions were performed in two steps, a predeposition, followed by a drive-in, during which the sources were removed. The drive-in is designed to allow antimony to diffuse to the desired depth within the silicon as well as to lower the sheet resistance by diffusing additional antimony into the silicon from a coating which forms during the predeposition. The predeposition duration was half-an-hour at 1250°C; the drive-in was an hour and a half at 1250°C, both under 5% oxygen/95% argon. Following diffusion, the sheet resistance of the silicon wafers was determined by a four-point probe, and the junction depth by secondary ion mass spectrometry, with an occasional verification by spreading sheet resistance. The sources were re-used until the sheet resistance rose significantly above 25 ohms/square. This was defined as the lifetime of the source.

Results

Figures 3(a),(b), and (c) show the sheet resistance as a function of doping cycle for sources containing 68% antimony trioxide by weight, 54.5% antimony trioxide, and 50% antimony trioxide. It will be seen that the higher the antimony trioxide content, the longer the lifetime. The longest lifetime (68% antimony trioxide) was 27 cycles. The weight loss versus doping cycle is shown in figures 4 (a),(b), and (c) corresponding to the same sources. Typically the sheet resistance was 10 ohms/square.

Figures 5(a) and (b) show the junction depth obtained by secondary ion mass spectrometry and by spreading sheet resistance for a source with 50% antimony trioxide. The agreement between the two techniques is very satisfactory. Typical junction depths were between 4 and 6 microns. Antimony surface concentrations were between 3×10^{19} and 8×10^{19}/cm^3.

Glass Transfer

In early diffusion experiments, oxygen was found to have a pronounced positive effect on source performance. The addition of oxygen to the diffusion-tube gas was found to cause a significant silica glass layer to build up on the silicon wafers during the predeposition. Without oxygen this did not occur, and only limited doping was accomplished. Another interesting result was that this coating was found to evaporate during the drive-in unless oxygen was added during this step as well.

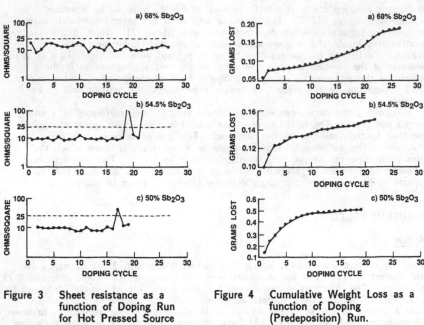

Figure 3   Sheet resistance as a
            function of Doping Run
            for Hot Pressed Source
            Material of Various
            Compositions.

Figure 4   Cumulative Weight Loss as a
            function of Doping
            (Predeposition) Run.

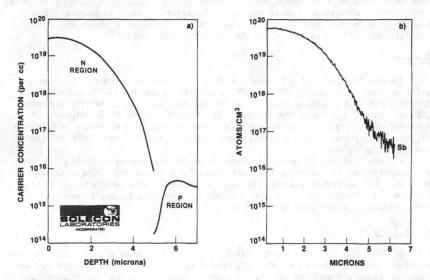

Figure 5   Junction Depth produced in Doped Silicon Wafer as measured
            by a) Spreading Sheet Resistance, and b) Secondary Ion Mass
            Spectrometry.   The Diffusion Source used was made from
            Material with 50% Antimony Trioxide.

Although some silica glass would be expected to form on the silicon wafer surface during the predeposition step via oxidation by gaseous oxygen, the glass layer formed during the predeposition was much thicker than would be expected from this mechanism alone. Additional silica may have been formed when antimony, vaporized from the source, and oxidized by the oxygen atmosphere, contacted the silicon, and was reduced to form silica and metallic antimony. Marshakov et al. [3] have shown that the kinetics of silicon oxidation by antimony oxide is almost a factor of one hundred faster than simple oxidation of silicon by gaseous oxygen, resulting in thicker glass layers.

In addition some glass was apparently transferred to the silicon from the source, rather than being produced by silicon wafer oxidation. To confirm this, an experiment was carried out using tantalum wafers in place of silicon wafers in a standard predeposition. After this procedure, the tantalum wafers were found to be coated with a silica glass very similar to that which was found on the silicon wafers during the standard predeposition step. This would account for at least part of the extra glass found on predeposited silicon wafers, the rest coming, perhaps, from an accelerated oxidation as previously described.

The fact that the glass coating forms only in the presence of oxygen and can be made to evaporate if the drive-in is carried out in an oxygen free environment may be due to the following set of reactions:

$$Si(s) + SiO_2(s) \rightarrow 2SiO(g)$$
$$SiO(g) + 1/2\ O_2(g) \rightarrow SiO_2(s)$$

During an oxygen-free drive-in, the first reaction may occur at the silica/silicon interface producing silicon monoxide gas which can evaporate through the coating or rupture the coating and cause it to flake off. Oxygen suppresses this evaporation by means of the second reaction. The coating would have trouble forming during predeposition in an oxygen-free environment since it would tend to evaporate almost as soon as it was deposited. This type of behavior is common in silica/silicon systems [4].

Maintaining a glass layer during drive-ins is important as it provides a barrier preventing antimony from escaping. Also the glass acts as an additional source of antimony during drive-in. The thickness of the glass layer produced on the silicon wafers during predeposition declines from run to run throughout the life of the sources until it becomes too thin to trap the antimony at the surface of the silicon during the drive-in. This point corresponds to the end of the life of the source. When the coating thickness has declined to about 1000 Å, the source has been depleted and the sheet resistance of the silicon after a 1250°C 1-1/2 hour drive-in exceeds 25 ohm/square.

In addition to maintenance of the glass layer during drive-ins, oxygen may enhance the diffusion of antimony from the glass into the silicon. From experiments involving diffusion into silicon of antimony from $Sb_xSi_yO_z$ glasses deposited by chemical vapor deposition, Pintchovski, et al. [5] concluded that oxygen in the ambient gas can cause a highly oxidized $Sb_xSi_yO_z$ film to form at the silicon/glass interface. This film creates a high concentration of antimony on the surface of the silicon due to the pile-up effect [6]. Pintchovski et al. [5] also found, however, that oxygen levels above 25% resulted in lower concentrations of antimony at the silicon surface itself. Their explanation of this phenomenon is that at high oxygen concentrations, the silicon surface oxidizes sufficiently rapidly that the resulting silica layer hinders the antimony diffusion.

CONCLUSIONS

It has been shown that high temperature synthesis of silicon and antimony oxide can be used to produce antimony planar diffusion source material. Diffusion sources made from such material dope silicon to acceptable values (sheet resistance ~ 10 Ω/square, junction depth ~ 4 μm), and have reasonable lifetimes. The lifetime increases as the antimony content in the sources increases. Glass transfer during doping is improved by the addition of oxygen to the diffusion tube gas.

REFERENCES

1. L. Kahlenberg and W. J. Trautmann. Trans. of Am. Electrochem. Soc. 39, 377 (1921)

2. Ya. A. Ugai, P. Fertsh, I. Ya. Miltova, V. Z. Anokhin, and T. A. Gadebskaya, Izvestiya Akademii Nauk SSSR, Neorganicheskie Materialy, 13 (9). 1611-1613, (1977)

3. I. K. Marshakov, V. Z. Anokhin, I. Ya. Miltova, S. S. Lavrushina, V. L. Gordin and Ya. A. Ugai. Zhurnal Fizicheskai Khimii, 50, 3094-3096 (1976) [Russian Journal of Physical Chemistry, 50 (12) 1840-1842 (1976)]

4. E. A. Gulbransen and S. A. Jansson. "Thermochemistry and Gas-Metal Reactions," in Oxidation of Metals and Alloys, editor D. L. Douglas. Proceedings of a Seminar held by American Society for Metals, October 1970.

5. F. Pintchovski, P. J. Tobin, M. Kottke and J. B. Price. J. Electrochem. Soc. 131 No. 8, 1875 (1984)

6. A. S. Grove, O. Leistiko, Jr., and C. T. Sah, J. Appl. Phys. 33, 2695 (1964).

Author Index

635

Subject Index

Chemical compounds are listed alphabetically at the beginning of
each letter according to the central metal atom. Abbreviations:
Me = CH_3 (methyl); Et = C_2H_5 (ethyl); Pr^i = CH_3CHCH_3 (isopropyl); Bu^i = $CH_2CH(CH_3)_2$ (isobutyl); t-Bu = $C(CH_3)_3$ (t-butyl);
allyl = CH_2CHCH_2.

Printed in the United States
By Bookmasters